W9-CRB-854

HANDBOOK OF

Chemical Technology and Pollution Control

HANDBOOK OF
Chemical Technology and Pollution Control

Martin B. Hocking
Department of Chemistry
University of Victoria
Victoria, British Columbia
Canada

ACADEMIC PRESS

San Diego   London   Boston   New York   Sydney   Tokyo   Toronto

Cover photographs: © 1993 by Eric Wunrow.

This book is printed on acid-free paper. ∞

Copyright © 1998 by ACADEMIC PRESS

All Rights Reserved.
No part of this publication may be reproduced or transmitted in any form or by any
means, electronic or mechanical, including photocopy, recording, or any information
storage and retrieval system, without permission in writing from the publisher.

Academic Press
a division of Harcourt Brace & Company
525 B Street, Suite 1900, San Diego, California 92101-4495, USA
http://www.apnet.com

Academic Press Limited
24-28 Oval Road, London NW1 7DX, UK
http://www.hbuk.co.uk/ap/

Library of Congress Card Catalog Number: 97-80796

International Standard Book Number: 0-12-350810-X

PRINTED IN THE UNITED STATES OF AMERICA
98 99 00 01 02 03 QW 9 8 7 6 5 4 3 2 1

■ CONTENTS

2 Air Quality Measurement and Effects of Pollution

3 Air Pollution Control Priorities and Methods

4 Water Quality Measurement

5 Raw Water Processing and Wastewater Treatment

6 Natural and Derived Sodium and Potassium Salts

7 Industrial Bases by Chemical Routes

8 Electrolytic Sodium Hydroxide, Chlorine, and Related Commodities

9 Sulfur and Sulfuric Acid

10 Phosphorus and Phosphoric Acid

11 Ammonia, Nitric Acid, and Their Derivatives

12 Aluminum and Compounds

13 Ore Enrichment and Smelting of Copper

14 Production of Iron and Steel

15 Production of Pulp and Paper

16 Fermentation and Other Microbiological Processes

17 Petroleum Production and Transport

18 Petroleum Refining

19 Petrochemicals

20 Condensation (Step-Growth) Polymer Theory

21 Commercial Polycondensation (Step-Growth) Polymers

22 Addition (Chain Reaction) Polymer Theory

23 Commercial Addition (Vinyl-Type) Polymers

*For Common Process Flowsheet Symbols and the Periodic Table of the Elements, see the inside
back cover and endpapers.*

PREFACE TO THE FIRST EDITION

This text of applied chemistry considers the interface between chemistry and chemical engineering, using examples of some of the important process industries. Integrated with this is detailed consideration of measures which may be taken for avoidance or control of potential emissions. This new emphasis in applied chemistry has been developed through eight years of experience gained from working in industry in the research, development, and environmental control fields, plus twelve years of teaching here using this approach. It is aimed primarily toward science and engineering students as well as toward environmentalists and practicing professionals with responsibilities or an interest in this interface.

By providing the appropriate process information back to back with emissions and control data, the potential for process fine-tuning is improved for both raw material efficiency and emission control objectives. This approach also emphasizes integral process changes rather than add-on units for emission control. Add-on units have their place, when rapid action on an urgent emission problem is required or when control simply is not feasible by process integral changes alone. Obviously, fundamental process changes for emission containment are best conceived at the design stage. However, at whatever stage process modifications are installed, this approach to control should appeal to the industrialist in particular, in that something more substantial than decreased emissions may be gained.

This book may also be used as a general source of information and further leads to the details of process chemistry or as a source of information relating to air and water pollution chemistry. Many references are cited to provide easy access to additional background material. The dominant sources cited may generally be recognized by the number of direct citations given in the chapter. Article titles are given with the citation for any anonymous material to aid in retrieval and consultation. Sources of further information on the subject of each chapter, but generally not cited in the text, are also given in a short Further Reading list immediately following the text. Trade names have been recognized by capitalization, when known, and sufficient detail is mentioned or referenced to each of these to enable them to be followed up, if desired. It would be appreciated if any unrecognized trade name usage is brought to the author's attention.

Martin B. Hocking
From *Modern Chemical Technology*
and Emission Control

■ PREFACE TO THE SECOND EDITION

The objectives that motivated the first edition, a unified treatment of the fields of industrial and environmental chemistry, have been maintained here. The result is intended to be of interest to senior students in applied chemistry, science, engineering, and environmental programs in universities and colleges, as well as to professionals and consultants employed in these fields.

This edition further develops, refines, and updates the earlier material by drawing on progress in these fields and by responding to comments from users of the first edition. Sections relating to air and water pollution assessment and theory have been expanded, chapters on petrochemical production and basic polymer theory and practice have been added, and the original material has been supplemented by new data. In addition, review questions have now been added to each chapter. These will be of interest primarily to students but could be of conceptual value to all users.

The new edition has been assembled to make it easy to use on any or all of three levels. The basic principles and theory of each process are discussed initially, followed by more recent refinements and developments of each process, and finally supplemented with material that relates to possible process losses and integral and end-of-pipe emission control measures. The user's interest can dictate the level of approach to the material in the book, from a survey of a selection of basic processes to an in-depth referral to one or more particular processes, as appropriate. Chemical reactions and quantitative assessment are emphasized throughout, using worked examples to aid understanding.

Extensive current and retrospective production and consumption data have been maintained and expanded from the first edition to give an idea of the scale and volume trends of particular processes and an indication of regional similarities and differences. This material also provides a basis for consideration of technological changes as these relate to changes in chemical processes. Specific mention should be made of the difficulties in providing recent information for Germany and the region encompassed by the former USSR because of political changes during this period.

The author would appreciate receiving any suggestions for improvement.

Martin B. Hocking

■ ACKNOWLEDGMENTS

ACKNOWLEDGMENTS TO THE FIRST EDITION

I am grateful to numerous contacts in industry and environmental laboratories who have willingly contributed and exchanged technical information included in this book. I would particularly like to thank the following people who have materially assisted in this way: B. R. Buchanan, Dow Chemical Inc.; W. Cary, Suncor; R. G. M. Cosgrove, Imperial Oil Enterprises Ltd.; F. G. Colladay, Morton Salt Co.; J. F. C. Dixon, Canadian Industries Ltd.; R. W. Ford, Dow Chemical Inc.; T. Gibson, B. C. Cement Co.; G. J. Gurnon, Alcan Smelters and Chemicals Ltd.; D. Hill, B. C. Forest Products; J. A. McCoubrey, Lambton Industrial Society; R. D. McInerney, Canadian Industries Ltd.; R. C. Merrett, Canoxy, Canadian Occidental Petroleum; S. E. Moschopedis, Alberta Research Council; J. C. Mueller, B. C. Research; J. A. Paquette, Kalium Chemicals; J. N. Pitts, Jr., Air Pollution Research Center, University of California; J. R. Prough, Kamyr Inc.; J. G. Sanderson, MacMillan-Bloedel Ltd.; A. D. Shendrikar, The Oil Shale Corp.; J. G. Speight, Exxon; A. Stelzig, Environmental Protection Service; H. E. Worster, MacMillan-Bloedel Ltd. They have been credited wherever possible through their own recent publications. These contacts are especially valuable because of the notorious slowness of new industrial practice to appear in print.

I also thank all of the following individuals, each of whom read sections of the text in manuscript form, and C. G. Carlson, who read all of it, for their valuable comments and suggestions that have contributed significantly to the authenticity of this presentation:

R. D. Barer, Metallurgical Division, Defence Research Establishment Pacific

G. Bonser, Husky Oil Limited

R. A. Brown, formerly of Shell Canada

M. J. R. Clark, Environmental Chemistry, Waste Management Branch, B. C. Government

H. Dotti, Mission Hill Vineyards

M. Kotthuri, Meteorology Section, Waste Management Branch, B. C. Government

J. Leja, Department of Mining and Mineral Process Engineering, University of British Columbia

L. J. Macaulay, Labatt Breweries of B. C. Ltd.

D. J. Maclaurin, formerly of MacMillan-Bloedel Ltd.

R. N. O'Brien, Department of Chemistry, University of Victoria

M. E. D. Raymont, Sulphur Development Institute of Canada

W. G. Wallace, Alcan Smelters and Chemicals Ltd.

R. F. Wilson, Dow Chemical Canada Inc.

M. D. Winning, Shell Canada Resources Ltd.

But without the support of the University of Victoria, the Department of Chemistry, and my family to work within, this book would never have been completed. I owe a debt of gratitude to the inexhaustible patience of my wife, who handled the whole of the initial inputting of the manuscript into the computer, corrected several drafts, and executed all of the original line drawings. Thanks also go to K. Hartman, who did the photographic work; to B. J. Hiscock and L. J. Proctor, who were unfailingly encouraging and helpful in use of the computer for manuscript preparation, even when the occasional seemingly hopeless scrambles occurred; and to L. G. Charron and M. Cormack, who completed the final manuscript.

Some of the line drawings and one photograph are borrowed courtesy of other publishers and authors, as acknowledged with each of these illustrations. To all of these I extend my thanks.

It would be tempting, since computer composition was extensively used, to blame any final errors on computer programming glitches. This may, occasionally, have been the case. Nevertheless, I must accept responsibility for them all. It would be appreciated if they were brought to my attention.

ACKNOWLEDGMENTS TO THE SECOND EDITION

Students using the first edition are thanked for providing useful feedback to improve the presentation in a general way and for testing the concepts of most of the problems. My former and present students in polymer chemistry have read and reviewed in detail the material of the new Chapters 19–23. They, several more casual readers, and Diana Hocking, who read the whole manuscript, are thanked for their contributions. Elizabeth Small, Carol Jenkins, Susanne Reiser, and Diana Hocking completed most of the typing, Devon Greenway provided bibliographic help, and Chad Beddie assisted with the proofreading. Ken Josephson and Ole Heggen executed the new graphics for preparation of the final manuscript.

Martin Hocking

1
BACKGROUND AND
TECHNICAL ASPECTS

*Take calculated risks. That is quite different from
being rash.*
                                        —*George S. Patton*

1.1. IMPORTANT GENERAL CHARACTERISTICS

The business niche occupied by the chemical industry is of primary importance
to the developed world in its ability to provide components of the food, clothing, transportation, accommodation, and employment enjoyed by modern hu-
manity. Most material goods are either chemical in origin or have involved
one or more chemicals during the course of their manufacture. In some cases
the chemical interactions involved in the generation of final products are relatively simple ones. In others, for instance, for the fabrication of some of the
more complex petrochemicals and drugs, more complicated and lengthy procedures are involved. But by far the bulk of all modern chemical processing
uses raw materials naturally occurring on or near the earth's crust, to produce
the commodities of interest.

Consider the sources of some of the common chemical raw materials and
relate these to the kinds of products that are accessible via one or two simple
chemical transformations in a typical chemical complex. Starting with just a
few simple components—air, water, salt (NaCl), and ethane—together with
an external source of energy, quite a range of finished products is possible
(Fig. 1.1). While it is unlikely that all of these will be produced at any one
location many will be, and all are based on commercially feasible and practiced processes [1]. Thus, a company that is basic in the electrolytic production
of chlorine and sodium hydroxide from salt can conveniently site itself on or
near natural salt beds, which can provide a secure source of this raw material.
This operation should preferably be located near a large source of freshwater,

1

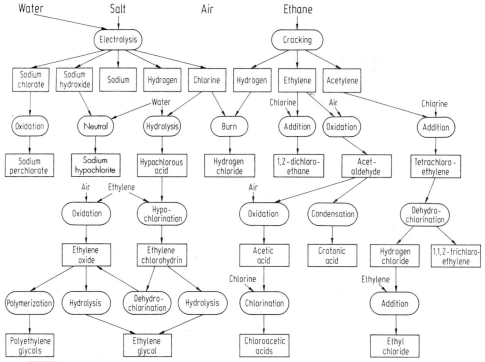

FIGURE 1.1 Flowchart of a hypothetical though credible chemical complex based on only air, water, salt, and ethane raw materials. Ellipses represent processes, rectangles indicate products.

such as a river or a lake, to provide for feedstock and cooling water requirements. Quite often an oil refinery forms a part of a matrix of companies which find it mutually advantageous to locate together. This can provide a supply of ethane, benzene, or other hydrocarbon feedstocks. In this manner all the simple raw material requirements of the complex can flow smoothly into the production of more than a dozen products for sale (Fig. 1.1).

A rapid rise in the numbers of chemicals produced commercially, and a steady growth in the uses and consumption of these chemicals historically (since the 1930–1940 period), has given the chemical industry a high growth rate relative to other industrial activities. In current dollars the average annual growth rate in the U.S.A. was about 11% per year in the 1940s and just over 14% per year through the 1970s, seldom dropping below 6% in the intervening period. Plastics and basic organic chemicals have generally been the stronger performing sectors of the chemical industry as far as growth rate is concerned. Basic inorganic chemicals production, a generally "mature" area of the industry, has shown slower growth. World chemical export growth has been strong too, having averaged just over a 17% annual growth rate during the 1968–1978 interval. But growth rates based on current dollar values, such as these are, fail to recognize the salutary influence of inflation. Using a constant value dollar, and smoothing the values over a 10-year running average basis gives the maximum for the real growth rate of about 9% per year occurring in 1959, tapering down to about 1–3% per year to 1990. The slowing of the real growth rate in recent years may be because the chemical industry generally is maturing. There may also have been a contribution over the more recent short term from the global business recession of that period.

Most of the machinery and containment vessels required for chemical processing are costly, in part because of the high degree of automation used by this industry. This means that the labor requirement is relatively low, based on the value of products. Put in another way, in the United States the investment in chemical plant per employee has amounted to about $30,000 per worker at the time when the average for all manufacturing stood at $14,000 per worker. In the U.K. this ratio of capital investment per employee in the chemical industry versus the investment by all manufacturing is very similar to the experience in the U.S.A. In 1963 these figures stood at 7000 and 3000 pounds, and in 1972, 17,000 and 7000 pounds, respectively [2].

Yet another way of considering the relationship of investment to the number of employees is in terms of the "value added per employee." The value added, which is defined as the market price of a good minus the cost of raw materials required to produce that good [3], is a measure of the worth of processing a chemical in terms of its new (usually greater) value after processing than before. When the gross increase in value of the products of a chemical complex is divided by the numbers of employees operating the complex, one arrives at a "value added per employee", one kind of productivity index. Considered on this basis the productivity of a worker in the chemical industry is at the high end of the range in comparison with the productivity of all workers within any particular country. There are, however, quite significant differences in relative productivity when the value added per employee of one country is compared with the same figure from other countries. In 1978, the value added for the U.S.A. stood at $58,820 per employee per year, as compared to a value of $17,800 for Spain, the extremes of the range among the countries compared in Table 1.1. This comparison also reflects the much higher investment per employee and the higher degree of automation generally used by American chemical companies versus their Spanish counterparts. The range of values given here is, however, also dependent on a number of other factors, such as scale and capacity usage rates, which have not yet been discussed. Relative positions may also change in a 10-year span, as is shown by Canada and West Germany, from differing investment rates and other factors.

The products of the chemical industry, because of the more or less steady stream of better performing replacement products, tend to have a high rate of obsolescence. During the 1950s and 1960s most of the sales volume increase noted by chemical companies for the period was from products developed in the preceding 15 years. To provide the steady stream of improved products required to maintain this record and still grow requires a significant commitment to research in order for a company to keep up with its competition. This requirement also provides much of the incentive for a chemical company to employ and provide facilities for the employment of chemists, engineers, biologists, and other professionals to ensure the continuing discovery and development of new products.

TABLE 1.1   Numbers Employed in Chemical Production, and Value Added per Employee for Selected Countries[a]

Country	Thousands employed, 1978	Value added per employee (US$)		Decade productivity growth factor
		1968	1978	
Austria	61	—	19,670	—
Belgium	62.6	8,410	37,100	4.4
Canada	84.7	18,020	39,130	2.2
France	305	9,540	32,540	3.4
Germany (West)	548	11,350	46,220	4.1
Italy	292	7,890	20,000	2.5
Japan	470	9,460	33,600	3.6
Netherlands	87.1	9,810	44,100	4.5
Norway	17.4	8,290	23,800	2.9
Spain	144.1	5,780	17,800	3.1
Sweden	39.7	12,180	36,500	3.0
United Kingdom	467	8,110	19,800	2.4
United States	1088	24,760	58,820	2.4

[a]Data from Cairns [4] and OECD [5].

From 2.5 to 3.5% of the value of sales of U.S. chemical companies is spent on research and development activity, about the same proportion of sales as spent by all industry. German companies tend to place a somewhat greater emphasis on research and development, and show an expenditure of 4 to 5% of sales in this activity. Drug (pharmaceutical) companies represent the portion of the chemical sector which spends the largest fraction of sales, about 6%, on research and development programs [6]. This is probably a reflection of both the generally higher rate of obsolescence of the products of this sector as compared to the chemical industry as a whole, as well as of the greater costs involved in bringing new, human use drugs to market, as opposed to new commodity chemicals.

1.2. TYPES AND SIGNIFICANCE OF INFORMATION

With the moderately high growth rate of the chemical industry and its high rate of obsolescence of both the products themselves and of the processes leading to an existing product, the competition in this industrial area is vigorous. Technological and market success of a company in this industry area is a composite of the financial resources, raw material position, capabilities and motivation of staff, and the information resources that the company has at its disposal. The last of these factors, the information resource, is a particularly important one for the chemical processing industry. Information, or "know-how," may be derived from prior experience of all kinds. It may be generated from self-funded and practiced research or process development. Or it may be purchased from appropriate other companies if it is already available in this form. Thus, sale of the results of research by a company, even if not used by that company to produce a product, may still produce an income for it in the form of licensing agreements, royalties per unit of product sold, and the like. In many ways this is a highly desirable component of a company's earnings since it does not require any capital investment or raw material and product inventories as is ordinarily required to generate an income from chemical processing.

Patents and the patenting system represent the orderly system of public documents used in most parts of the world to handle much of this kind of information. Patent protection is of interest to chemical as well as other companies. Patents must be applied for in each country in which protection is desired, since the subject of a patent may be practiced and the product sold without license in any country in which this precaution is not taken. "Composition of matter" patents, which relate primarily to newly discovered chemical compounds, are issued on successful application by an inventor (individual or company). Utility, i.e., some type of useful function of the compound must be demonstrated before a patent application of this type can be filed. In return this class of patent provides the best kind of protection for a new compound because the compound itself is protected from its sale by others, regardless of the synthetic route developed to produce it.

"Process" patents are used to protect a new process or refinements to an established process which is employed to produce an existing compound. This type of patent also provides useful protection against the commercial use by others of an improved, completely distinct process which may be developed by a company. The benefit provided by all types of process development may be lower costs achieved from higher conversion rates, or from better selectivity, more moderate operating conditions, and the like. In these ways it provides the company with an economic advantage to practice this improvement.

Other patent areas are used by chemical and other companies, such as those covering machines and registered designs, trademarks and symbols, and copyright, but these are generally less fundamental to the operations of chemical companies than the composition of matter and process patent areas [7]. Trademarks and symbols are generally of more importance for sales, since product recognition is a significant marketing factor. These patent office functions also have no expiry date, as long as the required annual maintenance fee is paid.

The patent, in printed form, comprises a brief description of the prior art (the narrow segment of technology) in areas related to the subject of the patent. Usually this is followed by a brief summary of what is being patented. A more detailed description of what is involved in the invention is then given, accompanied by descriptions of some detailed examples which illustrate the application of the invention. Usually at least one of the examples described is a description of an experiment which was actually carried out, but they need not all have been actually tested. Differentiation between actually tested examples and hypothetical examples described in the body of the patent is made on the basis of the tense used in the description. If it is described in the past tense, i.e., "was" is used throughout, then it is a description of a tested example. If it is given in the present tense then it describes a hypothetical example. To be able to differentiate the two types of examples is of particular interest to synthetic chemists, for example, who are likely to be more successful if they follow a procedure of a tested rather than a hypothetical example. The last, and most important part of a patent is the claims section. Here, numbered paragraphs, each of which by custom is written all in one sentence, cover the one or more novel areas to be protected by the patent in order of importance. In the case of any contest of the patent by other parties, these claims must be disproved in reverse order, i.e., the last and least important claim first followed by the others if the last claim is successfully contested.

The granting of a patent confers on the holder a time-limited monopoly in the country of issue, for a period of about 17 years, to cover the novel composition of matter or advance in the art that is claimed by the patent. During this time the company or individual may construct a plant using the patented principles, which may take 6 or 7 years. Once production has started a product can be marketed from this plant at a sufficiently high price that the research and development costs involved prior to patenting, as well as reasonable plant write-off expenses and the like, may be met. This stage of marketing can proceed without competition from others for the 10–11 years remaining from the original patent interval. Or a company may choose to license the technology and collect product royalties from another interested company. Or it may follow both options simultaneously, if it reasons that the market will be large enough to sustain both. For these reasons the patent system encourages a company to carry out its own research since it provides a reasonable prospect of the company being able to recover its early development costs while it is using the new art, protected from competition.

Seventeen years (20 years in European countries) from the date of issue of a patent, however, the subject matter of the patent comes into the public domain, that is, it becomes open to any other person or company who wishes to practice the art described in the patent and *sell* a product based on this technology. At this time, the price of the product will normally fall somewhat as the product starts to be produced competitively by others. But the originating company still has some production and marketing advantages from its longer experience in using the technology, from having one or more producing units which may be largely paid for by this time, and from having already developed some customer confidence and loyalty.

The new regulatory requirements which must be met before marketing new drugs and pesticides are now taking extremely long to satisfy, up to 7 years in some instances. This has increased the new product development costs, at the same time as decreasing the period of time available for monopoly marketing to allow recovery of development costs. Realization of this has led to moves in the U.K. and in the U.S.A. to extend the period of monopoly protection granted by the patent by the length of time required by a company to obtain regulatory clearance. These moves should at least encourage maintenance of the current level of research and development effort by companies even if it does not increase innovation.

In all cases patent protection for an idea is for a limited time, but even during the protected time the information in the patent becomes public knowledge. There may be some technological developments which a company wishes to keep completely to itself, or which are so close to prior art (already practiced) that there is some doubt of the likelihood of success of a patent application. Information falling into these categories may be simply filed in company records for reference purposes and not be patented or otherwise publicized at all. This type of know-how is termed "proprietary information," useful to the company but kept within the company only. Agreements signed by all new employees working for a company ensure that this proprietary information does not become public knowledge. In return for risking possible eventual leakage and use of this information by others, the company gains the advantageous use of the information in the meantime, it saves patenting costs (even if feasible), and it avoids the certainty of public disclosure on issuance of a patent covering this information. But the ideas involved are not directly protected from use by others, whether or not the knowledge is lost via "leaks" or via independent discovery by a second company working on the same common knowledge premises as the first company, hence the value of the patent system in providing this assurance of protection.

A second approach to decrease the impact of public disclosure when a patent is filed is to apply for many patents on closely related technologies simultaneously. Some will relate to the core technology for which protection is desired. The others serve as distractors to those who would wish to discover and explore the new technology competitively.

1.3. THE VALUE OF INTEGRATION

Integration, as means of consolidation by which a company may improve its competitive position, can take a number of forms. Vertical integration can be "forward" to carry an existing product of the company one or more stages closer to the final consumer. For instance, a company producing polyethylene resin may decide also to produce film from this resin, for sale, or it might decide to produce both film, and garbage bags from the film. By doing this, more "value-added" manufacturing stages are undertaken within the company. If these developments are compatible with the existing activities and markets of the company they can significantly enhance the profitability of its operations.

Vertical integration may also be "backward" in the sense that the company endeavors to improve its raw material position by new resource discoveries and acquisition, or by purchase of resource-based companies strong in the particular raw materials of interest. Thus, it can explore for oil, or purchase an oil refinery to put itself into a secure position for ethane and ethylene. Or it can purchase land overlying beds of sodium chloride or potassium chloride, or near sodium sulfate rich waters and develop these to use for the preparation of existing product lines. Either of these routes of backward integration can help to secure for a company an assured source of supply and stable raw material pricing, both helpful in strengthening the reliability of longer term profit projections.

Horizontal integration is a further type, in which the technological or information base of the company is applied to improve its competitive position in this and related areas. When a particular area of expertise has been discovered and developed this can be more fully exploited if a number of different product or service lines are put on the market using this technology. For instance, Procter & Gamble and Unilever have both capitalized on surfactant technology in their development of a range of washing and cleaning products. Surface active agents of different types have also been exploited by the Dow Chemical Company with its wide range of ion exchange resins, and cage structures by Union Carbide with its molecular sieve-based technology. It can be seen from these examples that judicious application of one or more of these forms of integration can significantly strengthen the market position and profitability of a chemicals based company.

1.4. THE ECONOMY OF SCALE

The size or scale of operation of a chemical processing unit is an important competitive factor since, as a general rule, a large-scale plant operating at full capacity can produce a lower unit cost product. This is the so-called "economy of scale" factor. How does this lower cost product from a larger plant arise? First, the labor cost per unit of product is lower for a very large than for a small plant. This comes about because proportionally fewer staff are required per unit of product to run a 1000 tonne/day plant than a 100 tonne/day plant. Secondly, the capital cost of the plant per unit of product is lower, if the plant is operating at full capacity.

Reduced labor costs result from the fact that if one person is required to control, for example, raw material flows into a reactor in a 100 tonne per day plant, in all likelihood they can still control these flows in a 1000 tonne/day plant. In fact an empirical expression has been derived by correlation of more than 50 types of chemical operations which, knowing the labor requirement for one size of plant, allows one to estimate with reasonable assurance the labor requirement for another capacity [8] (Eq. 1.1).

$$M = M_0(Q/Q_0)^n, \text{ where} \qquad\qquad 1.1$$

M is the labor requirement for plant capacity Q of interest,
M_0 is the known labor requirement for a plant capacity Q_0, and
n is the exponent factor, normally about 0.25, for the estimation of labor requirements.

If 16 staff are required to operate a 200 tonne/day sulfuric acid plant, this expression allows us to determine that only about 24 staff [$16 \times (1000/200)^{0.25}$] should be needed to operate a 1000 tonne/day plant. Thus, when operating at full capacity, the larger plant would only have three-tenths the labor charge of the smaller plant, per unit of product.

The lowered plant capital cost per unit of product comes about because of the relationship of capital costs of construction to plant capacity, which is an exponential, not a linear relation (Eq. 1.2).

$$\text{capital cost} \propto (\text{plant capacity})^{2/3} \qquad\qquad 1.2$$

The approximate size of the fractional exponent of this expression results from the fact that the cost to build a plant varies directly as the area (or weight) of metal used, resulting in a square exponent term [9]. At the same time the capacity of the various components of the processing units built increases in relation to the volume enclosed, or a cube root term. Hence, the logic of this approximate relationship.

In actual fact, a skilful design engineer is generally able to shave just a bit off this descriptively derived exponent, making capital cost relate to scale more closely in accord with Eq. 1.3 for whole chemical plants.

$$\text{capital cost} \propto (\text{plant capacity})^{0.60} \qquad\qquad 1.3$$

In order to use Eq. 1.3 to estimate the capital cost of a larger or smaller plant, when one knows the capital cost of a particular size of plant, one has to insert a proportionality constant (Eq. 1.4).

$$C = C_0(Q/Q_0)^n, \text{ where} \qquad\qquad 1.4$$

C is the capital cost for the production capacity Q of the plant to be determined,

C_0 is the known capital cost for production capacity Q_0, given in the same units as C, and

n is the scale exponent, which is usually in the 0.60 to 0.70 range for whole chemical plants.

Thus, if it is known that the capital cost for a 200 tonne/day sulfuric acid plant is $1.2 million ($1.2 mm) then using this relationship it is possible to estimate that the capital cost of an 1800 tonne/day plant will be somewhere in the range of $4.49 mm to $5.59 mm (Eq. 1.5).

$$
\begin{aligned}
C &= \$1.2 \text{ mm } (1800/200)^{0.60} & C &= \$1.2 \text{ mm } (1800/200)^{0.70} \\
&= \$1.2 \text{ mm } (3.7372) & &= \$1.2 \text{ mm } (4.6555) \qquad 1.5 \\
&= \$4.485 \text{ mm} & &= \$5.587 \text{ mm}
\end{aligned}
$$

From construction cost figures the actual capital cost of construction of an 1800 tonne/day sulfuric plant is about $5.4 mm, when taken at the time of these estimates. This figure agrees quite well with the two values estimated from the known cost of the smaller sized plant.

Of course, if one has recent capital cost information on two different sizes of plant for producing the same product, this can enable a closer capital cost estimate to be made by determination of the value of exponent n from the slope of the capital cost versus production volume line plotted on log–log axes for the two sizes of plant. For the particular example given, this experimentally determined exponent value would be 0.685. Note also that this capital cost estimation method is less reliable for plant sizes more than an order of magnitude larger or smaller than the plant size for which current costs are available [10].

From a comparison of the foregoing capital cost figures, it can be seen that nine times as much sulfuric acid can be made for a capital cost of only 3.7 to 4.7 times as much as that of a 200 tonne/day plant. Obviously if the large plant is operated at full capacity, the charge (or interest) on the capital which has to be carried by the product for sale by the larger plant is only about half (4.7/9.0) or even less than half (3.7/9.0) of the capital cost required to be borne by the 200 tonne/day plant, per unit of product.

To make the decision regarding the size of plant to build in any particular situation, careful consideration has to be given to product pricing, market size and elasticity, and market growth trends. Also it is a useful precaution to survey the immediate geographical area and public construction announcements for any other plans for a plant to produce the same product. The final decision should be based on a scale of operation which, within a period of 5 to 7 years, could reasonably be expected to be running at full capacity. That is, it should be possible to stimulate a sufficient market, within this period of time, to sell all of the product that the plant can produce. If the final size of plant built is too small, not only are sales restricted from inadequate production capacity but also the profit margin per unit of product is smaller than it potentially could have been if the product were being produced in a somewhat larger plant. If the final result is too large, and even after 10 years or so the plant is required to operate at only 30% of capacity to provide for the whole market, then the capital and frequently also labor costs per unit of product become higher than they would have been with say one-half or even one-quarter of the plant size. In this event, planning too optimistically can actually *decrease* the profitability of the operation. It is the significance of decisions such as these as to the financial health of a chemical company that justify the handsome salaries of its senior executives.

One remaining point to consider regarding scale is that the capital cost exponential factor of 0.60 to 0.70 relates to most whole plants. If considering individual processing units this factor can vary quite widely (Table 1.2). With a jaw crusher, for example, a unit with three times the capacity costs 3.7 times as much. Obviously here scaling up imposes greater capital costs per unit of product for a larger than for a smaller unit. But other associated costs may still be reduced. A steel vent stack of three times the height costs about three times as much, i.e., there is no capital cost economy of scale here, and these

TABLE 1.2   Typical Values for the Exponent Scale Factor and How These Relate to the Cost of the Unit for Particular Types of Chemical Processing Equipment[a]

Type of equipment	Typical value of exponent n	Cost factor for three times scale
Jaw crusher	1.2	3.74
Fractionating column, bubble cap	1.2	3.74
Steel stack	1.0	3.00
Fractionating column, sieve tray	0.86	2.57
Forced circulation evaporator	0.70	2.16
Shell and tube heat exchanger	0.65	2.04
Jacketed vessel evaporator	0.60	1.93
Stainless steel pressure reactor, 300 psi	0.56	1.85
Industrial boiler	0.50	1.73
Drum dryer, atmospheric pressure	0.40	1.55
Storage tank	0.30	1.39

[a]Exponent values for use with Eq. 1.4, $C = C_0(Q/Q_0)^n$, and selected from those of Peters and Timmerhaus [10] and Allen [11].

capital cost increases with height may still have to be borne by the plant. However, for very simple components of processing units such as storage tanks, the value of this exponent is small, about 0.30, which allows a tank of three times the capacity to be built for only about 1.4 times the price. Thus, a composite of the scale-up exponent factors for individual units averages out to the 0.60 to 0.70 range for a whole chemical plant.

1.5. CHEMICAL PROCESSING

The chemical side of the chemical process industries is concerned with the change of raw materials into products by means of *chemical conversions*. Since it is very seldom that a single reacted starting material gives only pure product it is also usually necessary to use physical separations such as crystallization, filtration, distillation, or phase separation to recover product(s) from the unreacted starting materials and by-products. By-products are materials other than product which are obtained from reacted starting materials. These physical separation processes are often called *unit operations* to distinguish these definable steps. Similar features of these may be compared from process to process, as distinct from the chemical conversion aspect of a process [12]. The combination of the chemical conversion step, with all of the unit operations (physical separations) that are required to recover the product of the chemical conversion, is collectively referred to as a *unit process*.

Unit processes may be carried out in single-use (dedicated) equipment, which is used solely for generating the particular product for which it was designed. Or they may be carried out in multiuse equipment that is used in sequence to produce first one product, followed by the production of one or more related products that have similar unit process requirements, in a series after this. Single-use equipment is invariably used for large-scale production, when 90% or so of full-time usage rates are required to obtain sufficient product to satisfy the market requirements. Multiuse equipment more often is chosen for small-scale production, and particularly for more complicated processes such as may be required for the manufacture of many drugs, dyes, and some specialty chemicals.

Proper materials of construction particularly with regard to strength and toughness, corrosion resistance, and cost must all be kept in mind at the design stage for construction of a new chemical plant. Early experiments during the conception of the process will have generally been conducted in laboratory glassware. Even though glass is almost universally corrosion resistant (and transparent, and thus useful in the lab) it is unduly fragile for most full-scale process use. Mild steel is used wherever possible, because of its lowest cost and ease of fabrication. But it is not, however, resistant to attack by many kinds of process fluids or gases. In these cases any of titanium, nickel, stainless steel, brass, Teflon, polyvinylchloride (PVC), wood, cement, and even glass (usually as a lining) among other materials may be used to construct components of a chemical plant. The final choice of construction material is based on a combination of experience and accelerated laboratory tests. Small coupons of the short-listed candidate materials are suspended in synthetic mixtures prepared to mimic those to be found in the process. These are then heated to simulate anticipated plant conditions. Preliminary tests will be followed by further tests during small-scale process test runs in a pilot plant, wherever possible. Even when the final full-scale plant is completed, there may still be recurring corrosion failures of a particular component which may require construction material changes even at this stage of development.

1.5.1. Types of Reactors

Industrial reactor types again can use the analogy between laboratory manipulations and a full-scale production plant. Very often in the laboratory a synthesis will be carried out by placing all the required reactants in the flask and then imposing the right conditions, heating, cooling, light, etc., on the contents to achieve the desired extent of reaction. At this stage the contents of the flask are emptied into another vessel for the product recovery steps to be carried out. Operating an industrial process in this fashion, which can be done, is termed a *batch process* or *batch operation*. Essentially this situation is obtained when all starting materials are placed in the reactor at the beginning of the reaction, and remain in the vessel until the reaction is over, when the contents are removed. This mode of operation is the one generally favored for smaller scale processes, for multiple use equipment, or for new and untried, or some types of more hazardous reactions.

On the other hand an industrial process may be operated in a *continuous mode,* rather than in a batch mode. To achieve this, either a single or a series of interconnected vessels may be used. The required raw materials are continuously fed into this vessel or vessels and the reaction products continuously removed so that the volume of material in the reactor stays constant as the reaction proceeds. The concentrations of starting materials and products in the reactor eventually reach a steady state. One or more tanks in series may be used to conduct the continuous process. Or a pipe or tube reactor may be used, in which case the reaction time is determined by the rate of flow of materials into the tube divided by the length of the pipe or tube.

Since, in general, the labor costs of operating a large-scale continuous process are lower than for a batch process, most large-scale industrial processes are eventually worked in a continuous mode [13]. However, because of the more complicated control equipment required for continuous operation, the capital cost of the plant is usually higher than for the same scale batch process. Thus, the final choice of the mode of operation to be used for a process will often depend on the relative cost of capital versus labor in the operating area in which the plant is to be constructed. Most developed countries opt for a high degree of automation and higher capital costs in new plant construction decisions. For Third World nations, however, where capital is generally scarce and labor is low cost and readily available, more manual and simpler batch-type operations will often be the most appropriate. A smaller scale of operation will generally be more than adequate to supply the market requirements in these economies. Also the more straightforward operating, maintenance, and repair operations for such a plant are more easily accomplished under these circumstances than would be possible with the more complex control systems required for a continuous reactor.

There are several common combinations within this broad division into batch and continuous types of reactor which use minor variations of the main theme. The simplest and least expensive of these subdivisions is represented by the straight batch reactor, which is frequently just a single stirred tank. All the raw materials are placed in the tank at the start of the process. There is no flow of materials into or out of the tank during the course of the reaction, i.e., the volume of the tank contents is fixed during the reaction (Table 1.3). There also usually will be some provision for heating or cooling of the reacting mixture, either via a metal jacket around the outside of the reactor or via coils placed inside the reactor, through which water, steam, or heat exchange fluid may be passed for temperature control. However, the temperature is not usually uniform in this situation since initially the concentrations and reaction rate of the two (or more) reactants are at a maximum, which will more or less rapidly taper off as the reaction proceeds. Thus, heat evolution (or uptake)

TABLE 1.3   A Qualitative Comparison of Some of the Main Configurations of Batch and Continuous Types of Liquid Phase Reactors

| Type of reactor | Illustration of concept | Composition with time | Uniformity of | |
			Composition within reactor[a]	Temperature throughout process[b]
a. Batch		no	yes	no
b. Semi-batch	feeds	no	yes	yes
c. Continuous stirred tank (CSTR) sometimes "tank flow reactor"	feeds / product	yes	yes	yes
d. Multistage CSTR	feeds / product	yes	partly	partly
e. Tubular flow, sometimes "pipe reactor" or "plug flow reactor"	feeds / product	yes	no	no

[a]Meaning the composition within the reactor at any particular point in time.
[b]Referring to temperature constancy during the whole of the reaction phase.

is going to be high initially and then gradually subside to coincide with a slowing of the reaction rate. At the end of the reaction the whole of the reactor contents is pumped out for product recovery.

A semi-batch reactor is a type of batch configuration used particularly for processes which employ very reactive starting materials. Only one reactant, plus solvent if required, is present in the reactor at the start of the reaction. The other reactant(s) is then added gradually to the first, while continuing stirring and controlling the temperature. By controlling the rate of addition of one reactant in this way, the temperature of the reacting mixture may be kept uniform as the reaction proceeds.

Continuous reactor configurations are generally favored for very large-scale industrial processes. If the process is required to produce only two million (2 mm) kg/year or less, generally the economics of construction will dictate that a batch process be used [14]. If, however, the process is called upon to produce 9 mm kg/year or more, there is usually a strong incentive to apply some type of continuous reactor configuration in the design of the production unit.

The stirred tank is used as the main element of the simplest type of continuous reactor, the continuous stirred tank reactor (CSTR). Continuity of the process is maintained by continuous metering in of the starting materials in the correct proportions, and continuous withdrawal of the stream containing the product from the same, well-stirred vessel. In this way the concentration and temperature gradients shown by simple batch reactors are entirely eliminated (Table 1.3). This type of continuous reactor is good for slow reactions, in particular, since it is a large, simple, and cheap unit to construct. However, reactors of the CSTR type are inefficient at large conversions [14]. For a process proceeding via first-order kinetics and requiring a 99% conversion a seven times larger reactor volume is needed than if only 50% conversion is desired.

A solution to the large volume requirement for high conversions in a single reactor is to use two or more CSTRs in series [14]. Use of two CSTRs in series allows the first reactor to operate at some intermediate degree of conversion, the product of which is then used as the feed to the second reactor to obtain the final extent of reaction desired (Fig. 1.2). From the diagrams it can be readily seen that the total reactor volume required to achieve the desired final degree of conversion is significantly reduced over the volume required to achieve the same degree of conversion in a single reactor. Carrying this idea further, it can also be seen that increasing the number of CSTRs operating in series to three or more units contributes further to the space–time yields and allows further reductions in reactor volume to be made to still obtain the same final degree of conversion. Therefore multiple CSTRs operating in series allow either a reduction in the total reactor volume used to obtain the same degree of conversion as with a single CSTR, or a higher degree of conversion for a given total reactor volume. In either case the engineering cost to achieve these changes is in the additional connecting piping and fluid and heat control systems required for multiple units, over a single reactor. This generally is the factor that limits the extent to which increased numbers of reactors are economic to use for improvement of process conversions.

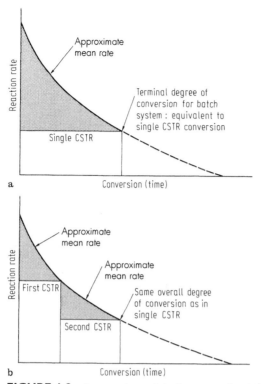

FIGURE 1.2 A comparison of the "space–time" yields, or saving in reactor volume achieved by carrying out a continuous process to the same degree of conversion in (a) a single stirred tank reactor versus (b) two CSTRs operating in series. (Adapted from Wynne [14], with permission.)

Taking the multiple CSTR concept to its logical extreme, a very large number of series-connected and small tank reactors can be likened to carrying out the same process in a very long, narrow-bore tube, referred to as a *tubular flow*, or *pipe reactor*. By placing sections or all of this tube in externally heated (or cooled) sections, any desired processing temperature of the fluid mixture flowing inside the pipe can be obtained. If a high flow rate is used, or if in-line mixers (streamline flow splitters employing an interacting series of baffles) are used, components of even immiscible mixtures may be made to interact intimately from turbulent flow as they move down the tube. So usually no external mixing is necessary to obtain good contact between the reacting materials moving through a pipe reactor. Turbulent flow also helps to ensure that good heat transfer occurs between all parts of the liquid flowing inside the tube, and the tube wall. If the heat transfer fluid or gas flowing outside the tube is also moving vigorously, the temperature difference between this external heat transfer medium and the contents of the tube is also kept to a desirable minimum.

The concentrations of starting materials flowing in a tube reactor decrease, and the concentration of product increases, as the mixture flows down the tube and the reaction proceeds. Thus reaction times for the raw materials flowing into a tube reactor can be calculated from the relation (distance from the inlet)/(reactant velocity). The induced turbulence in the tube occurs mostly in the cross-sectional dimension and very little along the length of the tube, i.e., there is little or no "backmixing," or mixing of newly entering raw materials, with raw materials that have already reacted for some time. This feature has led to the names *plug* or *plug flow* reactor as other descriptive synonyms for tubular flow or pipe reactor.

Most industrial processes using the interaction of fluids to obtain chemical changes can be classified into one, or sometimes more of the preceding five liquid reactor types. Variations on these themes are used for gas–gas, gas–liquid, or gas–solid reactions, but these parallel many of the processing ideas used for liquid–liquid reactors [15].

1.5.2. Fluid Flow through Pipes

To understand the mechanism of the turbulent mixing process occurring in pipe reactors we have to consider first some of the more general properties of fluid flow in pipes. Resistance to fluid flow in a pipe has two components, the viscous friction of the fluid itself within the pipe, which increases as the fluid viscosity increases, and the pressure differential caused by either a liquid level and vessel height difference or a pressure difference between the two vessels.

At relatively low fluid velocities, and particularly for a viscous fluid (where turbulence is damped) in a small pipe one will normally obtain streamline flow of the fluid within the pipe (Fig. 1.3a). Under these conditions the fluid is in a continuous state of shear with the fastest flow in the center of the pipe and low to zero flow right at the wall. The fluid velocity profile, along a longitudinal section of the pipe, is parabolic in shape.

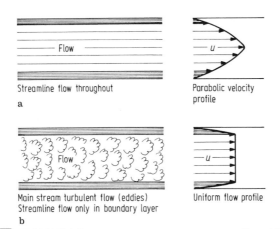

Streamline flow throughout

a

Parabolic velocity profile

Main stream turbulent flow (eddies)
Streamline flow only in boundary layer

b

Uniform flow profile

FIGURE 1.3 Fluid flow characteristics and profiles of fluid flow in pipes. (a) At low Reynolds numbers, where streamline flow is obtained throughout the cross section. (b) At high Reynolds numbers, where turbulent flow is obtained for most of pipe volume. Streamline flow is only obtained in a thin boundary layer adjacent to the pipe wall where the influence of the wall and viscous forces control turbulence. (Adapted from Raitt [16], with permission from The Nuffield Foundation.)

At high fluid velocities, and particularly for low viscosity fluids in large diameter pipes, small flow disturbances create eddies in the fluid stream which fill the whole of the cross-sectional area of the pipe (Fig. 1.3b). Only a residual boundary layer against the inside wall of the pipe will maintain streamline flow under these conditions. This turbulent condition is in fact the more usual one for fluid movement in pipes since smaller pipes, which cost much less, may be used when high fluid velocities are used [16]. The cost saving obtained by using smaller pipe generally far exceeds the small increase in pumping cost required to achieve the higher fluid velocities. There are also other reasons for this [16].

The development of turbulent flow depends on the ratio of viscous to inertial (density and velocity) forces, a ratio known as the Reynolds number, R_e (Eq. 1.6).

$$R_e = \frac{u\rho d}{\mu}, \text{ where} \qquad\qquad 1.6$$

u = fluid velocity,
ρ = fluid density,
d = pipe diameter, and
μ = fluid viscosity.

Either metric or English units, i.e., the g/cm sec, or the lb/ft sec system may be used for substitution, as long as the usage is kept consistent. In either case the result comes out to the same, dimensionless (unitless) Reynolds number. The lack of dimensionality of this value is expected for any pure ratio. Whenever the Reynolds number for fluid flow in a pipe exceeds about 2100, one obtains turbulent flow. However, the division between streamline and turbulent flow situations is also somewhat dependent on factors other than those included in the Reynolds number calculation, such as the proximity of bends and flow-obstructing fittings, and the surface roughness of the interior of the pipe. So normally a Reynolds number range is given for the dividing line between streamline and turbulent flow. If it is 2000 or less, this is indicative of a streamline flow situation. If it is 3000 or more, turbulence is expected [17].

Tubular or pipe reactors are designed to take advantage of this phenomenon to obtain good mixing. This means that relatively small bore tubes and relatively high flow rates are used for this type of continuous reactor. It should also be kept in mind that this dependence of good mixing on high flow rates may in some instances set a lower limit on the fraction of the design production rate at which the plant can operate. Turbulent flow of the raw materials in the pipe not only contributes good mixing, but also assists in maintaining good heat transfer conditions through the pipe wall separating the reactants flowing in the pipe from the jacketing fluid.

1.5.3. Controlling and Recording Instrumentation

Many kinds of sensors are needed to measure the process parameters important for effective operation of any type of chemical process. The principles of manual or automated process control require, first of all, an appropriate variable or variables which need to be measured—temperature, pressure, pH, viscosity, water content, etc.—in order to know the progress of the reaction or separation process. For the measured variable to be significant in the control of the process it must represent a control parameter, such as steam flow, pump speed, acid addition rate and the like which, when altered, will cause a response in the measured variable. Finally, there must be some actuating mechanism between the variable which is sensed and the process condition which requires adjustment, in order for the measured variable to be useful for the control of the process. In simple processes and where labor costs are low the actuation may be carried out by a person who reads a dial or gauge, decides whether the parameter is high, low, or within normal range, and if necessary adjusts a steam valve or pump power or cooling water to correct the condition. For automated plants the means of actuation may be a mechanical, pneumatic, electric, or hydraulic link between the sensor and the controlled parameter.

The amounts of materials fed to a chemical reaction are usually sensed by various types of flowmeters (Fig. 1.4). The proportions of raw materials reacting are known from metered flow rates, the altering of which, in turn, may be used to obtain the correct ratio of raw materials moving into the reactor. Control of liquid flow rates is usually achieved via valves. In the early days, this was accomplished by means of an on or off option. Today, flow control as well as many other process variables are designed to be proportional, that is, there is control of the degree to which a variable may be altered, not simply an all-or-nothing situation. Valves may be set to a variety of different flow rates, pump speeds may be altered, conveyors moving solids may be made to feed at different rates, and heat input as electrical energy or steam or cooling water flows may be varied at will. The development of proportional controls of this kind for more process variables has greatly improved the degree of refinement of process operation that is now possible. Occasionally raw material quantity measurement is carried out using measuring tanks or bins, with a float, sight tube, electrically activated tuning fork, or the like as level controls, or even mass measurement of the bin plus contents. These devices are more like the usual laboratory methods used for mass proportioning of reactions.

Actuation of process controls in response to the measured process variables has commonly been, and still is to a significant extent pneumatic (low pressure air) because of the reliability and inherent ignition safety aspect of this system. However, the greater ease of computer interfacing possible with electric or electronic actuation, and the now improved reliability and safety of these systems against ignition hazards have contributed to the growth in the use of these actuating methods. Increased use of computer-based technology for plant automation has stimulated this trend and at the same time has helped further to smooth process operations and improve yields and product quality while providing a savings in labor costs.

Automation of plant control using a computer to match ideal process parameters to the readings being taken from measured process variables allows close refinement of the operating process to the ideal conditions

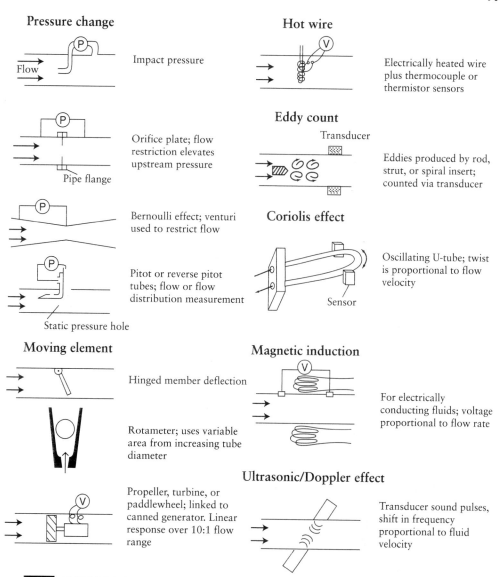

FIGURE 1.4 Some types of flow measuring devices used in chemical processing. (Adapted from McShane and Geil [18], with permission.)

(Fig. 1.5). Manually, a process reading may be compared to the ideal condition every hour or half hour as a reasonable operating procedure when being run under human control. However, under computer control it is possible to program the operating system so that 20 (or more) variables are monitored, compared to their ideal ranges, and actuators motivated to adjust process parameters, if necessary, every minute or at shorter or longer intervals as required. The computer is given override management of the main process loop. With the very short monitoring intervals that computer control makes

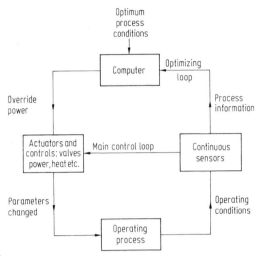

FIGURE 1.5   Outline of scheme used for computer control of a chemical processing unit. (Adapted from Raitt [16], with permission from The Nuffield Foundation.)

possible, process control is still not an easy task. However, the advantage gained is that no process using this system needs to deviate far from ideal operating conditions before parameter corrections are made.

The frequent monitoring capability of the computer has also provided a stimulus to the design and implementation of rugged, on-line process analyzers such as infrared and mass spectrometers, gas chromatographs, and the like that can provide quicker and more frequent composition information on the progress of the reaction than is possible from conventional laboratory analysis. Without on-line analysis, an interval of 2 or more hours would elapse for completion and reporting of a laboratory analysis. Only then could adjustment of appropriate process parameters and optimizing of the process condition be carried out. Under manual analysis conditions the extent of deviation of the process from ideal was sometimes quite significant, causing a reduction in yields and the quality of product.

Whether equipped with on-line analyzers or not, all chemical processes rely heavily on the information obtained from periodic manual laboratory analyses. These provide a means of checking the performance of the on-line analyzers. More importantly they provide quality control checks on all raw materials moving into the plant site and all products and waste streams (if required) moving out. Sampling frequency and methods of analysis used for raw materials will be geared to the size of the shipment lot and the method of delivery. Process and product quality monitoring will be carried out at a frequency depending on the operating stability of the process and the rigor of the quality control requirements (allowed impurity concentrations) of the product. A process that requires frequent checks and parameter adjustments to keep operating smoothly is, of course, going to need more frequent sampling and manual analysis to ensure consistent product quality.

A part of the analytical planning for operating a chemical complex is the setting up and maintenance of a "sample library," where analyzed samples from each tank car load or reactor lot may be stored for reference purposes. The retention time for these samples is set to exceed the probable delay between the time of product shipment and time of final consumption. A sample retention program of this kind enables the company to render rapid technical assistance to a customer who may be having difficulty with a particular batch of product. It also provides an independent means, by reanalysis of the sample, for the company to accept or reject claims regarding the quality of a product shipment. Thus, the whole area of proportional process control, analysis, sample retention, and careful record-keeping all comprise important integrated parts of an operating chemical complex necessary for the maintenance of product quality and customer satisfaction.

1.5.4. Costs of Operation

Since the primary purpose for the existence of a chemical company is to make a profit for its owners and shareholders, it is vital that it be possible to determine accurate operating costs so that product pricing and marketing can achieve this objective. Today, of course, this profit picture is complicated by the fact that it has to be achieved while other company obligations relating to the welfare of its employees, to safety, to environmental quality, and to the community and country of operation as a corporate good citizen are also met. It is of course possible to produce an uneconomic chemical commodity, such as when required under the exigencies of war or with particular political objectives. But this requires artificial inducements such as tax concessions, subsidies, or the economics of an organized economy to achieve this.

As a rough rule of thumb, the chemical process industries aim at an 8 to 10% after-tax profit (earnings), stated as percent of sales [19, 20]. In general, if the financial decision regarding construction of a new plant relates to a process for a new product, the economic projections required for construction to proceed will require a slightly higher profit, or higher rate of return on investment than this, for construction to proceed. If, however, the plant is to produce a standard chemical commodity that is already a large volume product (i.e., a less risky venture) then lower profit projections may still be acceptable. This is particularly true if the company is intending to use a substantial fraction of the product in its own operations, which is a favorable "value-added" practice known as providing a "captive market" for this product. However, these are just profit projections. The ever variable nature of business cycles places significant perturbations onto the realization of these projections so that actual, after-tax profit margins more usually range from 4 or 5% up to 12% in any particular year, and sometimes outside both extremes including a loss.

What should be remembered, however, is that a profitable company earns not only an income in its own right (part of which goes to the investors who put up the money to construct the plant), it also provides jobs and salaries to its employees who spend much of their earnings locally, stimulating further business activity. The company also usually pays local and federal taxes on a corporate basis and through income taxes paid by individual employees, together providing direct and indirect sources of income to different levels of government. Coupled with the sales taxes levied against many types of chemical commodities, because of this "multiplier effect," the local and country-wide benefit of a profitably operating commodity-based company is enormous.

To determine accurately the costs of operation for any particular process the chemical reactions being used must be known precisely. It is the effective stoichiometry, or quantities of product(s) which are to be expected from particular quantities of raw materials, which must be known. It is helpful to know something about the mechanism or chemical pathway to the materials being produced since this knowledge can be used to make process changes which can increase reaction rates or raise process yields. However, it also frequently happens that we have a process which operates profitably which produces salable product long before anything significant is known about the mechanism.

As examples of this, the early facilities to produce phenol by chlorobenzene hydrolysis and by cumene oxidation were both constructed when the stoichiometry demonstrated acceptable economics [21]. It was after the product had already been on the market for some years before anything was known about the respective mechanisms involved. Other secondary aspects of the process, such as the capital costs, heat requirements, electric power, labor, and water needs must also be known with some degree of certainty to determine the product pricing structure necessary for profitable operation.

1.5.5. Conversion and Yield

There are several important ways in which the description of the quantitative aspects of a reaction differ, using the same facts, when discussed in a research or academic context as compared to discussion in an applied, or industrial context. Probably the best way to understand these distinctions is to define the various terms used, employing a general example, Eq. 1.7.

A	→	B	+	A	+	by-products	
1.00 moles		0.60 mol		0.30 mol		0.10 mol	
starting		product		unreated		from reacted A	1.7
material				starting			
				material			

The yield which would be reported in a research or academic setting would be based on the yield definition given in the word Eq. 1.8.

$$\% \text{ research yield} = \frac{(\text{moles of B formed}) \times 100}{\begin{array}{c}(\text{theoretical moles of B which could} \\ \text{be formed from moles of A charged})\end{array}} \qquad 1.8$$

To calculate a yield from the information of Eq. 1.7 using this definition gives a value of 60% ($0.60/1.00 \times 100$) for the result to be reported using this system. This result takes no account of any starting A which did not react. This is in keeping with what is frequently the primary objective in the academic setting, the synthesized product of interest. The amount of any unreacted starting material in the residues from a reaction is seldom bothered with, and in fact is usually discarded with any by-products, etc., once the product of interest has been isolated.

In contrast to this, in an applied or industrial setting the definition of yield differs somewhat from that described above. Here, the amount of unreacted starting material remaining after a reaction is carried out is nearly as important as the amount of product obtained, and is taken care of in the yield definition used (Eq. 1.9).

$$\text{\% industrial yield (selectivity)} = \frac{(\text{moles B formed}) \times 100}{(\text{theoretical moles of B which could be formed from moles of A consumed})} \qquad 1.9$$

The reason for the commercial importance of this is that any unreacted starting material in an industrial process is usually recovered from product(s) and by-products. It is then recycled to the front end of the process, with the fresh starting material just entering the process (Fig. 1.6). This way the amount of fresh starting material required is decreased. Again, using the information example just given, the industrial yield works out to 85.7% [(0.60/(1.00 − 0.30) × 100] or, rounded, 86%. Thus, when taking into account the unreacted starting material a more favorable yield picture is presented.

However, to an applied chemist or an engineer at least one further piece of information related to the performance of the reaction is needed to estimate the quantitative results of a process, and that is the conversion. A good working definition of conversion is given by Eq. 1.10.

$$\text{\% conversion} = \frac{(\text{moles of main product}) \times 100}{(\text{moles of main product equivalent to the moles of main reactant charged})} \qquad 1.10$$

This takes into account the fact that not all reactions yield one mole of product for each mole of a starting material charged (placed in the reactor). Cracking reactions, for example, may yield 2 (or more) moles of product per mole

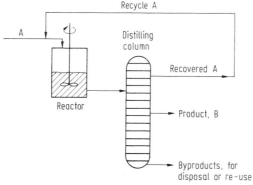

FIGURE 1.6 An illustration of the importance of recycle of recovered starting material in the industrial definition of yield.

of raw material, while condensation reactions and polymerizations can yield less (and sometimes very much less) than a mole of product per mole of raw material. Using this definition for the example at hand gives the working Eq. 1.11, which is useful for calculations.

$$\% \text{ conversion} = \frac{(\text{initial moles of A}) - (\text{final moles of A}) \times 100}{(\text{initial moles of A})} \qquad 1.11$$

Using this equation, a value of 70% [(1.00 − 0.30) × 100/(1.00)] is obtained for the conversion. What this means in practical terms is that 70% of the raw material started with is no longer starting material and has been converted to something. This number alone does not tell us how much of what was converted is product.

One needs to use the definitions of both the industrial conversion and the yield in order to determine how much of product B we can expect. This can be determined when having only the amount of starting A and the conversion and yield data for the process of interest. The industrial yield essentially amounts to the fraction of converted starting material which ends up as product, a sort of "selectivity" or "efficiency" term for the process. For this reason "selectivity" or "efficiency" are terms which are often used synonymously with industrial yield. As such, by multiplying the two fractions together one obtains the fractional yield of product to be expected from a batch process, or the "yield per pass" (yield on one passage of raw materials through the process) for a continuous process (Eq. 1.12).

$$\begin{array}{ccccc} (\text{fractional} & \times & (\text{fractional} & = & (\text{fractional product} \\ \text{industrial yield}) & & \text{conversion}) & & \text{recovery}) \\ 0.86 & & 0.70 & & 0.60 \end{array} \qquad 1.12$$

Again using the example, when the fractional product recovery is multiplied by the number of moles of A started with, one obtains a value of 0.60 [(0.70 × 0.86) × 1.00 mole of A] moles of B, as the amount of product to be expected from a batch operation, or from one pass of the raw material through a continuous process. This is in agreement with the quantities specified in the original example, and, it will be noted, is the same as the academic yield specified on a fractional basis. Thus we can write down the form of an additional relationship, specified in fractional terms, which is often useful in quantitative calculations which relate to industrial processes (Eq. 1.13).

$$(\text{industrial yield}) \times (\text{conversion}) = (\text{research yield}) \qquad 1.13$$

This can be further rearranged to another useful expression for calculations involving process efficiencies (Eq. 1.14).

$$\% \text{ selectivity} = \frac{\text{research yield (fractional)} \times 100}{(\text{fractional conversion})} \qquad 1.14$$

It is useful to consider the significance of having the ability to carry out yield and conversion manipulations. It is of course desirable to have any industrial process operate with high yields and high conversions. If both of these conditions prevail simultaneously, more product will be obtained from each passage of raw materials through a given size of reactor, and there will be less starting material to separate and recycle, than if this were not true. Many industrial processes do in fact operate with this favorable situation.

It is possible, however, to make a success of an industrial process which only achieves low conversions, as long as high yields are maintained. Very few industrial processes operate with industrial yields (selectivities) of less than 90%, and many operate with yields of 95% or better. Yet some of these, for example, the processes for the vapor phase hydration of ethylene to ethanol, and the ammonia synthesis reaction, have conversions which run in the 5 to 15% range and yet are operating on a large scale very competitively because both reliably achieve selectivities to the desired product which exceed 95%. If one only had research yield information about these processes, 4 to 5% and 15 to 20%, respectively, neither would appear to be promising candidates for commercialization. Thus, while it is desirable for an industrial process to obtain high conversions with high yields (selectivity), it is *vital* for a successful industrial process to have high yields.

1.5.6. Importance of Reaction Rate

Fast reactions, in general, are conducive to obtaining a large output from a relatively small volume of chemical processing equipment. For example the ammonia oxidation reaction, the first stage of producing nitric acid from ammonia, is essentially complete in 3×10^{-4} seconds at 750°C. This is sufficiently rapid that the catalytic burner required to do this occupies only about the volume of a file cabinet drawer for the production of some 250 tonnes of nitric acid daily. Except for the cost of the catalyst inventory, the fabrication cost of the ammonia burner itself is relatively low. Follow-up reactions for the process are much slower than this so that the volume of equipment required to contain these parts of the process are much larger and more costly (Chap. 11).

Ammonia oxidation represents a process with which it was realized, early in the design stage, that carrying out this step at 600–700°C instead of at near ambient temperatures speeded up the process sufficiently to allow large conversion volumes to take place in a relatively small reactor. This same philosophy is followed, wherever feasible, with all chemical processes, i.e., a reactor volume saving is a capital cost saving. But with some processes, such as the esterification of glycerin with nitric acid, technical complications exist that put an upper limit on feasible reaction temperatures. This effectively prevents the use of higher temperatures to increase the rate of the reaction. Consequently this process, under the normal operating temperature of about 5°C, has a 60- to 90-minute reaction time requirement. So to produce even 20 tonnes of nitroglycerin per day would require a batch reactor of 2 tonnes or so capacity, much larger than the ammonia oxidation unit required for a 250 tonne/day nitric acid plant.

These examples illustrate the principle that, wherever feasible, reaction conditions, catalysts, etc., are selected and developed such that the rate of a commercial process is maximized. In so doing the size of the processing units required for a given volume of production is reduced, in this way decreasing the costs of construction. Reducing the capital costs also reduces the capital charge per unit of product, decreasing the price required from the product to operate at a profit. In these ways improvement of the rate of a chemical process becomes a further contributing factor in the market competitiveness of the chemical industry.

1.6. CHEMICAL VOLUME PERSPECTIVES

The chemicals listed in Table 1.4 are presently produced on the largest scale and are examples of the so-called commodity, heavy, or bulk chemicals. Sodium chloride is not always classified as a *produced* chemical since most salt production is basically extractive in nature. Sulfur, too, is sometimes ranked with produced chemicals and sometimes with extracted minerals, depending on the origin of the sulfur. If one leaves these two chemicals aside, sulfuric acid emerges as the leading volume chemical product, both in the United States and worldwide. The availability of world production data for many chemicals is somewhat sporadic, but the world ranking for most of these lies near the U.S. ranking. American production of most of the chemicals on this list represents the largest single contribution to the world figure. U.S. ranking data can often be used to estimate both world rankings and world production levels when these data are not available directly.

TABLE 1.4 World and American Production of Large-Scale Chemicals, in Millions of Metric Tonnes[a]

| U.S. rank, 1995 | Chemical | United States | | World 1992–94 data |
		1980	1995	
1	Sulfuric acid	40.1	43.3	126.8
2	Nitrogen	15.5	30.9	
3	Oxygen	15.6	24.3	
4	Ethylene	12.5	21.3	
5	Lime	16.0	18.7	118
6	Ammonia	17.2	16.2	
7	Phosphoric acid	9.9	11.9	37.9[b]
8	Sodium hydroxide	10.3	11.9	
9	Propylene	6.2	11.7	
10	Chlorine	10.2	11.4	
11	Sodium carbonate	7.5	10.1	30.4
12	Methyl *t*-butyl ether	<0.1	8.0	
13	Ethylene dichloride	4.5	7.8	
14	Nitric acid	7.8	7.8	
15	Ammonium nitrate	7.8	7.3	
	Sugar	4.3[c]	7.0	112.1

[a]Data compiled from *Chemical and Engineering News* [22, 23] and *U.N. Statistical Yearbook* [24]. Figures for the United States can be used as a means of estimating world production. For many basic chemicals the ratio is 4–6 times American production.
[b]Stated as phosphate rock, for lack of phosphoric acid data.
[c]World 1980 sugar production was 92 million metric tonnes.

A number of other interesting observations can be made concerning these particular bulk chemicals. First, they are all not far removed, in terms of processing steps, from the natural raw materials from which they are derived. Virtually all of the oxygen and nitrogen and a significant proportion of the salt, sulfur, and sodium carbonate are all obtained relatively directly from natural sources. Also, these commodities interrelate quite closely to one another, in chemical terms. Thus sulfuric acid, largely produced from sulfur, is in turn used to produce phosphoric acid from phosphate rock. A large fraction of the nitrogen produced goes into the production of synthetic ammonia. Ammonia, in turn, is used for nitric acid production and also is combined with much of the nitric acid product in an acid–base reaction for the preparation of ammonium nitrate. Chlorine and sodium hydroxide are mostly obtained from the electrolysis of a solution of sodium chloride in water. It is interrelationships of this kind, coupled with very large world fertilizer markets for some of the secondary and tertiary products of these sequences, in particular ammonia, ammonium nitrate, and phosphoric acid (as salts), that keeps many of these chemicals on this large volume production list.

If one ranks all American chemical companies one obtains a list which includes a significant number of oil companies (Table 1.5). In fact this exercise shows that 6 of the 15 largest U.S. chemical companies are oil companies which are also producing chemicals.

Listing the world's chemical companies produces a similar picture (Table 1.6). The dominant positions of Germany, the United States, and the United Kingdom in chemicals production is evident from this ranking. Again oil companies show up on this list, even though chemicals represent a small fraction of their total sales. Shell, Exxon, and Elf Aquitaine had gross sales in 1995

TABLE 1.5 The Fifteen Largest Chemical Processing Companies in the United States Ranked on the Basis of Chemical Sales[a]

Overall rank, 1980	1995	Company	1995 Chemical sales (10^9 US$)
2	1	Dow Chemical	19.2
1	2	Du Pont	18.4
3	3	Exxon	11.7
6	4	Hoechst Celanese	7.4
5	5	Monsanto	7.2
n/a	6	General Electric	6.6
21	7	Mobil Oil	6.2
4	8	Union Carbide	5.9
n/a	9	Amoco	5.7
10	10	Occidental Petroleum	5.4
17	11	Eastman Chemical	5.0
39	12	BASF Corp.	4.8
7	13	Shell Oil	4.8
n/a	14	Huntsman Chemical	4.3
n/a	15	Arco Chemical	4.3

[a]Compiled from *Chemical and Engineering News* [22] and Peaff [25].

TABLE 1.6   World's 10 Largest Chemical Companies Based on 1995 Chemical Sales[a]

Rank, 1995	Company	Chemical business area	Annual chemical sales (10^9 US$)
1	BASF (Germany)	chemicals, plastics	22.0
2	Hoechst (Germany)	chemicals, dyes	21.7
3	Dow Chemical	chemicals, plastics	19.2
4	Bayer (Germany)	chemicals, drugs	18.8
5	DuPont (U.S.A.)	chemicals	18.4
6	Shell (U.K., Netherlans)	chemicals	15.4
7	ICI (U.K.)	chemicals, plastics	13.0
8	Exxon (U.S.A.)	chemicals	11.7
9	Elf Aquitaine (France)	chemicals	11.1
10	Formosa Plastics (Taiwan)	plastics	10.8

[a]Compiled from Layman [26].

of 150.6, 121.8, and 41.7 billion US$ respectively. Clearly oil and gas production and processing are the dominant business areas of these companies.

REVIEW QUESTIONS

1. What minimum flow rate in cm/sec is required in a 10-m length of open straight pipe of 2 cm inside diameter to obtain a uniformly mixed paint from two components of combined density 1100 kg/m^3 and 1.912 centipoise viscosity?

2. A 100 tonne/day electrolytic chlorine plant, complete with caustic soda facilities costs about 12 million dollars.
   (a) What would be the approximate capital cost of a 600 tonne/day facility?
   (b) What capacity chlorine plant could be built for $6 million?
   (c) If an operating staff of eight is are required to run a 100 tonne/day plant, what staffing would be needed for plants with capacities of 1000 tonne/day and 2500 tonne/day?

3. A tube reactor is to be used to contact an aqueous sugar solution with 5% by volume of a solvent of density 0.780 g/cm^3 for extraction. The aqueous solution has a density of 1.080 g/cm^3, and a viscosity of 1.201 centipoise.
   (a) If the tube to be used is 2 cm in diameter and the mixture of the aqueous solution and solvent is to flow at a combined velocity of 50 cm sec^{-1}, in the tube, would there be efficient contact (i.e., turbulent flow) between the sugar solution and the other liquid phase? Very briefly describe the criterion for your answer.
   (b) If the liquids to be contacted have a combined bulk density of 68 lb ft^{-3} and a viscosity of 7.923×10^{-4} lb ft^{-1} sec^{-1}, what flow velocity in feet per second would be required to ensure turbulent flow conditions in a 1-in. pipe (inside diameter)?

(c) What would be the daily rate (kg/24 hr) of sugar production in part (a) if the sugar concentration was 2% by weight and extraction efficiency was 99.9%?

4. The capital costs of 200 million kg/year and 500 million kg/year styrene plants are $18 million and $28 million.

(a) From a plot of log capital costs vs. log capacities, determine the particular value of the exponent n which could be used for more accurate estimation of costs for construction of other capacities of styrene plant.

(b) Using the exponent value determined in part (a), what would be the approximate capital cost of a 100 million kg/year plant?

(c) What would have been the estimated capital cost of a 100 million kg/year plant using the mean value for the exponent n of 0.60 and each one of the two known capital costs given in the preamble?

5. Naphthalene ($C_{10}H_8$) can be converted microbiologically to salicylic acid ($C_7H_6O_3$) on a 94%, weight for weight basis, in one pass. (CO_2 and H_2O are also produced.)

(a) At 100% conversion, what is the percent selectivity (industrial yield) of this process?

(b) What is the "academic" or research yield for this process?

(c) Air is 21% oxygen (take as mole ratio $O_2:4N_2$). Assuming 25% conversion of the oxygen content of the air in this process, what mass of air would be required to produce 1000 kg of salicylic acid? Ignore oxygen consumed.

FURTHER READING

American Chemical Society, "Chemical Abstracts," Columbus, OH, published since January, 1907. Chemistry, technology, and patent indices for all countries.

"Applied Science and Technology Index." H.W. Wilson Company, Bronx, NY, since 1958; formerly called The Industrial Arts Index, published since 1913.

"Business Periodicals Index." H.W. Wilson Company, New York, published since January, 1958. Current economic data.

"Chemical Marketing Reporter" (formerly Oil, Paint, and Drug Reporter). Schnell Publishing Co., NY, published in 2 volumes per year since October 1871, gives weekly chemical prices.

Library Association, "British Technology Index," London, published since January, 1962.

H.P. Meissner, "Processes and Systems in Industrial Chemistry," Prentice-Hall, Englewood Cliffs, NJ, 1971.

P.J.T. Morris, W.A. Campbell, and H.L Roberts, eds, "Milestones in 150 years of the Chemical Industry," The Royal Society of Chemistry, Cambridge, UK, 1991.

Patent Office, "Patent Office Record," Patent Office, Ottawa, Ontario, published since 1872. Also the parallel records of other countres of interest.

M.S. Peters and K.D. Timmerhaus, "Plant Design and Economics for Engineers," 3rd ed, McGraw-Hill, NY, 1980.

D.F. Rudd, G.J. Powers, and J.J. Siirola, "Process Synthesis," Prentice-Hall, Englewood Cliffs, NJ, 1973.

R.W. Thomas and P. Farago, "Industrial Chemistry," Heinemann, London, 1973.

U.S. Patent Office, "Official Gazette," Washington, DC, published weekly since 1872, replacing the previous "Patent Office Reports." Also the parallel records of other countries of interest.

REFERENCES

1. F.A. Lowenheim and M.K. Moran, "Faith, Keyes and Clark's Industrial Chemicals," 4th ed. Wiley-Interscience, New York, 1975.
2. M. Trowbridge, *Chem. Br.* **11**(1), 15, Jan. (1975).
3. H.S. Sloan and A.J. Zurcher, "Dictionary of Economics," 5th ed., 459, Barnes and Noble, New York, 1971.
4. A.C.H. Cairns, *R. Inst. Chem. Rev.* **2**(1), 41, Feb. (1969).
5. Organization for Economic Cooperation and Development, "The Chemical Industry," Paris, 1970, 1981.
6. Facts and Figures for Chemical R and D, *Chem. Eng. News* **60**(30), 38, July 26 (1982).
7. Most Patents Continue to be in Chemical Field, *Chem. Eng. News* **60**(14), 24, April 5 (1982).
8. F.P. O'Connell, *Chem. Eng. (N.Y.)* **69**(4), 150, Feb. 19 (1962).
9. J.P. Stern and E.S. Stern, "Petrochemicals Today," Edward Arnold, London, 1971.
10. M.S. Peters and K.D. Timmerhaus, "Plant Design and Economics for Chemical Engineers," 3rd ed. McGraw-Hill, New York, 1980.
11. D.H. Allen, *Chem. Ind. (London)*, Feb. 5, p. 98 (1977).
12. G.T. Austin, "Shreve's Chemical Process Industries," 5th ed., McGraw-Hill, New York, 1984.
13. K. Denbigh, "Chemical Reactor Theory," University Press, Cambridge, UK, 1965.
14. M.D. Wynne, Chemical Processing in Industry," Royal Institute of Chemistry, London, 1970.
15. H.S. Fogler, ed., "Chemical Reactors," ACS Symp. Ser. No. 167, American Chemical Society, Washington, DC, 1981.
16. J.G. Raitt, ed., "Chemical Engineering; a Special Study," Nuffield Foundation, Penguin, Harmondsworth, England, 1971.
17. R.H. Perry and D.W. Green, eds., "Perry's Chemical Engineer's Handbook," 6th ed., McGraw-Hill, New York, 1984.
18. J.L. McShane and F.G. Geil, *Res./Dev.* **26**(2), 30, Feb. (1975).
19. Facts and Figures for the Chemical Industry, *Chem. Eng. News* **60**(24), 31, June 14 (1982).
20. Facts and Figures for the Chemical Industry, *Chem. Eng. News.* **61**(24), 26, June 13 (1983).
21. P. Wiseman, "Industrial Organic Chemistry," Wiley-Interscience, New York, 1972.
22. Top 50 Chemical Products and Producers, *Chem. Eng. News* **59**(18), 35–42, May 4 (1981).
23. Facts and Figures for the Chemical Industry, *Chem. Eng. News* **74**(26), 38–79, June 24 (1996).
24. "United Nations Statistical Yearbook 1993," 40th ed., United Nations, New York, 1995.
25. G. Peaff, Dow Replaces DuPont to Lead Top 100 U.S. Producers, *Chem. Eng. News* **74**(19), 15–20, May 6 (1996).
26. P.L. Layman, Global Top 50 Chemical Producers Shift Rankings During Profitable 1995, *Chem. Eng. News* **74**(30), 29–31, July 22 (1996).

2

AIR QUALITY MEASUREMENT AND EFFECTS OF POLLUTION

We have first raised a dust and then complain we cannot see.
            —Bishop Berkeley (1685–1753)

2.1. SIGNIFICANCE OF HUMAN ACTIVITY ON ATMOSPHERIC QUALITY

In the early days of habitation of this planet, when the human population was small and its per capita consumption of energy was primarily as food (8400–12,600 kJ/day; 2000–3000 kcal/day), total human demands on the biosphere were consequently small. Early requirements of goods were minimal and quite close to direct (requiring little fashioning) so that this early society's total demands and wastes were easily assimilated by the biosphere, with little impact.

Today, with advances in technology providing an ever increasing range of goods and services, it has been estimated that the individual consumption of energy of all forms has risen some 100-fold from the requirements of primitive man. Of the 1 million kJ per person per day (230,000 kcal/person/day) that this consumption now represents, more than half, or 645,000 kJ (154,000 kcal), is estimated to be consumed by society's industrial, agricultural, and transportation needs. About a quarter is consumed per person through the medium of electricity generation and consumption, either for individual domestic use, or by the prorated industrial power consumption on their behalf. All of the major fields of human endeavor, high-technology agriculture, industrial production, and thermal electricity generation use combustion processes to provide a major fraction of their energy requirements. When this gross increase in per capita energy consumption is coupled to the global population growth of a thousand or more times the population of primitive societies, it is easy to see how the total demand placed on the biosphere by modern industrial societies has now become so significant.

Most of the world's population growth has taken place since the Middle Ages. Industrial development has accelerated since about 1850, so that most of the increase in demand for energy and materials has occurred during the period when these two component increases coincided. Human activities now contribute similar volumes of the minor gaseous constituents of the atmosphere as do natural processes. From a world fossil fuel consumption of about 5 billion tonnes per year, we now contribute about 1.5×10^{10} tonnes of carbon dioxide to the atmosphere annually, a significant fraction of the natural contribution, now estimated to be about 7×10^{10} tonnes/year. Of annual human atmospheric contributions of sulfur-containing gases of about 10^{10} tonnes, of oxidized and reduced nitrogen compounds of about 50 million tonnes, and of carbon monoxide of about 200 million tonnes, all are estimated to be of a similar order of magnitude to the natural contributions. The old jingle "The solution to pollution is dilution" no longer holds true, particularly for atmospheric discharges. There is just too large a total mass of atmospheric contaminants being discharged, and too small an atmosphere to accept these, to still be able to discharge at the same rate and obtain sufficient dilution. This is particularly the case when natural air movement is sluggish for any reason so that little pollutant mixing occurs. Under these circumstances discharged pollutants simply remain in the same local air mass, which greatly exaggerates the immediate effects of the discharge.

2.2. NATURAL CONTAMINANTS

Natural sources of many common air contaminants also make a contribution to this overall atmospheric pollutant loading. The seas contribute large masses of saltwater spray droplets to the air as a result of wave action. As the water evaporates from these droplets very fine particles of salts are left suspended in the moving air, one of the contributing factors to the sea "smell." It has been estimated that some 13 million tonnes of sulfate ion alone is contributed to the atmosphere annually from the oceans of the world.

Volatile organic compounds are contributed to the atmosphere by many forms of plant life, conifers such as cedar, pine, and eucalyptus, and herbs such as mint. Extensive tracts of such plants, such as the pine forests of New England or the eucalyptus forests of the Blue Mountains in Australia, contribute a large mass of terpenes to the air above them. Terpenes, plant products biosynthetically derived from isoprene, have a type formula $(C_5H_8)_n$, where n is based on the number of isoprene units in the compound (Eq. 2.1).

$$CH_2{=}C(CH_3){-}CH{=}CH_2$$

isoprene α-pinene   limonene   2.1

All terpenes have one or more reactive double bonds. Reactions of terpene vapors with sunlight and air generate a photochemically promoted blue haze over these forests when there is little air movement, hence the term Blue Mountains. Estimates of the global atmospheric contribution by plants of terpenes and oxygenated terpenes alone range from 2×10^8 to 10^9 tonnes per year.

The world's deserts contribute significant masses of dust and particulate to the atmosphere, some of which is transported a considerable distance. Meteoritic dust is estimated to be collected by the atmosphere to the extent of more than 900 tonnes annually. However, an active volcano is a far more significant particulate contributor. Some 90 million tonnes were estimated to have been discharged into the atmosphere by the eruption of Mount St. Helens in Washington State [1]. When several eruptions take place at one time the gross dust contribution to the global atmosphere is massive and long lived since it is forcefully injected as high as 40 km into the atmosphere. Therefore, the global influence of these events has more significance on atmospheric quality than more localized surface dusting events.

Sulfur dioxide and hydrogen sulfide from volcanic activity are also contributed to the atmosphere on the scale of 1 to 2 million tonnes per year. Other contaminating gases, as well as metal vapors, are also discharged in significant quantities during volcanic eruptions [2]. Vapors of the more volatile metallic elements such as mercury are also lost continuously over ore bodies. Sensitive vapor detection instruments can use these losses as a prospecting method to converge on the ore body using this mercury vapor "halo" [3].

Biological materials of many types do not represent a large mass of contamination but nevertheless comprise an important component of atmospheric pollution because of their potency. For example, very small quantities of the pollens from many types of wild flowers and grasses such as golden rod and ragweed, are sufficient to severely affect many people. Bacteria, viruses, and the living spores of some of the common molds can also cause problems, particularly when situations develop which promote a rapid, localized rise in numbers of the organism. When this occurs in seawater, such as near sewer outfalls, dispersal of microorganisms in the aerosols released by vigorous wave action can aggravate the problem [4].

Little can be done about controlling most of these natural sources of atmospheric contaminants. However, it is still important to catalog and quantify them to be able to quantitatively relate the importance of these components of atmospheric quality to the contributions from our own activities. Only in this way is it possible to compare the relative significance of the two sources of contaminants. This is a necessary preliminary to provide appropriate guidance for the modification of our own activities to decrease their negative impact on atmospheric quality where this is necessary.

2.3. CLASSIFICATION OF AIR POLLUTANTS

Air pollutants can be classified into one of three main categories based on their physical characteristics. By doing so, potential emissions may be grouped as to their appropriateness for one or more types of emission control devices, based on their mode of action.

The first of these, the coarse particles or particulate class of air pollutants, comprises solid particles or liquid droplets which have an average diameter greater than about 10 μm (10×10^{-3} mm). Particles or droplets of this class of air contaminant are large enough that they fall out of the air of their own accord, and more or less rapidly.

Air pollutants of the aerosol class can also comprise solid particles or liquid droplets, but are limited to a size range generally less than about 10 μm average diameter. The important consideration for this class is that the particles or droplets are small enough in size that there is a strong tendency for the substance to stay in suspension in air [5]. Thus for the powders of the denser solids, such as magnetite, the average particle size would have to be smaller than this 10 μm guideline for the particle to stay in suspension. A suspension of a finely divided solid in air is referred to as a "fume," and that of a finely divided liquid as a "fog," each term being only a more specific representation within the aerosol classification.

The gases comprise the third major classification of air pollutants, which includes any contaminant that is in the gaseous or vapor state. This comprises the more ordinary "permanent" gases, such as sulfur dioxide, hydrogen sulfide, nitric oxide (NO), nitrogen dioxide (NO_2), ozone, carbon monoxide, carbon dioxide (pollutant?), etc., as well as the less common ones such as hydrogen chloride, chlorine, tritium (^3_1H) and the like. It also includes mate-

TABLE 2.1 Gravitational Settling Velocity for Spheres of Unit Density in Air at 20°C[a]

Particle diameter (μm)	Terminal velocity (mm/sec)
0.1	8.5×10^{-4}
0.5	1.0×10^{-2}
1.0	3.5×10^{-2}
5.0	0.78
10	3.0
20	12
50[b]	72
100[c]	250

[a]Compiled from Barrett [6] and Spedding [7].

[b]Roughly equivalent to a powder which would pass through a No. 325 sieve (325-mesh, 54-μm average opening size) and be retained on a No. 400 sieve (400-mesh, 45-μm average opening size).

[c]Roughly equivalent to a powder which would pass through a No. 170 sieve (170-mesh, 103-μm average opening size) and be retained on a No. 200 sieve (200-mesh, 86-μm average opening size).

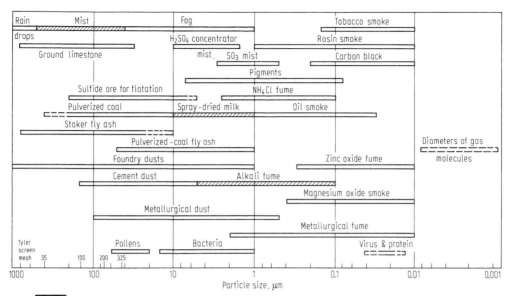

FIGURE 2.1 Typical particle size ranges of some types of precipitation, industrial and mineral processing streams and ambient air biologicals. (From Munger [8], with permission from McGraw-Hill.)

rials which are not ordinarily gases, such as hydrocarbon vapors, and volatile nonmetal or metal vapors (e.g., arsenic, mercury, zinc) when these are in the vapor state.

The dividing line between the particulate/aerosol classes, and the gaseous classes is clear enough because of the phase difference. But the position of the dividing line between the particulate and the aerosol classes, since it is based on whether or not a second phase stays in suspension in air, is far less obvious. However, consideration of the terminal velocities for particles of differing diameters helps to clarify this distinction. It can be seen from Table 2.1 that, for a substance with a density of 1 g/cm^3, a significant terminal velocity in air begins to be observed at a particle diameter of about 10 μm. This is basis of the approximate dividing line between these two classes. Figure 2.1 gives examples of typical particle size ranges for some common industrial and domestic substances.

2.4. PARTICULATE MEASUREMENT AND IDENTIFICATION

The particulate class of air pollutant, since it represents particles or droplets which more or less rapidly settle out of air, is also the easiest class to measure. If a source particulate determination is to be carried out, that is, if the particulates in the flue gases of a chimney or the exhaust gases of a vent stack are to be sampled, then special holes are required in the ductwork. Probes with associated equipment and a means of reaching the sampling holes with this equipment are necessary. More details of this procedure are given in connection with aerosol determination.

It is also possible, however, to gain a useful semi-quantitative record of source information by observation of the opacity of a plume. The Ringelmann system was set up by Maximilian Ringelmann in about 1890 to accomplish this for black smokes [9]. Ringelmann numbers 1, 2, 3, 4, and 5 represent 20, 40, 60, 80, and 100% opaque plumes, respectively. A circular card with shaded opacities corresponding to these Ringelmann numbers which surround a central viewing hole is available to assist in visual assessments. A trained observer is able to reproducibly estimate plume opacity to within half a Ringelmann number without this assistance. The obscuring properties of white smokes are specified simply on a percent opacity basis.

Another common type of particulate determination method uses particulate fallout from ambient air as a measure of the particle loading. This method can be as simple as a series of glass jars placed in suitable areas for which dustfall measurements are desired. The jars may be used dry, or may contain a liquid collecting agent to assist in trapping the fallout as it is collected. After a suitable interval, usually 30 days, the collected material is filtered (if wet collection was used), dried, and weighed. The weight obtained may be used to calculate a fallout value for each of the areas in which the collections were carried out. The result is specified as short tons per square mile per month, or more usually today as mg/m^2/day.

Essentially the same procedure is used with a plastic dustfall canister on a stanchion (upright stand) specially designed for the purpose, except that a few collection refinements are used to increase the reliability of the results. The stanchion, which places the top of the canister 4 ft above the base, is also sited a reasonable distance away from trees, buildings, or roof fixtures so as to minimize the interference of fallout by anomalous air currents. It is also usually fitted with a bird ring to discourage the contamination of the canister contents by birds resting on the edge of the canister itself. A further modification, a series of vertical spikes welded to the bird ring, is necessary in areas frequented by larger species such as gulls and crows, to prevent their roosting on the edge of the canister or the bird ring.

The amount of fallout normally obtained with either device in relatively undisturbed open country is often a surprise to those who have not previously worked in the area of air pollution studies. For example, if 100 mg of dustfall is collected during a 30-day period by a glass container having a 10-cm-diameter opening, this does not sound like much. But in conventional fallout units this amounts to 424 mg/m^2 day, or more impressively as 36.4 tons/(mile)^2 month (Eq. 2.2, 2.3),

$$(100 \text{ mg} \times 10{,}000 \text{ cm}^2/\text{m}^2) \div (3.1416 \times (5 \text{ cm})^2 \times 30 \text{ days}) \qquad 2.2$$

$$(0.10 \text{ g} \times (2.54 \text{ cm/in})^2 \div (12 \text{ in/ft})^2 \times (5280 \text{ ft/mile})^2) \\ \div (3.1416 \times (5 \text{ cm})^2 \times 9.07 \times 10^5 \text{ g/ton}) \qquad 2.3$$

actually quite a heavy fallout. Table 2.2 gives a range of some recent fallout values for the U.K. and Canada, to give a basis for comparison. The Canadian figures are somewhat high since they were obtained for an industrialized, relatively high population density area.

TABLE 2.2 Average Values for the Particulate Fallout Experienced by Areas of Differing Urban Activity in the United Kingdom and Canada[a]

Type of area	United Kingdom, 1962–1963 Particulate fallout		Canada, 1968 Particulate fallout	
	mg/m² day	ton/mi² month[b]	mg/m² day[c]	ton/mi² month
Industrial	159	13.6	233–350	20–30
High-density housing	116	9.95	—	—
Town center	112	9.60	210–256	18–22
Town park	90	7.7	—	—
Low-density housing	82	7.0	177–163	10–14
Suburbs, town edge	70	6.0	—	—
Open country	38	3.3	82–140	7–12

[a]Compiled from Barrett [6] and International Joint Commission [10].
[b]Calculated using the factor (tons/mi²/month) ÷ (mg/m²/day) = 0.08556.
[c]Calculated from the values in (tons/mi²/month) by multiplying by the factor 11.68 (1/0.08556).

An early Canadian particulate guideline stipulated that no discharge was to result in a fallout exceeding 20 tons per square mile per month for any area.

2.5. AEROSOL MEASUREMENT AND IDENTIFICATION

Since the aerosol class of air pollutants represent particles and droplets too small to fall out of their own accord, sampling and measurement of this class requires dynamic sampling equipment. Work has to be done on the gas to force it through the recovery or analyzing equipment so that the suspended matter can be captured. Again the analysis can be a source test, where a stack or waste vent is sampled directly, or it can be an ambient air survey, where the general condition of the outside air is determined.

Source testing requires the placing of a probe into the stack or waste vent through a sampling port in the side of the stack. A vacuum pump at the exhaust end of the sampling train provides the means for drawing the test gas sample into the train (Fig. 2.2). Also important for good source sampling technique is that the gas flow rate into the end of the probe is the same as the gas flow rate in the stack at the point being sampled, referred to as iso-kinetic sampling [12]. Otherwise some sorting of the particles sampled inevitably occurs, which can make the results unrepresentative and unreliable. Also different points in the cross section of the stack must be sampled to correct for any variation in particle concentrations across a section, a procedure referred to as "equal area sampling" [13]. This is necessary because, even in a long straight vertical duct, gas flow rates close to the wall of the duct will be lower due to frictional effects. This flow rate difference, in turn, affects the relative particle loadings in the gases moving at the center, and at the edge of the duct.

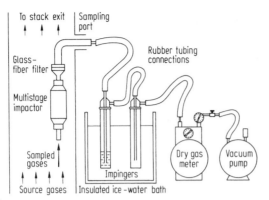

FIGURE 2.2   Simplified layout of an impactor sampling train. (From Pilat *et al.* [11], with permission.)

As the sampled gases flow into the first series of large jets in the impactor used for collection, large particles with significant momentum strike the impactor plate, many of which remain stuck there for later analysis (Fig. 2.3). Smaller particles with less momentum are diverted away from this plate, still carried by the diverted current of gases. As the gases proceed further into the impactor they reach smaller and smaller jet sizes forcing the gas to move at higher and higher velocities. As the gas velocities increase, smaller and smaller particles receive sufficient kinetic energy to impinge on the collector plates, and stick there. For some kinds of test the sticking tendency may be increased by application of a thin wipe of petroleum jelly. In this manner particles are collected roughly classified as to size. On completion of the test the impactor is carefully disassembled. The sorted collected material will give some information about the particle size distribution of the tested source, or the ambient air, if this is what is being sampled. The samples are also available for other tests, such as microscopic examination and wet chemical or instrumental tests if desired. This system does not permit good quantitative information to be obtained, unless it is fitted with a fine pore filter (membrane or the like) for final passage of the sampled gas. A filter would, however, impose gas flow rate restrictions onto the system, which in turn could limit its use for isokinetic sampling.

Filters are, however, separately useful in the sampling of ambient air for suspended matter. The high-volume ("high-vol") sampler is one of the commonest and simplest devices used for this purpose. An air turbine driven by a vacuum cleaner motor is used to provide the suction to pull an air sample of 2000 m^3 or more through a thick 12 $\times$ 15-cm filter sheet during a predetermined test period, usually 24 hours. A recording flowmeter is used to keep track of the volume of air filtered, since the flow rate gradually declines during the test period as material collected on the filter gradually impedes air flow. The quantities (flow rate) $\times$ (time) give the volume of air filtered, and the gross weight minus the tare weight of the filter, dried to standard conditions, gives the mass collected. A quantitative mass per unit volume result for ambient air is thus obtained, limited only at the lower end of the particle size

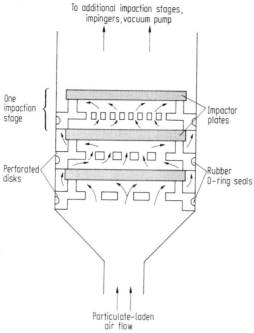

FIGURE 2.3 Enlarged longitudinal section of a portion of a cascade impactor for source testing. (From Pilat *et al.* [11], with permission.)

range by the pore size of the filter used. The large particles, plus the medium to larger size range of the aerosol classification will be collected. The result is usually specified in units of mass per unit volume, e.g., mg/m^3 or μg/m^3.

A refinement of the high-vol sampling technique enables separation of the collected sample into coarse (2.5 to 15 μm) and fine (<2.5 μm) particle fractions with the help of different Teflon filter elements. Ambient air sampling rates of about 1 m^3/hour are claimed for this device.

None of the preceding methods is very efficient at retaining particles of 1-μm diameter or less. For information on this particle size range, ultrafiltration on dense cellulose or molecular membrane filters of cellulose acetate (e.g., Millipore, Isopore) is necessary [14]. Cellulose is fine for qualitative filtration work but is difficult to use in quantitative studies because of weight variations caused by the variable extent of moisture adsorption. Cellulose acetate, which is much less affected by atmospheric moisture, is also close to 100% efficient at retaining particles larger than about 0.1 μm, and has the further advantage that the filter element can be used for direct examination under a light transmission microscope for size distribution or particle identification studies. The fine pore size of this filter requires a pressure differential of nearly an atmosphere to obtain adequate air filtration rates, and a good filter element support to enable the membrane to tolerate these pressures without rupture. Even using these conditions only about 0.03 m^3 of air per minute can be filtered through the standard 5-cm disk, one of the limitations of the method.

A Mine Safety Appliance development, an analytical scale electrostatic precipitator, provides one further technique for collection of fine aerosol material. This is particularly useful under low loading conditions. By prior taring of the removable glass tube of this device it is possible to determine particle loadings of as low as 10 μg/m^3 after the passage of 10 m^3 of air. Even larger volumes of air may easily be sampled with this device since there is negligible resistance to air flow through the glass collection tube of this design.

Identification of particles and aerosols can frequently be made using direct microscopic examination coupled with a knowledge of the collection circumstances. Particle features such as transparent/opaque, colorless/colored, rounded/angular, isotropic/anisotropic, density, and refractive index all may be determined in this way. When these properties are combined with the general visual features such as are cataloged in one of the particle atlases [15], this examination can very often be unequivocal for identification. Particles below about 0.2 μm in diameter, the limit for the best light microscopes, may be studied using electron microscopy. Various destructive wet chemical, or instrumental techniques such as ion chromatography, neutron activation analysis, or atomic absorption may also be used to provide confirmation of the identities provisionally established by the visual process. Details are available from a variety of sources [16].

2.6. ANALYSIS OF GASEOUS AIR POLLUTANTS

A wide variety of methods exists for the analysis of gaseous air pollutants within the realms of wet chemical, instrumental, and biological procedures. The analytical method of choice will depend on a number of factors among which the purpose (accuracy and number of results required), budget, equipment available, and level of training of available staff will all have a bearing. As an example of the spread of possibilities within these choices, an air (and water) pollution survey was conducted in the United Kingdom during 1972–1973 which was publicized by the *Sunday Times*. Hardware and methods literature were made available at cost through the Advisory Centre for Education, Cambridge. The unique aspect of this survey was that it was conducted predominantly by school children. Yet it produced a set of results that has been classed as one of the most comprehensive overviews of air quality yet conducted in the British Isles.

The chief difference between the two "particle classes" of air pollutants and the gaseous air pollutants, as far as analysis is concerned, is that for the latter normal filtration methods cannot be applied. Physical adsorption, such as on activated charcoal, or silica, or absorption into such matrices as silica gel, rubber, zeolites, or the like can be used to capture many gaseous air pollutants [17]. Recovery of the gases or vapors required for later analysis, or for recycle depending on the scale, is frequently achieved by warming the bed of adsorbent or absorbent while passing air or an inert gas through the bed. Sometimes this desorption is assisted by a pressure reduction. Wet absorption methods are also used, with a reagent dissolved in the liquid (usually water) to obtain reaction with and hence retain the pollutant of interest in the liquid as the air containing it is sparged through it. Sulfur dioxide, which is not well retained when contaminated air is sparged through water alone [18], is efficiently trapped when a solution of hydrogen peroxide or sodium hydroxide in water is used (Eq. 2.4–2.6).

$$SO_2 + H_2O \rightarrow H_2SO_3 \qquad\qquad 2.4$$

$$H_2SO_3 + H_2O_2 \rightarrow H_2SO_4 + H_2O \qquad\qquad 2.5$$

$$H_2SO_3 + 2\ NaOH \rightarrow Na_2SO_3 + 2\ H_2O \qquad\qquad 2.6$$

2.6.1. Concentration Units for Gases in Air

The concentration of a gas or vapor in air or any other gas phase, may be specified on a volume for volume basis, a weight for volume basis, or on a partial pressure basis. Each of these concentration units has one or more features which makes its use convenient and informative in particular situations.

Concentrations of one gas in another, specified on a volume for volume basis, are normally corrected to 25°C and 1 atm (760 mm Hg) pressure for reporting [19]. Under this system relatively high concentrations are specified in percent (%) so that 3% sulfur dioxide in air would correspond to 3 mL of sulfur dioxide mixed with 97 mL of air, both specified at 25°C and 1 atm. In more general terms this also corresponds to 3 parts by volume of sulfur dioxide in 100 parts by volume of the sulfur dioxide/air mixture, both specified in the same volume units. Lower concentrations are specified in smaller units, ppm for parts per million (1 in 10^6), ppb for parts per billion (1 in 10^9), and now that adequate sensitivity has been developed for the analysis of some air pollutants at these low concentrations occasionally even ppt for parts per trillion (1 in 10^{12}).

This volume for volume system of specifying gas concentrations in a gas is in essence dimensionless without units, i.e., it represents a pure ratio of quantities given in the same volume units which thus allows the units to divide out. To a chemist this system has the further advantage that comparisons made on a volume for volume basis are also on a molecular, or molar basis, i.e., a mole of any gas at normal temperature and pressure (NTP or STP; 0°C and 1 atm) occupies 22.41 L. This volume for volume comparison is true for most gases and vapors when existing as mixtures at or near ambient conditions of pressure and temperature, and only deviates significantly from this molar equivalency when the conditions (pressure and/or temperature) become extreme. But it must be remembered that volume for volume data are corrected to 25°C and 1 atm, under which conditions a molar volume corresponds to 24.5 (24.46) L (298°C/273°C × 22.41 L).

The behavior of this volume for volume concentration unit is also convenient for another reason. If the temperature or pressure is changed, there is no change in the concentration specified in volume for volume units, within reasonable limits. Any correction which may be necessary as a result of a minor variation of the measuring conditions is generally smaller than the experimental error of the analysis. Since many analyses of air samples are conducted at temperatures close to 25°C and normal atmospheric pressure, corrections are only occasionally needed. Because of the insensitivity of volume for volume units to changes in temperature or pressure these do not normally need to be specified when using these units for quoting concentrations of a gas in air or another gas.

Increasingly often, however, the concentration of a gas (or vapor) in air is specified on a weight per unit volume basis, common units for which are mg/L, mg/m^3, and μg/m^3:

$$1 \text{ mg/L} = 1000 \text{ mg/m}^3 = 10^6 \text{ μg/m}^3$$

Knowing the concentration of an air pollutant specified in this way allows easier determination of mass rates of emission, which are important for regulatory purposes and for exposure hazard calculations. But using this system, there is no direct molecular basis for comparison of the concentrations of any two gases, which makes it more difficult to visualize chemical relationships with this system. Also the concentration of a gas in a gas, specified in these units, does change with changes in temperature or pressure. This happens because the mass (or weight) of a minor component per unit volume in this system becomes smaller as the temperature of the gas mixture is raised, whereas the volume of the air (or other main component) in the mixture becomes larger. Both influences tend to make the specified concentration smaller with a rise in temperature. Since 0°C has been used as "standard temperature" by some authorities [20] and 25°C by others [21] for specifying weight per unit volume gas concentrations, to avoid ambiguity in specifying results in these units the temperature or the standard temperature basis used *must* also be specified. Data presented in this format are normally corrected to 1 atm pressure.

Since both volume for volume and weight for volume units are in common use for quoting air pollution results, it is necessary to be able to interconvert between these units. To obtain a value in mg/m^3 from a value in ppm one has to multiply the ppm value by the molecular weight of the component of interest in grams, and divide by 24.46, the molar volume at 25°C (Eq. 2.7).

$$\text{mg/m}^3 = \frac{1/10^6 \text{ (ppm)} \times \text{mol.wgt. (g/mol)} \times 10^3 \text{ (mg/g)}}{24.46 \text{ (L/mol)} \times 10^{-3} \text{ (m3/L)}} \qquad 2.7$$
$$= (\text{ppm} \times \text{mol.wgt. (g/mol)})/24.46 \text{ (L/mol), at 25°C}$$

The conversion process is similar for μg/m^3 and mg/L from ppm, again both at 25°C (Eq. 2.8, 2.9).

$$\text{μg/m}^3 = (\text{ppm} \times \text{mol.wgt. (g/mol)} \times 10^3)/24.46 \text{ (L/mol)} \qquad 2.8$$

$$\text{mg/L} = (\text{ppm} \times \text{mol.wgt. (g/mol)} \times 10^{-3})/24.46 \text{ (L/mol)} \qquad 2.9$$

For weight for unit volume values at temperatures other that 25°C the molar volume for the temperature of interest must be used in place of 24.46 L/mol.

The partial pressure system is also occasionally used to specify the concentration of a gas in a gas (e.g., Eq. 2.10).

$$\text{ppm of constituent} = \frac{(\text{partial pressure of constituent}) \times 10^6}{(\text{total barometric pressure})} \qquad 2.10$$

Common pressure units are used for the partial pressure of the component of interest and for the total barometric pressure readings, so that the units divide out. Thus, the partial pressure system is also dimensionless and in all important respects is equivalent to the volume per unit volume system for specifying concentrations of a gas in a gas. This system is of particular value, for instance, when making up synthetic mixtures of gases on a vacuum line, where the partial pressure of each component is known and can be used to determine the relative concentrations.

2.6.2. Wet Chemical Analysis of Gases

Sometimes the chemical capture reagent actually forms a part of the wet chemical analytical scheme to be used. For example, if hydrogen peroxide is used as the sulfur dioxide capture reagent, then titration using standard alkali gives the final sulfuric acid concentration obtained in the capture solution (Eq. 2.5). Relating the measured acid concentration back to the original volume of air passed through the absorbing solution then gives the sulfur dioxide concentration originally present in the air. The answer obtained by this method is subject to errors when other acidic or basic gases (e.g., NO, NO_2, NH_3) or aqueous aerosols containing sulfuric acid, nitric acid, ammonium hydroxide, or the like are present in the air being sampled. The influence of these interferences can be greatly decreased if the determination is carried out gravimetrically by the addition of barium chloride rather than by titration (Eq. 2.11).

$$H_2SO_4 + BaCl_2 \rightarrow 2\ HCl + BaSO_4\downarrow \qquad 2.11$$

When sodium hydroxide in water is used as the medium to trap sulfur dioxide it allows a separate result to be obtained for the sulfur dioxide content, independent of any sulfur trioxide or sulfuric acid present in the sampled gas. This distinction is not possible when using hydrogen peroxide. To obtain this separate result requires addition of an involatile acid to the collection medium after collection. This is followed by heating, or sparging of the acidified solution with inert gas while heating, to release the sulfur dioxide from solution (Eq. 2.12).

$$Na_2SO_3 + H_2SO_4 \rightarrow Na_2SO_4 + H_2O + SO_2\uparrow \qquad 2.12$$

Any sulfur trioxide forms sulfate which is involatile on acidification, unlike sulfite. Capture of the released sulfur dioxide in fresh aqueous base, followed by titration with standard iodine solution, then gives the concentration of sulfur dioxide present in this second solution (Eq. 2.13, 2.14).

$$SO_2 + 2\ NaOH \rightarrow Na_2SO_3 + H_2O \qquad 2.13$$

$$\underset{\text{brown}}{SO_3^{2-}} + \underset{\text{colorless}}{I_2} \rightarrow SO_3 + 2\ I^- \qquad 2.14$$

Similar wet chemical techniques have been developed for many other polluting gases, but are beyond the scope of this outline. The general features of wet chemical analysis are such that the investment in equipment and materials required is generally low, but the skill level required of the analyst, particularly for some of the tests, is high. While it is possible to streamline some of the wet chemical methods so as to enable many results to be obtained in a working day, these methods are generally slow relative to other alternatives. Comprehensive accounts of suitable methods are available from many sources [22].

One gas analysis technique borrows from the colorimetry methods of manual procedures and the technology of the determination is prepackaged in sealed adsorbent tubes. Draeger and now also the Gastec and the Matheson-Kitagawa systems all use glass tubes packed with a solid support coated with an appropriate colorimetric reagent for the gas of interest. Company literature assesses the coefficient of variation for the tubes with most readily discernable color changes as 10%, and for the less efficient tubes as 20 to 30% [23] (Eq. 2.15–2.17).

$$\text{standard deviation} = [(x_1^2 + x_2^2 + \dots + x_n^2)/(n - 1)]^{0.5}, \qquad 2.15$$

where x_1, x_2, $\dots$ x_n are the deviations of individual determinations from the mean of all determinations in the series being measured, and n represents the number of individual determinations in the series.

$$\text{mean deviation} = (x_1 + x_2 + \dots + x_n)/n, \qquad 2.16$$

where only the absolute values of the individual deviations are used, without regard to sign. It is also referred to as the average deviation.

$$\text{coefficient of variation} = \frac{(\text{standard deviation})}{(\text{mean deviation})} \qquad 2.17$$

This is certainly a sufficient level of precision for the occasional determination of the less common gases, and provides a much more convenient method for this purpose than most alternatives (Table 2.3). It should be noted that in the early days of the development of these systems a tube for carbon tetrachloride vapor was questioned as to its capability to achieve a precision within 50% of stated values [26]. Nevertheless, even this level of accuracy is still adequate for many regulatory and industrial hygiene requirements.

2.6.3. Instrumental Methods for Gas Analysis

Instrumental analysis of air samples can frequently be conducted directly on the air or gas sample, particularly when either the concentration of the contaminant is high or when the analytical method being used is sensitive. Otherwise, prior concentration of the component of interest is necessary. This may be accomplished by adsorption onto a substrate from low concentrations, to be later released at higher concentrations by heating, evacuation, or the like. Or the concentration step may be accomplished by absorption from the air, as fine bubbles discharged under the surface of an entrapping liquid. The sampling method will be selected on the basis of the type of information required from the sample and the instrumentation that is to be used for analysis.

Spectrophotometric methods, both ultraviolet and infrared, are useful for air analysis in a wide variety of formats. Either a solution or a gas sample

■ **TABLE 2.3 Some Examples of the Gases Which May Be Determined by Prepacked Colorimetric Tubes**[a]

| Gas or vapor | Measurement[b] range | |
	Minimum	Maximum
Ammonia	5 ppm	10%
Benzene vapor	5 ppm	420 ppm
Carbon monoxide	5 ppm	7%
Chlorine	0.2	500 ppm
Ethylene oxide	5	3.5%
Formaldehyde	0.5 ppm	40 ppm
Hydrogen sulfide	1 ppm	7%
Mercury vapor	0.1 mg/m^3	2 mg/m^3
Methane thiol (methyl mercaptan)	2 ppm	100 ppm
Nitrogen dioxide	0.5 ppm	50 ppm
Oxygen	5%	23%
Ozone	0.05 ppm	300 ppm
Perchloroethylene	5 ppm	1.4%
Phenol	5 ppm	—[c]
Sulfur dioxide	0.1 ppm	2000 ppm
Tetrahydrofuran	100 ppm	2500 ppm
Vinyl chloride	0.5 ppm	50 ppm
Xylene	5 ppm	1000 ppm

[a]Compiled from Leichnitz [23], Matheson-Kitagawa [24], and Gastec Corp. [25].

[b]Concentrations are specified by volume (see text). Wider concentration ranges are covered with two to five detector tubes of differing sensitivities.

[c]Not specified.

cell may be used. For good ultraviolet sensitivity the component of interest must have two or more double bonds present, preferably conjugated. Thus sulfur dioxide, acrolein (CH_2=CH–CHO), and benzene all give very good ultraviolet sensitivity, when measurements are taken at suitable wavelengths near their absorption maxima. In contrast acetone, phosgene, and gasoline all have direct atmospheric measurement detection limits of only about 5 ppm, too high to be useful for many situations. However, even if direct measurement sensitivities are poor, a colorimetric method coupled with ultraviolet spectrophotometry may be used to obtain good sensitivity. Ultraviolet absorption, for example, is significantly more sensitive for the purple complex formed by reaction of formaldehyde with chromotropic acid than it is for formaldehyde itself. In this way solution detection limits of 3 μM (0.1 mg/L) can be achieved (Eq. 2.18).

$$H_2CO \text{ (colorless)} + \text{chromotropic acid} \rightarrow \text{purple complex} \qquad 2.18$$

For a 10-L air sample this sensitivity corresponds to an ambient air detection limit of 0.1 mg/m^3.

Ultraviolet absorption by sulfur dioxide may also be usefully applied in the field for direct plume observation. In a novel technique developed for the observation of the colorless discharge of sulfur dioxide and water vapor from a natural gas cleaning plant vent stack, use of silica optics and a special film allow the clear photographic differentiation and observation of plume behavior from the ultraviolet absorption of the sulfur dioxide component [27, 28].

The infrared spectra of many air pollutants are generally more complex than their ultraviolet spectra, since the infrared usually reveals several absorption maxima. But for many single components the pattern of maxima obtained gives a "fingerprint," which can permit positive identification of the material at the same time as obtaining the concentration from the absorbance readings. All organic compounds absorb infrared energy although the intensity of the absorption varies greatly from compound to compound. This affects the relative sensitivity of the method for different compounds. However, recently developed multiple path cells compensate for this somewhat by providing effective absorption path lengths of 20 m or more in a complex but compact gas cell which uses internal reflections (Fig. 2.4). Typical infrared sensitivities quoted for this system are 1.2 ppm for carbon monoxide, 1.5 ppm and 0.08 ppm for nitric oxide (NO) and nitrous (N_2O) oxide, respectively, 0.1 ppm for sulfur dioxide, and 0.05 ppm for carbon tetrachloride. The particular wavelengths at which different compounds absorb also varies, a feature used in dispersive infrared instruments to identify the absorbing species. Some of these infrared instruments are conveniently portable and fitted for 12-V power, suitable for operating from an ordinary car battery. These field capabilities avoid many of the deterioration problems faced by sample containment systems required for field collection and then transport to the laboratory for analysis.

Rapid multiple scans, signal storage, plus Fourier transform capabilities have combined to push infrared detection limits to about 1/100 of the concentrations possible with the more conventional infrared instruments just discussed, but at about 10 times the cost [30].

The really significant spectroscopic developments that relate to air pollutant analysis are closely linked to progress in laser technology. The coherence and high resolution possible with a laser source, coupled with the high

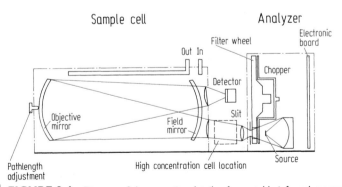

FIGURE 2.4 Diagram of the operating details of a portable infrared gas analyzer which uses a long path, multiple internal reflection gas cell. Adjusting the angle of the objective mirror alters the number of internal reflections obtained before the beam enters the detector. (From Wilks Scientific [29], courtesy of the Foxboro Company.)

power available with some configurations, makes inherently high sensitivities possible, particularly where the match between laser and absorbing frequencies is good. For example, it has been estimated that these methods enable sensitivities of the order of 1 ppb with nitrous oxide, using a 100-m path length. Even lower sensitivities are possible for more strongly absorbing gases [31]. With high source powers lasers may be used for remote sensing of air pollutants in real time, such as in smog situations, discharge plumes, or even for ambient air pollutant concentrations in open countryside [32].

Various laser instrument arrangements are possible depending on the particular remote sensing application (Fig. 2.5). The direct absorption mode (Fig. 2.5a) is the simplest, but suffers from the inconvenience that source and detector units must be sited separately, and aligned for each particular test situation. Retroreflection, either from a surface-silvered mirror or from a geographical feature such as a tree or rock (Fig. 2.5b) at least allows source and receiver to be operated from the same site. In the backscatter mode (Fig. 2.5c) the reflections from aerosol particles which have passed through the target air

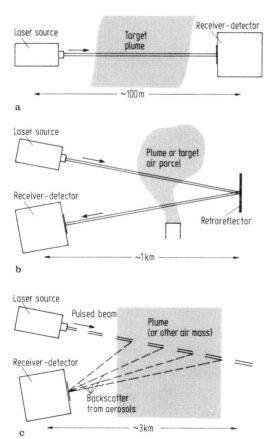

FIGURE 2.5 Some possible configurations of laser source and receiver detectors for use in remote sensing of air pollutants. (a) Simple direct absorption mode. (b) Retroreflective direct absorption mode. (c) Backscatter from aerosols mode.

sample are picked up by the receiver-detector operating at the same location. In this situation the laser is tuned to a high absorption frequency of interest and then to a low absorption frequency near this. The concentration of the component of interest may be computed from a comparison of the relative absorption signals, the basis of the so-called DIAL (differential absorption lidar) system [33].

One other nonspectroscopic instrumental technique which should be mentioned as being particularly appropriate to the analysis of gaseous air pollutants is gas chromatography, particularly when this is coupled to an efficient mass spectrometer. Gas chromatographic techniques allow efficient separation and detection of many common and less common air pollutants. With the more sensitive columns and an electron-capture or similar sensitivity detector it can give a measurable signal with as little as 10^{-12} g of separated component. But gas chromatography itself does not permit unequivocal identification of the separated component. Coupling a quadrupole mass spectrometer to the exit of a gas chromatograph allows both functions to be performed, the gas chromatograph to carry out separations of constituents, and the rapid acting mass spectrometer to identify them from their fragmentation patterns or molecular ions. This combination of instruments is able to provide rapid answers for the real time tracking of gases, vapors, and fumes released from a train derailment, or other emergencies which involve released chemicals [34].

Many other instrumental methods are also used for gas analysis, the details of which are available elsewhere [35]. Development, standardization, and calibration of the various methods used for particular pollutants of interest are continuing and ongoing processes.

Real-time measurements of gas concentration changes in plume chemistry can involve some ingenuity in delivery to instrumentation to the site of interest. Full sized, fixed wing aircraft or helicopters cause extensive mixing of the plume which makes the measured concentrations unrepresentative and at considerable risk to the aircraft and the stack. A small, unmanned radio-controlled aircraft carrying a compact gas chromatograph can take measurements without disturbing the plume and without risk. Freely rising helium-filled balloons can be used to lift instrumentation through the various layers of the atmosphere to obtain a snapshot of readings. A continuous profile of readings from one location may be obtained from a balloon (to 2 km), or a box-kite (to 10 km) tethered with Kevlar cord. Freely floating tetroons, which can be designed for neutral buoyancy with their instrumentation load, can give longer term results at an approximate altitude over a horizontal distance determined by wind speed and direction.

2.6.4. Biological Methods for Air Pollution Assessment

High concentrations of air pollutants are known to kill many annual plants and trees. This amounts to a biological indicator, albeit a coarse one. Less severe exposures can cause premature senescence (early leaf drop) of sensitive species of trees and shrubs [36]. For example, the aspen poplar, *Populus tremuloides* drops its leaves after exposure to as little as 0.34 ppm of sulfur dioxide for 1 hr. Early leaf drop slows tree growth, a measurable result which can be qualitatively determined by examining growth ring widths from tree cores. Several leaf drops in quick succession can kill trees or annual plants.

The olfactory membranes, as another kind of biological indicator, also provide an odor warning of air pollutant exposure to man and many animals. In some cases, for example in specially trained dogs, the sense of smell can rival the sensitivity of the most sophisticated instrumentation [37]. Since many toxic gases also possess an odor the avoidance response that this prompts can also help to reduce exposure.

Some examples of gases with which this warning effect is pronounced and some for which it is nonexistent are given in Table 2.4. It should be remembered that the sensitivity of the sense of smell can vary widely with individuals and also that olfactory membranes are subject to narcosis (anesthesia) by some of these gases, such as hydrogen sulfide and formaldehyde, which can remove the safety warning provided by odor for a gradual onset of exposure or for longer term exposures.

The lichens, as a group, can also be a generally useful, semi-quantitative indicator of pollution levels over time and space in both city and rural study areas [38]. Lichens represent a slow growing symbiosis between an alga, which photosynthetically manufactures carbohydrate, and a fungus, which aids the alga in water storage in an otherwise inhospitable site and uses the manufactured carbohydrate for its own metabolism and growth. Their very slow growth, and their high reliance on the air for most of the lichens' nutrient requirements, makes for efficient uptake and retention of many air pollutants. Hence, the level of air pollution more significantly affects the growth and reproduction of lichens than that of most other, faster growing forms of plants which rely on a soil substrate. Heavy metal fumes or other particulate air pollutant exposures also tend to accumulate to higher concentrations in lichens than in most other forms of plants.

TABLE 2.4   Examples of Extent of Warning Given from Relationship of Human Olfactory Thresholds of Some Common Gases to the Time-Weighted Average Industrial Hygiene Standards[a]

Gas or vapor	Olfactory threshold (OT)	Industrial hygiene standard[a] (HYG)	Warning effect, (HYG) ÷ (OT)
H_2S	25 ppb	10 ppm	400
O_3	<10 ppb	100 ppb	10
SO_2	0.3 ppm	2 ppm	7
NH_3	16 ppb	25 ppm	1.6
H_2CO	1 ppm	1 ppm	1
CO_2	1%	1%	1
CCl_4	540 ppm	2 ppm	0.004
CO	odorless	35 ppm	0

[a]No effect level for a healthy adult for time-weighted average exposures for an 8-hr working day over a 40-hr workweek.

In the hands of skilled botanists a lichen diversity and distribution study of an area can give useful long-term exposure information regarding continuous or intermittent fumigation by pollutants such as sulfur dioxide which would be difficult to obtain in any other way [39]. Even in the hands of the relatively uninitiated but careful observer it is possible to determine up to six or seven zones (levels) of relative air pollutant exposures by recording the diversity and types of lichens found growing in an area. Thus, in a relatively polluted area few or no healthy lichen varieties will be found. Those more resistant varieties that do occur in this exposed situation, such as species of *Pleurococcus,* will be recumbent forms which are closely bound to their support, a habit that minimizes exposure of the lichen to air pollutants. At the other extreme, in areas little exposed to air pollutants, both the species diversity and the number of specimens of each species found will be larger. There will also be a better representation of the more pollution sensitive foliose (leafy) forms such as species of the genera *Parmelia* and *Letharia,* or the fruiticose (shrubby) forms such as species of the genera *Usnea* or *Alectoria.* Biological methods based on lichen surveys such as this represent an important complement to the information obtained from wet chemical and instrumental analysis for determining the integrated levels of air pollution over time for an area.

2.7. EFFECTS OF AIR POLLUTANTS

Much has been said and written about the effects of air pollutants on plants and animals so that only a brief summary needs to be presented here. The health effects on man probably comprise the most direct, even if somewhat anthropocentric, concern with air pollutants. The primarily nuisance effect of the smell emanating from a fish packing plant represents an example at the minor end of the effects range. However, since the average person takes in some 22,000 breaths per day, amounting to some 14 to 16 kg (30 to 35 lb) of air per day, and this contacts the large area of efficient moist exchange membranes of the alveoli in the lungs even a relatively low concentration of many air pollutants is sufficient to have a noticeable effect. For this reason alone the air environment has to be the most important of the biospheric elements to be worth striving to maintain or improve in quality.

One logical approach which may be used to organize the consideration of air pollution problems and solutions is to start from the microscopic, or individual level with treatment of indoor air pollution, where the problems are of a scale more or less under the control of individuals or small groups of people. From these we can move to consideration of larger and larger scale macroscopic effects which become progressively more complex to understand and model. The scale and scope of these problems also become progressively more difficult for individuals, countries, or even the whole world to control. Topics include smogs, acid rain, arctic haze, climate effects, and damage to the stratospheric ozone layer as we progress up the scale of influence of these problems.

2.7.1. Indoor Air Pollution

Ever since people used natural shelters and lit poorly vented fires for warmth, to the building of small or large and complex facilities for living space, offices, factory operations, and warehousing, indoor air pollution to varying degrees has existed. For structures of all kinds in which involuntary air infiltration (drafts) and/or people and goods traffic, significant air exchange occurs and the quality of indoor air is seldom a problem. But with recent energy conservation concerns, tighter construction of new buildings and adjustments to the ventilation equipment of existing structures to provide lower levels of air exchange is in some instances sacrificing indoor air quality to save energy. Both in this area of stationary enclosed space and in the provision of fresh air in the transportation sector, for an example to the passenger cabins of modern jet aircraft, there are instances where the small savings in energy achieved to one stakeholder by these practices is being far exceeded by the larger system costs (general malaise, illnesses, absenteeism, and the like) [40].

We can use conditions common to the individual home or apartment to illustrate the concepts involved within a framework which most can relate to their own experiences. The degassing of composite woods such as chipboards, plywood, and hardboard as well as vapor loss from the more aromatic solid woods such as cedar and pine can contribute to the contaminants in home air. The finishes on any of these woods or on wall surfaces also make contributions, particularly in the period shortly after application. Carpeting and the many easy-care flooring materials together with soft furnishings also release small quantities of vapors, particularly when new. Human presence and activities in the residential space also add to these static sources. Oxygen is consumed and carbon dioxide, moisture, and heat are released by respiration in proportions which vary somewhat with the composition of foodstuffs being metabolized [40] (e.g., Eq. 2.19, 2.20).

$$(C_5H_{10}O_6)_n + {}^3/_2n\ O_2 \rightarrow 5n\ CO_2 + 5n\ H_2O$$
$$\text{starch}$$

2.19

$$(C_{51}H_{92}O_6)_n + 71\ O_2 \rightarrow 51\ CO_2 + 46\ H_2O$$
$$\text{typical fat}$$

2.20

Heat generated by a conscious person at rest is approximately 90 W. This means that excessive heating is probably the commonest discomfort factor that builds up in the air of some densely seated auditoria, but heating from this source is seldom a problem in private accommodation.

Entering the space, movement, and cleaning activities within the space raises dust levels, and cooking activities put moisture, heat (and with gas stoves, carbon dioxide and additional moisture), aerosol droplets of fat, etc., into the air. Some of the pollutants, for example the smell of baking bread, are much appreciated. But the U.S. Environmental Protection Agency has required the installation of control equipment for the volatile organic compounds released from some large commercial bakeries to help reduce contributions to photochemical smog. Unvented combustion units such as gas stoves and space heaters contribute their combustion gases to the indoor air, for which control is usually recommended via ventilation. Less pleasant cooking smells may be masked by room deodorizers or room ozonizers, which make their own contributions. As a final component, plants exposed to light undertake photosynthesis causing a net consumption of carbon dioxide (and some pollutants) and production of oxygen. But at night, metabolism continues in the plants consuming oxygen and producing carbon dioxide without the more than offsetting photosynthetic processes producing oxygen which occur during the day.

Various indoor air quality guidelines exist. To determine whether or not a guideline is being met one can use the ventilation rate formula to evaluate the situation (Eq. 2.21).

$$\text{allowable concentration of contaminant} = \frac{\text{rate of contaminant loss}}{\text{rate of dilution air supply}} \quad 2.21$$

If the allowable concentration is specified in volume for volume units, %, ppm, ppb, etc., then that is used (times the appropriate factor) on the left-hand side. In this instance volume per unit time in the same units should be used for both of the substitutions of the right-hand side. The ventilation rate equation is also valid for use when contaminant concentration allowed is specified in mass per unit volume units such as mg/L, mg/m^3, $\mu g/m^3$, etc. In this case mass per unit time units are used in the numerator and volume per unit time in the denominator of Eq. 2.21. To use Eq. 2.21 in either format does not require knowledge of the volume of enclosed space being ventilated. However, it does assume perfect uniformity of the gas mixture, and perfect mixing of the air with the contaminant.

Sometimes the rate of air exchange of an enclosed space is specified as air changes per hour (ACH, Eq. 2.22).

$$\text{air changes per hour} = \frac{\text{ventilation rate per hour}}{\text{volume of enclosed space}} \quad 2.22$$

In this instance one can use the ventilation formula structured to use the air changes per hour term (Eq. 2.23).

$$R = k_{ex}[C_i]v, \text{ where} \quad 2.23$$

R = rate of production of gas or vapor
K_{ex} = air changes per hour
C_i = interior equilibrium concentration
v = volume of enclosed space.

Use of this equation requires the same assumptions as Eq. 2.21. It is only valid if there is essentially zero concentration of the contaminant of interest in the outside air being used for ventilation.

If there *is* a significant concentration of the gas or vapor of interest in the outside air, for example carbon dioxide, then Eq. 2.24 rather than 2.23 must be used to account for this.

$$R + k_{ex}[C_o]v = k_{ex}[C_i]v \quad 2.24$$

where C_o is the concentration of gas or vapor in the outside air. Again the same assumptions as for Eq. 2.21 are necessary.

How are these equations used? Let us consider a propane-powered forklift truck that is producing 2.0 g/min carbon monoxide and operating in a warehouse. The industrial hygiene standard for carbon monoxide is 35 ppm. What minimum ventilation rate is required for the warehouse to ensure safe working conditions? To convert the mass of CO per minute to a volume, we can assume conditions of 1 atm pressure, 20°C.

$$v = \frac{2.0 \text{ g/min CO}}{28.01 \text{ g/mol}} \times \frac{0.0821 \text{ L atm K}^{-1} \text{ mol}^{-1} \times 293 \text{ K}}{1 \text{ atm}}$$
$$= 1.72 \text{ L/min CO}$$

Ventilation rate required for 35 ppm concentration is:

$$35 \text{ ppm} = \frac{1.72 \text{ L/min CO}}{x \text{ L/min air}}$$
$$\frac{35}{10^6} = \frac{1.72 \text{ L/min CO}}{x \text{ L/min air}}$$
$$x = 4.91 \times 10^4 \text{ L/min air, or } 49.1 \text{ m}^3/\text{min air}$$

The same ventilation would be required for a small or a large warehouse. A smaller warehouse would simply reach the equilibrium carbon monoxide concentration more quickly than the large one, assuming the same ventilation rates.

To convert the ventilation rate in m³/min to units of air changes per hour requires:

$$49.1 \text{ m}^3/\text{min} \times 60 \text{ min/hr} = 2948.6 \text{ m}^3/\text{hr}$$

For a nearly empty warehouse of $25 \times 40 \times 6$ m a ventilation rate of 2948.6 m³/hr requires:

$$\frac{2948.6 \text{ m}^3/\text{hr}}{20 \text{ m} \times 40 \text{ m} \times 6 \text{ m}} = 0.614 \text{ air changes per hour}$$

As an example of the use of Eq. 2.24, use the 1000 ppm carbon dioxide general guideline for interior air quality. The same propane-powered forklift produces 20 L/min carbon dioxide. What air change rate would be required to maintain the carbon dioxide guideline for the warehouse? Would it be the carbon monoxide or carbon dioxide requirement that would dictate the ventilation rate required? First we rearrange Eq. 2.24 to the form:

$$k_{ex} = \frac{R}{v([C_i] - [C_o])}$$

Taking the concentration of carbon dioxide in the outside air as 350 ppm (350×10^{-6}), and substituting:

$$k_{ex} = \frac{20 \text{ L/min} \times 10^{-3} \text{ m}^3/\text{L} \times 60 \text{ min/hr}}{4800 \text{ m}^3 \times ((1000 \times 10^{-6}) - (350 \times 10^{-6}))}$$
$$= 0.3846, \text{ or } 0.385 \text{ air changes per hour}$$

Comparing the 0.385 ACH required for carbon dioxide control with the 0.614 ACH required for safe control of carbon monoxide levels, it is clear that control of the carbon monoxide concentration determines the safe ventilation rate required.

2.7.2. Classical and Photochemical Smogs

Localized air pollution episodes tend to occur in areas subject to inversions. These stagnant air events will tend to occur in regions with low or no winds. A valley location or a plain bounded by mountains will also tend to increase the occurrences and persistence of inversions. In the normal daytime situation the sun warms the surface of the earth and the air mass immediately above it. Warming lowers the density of this surface air causing it to rise and in the process to mix with the upper air levels. This dilutes any pollutants discharged into the surface air.

At night, or when there is fog or another meteorological event which brings an inversion to the area, the situation changes. During the inversion, the air close to the surface 100 to 300 m is relatively more dense than the layers immediately above it, which creates a stable situation. There is little or no mixing of any pollutants which may be discharged into the surface air which causes a gradual buildup of these to uncomfortable or dangerous concentrations in this layer.

The elevated air pollutant levels which occur during localized classical smog episodes tend to severely impact on the chronically ill, the young, and the old. They can cause effects ranging from watering of the eyes and restricted breathing, to an aggravation of respiratory illnesses, and even to a noticeable rise in the death rate recorded for the affected area during the most severe episodes [41]. London, England, the Donora valley, Pennsylvania, and the Meuse valley, Belgium, are just a few of the documented older examples of occurrences of this last level of severity. A classical smog can also serve to drop the pH of wetted surfaces sufficiently to seriously damage statuary and stone or masonry buildings from a simple solution reaction of limestone or marble stonework with the acidic water (e.g., Eq. 2.25).

$$CaCO_3 + 2\,HNO_3 \rightarrow Ca(NO_3)_2 + H_2O + CO_2 \qquad\qquad 2.25$$
$$\text{insoluble} \qquad\qquad\qquad \text{soluble}$$

Today, two clearly differentiated types of smogs are recognized. In the classical variety of smog, represented by the pollution episodes mentioned above, the problem was caused by the accumulation of primary air pollutants such as sulfur dioxide, particulates, and carbon monoxide contributed by smoke, usually complicated by the presence of fog (Table 2.5). The fog also tended to slow air pollutant dispersal by cutting off sunlight, preventing the warming of air close to the earth's surface by the sun. The characteristics and time of occurrence of a classical smog differ markedly from the more recent phenomenon of photochemical smog which was first described for Los Angeles, but which is now also experienced by Tokyo, Mexico City, and other major centers with similar conditions [43].

TABLE 2.5   Distinguishing Features of Classical and Photochemical Smogs[a]

	Classical smog	Photochemical smog
Location example	London, 1950s	Los Angeles
Peak time of occurrence	winter	summer
Conditions	early a.m., 0 to 5°C, high humidity plus fog	around noon, 22–35°C, low humidity, clear sky
Atmospheric chemistry	primarily reducing, SO_2, particulates, carbon monoxide, moisture	oxidizing, nitrogen dioxide, ozone, peroxy-acetylnitrate
Human effects	chest, bronchial irritation	primarily eye irritation
Underlying causes	fog plus stable high, surface inversion, dispersal of primary pollutant emission is prevented, accumulate	sheltered basin, frequent stable highs, accumulation of secondary pollutants from photochemical oxidation of hydrocarbons

[a]Compiled from Williamson [42] and Kerr *et al.* [43].

In photochemical smog episodes, secondary air pollutants such as ozone, nitrogen dioxide, aldehydes, and peroxyacetylnitrate are formed as a result of the chemical interaction of the primary air pollutants, principally nitric oxide and hydrocarbon vapors, with sunlight and air (Fig. 2.6) [44]. This interpretation of the processes involved has been verified by smog chamber experiments (Fig. 2.7), and as the sensitivity of ambient air instrumentation has been improved has since been confirmed from field measurements [46]. In photochemical smog episodes it is the secondary pollutants that cause the severe eye irritation and upper respiratory effects felt by people and simultaneously damage plants.

The authorities of the London area were very successful in reducing emissions of sulfur dioxide and particulates sufficiently via passage of their Clean Air Act so that classical smogs are a thing of the past in this area. But the more frequent and intense sunshine experienced as a result of this improvement coupled with a rise in hydrocarbon concentrations from an increase in automobile traffic means that London, too, now experiences the irritating effects of photochemical smog occurrences [47].

Similar processes cause the attack of ironwork and other metalwork, and accelerate the decay of exposed wood. Even aluminum, which is highly favored for use in exterior metalwork because of its resistance to corrosion under ordinary conditions, is now showing the classic pits and erosion scars of corrosion when used in areas which have severe air pollution.

Initiation (at sunrise, with high reactive hydrocarbon concentration):

$$RH \text{ (hydrocarbon)} + O_2 \xrightarrow[\text{[O]}]{O_3, \text{ or}} R^{\cdot} + HOO^{\cdot}$$

$$HOO^{\cdot} + NO \rightarrow HO^{\cdot} + NO_2$$
$$NO + NO_2 + H_2O \rightarrow 2 \text{ HONO}$$
$$HOO^{\cdot} + NO_2 \rightarrow HONO + O_2$$
$$HONO \xrightarrow[290\text{--}400 \text{ nm}]{h\nu} HO^{\cdot} + NO$$

Propagation (converts reactive hydrocarbons to oxidized products):

$$HO^{\cdot} + RH \rightarrow H_2O_2 + R^{\cdot}$$
$$R^{\cdot} + O_2 \rightarrow ROO^{\cdot}$$
$$HOO^{\cdot} + RH \rightarrow H_2O_2 + R^{\cdot}$$

Termination (dominates at sundown):

$$R^{\cdot} + R^{\cdot} \rightarrow R\text{---}R$$
$$HO^{\cdot} + R^{\cdot} \rightarrow ROH$$

$$CH_3\text{---}\overset{\overset{\textstyle O}{\|}}{C}\text{---}OO^{\cdot} + NO_2 \rightarrow CH_3\text{---}\overset{\overset{\textstyle O}{\|}}{C}\text{---}OONO_2 \text{ (PAN)}$$

FIGURE 2.6 Generalized radical chain reactions for typical processes which occur during a photochemical smog episode.

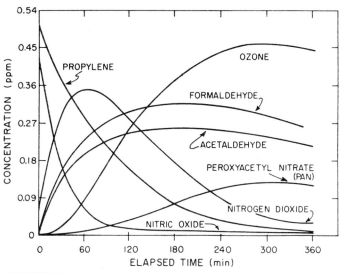

FIGURE 2.7 Results of a typical smog chamber experiment in which initial 0.5 ppm propylene, 0.45 ppm NO, and 0.05 ppm NO_2 in otherwise clean air is irradiated with artificial sunlight and analyzed at intervals. (From Lloyd et al. [45], with permission.)

2.7.3. Acid Rain

The normal pH of rain or melted snow in equilibrium with 360 ppm of atmospheric carbon dioxide is 5.6 (Eq. 2.26, 2.27).

$$H_2O + CO_2 \rightleftharpoons H_2CO_3 \qquad\qquad 2.26$$

$$H_2CO_3 + H_2O \rightleftharpoons H_3O^+ + HCO_3^- \qquad\qquad 2.27$$

When pHs of less than this value occur in precipitation above the freezing point it is referred to as acid rain.

This lowered pH is caused by rainout and washout of nitrate and sulfate from the atmosphere (Table 2.6) [48]. pHs of as low as 2.4, about the acidity of lemon juice or vinegar, have been measured for the rainwater of an individual storm in Pitlochry, Scotland. Also an annual pH value for precipitation of 3.78 has also been recorded for the Netherlands. It is evident that these acid pHs can cause structural damage to buildings in the manner just described for classical smogs. Lowered pH precipitation is also seriously affecting the biota of those lakes which have a limited carbonate–bicarbonate natural buffering capacity [50]. When the ionic exchange capacity of soils becomes exhausted it, too, can drop in pH to result in reduced growth, or actual die-back of forests. Liming of affected lakes to raise the pH of affected lakes into the normal range has been practiced successfully on an experimental scale. But this measure can obviously be only a local and temporary solution.

2.7.4. Arctic Haze

Related to smogs but only relatively recently noticed by the scientific community is the phenomenon of Arctic haze, a brownish turbidity occurring in the atmosphere of Arctic regions from late fall to March or April of each year. A suspended aerosol of primarily sulfates (2 μg/m^3), organic carbon (1 μg/m^3), and black carbon (0.3–0.5 μg/m^3) reduces the visibility in the region to

TABLE 2.6 Contributions of Sulfuric and Nitric Acids to the Acidity of Acid Rain[a]

Substance	Concentration (mg/L)	Contribution to total acidity[b]
H_2SO_4	5.10	57
HNO_3	4.40	39
NH_4^+	0.92	51
H_2CO_3	0.62	20
All others	ca. 0.4	ca. 12

[a]Data selected from Likens [9]. Results determined for a sample of rain of pH 4.01 collected at Ithaca, New York, in October 1975.

[b]Microequivalents per liter. Ammonium by titration to pH 9.0.

3 to 8 km, over an area of several thousand square kilometers and up to an altitude of about 3000 m during this period. The novelty here is that this represents an accumulation of pollutants, remote from the original points of discharge, which by the idiosyncracies of atmospheric movements and conditions resides over and adversely influences a site not responsible for the emissions.

This tendency of aerosols to reduce visibility during Arctic haze episodes is also a very general effect of atmospheric particles. It is caused by the scattering of light by particles or droplets in the aerosol size range. Apart from aesthetic considerations, high loadings of solid and liquid (fogs) atmospheric aerosols also influence flight conditions at airports, and have been implicated in anomalous rainfall patterns of the St. Louis area.

2.7.5. Human Effects of Particulate Exposure

Another important consideration related to atmospheric loadings and particulate size distributions concerns the potential for human effects on inhalation. The body's defenses in the upper respiratory system are adequate to trap more than 50% of particles larger than about 2 μm in diameter which are present in the air breathed. However, particles smaller than this are not efficiently captured by the upper respiratory tract. These "respirable particulates" penetrate the lungs to the alveolar level. Here, the more vigorous Brownian motion of the small aerosols increases their collision frequency with the moist walls of the alveoli, thus trapping a large proportion of them at these sites. This occurs in a region of the respiratory system poorly equipped to degrade or flush out accumulated material, particularly if the aerosol is inert or insoluble. Hence, any physiological effect of the presence of these foreign substances is aggravated, which tends to increase the incidence of respiratory illness experienced in areas that have high concentrations of polluting aerosols. The efficiencies of many types of air pollution control equipment for the lower aerosol size range is poor, an important consideration when the decision is made regarding process emission control options [51].

2.7.6. Climatic Effects

Proceeding from the local and regional effects of air pollutants to the global scale, there is no doubt that our escalating use of fossil fuels coupled with the widespread cutting of forests have contributed to a steady rise in the atmospheric carbon dioxide concentration, now averaging about 0.7 ppm per year (Fig. 2.8) [52]. Carbon dioxide is virtually transparent to the short-wavelength (ca. 1 μm) maximum in the incoming solar radiation, but has a substantial absorption band in the region of the longer wavelength infrared irradiation emanating from the relatively low surface temperature of the earth. Thus, an increase in the concentration of atmospheric carbon dioxide does not significantly affect the energy gained by the earth's surface from incoming solar radiation, because of the atmospheric absorption "window" in this region.

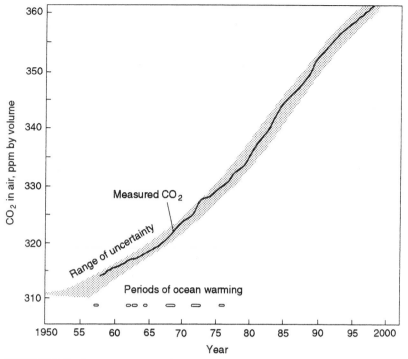

FIGURE 2.8 Past perspectives and present trends of atmospheric carbon dioxide concentrations at the Mauna Loa Observatory, Hawaii, compared with the recent periods of high solar flux and ocean warming intervals. (Assembled from the data of Lepkowski [53], Kerr [54], and extrapolated carbon dioxide concentration data of Keeling *et al.* [55].)

But the carbon dioxide and some of the other trace gases of the atmosphere absorb a significant fraction of the radiant energy in the outgoing long wavelengths, that would otherwise be lost (Table 2.7). This capture of infrared energy, dubbed the greenhouse effect, is what keeps the biosphere of the earth sufficiently warm to sustain life as we know it, and is therefore desirable. However, as the warming effect of this suite of gases becomes stronger it is predicted from current climate models that this warming effect will become incrementally greater. It is the extent and importance of this influence in relation to a number of other factors that is still not well understood.

Ice cores from Arctic and Antarctic ice caps have provided atmospheric carbon dioxide concentration data for periods of up to 160,000 years before the present. During most of this period, when mankind contributed little to the atmospheric carbon dioxide, there were nevertheless wide (180-300 ppm) fluctuations in concentrations [59, 60]. Climate studies have correlated warm periods with the periods of high atmospheric carbon dioxide, but the data are inadequately resolved to establish which caused which. Since other major factors are also known to influence climate it is possible that one or several of

TABLE 2.7 Concentrations, Trends, and Warming Effect of the Greenhouse Gases[a,b]

Gas	Present atmospheric concentration (by vol.)	Rate of increase per year (%)	Atmospheric lifetime	Contribution to Warming Effect[b,c]	Contribution to Warming Relative to CO_2
CO_2	350 ppm	1.2 ppm (0.35)	$\sim 500\ yr^d$	49%	1
CH_4	1.7 ppm	0.018 ppm (1.2)	7–10 yr	18%	20
$O_3 < 12$ km	0.02–0.1 ppm	variable	< few hr	see below	—
N_2O	310 ppb	0.58 ppb (0.019)	150 yr	6%	200
CFC 11 ($CFCl_3$)	230 ppt	8.9 ppt (0.056)	75 yr ⎱	14%	1000
CFC 12 (CF_2Cl_2)	380 ppt	5.0 ppt (0.018)	110 yr ⎰		
Methyl chloroform (CH_3CCl_3)	130 ppt	5.8 ppt (0.058)	8–10 yr	—	—
CCl_4	120 ppt	2.2 ppt (0.017)	—	$\sim 13\%$ (incl. O_3)	—
Halons	<100 ppt	—	25–110 yr	—	—

[a]Calculated and compiled from Hileman [57], Johnston [58], and Palmer [59].
[b]Water vapor and clouds are estimated to have twice the effect of all the gases listed here.
[c]Estimated in proportion to other greenhouse gases.
[d]Turnover rate is about 20% per year.

these exerted the primary influence and carbon dioxide and climate change simply tracked this.

Computer-based climate models are constantly being refined to try to provide more reliable predictions. Cloud cover and atmospheric moisture are currently estimated to have about twice the warming effect of the whole list of gases of Table 2.7, and this component is so far one of the more difficult ones to model. Many other very complex interactions are involved in climatic effects, among them lateral and vertical perturbations of ocean currents, changes in prevailing winds, periodic cycles in the earth's tilt and orbit, and changes in the solar flux. As the periods of these effects vary and the warming effect of some of these may be augmented by the warming effect or may be cancelled out by the cooling effect of others, the net effect is at best difficult to predict.

While the increase in average global temperatures that the greenhouse effect brings about may be only a Celsius degree or less, it could still cause a gradual decrease in water storage in the earth's ice caps. This in turn could raise the world's ocean levels, causing changes to low lying coastlines. Also, an increase in average global temperatures could increase the rate and extent of desertification of marginally arable land from changes in wind and rainfall patterns, reducing the already limited agricultural resource. Imperfect models predict the noticeable onset of some of these effects by the year 2050. Thus, the possible effects are sufficiently serious that we should try to start taking action that might decrease these effects now, at least with less drastic measures.

Chlorofluorocarbon production has recently been stopped for global warming and stratospheric ozone preservation reasons. Many nations are also involved in active discussion concerning possible measures to decrease carbon dioxide emissions. Examination of the annual atmospheric fluxes of carbon dioxide into and out of the atmosphere establishes that human activity probably contributes less than 5% of the total (Fig. 2.9). Consequently truly drastic

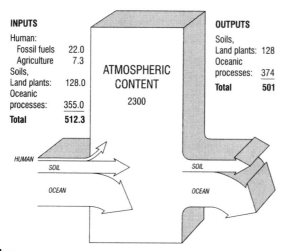

INPUTS

Human:
Fossil fuels   22.0
Agriculture   7.3
Soils,
Land plants: 128.0
Oceanic
processes: 355.0

Total   **512.3**

ATMOSPHERIC
CONTENT
2300

OUTPUTS

Soils,
Land plants: 128
Oceanic
processes: 374

Total   **501**

HUMAN
SOIL
OCEAN

SOIL
OCEAN

FIGURE 2.9 Estimated annual atmospheric fluxes of carbon dioxide into and out of the atmosphere in billions of metric tonnes carbon dioxide. (Compiled from the data of Bolin [61] and McHale [62].)

reductions in fossil fuel and biomass combustion would be necessary to have a significant effect. Preliminary experiments in which iron sulfate was added at the 2 nM level to otherwise fertile regions of the oceans doubled the phytoplankton growth [63]. This is a measure which might also help.

2.7.7. Stratospheric Ozone

The other potential global problem from increased atmospheric pollutant levels, and of chlorofluorocarbons, stable chlorocarbons, and fixed nitrogen in particular, concerns the destructive impact of these gases on the stratospheric concentrations of ozone [64, 65]. The ozone layer, consisting of ozone (O_3) concentrations of about 10^{11} to 10^{13} molecules/cm^3, exists largely in the stratosphere, 11 to 50 km above the earth's surface. Ozone formation in this region occurs by the interaction of molecular oxygen with sunlight, followed

Normal ozone formation:

$$O_2 + h\nu \rightarrow 2O$$
$$O + O_2 + M \rightarrow O_3 + M \text{ (energy sink)}$$

Ozone filter function:

$$O_3 + h\nu \rightarrow O_2 + O(^1D) \text{ (electronically excited oxygen)}$$

NO_x involvement:

$$NO + O_3 \rightarrow NO_2 + O_2$$
$$NO_2 + O \rightarrow NO + O_2 \text{ (a cycle)}$$

$$\overline{O_3 + O \rightarrow 2\,O_2 \text{ (net reaction)}}$$

Chlorofluorocarbon (CFC) involvement:
Primary (initiation)

$$CCl_3F + h\nu \rightarrow Cl + CCl_2F$$
(CFC 11)
$$CCl_2F_2 + h\nu \rightarrow Cl + CClF_2$$
(CFC 12)

Ozone destruction (propagation)

$$Cl + O_3 \rightarrow ClO + O_2$$
$$NO + O_3 \rightarrow NO_2 + O_2$$
$$ClO + NO_2 \rightarrow NO_3 + ClONO_2 \text{ (chemically stable)}$$
$$ClONO_2 + h\nu \rightarrow NO_3 + Cl$$
$$NO_3 + h\nu \rightarrow NO + O_2$$

$$\overline{O_3 + O_3 \rightarrow 3\,O_2 \text{ (net reaction)}}$$

FIGURE 2.10 Outline of representative normal processes and some interfering reactions which affect the equilibrium concentration of stratospheric ozone.

by the reaction of the atomic oxygen with further oxygen (Fig. 2.10). Its value to life on earth is the powerful filtering function that it passively performs on the short-wavelength UV-B (ca. 280-320 nm region) of sunlight before it reaches the earth's surface (Fig. 2.10). The equilibrium concentrations of ozone present are a resultant of these primary ozone formation and photolytic removal processes.

Chlorofluorocarbons (CFCs), when first produced as the working fluid for air conditioning and refrigeration systems, were valued for their safety and stability. But it is this very stability that has led to their contribution to damage of the ozone layer. Most gases and vapors discharged in the troposphere are destroyed or returned to the earth's surface in particulate form long before they reach the stratosphere. The high stability of the CFCs permits them to survive passage through the troposphere, to reach the stratosphere. Here, exposure to short-wavelength ultraviolet from the sun causes bond homolysis, releasing the destructive chlorine atoms. These chlorine atoms, in their turn, contribute many cycles of ozone removal reactions before ultimately forming a stable product which leads to their own removal.

It took some years to prove that this interference would cause a decrease in the equilibrium ozone concentration present [64, 65]. The net effect of a decrease in the ozone concentration was expected to cause an increase in the global exposure to short ultraviolet. This, in turn, was anticipated could cause a general increase in mutations in exposed species of the biosphere and more particularly in the incidence of human skin cancer (melanomas) [66]. Fortunately the downward trend in atmospheric CFC concentrations in recent years resulting from international regulations which curtailed production and use by the industry promises to minimize the influence of these effects [67]. Newer working fluids for cooling units which either contain hydrogen or contain no chlorine have been developed. These have much lower or zero ozone-destroying potential as compared to the earlier CFCs [68].

REVIEW QUESTIONS

1. (a) Calculate the gas or vapor concentrations, in milligrams per cubic meter (mg/m^3) at 25°C and normal atmospheric pressure, equivalent to 1.0 ppm (by volume) of ozone, nitrogen dioxide, sulfur dioxide, carbon monoxide, and mercury vapor.
   (b) To what concentration, in mg/m^3 also at 400°C, would a flue gas sulfur dioxide concentration of 1000 ppm at 400°C correspond? Assume 1 atm pressure and ideal gas behavior.
   (c) The flue gas from part (b), when cooled to 25°C, still contains about 1000 ppm sulfur dioxide. What would be the sulfur dioxide concentration of this gas now, in mg/m^3 at 25°C?
2. A sea level air pollutant analysis is carried out by passing 100 m^3 of air, measured at 0°C and 760 mm Hg, through an impinger containing 100 mL of 0.10 M aqueous sodium hydroxide. Subsequently, the entire contents of the impinger gave a titer to pH 8 of 35.6 mL of 0.100 M sulfuric acid.

(a) Assuming that sulfur dioxide is the only acid gas constituent in the air sampled that is titrated at the pH range given, what concentration of sulfur dioxide, in mg/m^3, would these results correspond to?

(b) When further acid was added, to bring the pH to about 4, bubbles were observed to form in the glass titration vessel. What gas are these bubbles likely to be?

(c) Consider an initially dry impinger tube, cooled in ice/water to chill the incoming air to 0°C for the analysis and placed ahead of the tube containing absorbing solution. The tube contains several milliliters of "water" at the end of the passage of the air sample. Might this observation affect the accuracy of the result quoted from part (a), and if so, how? What very simple test could you use to check this qualitatively?

3. A power station burns powdered anthracite coal which analyzes 88% C, 4% S, and 8% ash. If this facility used 20% more air than theoretically required to ensure complete combustion, and the sulfur was entirely converted to sulfur dioxide, what concentration of sulfur dioxide (ppm by volume) would be present in the flue gas if there were no emission controls? Assume air is 20% O_2, 80% N_2 by volume.

4. A certain old automobile uses 1 L of oil for every 500 km traveled. Assuming that oil contains 0.70 kg of carbon per liter, and that one-half the oil loss ended up as carbon monoxide, how many liters of carbon monoxide (at 0°C and 760 mm Hg pressure) would this car produce from the oil alone per 10,000 km of travel?

5. The average 1980 automobile discharges 1 g of nitric oxide (NO) per kilometer traveled. Assuming a gasoline (C_8H_{18}, density 0.9 g/cm^3) consumption of 10 km/L under stoichiometric combustion conditions, what volume concentration of nitric oxide would be discharged in the exhaust? Take air to be exactly 1 mol:4 mol, oxygen:nitrogen.

6. Coal is burned at the rate of about 21,000 tonnes per day in the Sundance Steam Electric Plant in Alberta, to produce 1725 MW of power.

(a) If the coal contains 0.6% sulfur and 11% ash, what masses of sulfur dioxide and ash would be produced daily?

(b) Assuming that the remainder of the coal is carbon and that combustion of the carbon and sulfur is accomplished with the stoichiometric amount of air, what volume concentration of sulfur dioxide and carbon dioxide would be present in the flue gases produced? Assume air composition is 1 O_2:4 N_2

7. (a) Briefly explain the significance of sunlight in the formation of photochemical smogs.

(b) Calculate the elemental percent composition by weight and the percent by weight of nitrogen dioxide in peroxyacetyl nitrate ($CH_3CO_3NO_2$), one of the more stable and irritating end products of this process.

8. In the Biosphere Experiment concluded on September 26, 1993, eight people stayed in a materially closed environment for 2 yr. Assume that each person consumed 500 g dry weight of starch $(C_6H_{10}O_5)_n$ daily during this experiment, and completely metabolized this component of their food to carbon dioxide and water.

   (a) What masses of carbon dioxide and water would the Biosphere closed system have to deal with annually from the human metabolism of starch?

   (b) Based on the assumptions above what annual mass of carbon dioxide would be produced annually by the current world population of 4×10^9 people with the same daily per capita starch metabolism as given above?

   (c) What fraction of the carbon dioxide produced by the annual global combustion of 11.4×10^9 tonnes of coal-equivalent fossil fuels (assume C content 85%) does the estimated human metabolic contribution represent?

9. The threshold limit value (TLV) for 1,2-dichloroethane (CH_2ClCH_2Cl) vapor in the air of work environments has been set at 100 ppm by volume.

   (a) What ventilation rate would be required to maintain safe working conditions in a dry cleaning establishment losing 1.0×10^{-3} m^3 of solvent vapor per minute, assuming uniform composition and perfect mixing of the air in the establishment?

   (b) Is the required ventilation rate affected by the size of the building? Explain.

   (c) Assuming 20°C and 1 atm conditions, what does 1×10^{-3} m^3 min^{-1} of dichloroethane vapor correspond to as a *mass* rate of emission?

   (d) What is a TLV of 100 ppm by volume for 1,2-dichloroethane equivalent to, in mg m^{-3} (at 25°C, 1 atm)?

   (e) The TLV for ethyl alcohol (CH_3CH_2OH) vapor is set at 1900 mg m^{-3}. What ventilation rate is required to maintain this standard in a distillery that is experiencing alcohol loss at the rate of 10 mg min^{-1}.

FURTHER READING

J.W. Berks, J.G. Calvert, and R.E. Sievers, eds., "The Chemistry of the Atmosphere: Its Impact on Global Change." American Chemical Society, Washington, DC, 1993.

P. Brimblecombe, "Air Composition and Chemistry," 2nd ed., Cambridge University Press, Cambridge, UK, 1996.

G.D. Clayton and F.E. Clayton, eds., "Patty's Industrial Hygiene and Toxicology," 3rd ed.,Vol. 1. Wiley, New York, 1978.

B.W. Ferry, M.S. Baddeley, and D.L. Hawkesworth, eds., "Air Pollution and Lichens," Athlone Press, London, 1973.

L.B. Lave, E.G. Seskin, and M.J. Chappie, "Air Pollution and Human Health." Johns Hopkins University Press, Baltimore, 1977.

H.B. Singh, ed., "Composition, Chemistry, and Climate of the Atmosphere." Van Nostrand-Reinhold, New York, 1995.

W.H. Smith, "Air Pollution and Forests: Interactions Between Air Contaminants and Forest Eco-
    systems." Springer-Verlag, New York, 1981.
State of the Environment Reporting, The State of Canada's Environment, Government of Canada,
    Ottawa, 1991.
A.C. Stern, ed., "Air Pollution," 3rd ed., Vol. 1. "Air Pollutants: Their Transformation and
    Transport," Academic Press, New York. 1976.
A.C. Stern, ed., "Air Pollution," 3rd ed., Vols. 2 and 5. Academic Press, New York, 1977.

REFERENCES

1. R. Findley, *Nat. Geogr.* **159**(1), 3, Jan. 1981.
2. M.B. Hocking and J.A. Jaworski, eds., "Effects of Mercury in the Canadian Environment."
    National Research Council, Ottawa, 1979.
3. S.H. Williston, *J. Geophys. Res.* **73**(22), 7051 (1968).
4. S.A. Berry and B.G. Notion, *Water Res.* **10**, 323 (1976).
5. "Atmospheric Aerosol: Source Air/Quality Relationships," ACS Symp. Ser. No. 167. Ameri-
    can Chemical Society, Washington, DC, 1981.
6. C.F. Barrett, *R. Inst. Chem. Rev.* **3**(2), 119. Oct. (1970).
7. D.J. Spedding, "Air Pollution." Clarendon Press, Oxford, 1974.
8. H.P. Munger, The Spectrum of Particle Size and its Relation to Air Pollution. In "Air Pol-
    lution" (L.C. McCabe, ed.), Chapter 16. McGraw-Hill, New York, 1952.
9. J.A. Dorsey and J.O. Burckle, *Chem. Eng. Prog.* **67**(8), 92, Aug. 1971.
10. "Joint Air Pollution Study of St. Clair-Detroit River Areas for International Joint Commission,
    Canada and the United States." International Joint Commission, Ottawa and Washington,
    DC, 1971.
11. M.J. Pilat, D.S. Ensor, and J.C. Bosch, *Atmos. Environ.* **4**, 671 (1970).
12. L. Svarovsky, *Chem. Ind. (London),* p. 626, Aug. 7 (1976).
13. A.W. Gnyp, S.J.W. Price, C.C. St. Pierre, and J. Steiner, *Water and Pollut. Control* **111**(7),
    40 (1973).
14. "Detection and Analysis of Particulate Contamination." Millipore Corp., Bedford, MA, 1966.
15. W.C. McCrone, R.G. Draftz, and J.G. Delly, "The Particle Atlas." Ann Arbor Sci. Publ., Ann
    Arbor, MI, 1971.
16. W. Strauss, ed., "Air Pollution Control," Part III: Measuring and Monitoring Air Pollutants,
    Wiley-Interscience, New York, 1978.
17. Tenax for Trapping, *Gas-Chromatogr. Newsl.* **21**(2), 8, June (1980).
18. M.B. Hocking and G.W. Lee, *Water, Air, Soil Pollut.* **8**, 255 (1977).
19. F.A. Patty, ed., "Industrial Hygiene and Toxicology," 2nd ed., Vol. 2. Interscience, New York,
    1963.
20. Committee on Environment Improvement, "Cleaning Our Environment: A Chemical Per-
    spective," 2nd ed., American Chemical Society, Washington, DC, 1978.
21. G.D. Clayton and F.E. Clayton, eds., "Patty's Industrial Hugiene and Toxicology," 3rd ed.,
    Vol. 2A. Wiley, New York, 1981.
22. M.B. Jacobs, "The Chemical Analysis of Air Pollutants." Interscience, New York, 1960.
23. K. Leichnitz, "Detector Tube Handbook," 3rd ed. Draegerwerk AG, Lubeck, W. Germany,
    1976.
24. "Matheson-Kitagawa Toxic Gas Detector System." Matheson, Lyndhurst, NJ, 1980.
25. "Gastec Precision Gas Detector System." Gastec Corp., Tokyo, ca. 1974.
26. R.M. Ash and J.R. Lynch, *J. Am. Ind. Hyg. Assoc.* **32**(8), 552 (1971).
27. J.J. Havlena, D. Hocking, and R.D. Rowe, "The Photography of Sulfur Recovery Plant
    Plumes," Proc. Alta Sulfur Gas Workshop II, Kananaskis, Alta., 1975, p. 30.
28. Photographing SO_2 Emissions, *Chem. Can.* **26**(11), 16, Dec. (1974).
29. "Quantitative Analysis with the Miran Gas Analyzer," Seminar. Wilks Scientific, South Nor-
    walk, CT, 1973.
30. J.N. Pitts, Jr., B.J. Finlayson-Pitts, and A.M. Winer, *Environ. Sci. Technol.* **11**(6), 568 (1977).
31. J.R. Alkins, *Anal. Chem.* **47**(8), 752A (1975).

32. W.F. Herget and W.D. Connor, *Environ. Sci. Technol.* **11**(10), 962 (1977).
33. E.D. Hinkley, *Environ. Sci. Technol.* **11**(6), 564 (1977).
34. D.A. Lane and B.A. Thomson, *J. Air Pollut. Control Assoc.* **31**, 122, Feb (1981).
35. R. Perry and R.M. Harrison, *Chem. Brit.* **12**(6), 185, June (1976).
36. D. Hocking and M.B. Hocking, *Environ. Pollut.* **13**, 57 (1977).
37. W. Summer, "Odour Pollution of Air." CRC Press, Cleveland, OH, 1971.
38. W.C. Denison and S.M. Carpenter, "A Guide to Air Quality Monitoring with Lichens." Lichen Technology, Corvallis, OR, 1973.
39. A.C. Skorepa and D.H. Vitt, "A Quantitative Study of Epiphytic Lichen Vegetation in Relation to SO_2 Pollution in Western Alberta," Inf. Rep. NOR-X-161. Northern Forest Research Centre, Edmonton, Alberta, 1976.
40. M.B. Hocking, Indoor Air Quality: Recommendations Relevant to Aircraft Passenger Cabins. (Submitted)
41. D.V. Bates, "A Citizen's Guide to Air Pollution." McGill-Queen's University Press, Montreal, 1972.
42. S.J. Williamson, "Fundamentals of Air Pollution." Addison-Wesley, Reading, MA, 1973.
43. J.A. Kerr, J.G. Calvert, and K.L. Demerjain, *Chem. Br.* **8**(6), 252, June (1972).
44. J.N. Pitts, Jr. and B.J. Finlayson, *Angew. Chem., Int. Ed. Engl.* **14**(1), 1 (1975).
45. A.C. Lloyd, W.P. Carter, and J.L. Sprung, *Calif. Air Environ.* **4**(3), 1, Spring (1974).
46. R.M. Harrison and C.D. Holman, *Chem. Br.* **18**(8), 563, Aug. 1982.
47. R.M. Harrison and C.D. Holman, *Atmos. Environ.* **13**, 1535 (1979).
48. E.B. Cowling, *Environ. Sci. Technol.* **16**(2), 110A (1982).
49. G.E. Likens, *Chem. Eng. News* **54**(48), 29, Nov. 22 (1976).
50. B. Hileman, *Environ. Sci. Technol.* **15**(10), 1119 (1981).
51. S.K. Friedlander, *Environ. Sci. Technol.* **7**(13), 1115 (1973).
52. C.S. Wong, *Science* **200**, 197 (1978).
53. W. Lepkowski, *Chem. Eng. News* **55**(42), 26, (1977).
54. R.A. Kerr, *Science* **269**, 633 (1995).
55. C.D. Keeling, R.B. Bacastow, A.F. Carter, S.C. Piper, T.P. Whorf, M. Heimann, W.G. Mook, and H. Roelloffzen. In "Aspects of Climate Variability in the Pacific and the Western Americas," Geophys. Monogr., Am. Geophys. Union, **55**, 1989.
56. V. Ramanathan, *Science* **240**, 293 (1988).
57. B. Hileman, *Chem. Eng. News* **67**(11), 25, Mar. 13 (1989).
58. H.S. Johnson, *Annu. Rev. Phys. Chem.* **35**, 481–505 (1984).
59. B.J. Palmer, *Environ. Letters* **5**(4), 249 (1973).
60. CO_2 Buildup's Effect on Climate Explored, *Chem. Eng. News* **55**(31), 18, Aug. 1 (1977).
61. B. Bolin, *Sci. American* **223**, 124 (1970).
62. J. McHale, "The Ecological Context," G. Braziller, New York, 1970.
63. R. Rawls, Ironing the Ocean, Iron, a Micronutrient for Marine Plants. *Chem. Eng. News* **74**(45), 40–43, Nov. 4 (1996).
64. M.J. Molina and F.S. Rowland, *Nature (London)* **249**, 810 (1974).
65. R.M. Baum, *Chem. Eng. News* **60**(37), 21, Sept. 13 (1982).
66. T.M. Sudgen and T.F. West, eds., "Chlorofluorocarbons in the Environment: The Aerosol Controversy." Society of Chemical Industry, Ellis Horwood, Chichester, UK, 1980.
67. R.A. Rasmussen, M.A.K. Kahlil, and R.W. Dalluge, *Science* **211**, 285 (1981).
68. L.E. Manzer, *Science* **249**, 31–35 (1990).

Oil refinery emergency flare for safe combustion of hydrocarbons during process upsets.

3
AIR POLLUTION CONTROL PRIORITIES AND METHODS

There is not always good cheer where the chimney smokes.

—Thomas Shelton, 1620

3.1. AIR POLLUTANT INVENTORIES

To apply the best pollution control strategies, it is important to start with an inventory of the sources, mass emission rates, and types of pollutants being discharged in the area of concern. This is accomplished by a detailed analysis of the emissions from each point source of the inventory area. Data obtained by actual measurements are used wherever possible. But where measured data are not available, chemical and engineering theory is applied to come up with the best possible emission estimates within the time frame required for the inventory. Moving sources, such as automobiles, aircraft, and the like are tallied by averages per unit, times the number of kilometers per unit, times the number of units to come up with good estimates for the contributions from these sources.

The final pollutant inventory obtained gives an overview of the total current emission picture for the study area. Table 3.1 presents this data for Los Angeles, where the inventory was conducted to assist in determining the origins or causes of the photochemical smog problem. For comparison, parallel data are presented for the whole of Canada which also has incipient smog problems in some of the major centers. The best foundation for meaningful planning of abatement measures is obtained from inventories of this kind. From the tabulated data it is evident that automobiles rather than industrial activity are the source of by far the largest mass of pollutants overall (Table 3.1). Road transport, consisting of multiple moving sources under the control of individuals makes emission abatement for this classification difficult to

TABLE 3.1 Air Emission Inventory for Los Angeles in November 1973 Compared on a Percentage Basis to the Annual Inventory for Canada for 1974 and 1985[a]

Emission source	Emissions to the atmosphere (tonnes/day)					Percent of total		
	Hydro-carbons	Nitrogen oxides	Sulfur dioxide	Carbon monoxide	Totals, by source	Los Angeles	Canada[b] 1974	1985
Automobiles	1750	445	27	9375	11,597	86.9	47.9	41.0
Organic solvent use	450				450	3.4		10.2
Oil refining	200	41	41	154	436	3.2	27.1	20.8
Chemicals production	50		60		110	0.8		
Combustion of fuels[c]	13	245	280	1	539	4.1	9.1	16.9
Miscellaneous	37	27	5	144	213	1.6	15.9	10.5
Totals, by pollutant	2500	758	413	9674	13,345	100.0	100.0	100.0

[a]From Acres [1], The Clean Air Act [2], and Environment Canada (available from http://www/ec.gc.ca/pdb/cacbk_e.html).
[b]Canadian air pollutant inventory totals for the listed categories for 1974 and 1985 was 25.8 million tonnes, respectively. In 1985 road vehicles produced 30.4%, 51.4% and 55% of the hydrocarbons, NO_x and CO, respectively.
[c]From stationary sources.

achieve. Nevertheless, by fundamental engine design changes coupled to accessory control units installed at the manufacturing stage, and by the application of emission control regulations which have to be met annually with vehicle license renewal, significant reductions have been achieved.

3.2. AUTOMOTIVE EMISSION CONTROL

As much as 70% of the hydrocarbons, 98% of the carbon monoxide, and 60% of the nitrogen oxides (NO_x), have originated from the operation of automobiles (Table 3.1). Hence, these are the principal automotive emissions of concern. The hydrocarbons and carbon monoxide arise from the fact that the most power for a given engine capacity (displacement) is obtained with enough air for only 70 to 80% complete combustion. Nitrogen oxides are formed from the interactions of atomic oxygen and atomic nitrogen, which are formed against hot metal surfaces at high temperatures, with the corresponding elements (Eq. 3.1, 3.2).

$$N_2 + [O] \rightarrow NO + [N] \qquad 3.1$$

$$O_2 + [N] \rightarrow NO + [O] \qquad 3.2$$

These emission problems are not easy to solve. To run with a leaner tuned engine (a higher air to fuel ratio) does obtain more complete fuel combustion and in so doing decreases the hydrocarbon and carbon monoxide concentrations in the exhaust, as would be expected. But at the same time the higher combustion temperatures obtained in the process raise the concentrations of nitrogen oxides obtained in the exhaust, and decrease the power output per liter of engine displacement [1] (Fig. 3.1). It also affects drivability, or smooth engine response, at various throttle settings.

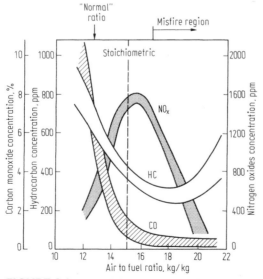

FIGURE 3.1 Plot of the approximate effect of various air to fuel ratios on the composition of automotive exhaust gas. Compiled from the data of Acres [1] and Stoker and Seager [3].

Early emission control methods were based on the use of a thermal reactor, for hydrocarbon and carbon monoxide oxidation, combined with exhaust gas recirculation (EGR) for reduction of nitrogen oxide emissions (Fig. 3.2a). Hydrocarbons and carbon monoxide in the hot exhaust fed to the reactor, once heated, were rapidly oxidized to carbon dioxide and water by the additional pumped air which was fed to the reactor (e.g., Eq. 3.3, 3.4).

$$HC \ (= hydrocarbon) + O_2 \rightarrow CO_2 + H_2O \qquad\qquad 3.3$$

$$2\ CO + O_2 \rightarrow 2\ CO_2 \qquad\qquad 3.4$$

Some of the relatively inert exhaust gas, before it entered the reactor, was used to dilute the air fed to the carburetor by some 15%, in so doing decreasing the peak combustion temperatures during normal engine operation. This effectively decreased the concentrations of nitrogen oxides formed. These combined measures achieved significant emission reductions (Table 3.2), but at the same time caused a noticeable increase in fuel consumption and loss of performance.

While thermal reactors have been used extensively, two-stage catalytic exhaust purification approached these emission problems in a different way. Here the first stage of control, NO_x reduction, is achieved catalytically using the reserve chemical reducing capacity of the residual hydrocarbons and carbon monoxide already present in the exhaust (Fig. 3.2b; e.g., Eq. 3.5–3.8).

$$2\ NO + 2\ CO \rightarrow N_2 + 2\ CO_2 \qquad\qquad 3.5$$

$$2\ NO_2 + 4\ CO \rightarrow N_2 + 4\ CO_2 \qquad\qquad 3.6$$

$$2\ NO + HC \ (= hydrocarbon) \rightarrow N_2 + CO + H_2O \qquad\qquad 3.7$$

$$2\ NO_2 + HC \rightarrow N_2 + CO_2 + H_2O \qquad\qquad 3.8$$

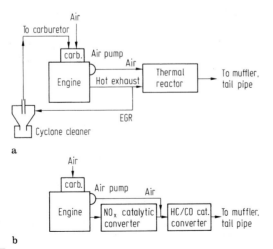

FIGURE 3.2 Block diagrams of two automotive emission control configurations. (a) Thermal reactor plus exhaust gas recirculation. (b) Catalytic exhaust purification.

■ TABLE 3.2   Typical Automotive Emissions Before and After Regulation, Related to U.S. Legislated Standards[a]

	Grams per mile				Gasoline evaporation (g/test)
	Carbon monoxide	Hydrocarbons	Nitrogen oxide	Particulates	
Before controls	80	11	4	0.3	60
1970 average	23	2.2	4–5	—	—
1975 standard	15	1.5	3.1	0.1	2
1980 standard	7.0	0.41	3.1	—	—
Thermal reactor + EGR[b]	4.5	0.2	1.3		
Catalytic + cyclone	0.85	0.11	1.65	0.1	1–2
1994 model year[c]	3.4	0.25	0.4		
2001, clean fuel fleets (proposed)[c]	3.4	0.075	0.2		

[a]Compiled from Acres [1], Mikita and Cantwell [4], *Chemical and Engineering News* [5], and Acres and Cooper [6].

[b]EGR, short for exhaust gas recirculation. Refers to the practice of recycling of about 10 to 15% of exhaust gases to the intake of the engine to decrease peak combustion temperatures.

[c]From new sequences under more severe test conditions, which includes measurements from a cold start.

All of these NO_x-reducing reactions are much more rapid and effective than the catalyzed direct redox reaction (Eq. 3.9),

$$2\,NO \xrightarrow{\text{cat.}} N_2 + O_2 \qquad\qquad 3.9$$

and hence convert most of the NO_x to innocuous gases. After NO_x reduction, secondary air injection and passage of the hot mixture into a second catalytic converter accomplishes complete oxidation of the residual hydrocarbons and carbon monoxide not used for NO_x reduction in the first converter (Eq. 3.7, 3.8). One or two grams of platinum or palladium finely distributed on a ceramic matrix provides the catalytic activity of both converters. This combination exhaust treatment achieves better control of the hydrocarbons and carbon monoxide, and almost as good control of nitrogen oxides as achieved with the thermal reactor, and at the same time restores much of the lost fuel economy and performance (Table 3.2). If a limited exhaust gas recirculation circuit is integrated with this dual converter system, the NO_x emission rate can be further improved.

Since lead-containing antiknock additives of leaded gasolines rapidly destroy the activity of the catalytic emission control systems, unleaded gasolines have to be used in automobiles fitted with this or related systems. However, since using this fuel modification also eliminates lead emissions this trend also has a highly favorable impact on the major source of lead particulate discharges to the atmosphere [7, 8]. Before unleaded gasolines were on the market it was common to find high lead levels in the core areas of busy cities and alongside freeways, a trend which is gradually reversing as the proportion of automobiles on the road equipped to burn unleaded gasolines increases.

While it is possible to reduce significantly particulate lead discharges from automobiles using leaded gasolines by employing cyclone-type particle collectors in the exhaust train [9], these methods only achieve an incremental reduction, not elimination, of lead dispersal from this source.

Today, to maintain or further improve the gaseous emission characteristics of internal combustion engines the trend is toward integral improvements in engine design, such as computer-controlled precision fuel injection and various stratified charge modifications which accomplish the chemistry of the add-on units but as an integral part of engine operation [10]. These developments are leading to more drivable, fuel efficient automobiles.

Combinations of these control measures have resulted in a decrease in ambient carbon monoxide concentrations, and in the severity and frequency of photochemical smog incidences, where they have been applied, even though there has been a concurrent increase in the number of automobiles on the road. Nevertheless, during severe and prolonged photochemical smog episodes it may be necessary to restrict or ban automobile use temporarily as a public safety measure, as was used in Mexico City during such an event in October–November 1996.

Electric automobiles may be a viable option. It is not clear that current technologies employing lead acid batteries charged by fossil-fueled power stations would necessarily result in improvement. However, methods which involve the refinement of higher charge density, longer lived batteries, or new innovations in fuel cell technology could improve the environmental merits of this alternative. Prototype electric automobiles which use solar cells for battery charging have been demonstrated, but the high costs of solar cells is likely to keep these experimental for the time being.

3.3. AIR POLLUTANT WEIGHTING

Accumulation of emission data on a mass basis is a required first step to assess the overall impact of air pollutant discharges of a region prior to decisions on

TABLE 3.3   Examples of Weighting Factors Used for Primary Air Pollutants

Contaminant	Weighting factors based on	
	California ambient air standards[a]	U.S. federal secondary air quality standards[b]
Carbon monoxide	1.00	1.00
Hydrocarbons	2.07	125
Sulfur oxides, SO_x	28.0	21.5
Nitrogen oxides, NO_x	77.8	22.4
Particulates	106.7	37.3
Oxidants	186.9	—

[a]Used as the basis of the proposed Pindex air pollutant rating system [11].
[b]Based primarily on federal standards with health effect considerations [12].

any necessary action. But for development of the most cost-effective control strategy for an area, the contributing pollutants should be considered on the basis of their relative significance in terms of health effects, smog occurrences, and the like, and not solely on a mass basis. Two of the various weighting factors which have been proposed to do this are given in Table 3.3. While the weightings of these two systems differ significantly from each other, particularly for hydrocarbon vapors, they both assign the lowest weighting to carbon monoxide [12, 13].

If the weighting factors based on California standards are used to recalculate the relative significance of the Los Angeles emission sources on this basis, transportation still remains as the largest single sector, but now only just larger than the combustion sources sector (Table 3.4). This treatment thus has the effect of redistributing the significance of emission control strategies among the transportation, oil refining and chemicals production, and combustion source sectors to obtain maximum effectiveness of ambient air improvement for this control area. It can be seen that a similar treatment of the Canadian data would give even higher emphasis to the need to control emissions from the industrial and power production sectors than found for Los Angeles.

3.4. METHODS AND LIMITATIONS OF AIR POLLUTANT DISPERSAL

Chimneys and vent stacks have been and still are popular for waste gas discharge from all types of stationary sources, large and small. Thermal power stations, smelters, refineries, and even domestic heating appliances all use discharge and dispersal methods to dispose of their waste gases. Serious problems can arise, however, if the source is a very large one (i.e., has a high mass rate of emission) discharging into a stable air mass, particularly in the absence of control of the more noxious components of its waste gas. With the occurrence of a long-term inversion a temporary shutdown of the facility may be necessary to protect public health.

Simple dispersal of the flue gases from large emission sources, when operated without emission controls, is less acceptable today because of the limited capability of the atmosphere of highly industrialized regions of the globe to accept further loading. Particulate matter discharged in this way only falls out again on the immediate area, and more or less rapidly. Discharged fumes, fogs, and mists too, although they are more widely dispersed, also eventually return to the earth's surface as they coagulate or coalesce into larger agglomerated particles, or are washed out of the atmosphere in precipitation. Discharged gases too, which have the best prospect of efficient dispersal, eventually return adsorbed onto, or reacted with other gases, particles, or water and also return in precipitation. So a higher chimney merely spreads the combined fallout from these processes over a wider area; it does not decrease the gross atmospheric loading or fallout rate.

Multiple high chimneys in a single area tend to produce plume overlap some distance downwind of the original discharge points. This effect serves

TABLE 3.4 Comparison of Actual Versus Weighted Daily Air Pollution Inventory for Los Angeles, November 1973[a]

Emission source	Total mass of gaseous primary air pollutants		Estimated mass of particulates and aerosols tonnes/day[b]		Weighted mass of primary pollutant emissions[c]	
	tonnes/day	% of total	Actual	Weighted	tonnes/day	% of total
Automobiles	11,597	86.9	74	7,896	56,271	41.5
Organic solvent use	450	3.4	0	0	930	0.7
Oil refining	436	3.2	123	13,124	19,714	14.6
Chemicals production	110	0.8				
Combustion of fuels	539	4.1	245	26,142	53,108	39.2
Miscellaneous	213	1.6	27	2,881	5,371	4.0
	13,345	100.0	469	50,043	135,394	100.0

[a]Weightings from California standards, Table 3.3. See also Babcock [11].
[b]Includes hydrocarbon vapors, nitrogen oxides, sulfur oxides, and carbon monoxide.
[c]Includes the primary gaseous air pollutants, plus particulates and aerosols.

to negate the reduction in ground level concentrations of air pollutants, the original objective of the high stack because of the reduced dilution by diffusion (Fig. 3.3a).

If stack dispersal is to be used, it is important that the stack be sited in such a way that adjacent buildings or natural features such as hills or gullies do not trap discharged waste gases close to the ground. Also to be avoided are backwash, a downdraft obtained on the leeward side of any large surface feature, or eddies which may occur in the same plane around the corners of obstructions. Any of these factors, if not allowed for, can cause plume impingement much closer to the point of discharge than anticipated, and hence cause highly elevated concentrations of the components of the discharged exhaust gases close to the ground.

A primary objective of stack discharge of waste gases is to obtain the minimum or zero elevation of the ground level concentrations of the discharged gases in the immediate area of the stack. The stack discharge point should be high enough that sufficient diffusion of the discharged gases occurs to make the concentrations of these gases acceptable at the point the diffusion cone intersects the ground. For any given mass emission rate, Eq. 3.10 may be used as a rough guide to the effect on the ground level concentration of a discharged gas which would be expected for different stack heights and wind velocities at the point of discharge.

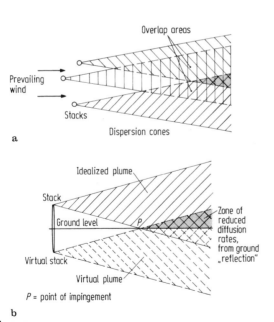

FIGURE 3.3 (a) Horizontal and (b) vertical sections of stack discharges showing how overlapping plumes and virtual stack effects increase ambient air concentrations of pollutants at ground level above that to be expected from simple diffusion cone calculations.

$$\text{ground level concentration} = K(m/uH^2), \text{where} \qquad 3.10$$

where   m = mass rate of emission
             u = wind velocity
             H = *effective* stack height.

In this expression the "effective stack height" corresponds to the physical stack height plus the plume rise. The plume rise is obtained from a combination of thermal (buoyant) rise and momentum rise, and the height component from this factor is defined as the vertical distance between the top of the stack and the point at which the center line of the plume levels off to within 10° of the horizontal. In the case of a warm or fan-forced discharge this plume rise component can add considerably to the actual stack height, significantly raising the effective height of the discharge. Most real situations, however, are far more complex than can be reasonably approximated using this expression, so that more complex expressions are usually required to accommodate the greater number of variables.

The point of impingement for particulates, or the location on the ground where discharge products of the stack are at a maximum, corresponds to about three or four times the effective stack height. For gases this corresponds to about 10 times the effective stack height. The influence of the "virtual stack effect" prevents further downward diffusional dilution of flue gases beyond the point of impingement because of the presence of the ground (Fig. 3.3b). Therefore, at points further from the stack than this the concentration of any discharged gases at ground level still continues to decrease, but is found to be roughly double what would be predicted from simple diffusion cone calculations.

The effective height of a chimney, or the efficiency of effluent dispersion may be improved by any of a number of measures. Addition of a jockey stack, a smaller diameter supplementary stack installed on top of an existing installation, can help. Or improvement may be gained by raising the flue gas temperature or velocity to obtain greater plume rise. Valley inversions may be avoided by running ductwork up a hillside for discharge via a stack at the top, taking advantage of both the increased height above the inversion layer and the higher wind speeds for more efficient dispersal. Diffusion and dispersal rates under stable air conditions may also be improved by the use of vortex rings or super high stacks. But ultimately, particularly with very large installations discharging in areas where inversions are frequent, containment measures of one sort or another in conjunction with stack discharge have to be considered.

3.5. AIR POLLUTION ABATEMENT BY CONTAINMENT

With the increasing dilution and dispersal burdens being placed on a finite volume of atmosphere it is becoming more and more clearly evident that containment or neutralization of air pollutants is required *before* discharge in order to prevent further deterioration in air quality. Air emission control measures which fall into this category may be precombustion or predispersal measures, which clean up a fuel or combustion process or modify a chemical process in such a way that the pollutant of concern is never dispersed into the exhaust gases. Or they may comprise measures taken to remove or neutralize the particulates, aerosols, and gases to be discharged *after* the combustion or chemical processing step. The choice of which action or combination of actions to be taken in any particular case depends on whether the process is already in operation or in the planning stage for construction, and the type of pre- or postcombustion emission control strategies which are available or may be readily developed. These options are each discussed separately.

3.5.1. Precombustion Removal Methods

If the technology is accessible and the cost of applying it is not too great, precombustion removal of potential air pollutants has a great deal of merit. At this stage not only is the offending substance (or substances) present in the highest concentration but it has not as yet been dispersed into the combustion or process exhaust gases, which should make recovery easier.

Consider, for example, the incombustibles in coal. These can contribute significantly to the particulate content of the flue gases produced on combustion, as well as to the bottom ash volume. Incombustibles are now removed before burning for more than half of the coal produced in North America. This is accomplished by using one or more of the following cleaning methods with coal that has been previously finely ground. It may be washed with water on riffle boards, during which the less dense coal particles are carried by the water stream while the more dense rock particles tend to sink and are captured in the cavities. In air jigs the powdered mixture is suspended on a bed of air in a fluidized form. In the jig the heavier rock particles tend to sink and may be drawn off the bottom, and the cleaned coal is drawn off the top. Or an actual liquid or fluidized dense solid may be used to obtain a more direct sink-float or froth flotation separation of the coal, of density about 1.5 g/cm^3, from the denser rock particles and other impurities.

In the process of the precombustion cleaning of powdered coal a significant proportion of the sulfur is also removed. Since about half of the sulfur compounds present in uncleaned coal resides in inorganic pyrite and sulfates and about half in organic sulfur compounds present in the coal itself, it is much of the inorganic sulfur that is removed by the particulate cleaning methods. This level of improvement can be further increased to some 80% or so of the total sulfur present by using various aqueous leaching chemicals such as ferric sulfate or sodium hydroxide, in the coal cleaning procedure. Extraction with an organic solvent, such as molten anthracene, *after* conventional coal cleaning [13] is also used (Eq. 3.11, 3.12).

Leaching:

$$FeS_2 + 8\,H_2O + 14\,Fe^{3+} \rightarrow 15\,Fe^{2+} + 2\,HSO_4^- + 14\,H^+ \qquad 3.11$$

Regeneration of leach solution:

$$2\,FeSO_4 + 1/2\,O_2 + H_2SO_4 \rightarrow Fe_2(SO_4)_3 + H_2O \qquad 3.12$$

Microbial methods of coal desulfurization have also been tested. By precombustion cleaning measures such as these the emission characteristics of coal-burning installations may be dramatically improved. Coal conversion to a liquid or gaseous fuel permits much greater improvement in the residual precombustion impurities present, but also at greater cost.

The same kinds of considerations apply in the precombustion emission control measures used for the nondistilled fractions of petroleum, the residual fuel oils, which are used as utility fuels. The residual fuel oils contain all of the involatiles of the original crude oil, concentrated by a factor of 6 to 10 times since the volatile fractions have been removed. As a result, these fuels possess a significant potential for particulate and sulfur oxides emission on combustion. Normal refinery crude oil desalting practice is capable of removing 95% or more of the particulate discharge problem before the crude is ever distilled. But to lower the sulfur dioxide emissions requires either use of only low sulfur crude oils for residual fuel formulation, or more complicated and expensive catalytic desulfurization processing of the residual feedstocks prior to formulation as a fuel [14].

Even the volatile petroleum fractions present in natural gas, principally methane, are not without their sulfur gas polluting potential if burned directly in the state in which it is obtained from the well head. But this is normally thoroughly desulfurized using an amine scrub before it is piped to industrial and domestic consumers as another precombustion cleaned fuel (Section 7.5).

3.6. POSTCOMBUSTION EMISSION CONTROL

3.6.1. Particulate and Aerosol Collection Theory

Of the two nongaseous air pollutant classifications, particulates and aerosols, the aerosols are much more difficult to control because they do not readily settle out of a gas stream of their own accord. The separation of aerosols from a waste gas stream is however greatly improved if the very small particles or droplets are agglomerated (coagulated or coalesced) to larger particles or drops, before removal is attempted. There are four primary means by which this may be achieved. Being aware of these contributing factors can provide a useful background to enable measures to be taken to accelerate the process, or make it more efficient.

Brownian agglomeration, an important contributing process to the natural removal of aerosols from the atmosphere, is the first of these (background, see Chap. 2). Microscopic particles of fumes and fogs are so small that the frequent collision of gas molecules with them causes them to move in an erratic path, commonly called Brownian motion. When this occurs with reasonably high concentrations of aerosol particles, i.e., high particle *numbers*, it results in frequent particle particle collisions. Particle collisions usually result in agglomeration by poorly understood mechanisms, forming loose clusters or chains with solid particles and larger drops with liquid droplets. The coagulation rate for any particular aerosol system is predictable to a significant extent (Eq. 3.13).

$$\text{Coagulation rate} = K_b c^2, \qquad\qquad 3.13$$

where K_b is the rate constant for Brownian coagulation of the particular type of particle, and c is the number of particles per cm^3. The coagulation rate is proportional to the square of the particle concentration. Thus, the number of particles per unit volume is the single most important factor affecting the rate of Brownian coagulation. Table 3.5 not only illustrates this aspect, but also gives the indication that concentrations of aerosols above about 10^6/cm^3 are unstable. Concentrations higher than this will rapidly drop to near this concentration range by agglomeration processes. As particles larger than ca. 5–10 μm form, i.e., in the particulate rather than the aerosol size range, they are removed by sedimentation. It is a combination of processes such as these which contribute to keeping the aerosol concentration in the earth's atmosphere in the range below about 10^4 to 10^5 particles/cm^3.

Turbulent coagulation occurs when particles in multiple flow line intersections collide, a process stimulated by turbulent flow or eddy conditions. The turbulent coagulation rate is again proportional to the square of the particle concentration (Eq. 3.14).

$$\text{Coagulation rate} = K_s c^2, \qquad\qquad 3.14$$

where K_s is the rate constant for turbulent coagulation. The magnitude of K_s, the proportionality constant for turbulent coagulation, is in the range of 10 to 4000 times K_b. Hence, turbulent coagulation can be used to accelerate greatly the natural agglomeration processes useful for emission control.

In natural situations, such as when burning wood out-of-doors, a smoke containing 10^6 particles/cm^3 may be generated. By a combination of Brownian and turbulent coagulation processes the average particle size in this smoke can increase from 0.2 to 0.4 μm before it has traveled 1000 m (Table 3.5).

TABLE 3.5 The Relationship of Coagulation Time to the Original Particle Concentration[a]

Initial number concentration of particles per cm^3	Time for number of particles to decrease to 1/10 initial concentration
10^{12}	0.03 sec
10^{11}	0.3 sec
10^{10}	3 sec
10^9	0.5 min
10^8	5.0 min
10^7	50 min
10^6	8.3 hr
10^5	17 hr
10^4	1.3 days

[a]Data calculated using the methods of the National Academy of Sciences [8].

Sonic agglomeration uses high-intensity sound waves to accelerate particle coagulation [16]. This procedure is based on the premise that particles of different sizes will tend to vibrate with different amplitudes, increasing the opportunities for collision and coagulation. Also, the particle concentration in the compression zone of a high-intensity sound wave will temporarily be raised, artificially increasing Brownian coagulation rates. Applying this technique involves the exposure of a particle-laden gas stream for a fraction of a second to a high-intensity siren or oscillation piston, operating at 300 to 500 Hz and 170 decibels in a sound insulated chamber. After exposure the particulate collection efficiency of downstream containment devices is significantly improved from the greater proportion of large particles produced by this process.

Electrostatic methods, the fourth class, may also be used to obtain efficient coagulation of particles of the aerosol size range and larger. By passing a particle-laden gas stream past the negative side of a high DC voltage corona (intense electric field), the particles become negatively charged. They are then efficiently attracted and coagulated against positively charged plates which are in proximity. Periodic rapping of the plates or a trickle of water is used to dislodge the large particles into collection hoppers. People involved with stack sampling of gases downstream of an operating electrostatic precipitator should remember that the 20,000 to 30,000 volts normally used is high enough to leave a hazardous residual charge in the very small fraction of particles not retained by the precipitator.

3.6.2. Particulate and Aerosol Collection Devices

Gas cleaning devices vary in their removal efficiencies, but almost invariably they are more efficient at removing particulate, rather than aerosol size range material. But even the least efficient device for removing small particle sizes, the gravity settling chamber (Fig. 3.4), still plays an important role in emission control. When the gas stream to be cleaned has only large particles present which can be effectively removed by the device, then this device alone represents an inexpensive method of control. When the particle size range to be collected is wide, and a dry cyclone is to be used for final particle collection, very large particles (100 to 200 μm) *should* be removed prior to the gas stream entering the cyclone. Otherwise the interior of the cyclone is subjected to very high abrasion rates.

Since the efficiency of gas cleaning devices is almost inevitably better for larger particles than for small and since the number of particles represented by any given mass goes up exponentially as the particle size goes down, mass removal, rather than particle removal, efficiencies have become the standard method of specifying air cleaning capabilities. In one example, a particulate control device with a 91% mass efficiency worked out to have only 0.09% particle count efficiency for the same hypothetical dust-laden gas sample. Thus, to be fully aware of the collection characteristics of an air cleaning device one must have both mass removal efficiency information, *plus* the particle size range over which the mass removal efficiency was measured. Typical

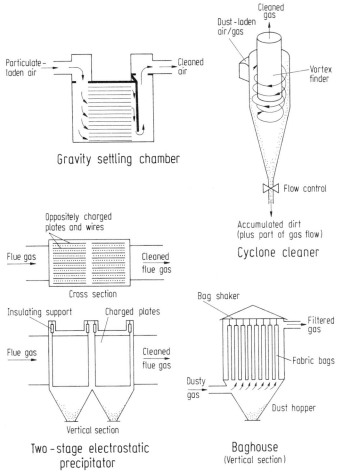

FIGURE 3.4 Simplified diagrams of some of the common types of particulate and aerosol emission control devices. Some of these are also useful for gas or vapor emission control; see text for details.

characteristics of a number of particle collection devices have been collected in Table 3.6.

The operating characteristics of some common particulate emission control units are illustrated in Fig. 3.4. Cyclones, which are extensively used for this purpose, are most efficient in small diameters which forces the entering gas stream through higher tangential velocities. It is high tangential velocities which most efficiently separate particles from the gas. To obtain the same high efficiencies with very large gas volumes a large number of small cyclones are connected in parallel, rather than using a single, very large cyclone.

Absolute filters, developed by the U.S. Atomic Energy Commission for containment of radioactive particles, are extremely efficient for retention of very fine particles but require considerable operating energy because of a large pressure drop across the filter element.

▇▇▇▇ **TABLE 3.6   Examples of Typical Collection Efficiencies for Some Common Types of Particulate and Aerosol Gas Cleaning Devices[a]**

Air cleaning device	Particle size range for 90% removal (μm)	Remarks
Gravity settling chamber	50–100	effective only at low gas velocities, <200 m/min
Baffle chamber	20–100	gas velocity ca. 300 m/min
Dry cyclone, 2–750 cm diameter,	10–50	gas velocity 300 to 1500 m/min, smaller are more efficient
Spray washer	5–20	gas velocity 70–100 m/min
Packed tower scrubber	1–5	gas velocities to 4000 m/min high resistance to gas flow (pressure drop)
Intensive rotary scrubber	0.1–1	high resistance to flow, energy intensive
Cloth filter	0.2–1	used in bag houses or continuous roll filters
Absolute filter	0.01–0.1	highly efficient but considerable resistance to flow
Electrostatic precipitator	0.01–10	must have low gas flow rates, very low pressure drop, very large units (>3000 m^3/min)
Activated charcoal	molecular	normally for control of gases, vapors, and odors; also retains particles, about as cloth filter

[a]Compiled from data of Mednikov [16], Williamson [17], Stern [18], and Patty [19].

Electrostatic precipitation is also very efficient for retention of very fine particles, as long as these are not highly electrically conductive. But the structural features of these units are such that they are only cost effective for particulate and aerosol emission control for very large volumes of gas, hence their typical use for the treatment of combustion gases of fossil-fueled power stations.

3.6.3. Hydrocarbon Emission Control

Adoption of automotive control devices has had the most significant effect on the gross hydrocarbon emissions for regulated areas, because of the large contribution to the total mass of hydrocarbons originating from this source (Tables 3.1, 3.2). But this measure on its own will not necessarily be effective for photochemical smog abatement since it is the reactive hydrocarbons, rather than total hydrocarbons, which are of greatest significance.

Measures such as the use of floating roofs for storage tanks which can reduce the ventilation losses of hydrocarbons from tank farms and potential losses from other points of oil refinery operations are discussed in Chapter 18.

In the dry cleaning business or in vapor degreasing plants, losses of hydrocarbon or chlorinated hydrocarbon vapors occur from the use and recovery of solvents. These losses can be prevented by the use of activated adsorbent systems to capture the organic vapors from the vents or hoods of these devices. Activated carbon, which is available in versions providing up to 1000 m^2/g (ca. 8 acres/oz) of adsorption area is one of the best adsorbents for this purpose because of its selectivity for organic vapors. Solvent present in high concentrations in air may also be recovered by cooling the air to liquid nitrogen temperatures to condense the solvent.

Air contaminated with low concentrations of organic vapors, which can also arise from a number of processes, may be economically controlled by using this air stream as boiler feed air. Any combustibles present are burned to provide fuel value, in the process destroying any hydrocarbons present before discharge [20]. This method is particularly appropriate when the concentration of organic vapors in air is too small for economic containment using adsorption or condensation.

The paint industry, with its high dependence on solvents and chemically active, film-forming components some of which come into the regulatory "reactive hydrocarbon" class, has also been hard pressed to meet air quality tests [21]. But by the further development of water-based coatings, even for many metal-finishing applications [22], and by innovations into such areas as high solids or even dry powder coating technologies the industry is decreasing its dependence on organic solvents for many finishing applications [23]. In turn these measures have decreased the emission of hydrocarbons when the coating is applied, and when combined with electrostatic application technology to better direct the coating particles or droplets onto the surface being coated these techniques simultaneously result in much less waste of coating solids as well.

3.6.4. Control of Sulfur Dioxide Emissions

Total sulfur dioxide discharges for the U.S.A. for 1981 amounted to 22.5 million tonnes. Combustion sources concerned with power generation produced about 60% of this, and a further 37% was evenly split between other stationary combustion sources (mainly space heating) and industrial processes. More than half of that originating from industrial processes, or about 11% of the total, arose from the smelting of sulfide minerals and metal refining activities. Only about 3% of the total originated with the transportation sector.

The extent of sulfur dioxide emission is of importance for the direct effect it has on the ambient air levels, for which guidelines have been laid down to protect public health (Table 3.7). However, the atmospheric half-life of discharged sulfur dioxide is estimated to be short, of the order of 3 days. This rapid return of sulfur dioxide to the earth's surface, both as sulfur dioxide itself and as its oxidized and hydrated products, is the reason for its significance as a contributor to the acidity of rainfall, about which comment has already been made. "Sulfurous acid," from hydrated sulfur dioxide, and sulfuric acid, formed from the oxidation of sulfurous acid or the hydration of sulfur trioxide, both contribute to low pH rain. Sulfur trioxide forms through the gas phase and heterogeneous oxidation of sulfur dioxide on particulate matter, which occurs to a significant extent both in the plumes downwind of large chimneys and in the ambient air.

The origins of two of the major sulfur dioxide discharges in Canada, combustion sources and sour gas plants, produce waste gas streams which

TABLE 3.7 Ambient Air Quality Standards for Canada and the U.S.A. in $\mu g/m^3$, and Parenthetically in ppm[a]

Pollutant	Averaging time	Primary standards (health related, or maximum acceptable)		Secondary standard (welfare related, or maximum desirable	
		Canada	U.S.A.	Canada	U.S.A.
Total suspended	annual[b]	70	75	60	60
particulate	24 hr	120	260	—	150
Sulfur dioxide	annual[c]	60(0.02)	80(0.03)	30(0.01)	—
	24 hr	300(0.11)	365(0.14)	150(0.06)	—
	3 hr	—	—	—	1300(0.50)
	1 hr	—	—	450(0.17)	—
Carbon monoxide	8 hr	15(13)	10(9)	6(5)	10(9)
	1 hr	35(30)	35(40)	15(13)	35(40)
Nitrogen dioxide	annual[c]	100(0.05)	100(0.053)	60(0.03)	100(0.053)
	24 hr	200(0.11)	—	—	—
Ozone	24 hr	50(0.025)	—	30(0.015)	—
	1 hr	160(0.08)	235(0.12)	100(0.05)	235(0.12)
Lead	3 month	—	1.5	—	1.5

[a]Compiled from EPA [24] and *Canadian Chemical Processing* [25].
[b]Annual geometric mean.
[c]Annual arithmetic mean.

contain only 0.15 to 0.50% sulfur dioxide. These are too low to be economically attractive for containment. Only with the smelter sources, which commonly produce roaster exhaust gas streams containing 2 to 5% sulfur dioxide and with modern equipment up to about 15%, is it anywhere near an economic proposition to capture the sulfur dioxide. This may either be sold as liquid sulfur dioxide, or may be converted to sulfuric acid for sale. The smelter grade is less pure than sulfuric acid from sulfur burning sources, and so fetches a much lower market price. However, it is still a useful product for such less demanding applications as metal pickling and the manufacture of fertilizers.

A number of solutions exist for decreasing the sulfur dioxide emissions from sources where the concentrations present are too low for economic containment. The choice in any particular situation depends on the existing ambient air quality (particularly with regard to public health) in the operating area, and the time frame within which abatement action has to be taken.

Temporary shutdown or a reduced level of operation may be required of a fossil-fueled power station or a smelter to avoid producing local, dangerously high ambient concentrations of sulfur dioxide during a severe, inversion-promoted smog situation. With experience temporary curtailment requirements of this kind may be anticipated. A power station, for instance, can arrange to have separate stockpiles of low-cost, relatively high sulfur content coal for periods of power generation during normal atmospheric conditions, and stockpiles of alternative more expensive lower sulfur coal, or even coke or natural gas (if suitably equipped), which may be burned during severe inversion episodes. With these choices available, costly and inconvenient emission curtailment by reduction of power production or temporary shut-down will be required less often.

Some emission reduction measures have longer lead times. These involve process modifications to avoid generating the sulfur dioxide-containing waste gases. For smelters this might mean adoption of hydrometallurgical technology, such as is now available for the processing of copper, nickel, and zinc, in order to bypass the older roasting methods of sulfur removal from ores [26]. For power stations this option might mean the installation of more rigorous precombustion sulfur removal techniques for their coal or residual fuel oil, or increased emphasis on natural gas firing, or selection of hydroelectric, nuclear, or the solar-related or wind power-based technologies in the utility's long-term plans for power generation.

A third choice to a company faced with a need to practice emission reduction is to remain with the existing front end process, which continues to produce a sulfur dioxide-containing waste gas stream, and move to some system which can effectively remove the sulfur dioxide from this waste gas before it is discharged. Many methods are available, each with features which may make one of them more attractive or suitable than the others for the sulfur dioxide removal requirements of the particular process (Table 3.8). Some of the factors to be considered in arriving at an appropriate choice are the waste gas volumes and sulfur dioxide concentrations which have to be treated and the degree of sulfur dioxide removal required, remembering that the trend is toward a continuing decrease in allowable discharges. Further considerations are the type of sulfur dioxide capture product which is produced by the process and the overall cost of it, considered with any product credit which may help to offset process costs. Finally, the type of controlled gas discharge required for the operating situation, i.e., warm or ambient temperature, moist or dry, etc., also has to be taken into account. Chemical details of the processes of Table 3.8 are outlined below.

Only one control procedure, ammonia injection, relies solely on gas phase reactions. The moist flue gas containing hydrated sulfur dioxide undergoes a gas phase, acid–base reaction to produce particulate solids (Eq. 3.15, 3.16).

$$SO_2 + H_2O + NH_3 \rightarrow \underset{\text{particulate}}{NH_4HSO_3} \qquad 3.15$$

$$SO_2 + H_2O + 2\ NH_3 + 1/2\ O_2 \underset{\text{catalysis}}{\rightarrow} (NH_4)_2SO_4 \qquad 3.16$$

The solid product, now a separate phase, can be readily captured by electrostatic precipitation or any other particulate collection device which is already normally in place for fly-ash control.

Heterogeneous reduction processes still involve the reaction of gases, but in these cases the reaction occurs in the presence of a suitable solid phase catalyst. Sulfur dioxide may be reduced to sulfur with hydrogen sulfide, if this is available, and the sulfur vapor condensed out of the gas stream by cooling, as in the second half of the Claus process (Eq. 3.17).

$$SO_2 + 2\ H_2S \xrightarrow{Fe_2O_3} 3\ S + 2\ H_2O \qquad 3.17$$

TABLE 3.8   The Chemical Details of Examples of Stack Gas Desulfurization Processes[a]

Reaction phases	Process examples	Regeneration method	Product(s)
Gas–gas	Homogeneous:		
	gaseous ammonia injection	none	$(NH_4)_2SO_3$,
	plus particle collection		$(NH_4)_2SO_4$
	Heterogeneous:		
	SO_2 reduction with H_2S + catalyst (Claus)	none	S
	SO_2 reduction with CH_4 + catalyst to H_2S, then amine scrub (e.g. Girbotol), or aqueous AQ (Stretford)[b] recovery	none air	conc. H_2S S
	SO_2 catalytic oxidation with air (chamber, or contact process), plus hydration	none	H_2SO_4
Gas–liquid	Absorption, plus chemical reaction:		
	by aqueous ammonia solution	none	$(NH_4)SO_3$, $(NH_4)_2SO_4$
	dimethylaniline solution (ASARCO)	heat	conc. SO_2
	aqueous sodium sulfite (Wellman-Lord)	heat	conc. SO_2, Na_2SO_4
	citrate process, U.S. Bureau of Mines	H_2S	S
	citrate process, Flakt-Boliden version	heat	conc. SO_2
	eutectic melt (Na_2CO_3/K_2CO_3), gives dry plume	water gas, (H_2, CO)	conc. H_2S
	limestone (or lime) slurry scrubbing,	none	$CaSO_3 \cdot nH_2O$,
	inexpensive, throw-away slurry product		$CaSO_4 \cdot 2H_2O$
Gas–solid	Physical adsorption:		
	SO_2 onto activated charcoal or alkalized alumina	heat	conc. SO_2
	Adsorption, plus chemical reaction:		
	powdered limestone injection, 20–60% efficient capture with particle collection (precipitator), waste product	none	dry $CaSO_3$ and $CaSO_4$
	CuO (or other metal oxides)	methane	conc. SO_2
	Chemical reaction:		
	SO_2 reduction in heated coal bed, Resox process	none	S

[a]Compiled from sources cited in the text, where further details are available, plus the reviews of Kaplan and Maxwell [27], Siddiqi and Tenini [28], and Engdahl and Rosenberg [29].
[b]AQ, short for anthraquinone disulfonic acids, see text for details.

Or the sulfur dioxide may be reduced catalytically with methane or other hydrocarbons to hydrogen sulfide. The hydrogen sulfide produced by this method is captured by amine scrubbing of the reduced gas stream (Eq. 3.18–3.20).

$$3\ SO_2 + 2\ CH_4 \xrightarrow{Al_2O_3} 2\ H_2S + 2\ CO_2 + 2\ H_2O + S \qquad 3.18$$

Absorption:

$$H_2S + H_2NCH_2CH_2OH \rightarrow HS^- + H_3\overset{+}{N}CH_2CH_2OH \qquad 3.19$$

Regeneration:

$$HS^- + H_3\overset{+}{N}CH_2CH_2OH \xrightarrow{heat} H_2S + H_2NCH_2CH_2OH \qquad 3.20$$

The high concentration of hydrogen sulfide obtained by heating the amine salt may then be easily and economically converted to elemental sulfur via the Claus process. The Stretford process, which may also be used for hydrogen sulfide capture, accomplishes both capture, using an aqueous mixture of 1,5-, and 1,8-disulfonic acids of 9,10-anthraquinone (AQ), and conversion of the hydrogen sulfide to sulfur in a single step [30] (Eq. 3.21).
Absorption–chemical conversion:

$$AQ{=}C{=}O + H_2S \rightarrow AQ{=}C(SH)(OH) \rightarrow AQ{=}C(H)(OH) + S \qquad 3.21$$

Regeneration of the quinone from the quinol is accomplished with air (Eq. 3.22).
Regeneration:

$$2\ AQ{=}C(H)(OH) + O_2 \rightarrow AQ{=}C{=}O + 2\ H_2O \qquad 3.22$$

Catalytic oxidation of sulfur dioxide with air, via the heterogeneous contact process or the homogeneous chamber process, also serves to improve the collection efficiency of the sulfur oxides. Sulfur trioxide has a very strong affinity for water, unlike sulfur dioxide, so that collection of sulfur trioxide by direct absorption into water is extremely efficient, and the product sulfuric acid is a salable commodity.

Sulfur dioxide containment by gas–liquid interactions can be as simple as flue, or process gas scrubbing with dilute ammonium hydroxide (Eq. 3.23, 3.24).

$$2\ NH_4OH + SO_2 \rightarrow (NH_4)_2SO_3 + H_2O \qquad 3.23$$

$$2\ NH_4OH + SO_2 + 1/2\ O_2 \rightarrow (NH_4)_2SO_4 + H_2O \qquad 3.24$$

The ammonium salt products crystallized from the concentrated spent scrubber liquors may be used as valuable constituents of fertilizer formulations. Or they may be first reduced to ammonia and hydrogen sulfide with natural gas, followed by conversion of the hydrogen sulfide to sulfur by a Claus-type sequence [31]. In this French-designed modification, the ammonia is recycled to the scrubbing circuit.

Similar acid–base chemistry is involved in the American Smelting and Refining Company's (ASARCO's) sulfur dioxide capture process using aqueous dimethylaniline (Eq. 3.25).

$$PhN(CH_3)_2 + SO_2 + H_2O \rightarrow Ph\overset{+}{N}(H)(CH_3)_2 + HSO_3^- \qquad 3.25$$

However, in this case the sulfurous acid-amine salt is heated to regenerate the much more expensive dimethylaniline solution for reuse and to obtain a concentrated sulfur dioxide gas stream. The more concentrated sulfur dioxide is now economic for acid production.

The Wellman-Lord process uses the effective sodium sulfite/sodium bisulfite equilibrium to capture sulfur dioxide from flue gases [32] (Eq. 3.26, 3.27).

Absorption:

$$SO_2 + H_2O + Na_2SO_3 \rightarrow 2\ NaHSO_3 \qquad 3.26$$

Regeneration:

$$2\ NaHSO_3 + heat \rightarrow SO_2 + H_2O + Na_2SO_3 \qquad 3.27$$

Most of the sodium bisulfite produced is converted back to sodium sulfite on regeneration, but some sodium sulfate inevitably forms from irreversible oxidation. This is crystallized out, dried, and sold as wood pulping chemical ("salt cake").

The citrate process, in which much development work has been invested by the U.S. Bureau of Mines and by Pfizer, Inc., uses an aqueous solution of citric acid to capture sulfur dioxide (Eq. 3.28, 3.29).

Ionization:

$$HO - C(CH_2CO_2H)_2CO_2H \rightarrow H_2Cit^- + H^+ \qquad 3.28$$
$$\text{citric acid (first ionization} = Cit^{-1})$$

Absorption:

$$H_2Cit^{-1} + HSO_3^- \rightarrow (HSO_3 \cdot H_2Cit)^{2-} \qquad 3.29$$

Regeneration of the uncomplexed citric acid at the same time as formation of an elemental sulfur product from the bisulfite anion is obtained by treating the absorption solution with hydrogen sulfide [33] (Eq. 3.30).

Regeneration:

$$(HSO_3 \cdot H_2Cit)^{2-} + H^+ + 2\ H_2S \rightarrow 3\ S + H_2Cit^- + 3\ H_2O \qquad 3.30$$

If hydrogen sulfide is not conveniently available from local process sources it may be produced on site for this requirement by reducing a part of the sulfur product of the process with methane (Eq. 3.31).

$$4\ S + CH_4 + 2\ H_2O \xrightarrow{\ Al_2O_3\ } 4\ H_2S + CO_2 \qquad 3.31$$

The Flakt-Boliden version of the citrate process uses thermal regeneration of the citrate absorbing solution to obtain stripped citrate solution and a stream of up to 90% sulfur dioxide at this stage. Recovery of a sulfur dioxide product gives flexibility to the final stage of processing as to whether liquid sulfur dioxide, sulfuric acid, or sulfur are obtained as the final product.

A further gas–liquid interaction process to a useful product is the sodium carbonate–potassium carbonate eutectic melt process. This operates at a temperature of about 425°C, well above the melting point of the eutectic [34]. Sulfur dioxide absorption takes place with loss of carbon dioxide (Eq. 3.32).
Absorption:

$$Na_2CO_3 + SO_2 \rightarrow Na_2SO_3 + CO_2 \qquad\qquad 3.32$$

The original eutectic is regenerated plus a hydrogen sulfide product obtained in a separate unit by treating the sodium sulfite with "water gas" (Eq. 3.33).
Regeneration:

$$Na_2SO_3 + CO + 2\ H_2 \rightarrow Na_2CO_3 + H_2S + H_2O \qquad 3.33$$

The hydrogen sulfide initial product can be subsequently converted to more useful sulfur via the Claus process. In return for the high operating temperatures required for this process, it gives a substantially dry plume, unlike the other gas-liquid interaction processes mentioned. This may be an important consideration if the process is to be operated in a fog susceptible area.

A finely powdered limestone or lime slurry in water used in a suitably designed scrubber is an effective and relatively low cost sulfur dioxide containment method (Eq. 3.34–3.37).

$$CaCO_3 + SO_2 \rightarrow CaSO_3 + CO_2 \qquad\qquad 3.34$$

$$CaCO_3 + SO_2 + 1/2\ O_2 \rightarrow CaSO_4 + CO_2 \qquad 3.35$$

$$Ca(OH)_2 + H_2O + SO_2 \rightarrow CaSO_3 \cdot 2\ H_2O \qquad 3.36$$

$$Ca(OH)_2 + H_2O + SO_2 + 1/2\ O_2 \rightarrow CaSO_4 \cdot 2\ H_2O \qquad 3.37$$

With either reagent, however, a throw-away product is obtained. Land has to be allocated for lagoon disposal of the spent scrubber slurries, or other systems have to be set up to handle the waste solid. A recent variation of this approach is to employ the alkalinity of fly ash itself, in a water slurry, as a means of capturing sulfur dioxide.

The simplest gas–solid containment systems conceptually are the direct adsorption ones. These accomplish adsorption on solids such as activated carbon, or alkalized alumina at relatively low temperatures and ordinary pressures [35]. In a separate unit, when the absorbent is regenerated by heating, a more concentrated sulfur dioxide stream is produced. The more concentrated stream is now more attractive as feed to an acid plant or for liquefaction or sulfur generation.

Among the simplest systems, at least for power stations that already burn powdered coal, is the use of powdered limestone injection. Essentially the same burner assembly is required, and the combustion temperature of the coal is sufficient to form calcium oxide (lime) from the limestone (Eq. 3.38).

$$CaCO_3 \rightarrow CaO + CO_2 \qquad\qquad 3.38$$

The alkaline lime, in a gas–solid phase reaction, reacts with sulfur dioxide in the combustion gases to form solid particles of calcium sulfite and calcium sulfate which are captured in electrostatic precipitators (Eq. 3.39, 3.40).

$$CaO + SO_2 \rightarrow CaSO_3 \qquad\qquad 3.39$$

$$CaO + SO_2 + 1/2\ O_2 \rightarrow CaSO_4 \qquad\qquad 3.40$$

The process is simple in concept, and to operate, but the stoichiometry of the gas-solid reaction is only about 20 to 60% of theory so that an excess of limestone is required for moderately efficient collection. This procedure also imposes a heavier solids handling load on the precipitators, a factor which has to be considered, and it, too, yields a throw-away product. It does, however, produce a dry plume, which may be an advantage in some situations.

Other gas–solid containment systems, in which a significant amount of the original development work was invested by Shell, are based on copper oxide on an alumina support . The absorption step of this system both oxidizes and traps the sulfur dioxide on the solid support as copper sulfate (Eq. 3.41).

Uptake:

$$CuO + SO_2 + 1/2\ O_2 \rightarrow CuSO_4 \qquad\qquad 3.41$$

Regeneration by methane reduction is conducted in a separate unit returning sulfur dioxide, but now in more useful high concentrations (Eq. 3.42).

Regeneration:

$$2\ CuSO_4 + CH_4 \rightarrow 2\ Cu + CO + 2\ H_2O + 2\ SO_2\uparrow \qquad 3.42$$

The finely divided copper on the solid support is rapidly reoxidized to cupric oxide when it comes into contact with the hot flue gases on its return to the absorption unit (Eq. 3.43).

$$Cu + 1/2\ O_2\ (hot\ flue\ gas) \rightarrow CuO \qquad\qquad 3.43$$

Ferric oxide on alumina has also been tested in a very similar process.

A final gas–solid interaction process involves chemical reaction for sulfur dioxide containment. A bed of crushed coal kept at 650 to 815°C is used to reduce a stream of concentrated sulfur dioxide passed through it, to elemental sulfur and carbon dioxide (Eq. 3.44).

$$SO_2 + C \rightarrow CO_2 + S \qquad\qquad 3.44$$

3.6.5. Control of Nitrogen Oxide Emissions

Discharges of nitrogen oxides in the U.S.A. in 1981 totaled 19.5 million metric tonnes, only slightly less than the total discharges of sulfur dioxide for the same year. But the transportation sector was much more significant for this classification, contributing about 8.4 million tonnes of this total. Stationary combustion sources, however, were still the most significant contributor with a total of 10.2 million tonnes, followed by industrial sources which totaled about 950,000 tonnes.

While both the primary and secondary ambient air quality standards for nitrogen oxides are somewhat higher than for sulfur dioxide the rationales behind the need for control measures are very similar, namely, health effects and their influence on the pH of precipitation (Table 3.7). In addition, however, the significance of the role of nitrogen oxides in photochemical smog episodes is a further factor in the matrix of information used to set up the ambient air quality standards.

The mechanism of formation of nitrogen oxides (NO_x) in flue gases or combustion processes generally is a composite of the relative concentrations of nitrogen and oxygen present and the combustion intensity (peak temperatures). Thus, small space heaters operating at lower temperatures typically produce combustion gases containing about 50 ppm NO_x, while large power plant boilers without abatement measures can produce flue gas concentrations of as high as 1500 ppm [37]. The formation reactions are relatively simple, involving a combination of atoms of oxygen with nitrogen and atoms of nitrogen with oxygen, both elements having finite concentrations present in atomic form at high combustion temperatures (Eq. 3.1, 3.2). Because of the higher concentration of atomic species present at high temperatures rather than low, and because of the very rapid reaction rates at high temperatures, the equilibrium of these reactions is established rapidly. A small further contribution to the nitrogen oxides in exhaust gases can come from the oxidation of nitrogen-containing organics (combined nitrogen) in the fuel.

To decrease nitrogen oxide emissions any one or a combination of measures may be used. In utility applications one of the simplest measures, which also serves to reduce average fuel combustion and hence the costs of power generation, is to take some care over the matching of combustion rate to load requirements. Fuel combustion at any rate greater than the heat removal (or load requirement) tends to produce excessive boiler and flue gas temperatures and to raise nitrogen dioxide concentrations.

More fundamental equipment modifications are necessary to adopt staged combustion, or larger flame volume changes as means of decreasing NO_x formation. Tests of systems that use a fuel-rich combustion core to produce reducing gases which partially reduce the nitrogen oxides which simultaneously form near the fuel-lean walls to nitrogen are also yielding promising results (e.g., Eq. 3.5–3.8, 3.45).

$$2\,NO_2 + CH_4 \rightarrow N_2 + 2\,H_2O + CO_2 \qquad\qquad 3.45$$

A second significant factor which can assist in decreasing nitrogen oxide formation is a reduction of the excess air used in combustion processes. This option requires the installation of oxygen analyzers for the continuous monitoring of the oxygen content of flue gases coupled to automatic controls or manual reset alarms, to ensure that the excess air used for combustion stays within predetermined limits. Changing from a practice of operating with a 15% excess of theoretical combustion air to a 2 to 3% excess can alone serve to decrease nitrogen oxide emissions by some 60%, and at the same time increase energy recovery efficiency.

Other kinds of process changes can also serve to decrease NO_x emissions by avoiding formation to a significant extent. Among these are recirculation of a fraction of the flue gas to decrease peak combustion temperatures, a measure which has also had some success in decreasing automotive NO_x emissions. Use of pure, or enriched oxygen instead of air for combustion can also help when used with flue gas recirculation, by decreasing the concentration of nitrogen available for oxidation. Use of fuel cells for power generation could eliminate utility-originated NO_x emission, but the pure gas requirement for present fuel cell technology makes this option prohibitively expensive except in special circumstances [39].

Postformation nitrogen oxide emission control measures include adsorption in packed beds, through which some decrease is observed [23]. But this method is not particularly effective. It is also not a very practical method either for utilities or for transportation sources. Ammonia or methane reduction of NO_x to elemental nitrogen is also an effective method, which is cost-effective for high concentration sources such as nitric acid plants (Chap. 11). Two-stage scrubbing, the first stage using water alone, and the second using aqueous urea, has also been proposed as an effective postformation NO_x control measure.

3.6.6. Carbon Dioxide Emission Abatement

Concern about current levels of carbon dioxide discharges stems from its contribution to global warming, as one of the greenhouse gases, and not from toxicity or acid rain considerations. To begin with let us put the need for reduction or control of carbon dioxide emissions into perspective. Current climate change models indicate that about two-thirds of the greenhouse effect is from the presence of moisture in the atmosphere, both in vapor form and as clouds. About one-third of the effect is from all the other greenhouse gases and about one-half of this, or one-sixth of the total effect, is from carbon dioxide. So one must bear in mind that other factors which are less under our control than carbon dioxide have a greater effect than carbon dioxide. It has been estimated that the total greenhouse warming effect from the suite of gases of Table 2.7 amounts to raising the global average temperature by about 30°C. A drop of average temperatures of as little as 5–6°C has been estimated was sufficient to bring on the ice ages. Therefore, the earth's biosphere is reliant on a rather delicate energy balance for mean temperatures as we know them. Venus, an example from our own solar system of an extreme greenhouse effect, has a drastically different atmosphere of predominantly carbon dioxide and water vapor shrouded in cloud. This gives Venus about 100 times the greenhouse warming from its atmosphere than does the earth, which produces mean surface temperatures of 480°C (750 K).

The earth's atmospheric reservoir of gases contains about 2.6 trillion (2.6×10^{12}) tonnes of carbon dioxide. The total annual flux of carbon dioxide into this reservoir is about 0.61 trillion tonnes, of which human activities contribute about 29 billion tonnes or 4.7% of the total. The rate of increase in concentration of atmospheric carbon dioxide has been 1.2 ppm/yr for a number of years now. Taking this rate of increase the annual amount of carbon dioxide it would take to do this is 7.7×10^9 tonnes/year (1.2 ppm/yr 2.6×10^{12} tonnes $\div$ 360 ppm current atmospheric concentration). What these approximations tell us is that 27% (7.7×10^9 t/yr $\div$ 29×10^9 t/yr $\times$ 100) of our present annual discharges of carbon dioxide would have to be curtailed, other things being equal, just to keep the atmospheric concentration of carbon dioxide constant at its current level of 360 ppm. Since natural sources of carbon dioxide contribute 20 times as much, annually, as human activities, a relatively minor perturbation of the natural sources may augment or defeat any changes in atmospheric concentration of carbon dioxide expected as a result of human curtailment activity.

Most of the human contribution of carbon dioxide to the atmosphere is from fossil fuel and biomass combustion. Since this combustion is for energy production, energy conservation measures in all forms would help. Better insulated buildings and housing, efficient mass transit, living close to one's workplace, increased efficiency of chemical processes, etc., can all contribute to this. Switching from carbon dioxide intensive fuels to less carbon dioxide intensive fuels can also help. Replacing coal by natural gas roughly halves the carbon dioxide produced per unit of energy (Table 3.9). The hydrogen listing here may be a bit misleading. It is only acting as a medium of energy transfer, like electricity. While hydrogen combustion itself does not produce carbon dioxide, most current methods of producing it are carbon dioxide intensive (Chap. 11). Increased use of all forms of geothermal, tidal, hydroelectric, and solar (wind, photovoltaic, etc.) energy can obviously help, although each of these alternatives also requires fossil fuels for construction. Nuclear power can help since it produces carbon dioxide-free energy (neglecting uranium mining, processing, reactor construction) but also has accompanying problems.

The other area for attention is to act on the reduced plant uptake side of the equation, which has occurred from massive deforestation of the last 1.5 centuries. It has been estimated that a rapidly growing rainforest can take up to 4–7 kg/m^2 year carbon dioxide, as compared to a typical crop uptake of 0.8–1.6 kg/m^2 year. Thus, vigorous reforestation can assist in improving the removal of carbon dioxide from the atmosphere. Attention to improved marine carbon dioxide uptake via enhanced plankton growth could have a significant effect, because of its wide distribution. Preliminary experiments have shown that provision of soluble iron can dramatically improve this process.

TABLE 3.9 Approximate Energy Equivalencies and Carbon Dioxide Production from Combustion of Different Fuels[a]

Fuel	BTU/lb	kWh/kg	kg CO_2/kg fuel	kg CO_2/kWh
Coal	13,000	8.39	3.22[b]	0.38[b]
Furnace Oil				
No. 5 equiv.	18,800	12.14	3.12	0.26
No. 2 equiv.	19,200	12.40	3.11	0.25
Gasoline	19,700	12.72	3.08	0.24
Propane	21,500	13.88	3.00	0.22
Natural gas	23,900	15.43	2.74	0.18
Methanol	9,500	6.13	1.37	0.22
Hydrogen	61,100	39.44	0.0[c]	0.0[c]

[a]Calculated from the known energy content and the stoichiometry of the respective combustion reactions.
[b]Also with a mean of 0.04 kg SO_2/kg fuel, or 0.005 kg SO_2/kWh energy.
[c]See text for explanation.

Promotion of an appropriate mix of these measures should be able to control the continued increase in atmospheric carbon dioxide.

REVIEW QUESTIONS

1. A coal-fired power station stack of an effective height of 20 m is discharging flue gases at a constant rate containing 1500 ppm sulfur dioxide and 8% carbon dioxide, both expressed on a volume for volume basis.
   (a) For a ground level concentration of sulfur dioxide found to be 20 ppm, 2000 m away when the wind speed was 25 km/hr, what would be the value of the constant in the equation:

$$\text{Ground level conc.} = K \cdot \frac{m}{uH^2}$$

   where m = mass rate of emission (or concentration, if invariant), u = wind velocity, and H = effective stack height?
   (b) What would be the expected concentration of carbon dioxide at the same point on the ground under the same conditions as in Part (a)?
   (c) What would be the expected ground level concentrations of both gases, 2000 m away, if the effective stack height was raised to 100 m which simultaneously raised the wind speed at the top of the stack to 35 km/hr?
2. To decrease carbon dioxide emissions to the atmosphere, it has been proposed that power generation be switched from No. 2 fuel oil (take as $C_{15} H_{32}$; 12.40 kWh/kg) to methane (CH_4; 15.43 kWh/kg) or methanol (CH_3OH; 6.13 kWh/kg).
   (a) How many kg of carbon dioxide would be produced on complete combustion of a kg of each of these fuels?
   (b) What mass of carbon dioxide would be produced per kWh of energy produced by each of these fuels?
   (c) What fuel choice(s) would produce even less carbon dioxide?
3. The national air quality objective for "suspended particulate matter" is 60 $\mu g/m^3$.
   (a) What would this correspond to in ng/L?
   (b) Can this 60 $\mu g/m^3$ be converted to ppm by volume? Explain your answer.
4. (a) Natural gas (take as 100% methane) is to be used to fuel a boiler. Taking air as exactly 1:4, mol O_2:mol N_2, what volume ratio of air to natural gas would theoretically be just sufficient to burn the methane completely?
   (b) What volume ratio of air to natural gas would be required to allow 10% excess air?
5. A resting adult flying in a commercial aircraft consumes 20.0 L hr^{-1} oxygen and produces 17.7 L hr^{-1} carbon dioxide, both at 20°C and 1 atm pressure.

(a) What ventilation rate would be required to maintain a carbon dioxide concentration of 1000 ppm in this passenger's cabin space, assuming 20°C and 1 atmosphere conditions and a carbon dioxide concentration of 350 ppm in the outside air?

(b) For an aircraft cabin air space of 1000 L per passenger, how many air changes per hour would be required to maintain 1000 ppm by volume carbon dioxide?

(c) What ventilation rate would be required at 30,000 ft altitude, where the captain is able to maintain conditions of 20°C and a pressure of 0.690 atm?

(d) What happens to the oxygen consumed that does not produce carbon dioxide? (Hint: Consider metabolism of carbohydrates.)

FURTHER READING

American Chemical Society, "Cleaning Our Environment: A Chemical Perspective," 2nd ed. Washington, DC, 1978.

R.G. Bond and C.P. Straub, eds., "CRC Handbook of Environmental Control," Vol. 1. CRC Press, Cleveland, OH ,1972.

H. Brauer and Y.B.G. Varma, "Air Pollution Control Equipment." Springer-Verlag, New York, 1981.

T.E. Graedel and B.R. Allenby, "Industrial Ecology." Prentice Hall, Upper Saddle River, NJ, 1995.

J.M. Marchello, "Control of Air Pollution Sources." Dekker, New York, 1976.

Organization for Economic Cooperation and Development, "Motor Vehicle Pollution: Reduction Strategies Beyond 2010." Paris, 1995.

A.C. Stern, ed., "Air Pollution," 2nd ed., Vol. 3, Academic Press, New York, 1968.

REFERENCES

1. G.J.K. Acres, *Chem. Ind. (London),* p. 905, Nov. 16 (1974).
2. The Canadian Environment, "The Clean Air Act, Annual Report 1977–78, Ottawa" (cited by M. Webb). W.B. Saunders Co., Can. Ltd., Toronto, 1980.
3. H.S. Stoker and S.L. Seager, "Environmental Chemistry: Air and Water Pollution." p. 20, Scott Foresman, Glenview, IL, 1972.
4. J.J. Mikita and E.N. Cantwell (Dupont), Exhaust Manifold Thermal Reactors—A Solution to the Automotive Emissions Problem. *68th Annu. Meet., Nat. Pet. Refiners Assoc.,* San Antonio, TX, April 5–8 (1970).
5. Auto Emissions Control Faces New Challenges, *Chem. Eng. News* **58**(11), 36 Mar. 17 (1980).
6. G.J.K. Acres and B.J. Cooper, *Platinum Met. Rev.* **16**(3), 74, July (1972).
7. M.E. Hilburn, *Chem. Soc. Rev.* **8**(1), 63 (1979).
8. "Airborne Lead in Perspective." National Academy of Sciences, Washington, DC, 1972.
9. K. Habibi, *Environ. Sci. Technol.* **7**(3), 223 (1973).
10. J. Daniels, *Recherche* **11**(114), 938 (1980).
11. L.R. Babcock, Jr., *J. Air Pollut. Control Assoc.* **20**(10), 653 (1970).
12. E.G. Walther, *J. Air Pollut. Control Assoc.* **22**(5), 353 (1972).
13. New Ways to Cope with Sulfur Oxides, *Hydrocarbon Process.* **51**(9), 19, Sept. (1972).
14. Desulfurization Refinery Capacities, *Environ Sci. Technol.* **7**(6), 494 (1973).
15. P.L. Magill, F.R. Holden, and C. Ackley, eds., "Air Pollution Handbook." McGraw-Hill, New York, 1956.

16. E.P. Mednikov, "Acoustic Coagulation and Precipitation of Aerosols" (translated by C.V. Larrick). Consultants Bureau, New York, 1965.

17. S.J. Williamson, "Fundamentals of Air Pollution." Addison-Wesley, Reading, MA, 1973.

18. A.C. Stern, ed., "Air Pollution," 2nd ed., Vol. 3. "Sources of Air Pollution and Their Control." Academic Press, New York, 1968.

19. F.A. Patty, ed., "Industrial Hygiene and Toxicology," 2nd ed., Vol. 1. Interscience, New York, 1958.

20. G.D. Arnold, *Chem. Ind. (London)*, p. 902, Nov. 16 (1974).

21. L.A. O'Neill, *Chem. Ind. (London)*, p. 464, June 5 (1976).

22. C.M. Hansen, *Ind. Eng. Chem., Prod. Res. Dev.* **16**(3), 266 (1977).

23. J. Cross, *Chem. Brit.* **17**(1), 24, Jan. (1981).

24. "National Air Quality and Emissions Trends Report, 1981," EPA 450/4-83-011. Research Triangle Park, NC, 1983.

25. Ottawa Sets Timetable for Cleanup, *Can. Chem. Process.* **58**(3), 19, March (1974).

26. R. Derry, *Chem. Ind. (London)*, p. 222, March 19 (1977).

27. N. Kaplan and M.A. Maxwell, *Chem. Eng. (N.Y.)* **84**, 127, Oct. 17 (1977).

28. A.A. Siddiqi and J.W. Tenini, *Hydrocarbon. Process.* **56**(10), 104, Oct. (1977).

29. R.B. Engdahl and H.S. Rosenberg, CHEMTECH. **8**(2), 118, Feb. (1978).

30. Hydrogen Sulphide Removal by the Stretford Liquid Purification Process, *Chem. Ind. (London)*, p. 883, May 19 (1962).

31. P. Bonnifay, R. Dutriau, S. Frankowiak, and A. Deschamps, *Chem. Eng. Prog.* **68**(8), 51, Aug. (1972).

32. B.H. Potter and T.L. Craig, *Chem. Eng. Prog.* **68**(8), 53, Aug. (1972).

33. F.S. Chalmers, *Hydrocarbon Process.* **53**(4), 75, April (1974).

34. SO$_2$ Removal, *Chem. Eng. News* **49**(6), 52, Feb. 8 (1971).

35. J.H. Russell, J.I. Paige, and D.L. Paulson, Evaluation of Some Solid Oxides as Sorbents of Sulfur Dioxide. *Rep. Invest.—U.S., Bur. Mines* **RI-7582** (1971).

36. F.M. Dautzenberg, J.E. Nader, and A.J.J. Ginneken, *Chem. Eng. Prog.* **67**(8), 86 Aug. (1971).

37. D.R. Bartz, *Calif. Air Environ.* **5**(1), 10, Fall (1974).

38. Low NO$_x$ Burner to be Demonstrated in Utah, *Chem. Eng. News* **59**(17), 16, April 27 (1981).

39. Zeroing in on Flue Gas NO$_x$, *Environ. Sci. Technol.* **3**(9), 808 (1969).

4
WATER QUALITY MEASUREMENT

*Water is H$_2$O, hydrogen two parts, oxygen one, but
there is also a third thing, that makes it water and
nobody knows what that is.*
—D. H. Lawrence (1885–1930)

4.1. WATER QUALITY AND SUPPLY OVERVIEW

Water is a vital commodity to industry for process feedstock (as a reacting raw material), as a solvent, and for cooling purposes, just as it is to us as individuals for all our personal water requirements. The global supply appears to be extensive, so much so that many people take it for granted. But when one considers that only 3% of this total resource exists as freshwater at any one time the concept of this as a globally limited and potentially exhaustible resource becomes more real (Table 4.1). When one further takes into account that roughly three-quarters of this 3% is frozen from immediate use by the ice caps and glaciers of the world, then an appreciation of the limited extent of the available global freshwater becomes more apparent.

If the total available surface freshwater supply of some 126,000 km^3, a quantity which excludes the ice caps and glaciers, was evenly distributed over the total nonfrozen land area of the earth, it would amount to only some 1.1 m in water depth. Addition of the net annual land-based precipitation, after evaporation, would add only a further 0.8 m in depth. The combined total certainly does not represent a limitless resource to serve the agricultural, industrial, and personal needs of a world population of about five billion.

Another aspect of the world's freshwater resource is that it is not at all uniformly distributed over the land surface, even in the form of lakes and water courses. This occurs partly from the uneven distribution of glaciers and their meltwaters, and partly from the wide disparities in rainfall over the

■ **TABLE 4.1   Estimated Distribution of the Global Water Resource**[a]

Location	Volume, 10^3 km^3	Fraction of total	Percent of fresh
Overall:			
Oceans[b]	1,319,000	97.2	—
Ice caps, glaciers	29,200	2.15	76.8
Freshwater on land or air	8,500	0.65	23.2
Total	1,358,000	100.00	100.0
Freshwater[c]:			
Antarctic ice cap	26,000	1.99	71.33
Greenland ice cap	21,200	0.15	5.57
Glaciers	26,000	0.02	0.56
Groundwater, to 4-km depth	8,360	0.62	22.17
Freshwater lakes	125	0.009	0.33
Rivers, average	1.25	0.0001	0.003
Atmosphere, average	13.0	0.001	0.03
Total	37,709	2.8	100.0
Precipitation, annual:			
Over oceans	320		
Over land area	100		
(less evaporation)	70		
Net, to land	30		

[a]Compiled from van der Leiden[1].
[b]Includes saline lakes, principally the Caspian Sea, of gross volume 10^4 km^3.
[c]Instantaneous values. Does not consider precipitation, which balances out evaporation, etc., on a long-term basis.

earth's surface. Total rainfalls of above 11 m are experienced in some years in the Mt. Waialeale area of Hawaii and at Cherrapunji, India, while less than 0.2 cm falls in the same period in Arica, Chile, or at Wadi Halfa in the Sudan [1]. This uneven distribution of source waters has already made it necessary to reuse much of the available natural supply. It has also stimulated the development of economical desalination techniques and the construction of large-scale desalination plants to enable brackish or marine sources of water to be used [2].

Multiple usage and reuse of available supplies, and sometimes poor liquid and general waste disposal practices have combined to severely degrade the quality of the surface waters of many parts of the world. As a recent example, in 1969 the Cuyahoga River, which flows into Lake Erie at Cleveland, Ohio, was declared a fire hazard from the accumulated loading of combustible organics floating on its surface [3]. Shortly thereafter it actually did catch fire and the resulting heat seriously damaged two steel railway bridges. Also, news reports from Mexico City in 1976 announced that the city had found it necessary to post "No Smoking" signs beside a promenade which ran alongside the Tlalnepantla River, motivated by similar risks. The Rhine River, as a European example, which flows through several different federal jurisdictions with sometimes confused responsibilities for water quality regulation, also has its share of degradation. Surface water quality trends such as these plus a contribution from aesthetics have led to a growing move from piped to bottled water for drinking purposes.

In Europe about 2% of the surface freshwater is in lakes whereas in North America, Asia, and Africa about 75% is in this form. Stationary bodies of water such as the Great Lakes, the Caspian and Black Seas, and Lake Victoria also have their quality problems even though they represent very large volumes of water. These problems are the consequence of the more limited or nonexistent water exchange rates possible in these surface water features, as compared to rivers. Limited water exchange rates, in turn, limit the contribution that dilution can make to the recovery of water quality from waste discharges.

Australia as the driest continent, at least partly because it is also the flattest, has different water supply problems again. Long periods of drought prevail for years over much of the interior, punctuated by brief periods of flooding from widespread shallow lakes. These reasons contribute to the coastal location of all major Australian cities.

The Mediterranean and Baltic Seas, like lakes, also have little external water exchange. The high level of industrial activity on their shores coupled with the sometimes poor emission control and waste disposal practices have combined to severely affect their water quality [4]. The very size of the main oceans provides a large buffer capacity for waste disposal before water quality is noticeably affected. But even these extended ultimate waste sinks are showing signs of degradation, particularly around busy ports and coastal industrial cities as well as in the region of major shipping lanes in the open ocean [5].

However, it is possible, with cooperation and a determined effort, to reverse the trends toward a deterioration in water quality in any of these areas. As an example, the Thames River through London, England, as recently as 1958 showed no fish at all in the 60-km stretch of estuary between Fulham and Gravesend. Only tubifex worms and eels, both quite resistant creatures, were found to be living in this region. This low species diversity was mainly a result of the high temperatures, a high biochemical oxygen demand (BOD), and a deficiency of dissolved oxygen. By 1982, however, the sweeping powers of the Thames Water Authority enabled control of most of the causes of this poor water quality, returning dissolved oxygen concentrations to an average of 50% of saturation. Many species of fish had by this time returned to this habitat, and plans were being made to restock the river with salmon. These aspects, together with the return of abundant waterfowl, are all signs of a dramatic improvement.

Contamination of the Great Lakes in North America was also clearly recognized in the late 1960s, as was evidenced by increased nutrient and pesticide levels, and decreased catches of commercial species of fish coinciding with increased populations of coarser (more pollutant tolerant) species. But fortunately this was recognized early enough by both the U.S.A. and Canada to enable a joint effort to bring about a noticeable improvement. Full restoration of former water quality in these lakes is likely to take longer, however, than it took with the Thames River because of the very much slower water exchange rates. Some 70 years would be required for the St. Lawrence River to drain the water of the Great Lakes system once.

4.2. WATER QUALITY CRITERIA AND THEIR MEASUREMENT

A surface water resource of very poor quality is easily recognized as such by sight by anyone, and frequently also by smell. The need for improvement is also self-evident. Even though these assessments are valid, they are subjective. They are not quantified or placed into categories appropriate to determine the steps necessary to obtain improvement. Also a body of water may appear to be "all right" to the senses but still be of poor quality for some kinds of uses. As examples of natural waters which superficially would appear pristine, groundwater supplies from springs and wells are known which are high in toxic arsenic, fluorides, or sulfides [6, 7], and recreational lakes in southern Norway and the Adirondacks (U.S.A.) are known which have pH's of 4 and less [8]. For these reasons it is useful to have a set of quantifiable criteria which may be used to measure water quality. The values obtained for each of these criteria allow a valid and quantitative relative assessment of the water quality of different sources to be made. They also permit identification of the appropriate action required to improve the quality of the surface water. If the surface water source is to be used as a municipal or industrial supply, the values obtained for these criteria also establish the complexity of the treatment measures necessary.

4.3. SPECIFYING CONCENTRATIONS IN WATER

There are six methods in use to specify the concentration of a substance in water, three of them common and three used less frequently. The simplest system, and probably the most widely used, is the weight fraction (Eq. 4.1). Using the weight fraction,

Weight fraction = $w_A/(w_A + w_B)$ 4.1
or, in dilute solutions, approximately = w_A/w_B (i.e., when w_A is small)

or mass fraction gives a dimensionless concentration term if the same mass units are used for the numerator and denominator. Typical values expressed in these terms are given in Table 4.2.

Another common unit system used for specifying aqueous concentrations is weight of constituent per volume of solution, for which common volume units are liters (L), or cubic meters (m^3). Again for dilute solutions or suspensions in water, which has a density of approximately 1 kg/L, these units would result in the same approximations given in Table 4.2.

Molarity is the third common unit used for aqueous solutions (Eq. 4.2).

Molarity = M = (moles of substance)/(L of solution)   4.2

It is only useful when the compositions of the constituents in water are known because to evaluate molarity requires dividing the mass of each constituent in grams (per liter of solution) by the respective molecular weights. This system is, therefore, more complicated to evaluate. But it is more meaningful in the chemical sense because reacting ratios of scrubber solutions can be more read-

▮▮▮ **TABLE 4.2 Common Weight Fraction Units for Specifying
Concentrations in Water**

Unit	Abbreviation	Factor times wt. fraction	Approximations[a]
Percent	%	100	
Parts per thousand	°/oo	1000	g/L
Parts per million	ppm	10^6	mg/L
Parts per billion[b]	ppb	10^9	µg/L
Parts per trillion[b]	ppt	10^{12}	ng/L
Parts per quadrillion[b]	ppq	10^{15}	pg/L

[a]Approximation is closer than experimental error for very dilute (<1 ppm) solutions.
[b]These terms may have different exponents for European data, e.g., ppb factor (a million million) is 10^{12}, ppt (a million million million) is 10^{18}, and ppq (million⁴) is 10^{24}.

ily calculated from molarity information than from weight fraction information.

Less commonly used systems are the volume fraction, mole fraction, and molality (Eq. 4.3–4.5).

$$\text{Volume fraction} = V_A / V_{(solution)}$$

A common difficulty is that $V_{(solution)} \neq V_A + V_B$ (4.3)

$$\text{Mole fraction of B} = \frac{\text{moles B}}{\text{moles A} + \text{moles B}}$$ (4.4)

$$\text{Molality} = \frac{\text{moles A}}{1000 \text{ g of solvent}}$$ (4.5)

Each of these systems has special applications which require their use, e.g., mole fraction in heats of mixing calculations, and molality for osmotic pressure calculations. But none of these is in common use for the evaluation or remediation of water pollution problems.

4.4. SUSPENDED SOLIDS

Insoluble matter in surface waters may be partly settleable, consisting of fairly large particles, or may be nonsettleable, falling into the colloidal size range of silts and clays. The dividing line by diameter between these two classifications differs with the density of the particle. For spherical particles of density 2 g/cm^3, diameters larger than about 100 µm (0.10 mm) settle out more or less quickly; diameters smaller than this are slower to settle (Table 4.3).

The suspended solids content of a water sample may by determined by filtering an appropriate measured volume through a tared, fine glass fiber filter mat, and then rinsing any dissolved salts from the mat with a small portion of distilled water. The solids collected are dried at a standard temperature, usually 100–105°C. The dry weight obtained is then related to the original

■ **TABLE 4.3 Terminal Settling Velocities versus Diameter of Sphere of Specific Gravity 2, in Water at 25°C**[a]

Particle diameter (μm)	Terminal velocity (mm/sec)	Characteristic description[b]
1000	100	coarse sand
500	60	coarse sand
200	30	coarse/fine sand
100	6	fine sand
50	1.5	fine sand
10	0.06	silt
5	1.5×10^{-3}	silt
2	2.2×10^{-4}	silt/clay
1	6.0×10^{-5}	clay

[a]Settling velocities estimated from Weast [9].
[b]Approximate dividing lines indicated by joint descriptions.

volume filtered to give the result [10]. Very fine sintered glass filters or well-prepared asbestos fiber pads (Gooch filters) may also be used, but paper filters perform poorly in this application [10]. Depending on the pore size of the filter medium used, this method gives a result which includes the settleable and much of the nonsettleable suspended solids present in the sample. The size distribution of the collected particles may be determined by passage of a suspension of the particles in a liquid through a Colter counter [11].

Centrifugation may also be used for suspended solids determination using the proper conical end centrifuge tubes. After centrifuging the sample the supernatant water is decanted from the precipitate, and the precipitate is rinsed with a small portion of distilled water to remove any dissolved salts. Centrifuging again, decanting, and drying the residue as before gives the suspended solids result. A smaller proportion of the nonsettleable suspended solid is retained by this method than by filtration, a factor which has to be considered when comparing results obtained by the two systems.

An Imhoff cone is used to determine the settleable solids content of treated waste waters. This cone-shaped measuring device, of 1-L capacity and made of transparent glass or plastic, is graduated down the side to units of tenths of a milliliter at its lower apex. The sample is placed in the cone and allowed to stand for a period of 1 hr. Then the volume of the solids layer settled is read directly from the graduations on the side.

In situ turbidity determinations, which are related to, but do not strictly correspond to suspended solids determinations, are carried out using a Secchi disk [12]. The disk, a circular plate of about 20 cm in diameter made of sheet steel or other metal, is painted with alternating black and white quadrants on its upper face. A rope or chain calibrated in meters is attached to the disk via an eye bolt at the center of the disk, so that the disk hangs horizontally. For a reading, the disk is lowered into the water until the black and white painted quadrants appear to be uniformly gray from the turbidity of the water, and the depth at which this occurs is recorded. It is then lowered a few meters more and then gradually raised again, noting the depth at which the black and white quadrants just become discernible again. The average of these two readings gives the turbidity result, which can range from 0.2 m or less for a silt-laden river in spring flood, to 25 m or more for an oligotrophic (geologically young) mountain lake.

4.5. DISSOLVED SOLIDS

Water quality and monitoring programs also have an interest in the dissolved solids or salts content of water systems. Among the monovalent cations, there is an interest in the concentrations of sodium, potassium, and ammonium ions (and neutral ammonia, which is in equilibrium). Among the polyvalent cations, calcium and magnesium are the main ones of concern because of their strong tendency to precipitate and form useless curds with natural soaps, and to form adhering deposits or scale on the walls of all kinds of water heating appliances, from domestic kettles to industrial boilers. Occasionally this category might also include iron, aluminum, or other polyvalent cations, since, if these are present, they can contribute to this tendency. This property common to polyvalent cations is referred to as hardness. Thus, hard waters, with a high polyvalent cation content, do not launder well with ordinary soaps, whereas soft waters do (Table 4.4). Soft waters also heat more cleanly in water heating appliances.

Among the principal anions of interest in water treatment programs are chloride, sulfate, carbonate, and fluoride. High chloride concentrations are of concern because of the tendency of chloride to accelerate the corrosion rates of pipelines and local water distribution systems. Sulfate at concentrations above 150 ppm can cause severe digestive upset, essentially the symptoms of diarrhea, especially in nonacclimatized people. The concentration of carbonate present has important consequences in relation to hardness, about which more is said later. Fluoride concentrations of ca. 1 ppm in water supplies are

TABLE 4.4 A Rough Guide to the Scale of Hardness of Natural Waters According to Content of Calcium Carbonate (or the Equivalent)[a]

Description	Hardness, as ppm (mg/L) of $CaCO_3$
Soft	0–50
Moderately soft	50–100
Slightly hard	100–150
Hard	200–300
Very hard	>300

[a]Data from Klein [10].

beneficial in reducing dental caries. Natural concentrations much above this, which do occur occasionally, can give rise to toxic symptoms [13].

The concentrations of the nutrient anions of interest in water monitoring programs include nitrogen, primarily as nitrate, NO_3^-, and nitrite, NO_2^-, and phosphorus, primarily as phosphate, PO_4^{3-}. Nitrite is also of independent interest because of its high toxicity, particularly to infants [14]. Since nitrate is also reduced to nitrite in the digestive system, the presence of either ion at concentrations above 10 mg/L in a water supply is cause for concern. During the 1945–1969 period in the United States 328 cases of methemoglobinemia (blue baby syndrome) and 39 fatalities were reported for infants from this cause [15]. Both groups of ions, but particularly phosphate, are limiting nutrients for the growth of algae and weeds in water systems. Hence, if the concentration of phosphate is allowed to rise above 0.015 mg/L (calculated as P) and nitrate above 0.3 mg/L (as N) in surface waters, then algal blooms (prolific algal growth) are a likely outcome [16, 17].

Specific conductance is a useful rough guide to the total dissolved ionic solids present in a water sample, and is a technique which also readily lends itself to the continuous monitoring of a river or waste stream for the total ion content (Fig. 4.1). It is also a simple method which can easily be used to check the accuracy of other types of analyses conducted for specific ions. Specific conductance is measured via a pair of carefully spaced platinum electrodes which are placed either directly in the stream to be measured or in a sample withdrawn from it. The water temperature should be 25°C, or the result corrected to this temperature. Voltages in the 12 to 14 range, and 60 to 1000 Hz AC are used, plus a Wheatstone bridge circuit to obtain a conductivity reading in (μmho/cm or (μS/cm (microsiemen/cm). The response obtained is linear with total ion content over a wide range of concentrations (Fig. 4.1). Examples of the conductance ranges and seasonal variation of some typical Canadian rivers are presented in Fig. 4.2. Rivers of both low and high dissolved solids, and with narrow or wide seasonal variations are clearly evident from these plots.

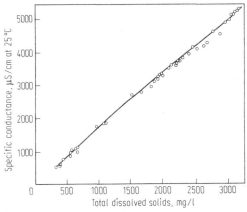

FIGURE 4.1   Total dissolved solids concentration (by analysis) versus specific conductance, shown by Gila River (Bylas, Arizona) water samples taken over a 1-year period. (Reprinted from Hem [18].)

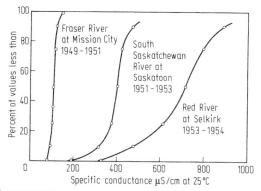

FIGURE 4.2 Distribution of the specific conductance readings for some Canadian rivers during 2-year data collection periods. (Data from Thomas [19].)

More specific and detailed information about the dissolved ionic solids present may be obtained through the application of conventional specific analytical techniques for the ions of interest. Complexation titration, for example, using a stable complexer such as EDTA (ethylenediaminetetraacetic acid) is useful for the determination of dissolved calcium, as well as other ions [20]. EDTA, being a hexacoordinate complexer, forms very stable $1:1$ complexes with many metal ions, including calcium. An 0.01 M standardized EDTA solution, usually as disodium EDTA, is first made up. Exactly 1.00 mL of this solution is equivalent to 400.8 (μg Ca, or 1000 (μg CaCO$_3$. Then titration of a sample containing calcium ion in the presence of a weakly complexing dye indicator gives displacement of the calcium from the indicator plus a corresponding change in color of the solution, which is used as the end point of the determination (Eq. 4.6, 4.7).

$$M^+ + dye \rightarrow M^+ \cdot dye \text{ (weak complex)} \qquad 4.6$$

$$M^+ \cdot dye + EDTA \rightarrow dye + M^+ \cdot EDTA \text{ (strong complex)} \atop \text{(colorless)} \qquad 4.7$$

Murexide, or Eriochrome blue-black R are suitable indicators for this analysis. A number of other metal ions may also be determined by related complexometric titrations.

Ion selective electrodes provide a simple and accurate method for the determination of many ions in solution. These have been developed using the same electrochemical principles as the pH electrode, which is basically an ion selective electrode specific for hydrogen ion [21]. Thus concentrations of Na$^+$, K$^+$, Ca^{2+}, and Pb^{2+} as examples of cations, and F$^-$, Cl$^-$, Br$^-$, I$^-$, S^{2-} and CN$^-$ as anions may all be measured using this method. With the appropriate measuring precautions and attention to possible interferences most can be determined at concentrations as low as 10^{-5} to 10^{-6} M, and lead for example down to 10^{-7} M.

Of the nutrient ions of current concern, "nitrogen" refers to "combined nitrogen," that is, nitrite, nitrate, ammonia or ammonium salts, and occasionally amines and the like. The dissolved element itself is not of concern from a nutrient standpoint, although the blue green algae are able to "fix" elemental nitrogen dissolved in water.

Nitrite concentrations are determined by, first, diazotization of sulfanilamide. The acidified waste sample to be analyzed is the source of the nitrous acid (Eq. 4.8).

When the diazotization is conducted in the presence of N-(1-naphthyl)-ethylenediamine, the reactive diazonium salt couples with the amine to form an intensely colored, reddish purple azo compound (Eq. 4.9).

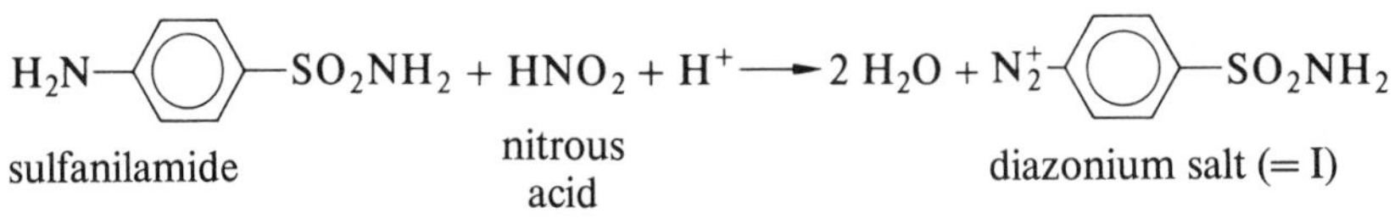

$$H_2N-\!\!\bigcirc\!\!-SO_2NH_2 + HNO_2 + H^+ \longrightarrow 2\,H_2O + N_2^+-\!\!\bigcirc\!\!-SO_2NH_2$$

sulfanilamide   nitrous acid   diazonium salt (= I)

4.8

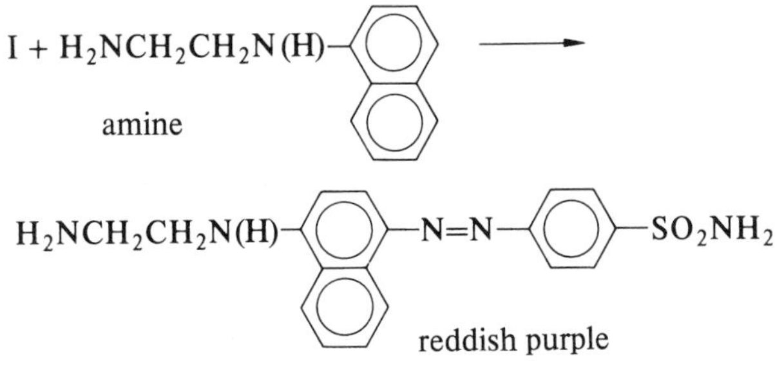

$$I + H_2NCH_2CH_2N(H)-\!\!\bigcirc\!\!\bigcirc \longrightarrow$$

amine

4.9

$$H_2NCH_2CH_2N(H)-\!\!\bigcirc\!\!\bigcirc\!\!-N=N-\!\!\bigcirc\!\!-SO_2NH_2$$

reddish purple

The concentration of azo compound obtained can be determined either colorimetrically, using a comparator, or spectrophotometrically. The nitrite detection limit of this method is 0.02 mg/L of nitrite, specified as N, which is equivalent to 0.02 ppm or 20 ppb N in fresh waters [20].

Nitrate is determined on a separate water sample by first reducing nitrate to nitrite with either hydrazine sulfate or with an easily oxidized metal such as cadmium (Eq. 4.10).

$$NO_3^- + Cd \rightarrow CdO + NO_2^- \qquad\qquad 4.10$$

The sensitive nitrite method is then used to determine total nitrite, which now will be equivalent to nitrite plus the original nitrate. Subtraction of the original nitrite from this total nitrite result gives the value for the nitrate concentration.

Reduced nitrogen, such as ammonia or ammonium salts, may be analyzed by spectrophotometric determination of the absorbance of the yellow-brown colloidal suspension formed on the addition of Nessler's reagent, which is potassium mercuri-iodide (Eq. 4.11).

$$2K_2HgI_4 + 2\,NH_3 \rightarrow NH_2Hg_2I_3 + 4\,KI + NH_4I \qquad\qquad 4.11$$
yellow-brown

If amines or amides are present, then the determination requires prior Kjeldahl decomposition of the amine to ammonium hydrogen sulfate using boiling sulfuric acid (Eq. 4.12, 4.13).

$$RCH_2NH_2 + H_2O + H_2SO_4 \rightarrow RCH_2OH + NH_4HSO_4 \qquad 4.12$$

$$RCONH_2 + H_2O + H_2SO_4 \rightarrow RCOOH + NH_4HSO_4 \qquad 4.13$$

Following this, the solution is made alkaline with sodium hydroxide and the ammonia content measured using Nessler's reagent. Again subtraction of the initial Nessler reagent reading yields an amine concentration independent of the original ammonia content for the water sample analyzed.

Phosphate concentrations may be measured gravimetrically, by weighing the yellow precipitate formed on the addition of a solution of ammonium molybdate to the water sample acidified with nitric acid (Eq. 4.14). The product stoichiometry for this process is certain, although the structure is not.

$$PO_4^{3-} + 12\ (NH_4)_2MoO_4 + 24\ HNO_3 \rightarrow$$
$$(NH_4)_3PO_4{\cdot}12MoO_3{\cdot}3H_2O + 9\ H_2O + 21\ NH_4NO_3 + 3\ NO_3^- \qquad 4.14$$

If higher sensitivities are desired then the ammonium phosphomolybdate product may be reduced with stannous chloride or ascorbic acid to give soluble molybdenum blue [20]. Spectrophotometric determination of the blue absorbance at 885 nm gives a detection limit of 0.003 mg/L phosphate, stated as phosphorus (3 ppb in freshwater) or 0.0092 mg/L as phosphate [20].

4.6. TOTAL SOLIDS OR RESIDUE ANALYSIS

Analysis of a water sample for total solids content (suspended and dissolved) requires evaporation of the water from a measured volume of sample, usually 1 L. For water samples the residue is usually dried at 180°C, to put calcium sulfate in the anhydrous state, and magnesium sulfate as the monohydrate, $MgSO_4 \cdot H_2O$, in the residue [10]. In any case, because of the dependence of the level of hydration of many salts on the drying temperature, and the different temperatures used, this should be stated with the results.

To determine the total solids content of river muds or sewage sludges, either of which may have a high organic content, lower drying temperatures of 100–105°C are normally used to reduce the risk of thermal decomposition of the organic content. For this reason, azeotropic drying is sometimes used for samples of this kind. When perchloroethylene is used as the azeotroping solvent for water removal from a weighed amount of mud or sludge, it drops the temperature required for water removal to 88.5°C, the boiling point of the perchloroethylene-water azeotrope. When azeotropic methods are employed the water removed from the sample is read from a graduated solvent/water separator, from which the solvent from the condensate of the process is returned to the sample flask.

4.7. DISSOLVED OXYGEN CONTENT

Oxygen, nitrogen (as N_2), carbon dioxide, and the other trace gases of the atmosphere are in a dynamic equilibrium between surface waters of the earth and the air. Gases in the surface layer of water move through (e.g., during river flow) or are largely restricted from moving through (e.g., in a thermally stratified lake) the water column below it, depending on the degree of mixing of the water phase. The solubility of oxygen in water is low, but its presence is significant to the existence of most kinds of water organisms, and to the perceived and actual water quality.

The solubility of any gas in a liquid with which it does not react is proportional to the pressure of that gas (i.e., the partial pressure) above the liquid (e.g., Eq. 4.15 for oxygen).

$$P_{O_2} = k_{O_2} X_{O_2} \qquad\qquad 4.15$$

where P_{O_2} is the partial pressure of oxygen above the water in atmospheres, k_{O_2} is the Henry's law constant, and X_{O_2} is the mole fraction of oxygen in the water phase. This solubility relation, also known as Henry's law, is quite closely followed for pressures not a large multiple removed from normal atmospheric pressure, and for any given temperature. Oxygen, which comprises about 21% (actual value 20.95%) of the atmosphere by volume, therefore contributes 0.21 atmosphere partial pressure above surface waters at sea level, and proportionally lower pressures at higher altitudes.

Gas solubility in a liquid is also inversely proportional to temperature (Eq. 4.16).

$$T = k_t / X_{O_2} \qquad\qquad 4.16$$

where k_t is the proportionality constant at absolute temperatures. This factor means that the discharge of warmed waste waters into a river or lake decreases the solubility of oxygen in that water, and thus tends to decrease the actual concentration of dissolved oxygen present.

The combined influence of temperature and pressure on the solubility of some common gases important in water quality studies is given in Table 4.5. A couple of interesting features to note are that oxygen, with a Henry's law constant of 1.91×10^7 mol/(mol mm Hg), is more than twice as soluble in water as nitrogen. Oxygen also has a very similar molar solubility to argon, a reflection of the diatomic nature of oxygen gas which has a similar molar mass as that of monatomic argon, 39.95. Also evident here is the rationale

▬▬▬ **TABLE 4.5 Henry's Law Constants for the Solubility of Some Common Dry Gases in Water**[a,b]

| Temperature (°C) | Henry's law constants/10^7 | | | | |
	Argon	Carbon dioxide	Helium	Nitrogen	Oxygen
0	1.65	0.0555	10.0	4.09	1.91
10	2.18	0.0788	10.5	4.87	2.48
20	2.58	0.108	10.9	5.75	2.95
30	3.02	0.139	11.1	6.68	3.52
40	3.49	0.173	11.0	7.60	4.14
50	3.76	0.217	10.9	8.20	4.50

[a]For the equation p = kX, where p = partial pressure of the gas in mm of Hg, X = mole fraction of the gas in water, and k = the Henry's law constant (e.g., 1.91×10^7 for oxygen at 0°C).
[b]Values selected from Mason [22].

for the use of oxygen-helium mixtures for deep sea diving [23]. The much lower solubility of helium in water, and hence in the blood, reduces the risk of gas bubble formation in the bloodstream on the return of a deep sea diver to ordinary atmospheric pressure. This can readily occur with nitrogen, giving the "bends" if ordinary air is used for diving under these circumstances.

A third factor influencing the solubility of oxygen or other gases in water is the presence of other solutes in the water. A high concentration of dissolved solids in the water phase decreases the solubility of oxygen. Other things being equal, therefore, marine waters tend to have a lower dissolved oxygen content than freshwaters.

The combined influence of all of these solubility factors on the actual oxygen content of ordinary surface waters is presented in Table 4.6. From this data it can be seen that the air-saturated oxygen content of seawater is normally about 20% less than the oxygen content of freshwater, at the same temperature. It can also be seen that the effect of elevated temperatures on solubility is such that at 30°C only about half the oxygen content is obtained in the water phase at saturation as at 0°C, for either freshwater or saltwater. Lakes at higher altitudes, exposed to lower atmospheric pressures and hence lower oxygen partial pressures, will tend to have lower concentrations of dissolved oxygen, other things being equal. Similarly the salt lakes which lie below sea level, such as the Dead Sea and Salton Sea, will tend to have higher concentrations of dissolved oxygen than would be predicted for similarly saline waters at sea level.

The Winkler test, or variations of it, is the standard wet chemical procedure for measuring the concentration of dissolved oxygen in water samples [20, 24, 25]. This test uses a standard 300-mL BOD sample bottle closed with a glass stopper which has a polished cone-shaped end to it to ensure that it is completely filled with the water sample during the test. This avoids inadvertent aeration of the sample during the test. Initially a white precipitate of manganous hydroxide is prepared in the BOD bottle (Eq. 4.17).

$$MnSO_4 + 2\ KOH \rightarrow Mn(OH)_2\downarrow + K_2SO_4$$
$$\text{white}$$

4.17

TABLE 4.6   Oxygen Content of Water in Equilibrium with Air Saturated with a Water Vapor at 760 torr (mm Hg)[a]

Temperature (°C)	Oxygen content (mg/L)		Water vapor pressure (mm Hg)
	Freshwater	Seawater 3.5% salinity	
0	14.6	11.22	4.58
10	11.3	8.75	9.21
20	9.2	7.17	17.54
30	7.6	6.10	31.82
40	6.6	5.13	55.32
50	5.6	—	92.51

[a]Compiled from Klein [10] and American Public Health Association [20].

Dissolved oxygen in the water sample being tested oxidizes the manganous hydroxide to a brown suspension of manganic hydroxide in the test solution (Eq. 4.18).

$$2 \ Mn(OH)_2 + O_2 \rightarrow 2 \ MnO(OH)_2\downarrow \qquad\qquad 4.18$$
$$\text{brown ppte}$$

If no brown coloration is observed at this stage of the test, there is no dissolved oxygen in the water sample being tested. If there is a brown coloration the amount of manganic hydroxide is accurately determined by first putting it into solution with sulfuric acid (Eq. 4.19).

$$MnO(OH)_2 + 2 \ H_2SO_4 \rightarrow Mn(SO_4)_2 + 3 \ H_2O \qquad\qquad 4.19$$

To determine the original levels of oxygen present, excess aqueous potassium iodide is added, releasing iodine equivalent to the manganic sulfate present in the most rapid reaction of this sequence (Eq. 4.20).

$$Mn(SO_4)_2 + 2 \ KI \rightarrow MnSO_4 + K_2SO_4 + I_2$$
$$\text{brown} \qquad\qquad 4.20$$
$$\text{solution}$$

The iodine generated is prevented from being lost as vapor by ensuring that an excess of potassium iodide is present, which forms the stable ionic complex, KI_3. The liberated iodine is measured by the familiar sodium thiosulfate titration, using starch as an indicator near the end point of the titration to sharpen its observation (Eq. 4.21).

$$I_2 + 2 \ S_2O_3^{2-} \rightarrow 2 \ I^- + S_4O_6^{2-}$$
$$\text{colorless} \qquad\qquad 4.21$$
$$\text{solution}$$

Each mole of liberated iodine is equivalent to a half mole of dissolved oxygen in the original water sample, so that each milliliter of 0.0250 M thiosulfate is equivalent to 200 µg of original dissolved oxygen content.

The Winkler test is reliable and accurate, to of the order of 20 µg/L for a 200-mL sample of ordinary clean surface waters, but is somewhat less accurate for sewage and industrial effluents where interfering substances may be present. It is also a primary determination method, which means that it may be used to calibrate or check other kinds of dissolved oxygen instrumentation such as dissolved oxygen meters based on the oxygen electrode. Improved designs of these instruments, such as those produced by the Yellowstone Instrument Company, provide rapid and convenient dissolved oxygen readings as long as care is taken to calibrate these, and that they are operated at or near calibration temperatures. Periodic performance checks should also be carried out [26]. Water samples of a range of dissolved oxygen levels may be readily prepared for this purpose. A 0 mg/L (ppm) test may be obtained by sparging a quantity of freshly boiled distilled water with nitrogen for 3 or 4 hr in a thermostat bath of the required test temperature. Sparging air through a water sample at 20°C for the same period would give 9.2 mg/L, and sparging pure oxygen through should give 44.0 mg/L at this temperature.

For meaningful short-term results, particularly with waters having prolific weed or algae growth, care must be taken to sample or measure the dissolved oxygen *in situ* at the same time of day, and note the ambient temperature. This is because photosynthetic production of oxygen toward the end of a sunny day can almost double the dissolved oxygen content of the water, even to well above saturation levels, as compared to the values in the very early morning after a cloudy day [27]. Metabolic consumption of oxygen under these circumstances can reduce the dissolved oxygen concentration to near zero.

Dissolved oxygen concentrations in surface waters below about 5 mg/L or 50% of saturation are generally unsatisfactory for a diverse biota. In fact, game (or sport) fish require more like 60 to 70% of saturation to do well. Lower dissolved oxygen concentrations than these tend to limit the habitat to the growth and reproduction of coarser species of fish. No dissolved oxygen eliminates the survival of gill breathers entirely, and discourages many other species so that only organisms capable of air breathing, such as eels and certain Tubifex worms, can survive in this situation.

The dissolved air, or more particularly dissolved nitrogen, in water can sometimes be too high for fish well-being, 110% of saturation already putting them at risk. If gill-breathers are exposed to waters supersaturated in air, their blood becomes supersaturated too, through gill action. Thus at 120% of saturation the excess nitrogen in the bloodstream, which is not metabolically consumed, comes out of solution behind the membranes of the eye sockets causing "pop-eye," and also behind the membranes of the fins, tail, and mouth causing gas blisters [28 and references cited therein]. These symptoms, collectively referred to as nitrogen narcosis, cause acute stress and when severe can kill the fish. Exposure to 120 to 140% of saturation for any length of time is sufficient to cause symptoms to appear, together with some fish mortality. Exposure to concentrations above 140% of saturation causes high mortalities.

Air supersaturation of natural waters can occur from the warming of a cold, air-saturated stream as it flows into a shallow reservoir, or from being used as a source of cooling water, which amounts to thermal causes [29]. It is also a common occurrence while releasing reservoir water into a plunge basin, at the bottom of which the pressure of air bubbles entrained in the water may be raised momentarily to 2 or more atmospheres from the pressure of the head of water over the bubbles. This increased pressure is sufficient to supersaturate the water, causing supersaturation problems here and downstream of the plunge basin. Plunge basin supersaturation can be avoided by the use of [flip-lip] or [flip-bucket] spillway designs which spread spilled water widely and as a fine spray to avoid carrying bubbles to depth.

From the point of view of disposal of oxidizable wastes to water, low concentrations of dissolved oxygen in a receiving body of water greatly exaggerate the impact of the waste discharge on water quality. Too low a waste dilution ratio on discharge can cause the dissolved oxygen content of the water course to drop to zero at any time. Discharge of any oxygen-consuming waste always causes a decrease in the dissolved oxygen content of the receiving lake or stream. But if this discharge coincides with low dissolved oxygen in the receiving waters the resultant drop in oxygen concentration will be greater and the recovery of the stream to normal oxygen concentrations will be less rapid than at higher initial oxygen concentrations. Also it is known that fish and other aquatic animals that are stressed by low dissolved oxygen concentrations are more susceptible on exposure to any additional stresses such as high temperatures, toxic substances, pH deviations, and the like.

4.8. RELATIVE ACIDITY AND ALKALINITY

The pH of a water supply, in conjunction with its dissolved solids content, is important because it relates to the corrosivity of the supply. Corrosivity is of concern in municipal water distribution systems, as well as for the feedwater of power boilers and process cooling water. pH's in the neutral to slightly alkaline range are generally preferred for these applications. Most aquatic creatures require a pH in the 5 to 8.5 range for optimal growth and reproduction, although they may survive for a time at pH's somewhat outside this range. Natural waters in the acid range, which can arise by absorption of acidic atmospheric gases (Chap. 2) or from the accumulation of humic acids on percolation through peaty soils, have the potential to mobilize elements of the rocks and soils through which they flow. In limestone, dolomites, and similar rock formations, calcium, magnesium, and other elements may be dissolved out, in the process increasing the pH but at the same time raising the hardness of the water (e.g., Eq. 4.22, 4.23).

$$CaCO_3 + H_2CO_3 \rightarrow Ca(HCO_3)_2 \qquad\qquad 4.22$$
$$\text{insoluble} \qquad\qquad\qquad \text{soluble}$$

$$MgCO_3 + 2\ HNO_3 \rightarrow Mg(NO_3)_2 + H_2O + CO_2 \qquad 4.23$$
$$\text{insoluble} \qquad\qquad\quad \text{soluble}$$

Industrial liquid waste streams, in general, have the greatest potential of any of the ordinary range of aqueous wastes to influence the pH of receiving waters, because of their sometimes significant deviation from neutral conditions and the large volumes commonly involved. Pulp mills, particularly those producing fully bleached kraft pulp, have some very acidic and some strongly alkaline waste streams generated by their processes. Drainage waters from coal mines and from metal mines working sulfide ores can also be quite acidic from the acid generated by the bacterial oxidation of sulfides (e.g., Eq. 4.24–4.26).

$$2\ FeS_2 + 7\ O_2 + 2\ H_2O \rightarrow 2\ FeSO_4 + 2\ H_2SO_4 \qquad 4.24$$

$$2\ FeS_2 + 15/2\ O_2 + H_2O \rightarrow Fe_2(SO_4)_3 + H_2SO_4 \qquad 4.25$$

$$Fe_2(SO_4)_3 + 6\ H_2O \rightarrow 2\ Fe(OH)_3 + 3\ H_2SO_4 \qquad 4.26$$

Chemical plants producing phosphatic fertilizers or phosphoric acid may also have quite acidic discharges, and ammonia plants or coal-coking plants quite alkaline streams, as further industrial examples.

Monitoring the pH of waste streams or surface waters is most frequently accomplished using a pH meter. An important preliminary for accurate readings is that the meter be calibrated using buffers at slightly higher and lower pH's than those of the samples being measured. Also the temperature calibration should be properly adjusted and the electrode given sufficient time in the sample being measured to come to thermal as well as pH equilibrium. Using a meter allows determination of pH free from any interferences due to color, colloidal matter, coarse turbidity, free chlorine, or the presence of other oxidants or reductants. However, pH measurement of distilled water or high-purity natural waters may give inaccurate readings from too low conductivities. This can be corrected by addition of a small amount of a neutral salt, such as 1 g/L potassium chloride, to raise the total ionic strength to about 0.1 M [30]. If the pH measurement is not being conducted *in situ*, it should be measured promptly after sampling to avoid errors due to gas exchange or biochemical processes that may alter the pH after collection. In this connection it should be noted that the pH of eutrophic waters can be highly time dependent from the influence of variable rates of metabolic activity on carbonate-bicarbonate buffer. Marine waters, and any other streams with a relatively high sodium ion content will require a sodium ion correction, especially if the measurement is being conducted at high pH.

For low-cost occasional pH measurement and for measurements in situations unfavorable to the placing of a fragile glass pH electrode, narrow-range single indicator and wide-range multiple indicator pH papers are convenient and sufficiently accurate for most purposes. Some of these include eight or more indicators and are capable of being read to 0.5 pH unit and some to 0.3 pH unit.

4.9. TOXIC SUBSTANCES

Here we consider a number of potential water contaminants which, for want of space for details, are grouped together. Included are such candidates as toxic heavy metals, pesticides, water and weed treatment chemicals, radioactive particles, and the like.

The relative toxicities to fish of some appropriate examples of toxic substances are given in Table 4.7. It should be remembered that aquatic toxicity is difficult to pin down to an exact value since it depends not only on the particular species and the ages and state of health of the exposed individuals, but also on the time exposed, water temperature, simultaneous presence of other contaminants, oxygen content, prior acclimatization, hardness, and other associated factors. On the whole, though, the high sensitivities of fish to the substances tabulated is the consequence of the good blood/water exchange processes necessary in gill breathing animals for their respiration. A parallel in human exposures is our high sensitivity to many types of atmospheric contaminants.

More than 20 metals or metalloids, Al, Sb, As, Ba, Be, Bi, Cd, Co, Cu, Ce, In, Pb, Hg, Mo, Ag, Te, Tl, Sn, Ti, W, U, and Zn are significant in industrial hygiene programs [32, 33]. Therefore water contamination by any of these could be potentially hazardous.

Some wet chemical sample preparation, such as pH or oxidation state adjustment, is normally required for most metal ion determinations. Then complexometric titration using EDTA, as already mentioned, or diphenylthio-

**TABLE 4.7 Approximate Acute Toxicities
of Some Common Potential Water
Contaminants to Freshwater Fish**[a]

Substance	Approximate lethal concentration [mg/L (ppm)]
Chlorine	0.05–0.2
Chloride	ca. 6000
Copper	0.05
Cyanide	0.04–0.10
DDT	0.02–0.10
Fluorides	2.6–6.0
Malathion	13.0
Mercury, Hg^{2+}	ca. 0.01
Phenols	1–10
Pentachlorophenoe[b]	5
Natural soap (in hard water)[c]	10–12[b] (900–1000)
Synthetic detergent	6–7
Toxaphene	0.01
Zinc, Zn^{2+}	0.15–0.60

[a]Exposure times and toxicity criteria vary. Compiled from Klein [14], Mason [22], and Ryckman *et al.* [31].
[b]Sodium oleate ($NaO_2C(CH_2)_{16}CH_3$), which is much less toxic in hard water, parenthesized value.
[c]Sodium dodecylbenzenesulfonate ($NaO_3S(CH_2)_{12}C_6H_5$).

carbazone ("dithazone"; Eq. 4.27) may be used for cadmium, copper, lead, mercury, or zinc determinations.

$$Ph—N{=}N—C(S)—NH—NH—Ph \qquad\qquad 4.27$$

With the latter reagent, the concentration of the colored complex obtained may be measured colorimetrically, by comparison with solutions made up from complexes of known concentrations of the metals of interest. This method is easily applied in the field. Or it may be determined somewhat more accurately, but at greater expense, using a spectrophotometer in the laboratory.

Other versatile and sensitive techniques for determination of metal ions in water samples are atomic absorption, which uses the attenuation of a beam of light of the appropriate wavelength by the atomized metal as the measure of concentration, and anodic stripping voltammetry, which is an electrochemical technique. Further details of all these procedures are available in standard texts ([20] and also see Further Readings).

Determination of pesticide and herbicide content, or the presence of other organics in surface waters, usually involves a preliminary concentration step by extraction with an immiscible organic solvent such as hexane or heptane. But the concentrations of these substances present are often so low that special highly purified grades of solvent, so-called "pesticide grades," are required to avoid analytical problems from contamination of the sample by traces of pesticide already present in the solvent. After extraction, the small volume of organic phase obtained is then dried and carefully concentrated. Analysis is often by injection of a small sample into a highly sensitive gas chromatograph e.g., [34, 35]. A series of peaks of differing retention times is obtained on a strip chart from which probable identities of the compounds represented by each peak may frequently be established by comparison with the retention times for a solution of standard reference materials separately injected into the chromatograph. More unequivocal and rapid peak identification may be obtained from a mass spectrometer coupled to the outlet of the gas chromatograph. The mass spectrometer can be used to establish accurate masses and fragmentation patterns of the constituents of each of the peaks obtained on the chromatograph trace.

Radioactive particles are generally present in surface waters at such low levels that a preliminary concentration step is necessary for determination. Then gross counting of the concentrated sample is carried out to give an overall measure of the radionuclide content. The result obtained can then be related back to the original sample volume. Related to this it should be mentioned that deliberate radiolabeling provides a safe and very sensitive tracer for the tracking of water flow patterns and the dispersion patterns of waste discharges [36].

4.10. MICRO-ORGANISMS

Knowledge of the types and populations of micro-organisms present in water is an important part of any water quality considerations. For this purpose micro-organisms can be conveniently grouped into two main classes. The indigenous, or naturally occurring organisms, which include many ambient types of bacteria, phytoplankton and zooplankton, algae, and diatoms, comprise one of these. This class is relatively harmless in water supplies. There are also the bacteria, viruses, and parasites which can arise in surface water from the excreta of warm-blooded animals or from human sewage contamination. This group includes examples which are harmless, i.e., they form a component of the normal human intestinal flora, and some examples which are pathogenic (disease-causing). Detection of the presence of any of this second group of organisms in water supplies should be taken as a warning of the risk of contamination by pathogens.

A survey of the micro-organism status of a water sample can be taken by means of a standard plate count. To do this, the sample and several dilutions are inoculated onto a nutrient agar medium in separate petri dishes (shallow plates with loosely, fitted lids). The spotted dishes are then incubated for a period of time, after which the spots of growth in each are counted. Very high micro-organism numbers in the water sampled tend to cause a merging of spots, making accurate counting impossible, and agglomerations or clumping of organisms tends to give lower counts. Both problems are minimized by suitable sample dilutions before inoculation of aliquots of each dilution onto the plates.

Since it is the degree of contamination by the bacteria of warm-blooded animals that is most significant from a water quality standpoint, this is measured by means of the differential ratio test [37] (Eq. 4.28).

$$\text{Differential ratio test} = (20°C \text{ count})/(37°C \text{ count})$$
If > 10, reasonably good supply   4.28
If < 10, indicative of contamination

The count at 20°C is taken for a 48-hr incubation of the plates at this temperature. The 37° count is taken for a 24-hr incubation at this temperature, to approximate the propagation conditions provided by warm-blooded animals. If the ratio obtained by the differential ratio test is above 10, it is taken as an indication of a reasonably good water supply.

Use of different nutrients in the petri dish, as well as different incubation conditions and various staining and slide-making techniques, permits positive identification of the particular micro-organisms collected [38]. Tests giving results comparable to the differential ratio test may also be carried out by using different plating out media for inoculation, or by membrane filtration techniques [39]. There is also a simple dipstick method which may be used for more qualitative bacterial monitoring, the Coli-Count water tester. This uses a filter paper matrix already factory-impregnated with dry nutrient and with a grid marked on it to facilitate counting after incubation. The dipstick is immersed into the water to be sampled, which both inoculates and activates the medium. Incubation for the required time in the sterile container provided then gives a convenient indication of microbiological water quality.

Viral analysis of waters is more difficult because of their small size, about 10 to 300 nm. Also viruses need susceptible living cells such as chick embryos or tissue cultures for cultivation and identification in the laboratory which makes them more difficult to work with. Nevertheless, viruses represent an important microbiological class for water and wastewater monitoring programs since serious diseases such as polio and hepatitis are transmitted in this way.

4.11. TEMPERATURE

Unnaturally affected surface water temperatures are most often caused by the warmed discharges from industrial process or thermal power station cooling requirements. This may, in itself, seem to be a relatively superficial parameter. However, water temperatures have a fundamental connection to several important water quality paramaters. Increasing the temperature of the surface waters of a lake, for example, increases the rate of evaporation of water from the lake (Fig. 4.3). The net water consumption caused by this increased evaporation rate from the cooling load of a 100-MW thermal power station has been estimated to equate to the consumptive load of a city of 100,000 people [40]. This increased evaporation also has the effect of increasing the concentrations of dissolved solids and nutrients in the water system [41]. These ef-

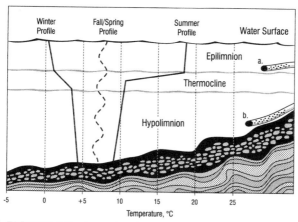

FIGURE 4.3 Seasonal, and warmed discharge effects on the change in temperature with depth of a lake. (a) Primarily surface effects resulting from a warm discharge into the epilimnion. (b) A larger mixing volume and normal temperature profile disruption obtained from a warmed discharge into the hypolimnion.

fects, in turn, tend to increase turbidities due to algal blooms and nuisance weed growth as a result of the increased nutrient concentrations. Higher average year-round temperatures also tend to decrease the average concentrations of dissolved oxygen, from the decreased solubility of oxygen at higher water temperatures.

High thermal loadings also affect the distribution of fish populations, since coarse fish such as suckers, carp, and catfish can better tolerate higher temperatures and lower dissolved oxygen concentrations than can game fish such as salmon, trout, or other commercially valuable species. But even for coarse fish, exposure to temperatures above 40°C for a brief period is usually fatal. Warm water temperatures can also favor year-round spawning of the coarser species of fish, which tends to further improve their competitive position in this habitat. These effects can cause coarse fish to gradually displace any preexisting population of sport fish.

Temperature measurement of surface waters using ordinary thermometers is relatively straightforward. Measurement at depth to determine thermal gradients is more difficult. Flip thermometers may be used, which are lowered to the depth of interest in the normal, vertical temperature-sensing position. At the depth of interest after waiting for a time for the temperature to register, the thermometer is flipped upside down. This breaks a slender portion of the column of mercury in the expansion capillary, preserving the temperature reading for examination at the surface. Thermocouples, which are more robust, may be used for continuous temperature measurement at the surface or at depth, a system which also lends itself to continuous automatic recording. A thermocouple, which uses the electrical potential generated by a pair of dissimilar metal junctions, commonly iron/constantan or chromel/alumel, gives either a millivolt reading, which can be translated to temperature using a table, or a direct temperature readout via solid-state circuitry. The fast response of this system facilitates the taking of a series of readings while gradually lowering the sensor through a single depth profile, or for monitoring the spread of warmed waters with distance.

Thermistors (thermal resistors), are semiconductor devices which have a very large drop in resistance with temperature, of the order of 20,000 ohms per degree over a 500°C temperature range. These may be used for temperature studies requiring quick response times. But for field overviews of a river or a lake receiving warmed discharges, aerial infrared photography (an aerial thermogram) or ERTS (Earth Resources Technology Satellites, Landsat and Seasat) studies can reveal more details more quickly than the majority of surface methods [42]. Satellite studies can also give an instantaneous thermal picture of very large areas of water, combined with the distributions of algae and the like, in a manner not possible by any other means.

4.12. OXYGEN DEMAND

Any dissolved oxidizable material present in water places a demand on the dissolved oxygen content of the water, as biochemical processes consume oxygen to utilize it. Therefore the oxygen demand level of surface waters or waste streams discharged into them has an important bearing on water quality and the concentration of dissolved oxygen that can be maintained in surface waters. The biochemical oxygen demand (BOD), the chemical oxygen demand (COD), and the total organic carbon content (TOC) tests are used to measure this parameter, so named because of the respective measurement methods used. Each of these tests is complementary to the others in defining the oxygen demand profile of a water sample. Each also has special situations in which it is a more useful or practical measure of oxygen demand, for one reason or another, than the other tests.

4.12.1. Biochemical Oxygen Demand

The biochemical oxygen demand, as the name suggests, is a measure of the biochemically oxidizable material present in the water sample expressed in terms of the oxygen required to consume it. More precisely, BOD is defined as the number of milligrams of oxygen taken up by a 1-L sample of water on incubation in the dark for 5 days at 20°C [10]. For freshwater samples 1 mg/L equates to 1 ppm (weight for weight) so both sets of units have been in common use. But to avoid potential ambiguity, since about 1970 the use of mg/L (or multiples of this) has been growing. In marine waters, because of density differences, it is more accurate to stick to mg/L units. The 5 subscript sometimes used with the BOD label refers to the 5-day test period, which is normally assumed. If the test period differs from this, it should be stated by the subscript used. The BOD test is the most lengthy of the oxygen demand tests to perform, and the answers obtained may be difficult to interpret. But it is also a useful test in that the result comes the closest to reproducing the natural oxidation and recovery conditions in a river or lake [43]. This rationale is also the reason for the selection of the standard test temperature of 20°C. The test can also be performed with relatively simple, inexpensive equipment requiring little operator time, even though there is a lengthy delay before the result is obtained.

There are three possible approaches to measurement of biochemical oxygen demands. The oxygen required to oxidize the organic matter in the water sample can come solely from the dissolved oxygen content of the sample itself or from added aerated dilution water, procedures referred to as direct and dilution techniques, respectively. Or the oxygen can come from a closed air space above the water sample to be analyzed, in which case the procedure is referred to as a manometric, or respirometer, method. It is difficult to measure the small volume of gas phase oxygen consumption obtained from a low BOD water sample. For this reason the manometric method is normally reserved for use with sewages and high oxygen demand industrial waste streams where the oxygen consumption will be sufficient to obtain a valid reading.

Using the direct technique, which is appropriate for ordinary clean river waters, requires complete filling of two 300-mL BOD bottles with the sampled water. The oxygen content of one is measured within 15 to 20 min of collection. The other bottle is incubated in the dark, to avoid photosynthetic contribution to the oxygen content of the water, for 5 days at 20°C. The extent of biochemical oxidation of any organics taking place is then determined by measuring the oxygen content of the second bottle. The difference between the measured oxygen contents of the first and second bottles is a direct measure of the BOD of the river sampled (Eq. 4.29).

Direct biochemical oxygen demand:

$$BOD_5, \text{mg/L} = DO_1 - DO_5, \text{where} \qquad 4.29$$

DO_1 is the dissolved oxygen content (mg/L) after 15 minutes, and DO_5 is the dissolved oxygen content after 5 days.

Since the oxygen content of freshwaters is ordinarily limited to a maximum of 9.2 mg/L (Table 4.5) and will usually be somewhat lower than this, BOD measurement by *filling* both BOD bottles with the sampled waters is limited to measurement of BODs which are less than the dissolved oxygen content of the water. For sewage and some types of industrial effluents, where BODs of from 100 up to 20,000 are not uncommon, the wastewater stream must be diluted with distilled aerated dilution water containing added nutrient salts before the BOD test is carried out. Two BOD bottles will be filled for each of several dilution ratios for a sample with an unknown BOD, so as to ensure that a test with at least one of the dilutions gives a consumption of about 50% of the initial dissolved oxygen content. Usually the right types of bacteria required will be present naturally. But if not, a small amount of treated municipal sewage (stored frozen) will also be seeded into each bottle to inoculate it. Again the dissolved oxygen of one bottle of each dilution will be measured immediately and of the other after 5 days in the dark at 20°C. The BOD of the sample from the test will be calculated using Eq. 4.30, or 4.31 if a sewage seed was used.

Biochemical oxygen demand with dilution:

$$BOD_5 = \frac{(DO_1 - DO_5) \times 300 \text{ mL}}{(\text{volume of sample per bottle, mL})} \qquad 4.30$$

Biochemical oxygen demand with dilution plus seed:

$$BOD_5 = \frac{(DO_1 - DO_5 - \text{seed correction}) \times 300 \text{ mL}}{(\text{volume of sample per bottle, mL})} \qquad 4.31$$

where the seed correction $= \dfrac{(DO_1 - DO_5)}{(\text{mL of seed per bottle})}$

which is determined in a separate test using only the seed plus dilution water in a BOD bottle.

Various refinements to these direct measurement techniques have been proposed. Substances such as glucose or potassium hydrogen phthalate, which are completely oxidized during the normal 5-day period of the BOD test, may be used as test substances to check the experimental technique of the BOD method used [20] (Eq. 4.32, 4.33).

$$C_6H_{12}O_6 + 6 \ O_2 \rightarrow 6 \ CO_2 + 6 \ H_2O \qquad 4.32$$

$$C_8H_5O_4K \ (= Ph(CO_2H)(CO_2K)) + 15/2 \ O_2 \rightarrow 8 \ CO_2 \atop + 2 \ H_2O + KOH \qquad 4.33$$

Parallel classifications of river water quality based on their BOD loadings and independently based on their dissolved oxygen content have been pro-

TABLE 4.8 Classification of River Quality Based on Dissolved Oxygen Content and Independently Based on Biochemical Oxygen Demand[a]

Dissolved oxygen (% of saturation)	River pollution status[b]	BOD loading (mg/L)[c]
90 or more	very clean	1 or less
ca. 90	clean	2
75 to 90	fairly clean	3
50 to 75	moderately polluted	5
25 to 50	heavily polluted	10 to 20
<25	severely polluted	20 or more

[a]Compiled from Klein [10,14] and Woodiwiss [44].

[b]Oxygen content of the river water is a result indpendent from the BOD result, even though it may be the first part of the two steps required for BOD measurement. While the two readings are not necessarily connected, frequently there will be a low oxygen saturation level coinciding with high BOD occurrences.

[c]Whole-river BOD loadings of at least 70 mg/L have been recorded for the Trent River [44], and exceeding 450 mg/L for the River Irwell (Radcliffe, U.K.) [14].

posed (Table 4.8). As a rough rule of thumb intended to maintain river quality a guideline has been suggested that no discharge to a river should bring the BOD of the river to more than 4 mg/L.

The manometric method of BOD determination measures the volume of oxygen uptake by a measured volume of the undiluted sample, placed in a brown glass bottle to prevent any photosynthetic influence (Fig. 4.4). After filling, the system is closed and the mercury manometer is set to a zero reading. Stirring is continued during the test to facilitate gas transfer across the air-water interface while bacterial attack of the organics present consumes oxygen and produces carbon dioxide (Eq. 4.34).

$$C(organics) + O_2 \rightarrow CO_2 \qquad 4.34$$

However, there would be little or no gas volume change without some means of taking up carbon dioxide as it is formed. This is accomplished by a wick, moistened with aqueous potassium hydroxide and placed in a stainless steel cup above the surface of the liquid (Eq. 4.35).

$$CO_2 + 2\,KOH \rightarrow K_2CO_3 + H_2O \qquad 4.35$$

The net oxygen uptake is read directly from a mercury manometer. This procedure makes it easy to take intermediate readings during the test to obtain a profile of the oxygen consumption during the 5 days, or for a longer period if desired. The manometer reading obtained after the elapse of 5 days' stirring is the BOD for the sample tested. Samples of a higher BOD than the range of the manometer would require dilution before loading the apparatus, to get a valid result within the capacity of the instrument.

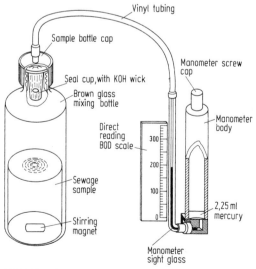

FIGURE 4.4 Diagram of a manometric apparatus for determination of biochemical oxygen demand [45]. (Courtesy of the Hach Company, Ames, IA.)

4.12.2. Chemical Oxygen Demand

In contrast to the BOD test, which is designed to measure the oxygen demand of only the biochemically oxidizable carbon compounds present in the sample, the COD test gives a close measure of the total oxygen demand of the sample. It is also a practical method to obtain a valid oxygen demand result when the sample contains toxic substances which cannot be easily neutralized. A standard BOD test conducted under these conditions would give a low or zero result from the toxicant action on the micro-organisms.

If strong chemical oxidants such as potassium permanganate [46] or potassium dichromate [20] are used under strongly acidic conditions, both the easily oxidized and the more resistant organic materials are converted to carbon dioxide and water (Eq. 4.36).

$$C_nH_aO_b + c\ Cr_2O_7^{2-} + 8c\ H^+ = n\ CO_2 + [(a + 8c)/2]\ H_2O$$
$$+ 2c\ Cr^{3+} \qquad\qquad 4.36$$
$$\text{where } c = 2/3n + a/6 - b/3$$

During the process any inorganic salts present are also converted to the oxidized forms. The result is specified as the mg/L of oxygen which would be consumed equivalent to the amount of chemical oxidant required (e.g., Eq. 4.37, 4.38).

$$3\ CH_2O + 2\ Cr_2O_7^{2-} + 16\ H^+ \rightarrow 3\ CO_2 + 11\ H_2O + 4\ Cr^{3+} \quad 4.37$$

$$3\ CH_2O + 3\ O_2 \rightarrow 3\ CO_2 + 3\ H_2O \qquad\qquad 4.38$$

Under the strong oxidizing conditions of the test most organic compounds give 95 to 100% of the theoretical oxygen consumption although some of the more stable aromatics, such as benzene and toluene, give lower oxidant consumptions than this. A micro semiautomated method of COD determination of surface and wastewaters has been developed which can help process large numbers of samples through this test.

4.12.3. Total Organic Carbon and Oxygen Demand Estimation

Total organic carbon determination of water samples is an instrumental method originally developed by the Dow Chemical Company [47, 48] and now sold commercially by Beckman under license from Dow. Either normal surface water samples or waste streams high in organics may be analyzed, and the result may be used to estimate oxygen demand from the TOC reading [49].

The procedure involves, first, acidification of the bulk sample to pH 2 with hydrochloric acid, and then sparging of the sample with nitrogen or helium to remove any carbon dioxide or any other inorganic carbon present (Eq. 4.39, 4.40).

$$Na_2CO_3 + 2\ HCl \rightarrow 2\ NaCl + H_2CO_3 \qquad\qquad 4.39$$

$$H_2CO_3 \rightarrow H_2O + CO_2\uparrow \qquad\qquad 4.40$$

A 10–30 μL sample is then injected into the instrument. Initially the water and any volatile organic carbon compounds present are vaporized at 200°C.

The volatilized carbon compounds are first all completely oxidized to carbon dioxide, and then the carbon dioxide is all catalytically converted to methane, in a hydrogen atmosphere (Eq. 4.41).

$$CO_2 + 4 H_2 \xrightarrow[Ni]{350°} CH_4 + 2 H_2O \qquad 4.41$$

The final step to obtain a volatile organic carbon reading is a quantitative analysis of the methane produced in a gas chromatograph using a sensitive flame ionization detector, which is unaffected by water. This instrumental configuration gives a very good linear response to the quantity of methane produced, which is the reason for the preliminary conversion of organic compounds to methane.

The TOC instrument also has provision for the analysis of refractory organic carbon, the more oxidation or volatilization resistant forms of carbon which may be present. This uses a vaporization and oxidation mode at 850°C, which gives a separate reading. The total organic carbon reading for the sample is the sum of the two-component readings (Eq. 4.42).

$$TOC = VOC + ROC, \qquad 4.42$$

where VOC is the volatile organic carbon and ROC is the refractory organic carbon. The result is specified in mg/L carbon. This may be readily translated into an estimate of oxygen demand for the carbon content *only*, by calculating the oxygen consumption required to convert this to carbon dioxide, i.e., by multiplying the TOC reading by $(32.00 \div 12.01)$ (Eq. 4.43).

$$\begin{array}{ccccc} C & + & O_2 & \rightarrow & CO_2 \\ 12.01 & & 32.00 & & 44.01 \end{array} \qquad 4.43$$

Each of these measures of oxygen demand will give somewhat different results because of the differences in the analytical methods used (Table 4.9). But each can generally be related to the others by a factor, depending on the particular kind of waste analyzed. Thus the COD for ordinary sewages, for example, is frequently about twice the BOD. The COD reading minus the BOD reading is sometimes referred to as the "refractory organic compounds," i.e., referring to the organics which are more resistant to oxidation.

4.13. BIOLOGICAL INDICATORS

By the number of different species represented (diversity) and the relative sensitivity to pollutants of the species found, much can be learned about the condition of fresh or marine waters. Sessile (more or less stationary) organisms are preferred for biological assessments, since they are unable to avoid discharged wastes. These can either be sampled by scooping up portions of the bottom material for examination and counting, or artificial substrates may be placed in the stream bed for colonization and later examined.

Using solely biological methods to assess water quality, if properly quantified against appropriate control stations or against surveys conducted prior

TABLE 4.9 Comparison of the Experimental and Theoretical Oxygen Demand with That Estimated from the Total Organic Carbon Reading for Some Representative Compounds in Water[a]

Compound	BOD (g O_2/g compound)		COD (g O_2/g compound)		TOC	
	5 day	20 day	$K_2Cr_2O_7$[b]	Theoretical	g C/g comp'd	g O_2/g comp'd[c]
Acetone	0.8	1.6	1.6–2.2	2.2	0.62	1.65
Adipic acid	0.6	1.2	1.3	1.42	0.49	1.31
Benzene	0.3–1.0	1.5–2.0	1.9	3.08	0.97	2.59
Chloroform	0	0	0.02	0.34	0.10	0.268
Ethanol	0.8–1.5	1.5–1.6	1.8–2.0	2.08	0.52	1.39
Heptane	1.9	—	0.06–0.2	3.52	0.84	2.24
Methanol	0.8–1.1	1.3	1.4–1.5	1.50	0.38	0.99
Phenol	1.4–1.9	1.9–2.1	2.4	2.38	0.77	2.04
Styrene	0.5–1.5	1.7–2.4	2.9	3.08	0.92	2.46

[a]Specified as g/g to assist understanding; i.e., a pure ratio, or dimensionless. Data compiled from Verschueren [50], and calculated from first principles.
[b]Or $KMnO_4$. Either may be used as the chemical oxident.
[c]That is, an estimate of the oxygen demand based on the TOC reading. Compare with the theoretical COD readings.

to discharge, is said to give a superior indication than either chemical or physical data alone. Several numerical methods for indexing pollution levels have been published.

REVIEW QUESTIONS

1. What are the significant factors which influence dissolved oxygen content of surface waters?
   (a) What are the wt/wt (ppm) and wt/vol (mg/L) oxygen concentrations for surface waters which contain 1.50×10^{-4} M dissolved oxygen?
   (b) Is this an acceptable oxygen concentration in surface waters for the respiration requirements of aquatic life?
2. Surface water temperature changes also significantly affect a number of other water quality parameters. What are they, and in which direction is each of these affected? Explain.
3. (a) Calculate the COD (chemical oxygen demand) for a refinery aqueous waste stream containing 900 mg/L phenol (C_6H_5OH).
   (b) Assuming that phenol is fully biodegradable, what would be the BOD_5 for this waste stream?
   (c) What instrumental TOC (total organic carbon) reading (mg/L) should be obtained?
   (d) What would be a good estimate of the oxygen demand based only on the TOC reading?
4. A BOD_5 test run on a sample of river water without dilution, using a standard 300-mL BOD bottle, gave an initial DO (dissolved oxygen) reading of 9.1 mg/L O_2 (15 min after sampling) and a final DO reading of 2.7 mg/L O_2.
   (a) What BOD would these results indicate for the river water?
   (b) Based on established guidelines, what appropriate comment could be made concerning the water quality of this river based on the BOD reading only?
5. A 5-mL sample of municipal sewage was placed in a standard BOD bottle (300 mL) and topped up with aerated dilution water. Initial dissolved oxygen was 4.7 mg/L, and final (after 5-day test period) was 1.3 mg/L.
   (a) What was the BOD of the sewage sampled?
   (b) Additional precautionary BOD runs of 1-, 2-, 10-, and 20-mL aliquots of sewage were also carried out in separate BOD bottles, each using the same aerated dilution water as given above. What final dissolved oxygen (after the test period) should be observed for each of these runs?
6. Potassium hydrogen phthalate ($C_8H_5O_4K$) has been suggested for use as a primary standard to check BOD determination procedures. A standard solution of this reagent is made up containing a weighed 50 mg/L. Assuming complete biochemical oxidation during the 5-day

test period, what BOD reading (mg/L) should one obtain for this
solution?

7. A washwater stream from a frozen french fries processing plant gives
a TOC (total organic carbon) reading of 1.1% (by weight) as it
enters the extended aeration facilities.

(a) Assuming full biodegradability of this primarily starch plus
hydrolyzed starch suspension, what would be the BOD of this
stream?

(b) Treated effluent from these facilities, analyzing 245 mg/L organic
carbon, is to be discharged into a municipal sewerage system. What
BOD [same assumptions as Part (a)] would this correspond to?

(c) The volume from this operation is to comprise, under maximum
flow conditions, 5% of the total effluent being treated by the
municipal sewage works. Is this likely to upset the sewage treatment
operation? If so, why? If not, why not?

FURTHER READING

M.L. Hitchman, "Measurement of Dissolved Oxygen." Wiley-Interscience, New York, 1978.

K.J.M. Kramer, ed., "Biomonitoring of Coastal Waters and Estuaries." CRC Press, Boca Raton,
FL, 1994.

W. Leithe, "Analysis of Organic Pollutants in Water and Wastewater" (translated by STS, Inc.).
Ann Arbor Sci. Publ., Ann Arbor, MI, 1973.

L.J. Thibodeaux, "Chemodynamics, Environmental Movement of Chemicals in Air, Water, and
Soil." Wiley, New York, 1979.

REFERENCES

1. F. van der Leiden, compiler-editor, "Water Resources of the World." Water Information
Center, Port Washington, NY, 1975.

2. D.G. Downing, R. Kunin, and F.X. Pollio, *Chem. Eng. Prog., Symp. Series* **64**(90), 126
(1968).

3. G. Young and J.P. Blair, *Nat. Geogr.* **138**(6), 743 (1970).

4. Meditteranean Conference, *Chem. Eng. News* **54**(8), 19, Feb. 23 (1976).

5. W. Marx, "The Frail Ocean." Coward McCann, New York, 1967.

6. S.K. Gupta and K.Y. Chen, *J. Water Pollut. Control Fed.* **50**(3), 493 (1978).

7. J. Cholak, *Arch. Ind. Health* **21**, 312 (1960).

8. D. O'Sullivan, *Chem. Eng. News* **54**(25), 15, June 14 (1976).

9. R.C. Weast, ed., "Handbook of Chemistry and Physics," 56th ed., CRC Press, Cleveland,
OH, 1975.

10. L. Klein, "River Pollution," Vol. 1. "Chemical Analysis," Butterworth, London, 1959.

11. G.P. Treweek and J.J. Morgan, *Environ. Sci. Technol.* **11**(7), 707 (1977).

12. The Aerial Photo—Water Quality Link, *Environ. Sci. Technol.* **10**(3), 229 (1976).

13. D. Rose and J.R. Marier, "Environmental Fluoride," Report No. 16081, National Research
Council, Ottawa, 1977.

14. L. Klein, "River Pollution", Vol. 2, "Causes and Effects," Butterworth, London, 1962.

15. "Cleaning Our Environment: A Chemical Perspective," 2nd ed., American Chemical Society,
Washington, DC, 1978.

16. C.N. Sawyer, *J. N. Engl. Water Works Assoc.* **61**, 109 (1947).

17. C.N. Sawyer, *Sewage Ind. Wastes* **26**, 317 (1954); cited by Klein [14], p. 125, and vol. III,
p. 159.

18. J.D. Hem, Study and Interpretation of the Chemical Characteristics of Natural Water. *Geol. Surv. Water-Supply Pap. (U.S.)* **1473**, Washington, DC (1959).

19. "Surface Water Quality in Canada—An Overview." Inland Waters Directorate, Water Quality Branch, Ottawa, 1977.

20. "Standard Methods for the Examination of Water and Wastewater," 13th ed., American Public Health Assoc., Washington, DC, 1971.

21. M.S. Frant, J.W. Ross, Jr., and J.M. Riseman, *Am. Lab.* **1**(1), 14, Jan. (1969).

22. C.F. Mason, "Biology of Freshwater Pollution." Longmans, London, 1981.

23. E.A. Hemmingsen, *Science* **167**, 1493 (1970).

24. L.W. Winkler, *Ber. Dtsch. Chem. Ges.* **21**, 2843 (1888).

25. W.R. Stagg, *J. Chem. Educ.* **49**(6), 427 (1972).

26. M.L. Hitchman, "Measurement of Dissolved Oxygen." Wiley-Interscience, New York, 1978.

27. G.J. Shroepfer, *Sewage Works J.* **14**, 1030 (1942); cited by Klein [10].

28. M.J.R. Clark, "Annotated Extracts of Some Papers Dealing with Various Aspects of Dissolved Atmospheric Gases." Water Resources Service, Victoria, B.C., 1977, plus Appendix 2, reference list for the above.

29. B.G. D'Aoust and M.J.R. Clark, *Trans. Am. Fish. Soc.* **109**, 708 (1980).

30. H.L. Youmans, *J. Chem. Educ.* **49**(6), 429 (1972).

31. D.W. Ryckman, A.V.S. Prabhakara Rao, and J.C. Buzzell, Jr., "Behavior of Organic Chemicals in the Aquatic Environment." Manufacturing Chemists Assoc., Washington, DC, 1966.

32. G.S. Fell, *Chem. Br.* **16**(6), 323, June (1980).

33. F.A. Patty, ed., "Industrial Hygiene and Toxicology," 2nd ed., Vol. 2. Interscience, New York, 1963.

34. F.W. Kawahara, *Environ. Sci. Technol.* **5**(3), 235 (1971).

35. K.W. Kawahara, *Anal. Chem.* **40**(11), 2073 (1968).

36. K.E. White, *Chem. Br.* **12**(12), 375, Dec. (1975).

37. P.V. Scarpino, *in* "Water and Water Pollution Handbook" (L.L. Ciaccio, ed.), vol. 2, p. 639, Dekker, New York, 1971.

38. R. Hare, "An Outline of Bacteriology and Immunity," 3rd ed., Longmans, London, 1967.

39. "Biological Analysis of Water and Wastewater." Millipore Corp., Bedford, MA, 1973.

40. T.L. Brown, "Energy and the Environment." Merrill, Columbus, OH, 1971.

41. R.J. Allan and D.J. Richards, "Effect of a Thermal Generating Station on Dissolved Solids and Heavy Metals in a Prairie Reservoir." Inland Waters Directorate, Regina, Sask., 1978.

42. T.M. Lillesand and R.W. Kiefer, "Remote Sensing and Image Interpretation," p. 393, 583, Wiley, New York, 1979.

43. T.J. McGhee, R.L. Torrens, and R.J. Smaus, *Water Sewage Works* **119**, 58, June (1972).

44. F.S. Woodiwiss, *Chem. Ind. (London)* No. 11, p. 443, (1964).

45. "Operating Manual, 6 Bottle Manometric Apparatus, Model 2173B," 3rd ed., Hach Company, Ames, IA, 1982.

46. D.W. Ryckman, A.V.S. Prabhakara Rao, and J.C. Buzzell, Jr., "Behavior of Organic Chemicals in the Aquatic Environment." Manufacturing Chemists Assoc., Washington, DC, 1966.

47. C.E. Van Hall, J. Safranko, and V.A. Stenger, *Anal. Chem.* **35**(3), 315 (1963).

48. C.E. Van Hall and V.A. Stenger, *Anal. Chem.* **39**(4), 503 (1967).

49. R.H. Jones, *Water Sewage Works* **119**(3), 72, Mar. (1972).

50. K. Verschueren, "Handbook of Environmental Data on Organic Chemicals." Van Nostrand-Reinhold, Toronto, 1977.

Activated sludge aeration stage for a small-scale secondary sewage treatment plant operating at Central Saanich, British Columbia.

5

RAW WATER PROCESSING AND WASTEWATER TREATMENT

It's no good throwing away dirty water until you get clean.

—V. Bridges, 1922

5.1. WATER QUALITY RELATED TO END USES

Water quality requirements differ according to the application for which the water is required. While the highest quality surface waters can generally be used for any freshwater need, many applications do not have as stringent requirements [1]. A detailed consideration of water quality needs by use and by parameter is a difficult and extensive exercise beyond the scope of this treatment [2]. Table 5.1 gives an appreciation of some of the considerations involved. This illustrates how the required levels of water quality parameters may change with the anticipated end use. Different water supplies may all have appropriate end uses. It should also be remembered that the exact requirements for any particular end use will vary with local conditions and with the government objectives. Setting of standards is a complicated exercise which requires consideration of direct factors such as toxicity and aesthetics (e.g., for recreational use). It also requires consideration of the effects of long-term consumption and use, the efficiency of raw water processing methods to remove impurities, the seasonal variations in supply characteristics, alternative supplies which may be available, and local (particularly upstream) wastewater disposal practices and the like, which involve the skills of many disciplines. Guidelines are available to assist in interpreting regulatory standards.

■ **TABLE 5.1   The Water Quality Requirements for Various Primary Classes of End Use**[a]

Agricultural Uses
Livestock: Low bacteria, <40/100 mL, and low concentrations of toxic substances (e.g., F^-)
Irrigation: Low dissolved solids, <500 mg/L, (to avoid increased soil salinity)
        Total bacteria: Allowable 100,000/100 mL; desirable <10,000/100 mL
        Low heavy metals

Fish, Aquatic Life, Wildlife Requirements
        Concentrations of toxic substances low
        pH near neutral, 6.5–8.5
        Low BOD, 1–2 mg/L or less
        High dissolved oxygen: cold, 6–7 mg/L at 15°C or less; warm, 4–5 mg/L at ca. 20°C
        Low temperature, turbidity, etc.

Industrial Uses
Cooling: Low hardness, <50 ppm of ($Mg^{2+} + Ca^{2+}$) (usually as SO_4^{2-} and CO_3^{2-})
        Low corrosivity
Food processing, brewing, and soft drinks
        As public drinking water, but F^- <1 ppm
Thermal power:
        Total dissolved solids <0.1 ppm (and lower for boiler pressures above 135 atm)

Public Recreational Requirements
        Free of color, odor, taste, and turbidity
        Total bacteria <1000/100 mL; coliforms <100/100 mL
        Low nutrients, to avoid nuisance algal growths

Public Drinking Water (treated)
        No bacteria
        Low nitrates, nitrites <10 ppm
        Very low pesticides, none >0.05 ppm
        Fluoride allowable to 2.4 ppm
        Toxic substances (metals, etc.) below criteria levels
        Total dissolved solids <500 ppm

---

[a]Condensed from McNeeley *et al.* [1] and Canadian Public Health Association [3].

5.2. TREATMENT OF MUNICIPAL WATER SUPPLIES

The primary physiological water requirement of an adult human not subjected to heat stress is only 2.5 to 3 L per day. Yet in households equipped with facilities for running water the actual daily per capita consumption for such uses as human waste transport, bathing, washing/cleaning, and garden irrigation is 95–380 L/day (25–100 U.S. gal/day) [4]. When the very large water demand of most industries is added to domestic requirements it can easily be seen why the facilities for the production of a safe and convenient water supply rank among the largest scale materials processing operations.

5.3. SIMPLE MUNICIPAL WATER TREATMENT

Depending on the raw water quality and the seasonal stability of its characteristics many municipalities are able to use a simple two-stage supply treatment involving preliminary filtration, followed by a disinfection step. Fast filtration methods use a pressure differential to force the raw water through a bed of finely granulated clean sand, crushed anthracite coal, or sometimes a mixture of media, to remove suspended solids present in the supply (e.g., see Fig. 5.1). As the filtration proceeds, accumulated solids in the filter bed cause a gradual slowing of the water flow rate through the filter. For periodic cleaning the filter is back-flushed using filtered water at a sufficiently high flow rate to lift the filter medium and free the accumulated algae, diatoms, silt, etc. The flushate is discarded to the sewer. This need for periodic filter cleaning requires a water treatment plant to have either two filters, one of which can continue operation while the other is in the back-flush mode, or a large enough holding basin for filtered water to provide for both back-flushing and normal filtered water requirements while a single filter is being cleaned.

There is also a "slow filtration" variant of filtration used by some water treatment plants, which combines both a physical separation of suspended solids plus some biological consumption or adsorption of undesirable dissolved substances in the water supply in a single-, two-, or three-stage unit [5]. Bacteria, algae, and diatoms that accumulate on the coarse sand layer of the filter bed metabolize nutrients, etc., from the supply, thereby removing them. This slow filtration system has been reported to be capable of removing up to 50% of the chlorinated pesticide content of the influent water [5]. Slow filtration may require a prior aeration step to ensure that the biochemical processes employed remain aerobic, since anaerobic operation can contribute bad odors or tastes to the supply in the form of sulfides and amines.

After suspended matter has been removed a disinfection step, usually with chlorine, is necessary to ensure that the supply is free of any viable pathogenic

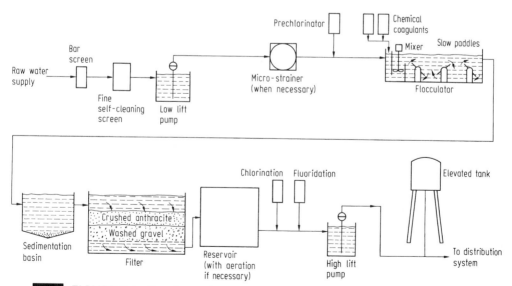

FIGURE 5.1 Flow diagram for an intermediate complexity municipal water treatment plant.

organisms. The active bactericide, undissociated hypochlorous acid, is formed immediately on contact of the chlorine gas with water (Eq. 5.1, 5.2).

$$Cl_2 + H_2O \rightarrow H^+ + Cl^- + HOCl \qquad 5.1$$

$$\underset{\text{active}}{HOCl} \rightarrow \underset{\text{relatively inactive}}{H^+ + ClO^-} \qquad 5.2$$

Sufficient gaseous chlorine is added to the water to leave a "residual chlorine" concentration of 0.1 to 0.2 mg/L, after the normal consumption of a part of the added chlorine in reactions with any dissolved or residual suspended matter in the supply has taken place [5]. A preferred contact time of 1 to 2 hr is recommended, but at least 20 to 30 min of contact time should be ensured before use. This residual chlorine content is necessary to maintain safe transmission of the treated water supply through a local piping system.

Since the dissociation constant for hypochlorous acid is very small, in neutral solutions undissociated hypochlorous acid is the dominant species of this equilibrium (Eq. 5.3).

$$K = \frac{[H^+][OCl^-]}{[HOCl]} = 2.95 \times 10^{-8} \text{ at } 18°C \qquad 5.3$$

Under acid conditions the higher concentration of hydrogen ion present tends to depress the hypochlorous acid dissociation reaction (Eq. 5.2) which consequently tends to raise the concentration of hypochlorous acid above that present in near neutral solutions (100% at pH 5 or lower). Thus, the effectiveness of chlorine disinfection is maintained or enhanced at low pH.

At a pH of 7.5, near neutral conditions, the hypochlorous acid will be roughly 50% dissociated and disinfection will still be quite effective. However, in disinfections of somewhat alkaline water supplies, the decreased hydrogen ion concentration depresses the concentration of undissociated hypochlorous acid; i.e., at pHs of 10 or higher, hypochlorous acid is present almost entirely (ca. 99.7%) as hypochlorite anion (e.g., see Section 15.5.2). This has the effect of decreasing the effectiveness of chlorine disinfection under these conditions [5], which can be remedied by pH adjustment of the supply before disinfection, by a higher chlorine dose rate, or by a longer contact time before water use.

When the time between disinfection and delivery at the householder's tap is long, such as when treatment is carried out at a reservoir some distance away from the consuming urban center, ammonia may be added at the same time as the chlorine. The more stable chloramines which form give a longer term residual chlorine to the supply than is possible with chlorine alone (Eq. 5.4–5.6).

$$HOCl + NH_3 \rightarrow NH_2Cl + H_2O \qquad 5.4$$

$$HOCl + NH_2Cl \rightarrow NHCl_2 + H_2O \qquad 5.5$$

$$HOCl + NHCl_2 \rightarrow NCl_3 + H_2O \qquad 5.6$$

However, chloramines also require a much longer disinfection period than chlorine to obtain the same effectiveness, and so are only used in situations like this.

Hypochlorites (e.g., household bleach) are also effective for disinfection of water supplies, and are an economic and convenient choice for smaller volume requirements such as for a single household, a hamlet, or for small campsites [6, 7]. With hypochlorite salts similar equilibria operate to rapidly establish a concentration of hypochlorous acid which corresponds to the amount of hypochlorite salt added (Eq. 5.7, Table 5.2).

$$Ca(OCl)_2 + 2\ H_2O \rightarrow Ca(OH)_2 + 2\ HOCl$$

Mol. wts.	143	18	74	52.5	5.7
Moles	1	2	1	2	
Mass ratios	143	36	74	105	

Thus, the "% available chlorine" from calcium hypochlorite, which by definition includes Cl_2, $HClO$, and OCl^- (but not Cl^-), is 99.2%, or nearly the same in effectiveness, on a weight for weight basis, as treating with elemental chlorine [8]. As long as the residual chlorine level used with hypochlorite salt disinfections is the same as with chlorine gas, there is no difference in the relative effectiveness of the two methods.

Chlorine is the most common disinfectant for public water supplies, and there is no doubt that its effectiveness, simplicity of application, and low cost have been important factors in the widespread adoption of chlorination as a significant public health measure. Questions have been raised about the presence of chloroform and other chlorinated organics, formed from chlorination of substances in public drinking water supplies [9, 10]. However mutagenic risks from this source are small relative to the health risks from not disinfecting the water supply [11].

Other methods have also been used for water disinfection, including the hypochlorites already mentioned, which would have less effect than

TABLE 5.2 "Available Chlorine" Content of Various Chlorine-Containing Water Disinfectants[a]

Disinfectant	Molecular weight	Moles of equivalent chlorine[b]	Percent by weight	
			Actual chlorine	Available chlorine[c]
Chlorine	70.9	1	100.0	100.0
Hypochlorous acid	52.5	1	67.7	135.0
Sodium hypochlorite	74.5	1	47.7	95.2
Calcium hypochlorite	143.0	2	49.6	99.2
Chloramine (NCl$_3$)	120.4	3	88.4	176.7[d]

[a] Data recalculated from White [8].

[b] The number of moles of hypochlorite (oxidizing chlorine) which could form on dissolving 1 mole of the disinfectant in water.

[c] This equates to the calculated weight of elemental chlorine required to produce the same amount of hypochlorous acid in water as the given disenfectant; e.g., for calcium hypochlorite

$$\frac{(70.9\ g/mol \times 2\ mol/mol \times 100)}{143.0\ g/mol} = 99.2\%$$

[d] Action is very slow.

chlorine in aqueous chlorinations. Chlorine dioxide (ClO_2) has a negligible chlorination tendency, and is also an effective disinfectant at high pH's, when chlorine and hypochlorites are relatively ineffective. Other water disinfectants such as ozone, iodine, ultraviolet irradiation, and gamma irradiation, avoid the use of chlorine entirely. Chlorine dioxide, favored in Germany, has an essentially similar mode of action to chlorine [12]. Ozone, which is favored in France, not only functions as a disinfectant but also serves to remove odor, taste, and color from its more general oxidizing properties, which is an important consideration with a poor-quality raw water supply [13]. However, it is not possible to maintain a disinfection residual in the water supply with ozone, which makes it necessary to back up the ozone treatment with a low chlorine dosage to provide this transmission safeguard [5]. Iodine, as crystals, or more conveniently as a tincture or iodine-releasing tablets, is also an effective field disinfection method but is too expensive for municipal water treatment. The remaining methods mentioned, although effective, have cost, maintenance, or lack of familiarity problems attached to their use.

5.4. MORE ELABORATE MUNICIPAL WATER TREATMENT METHODS

When there is little control over raw water quality, such as when the supply is near the mouth of a river flowing through a populous watershed, or when the natural supply from a groundwater source is poor, more complicated treatment methods are required (e.g., Fig. 5.1, or more complex). Following the preliminary treatment steps outlined below, filtration plus chlorination, or chlorination alone can be used to disinfect the finished supply. It is generally desirable, if not always economic, to remove excessive hardness (>150 ppm) from water supplies before distribution. Synthetic detergents have circumvented much of the problem from the use of carboxylate soaps in hard waters. This does not, however, help the aesthetics of personal bathing for which ordinary soaps are usually preferred. Hardness removal is also not particularly necessary from a health standpoint since, if anything, a degree of hardness in the water appears to reduce the risk of heart attack [14]. However, excessive hardness is still a serious problem in the scale buildup on water heating devices of all kinds, from kettles and domestic water heaters to commercial power boilers.

It is the carbon dioxide content of natural waters which contributes much of the temporary hardness, also referred to as bicarbonate hardness, into water supplies (Eq. 5.8, 5.9).

$$CO_2 + H_2O \rightarrow H_2CO_3 \qquad\qquad 5.8$$

$$H_2CO_3 + 2\ CaCO_3 \rightarrow 2\ Ca(HCO_3)_2 \qquad\qquad 5.9$$
$$\text{insoluble} \qquad\quad \text{soluble}$$

Temporary hardness may be removed by heating the water, which reverses the reaction which originally put the calcium and magnesium into solution (Eq. 5.10, 5.11).

$$Ca(HCO_3)_2 \underset{heat}{\rightarrow} CaCO_3\downarrow + H_2O + CO_2\uparrow \qquad 5.10$$

$$Mg(HCO_3)_2 \rightarrow MgCO_3\downarrow + H_2O + CO_2\uparrow \qquad 5.11$$

These equations also represent the processes which cause scale buildup from temporary hardness, and therefore are not of very great practical use. Oddly enough, however, addition of calcium in the form of slaked lime ($Ca(OH)_2$) can efficiently remove much of the temporary hardness of natural water supplies (Eq. 5.12–5.14).

$$Ca(HCO_3)_2 + Ca(OH)_2 \rightarrow 2\ CaCO_3\downarrow + 2\ H_2O \qquad 5.12$$

$$Mg(HCO_3)_2 + Ca(OH)_2 \rightarrow MgCO_3 + CaCO_3 + CaCO_3\downarrow + H_2O \quad 5.13$$

$$MgCO_3 + Ca(OH)_2 \rightarrow Mg(OH)_2\downarrow + CaCO_3\downarrow + H_2O \qquad 5.14$$

The second equivalent of calcium hydroxide is necessary to remove magnesium temporary hardness because the magnesium carbonate product obtained using one equivalent of calcium hydroxide is still somewhat soluble.

Neither calcium hydroxide nor heat is effective for removal of permanent hardness, that is, the hardness caused by the presence of the sulfates (or other anions), rather than the bicarbonates of calcium and magnesium [15]. However, treatment with sodium carbonate (soda ash) is effective for this purpose (Eq. 5.15).

$$Na_2CO_3 + CaSO_4 \rightarrow CaCO_3\downarrow + Na_2SO_4 \qquad 5.15$$

For effective removal of both temporary and permanent hardness, sodium hydroxide addition is required (Eq. 5.16).

$$Ca(HCO_3)_2 + 2\ NaOH \rightarrow CaCO_3\downarrow + Na_2CO_3 + 2\ H_2O \qquad 5.16$$

The sodium carbonate formed from the removal of temporary hardness is available to precipitate any calcium or magnesium salts present as permanent hardness. Thus, the usual water treatment practice for removal of hardness involves adding a mixture of quicklime (CaO) and sodium carbonate in accordance with the composition of the raw water, as determined by analysis. Frequently a coagulant such as alum ($Al_2(SO_4)_3 \cdot 14H_2O$) or ferrous sulfate will be added to help avoid the carryover of precipitated but still finely divided hardness in suspension.

The same steps may be effectively conducted at the household level to decrease the undesirable effects of a hard water supply by the addition of washing soda ($Na_2CO_3 \cdot 10H_2O$) or household ammonia to the laundry supply. Or a water softener which contains beads of a sulfonated polystyrene resin in the sodium form may be installed (Fig. 5.2). As hard water passes through, sodium ions are exchanged on a charge equivalent basis with any calcium, magnesium, or other polyvalent cations in the supply. In the process the household water is softened, and the resin content of the water softener gradually accumulates polyvalent cations until its capacity is exhausted. Regeneration of the resin requires passage of a saturated sodium chloride brine solution through the softener, to displace the polyvalent ions by sodium again.

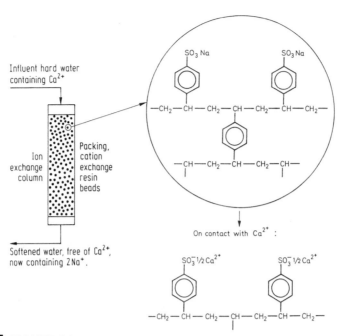

FIGURE 5.2   Details of operation of a household water softener.

During the regeneration step the outlet of the softener is connected to the sewer. After a short period of flushing with freshwater to remove the residual brine used for regeneration, the hardness removal operating cycle begins again.

Special techniques are required for the removal of other metal ions from raw water supplies, although lime treatment followed by flocculation serves to remove some heavy metal content by adsorption [5]. Excessive fluoride (i.e., >2.4 ppm in the supply), too, may be decreased by lime treatment. Fluoride precipitation is greatly improved by the presence of 50 mg/L or more of magnesium ion in the water at the time of treatment (Eq. 5.17).

$$2 \ F^- + Ca(OH)_2 \rightarrow \quad CaF_2\downarrow \quad + 2 \ OH^-$$
$$\text{adsorbed}$$
$$\text{on } Mg(OH)_2$$

5.17

Bone ash, and synthetic apatite, both consisting of essentially calcium phosphate [$Ca_3(PO_4)_2$], or a synthetic apatite, calcium hydroxide mixture, are also effective methods for fluoride removal because of the affinity of fluoride for these phosphate salts.

A taste or smell in the water supply may be removed by single or double aeration, if the causative agent is volatile. Otherwise adsorption on activated carbon, or chemical oxidation with ozone or chlorine, or both methods used in series may be required to control the problem. Fortunately, many organic toxic substances which may be present, such as pesticides, are also efficiently adsorbed by activated carbon and may be removed in this way. Lime treatment, if not already practiced for other reasons, is also efficient at removing dissolved colored substances, such as humic acids, when these are present in the supply. However, sometimes liming may have to be accompanied by activated carbon adsorption or coagulation plus filtration for effective color removal.

5.5. MUNICIPAL WATER BY DESALINATION

Coastal communities in arid areas may need to undertake more extensive and usually somewhat more costly methods to produce a potable supply from seawater or brackish supplies. The dissolved solids content of seawater is roughly 3.5%, 35°/oo (i.e., 35 "parts per thousand") or 35,000 mg/L. This needs to be reduced to about 1% of this figure to meet the dissolved solids requirement of a potable water supply. Brackish waters which have a lower salt concentration than seawater are a preferred raw water source when available, since the energy costs for salt removal are less.

Distillation can accomplish salt removal with reasonable thermal efficiencies if it is carried out under reduced pressure using several stages in series [16] (Fig. 5.3). Up to six stages have been found to be economic for maximum energy utilization at current energy costs. More than 95% of the world's existing desalination plants in 1969, representing a total capacity of nearly a million m³/day (240 million U.S. gal/day), utilize some form of distillation for saline water conversion. The largest capacity plants are installed in Kuwait and Oman, the coastal arid states of the U.S.A., the U.S.S.R., and the Netherlands.

Membrane processes of the reverse osmosis (hyperfiltration) or electrodialysis types are also used to a smaller extent by some largescale facilities (Fig. 5.4). Reverse osmosis units operate by using high brackish water or seawater charging pressures on one side of a semipermeable membrane, sufficient to exceed the osmotic pressure of the deionized water on the product side of the membrane. Thus, water free of ions permeates through the

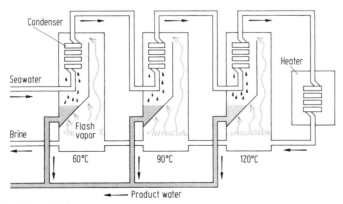

FIGURE 5.3   Schematic of the mode of operation of a multistage flash distillation unit. (Modified from Pryde [17], reprinted courtesy of Cummings Publishing Co.)

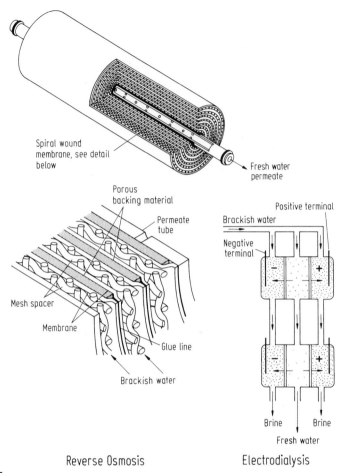

Spiral wound
membrane, see detail
below

Fresh water
permeate

Porous
backing material

Permeate
tube

Positive terminal

Brackish water

Negative
terminal

Mesh spacer

Membrane

Glue line

Brackish water

Brine   Brine

Fresh water

Reverse Osmosis   Electrodialysis

FIGURE 5.4   Principles of operation of two types of membrane water desalination units, reverse osmosis, and electrodialysis. (From Pryde [17], reprinted courtesy of Cummings Publishing Co.)

membrane at a rate proportional to the applied pressure (Table 5.3). Desalination by electrodialysis again uses membranes, but this time of a type permeable to ions. The driving force for ion removal in this instance is derived from the opposing electrochemical potentials on either side of the membrane. Electrochemical removal of ions is continued in several stages of electrodialysis cells in series until the salt concentration is brought down to the required level.

Differential freezing methods may prove to be particularly useful in water-short polar regions, where the energy costs for freezing would be low, but may also be utilized in more common desalination sites in hot climates. Reducing the brine temperature to below freezing point generates substantially salt-free ice crystals which are separated from the brine mother liquor (Fig. 5.5). For efficient energy utilization the heat removal capability of the thawing ice may be used to prechill incoming brine, prior to freezing.

TABLE 5.3   A Comparison of the Performance Characteristics of Three Commerical Reverse Osmosis Membrane Configurations[a]

Characteristic	Hollow fiber cellulose acetate	Hollow polyamide	Spiral wound sheets of cellulose acetate
Mandatory pretreatment	pH control	softener or pH	pH control
pH range (ideal)	4–7.5 (5.5)	4–11 (4–11)	3–6 (5.5)
Prefiltration, μm	5	5	25
Maximum Cl_2, ppm	1.0	0.1	1.0
Maximum temperature, °C	35	35	35
Normal operating pressure	2700 kPa (400 psig)	2700 kPa (400 psig)	2700–4100 kPa (400–600 psig)
Max. product back pressure	550 kPa (80 psig)	275–340 kPa (40–50 psig)	275–340kPa (40–50 psig)
Rated TDS level[b]	2000 ppm	2000 ppm	2000 ppm
TDS rejection rates[b]	90–95%	90–95%	90–98%
Water recovery	75%	50–75%	22–75%
Max. suspended solids	1.0 JTU[c]	1.0 JTU[c]	1.0 JTU[c]
Ease of cleaning	good	poor	very good
Surface water performance	good	poor	excellent
Well water performance	excellent	good	excellent
Packing density, m^2 membrane/m^3 module	29,500	—	660
Flux, L/m^2/day (pressure)	7.3 (4100 kPa)	—	730 (5500 kPa)
Flux density, L/m^3 module/day	220,000	—	480,000

[a]Properties compiled from McBain [18].
[b]TDS = total dissolved solids. Newer membranes can handle up to 35,000 ppm, i.e., seawater.
[c]Jackson turbidity units, on an arbitrary scale [19].

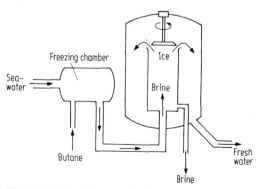

FIGURE 5.5   Schematic of differential freezing as a method for salt removal from seawater. (From Pryde [17], reprinted courtesy of Cummings Publishing Co.)

5.6. WATER QUALITY REQUIREMENTS OF INDUSTRY

Most industrial cooling water requirements may be met by the usual characteristics of the surface waters available, as long as the hardness and the concentrations of chloride or other substances which may cause corrosion are low. Bacteriological, color, dissolved oxygen, etc., requirements for this use are minimal. The food processing industry is one exception to this generalization, however, in that for the cooling of processed foods, either in bulk or in the final packages of bottles or cans, potable water quality is required. This is to avoid any risk of contamination in the event of leakage. Even the cooling of packaged products after pasteurization requires this quality since a pinhole leak plus the reduced interior pressure produced on cooling could cause coolant, and hence organism, leakage into the package.

A potable quality water supply or better is also required for the soft drinks and alcoholic beverage industries, for medical and pharmaceutical applications, and for the water needs of many types of research institutions. The beverage industry has a fluoride requirement of <1 mg/L, lower than that required for ordinary potable water supplies, probably to accommodate possible higher individual consumption rates. Deionized or distilled grades of water quality are required by and are sufficient for the majority of the other applications areas mentioned.

Among the most stringent of all water quality standards are those set for the feedwater requirements of modern high-pressure steam boilers [20]. Even by 1970 high-pressure boiler operation required direct conversion of feedwater to superheated (dry) steam at 160 atm (2400 psi) in the boiler tubes, so that any dissolved solids present would be deposited and gradually accumulated in the tubes, contributing to eventual failure. Consequently, this type of boiler requires water with a total impurity level of 0.03 mg/L (0.03 ppm, or 0.000003%), or less. Water of this standard amounts to the highest purity of any commercial chemical, more pure than the best grades of analytical reagents. An objective has just recently been made to raise even this high boiler water standard up to a requirement of 0.01 mg/L (10 ppb) total impurities.

The boiler feedwater quality requirements equate to or exceed the purity requirements of the electronics industry, which currently has among the more

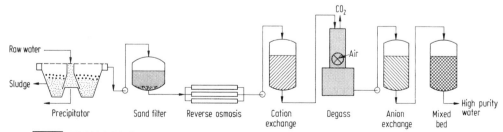

FIGURE 5.6 Schematic flowsheet of the water treatment system used by the electronics industry to obtain the required level of finished water quality (see text). (From Arden [20], reprinted courtesy of the Royal Society of Chemistry.)

stringent requirements. The following are among the basic minimum specifications: neutral salts, 0.02 mg/L; total dissolved solids, 0.02 mg/L; free base or mineral or organic acid, nil; and a conductivity of 0.05 μS/cm or better. The detailed processing steps necessary to meet these standards on a continuous basis requires a composite of all the best existing water purification technology wrapped up in a single processing train (Fig. 5.6).

5.7. TREATMENT OF MUNICIPAL WASTEWATERS

The principal chemical and physical characteristics of the domestic or sanitary sewage effluent must first be known in order to be able to decide the most efficient treatment and disposal options (Table 5.4). While the tabulated list of impurities appears long it should be remembered that the total impurities present still constitute less than 1% of the sewage volume to be treated. From a chemical standpoint, municipal sewage thus represents a dilute solution and suspension of a heterogeneous mixture of constituents—not an easy stream to treat. Nevertheless, since untreated domestic sewage represents about one-quarter of the total biochemical oxygen demand (BOD) waste loading to surface waters, at least for the United States, the treatment and/or disposal options selected can have a significant influence on surface water quality [23].

TABLE 5.4 Approximate Values for the Principal Physical and Chemical Characteristics of Municipal Waste Waters[a]

Characteristic	Domestic sewage			Urban storm waters
Volume	70–200 % of supply (from infiltration)			0–5000 + % of supply
Temperature	1–2° above supply			ambient
BOD	ca. 100–300 mg/L			1–>700 mg/L
COD	ca. 2 × BOD			5–3100 mg/L
Micro-organisms, (as bacteria, viruses etc., no. per 100 mL)	ca. 100–3000 × 10^6 (mostly nonpathogenic)			200, to 146 × 10^6
Total solids: mg/L	655			450–14600
Solids distribution:	Inorganic	Organic	BOD[b]	
Suspended[c]	65	170	110	2–11300
Dissolved[d]	210	210	30	450–3300[e]
Totals	275	380	140	

[a]Compiled from Klein [21], American Chemical Society [22, 23], and Field and Fan [24].
[b]Distribution of biochemical oxygen demand (BOD) between the suspended and dissolved matter of sewage with a total BOD of 140 mg/L.
[c]Suspended solids by volume, for domestic sewage only 1/3 nonsetteable (< 50 μm); 2/3 setteable (> 100 μm diam.)
[d]Dissolved solids for domestic sewage only Approx. twice municipal supply (dry weather); N 50 mg/L (mostly NH_3 and urea); P 30 mg/L (mostly as PO_4^{3-}); plus NaCl, etc.
[e]At high end of range found particularly in areas using highway deicing chemicals ($CaCl_2$, NaCl, etc.).

5.8. DISCHARGE REQUIREMENTS AND POSTDISCHARGE REMEDIES

Discharge of untreated sewage of the properties outlined, if permitted at all, could only be permitted if two composite requirements are met: (1) There is sufficient dilution by the receiving stream and (2) there is good mixing with this stream on discharge. The dilution requirement means that the gross BOD loading of the river should not be raised to more than 2 mg/L, and the dissolved oxygen should not be depressed to below about 4 mg/L (5 mg/L Alberta; 3 mg/L, New York), in order to meet the criteria of a clean river. To discharge raw sewage of an average BOD of about 200 mg/L demands a river flow of more than 100 times the sewage volume at low river flow periods, plus no other discharges in close proximity, i.e., a predischarge river BOD of 1 mg/L or less, to meet the dilution requirement alone. The second requirement, good mixing, is also difficult to achieve and maintain even for the common winding river courses [25]. So there are very few situations where the requirements for raw sewage discharge to rivers could be compatible with these criteria, even if it were aesthetically acceptable.

Similar considerations apply to the discharge of untreated domestic effluent to the sea [26]. Raw sewage discharge into shallow, poorly mixed saltwaters can give the same problems with odorous anaerobic decomposition and microbial contamination on the shoreline and in the surrounding air of the discharge area, as would be experienced with a river or a lake [27].

It is possible to remedy temporarily depressed oxygen levels or high BODs in a river or lake. This may occur for short periods during times of stream low flow or process upsets, even with normally adequate waste treatment. Electrically driven aerators can be placed on floats across the stream, such as has been used for the Lippe River, Germany. Using air of oxygen partial pressure 0.21 atm this system was found capable of boosting the dissolved oxygen content by about 1 kg/900 watts/hour (2 lb/horsepower hour) in the Upper Passiac River, New Jersey [28]. A similar arrangement has also been used temporarily in the North Saskatchewan River just below Prince Albert, Saskatchewan.

Alternatively, to take advantage of the increased oxygen exchange rates provided by increased air pressure, the whole of the low flow river water may be passed through a U-tube bored 14 m or more beneath the river bed. Passage of air bubbles into the downflow side boosts the oxygen partial pressure in the bubbles about 2.5 times to about 0.50 atm, greatly accelerating oxygen dissolution rates (Fig. 5.7). For the Red Deer River in Alberta, this procedure more than doubles the dissolved oxygen content when it is used at times of low flow. This is particularly useful in winter when ice cover prevents reaeration. When the river flow is high, and aeration therefore unnecessary, the bulk of the water flows over the weir bypassing the U-tube.

Aeration using pure oxygen, i.e., providing an oxygen partial pressure of 1 atm (plus) has been proposed for the rejuvenation of highly eutrophic lakes by Union Carbide, a producer of industrial gases. Certainly this system would provide excellent oxygen transfer conditions, but it would be likely to be unduly expensive except for experimental or other special cases.

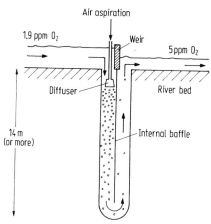

FIGURE 5.7 Automated U-tube river aeration, at times of low flow. A 1.3-m head drives the whole flow through the U-tube, which in practice can consist of a single or several parallel boreholes with an internal baffle [29].

5.9. STREAM ASSIMILATORY CAPACITIES

Adequate sewage treatment is necessary to minimize the frequency of these less efficient, postdischarge solutions to improve river water quality. The degree of treatment required is significantly dependent on the capacity of the receiving body of water to accommodate oxidizable wastes. In general this will be higher for a river which has continuous water turnover than for a lake. For the same gross water flow a shallow, fast-flowing river will have a higher assimilatory capacity than a deep, slow-flowing river. This is because the initial content of dissolved oxygen is likely to be higher, and the oxygen exchange rate with the air will also be more rapid for the shallow river. By the same token riffles, or the shallow fast-flowing stretches of a river, have two or three times the reaeration capacity of deeper, slow-moving pools. These considerations have led to the tabulation of stream assimilatory capacities as related to their volume of flow, depth, and oxygen exchange rates (Table 5.5).

Knowing the approximate assimilatory capacity of a stream, which equates to its recovery or reaeration rate, allows an estimate of the acceptable BOD loadings which could be discharged to the river without severely degrading river quality. Assuming a discharge of 2.4×10^6 L/day of untreated sewage of 200 mg/L average BOD, a volume which could reasonably arise from a small town of 24,000 inhabitants, this would equate to a daily BOD loading of about 480 kg/day (Eq. 5.18).

$$200 \text{ mg/L} \times 2.4 \times 10^6 \text{ L/day} = 4.8 \times 10^8 \text{ mg/day} \qquad 5.18$$

Discharge of this volume of oxygen demand into a Class I river, if evenly distributed through the day, would result in a river BOD loading of 14 mg/L (Eq. 5.19).

■■■■ TABLE 5.5   Waste Assimilatory Capacity Related to River Size[a]

Stream class	Common examples[b]	Assimilatory capacity[c]	Average depth (m)	Average flow (m³/sec)
I	Numerous local	0.023	0.17	0.4
II	Allegheny, Kansas, N. and S. Saskatchewan, Rhone, Po	0.56	1.5	300–1500
III	Mississippi, St. Lawrence, Orinoco, Ganges, Yenisei	2.90	13.7	20,000

[a]Compiled mostly from the data of Todd [4] and Canadian National Committee [30].
[b]Based on flow rates of first example of each class.
[c]Round number estimates in tonnes BOD per kilometer per unit oxygen deficiency for the physical data as given. Originally stated in units of tons BOD/mile/unit oxygen demand: Class I, 0.04; Class II, 1.0; Class III, 5.2.

$$\frac{4.8 \text{ m}^3/\text{sec} \times 10^8 \text{ mg/day}}{0.4 \text{ m}^3/\text{sec} \times 86{,}400 \text{ sec/day} \times 1000 \text{ L/m}^3} = 13.9 \text{ mg/L BOD} \qquad 5.19$$

This is a very heavy loading sufficient to severely degrade river quality, even if good mixing is achieved. However, if the BOD of this sewage was decreased to 20 mg/L by suitable treatment before discharge, i.e., to one-tenth of the raw BOD values, then discharge could occur without severe river degradation.

Another approach to this problem is to use a figure for the oxygen demand loading per capita, estimates of which range from 54 to 115 g per day [23]. For the community of 24,000 this method gives a range for the gross BOD discharge of 1296 to 2760 kg/day for untreated sewage, much higher than the estimate based on the previous method. If discharges on this scale were actually practiced to a Class I river this would correspond to very severe river degradation from gross BOD loadings of 37.5 to 80 mg/L.

From Table 5.5, the background calculations allow a quick estimate for a Class I river that 23 kg of BOD discharge would cause a drop in river BOD of 1 mg/L, one kilometer below the point of discharge. It can also be seen that although the flow volume of a Class III river is of the order of 50,000 times greater than a Class I river, the assimilatory capacity is only 130 times more. This fact is a reflection of the much slower reaeration rate of a deep, slow-flowing river with vertical stratification as compared to a shallow, fast-moving river. It also is a reminder that just because one river has 10 times the average flow volume of another does not necessarily mean that it can accept 10 times the oxidizable waste volume without degradation: hence the need for sewage treatment.

5.10. PRIMARY AND SECONDARY SEWAGE TREATMENT

Methods for municipal sewage treatment may be conveniently grouped into three categories according to the stage of wastewater processing and the types of treatment used since these two aspects generally coincide quite closely. The primary treatment step applies basically mechanical processes, amounting to physical separation procedures to sewage cleanup. Secondary sewage treatment uses mainly biological methods to remove and consume wastes by a variety of configurations, and provides a further significant degree of improvement of effluent quality over that obtained by primary treatment alone. Tertiary treatment, sometimes called advanced treatment, involves one or more of biological, chemically based, or physical separation processes for a final refinement of effluent quality. The level of treatment required in any particular situation is decided on the basis of the population loadings to be served by the system and the assimilatory capacity of the receiving body of water. If this is a river, the level of treatment required has to reflect the times of lowest flows, periods of ice cover, and whether it is to be used for recreational or other purposes. A plot of the interrelationships between the dissolved oxygen and biochemical oxygen demand helps to make the influence of these considerations clearer (Fig. 5.8).

Primary treatment involves the use of a combination of one or more of the following types of processing units (Fig. 5.8). First the effluent is screened by self-cleaning bar screens of 2 to 5 cm spacing to remove any coarse material, followed by a set of finer screens which are also usually self-cleaning. These preliminaries serve to protect the downstream pumps and sludge handling equipment from fine gravel and any other large suspended matter which often enters combined sewerage systems used to collect both sanitary and storm runoff waters in the same piping. After screening, the waste stream passes through a comminutor, an electrically driven fragmenter, which chops any residual coarse material ready for further processing. Following comminution the wastewaters are passed into quiescent channels with underflow discharge which remove floatables, grease, hair, plastics, scum, etc. The underflow discharge from these channels feeds into a large sedimentation tank (clarifier), which has a retention time of a few hours. The supernatant overflow from this tank comprises primary treated sewage effluent. These

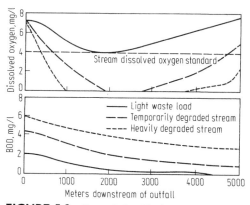

FIGURE 5.8 Plots of plausible degrees of degradation and recovery rates for oxygen-consuming discharges to a stream which heavily degrades, temporarily degrades, or stays within reasonable guidelines to maintain stream quality.

processing steps achieve about a 60% reduction in suspended solids content and a 30% reduction in oxygen demand over the influent sewage, plus a smaller improvement in other parameters, as summarized in Table 5.6. Sludge collected from the bottom of the primary settling tank, which at this stage contains only 2.5–5% solids, has to be dewatered and further treated for disposal. Details of disposal options are discussed later.

Biological consumption and immobilization, and further coagulation plus settling of residual wastes in the sewage stream are the objectives of secondary treatment. Greatly enhanced concentrations of micro-organisms are employed to bring about accelerated biochemical and biological consumption and destruction of residual wastes, aided by a vigorous supply of air (oxygen). By these measures the BOD and suspended solids are both decreased to about 1/10 of influent values and other qualities of the effluent are also improved somewhat over that provided by only primary treatment (Table 5.6). With the augmented micro-organism populations and artificial aeration the treatment time required is reduced to a matter of hours, instead of the several days that the same processes would take in a natural setting. Thus the effluent improvement may be conducted in much more compact equipment than would otherwise be possible. The detailed methods used to achieve this rapid waste utilization by micro-organisms vary widely with country and the local conditions which have to be met [33].

In North America, and to a lesser extent in Europe, activated sludge plants or variations of this process are common for secondary sewage treatment, as shown schematically in Fig. 5.9. High bacterial populations are maintained by the return of settled sludge and artificial aeration is maintained by compressed air or "brush" aerators to stimulate a high level of aerobic micro-

---

◼◼◼ TABLE 5.6 Approximate Cumulative Waste Removal Efficiencies of Various Sewage Treatment Procedures, in Percent[a]

	Primary	Primary + secondary	Shallow lagoons	Chemical coagulation	Reverse osmosis or electrodialysis
BOD	35%	90%	95%	95%	95%
COD	30	80	90+	85	95
Refractory organics	20	60	85	80	90
Suspended solids	60	90	95	95	95+
Total N	20	50	85	60	90–95
Total P	10	30	85	85	90–95
Dissolved minerals	1–2	5	10	10	50–90
Incremental cost, US$ per 100 m^3 [a,c]	0.80–1.05	1.85–4.00	1.60–3.20	3.20–6.60	33–100
Bacteria:	35%	90%	99%		
Postdisinfection[b]	90%	99%	99.9+		

[a]Data from American Chemical Soceity [23], Babbitt and Baumann [31], and Weinberger *et al.* [32]. Costs are approximate, varies greatly with scale.
[b]Approximate values postdisinfection, usually by chlorination [31].
[c]100 m^3 equates to roughly 22,000 imperial gallons or 26,400 U.S. gallons

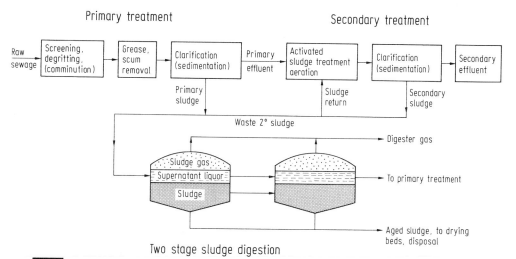

FIGURE 5.9 Schematic flow diagram for the principal steps involved in municipal sewage treatment to the secondary level. (A composite of components given by the American Chemical Society [23] and Field and Fan [34].)

bial activity in this tankage. By this means both dissolved and suspended matter is rapidly converted to settleable bacterial biomass. Biomass plus absorbed and adsorbed impurities are removed in a secondary sedimentation tank to produce a supernatant sewage effluent treated to the secondary level, of BOD 20–30 mg/L, plus a residual secondary sludge (Table 5.6). The Unox and Oxitron systems are secondary treatment variants promoted by Union Carbide, which use oxygen instead of air for aeration to obtain accelerated rates of waste treatment [35]. By the use of oxygen instead of air the capacity of a sewage treatment plant may be increased without an increase in size.

Other variants of this procedure to obtain efficient microbiological action include the trickling filter, common in Europe, and the biodisk, or rotating biological contactor (RBC), which are particularly suitable for smaller installations [36]. Both of these systems use efficient air exchange on a microbial support medium, rather than using compressed air, to obtain rapid aerobic consumption of wastes in the effluent. Where land is available and inexpensive, and the volumes of sewage to be treated are not unduly large, the Dutch oxidation ditch also provides a technologically simple method of obtaining roughly equivalent effluent quality. Where land is scarce, on the other hand, there is also the deep shaft treatment method, originally developed by I.C.I. (U.K.) which accomplishes the same type of improvement in a bore hole of 0.3 to 10 m in diameter and 150–300 m deep [37]. The pressure of the head of water is said to boost aeration rates to a factor of 10 higher than is possible with a regular activated sludge plant, and in the process halves the volume of sludge produced. Apparently the working bacteria, with their very high surface to volume ratio and hence high intracellular-extracellular gas exchange rates, are not afflicted with the "bends" under these conditions.

Bacterial counts in the supernatant liquor from a secondary treatment plant are greatly decreased, to about 10% of the influent raw sewage count, but are still high. Therefore, if secondary treatment produces adequate improvement in effluent quality it is normally disinfected by a fairly heavy dose of chlorine (ca. 10 ppm) before discharge, causing a further reduction to about 1% of the influent bacterial counts [32]. Each 1 ppm of postchlorination also tends to decrease the BODs of the effluent by about 2 ppm from the oxidizing action of the chlorine, which is estimated to consume 50 to 80% of the chlorine added. The higher chlorine dose rates used for sewage disinfection, together with the higher concentrations of a variety of organics also present in these waters, represent a more significant source of chlorinated organics to surface waters than that contributed from the disinfection of municipal water supplies. Gamma irradiation and bromine chloride are other effective measures which have been proposed for disinfection of treated sewage effluent [38]. At least the first of these disinfection methods does not contribute any halogenated organics, or additional toxicity to the effluent which may require neutralization before discharge.

If the final chlorine-disinfected effluent is to be discharged to a sensitive watercourse, dechlorination may be necessary to avoid problems from the toxicity of the residual chlorine. After a 20- to 30-min holding time for completion of the disinfection to take place, dechlorination is accomplished by converting chlorine to chloride with a reducing agent such as sulfur dioxide followed by pH neutralization with lime (Eq. 5.20, 5.21).

$$HOCl + SO_2 + H_2O \rightarrow HCl + H_2SO_4 \qquad 5.20$$

$$2\ Ca(OH)_2 + 2\ HCl + H_2SO_4 \rightarrow CaCl_2 + CaSO_4 + 4\ H_2O \qquad 5.21$$

It should be realized that this dechlorination step, while it does decrease chlorine-induced toxicity, does nothing to remove any chlorinated organics which may also have formed during the disinfection step.

One of the outstanding problems of this sequence of sewage treatment is the wide variability of influent volumes which result from the use of combined sewerage systems. Combined systems are those in which domestic sewage and storm runoff are collected in the same piping system as an installation economy. This type of system may at times have outflows of 50 or more times normal domestic flows, far exceeding the throughput capacity of the treatment plant. To deal with this situation, overflow points are provided at critical junctions in the collection system to release combined sewage flows directly to surface waters at the time of a storm. Towards the end of a prolonged storm these outflows become relatively dilute, low-toxicity wastes for which direct discharge does not constitute a serious problem. However, the initial storm flows from such a system can have characteristics significantly worse in terms of its effect on discharge to surface waters, than those of ordinary untreated domestic sewage (Table 5.4). This happens from the street accumulation of wastes such as soil, vehicle residues, animal droppings, pesticides, and fertilizers during the preceding dry spell, as well as from material sedimented out of domestic sewage itself during quiescent flow periods, which will all be flushed out of the system in the initial period of a vigorous rainstorm.

Solutions to these problems of combined sewerage systems include gradual separation of the two systems, as the construction costs become budgetarily feasible, plus provision of separate storm water treatment facilities. Initial storm flows may then be retained for treatment while allowing discharge of later, cleaner flows. This treatment philosophy can also be applied to combined sewerage systems by provision of initial storm water storage using the excess storage capacity within the ducting itself, simply by raising the regulation levels of overflows. Or system storage plus holding basins at the treatment facility can be combined to retain most of the highly contaminated initial storm flows for later gradual treatment, combined with the discharge of later flows from a prolonged storm, if necessary, when holding capacity is exceeded.

5.11. TERTIARY OR ADVANCED SEWAGE TREATMENT

If the effluent quality produced by secondary treatment is still not adequate, as may well be the case for discharges to a protected watershed or to a small stream, then one or more of the tertiary treatment options may have to be employed to follow the primary and secondary steps. Tertiary treatment, sometimes referred to as advanced treatment, covers a variety of methods which may be used either singly or in combination for the upgrading of secondary effluent quality [39].

One or more shallow lagoons in series having a total secondary effluent holding time of the order of days can alone accomplish significant further improvement. The shallowness requirement ensures good sunlight penetration and oxygen exchange to the bed of the lagoon, assuring the continued action of aerobic bacterial as well as photosynthetic waste utilization. Element uptake by the rooted and free-floating plants and bacterial action in the lagoons continues to remove nutrient ions plus C, H, O, and S compounds in the secondary effluent, albeit at a slower rate (Table 5.6). This is why a longer holding time is required with this step than for primary or secondary treatment. It also provides sufficient time to further decrease the suspended solids content by a combination of biological coagulation processes and settling. Bacterial counts are also reduced by a factor of about 10 by a combination of soil adsorption and coagulation/settling [40]. Tertiary lagoons, however, while simple in concept, do have a significant land requirement for their use.

Chemical methods of effluent improvement, being generally fast acting, have a smaller treatment volume and land requirement but have chemical costs attached to their use. Coagulation of suspended solids may be accomplished using an inorganic coagulant, such as alum. Chemically, alum is $K_2SO_4 \cdot Al_2(SO_4)_3 \cdot 24H_2O$, but since it is the aluminum ion of this double salt which is the active coagulant, "alum" for sewage treatment normally refers to aluminum sulfate itself, $Al_2(SO_4)_3$. Alum treatment removes much of the residual colloidal matter by promoting agglomeration to larger particles which then settle. It also can remove 90–95% of the dissolved phosphate, dropping this from 35–55 mg/L to 0.5 mg/L or less [41] (Eq. 5.22, 5.23).

$$Al^{3+} + \text{negatively charged colloids} \rightarrow \text{settleable floc} \qquad 5.22$$

$$Al_2(SO_4)_3 + 2\ Na_3PO_4 \rightarrow 2\ AlPO_4\downarrow + 3\ Na_2SO_4 \qquad 5.23$$
$$\text{soluble} \qquad \text{insoluble}$$

However, alum dosages of 200–300 mg/L are required to achieve this, which contributes a significant inorganic content to the sludge. This sludge mineral loading may be a nuisance in some disposal systems.

Lime, employed at similar dosages as alum, has about the same phosphate removal efficiency (Eq. 5.24).

$$3\ Ca(OH)_2 + 2\ PO_4^{3-} \rightarrow Ca_3(PO_4)_2\downarrow + 6\ OH^- \qquad 5.24$$

Lime treatment is generally lower in cost than alum treatment, but it still has the attendant disadvantage of adding significant inorganic (and hence incombustible) mass to the sludge disposal system.

Some commercial organic polyelectrolytes are also efficient for colloid flocculation. These, being combustible, may be readily burned along with the sludge. They are also active at dosages of only 1–2 ppm, much lower than required with the inorganic flocculants, which also helps in this respect. Thus, even though they are more expensive than inorganic flocculants on a weight for weight basis, they are competitive on a treated sewage volume basis. If used in conjunction with ferrous chloride or sulfate, which are by-products of the steel pickling industry, they too are capable of decreasing the phosphate concentration by some 80% [42] (e.g., Eq. 5.25).

$$3\ FeCl_2 + 2\ Na_3PO_4 \rightarrow Fe_3(PO_4)_2\cdot8H_2O\downarrow + 6\ NaCl$$
$$\text{vivianite} \qquad\qquad\qquad 5.25$$
$$\text{(finely divided)}$$

Subsequent settling removes the flocculated insoluble vivianite which forms, as the second step of this two-stage chemical treatment.

Phosphate may also be removed from treated effluent by passage over aluminum electrodes carrying an AC current [43]. With high effluent flow rates and low treating currents, about 1.4 g of aluminum is consumed per gram of phosphate removed.

If nitrogen as well as phosphate removal is required, combinations of chemical and biological action are required. Secondary effluent treated with lime to decrease the pH and then subjected to anaerobic digestion effectively converts up to 90% of the combined nitrogen compounds present to ammonia (Eq. 5.26).

$$NO_3^-, NO_2^-\ \text{(anaerobic digestion)} \rightarrow NH_3\ \text{in solution} \qquad 5.26$$

The ammonia, still in solution, is air-stripped from the effluent before discharge in a procedure that was adopted for the treatment plant in the sensitive Lake Tahoe, California, watershed.

Alternatively, removal of combined nitrogen may be accomplished by denitrification of nitrate and nitrite using appropriate bacteria in the presence of a carbon source such as methanol (Eq. 5.27).

$$6 \ NO_3^- + 5 \ CH_3OH + 6 \ H^+ \rightarrow 3 \ N_2 + 5 \ CO_2 + 13 \ H_2O \qquad 5.27$$

With this option the nitrogen gas end product altogether gets rid of the combined nitrogen originally present in the effluent from the watershed, unlike the possible return in subsequent precipitation of a fraction of the ammonia produced by anaerobic digestion.

Finally, removal of other dissolved salts from the effluent, which may be necessary for water reuse in some water conservation programs, may best be accomplished using reverse osmosis or electrodialysis [44]. Ion exchange may also be used [45]. More information concerning the operating details of these processes is given with the discussion of municipal water treatment.

For some water reuse applications activated carbon may be employed to adsorb 90 to 98% of any residual degradation-resistant organics when necessary. The resultant effluent quality is certainly adequate for many industrial coolant or irrigation applications, and could even be reused for potable purposes in emergency situations after minimal further treatment such as reaeration and chlorination [46]. However, not enough is known about the potential for accumulation of trace toxins to recommend this procedure for long-term potable water use.

5.12. SLUDGE HANDLING AND DISPOSAL

The large volumes of sludge produced during sewage treatment and its low solids content pose a secondary environmental problem, its disposal. Frequently the costs of sludge treatment and ultimate disposal or destruction equal all the other costs of sewage treatment [47]. Primary sludges, of 2.5–5% solids, and the sludges from trickling filters and activated sludge plants of 0.5–5% and 0.5–1% solids, respectively, are particularly difficult to dewater because much of the water is tied up intracellularly by micro-organisms and in the sedimented flocs, both of which retain water tenaciously. In terms of disposal weights, roughly 1 tonne dry weight of sludge is produced daily for each 10,000 population (domestic wastes or industrial equivalent) connected to the sewerage system.

Anaerobic sludge digestion at 30 to 35°C reduces sludge volume by about 90% while producing a medium-grade fuel gas of methane and carbon dioxide, in the proportions of roughly 2 CH_4:CO_2 (e.g., Eq. 5.28–5.31, Fig. 5.9).

$$CO_2 + 4 \ H_2 \rightarrow CH_4 + 2 \ H_2O \qquad\qquad 5.28$$

$$4 \ HCOOH \rightarrow CH_4 + 3 \ CO_2 + 2 \ H_2O \qquad\qquad 5.29$$

$$CH_3COOH \rightarrow CH_4 + CO_2 \qquad\qquad 5.30$$

$$2 \ CH_3CH_2OH \rightarrow 3 \ CH_4 + CO_2 \qquad\qquad 5.31$$

The gas produced may be usefully used for energy requirements in the sewage treatment plant itself, or it may be relatively easily upgraded by carbon dioxide removal (water washing) and added to a town gas distribution system [48]. Small volumes of digested sludge may be disposed of by plowing into land, though safety questions remain related to residual pathogens and possible toxic heavy metal accumulations, particularly in the sludges from heavy industrial areas, which may be contributed to the land in this way.

Sludge may be disinfected by chlorination, accomplished by the direct addition of 2–4 mg/L chlorine, or by sludge electrolysis which produces chlorine *in situ* from the residual sodium chloride present in the wet sludge [49]. Gamma irradiation has also been used for sludge disinfection, and various heat treatments are also effective for this purpose [50]. Any of these disinfection alternatives improves the safety and appropriateness of land disposal of sludges, provided that this is to a level area removed from any surface water courses, that the heavy metal content of the sludge is low, and that the sludge application to land is not too frequent or heavy. For the year following application to agricultural land the site should be clearly marked, and the land used only for pasture, fallow, or forage crops, not for produce or dairy cattle. This sludge disposal option has been usefully applied to assist in the reclamation of strip-mined areas. The sludge may also be composted with the organic content of municipal waste or wood chips to produce a useful soil conditioner. Or dried, alone, it can be more economically shipped greater distances and marketed as a low analysis fertilizer, as is promoted by the city of Milwaukee, with their Milorganite.

If none of the disposal options above or their variations can be practiced, then wet or dry incineration are the only alternatives. Wet oxidation with air at 300–350°C, the Zimmerman process, yields liquid effluent containing a much reduced volume of easily settled ash [51]. Dry incineration in multiple hearth units uses much of the heat generated by combustion to complete the drying of the entering sludge [52]. The final product, a sterile ash of 3–5% of the original dewatered volume, comprises a much smaller volume for discarding to landfill or the like than the original sludge. One should be aware of the possibility of metal elution from the ash in the landfill, and take appropriate precautions. Or the sludge can be more thoroughly dewatered by filtration or centrifugation techniques, combined with milled coal to form briquets as developed by BASF, and then burned in a coal-fired power station for energy recovery from the composite fuel [53].

5.13. INDUSTRIAL LIQUID WASTE DISPOSAL

Industrial liquid wastes can only be considered for discharge into municipal sewerage systems if they meet certain strict criteria, because of frequent incompatibility problems. They may have very high BODs, in the tens of thousands instead of 100 to 300 mg/L, corresponding to carbon loadings of 1% or more. Oxygen demands that are this high in liquid wastes fed to a municipal sewage treatment plant can seriously overload the BOD removal capabilities. Industrial waste streams may contain toxic constituents such as cyanide ion, heavy metals, or toxic organics which could kill the active micro-organisms employed for secondary sewage treatment. They may also contain dissolved or immiscible solvents which are not only hazardous to the sewage treatment plant but also to the sewage collection system itself. An accident with released hexane, for instance, left the main street of Louisville, Kentucky, pock-marked with 7-m craters as a result of a sewer explosion a few years ago. Therefore, many kinds of industrial liquid wastes require at least some kind of preliminary treatment before acceptance by a municipal system and many are not acceptable to a municipal system under any circumstances.

With any by-product or waste stream, utilization as far as possible, is a far more attractive proposition than paying for ultimate disposal [54]. For instance, diphenyl ether is obtained as a by-product of the chlorobenzene to phenol process roughly according to the stoichiometry of Eq. 5.32 and is not a direct objective of the process.

$$20 \ C_6H_5Cl + 20 \ NaOH \rightarrow 18 \ C_6H_5OH \qquad \qquad 5.32$$
$$+ \ (C_6H_5)_2O + 20 \ NaCl + H_2O$$

Diphenyl ether production may be decreased by recycling a part of it back into the process, but markets have also been developed for this by-product as a component of high-temperature heat transfer fluids, brake fluid formulations, and the like. In these applications diphenyl ether earns a profit rather than imposing a disposal cost burden on the process.

As other examples of this philosophy, the anhydrous hydrogen chloride in excess of market requirements produced during chlorinations of hydrocarbons to halocarbon solvents, or during the manufacture of toluene diisocyanate, may be usefully employed in the production of phosphoric acid from phosphate rock. Fluorides associated with the production of phosphoric acid and phosphatic fertilizers may be used to manufacture synthetic cryolite (Na_3AlF_6), useful in aluminum smelting, or fluorspar, valuable in iron and steel metallurgy. In turn the iron-containing red muds from alumina purification prior to smelting can be used to prepare chemicals useful in water and sewage treatment. Aluminum itself can also be economically recovered from fly ash [55].

A large number of examples of these kinds of utility can be found for process by-products. However, each new process developed and started up can pose a new by-product problem to test the ingenuity of the developers to come up with a use for it, rather than generate a waste disposal problem. Thoughts (of this kind are behind the "Product Stewardship" ideas put forward within the Dow Chemical Company some time ago to promote this concept early in the formative stage of new process ideas, and have recently been adopted by the Chemical Specialties Manufacturers Association.

5.13.1. Aqueous Wastes with High Suspended Solids

Liquid waste streams with a high suspended solids content can be cleaned up by solids removal in clarifiers, thickeners, and liquid cyclones and by accelerated settling by inclined "Chevron" settlers or the like [56, 57]. For waste streams with very finely divided solids in suspension (i.e., less than about 100 μm) dissolved air flotation techniques have been shown to be more efficient than methods employing sedimentation. Final dewatering of the sludges obtained may be carried out on a continuous filter or a centrifuge. Water clarified in these ways can be accepted for more potential options of reuse or final disposal than untreated water, and the separated solids may be burned or discarded to landfill, as appropriate [58].

5.13.2. Aqueous Wastes Containing an Immiscible Liquid

Liquid waste streams containing an insoluble liquid, such as can arise from extraction processes, from steam ejectors operating on solvent distillation systems, or from the loss of heat exchange fluid from a heat exchanger, should be phase-separated before final disposal measures are undertaken. A simple settler, or a unit such as an API separator can be used to accomplish this step. When the initial separator is coupled to an entrained or dissolved air flotation unit, the concentration of residual organics can be further reduced [59]. The recovered organics can be returned to their processing origins, via a further cleanup if required, and the water phase more safely discarded.

5.13.3. Heated Effluent Discharges

Various options exist to deal with high thermal loads to water [60]. If the cooling water supply is plentiful the impact of the heated discharge can be decreased by reducing the design temperature rise of the cooling water flow. This can be accomplished within limits by increasing the cooling water flow rate to existing heat exchangers, or by combining this step with the installation of larger heat exchangers. If limited water availability precludes this option, the cooling water may be reused by passage through a cooling pond or through convective or forced air circulation evaporative cooling towers (Fig. 5.10). When heated effluent is available near an urban center it can be used for community heating requirements, with mutual benefit. Otherwise it can be employed for the warming of greenhouses or soil heating, or for accelerated fish rearing. If water supplies for this purpose are very limited, indirect cooling towers, not employing evaporation, or air cooling options may have to be adopted [61]. These options may also be necessary in the event of severe fogging in the vicinity of evaporative cooling towers, caused by the very large water vapor discharges from such installations.

5.13.4. Aqueous Waste Streams with a High Oxygen Demand

Aqueous waste streams with BODs of 5000 to 20,000 but which are essentially nontoxic, such as those produced by the food processing industry or by a distillery, may be efficiently treated by extended aeration in ponds or lagoons, or in deep shaft aeration units [62]. With added nutrients in the form of ammonium phosphate to promote bacterial waste utilization, BODs may be decreased by 90–95% [52]. Spray irrigation of this type of waste stream

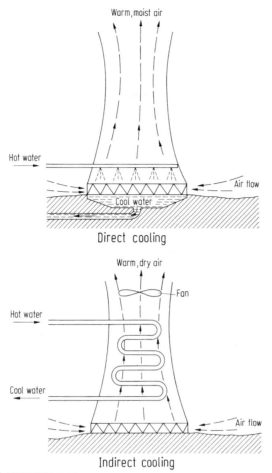

FIGURE 5.10 Diagrams of direct (evaporative) and indirect cooling towers designed to shed heat loads to air rather than surface waters.

with or without prior BOD reduction has also been advocated, in which case infiltration to the soil and the activity of soil bacteria are relied on to deal with the oxygen demand. If this disposal method is used it is important not to overload the land disposal area to avoid placing the soil microbial system under anaerobic conditions.

5.13.5. Highly Colored Wastewaters

Activated carbon has been tested for the cleanup of highly colored waste streams such as those resulting from dyeing operations [63]. Coal carbons were 99% effective for dye removal and could accept dye loadings as high as 0.40 kg dye/kg carbon. Lignite carbon was also useful, but less effective.

5.13.6. Fluid and Solid Combustible Wastes

Combustible wastes which are fairly fluid at ordinary temperatures may be simply atomized and burned for disposal. But the combustion units to handle organics containing a heteroatom such as nitrogen or chlorine must be designed to handle more corrosive than normal combustion gases to avoid corrosion problems in the combustion chamber, and to have a water scrubber or other type of exhaust gas emission controls integral with the unit. Very viscous or solid combustible wastes are normally disposed of by direct burning in fluidized bed or multiple hearth combustors [58, 64] (Fig. 5.11).

Aqueous solutions of combustible organics such as phenols, alcohols, and lower ketones can also be burned for disposal, but proper combustion may have to be assisted by co-combustion with some external fuel. If, however, fairly fluid combustible wastes are burned at the same time as these aqueous wastes containing combustibles in special burners designed to accommodate dual fuels, both may be disposed of simultaneously without having to use purchased fuel to assist the process.

5.13.7. Neutralization and Volume Reduction of Intractible Waste Streams

Noncombustible or toxic liquid wastes that do not lend themselves to disposal by any of the means just outlined are the most difficult to dispose of safely. With these kinds of wastes in particular, it is in the producer's best interests

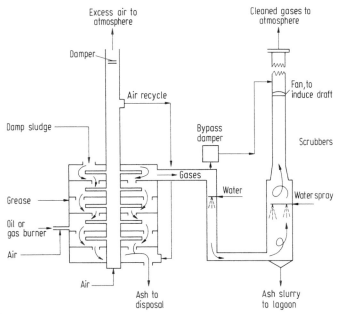

FIGURE 5.11   Multiple hearth combustion unit with emission controls, for the disposal of sludges and combustible solid wastes. Rakes around air preheat tube at center of multiple hearth unit rotate slowly around tube to move initial sludges or solids across each tray and down, eventually to exit as ash from the base of the unit. (Adapted from Sebastian and Cardinal [58]. Excerpted by special permission from CHEMICAL ENGINEERING [Oct. 1968]. Copyright © 1968, by McGraw-Hill, Inc., New York, NY 10020.)

to minimize ultimate disposal costs. Thus, acidic pickling plant wastes, which will contain iron plus unused sulfuric or hydrochloric acids, may be used to precipitate phosphate from secondary sewage effluent (e.g., Eq. 5.25). Or they may be blended with aqueous alkaline wastes containing phenolate and unexpended sodium hydroxide to neutralize the pH extremes of both streams.

Wastes containing cyanide ion and various heavy metals may be at least partially detoxified (cyanide neutralized) by combining these wastes in proper proportions with the toxic waste streams from chlorohydrin-based ethylene oxide or propylene oxide manufacture, which will usually contain unexpended hypochlorite [65] (Eq. 5.33).

$$2 \text{ NaCN} + \text{Ca(OCl)}_2 \rightarrow 2 \text{ NaCNO} + \text{CaCl}_2 \qquad 5.33$$

At higher ratios of oxidant to cyanide, complete conversion to carbon dioxide and nitrogen can be achieved [66] (Eq. 5.34).

$$4 \text{ NaCNO} + 3 \text{ Ca(OCl)}_2 + 2 \text{ H}_2\text{O} \rightarrow \qquad 5.34$$
$$\text{N}_2 + 4 \text{ NaCl} + \text{CaCl}_2 + 2 \text{ Ca(HCO}_3)_2$$

Since a slight excess of chlorine is usually employed to obtain cyanide destruction, the waste stream may still be quite toxic from the residual chlorine present. Dechlorination using sulfur dioxide effectively reduces any residual chlorine to chloride, removing the toxicity from this source before discharge (Eq. 5.20).

The metals content of an aqueous waste stream can be substantially decreased by complexation or adsorption methods and the recovered metal used to offset a part of the treatment costs [67]. Unexpended hypochlorite can also be neutralized by contacting this stream, in the correct proportions, with spent scrubber liquor from sulfur dioxide emission control systems (Eq. 5.35).

$$\text{Ca(OCl)}_2 + 2 \text{ CaSO}_3 \rightarrow \text{CaCl}_2 + 2 \text{ CaSO}_4 \qquad 5.35$$

Decreasing the volume of waste streams containing a high concentration of dissolved salts by such means as ion exchange or reverse osmosis can also help reduce the ultimate disposal costs of the solutions of salts produced by these neutralization reactions.

Since ingenuity in dealing with these kinds of wastes frequently requires good communication between companies in widely differing business areas, participation in a waste trading exchange [68, 69] or the setting up of an independent waste disposal company [70, 71] can be useful options. These clearing houses can then contract workable price schedules for proper disposal of various classes of wastes, and by so doing are set up to maximize the benefit of neutralization interactions or codisposal options between the various kinds of wastes handled.

5.13.8. Ultimate Destruction or Disposal of Hazardous Wastes

Incineration at the high continuous operating temperatures of a cement kiln [72] or by large ocean-going incinerators such as the Chemical Waste Management, Inc., and Bayer vessels, *Vulcanus* and *Vesta,* operating at sea, are two ultimate disposal options for more than usually hazardous wastes. Chemical fixation into absorbent solids by proprietary processes such as those tradenamed Sealosafe, Chemfix, or Poz-o-tec, and then use of the fixed product as landfill or as foundation material is a further alternative [73]. Direct landfilling and acceptance of long-term sterilization of the land with proper management, recording, and site posting has also been used for hazardous waste disposal [71]. A clay lining to the disposal site does decrease the transmission of larger molecules through the soil, but is still quite permeable to smaller organics [74]. More rigorous sheet or cement linings plus a drainage system under the lining may be called for in some circumstances.

Deep well disposal, generally to a brine aquifer at depth, has also been used, particularly for waste brines (primarily $CaCl_2$, $NaCl$, etc.). To use this method without damage to groundwater resources requires careful attention to the basic requirements. These are that the wells be located in a seismically stable region, and that the discharge take place via a two casing (double pipe lined) well to a porous stratum capped by one or more impermeable strata to ensure retention of the waste at the level discarded. Drilling the well itself is expensive, US$30-$100 per meter, and there are also a number of important precautions to be taken in the well installation, such as prior removal of suspended solids and low injection pressures, to ensure safe operation [75] (Fig. 5.12). So deep well disposal is not an inexpensive disposal method, although it does have some virtues, e.g., for brine disposal into a brine aquifer. The long-term costs of poorly operated deep well disposal installations if groundwater contamination should occur are so great that licensing of these operations should be required by more jurisdictions, as it is by Ontario. The

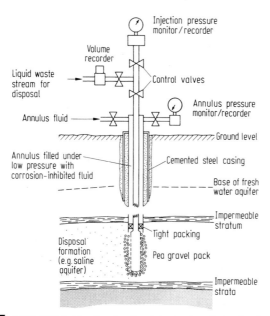

FIGURE 5.12 Details of an injection installation for deep well disposal of liquid wastes.

potential for contamination of the groundwater resource is by no means limited to deep well disposal accidents, however, but can also occur from bacteria and viruses in groundwater recharge operations using treated sewage [76] or from toxins leached from poorly placed or poorly operated municipal and industrial waste dumps [77, 78].

REVIEW QUESTIONS

1. (a) Where does the dividing "line" come between particle diameters which settle relatively rapidly and particle diameters which settle slowly in water? How does this dividing line compare with the parallel dividing line for particles in air or other gases?
   (b) Briefly, what effect do you expect the above comparison to have on the relative ease of removal of particles from wastewater streams as compared to gas streams?

2. Stearic acid, $C_{17}H_{35}COOH$, may be used to provide a monolayer to reduce the rate of water loss by evaporation from the surface of a reservoir. The molecule floats at the surface with the hydrophobic hydrocarbon chain upward and the hydrophilic carboxylic acid downward, and in so doing covers roughly a circular area of radius 2.82 Å (0.282 nm) per molecule. How many kilograms of stearic acid would be required to form a solid monolayer over the surface of the Elk Lake reservoir, which covers an area of 289 acres? (1 acre = 4047 m^2)

3. (a) When chlorine is added to water of initial pH 5 for disinfection, all of the active hypochlorous acid formed remains undissociated. What is the percent weight ratio, hypochlorous acid formed to initially added chlorine, under these circumstances?
   (b) If calcium hypochlorite $(Ca(OCl)_2)$, for convenience, is used for water disinfection instead of chlorine, what percent ratio of hypochlorous acid to the initially added weight of calcium hypochlorite could optimally be formed?
   (c) Considering your answer to part (b), what rate of anhydrous calcium hypochlorite would have to be added to water to equate to the effectiveness of treating with 2 ppm of elemental chlorine?
   (d) What dose rate, in mg/L, would have to be used for a calcium hypochlorite bleach solution of $5\frac{1}{4}\%$ available chlorine (by weight) to equate to treating with 2 ppm of elemental chlorine?
   (e) Addition of potassium hypochlorite to water of pH 5 for disinfection produces what percent weight ratio, hypochlorous acid to initially added potassium hypochlorite, using this disinfecting agent?
   (f) Explain at least two factors which would favor use of sodium or calcium hypochlorite over potassium hypochlorite for water disinfection.

4. Give two reasons why sparging a refinery wastewater stream with small air bubbles (froth flotation) should provide such an effective

way of removing further hydrocarbon impurities in a refinery wastewater stream, even after phase separation in an API separator.

5. Chlorine and calcium hydroxide (slaked lime), used to ensure high pH conditions to avoid generation of extremely toxic hydrogen cyanide gas, are used in combination to chemically neutralize waste aqueous solutions of sodium cyanide. In this way the much less toxic cyanate ion (CNO^-) is produced for discharge by oxidation of the cyanide ion.

    (a) Write the equations which would govern the chemical reactions required to carry out this neutralization.

    (b) What would be the required stoichiometric amounts of chlorine and calcium hydroxide to neutralize 1000 L of effluent containing 100 mg/L sodium cyanide?

6. Sulfide ion (S^{2-}), which is toxic, may be present in the wastewaters from froth flotation of sulfide ores or from refinery operations. It may be neutralized by the addition of the stoichiometric (not excess) amount of chlorine to oxidize sulfide to sulfate.

    (a) Give the equations which describe the chemistry of this neutralization.

    (b) What rate of chlorine addition (g/hr) would be required to neutralize an alkaline waste stream of 35 L/hr containing 1200 mg/L sulfide?

    (c) What problems, apart from excessive reagent cost, might the operator of such a neutralization plant face on the addition of a large excess of chlorine to the sulfide-containing waste stream?

7. With a 2.5:1 mole ratio of sodium hypochlorite to potassium cyanide it is possible to convert the toxic cyanide ion all the way to harmless nitrogen gas:

$$2 \; KCN + 5 \; NaOCl + H_2O \rightarrow N_2 + 5 \; NaCl + 2 \; KHCO_3$$

What amounts of sodium hydroxide and chlorine would be required to neutralize 100 L of effluent containing 30 mg/L potassium cyanide in this way?

8. Seawater at 25°C, assumed to contain only NaCl at 3.00% by weight (density 1.0197 g/mL), is to be desalted to 10% by volume of a freshwater permeate still containing ca. 200 ppm by weight salt.

    (a) What would be the approximate osmotic pressure (Π) of the raw seawater used (i.e., the pressure which would have to be exceeded to produce desalinated water)?

$$\Pi = i[S]RT \text{ (approx.), where}$$

    i = no. of ions (species) produced on dissolving solute in water
    $[S]$ = molarity of solute
    R = 0.08206 L atm/deg mol
    T = absolute temperature, K.

    (b) What would be the molar concentration of salt in the rejected brine (i.e., in the 90% that is rejected), and the approximate osmotic pressure of the reject?

(c) If 50% of the raw seawater was to be retained as freshwater permeate, what would be the salt concentration, and the approximate osmotic pressure of the rejected brine?

(d) Give two reasons why 10% freshwater retention may be more practical for routine reverse osmosis operations than 50% retention.

9. (a) What absolute air pressure would just be sufficient to allow aeration by compressed air fed to a perforated pipe laid at the 10-m-deep dredged floor of a stagnant freshwater harbor. (Assume water density 1.00 g/mL.)

(b) For a given volume of air injected, would small or large bubbles give better oxygen absorption. Why or why not?

(c) For an optimum bubble size in the same aeration system with both gases, what volume of pure oxygen would be theoretically required to obtain equivalent dissolved oxygen in the harbor waters as with 1000 L of air under the same conditions?

(d) Why might oxygen bubbles be completely absorbed, under microbubble conditions, when air microbubbles continue to break the surface?

FURTHER READING

American Water Works Association, "Water Quality and Treatment: A Handbook of Public Water Supplies," 3rd ed., McGraw-Hill, New York, 1971.

L.O. Benefield, J.F. Judkins, and B.L. Weand, "Process Chemistry for Water and Wastewater Treatment." Prentice-Hall, Englewood Cliffs, NJ, 1982.

A.V. Bridgewater and C.J. Mumford, "Waste Recycling and Pollution Control Handbook." Van Nostrand-Reinhold, New York, 1980.

D.W. Sundstrom and H.E. Klei, "Wastewater Treatment." Prentice-Hall, Englewood Cliffs, NJ, 1979.

N.C. Thanh and A.K. Biswas, eds., "Environmentally-sound Water Management." Oxford University Press, New York, 1990.

T.J. Tofflemire and F.E. Van Alstyne, Land Disposal of Wastewater. *J. Water Pollut. Control Fed.* 46(6), 1201 (1974).

P.A. Vesilind, "Treatment and Disposal of Wastewater Sludges." Ann Arbor Sci. Publ., Ann Arbor, MI, 1975.

REFERENCES

1. R.N. McNeeley, V.P. Neimanis, and L. Dwyer, "Water Quality Source Book: A Guide to Water Quality Parameters." Inland Waters Directorate, Water Quality Branch, Ottawa, 1979.
2. J.E. McKee and H.W. Wolf, "Water Quality Criteria," 2nd ed., Publ. No. 3-A. California State Water Quality Control Board, Sacramento, CA, 1971.
3. "Canadian Drinking Water Standards and Objectives 1968," Joint Committee on Drinking Water Standards, Advisory Committee on Public Health Engineering, and Canadian Public Health Assoc., Ottawa, 1969.
4. D.K. Todd, "The Water Encyclopedia." Water Information Center, Port Washington, NY, 1970.
5. "Water Treatment Handbook" (translated from the French by D.F. Long and Co.). Degremont Co., London, 1973.
6. A Refresher on Sodium Hypochlorite, *Water Pollut. Control* 115(1), 11, Jan. (1977).
7. M.B. Hocking, *Chem. Eng. News* 61(1), 50, Jan. 3 (1983).

8. G.C. White, "Handbook of Chlorination," p. 189. Van Nostrand-Reinhold, New York, 1972.

9. J.G. Smith, *J. Chem. Educ.* **52**(10), 656 (1975).

10. B.G. Oliver and J.H. Carey, *Environ Sci. Technol.* **11**(9), 893 (1977).

11. Evidence that Chlorination May Form Mutagens, *Chem. Eng. News* **56**(3), 33, March 27 (1978).

12. E.J. Calabrese, G.S. Moore, and R.W. Tuthill, *J. Environ. Health* **41**(1), 26, July/Aug. (1978).

13. M. Peleg, *Water Res.* **10**, 361 (1976).

14. Calcium May be Key to Health, *Chem. Eng. News* **55**(16), 17, April 18 (1977).

15. R.A. Horne, "The Chemistry of Our Environment." Wiley-Interscience, New York, 1978.

16. A. Porteous, "Saline Water Distillation Processes." Longmans, London, 1975.

17. L.T. Pryde, "Environmental Chemistry: An Introduction." Cummings, Menlo Park, CA, 1973.

18. D. McBain, *Chem. Brit.* **12**(9), 281, Sept. (1976).

19. "Standard Methods for the Examination of Water and Wastewater," 13th ed., American Public Health Assoc., Washington, DC, 1971.

20. T.V. Arden, *Chem. Brit.* **12**(9), 285, Sept. (1976).

21. L. Klein, "River Pollution," Vol. 2. Butterworth, London, 1962.

22. "Cleaning Our Environment: A Chemical Perspective," 2nd ed., American Chemical Society, Washington, DC, 1978.

23. "Cleaning Our Environment: The Chemical Basis for Action." American Chemical Society, Washington, DC, 1969.

24. R. Field and C.-Y. Fan, *J. Environ. Eng. Div.(Am. Soc. Civ. Eng.)* **107**, 171, Feb. (1981).

25. C.G. Patterson and J.R. Nursall, *Water Res.* **9**, 425 (1975).

26. D.V. Ellis, "Sewage Disposal to the Sea," West. Geogr. Ser. Vol. 12, p.289. University of Victoria, Victoria, B.C., 1977.

27. S.A. Berry and B.G. Notion, *Water Res.* **10**, 323 (1976).

28. W. Whipple, Jr., *Chem. Eng. Prog. Symp. Ser.* **65**, No. 97, 75 (1969).

29. A First for Alberta, Aeration by U-tube, *Water Pollut. Control* **110**(4), 42, April (1972).

30. "Discharge of Selected Rivers of Canada." The Secretariat, Canadian National Committee for the International Hydrological Decade, Ottawa, 1972.

31. H.E. Babbit and E.R. Baumann, "Sewerage and Sewage Treatment," 8th ed., Wiley, New York, 1958.

32. L.W. Weinberger, D.G. Stephan, and F.M. Middleton, *Ann. N. Y. Acad. Sci.* **136**, 131, July 8 (1966).

33. D.L. Smith and R.V. Daigh, *J. Water Pollut. Control Fed.* **53**(8), 1272 (1981).

34. "Introduction to Popular Treatment Methods for Municipal Wastes and Water Supplies," Ontario Water Resources Commission, Toronto, ca. 1972.

35. Waste Treatment is Undergoing Rapid Change, *Can. Chem. Process.* **61**(11), 4, Nov. (1977).

36. J.A. Chittenden and W.J. Wells, Jr., *J. Water Pollut. Control Fed.* **43**(5), 746 (1971).

37. O.C. Collins and M.D. Elder, *J. Inst. Water Pollut. Control* **79**(2), 272 (1980).

38. K.L. Murphy, *Water Pollut. Control* **112**(4), 24, April (1974).

39. F.B. DeWalle, W.G. Light, and E.S.K. Chian, *Environ. Sci. Technol.* **16**, 741 (1982).

40. D.W. Hendriks, F.J. Post, and D.R. Khairnar, *Water, Air, Soil Pollut.* **12**, 219 (1979).

41. R.E. Finger, *J. Water Pollut. Control Fed.* **45**(8), 1654 (1973).

42. Fe Wastes Help Remove Phosphates, *Can. Chem. Process.* **58**(6), 44, May (1974).

43. L.A. Campbell and A.J. Horton, *Water and Pollut. Control* **110**(3), 28, March (1972).

44. Reverse Osmosis Studied for Sewage Treatment, *Water Pollut. Control* **112**(4), 15, April (1974).

45. R. Kunin and D.G. Downing, *Chem. Eng.(N.Y.)*, **78**, 67, June 28 (1971).

46. E.J. Middlebrooks, ed., "Water Reuse." Ann Arbor Sci. Publ., Ann Arbor, MI, 1982.

47. J.L. Jones, D.C. Baumberger, Jr., F.M. Lewis, and J. Jacknow, *Environ. Sci. Technol.* **11**(8), 968 (1977).

48. D.W. Osborne, *J. Proc. Inst. Sewage Purif.* **3**, 195 (1962); cited by L. Klein, "River Pollution," Vol. 3. Butterworth, London, 1966.

49. Spinoff Process Aids Sewage Treatment, *Environ. Sci. Technol.* **5**(9), 756 (1971).

50. Porteous Unit Ready For Start-up, *Environ. Sci. Technol.* **2**(11), 1068 (1968).

51. Wet Air Oxidation Processes, *Chem. Can.* **29**(5), 15 (1977).

52. H.F. Lund, ed., "Industrial Pollution Control Handbook." McGraw-Hill, New York, 1971.

53. Energy from Sludge, *Chem. Br.* **19**(2), 98, Feb. (1983).

54. M. Campbell and W. Glenn, "Profit from Pollution Prevention." Pollution Probe, Toronto, 1982.

55. Process Extracts Aluminum from Flyash, *Chem. Eng. News* **60**(16), 28 (1983).

56. W. Graham and R. Lama, *Can. J. Chem. Eng.* **41**, 162, Aug. (1963).

57. M.B. Hocking and G.W. Lee, *Fuel* **56**(7), 325 (1977).

58. F.P. Sebastian and P.J. Cardinal, Jr., *Chem. Eng. (N.Y.)* **75**, 112, Oct. 14 (1968).

59. A.E. Franzen, V.G. Skogan, and J.F. Grutsch, *Chem. Eng. Prog.* **68**(8), 65, Aug. (1972).

60. F.L. Parker and P.A. Krenkel, "Engineering Aspects of Thermal Pollution." Vanderbilt University Press, Nashville, TN, 1969.

61. P.T. Doyle and G.J. Benkly, *Hydrocarbon Process.* **52**(7), 81, July (1973).

62. CIL to Test New Sewage Unit, *Can. Chem. Process.* **59**(10), 32, Oct. (1975).

63. P.B. Dejohn and R.A. Hutchins, *Text. Chem. Color.* **8**(4), 34, April (1976).

64. T.J. Chessell, *Can. Chem. Process.* **62**(6), 42, June (1978).

65. R.F. Curran, *Chem. Eng. Prog., Symp. Ser.* **64**, No. 90, 162 (1968).

66. G.R. Mapstone and B.R. Thorne, *J. Appl. Chem. Biotechnol.* **28**(2), 135, Feb. (1978.)

67. A.B. Wheatland, C. Gledhill, and J.V. O'Gorman, *Chem. Ind. (London)*, p. 632, Aug. 2 (1975).

68. Waste Exchange Achieves Initial Target, *Can. Chem. Process.* **63**(2), 12, Feb. (1979).

69. Canadian Recycling Market, biweekly published by Corpus, Don Mills, Ontario.

70. A.K. Coleman, *Chem. Ind. (London)*, p. 534, July 5 (1975).

71. D.C. Wilson, *Chem. Br.* **18**(10), 720 (1982).

72. Cement Kilns Can Destroy Toxic Chlorinated Compounds, *Can. Chem. Process.* **61**(5), 6, May (1977).

73. Waste Solidification, *Chem. Br.* **10**(3), 105, March (1974).

74. W.J. Green, G.F. Lee, and R.A. Jones, *J. Water Pollut. Control Fed.* **53**(8), 1348 (1981).

75. R.O. Van Everdingen and R.A. Freeze, "Subsurface Disposal of Waste in Canada," Tech. Bull. No. 49. Department of the Environment, Ottawa, 1971.

76. B.H. Keswick and C.P. Gerba, *Environ. Sci. Technol.* **14**(11), 1290 (1980).

77. J.R. McBride, E.M. Donaldson, and G. Derkson, *Bull. Environ. Contam. Toxiol.* **23**, 806 (1979).

78. Waste Sites Pose Risk to Water Supply, *Chem. Eng. News* **58**(40), 6, Oct. 6 (1980).

Cement dissolving pad for producing a saturated aqueous solution of sodium chloride from solar salt. The conveyor for loading the pad is visible at the top right.

6

NATURAL AND DERIVED SODIUM AND POTASSIUM SALTS

Ye are the salt of the earth: but if the salt have lost
its savour, wherewith shall it be salted?
                                   —*Matthew, ca. 65* A.D.

6.1. SODIUM CHLORIDE

Sodium chloride, or common salt, is among the earliest of the chemical commodities produced. Its production was prompted by its essential requirement in the diet and the scattered accessibility of land-based supplies. The word salary is derived from the Roman "salarium", which was a monetary payment given to soldiers for salt purchase to replace the original salt issue. While the initial production and harvesting of sodium chloride was from dietary interests, today this application represents less than 5% of the consumption, and uses as a chemical intermediate far exceed this (Table 6.1). The wide availability of sodium chloride has contributed to the derivation of nearly all compounds containing sodium or chlorine from this salt, and to the establishment of many large industrial chemical operations adjacent to major salt deposits. Three general methods are in common use for the recovery of sodium chloride, which in combination were employed for the worldwide production of 180 million tonnes of this commodity in 1994 and slightly more in 1990 (Table 6.2).

6.1.1. Solar Salt

Outdoor recovery of sodium chloride by evaporation of seawater or natural brines is carried out in areas which have a high evaporation rate and low rainfall. Slow evaporation rates in temperate countries restrict the use of natural evaporation to only a small fraction of the production in these areas.

167

TABLE 6.1   Sodium Chloride Use Profile for Canada and U.S.A., 1976[a]

	U.S.A.		Canada	
Application	Thousand tonnes	% of total	Thousand tonnes	% of total
Highway snow and ice control	8,101	20.1	2,224	49.3
Chlorine and sodium hydroxide production	19,003	47.2	1,348	39.9
Sodium carbonate (soda ash)	3,684	9.1	616	13.6
All other chemicals	1742	4.3	—	—
Fishing industry	—	—	97	2.1
Food processing	1,080	2.7	91	2.0
Livestock feeds	1,746	4.3	47	1.0
Meatpacking, tanning	524	1.3	49	1.1
Pulp and paper	193	0.5	38	0.8
Table salt	905	2.2	—	—
Textile and dyeing	185	0.5	3	0.1
Miscellaneous uses	3,146	7.8	3	0.1
Total	40,311	100.0	4,516	100.0

[a]Calculated from data in *Minerals Yearbooks* [1,2], and the *U.N. Statistical Yearbook* [3].

However, the low capital, energy, and labor costs of this process still make this a dominant and attractive method for world salt recovery. The fraction of the world's salt which is produced by solar evaporation is about 45%. Some examples of contributions to this total by various countries are Namibia (Southwest Africa), 100%; Ethiopia, 88%; Portugal, 42%; Spain, 41%; France, 22%; Italy, 15%; Germany, 11%; U.S.A., 4%; Canada, 0% [5].

Feed brines for solar salt recovery may be derived from seawater having a salinity about 3.5% or from one of the enclosed seas or natural salt lakes of higher salinities (Table 6.3). The Caspian Sea, salts of which are about 25% lower in chloride and about three times higher in sulfate than ordinary sea salt, is not used for solar salt production because of its much lower salinity (1.3%) and its temperate location. The brine is moved, with tidal assistance if feasible, into large shallow initial evaporation ponds where solar evaporation is allowed to proceed until it reaches a specific gravity of about 1.21 g/cm. It is then pumped to lime ponds where further evaporation, sometimes with the absorptive assistance of added dyes, causes crystallization and precipitation of most of the calcium sulfate (for phase diagrams, see [8]). When the brine density reaches about 1.24, gates are opened to allow gravity flow of the brine to the harvesting or crystallizing ponds where sodium chloride crystallization is allowed to proceed until the brine density reaches 1.25 to 1.29, when some 75% of the sodium chloride will have crystallized (Table 6.4). The mother liquor, or bittern, is then run off, and the residual crystallized salt scraped into long piles or windrows to drain. This product is frequently marketed in moist form (8 to 10% water) [11]. Simple drying, either by nat-

■ **TABLE 6.2 Production of the Major World Salt (NaCl) Producers**[a]

	Thousands of metric tonnes			
	1976	1980	1990	1994
U.S.A.	40,114	36,607	37,000	39,800
China	30,000	17,280	20,000	29,700
U.S.S.R.	14,000	14,500	14,700	n/a
United Kingdom	7,900	6,586	6,430	5,700
Germany	7,495[b]	12,970[b]	15,700	12,700
France	6,416	7,100	6,610	5,440
Canada	6,398	7,029	11,300	11,500
Poland	5,470	3,357	4,060	3,800
Australia	5,350	5,315	7,230	7,800
Mexico	4,591	5,990	7,140	7,460
India	4,480	7,262	9,500	9,500
Italy	4,012	5,270	4,430	3,100
Other Countries	31,199	35,488	38,900	43,500
Total	167,425	164,754	183,000	180,000

[a]From 1982 and 1994 *Canadian and U.S. Minerals Yearbooks* [2,4].
[b]For the former West Germany.

ural or artificial means, produces an industrial-grade solar salt consisting of about 95% sodium chloride. Yields range from about 50 tonnes per hectare per year (or about 20 tonnes per acre) in the San Francisco Bay area [12], 150 to 175 tonnes per hectare per year for the May to September season from the Great Salt Lake, and 12 to 185 or more from seawater for areas with higher evaporation rates [13].

To obtain a food or dairy grade of product the solar salt is redissolved and the solution evaporated to induce crystallization, which greatly reduces the calcium sulfate concentration. Pure salt is hygroscopic and will tend to cake with changes in relative humidity. To avoid this, salt crystals are coated with 0.5 to 2% of finely powdered hydrated calcium silicate, magnesium carbonate, or tricalcium phosphate to give the crystals free-flowing characteristics. Iodized salt for table use would have, in addition, 0.01% potassium iodide plus stabilizers added to the product. Cattle blocks comprise highly densified dyed aggregations of salt, to which various micronutrients are added according to local soil deficiencies of the marketing area.

The bittern from solar salt production containing 300–400 g/L dissolved solids, and now relatively enriched in the less concentrated salts, may be either discarded or further worked to recover other elements of value. Brine from the Great Salt Lake, for instance, is processed for magnesium chloride hexahydrate recovery [10] for later conversion to metallic magnesium [11]. Dead Sea brines, which are processed primarily for potassium chloride (potash), are also worked for sodium and magnesium chlorides and derived products such as bromine and hydrochloric acid [14] (Sections 6.2.2 and 8.6).

TABLE 6.3. Salinities and Compositions, in Percent by Weight of Dissolved Salts, for Some Major Brine Sources of Solar Salt

	Seawater[a]	Red Sea[b]		Searles Lake, California[c]		Dead Sea[a,d]		Great Salt Lake,[e] Utah
		Surface	Below 2000 m	Upper	Below 40 m	Upper	Below 100 m	
Salinity, %:	1 to 5	3.6–4.1	25.6	34.83	34.60	19.5–25	26.7	13.8–27.7
Average	3.0	3.8	—	—	—	—	—	25.0
Varies with	place, depth	depth				time, depth		
Density, g/mL	1.025		—	1.30	1.31	1.17	1.24	1.10–1.22
NaCl	77.8	77.8	89.1	46.66	46.97	31.3	40.2	84.4
$MgCl_2$	10.9	10.9	—	—	—	48.4	51.5	3.8
$MgSO_4$	4.7	4.7	trace	—	—	—	—	8.4
$CaSO_4$	3.6	3.6	trace	—	—	0.56	0.02	—
KCl	2.5	2.5	1.6	13.92	8.67	4.35	4.69	3.1
$CaCO_3$	0.34	0.34	—	—	—	—	—	0.3
$MgBr_2$	0.22	0.22	2.4	—	—	1.94	2.29	—
$CaCl_2$	—	—	5.3	—	—	13.50	14.33	—
Na_2SO_4	—	—	—	20.67	19.51	—	—	—
Na_2CO_3	—	—	—	13.35	18.35	—	—	—
Other	—	—	trace[f]	1.48[g]	1.52[g]	—	—	traces[h]
Total salts	8.44×10^{16}[i]	—	—	—	—	1.15×10^{10}[j]		5.44×10^{9}

[a]Calculated from data in *Encyclopaedia Brittanica* [5].
[b]Calculated from data of Degens and Ross [6].
[c]Data from Shreve and Brink [7]. Composition of the nearly dry Searles Lake salts is very similar to that of nearby Owens Lake, from which it was separated by evaporation [8].
[d]Calculated data [3] agree quite closely with the analyses reported by Epstein.
[e]Calculated from data of Rankama and Sahama [8], Hutchinson [9], and Kirk-Othmer [10]. Traces of magnesium carbonate and other more minor constituents are also present.
[f]Predominantly of other sulfates.
[g]Includes predominantly $Na_2B_4O_7$, Na_3PO_4, and NaBr.
[h]Includes ionic strontium and borate.
[i]Corresponds to a volume of 1.87×10^7 km^3.
[j]It has been estimated that a total of some 910,000 tonnes of salts per year is entering via the Jordan River.

◼︎ **TABLE 6.4   Properties of Sodium Chloride
and Aqueous Solutions**

Density:	2.165 g/cm^3
Melting point:	800.8°C
Boiling point:	1465°
Solubility:	35.7 g per 100 g water at 0°
	39.8 g per 100 g water at 100°

Saturated sodium chlorine brine has a boiling point of
108.7°, and contains 28.41% NaCl. At 25°C it has a
specific gravity of 1.1978, and contains 26.48° NaCl.

6.1.2. Sodium Chloride by Conventional Mining

By the evaporation in geologic history of large inland seas or land-locked lakes
of large drainage basins, extensive subterranean deposits of sodium chloride
and other salts in layers of cumulative thicknesses as great as 400 m have
been laid down in many parts of the world. Deposits less than 500 to 600 m
below the surface, in well-consolidated strata, are usually economic to work
by conventional mining techniques. In Canada these accessible deposits
amount to about 74% of the total, and in the U.S.A. about 35% of the total [1,2].

The extent and quality of the deposit is determined from a pattern of core
drillings (samples of the subterranean rock brought to the surface) which are
obtained for the specific area of interest. Access to the horizontal or near
horizontally bedded salt is obtained by sinking a vertical shaft of about 5 m
in diameter. This is then mined by undercutting, drilling, blasting and underground haulage to move the rock salt to the shaft and then lift it to the surface.
Deposits of this type are quite dense so that underground cavities can range
up to 15 m in each dimension. However stout supporting pillars of 20 to 60
m^2 are required at regular intervals because of the plastic (flows under pressure) nature of sodium chloride. In this way salt recoveries from the conventional mining operation may range from 25 to 40% and rarely as high as
60% of that present in the deposit.

Crushing and screening of the mined salt to size may either be carried out
underground or at the surface. Mechanical and electronic sorting then gives
an unrefined but beneficiated product of better than 95% NaCl (occasionally
better than 99%). Most of the Canadian and American mined rock salt is
employed as mined in highway deicing, although a further significant fraction
of the American rock salt is consumed in the production of chlorine, sodium
hydroxide, and other chemicals (Table 6.1). For the small quantities of higher
purity sodium chloride made from this raw material the beneficiated material
is dissolved in water, concentrated, and crystallized. The purified sodium chloride crystals are filtered and dried. Further finishing steps for marketing are
as described in Section 6.1.1.

6.1.3. Solution Mining of Sodium Chloride

To recover salt from underground deposits located more than about 600 m below the surface, or with less well-consolidated intervening strata, solution mining is usually the method of choice. This method provides the dominant source of brine for Canadian chloralkali operations. Low production rates may be accomplished by sinking a single drilled well and fitting this with a casing and a smaller bore concentric tube fitted inside this. Tubes are usually made of fiberglass-reinforced plastic to avoid corrosion problems. Pumping water down one tube displaces brine up the other. For larger rates of production two (or more) boreholes will normally be sunk into the salt formation, and water will be pumped at high rate and pressure into one to introduce an expansion fracture connecting the two wells, a process called "fracturing." These fractures allow water passage between the wells and hence brine recovery from one of them.

Fields of wells in Canada recover brine from deposits as deep as 2000 m, and may have as many as 20 wells for saturated brine production from a single deposit. At intervals, new wells will be bored some distance removed from both the initial solution mining site and surface producing facilities. This procedure leaves intervening support columns underground to reduce the risk of subsidence. Saturated brine emerging from the underground cavity will contain 26% by weight sodium chloride, small amounts (0.01 to 0.1%) of calcium and magnesium salts and possibly traces of ferrous and sulfide ions. Insoluble matter will have largely remained underground. This brine will be suitable as obtained, or it may require some specific chemical treatment to purify it before use (see Section 8.1).

To obtain a high-purity crystalline salt product from this solution, it is aerated to remove most of the sulfide as hydrogen sulfide and then chlorinated to oxidize any residual sulfide and ferrous ions (Eq. 6.1–6.3).

$$Cl_2 + H_2O \rightarrow HCl + HClO \qquad\qquad 6.1$$

$$HCl + HClO + 2\ FeCl_2 \rightarrow H_2O + 2\ FeCl_3 \qquad\qquad 6.2$$

$$4\ HClO + S^{2-} \rightarrow SO_4^{2-} + 4\ H^+ + 4\ Cl^- \qquad\qquad 6.3$$

Calcium, magnesium, and ferric ions are then removed as their hydroxides by treating the brine with small amounts of sodium carbonate (soda ash) and sodium hydroxide (caustic soda) and filtering or allowing the mixture to settle (Eq. 6.4–6.6) [12].

$$CaCl_2 + Na_2CO_3 \rightarrow CaCO_3\downarrow + 2\ NaCl \qquad\qquad 6.4$$

$$MgCO_3 + 2\ NaOH \rightarrow Mg(OH)_2\downarrow + Na_2CO_3 \qquad\qquad 6.5$$

$$FeCl_3 + 3\ NaOH \rightarrow Fe(OH)_3\downarrow + 3\ NaCl \qquad\qquad 6.6$$

Water is then removed from the treated brine by multiple-effect vacuum evaporation with low-pressure steam providing the heat for the first effect (Fig. 6.1).

To evaporate water from brine by heating with steam, one can only obtain somewhat less water evaporation than the mass of steam consumed (a heat

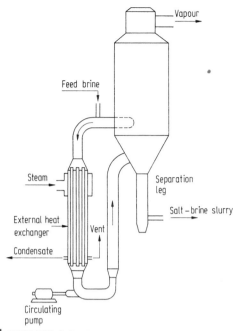

FIGURE 6.1 Forced circulation evaporator such as that used for simultaneous brine evaporation-crystallization. Mechanical movement of brine past the heat exchange surface avoids decreased heat transfer efficiency from crystallization on this surface. Constructed of Monel or Monel-clad steel for parts contacting brine. (Reprinted from Kirk-Othmer [10], with permission.)

transfer efficiency limitation). However, application of a partial vacuum to the heated brine in the first stage and the use of hot water vapor from this stage to heat another stage and so on allows far more efficient use of steam and the evaporation of progressively more water for each tonne of primary steam used as the number of stages is increased (Fig. 6.2). Further thermal efficiencies are introduced by countercurrent movement of brine and steam (Chap. 1) and by use of the still warm exit brine to prewarm incoming brine. In combination, these measures can give evaporation efficiencies of the order of 1.75, 2.5, 3.2, and 4.0 tonnes of water per tonne of steam from two, three, four, and five stages of evaporation, respectively [15]. Capital costs increase with the number of stages, and the percent improvement obtained with each additional effect decreases so that the maximum practical number of effects is 12. Partial vacuum for the last effect (stage of evaporation) is economically provided without moving mechanical devices by the use of cooling water with a barometric condenser, and a steam or water ejector for removal of noncondensible gases (Fig. 6.3).

A stream of a saturated brine suspension of crystals is continuously withdrawn from each evaporator-crystallizer, and the salt crystals separated on a continuous rotary filter with return of the brine to the evaporator. Much of the salt may be marketed in moist condition, or it may be passed through a drier countercurrently to heated air to give "vacuum salt" of typically 99.8 to 99.9% purity [17].

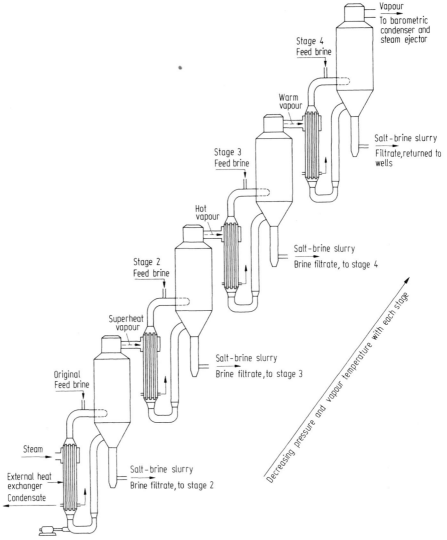

FIGURE 6.2   Quadruple effect evaporator-crystallizer. Pressure and boiling temperature decrease from left to right, temperature decreases about 20 to 30°C across the four effects. Feed brines to each effect may be directly from brine wells, or may be fed cascade fashion from the filtrate to the salt brine slurry of the previous effect, or a combination of these methods. (Adapted from Kirk-Othmer [10], with permission.)

There must be some net throughput of brine to waste through the evaporator circuit to avoid gradual buildup of impurities. In some operations, in fact, when the feed brine is of consistently high purity the bleed rate of brine mother liquor can be controlled at an appropriate level by continuously keeping track of calcium and magnesium concentrations, to give a finished sodium chloride purity of 99.5% or better, without requiring brine pretreatment [12].

The crystal size of vacuum salt is consistently quite small because of the high evaporation rates and turbulence in the evaporator crystallizers which makes this product less suitable for some uses. For this reason there is still a

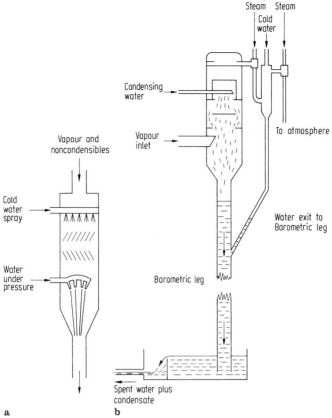

Steam   Steam

Cold water

Condensing water

Vapour inlet

To atmosphere

Vapour and noncondensibles

Cold water spray

Water under pressure

Water exit to Barometric leg

Barometric leg

Spent water plus condensate

a   b

FIGURE 6.3   (a) Detail of a water ejector using a supply pressure of 1.4–1.7 × 10⁵ kilopascals (gauge pressure; 20 to 25 psig) or higher. (b) Diagram of a barometric condenser plus steam-driven ejector to remove and condense vapor and provide the reduced pressures required for the last stage of evaporation. With steam pressures of 2.8–3.5 × 10⁵ kilopascals (gauge pressure; 40 to 50 psig) or higher and six stages of ejection absolute pressures in the range of 0.005 to 0.50 mm Hg are available. (See [16] for further information.)

small segment (ca. 1%) of the brine evaporation business conducted at atmospheric pressure in open pans, the "Grainer" process. Here, after special treatment to remove calcium sulfate, the brine evaporation takes place in a quiescent manner at 95°C giving large crystals of high surface area and purity suitable for the dairy industry. This process is, however, more extravagant of heat than the vacuum process. Alternatively, for the large dense crystals required for the regeneration of ion-exchange water softeners, quantities of vacuum salt may simply be heated to about 815°C and the melt cooled and broken up for this purpose.

6.1.4. New Developments in Sodium Chloride Recovery

As pressure on existing freshwater supplies tightens, seawater desalination plants, using multieffect vacuum distillation or reverse osmosis, are required in increasing numbers for provision of freshwater. The residual evaporated brines from these plants contain much higher salt concentrations than ordinary seawater and this is also obtained near potential salt markets. These features are likely to encourage further development of vacuum or solar evaporation salt recovery operations to work these brines in proximity to the desalination plants. In this way various salts may be more profitably recovered from these artificially enriched seawaters for reasons similar to the present incentives to use the rich natural brine sources for sodium chloride production (Table 6.3). Similar energy savings should be obtained.

6.2. POTASSIUM CHLORIDE

Potash, as potassium chloride or any other potassium salts are collectively referred to in industry, is the name derived from the methods originally used for recovery of potassium carbonate from wood ashes. About 1 tonne of crude potassium carbonate was recovered from the water leachate of the ash generated from the burning of some 400 tonnes (more than 200 cords) of hardwood [18]. Perhaps because of the availability of a large wood supply, Canada was the world's largest potash exporter during much of the first half of the nineteenth century.

At this time purified potassium carbonate was primarily of value as an ingredient in glass making. The discovery of mineral potassium chloride at Stassfurt, Germany, in 1852 rapidly provided competition to the wood ash leachate industry since this mineral could be readily converted to potassium carbonate by the Leblanc process which was already in commercial scale operation (see Chap. 7). The mineral potash is also more specifically referred to as "muriate of potash" (muriatic acid = hydrochloric acid) for potassium chloride, and "sulfate of potash" for potassium sulfate in reference to these particular salts.

6.2.1. Potassium Chloride Production and Use Pattern

Prior to 1960, world production of potassium chloride was dominated by the U.S.S.R., U.S.A., East and West Germany, and France (Table 6.5). But since the incidental discovery of potash mineralization in Saskatchewan during oil prospecting in 1943, and the first commercial production there in 1962, Canada became the world's leading producer in 1968 and 1969. Canada is at present the leading exporter and stands second after the U.S.S.R. in potash production. World production volume doubled in the 10 years from 1961 to 1971 from 9.82 to 19.1 million tonnes of K_2O equivalent, and rose to 25.8 million tonnes in 1976. Since then, however, production growth appears to have levelled, with only 25.3 million tonnes produced in 1994. At 1976 levels of production, present Canadian reserves of 107×10^9 tonnes of potassium chloride (67×10^9 tonnes, K_2O equivalent), primarily in sylvinite and carnallite minerals, would be sufficient to supply the world demand for some 2600 years. West German and Russian reserves appear to be of a similar order of magnitude [20].

Some 95% of the potassium chloride is consumed directly in fertilizer applications. Most of the rest is used to make potassium hydroxide (see

TABLE 6.5   Production of Potash by the Major World Producers Given as the Mass of K$_2$O Equivalent (see text)^{a,b}

	1976		1994	
	Thousand tonnes	% of total	Thousand tonnes	% of total
Canada	5,000	19.4	6,990	27.7
East Germany	3,175	12.3	2,650^c	10.5
France	1,606	6.2	870	3.4
Israel	680	2.6	1,310	5.2
China	490	1.9	29	0.11
Spain	530	2.1	686	2.7
U.S.A.	4,010	15.6	1.400	5.5
U.S.S.R.	8,260	32.0	8,000^c	35.6
Germany	2,032^d	7.9	2,310^c	9.1
World Total	25,783	100.0	25,274	99.9

^{a}Data from 1976 and 1994 *Minerals Yearbooks* [1, 19].
^{b}Dominant product is potassium chloride, e.g., for Canada 99% of the potash produced, but small amounts of potassium sulfate and potassium nitrate are also produced and are included in these figures.
^{c}1990 production.
^{d}Production of the former West Germany.

Chap. 8) which is ultimately employed to manufacture special types of glass. It is also used for the saponification of fats to produce liquid soaps, among other relatively small-scale uses.

6.2.2. Potassium Chloride Recovery from Natural Brines

While it would seem feasible to use seawater bitterns from solar salt facilities (Section 6.1.1) as the raw material for potassium chloride recovery, there are as yet no commercial processes operating on this basis [21]. Virtually all present production facilities use natural brines from sources already relatively richer in potassium than raw seawater (Table 6.3), and all have to cope somehow with the solubility differentials with temperature (Table 6.6). In general

TABLE 6.6   Change of Aqueous Solubility of Potassium Sulfate with Temperature^a

Temperature (°C)	Solubility (g/100 mL)
0	7.35
20	11.11
40	14.76
60	18.17
80	21.40
100	24.10

^{a}Data selected from *Chemical Engineers' Handbook* [22].

it is expected that mixtures of salts, such as in seawater, will tend to come out of solution in the reverse order of their solubilities, though this will also be strongly affected by their initial relative concentrations and somewhat by common ion effects. Thus, a precipitation sequence something like calcium carbonate > gypsum ($CaSO_4 \cdot 2H_2O$) > sodium chloride > magnesium sulfate ($MgSO_4 \cdot 6H_2O$) > magnesium chloride ($MgCl_2 \cdot 6H_2O$) > potassium chloride > sodium bromide would normally be expected. For this reason recovery of the less concentrated minerals, particularly if they are more soluble than sodium chloride, requires progressively more complex processing than is required for sodium chloride recovery.

Searles Lake, California, is a remnant of a former salt lake which is now worked for minerals. It consists primarily of beds of salts containing brine in the interstices, the brine lying from 15 cm below to 15 cm above the beds of salts depending on season and extent of flooding [23]. By comparison with seawater these brines contain far higher concentrations of sodium and potassium ions, much of it present as the sulfate, not chloride, and negligible concentrations of calcium and magnesium (Table 6.3). Initial triple-effect evaporation allows separation of mainly crystals of sodium chloride and Burkeite ($Na_2CO_3 \cdot 2Na_2SO_4$) with a small amount of dilithium sodium phosphate. With a complex series of subsequent steps, sodium sulfate (Glauber's salt) and sodium carbonate (soda ash) are recovered, and the lithium salt is converted to lithium carbonate and phosphoric acid for recovery of both components. Brines from the initial evaporation are then rapidly chilled (by cooling water and reduced pressure) to bring down potassium chloride crystals on a scale of some 675 tonnes/day, and leaving a supercooled solution of sodium borate. Seed crystals and more prolonged crystallization time produce crystalline sodium borate as a further product. Full details have been described [13].

Solar evaporation, from primary and secondary ponds of 100 and 30 km^2 in extent, the initial stage for potassium chloride recovery from Dead Sea brines [21]. The smaller number of constituent ions present in these waters significantly simplifies salts recovery, and the fact that they contain nearly twice the relative potassium chloride concentration of seawater also improves profitability. Developed from a process which was first operated in 1931, evaporation in the first pond reduces the volume of the brine to about one-half of the initial volume and brings down much of the sodium chloride together with a small amount of calcium sulfate (Fig. 6.4). The concentrated brines are then transferred to the secondary pond where evaporation of a further 20% of the water causes carnallite ($KCl \cdot MgCl_2 \cdot 6H_2O$) and some further sodium chloride to crystallize out. With care a 95% potassium chloride product on a scale of some 910,000 tonnes/year is obtained either by countercurrent extraction of the carnallite with brines, or by hot extraction of potassium chloride from the sylvinite matrix followed by fractional crystallization for its eventual recovery [14].

The "end brines" from the secondary evaporation pond, consisting largely of concentrated calcium and magnesium chlorides, are mostly returned to the sea. However, some 73,000 tonnes of sodium chloride, 900 tonnes of magnesium chloride and with some chemical conversion 23,000 tonnes of magnesia (MgO), 91,000 tonnes of hydrochloric acid, and 18,000 tonnes

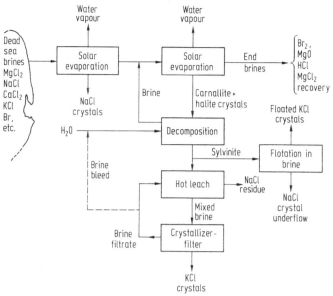

FIGURE 6.4 Flow sheet for potassium chloride recovery from Dead Sea brines.

of bromine are also obtained annually from these brines. Magnesium chloride is obtained from further processing of the sodium chloride residual bittern.

Potassium is also recovered from the Great Salt Lake, Utah, but mainly as the sulfate [11].

6.2.3. Potassium Chloride Recovery by Conventional Mining and Froth Flotation

Dominant minerals which occur in potash-bearing deposits are halite (NaCl), sylvite (KCl), and carnallite (KCl·MgCl$_2$.6H$_2$O), in which the two salts appear as separate crystals ranging in size from a few microns (μm) to 25 mm. Halite and sylvite minerals also occur together as physical mixtures known as sylvinite (nNaCl·KCl; n normally about 2). Sylvite and carnallite occur both separately and in closely associated mixtures with one another. All of these potash-bearing combinations also contain varying amounts of insolubles such as clays, anhydrite (CaSO$_4$), dolomite (CaCO$_3$·MgCO$_3$), and calcite (CaCO$_3$) which variously affect potash beneficiation steps.

Recovery of sylvinite from the shallower deposits of about 1000 m depth at the northern margin of the sloping 30-m-thick deposits in Saskatchewan, Canada, is entirely by the shaft and room method. One natural obstacle to overcome was an underground 60-m-thick bed of poorly consolidated, water-saturated sand (Blairmore, and other formations) lying below the 380-m level which introduced considerable access difficulty. This was eventually solved by prolonged chilling of these formations with lithium chloride brine at −46°C after which the frozen formation was then safely penetrated by rapid excavation and placement of watertight caissons through the unstable formation before it thawed. Apart from the access difficulties for this particular deposit, conventional mining of sylvinite is conducted in a parallel manner to the mining of salt (Section 6.1.2).

The high grade of sylvinite obtained from deposits in Saskatchewan (about 40% KCl; n about 2) favored the application of the differential surface activity between the finely ground potassium and sodium chlorides present to effect a separation of the two minerals, rather than the considerably more energy-intensive solution of using fractional crystallization technology [12]. This process marks an unusual application of froth flotation technology, commonly applied to the concentration of metal ores (Chap. 13), and here employed in the beneficiation of a water- soluble mineral. Mined sylvinite lumps are first dry crushed in gyratory or roll crushers and then ground to a fine pulp in saturated brine in a ball mill (Fig. 6.5). After classification in a cyclone, the coarse material is returned for regrinding and the fine pulp is deslimed (fine clays removed) by rigorous agitation with brine and removal of the clay-laden brine fraction. Brine is recycled after clay removal in thickeners (Fig. 6.6). The cleaned pulp is then treated with guar gum, a suppressant which coats any remaining clay to prevent adsorption of the amine collector by clays. For each tonne of ore processed, 100 g of a tallow amine (a long alkyl chain, primary amine) is added to induce attachment of the potassium chloride particles to air bubbles, which does not affect the surface properties of sodium chloride. Also 110 g/tonne of a polyethylene glycol is added to the mixture to encourage formation of a stable froth. An anionic (fatty acid salt) collector may be used if it is desired to reverse the process and float sodium chloride to give a potassium chloride underflow [18].

Froth flotation of the surface-sensitized pulp by vigorous aeration and agitation in saturated brine (density ca. 1.18 g/mL), first in a series of

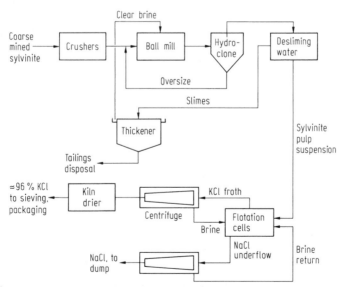

FIGURE 6.5   Flow diagram of the separation of fertilizer grade potassium chloride (sylvite) from sylvinite by froth flotation in saturated brine.

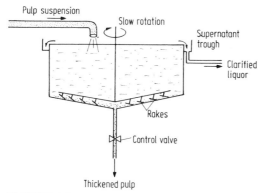

FIGURE 6.6 Diagram of a single deck mechanical thickener for producing a clear supernatant and a solids enriched slurry from a pulp suspension in water. Slow-moving rakes transport sedimented solids toward thickened pulp outlet, and an exit valve for this stream allows control of the solid/liquid ratio.

"rougher" cells with further refinement in "cleaner" units, produces a stable froth consisting almost entirely of beneficiated potassium chloride and an underflow of sodium chloride (particle densities of 1.984 and 2.165 g/mL) respectively. Potassium chloride recoveries from sylvinite by this procedure are 90 to 95% complete [18].

Flotation streams are individually centrifuged (Fig. 6.7) for separate recovery of the potassium chloride and sodium chloride and the brine is returned, usually via a thickener, to the flotation circuit for reuse. At present the sodium chloride is simply dumped. Potassium chloride is dried in gas-fired rotary kilns, and then sized by sieving into one of the four commercial grades: Granular, Coarse, Standard, and Soluble, all of which fall into the range of 60.2 to 60.7% K_2O equivalent purities. Potash dust from the drying operation is captured, dissolved in hot brine, and chilled under vacuum to cause purified potassium chloride crystals to form. After filtration and drying, this produces a further Soluble grade of 62.5% K_2O purity. Re-solution of this material

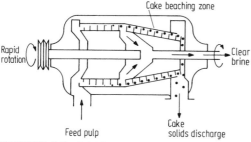

FIGURE 6.7 Solid bowl continuous centrifuge for separation of a solid suspension from a liquid. With sodium chloride or potassium chloride slurries in brine it is capable of producing a dry cake of 92 to 99% solids.

and crystallization from water again gives Refined, 99.9% pure potassium chloride, which is used primarily in chemical (not fertilizer) applications.

The bulk of the fertilizer market demand is for the two coarse grades, necessitating some compaction of the finer grades to produce these. Passage of the finer Standard and Soluble grades of material through closely spaced steel rollers under pressure forms a fused sheet of potassium chloride, because of the plastic nature of this salt. This fused sheet is then broken up and rescreened to raise the proportion of Granular and Coarse products produced.

Potassium chloride as a component in fertilizers is specified in terms of its "K_2O equivalent." On this basis 100% pure potassium chloride is equated to 63.18% K_2O equivalent. The approximately 96% pure potassium chloride product of the flotation separation is thus equivalent to 60.7% K_2O (see Chap. 9). Chemical grades correspond to the once-crystallized, soluble product of about 99.5% KCl, and 99.95% KCl which is the Refined, twice crystallized material. The price differential, 60 to 62.4% K_2O equivalent at 108 to 112 US\$ per tonne, and 99.95% KCl at 115 to 137 US\$ per tonne (1995 prices [24], range depends on quantities and contracts) is sufficient to cover the cost of the additional processing for the small amount of chemical grade potash produced. Together with the decreased environmental impact which results from dust containment, which is the source of the crystallization feed, these steps are normally considered to be worthwhile by most operators.

6.2.4. Solution Mining of Potassium Chloride

Recent innovative technology has been successfully applied to the preferential solution mining of potassium chloride from the sylvinite deposits 215 m thick and lying some 1600 m below the surface situated toward the southern extremity of the bed underlying much of southern Saskatchewan. These beds would at best be only marginally profitable to mine by conventional means. This unique operation takes advantage of the differential which exists between the hot and cold water solubilities of sodium chloride and potassium chloride (Table 6.7), to recover a far larger proportion of the potassium chloride than the sodium chloride from the deposit [25]. Present capacity of this particular operation is 1.35 million tonnes/year [26].

TABLE 6.7 Aqueous Hot and Cold Solubility Differentials Between Potassium Chloride and Sodium Chloride[a]

Temperature (°C)	Solubility (g/100 mL H_2O)	
	Potassium chloride	Sodium chloride
0	27.6	35.7
20	34.7	36.0
40	40.0	36.6
60	45.5	37.3
80	51.1	38.4
100	56.7	39.8

[a]Data selected from *Chemical Engineers' Handbook* [22].

The feature of this process which distinguishes it from conventional solution mining for sodium chloride lies in its application of a heated, mixed sodium chloride/potassium chloride brine, rather than freshwater at ambient temperature for salt dissolution. The warm extracting solution is pumped into the ore body via a concentric pair of pipes placed in a drill hole such that the reservoir brine temperature is maintained at 45°C or higher (Fig. 6.8). Natural formation temperatures aid in maintaining these temperatures in a large reservoir. By this means, some sodium chloride but significantly more potassium chloride is dissolved from the formation and brought to the surface via the "up-brine" piping for potassium chloride recovery. Any suspended sediment is removed from the mixed brine in thickeners plus a filtration circuit. Following this it is concentrated to 99% saturation at 100°C in two sets of quadruple-effect vacuum evaporators operated in a series-parallel arrangement (Fig. 6.1). Sodium chloride and any precipitated calcium sulfate are removed, hot, in a Bird solid bowl centrifuge (Fig. 6.7). Chilling the saturated, mixed brine from the centrifuges by a combination of vacuum and cooling water then produces a suspension of potassium chloride crystals in a predominantly sodium chloride brine. The crystals are separated on horizontal table filters and dried by countercurrent application of the clean flue gases produced from gas-fired steam generation, in a fluid bed drier (Fig. 6.9). This product, 99+% potassium chloride (62.5% K_2O equivalent), is somewhat higher purity than the crude froth flotation product.

Part of the centrifuged sodium chloride is processed for table and cattle block salt production. Unused sodium chloride is slurried to a 162-hectare (1.62-km^2) impounding basin, where wet storage minimizes windage losses. The bulk of the mother liquor, together with a small amount of make-up water, is reheated and returned to the formation for resaturation with potassium chloride. Any wastewater has been disposed of underground to a water bearing formation at the 1200–1500 m level. Dust from drying operations and the Tyler-Hummer screens, which are used for grade classification, is captured in water scrubbers. The brines produced by dust capture are returned to process streams for salts recovery.

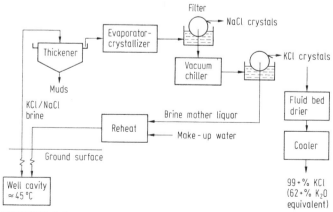

FIGURE 6.8 Potassium chloride from subterranean sylvinite by solution mining.

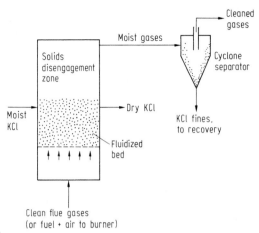

FIGURE 6.9 Fluidized bed drier. Indirectly heated air rather than the combustion gases from burning natural gas may also be used for bed fluidization, for cleaner drying action.

6.2.5. Environmental Aspects of Sodium and Potassium Chloride Recovery

Processes which recover sodium or potassium chlorides from natural brines originating from the ocean or salt lakes probably have the least salination impact on surrounding lands or water. Since many of these operations use solar evaporation they also have a low external energy requirement. The ratio of reserves to annual production rates, even for the salt lakes, is so large that there is not likely to be any noticeable salinity decrease for many years. Estimates for the Dead Sea indicate that the present production rate exceeds the rate of influx by only some 23%. However, these types of operations do not have proximity to all major markets. This provides the economic incentive to recover both sodium and potassium salts from inland deposits as well as from the obvious saline water sources.

Purification of mined inland sodium chloride has created problems with the disposal of the waste mother liquors obtained from the purification process. Deep well disposal has been practised, frequently into a deep brine aquifer to minimize subterranean impact [27]. River discharge has also been employed, but with growing demands on this freshwater resource and because of the significant negative impact of a salt discharge on many other freshwater uses, this method is now more strictly regulated.

For the potash recovery operations in Saskatchewan significant concerns exist regarding the sodium chloride stockpiles and waste disposal basins, which are presently accumulating sodium chloride at somewhere near the annual 4 million tonnes/year of potassium chloride production [28]. Precautions are being taken to avoid significant contamination of the limited surface and aquifer freshwater courses of the area by such methods as surrounding the waste disposal basins with impermeable clay dikes [29]. Aeolian salt contamination of the surrounding soils is, however, being felt to some degree despite windage control measures such as stockpile wetting and polymeric spray coatings. This may be affecting salt-sensitive plant life in the vicinity [30].

"No particular health hazards have been demonstrated . . ." to workers, from potassium chloride dust in mines and mills, although a comprehensive baseline study of these operations has not been conducted [31].

6.2.6. New Developments in Potassium Chloride Recovery

Recovery of potassium salts from natural brines using solar evaporation ponds has been made less dependent on the weather at Great Lakes Minerals and Chemicals Corporation by a chloride to sulfate conversion process followed by recovery of the sulfate [1]. This is being practiced by Great Lakes Minerals and Chemicals Corporation [1,10]. Highly absorbing dyes, particularly when added to the brine at later stages of production, may also allow significant increases in production from the same evaporation area by improved efficiency of solar energy absorption. Artificial sylvinite, formed by evaporation of marine waters, has been found to be less subject to potassium chloride and sodium chloride crystal agglomeration if a small amount of a crystal habit modifier such as sodium ferrocyanide is added during the carnallite decomposition stages [19]. In this way, flotation recoveries of potassium chloride are markedly enhanced.

Improvements in the processing of mined potash ores have also been achieved, principally in the methods by which clay brought up from the deposit. Clay in the pulped ores causes both higher consumption of collector reagents (by adsorption) and, simultaneously, reduced sylvite recoveries. The clay may be advantageously separated from the pulped ore prior to the addition of the amine collector by a desliming decantation, or a gentle preflotation collection of the clays in a clay-laden froth. The brine from the clay/brine froth is readily recycled to the flotation circuit after flocculation of the clay with organic reagents followed by removal of the floc in a thickener [18].

Collector adsorption on any residual clays present has been found to be markedly decreased by employing various triethanolamine derivatives (molecular weight < 5000) as blocking agents for the clay active sites, in addition to the usual starches or guar gums which are traditionally used for this purpose. Combining this treatment with the employment of straight-chain (not branched) fatty amines (and not amine salts) as the potassium chloride collector ensures improved recoveries. To provide a closer particle size match between the crude product obtained and market demands, it has been proposed that more emphasis be placed on straight pneumatic flotation assisted by the presence of more viscous reinforcing collector oils, ideally of the butylpolyoxypropylene type $C_4H_9 - (O - CH_2 - CH(CH_3))_n - OH$, n = 2 to 5, to allow the flotation separation of potash of larger average particle size. Adoption of this procedure might permit abandonment of roller fusion of potash fines and enable direct production of a size distribution more closely matching market demands.

Modified fracturing techniques and multiple well sitings have been suggested as a means of boosting ultimate potassium chloride recoveries from solution mining operations to the 60–70% range of in-place potash (versus ca. 40% experience for conventional shaft mining). Improvements may also be achieved by employing an oil blanket, with or without a gas head, and solvent density stratification to allow greater control of zones of dissolution [32]. Controlling brine inlet and exit pressures at the same time enables achievement of calculated overburden support from the cavity fluids which thus may also allow solution mining of deeper deposits than previously considered possible. Cavity interconnections may be detected by dye injections into down-brine wells to assist in avoiding or controlling subsidence. Occasionally it is found that further solution of sylvite is prevented by a coating of sodium chloride crystallized out on the walls of the cavern. However, it has been found that addition of about 100 ppm of manganese ion (or any of a number of other transition metal ions) prevents sodium chloride recrystallization and deposition in the well, and hence avoids the blocking of further dissolution of sylvite.

Relevant to the potash recovery part of solution mining, suggestions to employ high ammonia concentrations in water to achieve greater sodium chloride/potassium chloride selectivity in solubility do not appear to have offered sufficiently attractive improvements to be commercially adopted. Overall, however, solution mining of potash has matured sufficiently from the pioneering work of Kalium Chemicals at Belle Plaine, Saskatchewan [33], to encourage adoption of this technology by at least two operations which were formerly based on conventional mining [34]. These recently modified processes are both advantageously sited to favor solar evaporation rather than employing multiple-effect evaporator-crystallizers for potash recovery. One of these makes use of 162 hectares (1.62 km^2) of PVC-lined evaporation ponds. In this way they achieve significant energy economies in the design production rate of 236,000 tonnes of potassium chloride per year [29]. Good pond solution containment is vital to the economic success of this method of salt purification and recovery [10]. Experiments with solar absorbing, nonemitting dyes at these operations, and the application of mechanical spray systems such as has been proposed to reduce brine disposal problems, may lead to further boosts in production from the existing evaporation area.

6.3. SODIUM SULFATE

Synthetic or manufactured sodium sulfate was originally produced in Canada in the late 1800s [35], as a by-product of the Mannheim furnace method for the production of hydrochloric acid [36]. This process consists of an acidulation step, where sodium chloride is treated with sulfuric acid at high temperatures, followed by a purification step where the crude sodium sulfate is recrystallized from water (Eq. 6.7, 6.8).

$$2\,NaCl + H_2SO_4 \xrightarrow[600–800°C]{} 2\,HCl + Na_2SO_4 \quad\quad 6.7$$
$$\text{crude salt cake}$$
$$\text{(ca. 98\% on NaCl)}$$

$$Na_2SO_4 + 10\,H_2O \rightarrow Na_2SO_4 \cdot 10H_2O \quad\quad 6.8$$
$$\text{Glauber's salt (ca. 98\% recovery)}$$

This chemistry is the basis of one operating plant in the U.S.A. It is still a significant source of synthetic sodium sulfate in Europe, and is the origin of the term "salt cake" which is used synonymously with sodium sulfate to specify industrial grade material.

6.3.1. Production and Use Pattern for Sodium Sulfate

World production of sodium sulfate has averaged about 4 million tonnes/year for at least 25 years, which is approximately one-sixth the scale of potassium chloride production (Table 6.8). Very nearly half of this is from sodium sulfate recovered from natural sources, generally from captive lake basins in areas with high evaporation rates or from aquifers with a high dissolved sodium sulfate content, and the remainder is recovered as by-product material from other industrial processes. The major use for sodium sulfate is as make-up for the chemical losses incurred in the kraft process for the production of pulp and paper, some 79% of the total supply in Canada and 70% in the United States in 1976. Further significant quantities, 14% and 20% in these two countries respectively, are employed as builders in the formulation of synthetic detergents followed by about 4 and 10%, respectively, as ingredients in glass making.

6.3.2. Recovery from Natural Brines

In recent years, only about one-half of the sodium sulfate in the United States has been produced from natural sources (e.g., see Table 6.3), while almost all the Canadian product is obtained in this way. Some plants use a floating dredge to mine lake bottom crystal beds of the minerals mirabilite ($Na_2SO_4 \cdot 10H_2O$), laid down by successive natural seasons of evaporation and chilling, or thenardite (Na_2SO_4), which tends to crystallize out in the presence of significant concentrations of sodium chloride [32]. Processing of natural

TABLE 6.8 Major World Producers of Natural and Synthetic Sodium Sulfate, Including Salt Cake and Detergent Grades[a]

	Thousands of tonnes		
	1972	1980	1994
Belgium	—	249	250
Canada	460	454	312
Germany	300[b]	202[b]	110
Japan	410	290	220
Mexico	128	400	500
Spain	210	309	600
U.S.A.	1,205	1,142	165[c]
U.S.S.R.	420	603	220[d]
Other countries	627	854	1,483
World total	3,760	4,502	3,860

[a]Calculated from *Minerals Yearbooks* [1,19].
[b]For the former West Germany.
[c]Estimate.
[d]For 1991.

salts recovered in this way in its simplest form involves dehydration by melting the decahydrate (melting point 32.4°C) and evaporation of the water of hydration in a submerged combustion unit. More than one-half of the weight of the crude product is water. Kiln drying of the wet crystal mass at higher temperatures then produces a "salt cake" grade of product, with a minimum 97% Na_2SO_4 content, suitable for kraft process use and selling at about $90 per tonne in 1995 [22].

For a "detergent grade" product, brines may be ponded for natural evaporation or may be put through multiple-effect evaporators for concentration, and then chilled to 0°C or below to induce crystallization of the decahydrate, Glauber's salt. More frequently, however, by-product sodium sulfate, which is frequently recovered from process liquors by a crystallization step, is produced on a sufficient scale and purity (up to 99.77% Na_2SO_4) to supply the small detergent grade market at $100 to $139 per tonne in 1995 [22].

6.3.3. By-Product Sodium Sulfate

By-product sources include that material available from the Mannheim furnace process for the production of hydrogen chloride which directly yields by-product sodium sulfate. This may be sold in the form obtained as a "salt cake" product, or it may be recrystallized after neutralization and removal of insoluble matter to give a detergent grade product. Careful recovery of the sodium sulfate solution generated from rayon spinning (1.1 kg of sodium sulfate is obtained for each kilogram of rayon spun) gives Glauber's salt on a scale second only to natural brines, and a mother liquor still containing some sodium sulfate which may be returned to the spinning process. Dehydration of the Glauber's salt from this source yields a detergent grade sodium sulfate.

At present, the Hargreaves-Robinson process to produce hydrogen chloride is also providing a significant proportion of the by-product sodium sulfate in Europe, and at least one plant in the U.S.A. still uses this process. Air, steam, and sulfur dioxide are passed over a heated bed of porous salt granules causing a heterogeneous reaction which yields hydrogen chloride and sodium sulfate (Eq. 6.9).

$$4 \ NaCl + 2 \ SO_2 + O_2 + 2 \ H_2O \rightarrow 2 \ Na_2SO_4 + 4 \ HCl$$
$$93 \text{ to } 98\% \text{ yields}$$

6.9

The crude salt cake by-product may be purified for detergent markets in a manner similar to that used for the crude Mannheim furnace product.

REVIEW QUESTIONS

1. (a) What are the virtues and the restrictions that relate to sodium chloride recovery from natural sources?
   (b) How do these considerations affect the selection of the best sodium chloride recovery method for a recovery plant located at any particular latitude?

2. (a) How is it possible to practice froth flotation in water as a means of separation of water-soluble minerals?

(b) What measures can be used to improve the ability of froth flotation to differentiate between the minerals one wishes to separate?

3. What factor is used to give preferential recovery of potassium chloride from a solution mined underground mineral mixture of potassium chloride and sodium chloride, and how does this operate?

4. (a) What masses of anhydrous hydrogen chloride and crude salt cake are theoretically possible on the acidulation of 1.000 tonne of sodium chloride in the Mannheim process?

(b) With a 100% conversion, and a selectivity of 96% to HCl, and 98% to crude sodium sulfate (both based on NaCl), what actual masses of the two products would be obtained?

(c) Is this a hydrogen chloride process or a sodium sulfate process?

FURTHER READING

J.B. Davis and D.A. Shock, Solution Mining of Thin Bedded Potash. *Min. Eng. (Littleton, Colo.)* **22**, 106, July (1970).

"Potassium Chloride." Environment Canada, Environmental Protection Service, Ottawa, 1985.

J.B. Mitchell, Three Ways to Process Potash. *Min. Eng. (Littleton, Colo.)* **22**, 60, March (1970).

Potash Production Moves to Lower Grade Ores, *Eng. Min. J.* **176**, 166, June 1975.

S. Serata and W.G. Schultz, Application of Stress Control in Deep Potash Mines. *Min. Congr. J.* **58**, 36, November (1972).

R.B. Tippin, Potash Flotation Handles Variable Feed. *Chem. Eng.* **84**, 73, July 18 (1977).

Vinyl-lined Ponds, Key to Solution Mining of Potash, *Civ. Eng. (N.Y.)* **42**, 98, June (1972).

REFERENCES

1. "1976 Minerals Yearbook," Vol. I, p. 1136, 1156, 1184, 1190. U.S. Dept. of the Interior, Bureau of Mines, Washington, DC, 1978.

2. "Canadian Minerals Yearbook 1977," p. 382 to 388. Energy, Mines and Resources Canada, Minister of Supply and Services, Hull, 1979,

3. "United Nations Statistical Yearbook." 29th ed., United Nations, New York, 1977.

4. "Minerals Yearbook 1980," Vol. I, p.763. U.S. Dept. of the Interior, Bureau of Mines, Washington, DC, 1981.

5. "Encyclopaedia Brittannica," 15th ed., Vol. 16, p. 192. Macropaedia, Chicago, 1974.

6. E.T. Degens and D.A. Ross, *Sci. Amer.* **222**(4), 32 (1970).

7. R.N. Shreve and J.A. Brink, Jr., "Chemical Process Industries," 4th ed., p. 266. McGraw-Hill, Toronto, 1977.

8. K. Rankama and T.G. Sahama, "Geochemistry," p. 283-287. University of Chicago Press, Chicago, 1950.

9. G.E. Hutchinson, "A Treatise on Limnology," Vol. 1, p. 569. Wiley, New York, 1957.

10. "Kirk-Othmer's Encyclopedia of Chemical Technology," 2nd ed., suppl. p. 438, 444. Interscience, Toronto, 1968.

11. Chemicals Output Rises, *Chem. Eng. News* **55**(11), 10, March 14 (1977).

12. F.A. Lowenheim and M.K. Moran, "Faith, Keyes and Clark's Industrial Chemicals," 4th ed., p. 723. Wiley-Interscience, Toronto, 1975.

13. R.M. Stephenson, "Introduction to the Chemical Process Industries," p. 38. Reinhold, New York, 1966.

14. J.A. Epstein, *Hydrometallurgy*, **2**, 1 (1976).

15. B.G. Reuben and M.L. Burstall, "The Chemical Economy," p.149. Longmans, London, 1973.

16. W.J. Meade, ed., "The Encyclopedia of Chemical Process Equipment, p. 329. Reinhold, New York, 1964.

17. R.W. Thomas and P.J. Farago, "Industrial Chemistry," p.79. Heinemann, Toronto, 1973.

18. C.J. Warrington and B.T. Newbold, "Chemical Canada," p. 5. Chemical Institute of Canada, Ottawa, 1970.

19. "Minerals Yearbook 1994," Vol. I, U.S. Dept. of the Interior, Bureau of Mines, Washington, DC, 1994 (and earlier).

20. V.A. Arsentiev and J. Leja, *Can. Min. Metall. Bull.* **70A**, 154, March (1977).

21. J.A. Epstein, *Chem. Ind. (London)* **55B**, 572 (1977).

22. R.H. Perry, ed., "Chemical Engineers Handbook," 4th ed., pp. 3–92, 3–93. McGraw-Hill, Toronto, 1969.

23. R.W. Mumford, *Ind. Eng. Chem.* **30**, 872 (1938).

24. Prices Calculated in U.S. Dollars per Tonne of KCl, Based on "Unit K_2O Equivalent," as quoted in "Chemical Marketing Reporter," p. 31, 32. Schnell Pub. Co., New York, 1995.

25. J.B. Dahms and B.P. Edmonds, Can. Pat. 672,308 (to Pittsburgh Plate Glass Co.) (1963).

26. Potash Product Profile, *Can. Chem. Process.* **60**(3), 58, March (1976).

27. H.F. Lund, ed., "Industrial Pollution Control Handbook," p. 7–34. McGraw-Hill, Toronto, 1971.

28. J.A. Vonhof, "A Hydrogeological and Hydrochemical Investigation of the Waste Disposal Basin at IMCC (Canada) Ltd." K2 Potash Plant, Esterhazy, Saskatchewan, National Hydrology Research Institute, Ottawa, 1980.

29. A. Vandenberg, "An Unusual Pump Test Near Esterhazy, Saskatchewan," Tech. Bull. No. 102. Inland Waters Directorate, Water Resources Branch, Ottawa, 1978.

30. R.S. Edwards, The Effects of Air-borne Sodium Chloride and Other Salts of Marine Origin on Plants in Wales. In "Proceedings of the 1st European Congress on Air Pollution Effects on Plants and Animals," p. 99. Centre for Agric. Publ. and Documentation, Wageningen, 1969.

31. John Markham, cited in *Chem. Can.* **27**(9), 14, Oct. (1975).

32. H.H. Werner, Can. Pat. 838,477 (1970).

33. First Solution-mined Potash, *Chem. Eng. (N.Y.)* **71**(22), 84, Oct. 26 (1964).

34. R.L. Curfman, *Min. Congr. J.* **60**, 32, March (1974).

35. C.J.S.Warrington and R.V.V. Nichols, "A History of Chemistry in Canada," p. 213. Pitman, Toronto, 1949.

36. F.A. Lowenheim and M.K. Moran, "Faith, Keyes and Clark's Industrial Chemicals," 4th ed., p. 762. Wiley-Interscience, Toronto, 1975.

7

INDUSTRIAL BASES
BY CHEMICAL ROUTES

From the lime-kiln into the coals.
      —*Quintus Septimius Tertullian, ca. 150* A.D.

I sometimes dig for buttered rolls
Or set limed twigs for crabs; . . .
      —*Lewis Carroll (1832–1898)*

7.1. CALCIUM CARBONATE

The dominant source of calcium carbonate is limestone, the most widely used of all rocks. This often occurs in nature with traces of clay, silica, and other minerals which may interfere with some applications. However, high calcium limestone can consist of 95% or better calcium carbonate. White marble, a metamorphosed form of pure limestone with a closely packed crystal structure, is also chemically suitable, but is usually of higher value for other applications than as a chemical feedstock. Dolomitic limestones consist of calcium and magnesium carbonates present in a near one-to-one molar basis, though this ratio can vary widely in ordinary dolomites. For some applications at least, the presence of the magnesium carbonate is not a handicap to the use of this calcium base component of the dolomite. The principal remaining natural sources of calcium carbonate more closely reflect their biogenetic origin. These are chalk, which comprises the shells of microscopic marine organisms, bivalve shells, which are accessible in sufficient quantities on the shores of the Gulf of Mexico to be employed as an industrial feedstock, and coral, which consists of the massive, sub-marine fused skeletons of multiple stationary organisms.

Both to obtain higher purity and higher specific bulk density grades of calcium carbonate, and as an expedient to a salable product from the waste streams of some processes, synthetic calcium carbonate is also produced on a large scale [1]. As early as 1850 J. & E. Sturge Ltd. of Birmingham, England was treating a calcium chloride Solvay waste stream with sodium or ammonium carbonates to produce a high grade of calcium carbonate (Eq. 7.1, 7.2).

$$CaCl_2 + Na_2CO_3 \rightarrow 2\,NaCl + CaCO_3 \downarrow \qquad\qquad 7.1$$

$$CaCl_2 + (NH_4)_2CO_3 \rightarrow 2\,NH_4Cl + CaCO_3 \downarrow \qquad\qquad 7.2$$

In the U.S.A., certainly by 1900, not only Solvay by-product calcium chloride, but also causticizer sludges and direct carbonation of slaked lime were used to produce fine grades of calcium carbonate (Eq. 7.3, 7.4).

$$Ca(OH)_2 + Na_2CO_3 \rightarrow 2\,NaOH + CaCO_3 \downarrow \qquad\qquad 7.3$$

$$Ca(OH)_2 + CO_2 \rightarrow H_2O + CaCO_3 \qquad\qquad 7.4$$

Industrial uses of natural calcium carbonate include direct, very large scale consumption in the cement, iron and steel, and other metal refining industries as well as the Solvay ammonia soda process. It is also used directly or indirectly in the manufacture of at least 150 other chemicals. In recent years, direct or indirect use in emission control measures for combustion processes has taken a growing fraction of limestone production. The higher purities and more tailored properties of the synthetic grades of calcium carbonate both command a higher price and for this reason limit the appropriate applications somewhat. More than 75% of U.S. synthetic calcium carbonate is applied as a surface coating to paper to impart a smoother, whiter surface. Smaller amounts are used in paints, to give flatness and low gloss, in rubber, as a reinforcing agent, in plastics, as a filler, and in many other consumer items such as foods, cosmetics, toiletries, pharmaceuticals, and the like.

7.2. CALCIUM OXIDE

Calcium oxide, burned lime, or simply lime as it is known in industry, is obtained by heating calcium carbonate to a temperature of 900 to 1000°C to cause loss of carbon dioxide (Eq. 7.5).

$$CaCO_{3(s)} \xrightarrow{\;900-1000°\;} CaO_{(s)} + CO_{2(g)}$$
$$\quad 100.09 \qquad\qquad 56.08 \quad 44.01 \text{ mol. wt.} \qquad\qquad 7.5$$
$$\Delta H = +177.8 \text{ kJ } (+42.5 \text{ kcal})$$

TABLE 7.1   Calcium Carbonate Dissociation Pressures[a]

Temperature (°C)	Pressure (mm Hg)	Temperature (°C)	Pressure (mm Hg)
500	0.1	852	381
550	0.41	898	760
605	2.3	950	1490
701	23.0	1082.5	6758
749	72	1157.7	14202
800	183	1241	29711

[a]Compiled from the data of Kirk-Othmer [1] and Weast [2].

A temperature below this range slows the rate of carbon dioxide loss sufficiently to leave some unburned limestone in the product (Table 7.1). It also fails to confer any significant energy saving if the limestone retention time has to be prolonged to ensure complete reaction, because of the greater heat loss experienced for the longer calcination period. Also, when a moderate-size range of crushed limestone, rather than finely ground material, is used to reduce dust carryover in the kiln, a certain amount of additional excess carbon dioxide partial pressure must be developed in the larger granules to ensure carbon dioxide diffusion to the outside of the granule during the hot period.

In Canada, and probably elsewhere in the early nineteenth century, burned lime manufacture was carried out as a batch process using pot kilns heated with wood or coal [3]. These produced 6 to 27 tonnes per cycle, as did the draw kilns of the late nineteenth century, which were designed to use fuel more efficiently. Today a variety of vertical flow kilns are in common use, these mainly designed to operate with lower energy requirements than the early models [4]. One further feature of some of these newer designs is the separation of the function of gas flow required for combustion from the limestone heat requirement, thus obtaining a gas stream from above the limestone as high in carbon dioxide concentration as possible. This is a desirable feature, for example, when the process is operated as a component of the Solvay ammonia-soda process (Section 7.4).

Probably the most broadly useful design of a modern lime kiln, since it allows a feed of lumps or fines of limestone or marble, or wet or dry calcium carbonate sludges, is the rotary horizontal kiln (Fig. 7.1). The main component of this calcination unit is a 2.5- to 3.5-m-diameter by 45- to 130-m-long firebrick-lined inclined steel tube. Heat is applied to the lower end of this via oil, gas, or coal burners [5]. The feed to be calcined is fed in at the top end and slow rotation of the tube on its axis gradually moves the feed down the tube, as it tumbles countercurrent to the hot combustion gases. In this way, wet feed is dried in the first few meters of travel. Further down the tube, for wet or dry feeds, carbon dioxide loss begins as the temperature of the feed is raised. By the time the solid charge reaches the lower, fired end of the kiln it

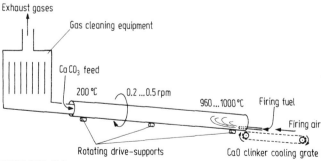

FIGURE 7.1 Details of operation of a rotary horizontal lime kiln. Temperature of calcium carbonate charge rises as it moves down the kiln. The hot calcium oxide, as it cools, is used to simultaneously preheat the firing air. Fuel may be natural gas or oil.

reaches temperatures of 900 to 1000°C and carbon dioxide evolution is virtually complete. Normally the temperature of the lower end of the kiln is not allowed to go much above this as it reduces the life of the kiln lining. It also adversely affects the crystal structure of the lime product since it produces a "dead-burned" or "overburned" lime. This overheated lime is difficult to slake to convert it to calcium hydroxide. It also raises fuel costs unnecessarily. The calcined product is normally packed in airtight sheet steel barrels or bulk containers for shipment in anhydrous form.

With a specific heat of about 0.90 J/g °C at 100°C (0.214 cal/g °C; varies with temperature) calcium carbonate requires some 7.92×10^5 kJ/tonne of sensible heat simply to bring the salt up to the final kiln temperature.

$$0.90 \text{ J/g °C} \times 10^6 \text{ g/tonne} \times (900 - 20)°C = 7.92 \times 10^8 \text{ J/tonne}$$

The endotherm of the reaction consumes a further 1.78×10^6 kJ/tonne, which, when combined with the sensible heat requirement and an allowance of a further 52×10^5 kJ/tonne for imperfect heat transfer efficiencies, gives a total heat requirement for this process of about 2.8×10^6 kJ/tonne.

$$177.8 \text{ kJ/mol} \times \frac{10^6 \text{ g/tonne}}{100.09 \text{ g/mol}} = 1.776 \times 10^6 \text{ kJ/tonne}$$

Production estimates for different fuels range from 2.5 to 3.4 kg of lime/kg of coke [5] or coal [6], approximately equivalent to 4.5 kg/kg of Bunker C fuel oil [7].

Improved energy efficiencies are obtained by using "recuperator chains" loosely slung inside the kiln to aid in heat transfer from hot gases to the drying and reacting solids, and by using the heated air obtained from lime cooling as the firing air for the kiln. Some other kiln designs provide better heat transfer and recovery than the rotary kiln but have certain other restrictions in the form of suitable feeds, etc., which do not affect the rotary horizontal kiln [4]. A recent survey examines kiln operating problems and suggests solutions [8]. A potential energy gain is possible from the use of oxygen-enriched air, which is accessible through one of the new low-cost enrichment techniques [9]. Using oxygen enrichment decreases the volume of associated nitrogen which has to be heated. Apart from the much greater additional expense, there would be little point in utilizing pure oxygen since one of the functions of gas throughput is to displace the dissociated carbon dioxide.

7.2.1. Lime Kiln Emission Control

Environmental concerns of lime kiln operation relate primarily to exhaust gas dust control and are usually solved by water scrubbing. The slaked lime (calcium hydroxide) produced by the scrubber can be otherwise employed as a base in the operations of a chemical complex, or sold. Electrostatic precipitation of precooled gases is also used. Occasionally both control measures are employed in series. The dissociated carbon dioxide discharged is not ordinarily regarded as a pollutant. The amount discharged from this source is far exceeded by the volumes of carbon dioxide discharged from fossil fuel combustion. Lime kilns associated with Solvay ammonia-soda plants may be able to usefully recycle a part of the dissociated carbon dioxide, particularly if the concentrations are raised by the use of oxygen-enriched combustion air. Recently interest has been shown in the collection of carbon dioxide from industrial sources such as lime kilns for use in enhanced oil recovery operations.

7.2.2. Uses of Calcium Oxide

The extensive industrial uses of calcium oxide (lime) are a reflection of both its low cost and its chemical properties. It is consumed on a very large scale by both the cement and the iron and steel industry. In both cases *in situ* lime preparation from limestone by heat is an integral part of the process. The metals smelting industry also employs lime in a general way as a basic refractory lining material. This basicity is also employed in processes for the heterogeneous absorption of acid gases produced from combustion or metallurgical roasting processes. The strong affinity of lime for water is applied to the industrial drying of alcohols as, for example, to remove water from the glycerin obtained from the saponification of fats, or from the 5% water azeotrope of ethanol. Most of the remaining uses are indirect since they employ slaked lime (calcium hydroxide), the product of the reaction of lime with water.

7.3. CALCIUM HYDROXIDE

Calcium oxide (burned lime) can be slaked with a theoretical amount of water to form calcium hydroxide (slaked lime) as a white powder, but this must be conducted with care because the highly exothermic reaction may produce sufficiently high temperatures that it could slow the rate of hydration [5] (Eq. 7.6).

$$CaO + H_2O \rightarrow Ca(OH)_2 \qquad \Delta H = -64.9 \text{ kJ } (-15.5 \text{ kcal}) \qquad 7.6$$

This process is sufficiently exothermic that it is dangerous to store large quantities of calcium oxide in combustible bags or even in non-fireproof buildings if there is any risk of contact of the burned lime with water or air, since even moist air is sufficient to slake burned lime. Calculation can demonstrate that the high heat of hydration coupled with the relatively low specific heats (CaO = 0.226 cal/g; Ca(OH)$_2$ = 0.286 cal/g [7]) is theoretically capable of generating a temperature rise of 730°C above ambient, more than enough to ignite combustible materials. This is apart from the explosive risks connected with possible sudden steam generation in an enclosed space.

Other variables in the preparation of slaked lime powder by the addition of the theoretical amount of water to lime are the original calcining temperature used and the magnesia (MgO) content. Very high calcining temperatures, which produce a "dead-burned" lime, can have a retarding effect on the hydration reaction. The presence of any significant concentrations of magnesia, such as may be obtained from the calcination of a dolomitic lime-stone (a "dolime") also shows this effect. However, the hydration retardation shown by burned dolimes is actually regarded as a working advantage in its application as a plaster.

Both continuous rotary and batch hydrators are used commercially to produce calcium hydroxide (slaked lime) powder. The continuous version uses a gently inclined, slowly rotating steel cylinder of about 1 by 6–7 m in length. The calcium oxide is fed into the upper end and a correctly proportioned water spray is added at the same point followed by tumbling in the cylinder to produce a uniform product. Open trough-type batch hydrators, in which the reacting components are mechanically combined, provide a greater degree of control of hydration rate and temperature. By this method there are some improvements in the quality of the product for certain uses, but the properties are more variable. An "explosion process," in which the hydration is con-ducted in a pressure vessel, produces a slaked lime with better flow charac-teristics [10.] The smaller mean particle size obtained by this method yields a product more suitable for uses where this is important, such as for filters.

The same exotherm is obtained when lime is slaked with a liberal amount of water to produce a slurry or suspension of calcium hydroxide in water. However, coping with this is less of a problem because of the high heat ca-pacity of the excess water present. In the preparation of cement mixes, mor-tars, and plasters the low solubility of slaked lime in water (1.84 g/L at 0°C; 0.77 g/L at 100°C) does not pose any problems. However, this factor does have to be considered in the applications of calcium hydroxide when used as an industrial base.

7.3.1. Uses of Calcium Hydroxide

Some of the uses of slaked lime are direct, in this form, and some are indirect via calcium oxide addition to water or to an aqueous medium. These uses range from its employment as a component of insecticides, to its value as a component of water treatment programs. The venerable but still viable use as an active pesticide component in lime-sulfur sprays and its substrate utility in many insecticidal dusts are also important. The chemical basicity of this com-modity is used in its medicinal applications as an antacid. This property also prompts its use for pH adjustment in animal food formulations, and for the neutralization of overly acid soils. Large quantities are employed as a slurry in water to absorb acid contaminants from flue gases and from roaster off-gases of small smelters. It is also used in papermaking. In the sulfite process it is used directly to provide the active cation of the pulping liquor (Chap. 15). In the kraft process it is used indirectly for the regeneration of the sodium hydroxide component of the pulping liquor from sodium carbonate recovered from the combustion of spent pulping liquor.

Calcium hydroxide frequently performs a dual role in water treatment. First, it aids the removal of colloidal suspended solids by encouraging coag-ulation and settling. Second, it reacts with any dissolved calcium, magnesium or other cations which are present as their bicarbonates to precipitate these ions and in this way softens the water supply for domestic and industrial uses (Eq. 7.7–7.9).

$$Ca(OH)_2 + Ca(HCO_3)_2 \rightarrow 2\ CaCO_3\downarrow + 2\ H_2O \qquad 7.7$$

$$Ca(OH)_2 + Mg(HCO_3)_2 \rightarrow MgCO_3 + CaCO_3\downarrow + 2\ H_2O \qquad 7.8$$

$$Ca(OH)_2 + MgCO_3 \rightarrow Mg(OH)_2\downarrow + CaCO_3 \qquad 7.9$$

7.4. NATURAL AND SYNTHETIC SODIUM CARBONATE

Up until the end of the eighteenth century the only commercially available sodium carbonate was a low grade product containing only 5–20% of this salt. It was obtained from the burning of kelp or barilla (a stout, berry-bearing Spanish shrub), followed by extraction of the ashes with water [11]. Early in the nineteenth century the availability of sodium carbonate was enhanced by the introduction of the Le Blanc process, which both increased the potential supply and produced a higher analysis product (Eq. 7.10, 7.11).

$$2\ NaCl + H_2SO_4 \rightarrow Na_2SO_4 + 2\ HCl \qquad 7.10$$

$$Na_2SO_4 + 2\ C + CaCO_3 \xrightarrow{heat} CaS + Na_2CO_3 + 2\ CO_2 \qquad 7.11$$

The early versions of this process were significant contributors to local air pollution since about 3/4 of a tonne of hydrogen chloride was discharged for each tonne of sodium carbonate produced. Fortunately the scale of operations at that time was small. In 1836 it was discovered that the acid discharge could be controlled by absorption in water, but it was 1863 before this control measure was made compulsory [11]. The other negative feature of LeBlanc process operations at that time was the large piles of waste calcium sulfide and calcium oxysulfide which were generated and discarded nearby, reclamation of which was very difficult.

The ammonia-soda process to sodium carbonate, which is still practiced on a large scale today, was known in 1822 [11] (Eq. 7.12–7.15).

$$NH_3 + H_2O \rightarrow NH_4OH \qquad 7.12$$

$$NH_4OH + CO_2 \rightarrow NH_4HCO_3 \qquad 7.13$$

$$NH_4HCO_3 + NaCl \rightleftharpoons NaHCO_3\downarrow + NH_4Cl \qquad 7.14$$

$$2\ NaHCO_3 \xrightarrow{175°} Na_2CO_3 + CO_2 + H_2O \qquad 7.15$$

However, this early version had no ammonia recovery phase so that the cost of the ammonia required to work this process was prohibitive. For this reason the process was not commercialized until 1863 when E. Solvay devised a method which employed a suspension of slaked lime in water to capture ammonia (Eq. 7.16).

$$2\ NH_4Cl + Ca(OH)_2 \rightarrow 2\ NH_3\downarrow + CaCl_2 + H_2O \qquad 7.16$$

Modern practice of the Solvay ammonia-soda process involves initial successive saturation of a purified saturated solution of sodium chloride in water, first with ammonia gas and then with carbon dioxide. The carbonation step is used to help drive reaction 7.14 to about 73% to the right. Sodium bicarbonate, being the least soluble salt component of this mixed solution, is precipitated and filtered off. It is then dried and calcined under mild conditions in a rotary kiln to yield the sodium carbonate product (Table 7.2). Conducting the calcination in a sealed rotary kiln enables recovery of the liberated carbon dioxide (Eq. 7.15) which provides some of that required for the primary carbonation (Fig. 7.2). The remaining carbon dioxide is obtained from the operation of a separate lime kiln which produces calcium oxide. The calcium oxide (burned lime) is required to produce calcium hydroxide for ammonia recovery, and also for carbon dioxide recovery from boiler flue gases as well, if necessary.

The ammonia recovery system of modern configurations of Solvay plants is quite efficient. Only about 3 kg of ammonia make-up is required to maintain the 300 kg or so of absorbed ammonia which is needed for each tonne of sodium carbonate produced [13]. The development of this high ammonia recovery efficiency was the feature that gradually spelled the end of the commercial Le Blanc sodium carbonate plants, the last of which closed down in the 1920 to 1930 period [11].

The crude sodium carbonate from the Solvay process, while of quite good purity, is a low-density fluffy material of bulk density of about 0.5 g/cm^3, and referred to as "light soda ash." The density may be increased to a bulk density of about 1.0 g/cm^3 required for some uses by conversion to the monohydrate, $Na_2CO_3 \cdot H_2O$. This is effected by tumbling the anhydrous light ash with the appropriate quantity of water and then drying at <100°C. Sal soda or washing soda, which is the decahydrate, $Na_2CO_3 \cdot 10H_2O$, is generally obtained by crystallization of the light ash from water and drying at ambient temperatures (<35.5°C).

Oddly enough the sodium bicarbonate ($NaHCO_3$) initial product of the Solvay process is not the article of commerce principally because of impurities, in particular ammonia, that are present. Instead the relatively much smaller market for the bicarbonate is satisfied by the carbonation of a saturated solution of sodium carbonate, the final Solvay product, and recovery of the

TABLE 7.2 Quality of Typical Sodium Carbonate from the Ammonia-Soda (Solvay) Process[a]

Component	Analysis (%)
Na_2CO_3	99.50 (58% Na_2O)
NaCl	0.20 to 0.25
Na_2SO_4	0.02
Insolubles	0.02
Water	balance

[a]Data selected from Kent [12].

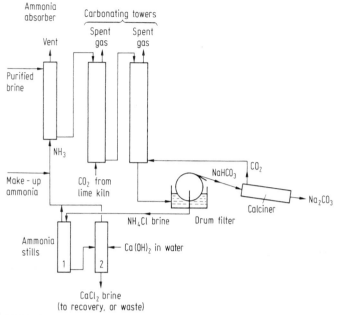

FIGURE 7.2 Flowsheet for the ammonia-soda (Solvay) process to provide sodium carbonate. Units numbered 1 and 2 are ammonia stills, the first being used to recover free ammonia and the second to recover that produced by the addition of slaked lime (Eq. 7.16).

precipitated, much less soluble sodium bicarbonate by filtration [4] (Eq. 7.17). This product is the household chemical "baking soda."

$$Na_2CO_3 + CO_2 + H_2O \rightarrow 2\ NaHCO_3 \qquad\qquad 7.17$$

The same source is also used to supply the sodium bicarbonate used for the manufacture of baking powder, for the preparation of carbonated mineral waters, and also as the filling for one type of dry powder fire extinguisher.

7.4.1. Environmental Aspects of Sodium Carbonate Production

An area of difficulty of present-day Solvay operations is related not so much to the technology of the process but to the basic chemistry, which determines the coproduction of more than a tonne of calcium chloride for each tonne of the product sodium carbonate obtained. For producers operating in areas with well-developed dissipative markets for calcium chloride in applications such as high early strength cement, winter ice and snow removal for highways, or dust control on unimproved roads, the coproduct can be sold at reasonable prices. Nondissipative uses of calcium chloride such as employment as the dissolved salt of an aqueous heat transfer fluid in artificial ice rinks and the like or as a component of solar energy storage systems [14], do not at the moment consume large amounts of the salt. Because of this, some producers are faced with high brine disposal costs since the waste, low-pH calcium chloride brines cannot be discharged to surface freshwater courses without creating significant pollution problems. Some processors have adopted calcium chloride brine discharge into underground brine aquifers as a disposal method.

Brine disposal problems can be avoided by proposed modifications to the Solvay process which employ magnesium hydroxide for ammonia recovery [13]. The magnesium chloride formed in the process could be recycled and hydrogen chloride obtained as a coproduct by contacting the magnesium chloride with superheated steam (Eq. 7.18, 7.19). However, this process has not as yet been commercially proven.

$$MgCl_2 + H_2O \xrightarrow[\text{heat}]{} MgO + 2\ HCl \qquad\qquad 7.18$$

$$MgO + H_2O \rightarrow Mg(OH)_2 \qquad\qquad 7.19$$

A dominant alternate source of sodium carbonate, at least in the United States, where large reserves of natural brines containing sodium carbonate are accessible, has been to process these for sodium carbonate recovery. The calcium chloride disposal problems of the Solvay process and the much lower capital cost of a natural sodium carbonate plant (about one-half the cost of a similarly sized Solvay plant) have contributed to this shift [13]. These factors have stimulated a strong trend in the U.S.A. toward natural sources in recent years (Table 7.3). Only two other countries, Chad and Kenya, are reported to be recovering natural sodium carbonate, and these were operating on the scale of about 10,800 and 119,000 metric tonnes annually, in 1977 (calc. from [15]). All other world producers still rely heavily on the ammonia-soda process [13,15] (Table 7.4).

An expedient but not large-scale additional source of sodium carbonate is via the carbonation of sodium hydroxide, as it is formed in an electrolytic cell, or separately, later. The precipitated sodium hydrogen carbonate is calcined to obtain sodium carbonate from the initial product in a manner very similar to the last step of the Solvay process (Eq. 7.15). This is not much practiced because of high cost.

TABLE 7.3   Growth of Natural Sodium Carbonate Production at the Expense of Synthetic Material in the United States[a]

| | Sodium carbonate produced (thousands of tonnes) | | Natural, as |
Year	Manufactured	Natural	percentage of total
1940	2744[b]	118	4.1
1950	3621	318	8.1
1960	4134	734	15
1970	4004	2439	38
1978	1360	6160	82
1994	—	9320	100

[a]Calculated from the data of Minerals Yearbooks [15].
[b]A further 22,680 tonnes was made in this year by the carbonation of electrolytic sodium hydroxide.

TABLE 7.4 **Major World Producers of Sodium Carbonate (thousands of metric tonnes)**[a]

	1960	1970	1979	1994
Bulgaria	128	300	1498	550
China, Mainland	—	—	1785	5680
France	848	1419	1355	1200
Germany	1117[b]	1334[b]	1180[b]	2290[c]
India	145	446	610	1500
Italy	—	662	725	500
Japan	516	1230	1180	1060
Poland	522	644	662	950
Romania	180	582	NA	450
U.S.S.R.	1793	3485	5350	4360[c]
United Kingdom	—	—	1630	1000
United States[a]	4135	3859	NA	9320
Total World[d]	11,480	16,610		30,400

[a]Does not include natural sodium carbonate. Data calculated from Minerals Yearbooks [15] and from U.N. Statistical Yearbooks [16].
[b]For the former West Germany.
[c]For 1990.
[d]Natural sodium carbonate represents all of the U.S. total. See Table 7.3.

7.4.2. Uses of Sodium Carbonate

Nearly half of the sodium carbonate produced in the U.S.A., whether from synthetic or natural sources, is consumed by the glass industry. A further quarter of the product goes into general chemicals production. The detergent industry, pulp and paper, and water treatment processes each consume somewhat less than 10% of the total, although there are also industries, such as kraft paper production, which produce and consume large quantities of sodium carbonate internally [15]. For this reason the figures for this captive production by the pulp industry do not appear in the consumption data reported.

7.5. SODIUM HYDROXIDE BY CAUSTICIZATION

Prior to 1850 sodium hydroxide was not available commercially. Processes such as soap making which required sodium hydroxide (caustic soda) had to obtain this by causticization of a solution of purchased sodium carbonate with lime (Eq. 7.20).

$$Na_2CO_{3(aq)} + Ca(OH)_{2(aq)} \rightleftharpoons 2\ NaOH_{(aq)} + CaCO_{3(s)}$$
$$\Delta H = -8.4\ kJ\ (-2.0\ kcal) \qquad 7.20$$

Sodium hydroxide, which first became commercially available in the early 1850s, was also made in this way until about 1890 when the first electrolytic product started to be produced [11]. Despite the advantages of electrolytic production of caustic soda, the causticization or "lime-soda" process dominated the production of trade sodium hydroxide in the United States until at least 1940 and is still of commercial significance in the regeneration of waste pulping liquors in the manufacture of pulp and paper by the kraft process [17]. It is also used for the production of sodium hydroxide without chlorine at times when the market demand ratio of sodium hydroxide to chlorine exceeds the stoichiometric ratio available from electrolysis of sodium chloride solutions.

The chemical change involved in this process depends for its success on the low solubility of calcium carbonate, which is also the feature which lends suitability of this process for small-scale low capital production of sodium hydroxide by batch processes. For batch operation, several of the functions such as slaking, mixing, and settling, may be carried out in the same wooden (or steel) vessel unlike the separate units required by the continuous process. A further chemical feature important to the recycle of the spent lime of this process is the relatively easier thermal loss of carbon dioxide from calcium carbonate than from sodium carbonate. It might be expected that, because sodium bicarbonate ($NaHCO_3$) may be calcined at 175°C to obtain carbon dioxide loss, sodium hydroxide may be made by calcining sodium carbonate at a somewhat higher temperature followed by hydration of the resulting sodium oxide (Eq. 7.21, 7.22).

$$Na_2CO_3 \xrightarrow[\text{heat}]{} Na_2O + CO_2 \qquad\qquad 7.21$$

$$Na_2O + H_2O \rightarrow 2\ NaOH \qquad\qquad 7.22$$

While this series of reactions is possible, very much higher temperatures are required for sodium carbonate dissociation to sodium oxide and carbon dioxide, than for calcium carbonate dissociation. Hot, concentrated sodium hydroxide is also very corrosive. Therefore it is not nearly as easy to use this sequence of reactions as it is to go through the same sequence with calcium carbonate, which has a much higher decomposition pressure at all tempera-

TABLE 7.5   Carbon Dioxide Dissociation Pressures of Sodium Hydrogen Carbonate and Calcium and Sodium Carbonates[a]

NaHCO$_3$			CaCO$_3$	Na$_2$CO$_3$
Temperature (°C)	Pressure (mm Hg)	Temperature (°C)	Pressure (mm Hg)	Pressure (mm Hg)
30	6.2	700	22.6	0.99
50	30.0	820	237	3.0
70	120.4	880	595	9.9
90	414.3	990	3,324	12.0
100	731.1	1,100	8,490	21
110	1,252.6	1,150	13,440	28
115.5	1,654.6	1,180	18,068	38
		1,200	21,534	41

[a]Dissociation pressures for sodium salts calculated from International Critical Tables [18], and for calcium carbonate obtained by interpolation of data from Weast [19].

tures (Table 7.5). This dissociation can then be followed by the slaking and the causticization reactions (Eq. 7.6, 7.20).

A hot (80–90°C) 20% solution of sodium carbonate is stirred with a slight stoichiometric excess of a slurry of calcium hydroxide in water ("milk of lime"). Since this reaction is only slightly exothermic (Eq. 7.2) the equilibrium is not significantly affected by operation of the reaction hot. The reaction is also faster and produces larger, more readily precipitated particles of calcium carbonate at high rather than at ambient temperatures. An additional advantage gained by operation at temperatures near the boiling point of water is obtained from the higher concentrations of sodium carbonate which may be employed (Table 7.6). Starting with high concentrations of sodium carbonate raises the concentration of sodium hydroxide obtained directly from the reaction to about 12%, sufficient for direct utilization in pulping liquor preparation [21]. It also decreases evaporation costs for applications which demand higher sodium hydroxide concentrations.

Operation of a causticizer at 80°C or higher is not without an attendant drawback, however. While the solubility product for calcium carbonate (Eq. 7.23) is low in pure water, it is higher in water containing sodium hydroxide which, in effect, displaces the equilibrium of the formation reaction somewhat to the left.

$$K_{sp} \text{ (CaCO}_3\text{): } [Ca^{2+}][CO_3^{2-}] = 0.87 \times 10^{-8} \text{ at 25°C} \qquad 7.23$$

$$K_{sp} \text{ (Ca(OH)}_2\text{): } [Ca^{2+}][OH^-]^2 = 1.3 \times 10^{-6} \text{ at 25°C} \qquad 7.24$$

In practical terms this means that at initial 2, 5, 10, 15, and 20% sodium carbonate concentrations one obtains approximately 99.4, 98.9, 96.5, 91.5, and 83.5% conversions of the sodium carbonate to sodium hydroxide, respectively [22]. Industrial yields on sodium carbonate for a lime recycle

TABLE 7.6 Solubilities of Sodium Carbonate and Sodium Hydrogen Carbonate in Water[a]

Temperature (°C)	Solubility (wt %)	
	Sodium carbonate	Sodium hydrogen carbonate
0	6.5	6.5
10	11.5	7.5
20	17.7	8.8
25	27.8	9.4
30	31.4	10.0
40	—	11.3
42	32.7	—
50	—	12.6
60	—	13.8
70	31.7	—
80	31.4	—
90	—	—
100	31.1	19.1

[a]Data from Kirk-Othmer [1] and Stephen and Stephen [20].

process on the above bases amount to about 90% [6] giving a yield per batch, or yield per pass for a continuous process of about 75% (0.835 × 0.90) when using a 20% initial sodium carbonate concentration.

The situation described above, showing gradually diminishing sodium carbonate conversions for increasing concentrations, is the concern that limits the practical sodium carbonate concentrations used in this process to well below what would be technically feasible. An empirical expression which may permit estimation of sodium carbonate conversions from the known solubility products and the initial sodium carbonate concentrations has been derived [23], but has been found to be unreliable [24]. This lack of rigor is partly because the solubility of calcium hydroxide is too high to obtain a solubility product constant, but also so low that only partial ionization precludes the application of ionic equilibria to the determination of ion concentrations. Thus, this expression would predict higher sodium carbonate conversions for higher excess ratios of slaked lime to sodium carbonate whereas in practice this has been found to make very little difference [22].

The insoluble calcium carbonate is separated from the sodium hydroxide solution by simple settling in a batch process, or by settling in a thickener (Fig. 6.6) followed by filtration of the more concentrated slurry from the thickener in a continuous process (Fig. 7.3). Continuous filtration, as used in this process, involves aqueous sodium hydroxide removal from a sludge by drawing this solution through a filter element fastened to the periphery of a slowly rotating drum. The driving force for fluid flow through the filter is the reduced pressure maintained in the interior of the drum (Fig. 7.4). Normally in the larger scale causticization processes the calcium carbonate filter cake, or "lime muds", will be calcined to obtain calcium oxide. After slaking, this regains the calcium hydroxide required for causticization (see Section 7.3).

The crude sodium hydroxide solution is obtained at a concentration of about 12%. This concentration may be suitable for a commercial product if

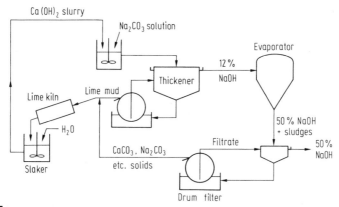

FIGURE 7.3 Flowsheet giving the details of a continuous causticization process for the production of sodium hydroxide. Slaker water is normally derived from lime mud washing. The process may also be operated in a batch mode requiring a smaller number of process vessels, and with or without lime recycle.

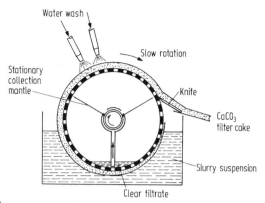

FIGURE 7.4 Cross section through the drum and slurry trough of a drum filter, used for the continuous separation of a solid from a liquid suspension. This version includes a water wash for scavenging the last traces of a solution of value. Sodium hydroxide washings are retained in the process by collecting them separately in the filter and using this to dissolve fresh sodium carbonate, and to slake burned burned lime. (Modified from Codd *et al.* [25] with permission.)

it is to be used close by. This might be the case, for example, to prepare the white liquor for the chemical pulping of wood, or for soap making. At this stage the crude sodium hydroxide will contain 2–3% impurities, mainly sodium carbonate with traces of sodium chloride.

Shipping costs of the sodium hydroxide are greatly reduced for distant markets if most (or all) of the water is removed first. Evaporation is carried out either batchwise, in corrosion-resistant close-grained cast iron pots over an open flame, or in continuous caustic evaporators heated with an industrial heat exchange fluid (Section 8.1). When 50% sodium hydroxide is cooled, sodium carbonate and sodium chloride residues are less soluble and crystallize out. These salts are removed from the 50% product by settling or filtration. For distant shipping, or for small scale customers, all the water may be removed from the solution. The strong affinity of sodium hydroxide for water requires temperatures of 500 to 600°C to accomplish this, leaving molten sodium hydroxide containing less than 1% water [26]. If this sodium hydroxide is manufactured from Solvay sodium carbonate it would give roughly the following analysis [20]:

$$NaOH, 98.62\%; Na_2CO_3, 0.90\%; NaCl, 0.30\%; Na_2SO_4, 0.18\%$$

Detailed use profiles for sodium hydroxide are given in Chapter 8.

7.5.1. Emission Control of the Causticization Process

Emission problems in a causticization operation arise chiefly in the area of dust control of the exit gases from the lime kiln since the water circuit is virtually self-contained. Effective containment is obtained by the use of scrubbers which achieve some 99% by mass dust removal from the exhaust gases. Spent scrubber liquor may be returned to the causticization circuit for recycle. Using spent scrubber liquor either for the slaking of lime, or to prepare fresh sodium carbonate solutions for causticization, avoids creating a water emis-sion problem from this aqueous waste stream. It also improves the raw ma-terial balance of the process.

Some facilities use electrostatic precipitation for lime kiln dust control, which avoids generation of a wastewater stream. The captured dust can be recycled to the slaker to avoid dust disposal costs and assisting the material balance of the operation.

REVIEW QUESTIONS

1. The Solvay process to produce sodium carbonate ordinarily operates with about 90% conversion and 75% yield (industrial definition) on sodium chloride.

    (a) Give the molar quantities of sodium carbonate, sodium chloride, and miscellaneous sodium compounds (by-products, $NaHCO_3$, etc.) to be expected from one passage (i.e., batch basis) of 2 mol of sodium chloride through this continuous process.

    (b) What molar quantities of sodium carbonate, sodium chloride, and sodium-containing by-products (expressed as moles of Na^+) would be expected if the unreacted sodium chloride was recycled to extinction (i.e., on an annual basis)?

2. The sole Canadian Solvay plant produced 400,000 tonnes (1 tonne = 1000 kg) of anhydrous sodium carbonate in 1975.

    (a) What quantity of sodium chloride, in tonnes, would theoretically be required to produce this amount of sodium carbonate, assuming quantitative reactions?

    (b) Current prices for limestone (for producing the calcium hydroxide, salt, and sodium carbonate monohydrate ($Na_2CO_3 \cdot H_2O$) are 22, 31, and 62 dollars per tonne, respectively. What gross (*before* deduction of operating costs, labor, etc.) annual profit should this operation be currently making, assuming quantitative reactions?

3. Batch causticization of 8000 L of a hot, 2.0 M solution of sodium carbonate is accomplished by the addition of 10% excess preslaked calcium hydroxide as a slurry in 2000 L of water. Under these conditions conversion and yield on sodium carbonate are 95% and 90%, respectively.

    (a) What would be the conversion and yield on calcium hydroxide?

    (b) Assuming that no residual water is lost with the precipitated calcium carbonate as it is removed, what resultant molar concentrations of sodium hydroxide and sodium carbonate would be obtained?

    (c) Evaporation in stages, with intermediate removal of impurities (with no loss of the associated sodium hydroxide), should yield what mass of 100% sodium hydroxide?

4. Sodium hydroxide solution, 50% by weight, is available commercially at $225 per tonne of solution.

    (a) What theoretical profit or loss, on a raw material cost basis only, would be incurred to produce each tonne of anhydrous sodium carbonate from this sodium hydroxide and sell the product at $210 per tonne? Carbon dioxide for carbonation is to be obtained from flue gas; assume essentially zero cost.

(b) Sodium carbonate monohydrate and sodium carbonate decahydrate are also commercial grades, selling at $290 and $430 per tonne, respectively. Give the relative dollar profit margins per tonne of contained Na_2CO_3 for each of these grades, neglecting any costs other than those outlined in part (a). Which of the three grades would be the most profitable?

FURTHER READING

Environmental Protection Service, "Sodium Hydroxide." Technical Services Branch, Ottawa, 1984.

R.D.A. Woode, Sodium Carbonate. In "Industrial Inorganic Chemicals: Production and Uses" (R. Thompson, ed.) pp.123–148. The Royal Society of Chemistry, Cambridge, UK, 1995.

REFERENCES

1. "Kirk-Othmer Encyclopedia of Chemical Technology," 2nd ed., Vol. 4, pp. 2,7. Wiley, New York, 1964.
2. R.C. Weast, ed., "Handbook of Chemistry and Physics," 51st ed., pp. F-64, B-78. Chem. Rubber Pub. Co., Cleveland, OH, 1970.
3. C. J. S. Warrington and R. V. V. Nichols, "A History of Chemistry in Canada," p. 96. Sir Isaac Pitman and Sons, Toronto, 1949.
4. R. N. Shreve and J. A. Brink, Jr., "Chemical Process Industries," 4th ed., pp. 166, 211. McGraw-Hill, Toronto, 1977.
5. R. W. Thomas and P. Farago, "Industrial Chemistry," p. 63. Heinemann, Toronto, 1973.
6. W. L. Faith, D. B. Keyes, and R. L. Clark, "Industrial Chemicals," 3rd ed., pp. 483, 694. Wiley, New York, 1965.
7. R. H. Perry, C. H. Chilton, and S. D. Kirkpatrick, eds., "Chemical Engineers' Handbook," pp. 9-6, 3-116. McGraw-Hill, Toronto, 1963.
8. E. Notidis and H. Tran, Survey of Lime Kiln Operation and Ringing Problems. *TAPPI J.* 76(5), 125–131, May (1993).
9. R. N. O'Brien and W. F. Hyslop, Oxygen Enhancement of the Air. Br. Pat. Appl. 8,003,884, (1980).
10. Corson (of Lime Hydration Process) Honoured, *Chem. Trade J.* **156**, 414 (1965).
11. F. S. Taylor, "A History of Industrial Chemistry," Reprint ed., p. 182. Arno Press, New York, 1972.
12. J. A. Kent, ed., "Riegel's Industrial Chemistry," p. 129. Reinhold, New York, 1962.
13. F. A. Lowenheim and M. K. Moran, "Faith, Keyes and Clark's Industrial Chemicals," 4th ed., pp. 706, 714. Wiley-Interscience, Toronto, 1975.
14. G. A. Lane, J. S. Best, E. C. Clarke, D. N. Glew, G. C. Karris, S. W. Quigley, and H. E. Rossow, "Solar Energy Subsystems Employing Isothermal Heat Sink Materials," ERDA Contract No. NSF-C906. The Dow Chemical Company, Midland, MI, 1976.
15. "Minerals Yearbook, Metals and Minerals," Vol. I, 1994. U.S. Dept. of the Interior, Bureau of Mines, Washington, DC, 1996, and earlier editions.
16. "United Nations Statistical Yearbook 1993," 40th ed. United Nations, New York, 1995, and earlier editions.
17. R. N. Shreve, "The Chemical Process Industries," p. 274. McGraw-Hill, New York, 1945.
18. E. W. Washburn, ed., "International Critical Tables," Vol. VII, p. 305. McGraw-Hill, New York, 1930.

19. R. C. Weast, ed., "Handbook of Chemistry and Physics," 56th ed., p. F-86. Chem. Rubber Publ. Co., Cleveland, OH, 1975.

20. H. Stephen and T. Stephen, eds., "Solubilities of Inorganic and Organic Compounds," Vol. I, page 115. Pergamon, New York, 1963.

21. J. A. Kent, ed., "Riegel's Industrial Chemistry," p. 158. Reinhold, New York, 1962.

22. J. C. Olsen and O. G. Direnga, *Ind. Eng. Chem.* **33,** 204 (1941).

23. L. F. Goodwin, *J. Soc. Chem. Ind.* **45,** 360T–361T (1926).

24. R. M. Stephenson, "Introduction to the Chemical Process Industries," p. 56. Reinhold, New York, 1966.

25. L.W. Codd, *et al.,* eds., "Chemical Technology: An Encyclopaedic Treatment," Vol. 1, p. 23. Barnes and Noble, New York, 1968.

26. F. A. Lowenheim and M. K. Moran, "Faith, Keyes and Clark's Industrial Chemicals," 4th ed., p. 741. Wiley-Interscience, Toronto, 1975.

8

ELECTROLYTIC SODIUM HYDROXIDE, CHLORINE, AND RELATED COMMODITIES

> . . . apparatus for the production of Chloride gas
> and of metallic Sodium through the agency of dy-
> namic electricity.
> —A. L. Nolf, 1882

8.1. ELECTROCHEMICAL BACKGROUND AND BRINE PRETREATMENT

All electrolytic routes to sodium hydroxide and chlorine from sodium chloride brine have to contend with keeping the highly reactive products separated. Sodium and chlorine products would react vigorously together to return starting salt (Eq. 8.1, 8.2).

$$Na^+ + 1\ e^- \rightarrow Na\ (or, with\ water \rightarrow NaOH + 1/2\ H_2) \qquad 8.1$$

$$Cl^- - 1\ e^- \rightarrow 1/2\ Cl_2 \qquad 8.2$$

A sodium hydroxide product, without separation, would react vigorously with chlorine to give sodium chloride and sodium hypochlorite (Eq. 8.3, 8.4), so that similar precautions are required with either option.

$$2\ Na + Cl_2 \rightarrow 2\ NaCl \qquad 8.3$$

$$2\ NaOH + Cl_2 \rightarrow NaCl + NaClO + H_2O \qquad 8.4$$

Two main procedures have for many years dominated the methods used to keep these electrolytic products apart. One involves the separation of the electrochemical cell into two compartments by a porous vertical diaphragm which permits the passage of ions, but keeps the products separated. The other employs a flowing mercury cathode to continuously carry away sodium, as an amalgam, from the brine and chlorine streams of the cell. Diaphragm cells, with which efficient production can be maintained using natural brines (brines solution mined from underground deposits) account for some three-quarters of North American chloralkali production. Mercury cells, which require a higher degree of brine purity for efficient operation, are used by the majority of European chloralkali producers who mostly use vacuum pan or recrystallized salt to make up the electrolyte feed. Two other types of solution cells have been employed in the past, the Billiter, or bell-jar type cell, in which chlorine was drawn off by a bell-shaped housing suspended in the brine and surrounding the anode, and the horizontal diaphragm cell, which is still employed in some European installations [1,2].

The initial stages of brine pretreatment are quite similar for both the diaphragm and the mercury cell. If calcium sulfate is present, removal requires the addition of a barium carbonate suspension (Eq. 8.5) or if present as the chloride it is removed by the addition of sodium carbonate (Eq. 8.5, 8.6).

$$CaSO_4 + BaCO_3 \rightarrow CaCO_3\downarrow + BaSO_4\downarrow \qquad 8.5$$

$$CaCl_2 + Na_2CO_3 \rightarrow CaCO_3\downarrow + 2\,NaCl \qquad 8.6$$

Magnesium ion requires a higher pH for effective removal, which is accomplished by the addition of small amounts of sodium hydroxide (Eq. 8.7).

$$MgCl_2 + 2\,NaOH \rightarrow Mg(OH)_2\downarrow + 2\,NaCl \qquad 8.7$$

The bulk of the precipitated salts is removed by settling, either in the treating unit itself or in a separate clarifier. This is followed by a polishing filtration to give a brine containing 5 ppm or less of Ca^{2+} and Mg^{2+}. Acidification with hydrochloric acid to about pH 2 produces a brine which is then suitable as feed for a diaphragm cell.

For mercury cell operation brine purity requirements are more stringent, particularly with regard to traces of transition metal ions which can cause severe process upset by lowering the hydrogen overvoltage on mercury. Normally much of the control required will be achieved by simple adsorption of interfering ions onto the precipitated calcium and magnesium flocs. Iron is simultaneously precipitated as the hydroxide (Eq. 8.8).

$$FeCl_3 + 3\,NaOH \rightarrow Fe(OH)_3\downarrow + 3\,NaCl \qquad 8.8$$

If this is not sufficient, the general practice is to further increase the volume of precipitate by deliberate addition of calcium and/or iron salts until control of vanadium, chromium, manganese, and iron to 0.5 ppm or less is achieved. To further increase the hydrogen overvoltage on mercury the feed to mercury cells is also acidified with hydrochloric acid to pH 2 to 3 prior to use. From this point on details of the two processes differ and will be treated separately.

8.2. BRINE ELECTROLYSIS IN DIAPHRAGM CELLS

The first electrolytic production of chlorine was by the electrolysis of a potassium chloride brine with the coproduction of potassium hydroxide by the Griesheim Company, in Germany, in 1888 [1]. Sodium hydroxide and chlorine were first commercially produced electrolytically in 1891 in Frankfurt,

Germany, using technology advanced in the patents of Mathes and Weber and in 1892, and in the United States at Rumford Falls, Maine, by the Electrochemical Company. Both of the early cells used an operating procedure which amounted to a scaled-up version of a laboratory electrolyzer (Fig. 8.1a) and were operated in batch modes.

The cell was loaded with saturated potassium chloride brine, indirectly heated to 80–90°C with steam pipes, and then a current of about 3500 A was passed through the cell for a period of about 3 days. During this time the chlorine and hydrogen produced were collected, and the potassium hydroxide in solution rose to about 7% (Eq. 8.9, 8.10).

$$\text{Anode:} \quad 2\ Cl^- - 2\ e^- \rightarrow Cl_2 \qquad\qquad 8.9$$

$$\text{Cathode:} \quad 2\ H^+ + 2\ e^- \rightarrow H_2 \qquad\qquad 8.10$$

This brought about accumulation of potassium ions and hydroxide ions in the electrolyte. The electrolyte was then pumped out and processed to recover the potassium hydroxide, and the cell was recharged with fresh potassium chloride and water to repeat the whole process on a batch basis.

The Griesheim method was successful and represented a significant advance in simplicity over the earlier chemically based causticization procedures used to produce potassium hydroxide and chlorine. At the same time, it was slow, and it made for inefficient material handling with the frequent necessity for emptying and filling of cells. It also had a low electrical efficiency, partly because solution concentrations were not optimum and partly because of side reactions caused by diffusion of the products through the diaphragm. These side reactions consumed both of the electrolytic products, in the process producing sodium chlorate (Eq. 8.11–8.15).

$$2\ OH^- + Cl_2 \rightarrow 2\ HClO \qquad\qquad 8.11$$

$$HClO + H_2O \rightleftharpoons H_3O^+ + ClO^- \qquad\qquad 8.12$$

$$2\ ClO^- \rightarrow ClO_2^- + Cl^- \qquad\qquad 8.13$$

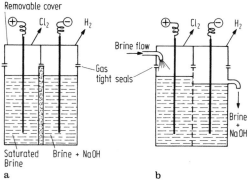

FIGURE 8.1 (a) Schematic diagram of a Griesheim-type cell which operated on a batch basis. The product was pumped out after electrolysis for about 3 days. (b) Schematic diagram showing the LeSueur percolating diaphragm which changed the mode of cell operation to a continuous basis and decreased the concentration of impurities in the sodium hydroxide product.

$$ClO_2^- + ClO^- \rightarrow ClO_3^- + Cl^- \qquad\qquad 8.14$$

$$Na^+ + ClO_3^- \rightleftharpoons NaClO_3 \qquad\qquad 8.15$$

Many of these difficulties were overcome by the Rumford Falls, Maine, chor-alkali operation in 1892 by incorporating improvements developed by E. A. LeSueur of Ottawa, Canada, just prior to this [3]. These innovations centered on a percolating diaphragm which allowed the continuous passage of both ions and brine solution from the anode to the cathode compartment [4] (Fig. 8.1b). To obtain brine flow in the desired direction, a tube fed saturated brine continuously into the anode compartment. An outlet was placed at the desired cathode compartment solution level, lower than the anode compartment level, which allowed for the continuous removal of nonelectrolyzed brine and the caustic product. These measures jointly contributed to a positive pressure in the anode compartment and a continuous brine flow through the diaphragm to the cathode compartment. In this way, back-diffusion of hydroxide ions through the diaphragm was largely prevented, which both avoided consumption of products and formation of impurities, and also achieved continuous rather than batch operation.

These features, the percolating diaphragm together with continuous brine entry and brine and caustic removal, are incorporated into virtually all modern diaphragm cell designs. This includes the North American Hooker, Diamond Alkali, and Dow filter press designs, which all use vertical diaphragms, as well as the European horizontal diaphragm designs such as the Billiter cell and the I.G. Farben-industrie modification, which introduced large flutes to the diaphragm and cathode for increased capacity [5.] A particular feature of interest with the horizontal diaphragm is the manner in which hydroxide ion back-diffusion was further minimized by maintaining a hydrogen gas head between the catholyte (cathode compartment electrolyte) and the diaphragm.

The percolating diaphragm of most modern cells consists of a thick asbestos fiber pad, supported on a crimped heavy iron wire mesh or punched steel plate located on the cathode side of the diaphragm. In this way the wire diaphragm support also serves as the cathodic electrode which places it as close as possible to the anode for electrical efficiency while remaining on the cathode side of the diaphragm (Fig. 8.2). The iron does not corrode under these severe conditions (contact with the hot saturated brine) because it is made electrochemically cathodic (cathodic protection) from its function in the cell. In this way it continuously supplies electrons to the solution such that the metal itself is not electrochemically scavenged to do so (Eq. 8.16).

$$Fe_{(s)} - 3\ e^- \rightarrow Fe_{(aq)}^{3+} \qquad\qquad 8.16$$

In this severe service, however, even the durable asbestos fiber pad of which the diaphragm is composed has to be replaced at intervals ranging from 1 to 3 yr. During this time, graphite particles from the anode and insoluble impurities in the brine gradually accumulate on the anode side and into the pad of the diaphragm, reducing its permeability. Also, the passage of electrical power through the diaphragm causes the asbestos fibers of the pad to swell. To maintain the chlorine and caustic production rate under these conditions, the brine level in the anode compartment has to rise, which raises the

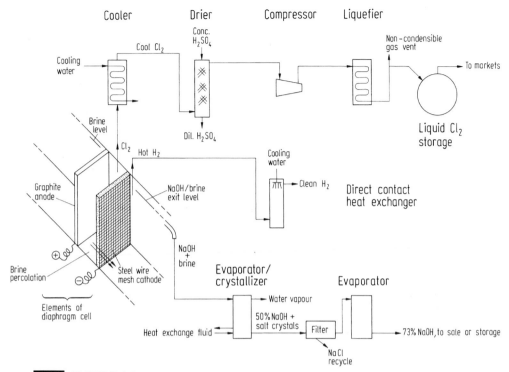

FIGURE 8.2 Diagram of the components of a commercial asbestos diaphragm cell, and flowsheet for the electrolytic production of chlorine and sodium hydroxide. All commercial cells employ a multiple series of anodes and diaphragms in a single brine container, only one of which is shown here. The asbestos layer is located on the anode side of the steel wire grid cathode, and located relatively much closer than shown here to minimize operating voltage.

differential pressure across the diaphragm and in this way maintains the brine flow rate through it. During this time the operating voltage is also raised to offset the decreased ion mobility. Diaphragm permeability can be improved several times, while in service, by a weak acid wash which serves to remove "hardness." But eventually there is no further adjustment possible and the cell is shut down and dismantled for overhauling. The spent asbestos is removed from the iron grid with high-pressure water jets, and the water slurry is collected in a disposal pit. New asbestos is then slurried in brine, placed in the anode compartment, and by means of a vacuum on the hydrogen exit of the cell it is applied in an even coating to the cathode grids. After fitting new anodes, if necessary, the cell is then reassembled. It may then be placed back in service.

The saturated (about 300 g/L NaCl) pretreated brine is preheated to about 90°C and fed to the anode compartment as a broken stream, *broken* to avoid electrical losses. Preheating the brine both increases cell conductivity by increasing ion mobility from the lower solution viscosities resulting, and decreases chlorine solubility in the brine. In this way, less dissolved chlorine is carried through the diaphragm in the percolating brine which contributes to higher cell current efficiencies and lower concentrations of impurities in the cathode exit brine (Eq. 8.11–8.15). It also tends, however, to result in increased graphite wear rates.

Anode electrode reactions produce chlorine gas which is removed from the vapor space above the anode compartment by ducting made of polyvinyl chloride (PVC) or other chemically resistant material, under slight negative pressure (Eq. 8.17–8.19).

$$Cl^-_{(aq)} \rightarrow Cl^\circ_{(aq)} + 1\ e^- \qquad\qquad 8.17$$

$$2\ Cl^\circ_{(aq)} \rightarrow Cl_{2(aq)} \qquad\qquad 8.18$$

$$Cl_{2(aq)} \rightarrow Cl_{2(g)} \qquad\qquad 8.19$$

High brine temperatures assist removal of chlorine from the brine and the slight negative pressure both aids in chlorine removal and provides a safeguard in the event of any minor leaks in the chlorine collection system.

At the cathode, which is the crimped iron wire grid supporting the diaphragm, electrons are added to water to form hydrogen gas and hydroxide ions (Eq. 8.20–8.22).

$$H_2O_{(aq)} + 1\ e^- \rightarrow H^\circ + OH^- \qquad\qquad 8.20$$

$$2\ H^\circ_{(aq)} \rightarrow H_{2(aq)} \qquad\qquad 8.21$$

$$H_{2(aq)} \rightarrow H_{2(g)} \qquad\qquad 8.22$$

The hydrogen gas is removed under slight positive pressure. It might be thought reasonable to expect that it would be sodium ions which are electrolyzed, not water, since these are also present at high concentrations [6.] Sodium, if formed, would react with water under these conditions to yield sodium hydroxide and hydrogen gas, the observed products. But examination of the relevant electrochemical data shows that this is not so. The deposition potential for hydrogen is -0.828 V, which together with the -0.30 V hydrogen overpotential on iron gives a total of -1.128 V required for hydrogen generation. The single electrochemical potential for sodium deposition is -2.71 V, more than twice that required for hydrogen deposition [7]. Therefore, hydrogen and hydroxide ion are the direct, not indirect, cathode electrochemical products. So the overall products of brine electrolysis in a diaphragm cell are sodium hydroxide, chlorine, and hydrogen (Eq. 8.23).

$$NaCl + H_2O \xrightarrow{\text{electrol.}} NaOH + 1/2\ Cl_2 + 1/2\ H_2 \qquad\qquad 8.23$$

In practice, brine and current flows to a diaphragm cell are synchronized to only electrolyze about 30–40% of the sodium chloride present in the feed brine solution. To electrolyze a larger fraction of the sodium chloride present than this would appear to be advantageous in that less salt would have to be removed from the cell effluent to produce a grade of sodium hydroxide suitable for sale. However, to do this would require an increased current flow for the same rate of brine flow through the diaphragm which would, in turn, necessitate a higher operating voltage. The higher voltage requirement would result in higher electrical power cost for each tonne of product made. Or, alternatively, it would require that the current flow be maintained at its original value while the brine flow rate is decreased. Decreasing the brine flow would increase hydroxide ion back-diffusion through the diaphragm with two undesirable consequences. Loss of both hydroxide ion and chlorine cell products would occur, resulting in an impurity in the sodium hydroxide (Eq. 8.11–8.15). Also the graphite consumption would increase from the increased rate of reaction of the anode material with oxygen from the electrolyzed hydroxide ion (Eq. 8.24–8.25).

$$2 \; OH^- \rightarrow H_2O + 1/2 \; O_2 + 2 \; e^- \qquad \qquad 8.24$$

$$2 \; C + O_2 \rightarrow 2 \; CO\uparrow \qquad C + O \rightarrow CO\uparrow \qquad \qquad 8.25$$
$$CO + O \rightarrow CO_2\uparrow \qquad 2 \; CO + O_2 \rightarrow 2 \; CO_2\uparrow$$

Even with only about 30–40% electrolysis of the brine, anode life is limited to about 2–3 yr from graphite consumption at the rate of 2–3 kg/tonne of chlorine.

8.2.1. Purification of Diaphragm Cell Products

Chlorine and hydrogen as obtained directly from the diaphragm cell are both hot and wet, and the 8 to 12% sodium hydroxide liquid cell effluent still contains 12 to 19% dissolved (and "undecomposed") sodium chloride. A few applications of these products can use them in the form obtained directly from the cell. For example, propylene chlorohydrin may be ring-closed to propylene oxide using the crude sodium hydroxide solution still containing sodium chloride (Eq. 8.26). However, the majority of uses require at least a simple cleanup procedure of the crude products prior to use.

$$\underset{\displaystyle CH_3 - CH - CH_2}{\overset{\displaystyle Cl \qquad OH}{\mid \qquad \mid}} + NaOH/NaCl \rightarrow \underset{\displaystyle CH_3 - CH - CH_3}{\overset{\displaystyle O}{\diagup \; \diagdown}} + 2 \; NaCl$$
$$8.26$$

Elemental chlorine under ordinary conditions is a heavy (dense) greenish-yellow gas, which may be maintained in the liquid form at temperatures below −35°C, its atmospheric pressure boiling point. Important considerations for its collection and purification are that it is highly toxic (hence the slight negative pressure for collection) and it is highly reactive with many ordinary metals, especially when wet, and with organic substances generally. It is also slightly soluble in water, with which it partially reacts. The first step in cleanup is indirect cooling in process water-cooled heat exchangers constructed of polyvinyl chloride, titanium, or fiberglass-reinforced plastic which removes much of the water vapor and entrained brine droplets from the chlorine. It is then contacted countercurrently to concentrated sulfuric acid, which has a strong dehydrating action. This takes place in one or more fiberglass-reinforced plastic towers packed with chemical stoneware and placed in series. The packing is used to provide a large surface area for contact of the chlorine with the sulfuric acid. Thus the acid consumed is only 2 to 5 kg/tonne of chlorine. Once dried, steel piping may safely be used to move chlorine. However, its reactivity must still be kept in mind during compression and liquefaction of the gas since chlorine can spontaneously react with iron at temperatures above 150°C. Care must also be taken to ensure maintenance of anhydrous conditions during these steps.

Much of the chlorine produced will be simply compressed and piped if it is to be consumed locally, either by the same company producing it ("captive consumption") or by another. But if it has to be shipped any distance, chlorine will normally be liquefied by compression and cooling to minimize storage and shipping volumes. Early compressors were reciprocating, and operated in a similar fashion to a piston engine but with graphite piston rings and using no cylinder lubricant. Larger facilities today use several stages of centrifugal turbine compressors [8]. The compressed hot gases are first air or water cooled and then further chilled either by Joule-Thomson effect cooling obtained by expansion of a portion of the cooled compressed gas, or occasionally by using liquid ammonia or other refrigerant in a separate circuit. Liquefaction may be achieved at 700 to 1200 kPa (about 7–12 atm) pressure at 18°C or at 100 kPa at −40°C, just below the atmospheric pressure boiling point. Most producers settle for intermediate liquefaction conditions of about 200 kPa and 4°C [9].

Small amounts of noncondensible gases, called "blow gas" or "sniff gas," which consists of traces of hydrogen, oxygen, carbon dioxide, and air, accumulate during liquefaction. This gas, which also contains some residual chlorine, is vented through an aqueous sodium hydroxide or sodium hypochlorite stream to capture the traces of chlorine before discharge. In mercury cell plants, the aqueous sodium hypochlorite stream produced in this way may be used for mercury recovery. Hydrogen gas, which tends to get more concentrated in the chlorine gas stream as liquefaction proceeds, must not be allowed to rise to more than 5% in the liquefaction or venting systems or there is risk of an explosion from the sudden reaction of chlorine with it.

The primary hydrogen stream from the cell is also hot and wet. Since hydrogen has a low solubility in water it is usually cooled and cleaned of chlorine, salt, etc., by a direct, countercurrent contact with a cool spray of dilute sodium hydroxide in water. Clean hydrogen exits from the top of the cooler through baffles or plastic mesh placed so as to remove entrained water droplets.

Crude 10% sodium hydroxide containing sodium chloride is purified in a similar manner to the product of the causticization process. The water is evaporated in nickel or nickel-clad steel (to reduce corrosion) multiple-effect evaporators to about 50% sodium hydroxide concentation. At this concentration sodium chloride is only about 1% soluble (2%, on a dry basis) in the more concentrated caustic so that the bulk of it crystallizes out and is filtered off. This quite pure sodium chloride is recycled to the cells. For many purposes, such as for pulp and paper production, this purity of 50% sodium hydroxide is quite acceptable. If higher purities are required, sodium hydroxide may be separated from residual water and salt by chilling to the double hydrate crystals $NaOH \cdot 2H_2O$, m.p. about 6°C, or as $NaOH \cdot 3.5H_2O$, with a m.p. of about 3°C, or by counter-current extraction [10]. Sodium hydroxide is obtained which contains only 2–3 ppm sodium chloride, equivalent to the purity of the mercury cell product ("rayon grade") [11]. Concentrations of 3 and 100% sodium hydroxde (see details, Section 7.5) are also marketed.

8.3 BRINE ELECTROLYSIS IN CHLORATE CELLS

A cell similar in principle to the diaphragm cell described above but operated without a diaphragm gives the same initial products but allows these to react (Eq. 8.23, 8.11–8.15). Thus the initial electrochemical products ultimately form sodium chlorate as the final cell product from aqueous sodium chloride electrolysis. The processes involved are as follows. Part of the product formation involves the "chemical chlorate" formation just outlined (Eq. 8.11–8.15), and part of it forms from "electrolytic chlorate" formation (Eq. 8.27).

$$6 \; ClO^- + 3 \; H_2O - 6 \; e^- \rightarrow 2 \; ClO_3^- + 6 \; H^+ + 4 \; Cl^- + 3/2 \; O_2 \qquad 8.27$$

So the overall process may be summarized as given in Eq. 8.28 [12].

$$NaCl_{(aq)} + 3 \; H_2O \xrightarrow[\text{6 Faradays}]{\text{electrolysis}} NaClO_{3(aq)} + 3 \; H_2 \qquad 8.28$$

Hydrogen, the primary gaseous product, is used to drive the brine solution to be electrolyzed through the chlorate cell and up to the vessel which serves as the chemical chlorate reactor. Normally chlorate cells are operated with added sodium dichromate inhibitor present in the brine to prevent corrosion which could otherwise be caused by hypochlorite ion.

Theoretical and actual operating voltages are 1.71 and about 3.5 V, respectively, and current efficiency is about 94%. These electrochemical parameters give a power consumption of 4950–6050 kWh per tonne of sodium chlorate for these cells [13].

The major market for sodium chlorate is for the preparation of the chlorine dioxide used for bleaching of wood pulp (Chap. 15). This market has jumped by an order of magnitude in the last 10 years with the growth in the need to replace part or all of the chlorine previously used for this purpose by chlorine dioxide. Other uses of sodium chlorate are as a weed killer, for the making of matches and fireworks, and for the tanning of hides. A small market exists for sodium (and potassium) perchlorate, made by electrolysis of sodium chlorate using platinum electrodes (Eq. 8.29).

$$4 \; NaClO_3 \xrightarrow{\text{electrol.}} 3 \; NaClO_4 + NaCl \qquad 8.29$$

8.4. ELECTROCHEMICAL ASPECTS OF BRINE ELECTROLYSIS

The overall chemical changes involved in the electrolysis of aqueous sodium choride can be expressed as given in Eq. 8.30.

$$NaCl_{(aq)} + H_2O_{(\ell)} \xrightarrow{\text{electrol.}} NaOH_{(aq)} + 1/2 \; H_{2(g)} + 1/2 \; Cl_{2(g)} \qquad 8.30$$

To enable thermodynamic insight into the overall energetics of this process, it may be broken down to its component enthalpy changes as follows (Eq. 8.31–8.33).

	$\Delta°H$	
$Na_{(s)} + 1/2\ Cl_{2(g)} \rightarrow NaCl_{(aq)}$	-407 kJ (-97.0 kcal)	8.31
$H_{2\ (g)} + 1/2\ O_{2(g)} \rightarrow H_2O_{(\ell)}$	-286 kJ (-68.3 kcal)	8.32
$Na_{(s)} + 1/2\ O_{2(g)} + 1/2\ H_{2(g)} \rightarrow NaOH_{(aq)}$	-469 kJ (-112 kcal)	8.33

To obtain a net enthalpy change for the process of Eq. 8.30, the signs of the first two processes have to be changed to correspond to these in the opposite direction to that given. Then summing the three components ($+\ 407 + 286 - 469$) kJ gives a net enthalpy change of $+224$ kJ, i.e., that it is an endothermic process.

The net enthalpy change of a process may be used to obtain the theoretical voltage required for electrochemical equilibrium by using the Gibbs-Helmholtz equation [13]. It also can be summed for the two component electrochemical changes taking place (Eq. 8.34, 8.35).

$2\ Cl^- \rightarrow Cl_2 + 2\ e^-$	-1.360 V	8.34
$2\ H_2O + 2\ e^- \rightarrow H_2 + 2\ HO^-$	-0.828 V	8.35

In both cases a value of close to 2.20 V is obtained, which is defined as the "theoretical cell voltage" or "theoretical decomposition voltage." Actual operating voltage is 3.7–4.0 V. Factors such as the finite spacing required between electrodes, the cathode overpotential requirement for deposition of hydrogen, the resistance of the hot brine, and the resistance to ion diffusion imposed by the presence of the diaphragm (and other minor components) all contribute to this higher than theory operating voltage. The ratio of the theoretical cell voltage to the actual operating voltage is defined as the "voltage efficiency" of the cell. Expressed as a percentage ($\times$ 100) the voltage efficiency normally lies in the 55 to 60% range. Voltage efficiencies usually drop somewhat with increased production rates, as the anodes wear (increasing electrode separation) or as diaphragm permeability decreases.

The current consumption, or electrical flow rate in amperes required to operate the cell is the factor that directly relates to the amount of electrochemical product collected. One Faraday (1 mol of electrons, after the original discoverer) of electricity is required to deposit 1 g equivalent weight of a substance in an electrochemical reaction (Eq. 8.36).

$$1\ Faraday = 96,494\ coulombs/(g\ equiv.\ wt.)\qquad or$$
$$1\ Faraday = 96,494\ ampere\ sec/(g\ equiv.\ wt.)\qquad 8.36$$
$$since\ 1\ coulomb\ (C) = 1\ ampere\ second\ (A \cdot sec)$$

So chloride, being a monovalent ion, would theoretically require 3.15×10^4 amperes to be converted to chlorine at the rate of 1 tonne per day (Eq. 8.37).

$$\frac{96,494\ A \cdot sec/(g\ equiv.\ wt.)}{(24 \times 60 \times 60)\ sec/day} \times \frac{10^6\ g/tonne}{35.453\ g/(g\ equiv.\ wt.)} \qquad 8.37$$
$$= 3.1502 \times 10^4\ ampere\ days/tonne\ chlorine$$

Side reactions from hydroxide ion back-diffusion across the diaphragm and from the presence of some dissolved chlorine in the brine percolating from the anode to the cathode compartment, as well as the production of small amounts of oxygen at the anode, all consume electrical power without contributing to the primary products of the cell. These factors reduce the cell efficiency to somewhat less than theory, so that about 32.5 kA is actually required for a daily tonne of chlorine. The ratio of the theoretical current requirement to the actual current consumed is called the "current efficiency", and ($\times$ 100) is thus about 92–96%. By convention, the current efficiency is normally understood to refer to the cathode current efficiency, which for a diaphragm chloralkali cell is nearly the same, though marginally higher than the anode current efficiency. Ordinarily, cell operators expect current efficiencies to fall slightly with increasing decomposition efficiencies of the cell (decreasing salt to sodium hydroxide ratios in effluent brine) for the reasons outlined earlier.

Current density refers to the total current flow in kiloamperes divided by the anode electrode area in square meters, expressed as kA/m^2. High current densities are desirable, particularly for electrochemical processes which yield unstable products. With current densities of 2–3 kA/m^2 electrolytic products of the diaphragm cell are rapidly moved from the site of formation which decreases side reactions and maximizes current efficiencies [14]. High current densities, however, do increase heat generation, anode wear, and the operating voltage so that lower current densities (and more cells) are better if the cells can be made cheaply.

The overall "energy efficiency" of the cell is the voltage efficiency times the current efficiency and relates to the gross *power* consumption for any given quantity of product, e.g., a tonne of chlorine. By lowering the energy efficiency an increased operating voltage directly increases the power cost per tonne of product. For example, the mean power requirement for chlorine production in a diaphragm cell amounts to some 2700–2900 kWh/tonne. But if the cell is operated below capacity, at 3.2 V, it would only require some 2496 kWh/tonne (3.2 $\times$ 32.5 $\times$ 24 hr). Operating the cell at near the end of its diaphragm life and/or pushing the cell to produce somewhat beyond its rated capacity by running at 4.2 V would require *3276* kWh/tonne, an additional 31% more power (and power cost) for the same mass of chlorine as at 3.2 V. Since the power costs some 60 to 70% of the total chlorine production costs, these voltage increments add significantly to operating cost [15]. Nevertheless, for temporarily slack or tight markets operating cells at 5 to 10% below or above rated capacities (or actually taking cells out of service) may be used to adjust production rate to match demand.

Finally, the electrochemical "decomposition efficiency" is defined as the equivalents produced divided by the equivalents charged (or entering) the electrochemical cell (Eq. 8.38).

$$\text{Decomposition efficiency} = \frac{\text{(equivalents produced)}}{\text{(equivalents charged)}}, \quad \text{or} \quad \quad 8.38$$
$$= \frac{\text{(moles electrolyzed)}}{\text{(moles fed to cell)}}$$

TABLE 8.1   Typical Operating Characteristics of Relatively Small and Large Diaphragm Chloralkali Cells

	Rated current flow (A)	
	30,000[a]	**150,000**[b]
Current efficiency	96.5	96
Current density[c], kA/m^2	1.29	2.7
Operating voltage[c] (nominal)	3.82	3.48
Operating temperature, °C	90	85–90
Cell effluent:		
Temperature, °C	95	—
% NaOH	10.5	10.5
NaClO$_3$, g/L	0.07	0.05–0.25
Cell output, tonnes per day:		
Chlorine	0.92	4.60
NaOH	1.03	5.19
Power,[c] kWh/tonne Cl$_2$	3000	2740

[a]Data calculated from Sconce [1].
[b]Data selected from Puschaver [15] and Thompson [16].
[c]The improved performance of the larger cell is at least partly the consequence of the dimensionally stable anodes used (see Section 8.5).

This quantity relates to the proportion of the salt present in the incoming solution which is electrolyzed on its passage through the cell. For the reasons previously discussed, for a diaphragm cell this electrolyzed fraction is about 0.30 to 0.40, or ($\times$ 100) 30 to 40%. Since this represents a mole (not weight) fraction electrolyzed, the crude 10–12% sodium hydroxide in water produced also contains 12–15% by weight dissolved sodium chloride.

The interrelationships between these various operating and electrolytic characteristics for diaphragm cells are given in Table 8.1.

8.5. BRINE ELECTROLYSIS IN MERCURY CELLS

Use of the knowledge that an electrolytic cell employing mercury as a cathode would cause sodium deposition in the mercury, rather than hydrogen generation, was made simultaneously in 1892 by H. Y. Castner in the U.S.A, and by Karl Kellner in Austria, each independently of the other [1]. Their designs employed a rocking cell to move mercury from the sodium amalgam (sodium/mercury alloy or solution of sodium in mercury) forming side of the cell to the sodium amalgam hydrolysis side (Fig. 8.3). This design became a commercial success after further development by Solvay et Cie. This method of brine electrolysis gave no direct contact between the brine and the sodium hydroxide streams, so it was possible to produce a high-purity 50% sodium hydroxide solution directly. No concentration, sodium chloride crystallization, etc., was required. This provided such clear advantages for the high-purity sodium hydroxide market that Castner-Kellner rocking cells were used at an

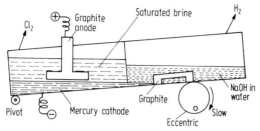

FIGURE 8.3 Vertical section of a Castner-Kellner rocking mercury cell which demonstrates the principle of operation. The cell had provision for brine entry and removal of depleted brine from the left compartment, and for purified water entry and removal of sodium hydroxide from the right compartment. The eccentric revolved slowly to move the mercury on a regular basis between the two compartments. A cell divider extended close enough to the bottom of the cell to maintain a mercury seal which prevented any mixture of the water phases of the two compartments.

Olin-Mathieson chloralkali operation at Niagara Falls until as recently as 1960, when they converted to a new mercury cell design. These newer designs now use variations on the theme of a long shallow sloping flat-bottomed trough in which the electrolysis is conducted. Pumps are used to circulate mercury continuously, as a thin film flowing on the bottom of the trough.

Pretreatment of brine to be used as feed for a mercury cell is the same as for diaphragm cell feed plus precautions to ensure that molybdenum, chromium, and vanadium ions are maintained at levels below 1 μg/L (1 ppb) and that (magnesium + iron) salts are kept below 1 mg/L (1 ppm) to avoid reducing the hydrogen overvoltage on mercury relative to the sodium deposition voltage. If one or more of these ions exceeds these limits, then hydrogen gas evolution in the electrolyzer (rather than sodium deposition in mercury) can become significant. This hydrogen contamination of the chlorine introduces an explosion hazard at the time the chlorine is compressed, especially if compression is accompanied by liquefaction, which would raise the concentration of hydrogen (Eq. 8.39).

$$H_{2(g)} + Cl_{2(g)} \rightarrow 2\ HCl_{(g)} \qquad \Delta H = -92.3 \text{ kJ/mol } (-22.1 \text{ kcal/mol}) \qquad 8.39$$

Generally these interfering ions in the brine are removed to sufficiently low concentrations by simple adsorption onto the colloidal precipitates of calcium carbonate, magnesium carbonate, and barium sulfate produced by conventional brine pretreatment. The only additional step required is acidification of the brine to about pH 4 with hydrochloric acid. This helps to maintain a high hydrogen overvoltage on mercury in order to minimize the formation of hydrogen in the electrolyzer (Table 8.2). The brine is then preheated to about 65°C before passage to the electrolyzer, for the same reasons as heating of the diaphragm cell feed.

Modern mercury cells comprise two key parts, the electrolyzer and the decomposer (Fig. 8.4). The electrolyzer consists of a slightly inclined (about 1 in 100) ebonite (a hard rubber) lined steel trough with an unlined, smoothly machined steel floor. This steel floor is connected to the negative side of a DC potential, and hence so is the mercury which covers it, which makes it the

TABLE 8.2   A Comparison of Mercury Cell Gas Composition for Graphite versus Dimensionally Stable Anodes[a]

	Graphite anode (%)	Dimensionally stable anode[b] (%)
Chlorine	97.5–98.5	97.5–98.5
Hydrogen	0.4–0.6	0.4–0.6
Oxygen	0.1–0.3	0.2–0.5
Nitrogen	0.3–0.7	0.3–0.7
Carbon dioxide	0.4–0.9	0.3–0.7

[a]Data from Puschaver [15].
[b]For details, see Section 8.7.

cathode. The upper anode elements of the electrolyzer consist of graphite blocks with spaces between them to allow the escape of chlorine. They are suspended with a 3- to 4-mm gap between the mercury film and the graphite. Both saturated brine and mercury enter at the top of the electrolyzer and move cocurrently through the electrolyzer while a high current flow is maintained. In its passage through the electrolyzer the sodium chloride concentration in the brine is decreased by electrolysis from an initial 300 g/L or so, to 260–280 g/L. In the process chlorine gas is generated and collected (Eq. 8.17–8.19), and a solution of 0.3 to 3% sodium metal in the mercury stream is produced (Eq. 8.40).

$$Na^+ + Hg + 1\ e^- \rightarrow Na_{(Hg)} \qquad\qquad 8.40$$

Since the electrolyzed fraction of the sodium chloride feed concentration is much lower than for the diaphragm cell, the decomposition efficiencies are correspondingly less, about 7 to 14%. The mercury cell is operated in this way to keep the electrolyzer operating voltage down.

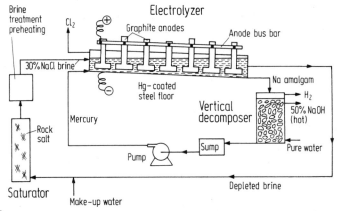

FIGURE 8.4   Mercury cell process for the electrolytic production of chlorine and sodium hydroxide. Packing in the vertical decomposer is graphite lumps. For treatment of cell products, see Fig. 8.2.

The equilibrium electrochemical potential, or theoretical voltage requirement for the electrolyzer reaction (Eq. 8.41) is 3.13 V, significantly higher than the 2.20 V required by the diaphragm cell.

$$NaCl_{(aq)} + Hg_{(\ell)} \xrightarrow{\text{electrol.}} Na_{(Hg)(\ell)} + 1/2 \ Cl_{2(g)} \qquad 8.41$$

However, because of better internal cell conductivities the operating potential is only about 4.3 V, which gives a voltage efficiency of about 70%.

The sodium amalgam cathode product of the electrolyzer of the mercury cell is the chief difference between this and the diaphragm cell. The production of sodium hydroxide from this stream uses a separate set of electrochemical reactions conducted in a decomposer or denuder. This unit is usually located below the electrolyzer to allow gravity feed of the sodium amalgam from the electrolyzer to the decomposer (Fig. 8.4). Deionized water is fed into the bottom of the decomposer to provide countercurrent flows of sodium amalgam and water. Mercury, stripped (or denuded) of sodium, is continuously drawn off the bottom of the decomposer, and a hot solution of 50 to 70% sodium hydroxide in water, plus hydrogen gas, from the top.

The decomposer is packed with graphite lumps or plates which are in direct contact with each other and with the metal shell of the decomposer. These provide an electrochemical short circuit for the potential generated by the sodium solution reaction. Thus, in this unit the sodium amalgam becomes the anode, electrons being given up by the sodium to the mercury matrix (Eq. 8.42) in the process of going into solution in the water phase.

$$Na_{(Hg)} \rightarrow Na^+ + Hg + 1 \ e^- \qquad 8.42$$

At the same time hydrogen evolution occurs at the graphite surfaces which provide the electrons for this process by the direct contact with the mercury phase, which simultaneously generates hydroxide ions in solution (Eq. 8.43).

$$H_2O + 1 \ e^- \rightarrow OH^- + 1/2 \ H_2 \qquad 8.43$$

There have been many attempts to utilize the approximately 1.8 V generated by the electrochemical reactions of the decomposer. However, it has not been found possible to do this *and* to maintain high sodium hydroxide concentrations and low residual sodium in the stripped mercury. The sodium hydroxide product obtained from the decomposer of a mercury cell is very pure, containing 0.001% or less sodium chloride. This product is referred to as "rayon grade caustic" because the high purity and low sodium chloride content makes it particularly suitable for rayon manufacture. This high purity is achieved without the special purification steps required for the diaphragm cell product.

Typical operating characteristics of mercury cells, which are capable of much higher production rates per cell than the largest commercially available diaphragm cells, are given in Table 8.3. Current efficiencies of the two types of cells are approximately the same, so that the approximately 20% additional power requirement per tonne of chlorine of the mercury cell results almost entirely from the higher operating voltage. This is not regarded as a significant penalty, since there is considerable energy saving in the areas of caustic (sodium hydroxide) evaporation and purification with the mercury cell.

■ **TABLE 8.3 Typical Operating Characteristics of Small and Large Mercury Chloralkali Cells**

	Rated current flow (A)	
	30,000[a]	400,000[b]
Current efficiency	95	96
Current density,[c] kA/m^2	4.8	12
Operating voltage[c] (nominal)	4.5	4.32
Cell output, tonnes per day:		
Chlorine	0.95	12.2
NaOH	1.08	13.8
Power, kWh/tonne Cl_2	3580	3400
Mercury:		
Inventory, kg	1410	—
Loss rate, g/tonne Cl_2	250[d]	—
Graphite consumption, kg/tonne Cl_2	2.6	
Electrode gap, mm	—	3–4

[a]Data calculated from Sconce [1].
[b]Data selected from Warrington and Newbold [4].
[c]Improved operating characteristics of the larger cell achieved by the use of dimensionally stable anodes (see Section 8.7).
[d]Current cells show mercury loss rates of 1 g or less per tonne of chlorine.

However, if the impure direct caustic product of the diaphragm cell can be used right at the plant site, there is a significant power saving to be gained. This is one of the reasons why some chloralkali producers operate both types of cells, so that the lower cost, lower grade product may be economically used localwhere feasible. Mercury loss in the products and waste streams of the mercury cell has also become a significant negative factor now in the choice of the chloralkali process used in many cases, as has the wild swings in mercury prices.

8.6. EMISSION CONTROL

8.6.1. Emission Control Aspects of Mercury Cells

Any of the products of brine electrolysis, chlorine, sodium hydroxide, and hydrogen can in themselves be hazardous if released. However, releases of these materials largely result from process upsets or accidental breakdowns, which may be minimized by the construction of fail-safe plants, and by safe transport and storage practices. Probably of greater long-term concern is the mercury loss experienced through the process streams of a mercury cell chloralkali operation. These losses can also carry over to the products of the diaphragm cell, even though this does not use mercury, if a common brine well or common salt dissolver is used for both sets of cells.

Precautions have always been practiced to contain the mercury (Table 8.3) of operating mercury cells, primarily for the value of the metal itself. The steel flasks of the metal are also covered with a layer of water to reduce mercury vapor loss during shipping and storage. However, in 1970 it was demonstrated that mercury ions and the free metal could be converted by natural processes to the far more toxic forms of mercury, methylmercury salts and dimethyl mercury even under water [17,18]. Industrialists, toxicologists, and legislators alike were alarmed by this discovery, which led to the rapid installation of control measures to drastically reduce mercury loss rates in Europe, Japan, and North America [19] (Table 8.4).

Electrolyzer mercury losses occur in the spent brine as finely divided mercury droplets, as dissolved mercuric chloride, and as the stable tetrachloromercury complex ($HgCl_4^{2-}$). Droplet entrainment in the brine stream occurs both directly, from the use of a high-speed centrifugal pump to return mercury from the decomposer to the upper end of the electrolyzer, and indirectly from the interaction of the two flowing liquids. Elemental chlorine present in the electrolyzing brine produces oxidizing conditions in the brine. Any contact of this hot, low-pH layer with finely divided mercury droplets oxidizes the mercury to mercuric chloride and tetrachloromercury dianion, which dissolve in the brine (Eq. 8.44, 8.45).

$$Hg^\circ + NaClO + H_2O + NaCl \rightarrow HgCl_2 + 2\,NaOH \qquad 8.44$$

$$Hg^\circ + NaClO + H_2O + 3\,NaCl \rightarrow HgCl_4^{2-} + 2\,Na^+ + 2\,NaOH \qquad 8.45$$

Total mercury in the spent brine may be 3–5 µg/g (ppm) amounting to a potential mercury loss rate of 2–15 g/tonne chlorine produced. However, since the depleted brine normally flows in a closed circuit either to underground wells or to a dissolver, this does not result in a mercury loss to the biosphere. Brine losses which occur during process upsets and brine pretreatment sludges, however, must both be treated before discharge to avoid mercury losses.

If diaphragm and mercury cells were run on the same dissolver or brine wells, operators still had to cope with possible mercury losses in the diaphragm cell products since the resaturated brine from either source still contained from 0.2–1 µg/g (0.2-1 ppm) mercury. Some settling occurs during resaturation, reducing the mercury concentration only slightly from that leaving the cell. This problem could be solved by an additional sulfide treatment step for the brine pretreatment for the diaphragm cells (Eq. 8.46, 8.47) [21,22].

$$S_n^{2-} + Hg^\circ \rightarrow HgS{\downarrow} + S_{n-1}^{2-} \qquad 8.46$$

$$Na_2S + Hg^{2+} \rightarrow HgS{\downarrow} + 2\,Na^+ \qquad 8.47$$

Care was required, however, to avoid treating the brine with excess sulfide, a natural tendency, or re-solution of mercuric sulfide would occur (Eq. 8.48).

$$HgS_{(s)} + S^{2-} \rightarrow HgS_{2(aq)}^{2-} \qquad 8.48$$

Dechlorination of the brine by sparging with air was also necessary, sometimes also with a reductant such as hydrazine prior to sulfide treatment. Otherwise precipitate reoxidation and resolution would occur (Eq. 8.49).

$$HgS + 4\,ClO^- \rightarrow HgCl_4^{2-} + SO_4^{2-} \qquad 8.49$$

TABLE 8.4 Grams of Mercury Lost per Tonne of Chlorine Produced by Mercury Cell Electrolysis Plants[a]

Source of loss	Canadian practice prior to 1970	Canada 1971	Sweden 1969	Sweden Stenungsund	Britain 1976	Canada 1974	Japan 1974
In products:							
Chlorine	0.05	0.05	<0.5	Slight	1.4	0.6	0.024
Caustic soda	1–11.5	1.75	1.0	0.4–1.0	—	—	—
To sewer:							
Hydrogen coolers	0–100	—	—	—	—	—	—
Basement drains	15–150	0.15–1.25	<5.0	0.5	18	—	0.00
Brine sludges	1–25	—	<2.0	0.11	—	0.43	—
To atmosphere:							
End box vents	1–10	1.25	not given	—	10–25	—	—
Hydrogen	10–200	1.25	1.0	<0.5	3.6	5.58	0.76
Ventilation	1–10	2.5–5.0	variable	ca. 1.0	1.5–2.5	—	—
Losses not accounted for:	15	—	—	—	0.2	6.50	8.68
Solid wastes:							
From brine system	—	1.25–5.0	—	0.2–0.6	0.82	—	—
Caustic filter, cell room muds, effluent treatment	—	1.25	—	—	—	3.00	—
Totals, g/tonne Cl₂:							
Range	32–510	9.45–16.8	ca. 9.5	2.51–5.62	34.6–51.3	ca. 14.0	12.5–116
Average experience	196						50.1 (5 y)
Prior experience	—		150–200	new facility	250		new facility
Other comparisons:	300, U.S. 50–250, world	0.77–3.58[b] (1972)	—	—	—	—	—

[a]From Hocking and Jaworski [20] with permission.
[b]In liquid effluents only.

A more permanent and less troublesome alternative was simply to separate the diaphragm and mercury cell resaturation systems by installing an additional dissolver or by drilling additional brine wells.

Mercury droplets, mercuric chloride, and mercury vapor are present in the hot moist chlorine gas produced by the electrolyzer. Carryover of entrained mercury in the chlorine is decreased by passage through a "demister," a labyrinth filter of titanium ribbon. Any mercury vapor present is condensed along with water vapor in an indirect, process water-cooled heat exchanger. The condensate, which contains 0.1–0.2 μg/g mercury, may be recycled to the brine circuit or charged to a waste brine treatment system prior to discharge [19].

Mercury loss in the decomposer section occurs as finely divided metal droplets in the sodium hydroxide stream as a result of the vigorous interaction between the deionized water and sodium amalgam streams promoted to ensure as complete sodium removal from the mercury as feasible. Mercury concentrations in the range of 0.5–10 μg/g (0.5–10 ppm) in the sodium hydroxide as it leaves the decomposer represent a mercury loss rate of 1–20 g/tonne of chlorine produced. This loss generally occurs indirectly through processes which use the sodium hydroxide, such as pulp and paper production. Efficient centrifuging and filtration on porous carbon or other caustic resistant media (occasionally both) are procedures capable of reducing the mercury concentrations in 50% sodium hydroxde to about 0.1 μg/g.

A far more significant mercury loss used to occur from the hydrogen stream of the decomposer. Hydrogen has a relatively low water solubility, so it used to be efficiently cooled and washed by direct countercurrent contact with a water shower. It left the decomposer hot and nearly saturated in mercury vapor, i.e., at 60°C it contained about 300 mg/m^3 Hg in hydrogen. This loss rate represented about 92 g of mercury per tonne of chlorine. When this hydrogen contacted the cooling water most of the mercury condensed into the water phase which produced a water stream high in suspended mercury. The hydrogen stream, too, still contained sufficient mercury to represent a potential loss of 2–11 g/tonne chlorine produced. Current practice dictates that hydrogen be cooled indirectly, in two stages, sometimes including prior compression to 2–3 atm to improve mercury recovery. The first stage uses process water to cool to about 20°C followed by a stage using a refrigerant such as ammonia, liquid propane, or a Freon to chill the hydrogen to −10 to −20°C. With good mercury mist control, these measures bring the mercury content down to <1 mg/m^3 and the loss rate in hydrogen down to <1 g/tonne chlorine.

Sludges result from the pretreatment of resaturated brine for removal of impurities, and from brine to be discharged, occasionally necessary because of water buildup in the brine circuit. These sludges contain 8–15 mg/g (dry basis) mercury as a complex mixture of compounds. To recover the mercury, most of the water is removed, and then the sludge is resuspended in aqueous sodium hypochlorite. The hypochlorite oxidizes the sulfide and any elemental mercury present (Eq. 8.44, 8.49) to produce a concentrated aqueous stream of dissolved mercury salts. Insoluble components are then removed by filtration, and the solution is then returned to the brine circuit. When this reaches the electrolyzer, electrolytic reduction recovers the dissolved mercury present (Eq. 8.50).

$$Hg^{2+} + 2\ e^- \rightarrow Hg^\circ \qquad\qquad 8.50$$

This is a neat example of the use of ingenuity to recycle mercury salts present in aqueous streams.

The residual solid filtered from the hypochlorite dissolving solution now contains less then 50 $\mu g/g$ mercury. This is discarded by blending it with some inert diluent material such as sand and then burial. Disposal areas are sealed using plastic film or layers of impermeable clay for containment, and have a ring of drainage tile placed around the perimeter with provision for treatment of any collected eluate. Addition of a chemical reductant such as zinc followed by vacuum distillation of the sludges for recovery of the mercury has also been used [23].

Mercury vaporization losses to the cell room air can amount to 1–5 g/tonne chlorine [24]. Therefore, ventilation is important to ensure safe working conditions, which require that the mercury vapor concentration be kept below 0.05 mg/m^3. This is achieved by tight cell construction, localized hoods and venting of critical cell areas, and cell room ventilation rates of six to eight air changes per hour. Mercury cell chloralkali plants in moderate climates are able to operate their cells outside which avoids these ventilating problems.

The provisions outlined above, together with good housekeeping practices, provide a safe working environment. However, the ventilation requirements cause a noticeable elevation of the mercury concentration in the air and in the precipitation in the immediate area of an operating mercury cell chloralkali plant [25,26]. It would be possible to employ the chemistry of Eqs. 8.44 and 8.45 in massively sized scrubbers to treat the ventilation air and reduce discharges. But the very high installation costs have meant that this measure has never been adopted. So the best overall compromise has been to maintain vapor control at the cells as tightly as possible.

Complications of mercury containment have convinced the North American chlorine and caustic producers to build new facilities, or at the time of a major overhaul to convert existing facilities to diaphragm cell technology [27,28]. Nearly 60% of U.S. chloralkali production was by mercury cells in 1970, whereas by 1977 it had dropped almost half of this value. In Japan mercury cell chloralkali production has been phased out entirely. In Europe these complications have not had such a dramatic effect on the cell choice since a large proportion of the total chlorine is still produced by mercury cells.

8.6.2. Emission Control Aspects of Diaphragm Cells

Conventional diaphragm cells also have potential losses from operating materials. Fiber dispersion into the cell products occurs from the asbestos diaphragm, and steps have been taken to monitor the degree of hazard [29]. Cell streams are all generally either in closed circuits, or are fed as components to following processes so that the risk of outside asbestos dispersal from these sources during normal operation is low. During diaphragm changes, however, when the fiber is dispersed in water or brine for removal, precautions are taken to avoid both personnel exposure and dispersal of waste fiber on final disposal. Little is known about the hazards of oral ingestion of asbestos fiber.

Another emission control aspect of diaphragm cell operation concerns the use of the crude cell product, still containing sodium chloride, to carry out base-catalyzed reactions such as ring closure of propylene chlorohydrin (Eq. 8.29) or hydrolysis of chlorobenzene (Eq. 8.51).

$$C_6H_5Cl + NaOH + H_2O \rightarrow C_6H_5OH + NaCl + H_2O \qquad 8.51$$

It is desirable that the water phase from these reactions, now high in sodium chloride, be reused in chloralkali cells, but the traces of organics present after use interfere with efficient cell operation. By providing the right metabolic conditions in ponds and seeding this waste stream with a heterogeneous bacterial population using sewage sludge it has been possible to remove the organic constituents in what are referred to as "Bio-ox" units [30]. After clarification, the brine may be resaturated and fed to electrolysis cells without causing problems. The discharge alternative, rather than bio-oxidation and recycle, would discharge a waste stream with both a high oxygen demand and ionic loading to any receiving water body.

8.7. NEW DEVELOPMENTS IN BRINE ELECTROLYSIS

Dimensionally stable anodes (DSAs) made of a corrosion-resistant titanium screening have been adopted by both mercury cell and diaphragm cell operators. A baked-on rutile titanium dioxide paste is used with these electrodes to decrease the electrochemical overvoltage and to improve practical operating current densities. While significantly more expensive than graphite initially, this type of anode has 30 times the conductivity of graphite and allows the maintenance of closer tolerances on the anode, mercury film spacing in the operation of mercury cells, resulting in savings of both power and operator time [15]. For diaphragm cells these electrodes mean that the cell overhaul time is dictated more by the asbestos diaphragm life rather than by the 3- to 4-yr graphite anode life that used to be experienced. Not only is anode life lengthened, but because dimensionally stable anodes do not continually shed particles as graphite anodes do, even the percolation lifetime of the asbestos diaphragm is extended somewhat. However, occasional problems with titanium corrosion do occur [31]. A further advantage, particulary with diaphragm cells, is a decrease in the carbon dioxide content of the chlorine (Tables 8.2, 8.5) and a significant (2 to 5%) increase in chlorine yields.

A further experimental improvement suggests the use of *tert*-butyl hydroperoxide as a diaphragm cell cathode depolarizer [32]. This cell would operate at only about 3 V, and would produce *tert*-butanol instead of hydrogen in a much more compact cell design.

Even highly resistant asbestos fiber is degraded and the fiber length shortened somewhat with time during normal diaphragm cell operation. Replacing the asbestos with a more inert synthetic polymer fiber allows longer periods of operation without cell overhaul. Teflon (polytetrafluoroethylene) fiber has

TABLE 8.5   A Comparison of Diaphragm Cell Gas Composition for Graphite versus Dimensionally Stable Anodes[a]

	Graphite anode (%)	Dimensionally stable anode (%)
Chlorine	96.5–98.0	96.5–98.0
Hydrogen	0.1–0.4	0.1–0.4
Oxygen	0.5–1.5[a]	1.0–3.0[b]
Nitrogen	0.1–0.5[a]	0.1–0.5[b]
Carbon dioxide	1–2	0.1–0.3[c]

[a]Data from Puschaver [15].
[b]From air.
[c]From brine.

been successfully used to maintain diaphragm porosity and percolation capacity longer than is possible with asbestos.

Conventional diaphragm cell technology still has a significant disadvantage over mercury cell technology in that the former produces a relatively dilute sodium hydroxide product mixed with sodium chloride, as opposed to the pure sodium hydroxide solution obtained directly from the mercury cell at high concentrations. New ion-selective membrane cells, already in use in the industry to some extent, substantially remove this handicap [33]. These cells use a polytetrafluoroethylene, ion permeable membrane through which no percolation occurs, to replace the usual nondiscriminating asbestos (or synthetic polymer) fiber brine percolation diaphragm (Fig. 8.5). By using carboxyl or sulfonic acid surface groups on the membrane it becomes selectively permeable to sodium ions and rejects chloride ions [34]. In this way it becomes possible to produce a sodium hydroxide stream low in sodium chloride.

As the saturated brine passes through the anode compartment of a membrane cell it becomes depleted in sodium chloride through the formation

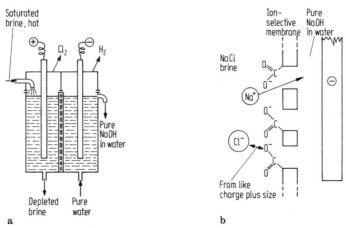

FIGURE 8.5   (a) Schematic diagram showing a vertical section of a membrane cell for the electrolysis of brine. (b) Details of a section of membrane illustrating the ion selection mechanism.

of chlorine gas (Eqs. 8.17–8.19) and through diffusion of sodium ions through the microporous membrane. The negatively charged groups on the surface of the membrane sheet prevent both the forward diffusion of chloride ion and the backward diffusion of hydroxyl ion. Purified water added to the cathode compartment is partially electrolyzed to hydrogen gas and hydroxide ions which, when combined with the diffused sodium ions, gives the sodium hydroxide solution cathode product. Since there is neither chloride diffusion nor brine percolation through the diaphragm the product is a nearly pure sodium hydroxide solution in water. Depending on current and water flow rates concentrations of sodium hydroxide of around 15% are common, and concentrations of up to 28% occasionally have been obtained from continuously operating membrane cells. This caustic contains only about 50 μg/g sodium chloride, as compared to about 30 μg/g in 50% sodium hydroxide from mercury cells. Higher sodium hydroxide ion concentrations from membrane cells lead to decreased current efficiencies from hydroxide ion back-diffusion, the reason for the present sodium hydroxide concentration limitation of these cells. Other remaining technical concerns with membrane cells relate to the generally somewhat lower current efficiencies and to short membrane lifetimes. At present this is limited to 2 to 3 yr of operation when this is coupled to much more careful brine pretreatment than is required for conventional asbestos diaphragm cells. A combination of mercury cell and membrane cell technologies has recently been tested for commercial feasibility [35]. The economics of the three primary chloralkali technologies have also been reviewed [36].

A more drastic realignment of diaphragm cell technology which accomplishes production of sodium hydroxide free of chloride from a sodium chloride/zinc chloride melt has been successfully tested in experimental prototypes [37]. In this system a salt mixture is used to obtain an eutectic depression of melting point to about 330°C rather than the 801°C required to melt pure sodium chloride [38]. Electrolytic ion mobility is thus obtained from the melt rather than from a solution of the salt. This anolyte melt is separated from the cathode compartment by a sodium ion permeable, β-alumina ceramic sheet (Fig. 8.6). Passage of an electric current through the cell requires a potential of about 5 V and produces dry chlorine. This cell uses a graphite anode,

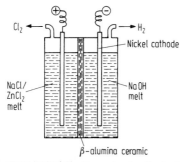

FIGURE 8.6 Operating details of a β-alumina ceramic cell for the production of chlorine, hydrogen, and nearly anhydrous sodium hydroxide. The β-alumina ceramic divider serves an ion-selective membrane function to the sodium chloride/zinc chloride analyte melt.

a sodium hydroxide melt (containing a small amount of water) electrolyte, and a pure nickel cathode. The much higher temperatures required for operation of this cell are maintained directly from internal cell resistance. When the smaller surface to volume ratio is considered, together with the larger electrical current flow of full-size commercial cells, this seems to be a feasible prospect. Commercial success of this cell rests on developments with the β-alumina ceramic to improve lifetime and ion flow capabilities, and on the completion of larger scale tests where voltage and current efficiencies may be determined more accurately at industrial rates of production.

8.8. BALANCING CHLORINE AND SODIUM HYDROXIDE PRODUCTION

Production figures for the major producers of sodium hydroxide are given in Table 8.6. From the growth of sodium hydroxide production on a country-by-country and a worldwide basis it can be seen that the volume of this commodity chemical slightly more than doubled every 10 years from 1950 to 1970. Since 1970 it appears that this steady growth has slowed significantly, so that recent decades only show a 30–50% increase. This is probably the result of the commodity reaching "mature" status, worldwide business recessions, and increased energy costs during this period.

Electrolytic production of sodium hydroxide and chlorine from sodium chloride solutions so heavily dominates the supply of these chemicals now that the production of chlorine can be quite closely approximated by multiplying the sodium hydroxide figures by the fraction calculated from the theoretical ratio of chlorine to sodium hydroxide (70.906:79.996) of 0.886: 1.000. For the U.S. production and marketing area for example, the actual production ratio of chlorine to sodium hydroxide is 0.953 to 1.000, quite close to this theoretical ratio.

The use pattern for both sodium hydroxide and chlorine is wide, reflecting the broad importance of these commodities to the chemical and other industries (Table 8.7). Most important for the demand of both commodities is the

TABLE 8.6 Major World Producers of Sodium Hydroxide[a]

	Thousands of metric tonnes					
	1950	1960	1970	1980	1990	1995
Canada	134	339	860	1,459	1,651	1,182
France	242	263	1,094	1,325	1,426	1,497
Italy	159	426	999	—	1,111	1,206
Japan	195	843	2,606	3,159	3,917	—
U.K.	—	—	—	967	1,002	773
U.S.A.	2,278	4,510	9,200	10,025	10,917	11,883
Germany	33[b]	776[b]	1,682[b]	3,187[b]	3,268	3,511

[a]Data selected from U.N. Statistical Yearbooks [39] and Facts and Figures Reports [40].
[b]Production of the former West Germany.

■■■■ **TABLE 8.7   Use Profile for Sodium Hydroxide and Chlorine in the United States**[a]

Sodium hydroxide		Chlorine	
Application	% of total	Application	% of total
Organic chemicals	34	Vinyl chloride	17
Inorganic chemicals	9	Propylene oxide	10
Petroleum refining	5	Methylene chloride	3
Food processing	1	$Cl_xC_2H_{6-x}$	14
Pulp and paper	15	Fluorocarbons	8
Soap and detergents	6	Other organic chemicals	10
Alumina from bauxite	5	Inorganic chemicals	11
Rayon and cellulose	3	Pulp and paper	13
Textiles	3	Water and wastewater treatment	5
Other, miscellaneous	7	Other miscellaneous	9
	100		100

[a]Based on data from *Chemical and Engineering News* [40] and Dickert [8].

manufacture of organic chemicals, which consume about one-third of the sodium hydroxide and almost two-thirds of the chlorine produced. The pulping of wood for paper and the bleaching steps for pulp brightening together consume about one-seventh of the sodium hydroxide and chlorine. The production of inorganic chemicals altogether accounts for only about one-tenth of each product. Water treatment and sewage disinfection, traditional uses of chlorine, only account for about 5% of the total produced.

Chlorine and sodium hydroxide production by the electrolysis of brine solutions necessarily locks the ratio of the two products to the theoretical ratio of the process. When the market for sodium hydroxide exceeds the market for chlorine, the causticization of sodium carbonate to sodium hydroxide (Section 7.3) may be used by some suppliers and consumers to supplement the available sodium hydroxide without producing large amounts of excess chlorine. Another expedient for large-scale chloralkali producers faced with this situation is to stimulate the chlorinated solvent or hydrochloric acid markets in an attempt to increase the consumption of chlorine to restore the balance. But these measures are not usually sufficiently rapid to respond over the short term, unless the solvent plant, too, is operated by the chloralkali producer.

Occasionally the market will swing the other way leaving chlorine in short supply. Fused sodium chloride is commercially electrolyzed in Down's cells to give chlorine and metallic sodium as the cell products [13]. Sodium production in the U.S.A. has averaged 135,000 to 150,000 metric tonnes annually since 1968, which represents about a 2–3% contribution to the chlorine supply from this source. In the U.K. it is estimated that as much as 10% of the available chlorine arises from Down's cell technology. Potassium chloride solutions are also electrolyzed for relatively small volumes of commercial potassium hydroxide, but the contribution to the chlorine supply from this source is even less than from fused sodium chloride electrolysis.

Chlorine has also been made commercially until very recent years by the chemical oxidation of sodium chloride with nitric acid, the Salt process [1]. Nitrate is used to oxidize chloride to chlorine and in the process is converted to nitric oxide (Eq. 8.52).

$$6 \text{ NaCl} + 12 \text{ HNO}_3 + 2 \text{ Na}_2\text{CO}_3 \rightarrow 3 \text{ Cl}_2 + 10 \text{ NaNO}_3 \qquad 8.52$$
$$+ 2 \text{ CO}_2 + 2 \text{ NO} + 6 \text{ H}_2\text{O}$$

The coproduced sodium nitrate could be utilized and sold as a fertilizer constituent to offset the cost of chlorine production by this route.

Routes to chlorine from hydrogen chloride oxidation have been refined from the original commercial ventures of Weldon which employed manganese dioxide (Eq. 8.53). The Weldon process permitted a maximum recovery of 50% of the chlorine in the hydrogen chloride consumed.

$$4 \text{ HCl} + \text{MnO}_2 \rightarrow \text{MnCl}_2 + \text{Cl}_2 + \text{H}_2\text{O} \qquad 8.53$$

Measures were taken to recycle the manganese (II) chloride, with mixed success [41]. The Deacon process solved these problems by replacing manganese dioxide by air as the oxidizer (Eq. 8.54).

$$4 \text{ HCl} + \text{O}_2 \underset{\text{cooling}}{\overset{\text{Cu/heat}}{\rightleftharpoons}} 2 \text{ Cl}_2 + 2 \text{ H}_2\text{O} \qquad 8.54$$

Further tuning of the Deacon process resulted in the development of the Kel Chlor variant, which employs concentrated sulfuric acid as a dehydrating agent. Dehydration greatly assists the process by driving the equilibrium of Eq. 8.54 to the right [42]. The chlorine product of Eq. 8.54 may also be efficiently removed by olefin, which has the same net effect on the equilibrium as water removal by sulfuric acid (Eq. 8.55).

$$2 \text{ HCl} + \text{CH}_2{=}\text{CH}_2 + 1/2 \text{ O}_2 \xrightarrow[\text{catalyst}]{\text{Cu}} \text{CH}_2\text{Cl}{-}\text{CH}_2\text{Cl} + \text{H}_2\text{O} \qquad 8.55$$

This version is known as oxychlorination, and 1,2-dichloroethane, the chlorination product, is denser than water, and insoluble in water so is readily separated [43].

The steps outlined above may also be carried out separately by electrolysis of a hydrochloric acid solution, or hydrogen chloride gas in a fused eutectic of lithium chloride and potassium chloride to obtain hydrogen and chlorine [44]. The chlorine can then be used for chlorination and the hydrogen chloride produced recycled (e.g., Eqs. 8.56–8.58).

$$\text{CH}_4 + \text{Cl}_2 \rightarrow \text{CH}_2\text{Cl}_2 + 2 \text{ HCl} \qquad 8.56$$

$$\text{CH}_2{=}\text{CH}_2 + \text{Cl}_2 \rightarrow \text{CH}_2\text{Cl}{-}\text{CH}_2\text{Cl} \qquad 8.57$$

$$\text{CH}_2\text{Cl}{-}\text{CH}_2\text{Cl} \rightarrow \text{CH}_2{=}\text{CHCl} + \text{HCl} \qquad 8.58$$
$$\text{vinyl chloride}$$

Another expedient to improve the balance in the supply of chlorine and caustic is to increase the consumption of sodium hydroxide by diverting part or all of this product to production of sodium carbonate (soda ash) [45]. This process reacts sodium hydroxide with the carbon dioxide of clean flue gas (Eq. 8.59).

$$2\ NaOH + CO_2 + H_2O \rightarrow Na_2CO_3 + 2\ H_2O \qquad 8.59$$

It also simplifies shipment of the product in a dry form. As a general rule, however, recovery of sodium carbonate from naturally occurring trona is less expensive. In one or more of these ways, a balance in the supply and demand of these commodities may be achieved.

The hydrogen coproduct from the cells of large chloralkali facilities may be used to contribute to that required for ammonia or methanol production (Eq. 8.60, 8.61).

$$N_2 + 3\ H_2 \rightarrow 2\ NH_3 \qquad 8.60$$

$$CO + 2\ H_2 \rightarrow CH_3OH \qquad 8.61$$

If the chloralkali plant is operating in the vicinity of an oil refinery it can provide a part of the hydrogen gas requirements of hydrocracking or various hydrotreating processes used in refining. Hydrogen may also be burned in chlorine to profitably produce hydrochloric acid, either for captive use or for sale [46] (Eq. 8.62).

$$1/2\ H_{2(g)} + 1/2\ Cl_{2(g)} \rightarrow HCl_{(g)} \qquad \Delta(H = -92.3\ kJ\ (-22.1\ kcal) \qquad 8.62$$

When the chlorine market is slack this option can also usefully consume chlorine, if the hydrogen chloride can be used or sold. However, since the masses of hydrogen produced in all except the largest facilities are relatively minor, many of the smaller plants simply burn the hydrogen in air to produce energy to assist in steam raising. For a diaphragm cell plant producing 50% sodium hydroxide this option can reduce the external fuel requirement by as much as one-third.

REVIEW QUESTIONS

1. Briefly, how do the brine pretreatment steps required for diaphragm cell electrolysis and mercury cell electrolysis compare, and what are the reasons for these differences?
2. One thousand liters per hour of a 10% sodium hydroxide solution (wt. of solute/wt. of solution; density 1.116) is to be concentrated to 50% (density 1.540) in a single effect evaporator operating at a reduced pressure of 40 mm Hg. Atmospheric pressure at the site of operation is 760 mm Hg.
(a) How many kilojoules per hour would be required to evaporate at this rate under the conditions outlined, with caustic solution entry and exit at 20°C, and condensate discharge also at 20°C? Assume perfect heat transfer efficiencies.
(b) If the overall heat transfer efficiency, fuel oil combustion → steam raising → evaporation from caustic solution is only 25%,

how many liters of fuel oil would need to be burned each hour to provide this energy? (No. 2 fuel oil $\equiv$ 140,000 Btu per Imperial gallon; 1 Btu $\equiv$ 1054 joules.)

3. (a) What current flow would theoretically be required to electrolytically produce chlorine from aqueous brine at the rate of 1 tonne (1 Mg) per day?

(b) Assuming 95% current efficiency for a chloralkali cell operating at 4.1 V, what would be the number of kilowatt-hours required per tonne of chlorine produced?

(c) What masses of sodium hydroxide (100% basis) and hydrogen would be coproduced along with the 1 tonne of chlorine?

(d) Commercial prices for the cell products are currently chlorine, $159 per tonne; and 50% (by weight) sodium hydroxide, $225 per tonne of solution. For a power cost of 5.5¢/kWh, what proportion of the sales value of these two products is paid for the electric power to produce them?

4. The rate of production of chlorine gas by electrochemical cells is sometimes determined by addition of oxygen to the chlorine stream at a known rate, and accurate determination of the oxygen concentration at a well mixed point downstream of the point of addition. To the chlorine produced by diaphragm cells, which already contains 0.8% by volume oxygen, is added additional oxygen at the rate of 1 kg $\cdot$ min^{-1}. Analysis prior to compression (20?C, 1 atm) indicates a resultant oxygen concentration in the chlorine of 3.1%.

(a) What is the rate of chlorine production, in kg $\cdot$ hour^{-1}?

(b) What mass of sodium hydroxide (kg, 100% basis) will be produced during the same interval?

(c) Give two origins for the source of the small percentage of oxygen already present in the chlorine cell product and explain how this arises. Give equations if reactions are proposed.

5. Sodium hydroxide solution, 50% by weight, is available commercially at $225 per tonne of solution.

(a) What theoretical profit or loss, on a raw material cost basis only, would be incurred to produce each tonne of anhydrous sodium carbonate from this sodium hydroxide and sell the product at $210 per tonne? Carbon dioxide for carbonation is to be obtained from flue gas. Assume essentially zero cost.

(b) Sodium carbonate monohydrate and sodium carbonate decahydrate are also commercial grades, selling at $290 and $430 per tonne, respectively. Give the relative dollar profit margins per tonne of contained Na_2CO_3 for each of these grades, neglecting any costs other than outlined in part (a). Which of the three grades would be the most profitable?

6. Four rail tank cars filled with 340 tons of liquid chlorine (boiling point is $-34.6°C$) fell from a barge carrying them in Malaspina Strait, British Columbia, in February 1975. The vapor pressure of chlorine is given by the expression:

$$\log_{10} P = \left[\frac{-0.2185A}{K}\right] + B \quad \text{where } P = \text{pressure in torr (mm Hg)},$$
$$A = 5180.4,$$
$$B = 7.5499,$$
$$\text{and } K = \text{absolute temperature.}$$

Calculate the critical depth in meters below which any chlorine escaping from the gradually corroding cars would do so as a liquid. Temperature and density of the seawater in the area are 10°C and 1.025 g/ml, respectively.

7. (a) In a small-scale solvent plant 40 tonnes of carbon tetrachloride and 10 tonnes of chloroform are produced daily, based on methane chlorination. Methyl chloride and methylene chloride are recycled. What daily mass of hydrogen chloride (in tonnes) would be coproduced by this process?
(b) Give the equations and calculate the mass ratios, hydrogen chloride to chlorine, theoretically possible from *each* of the three chemical processes for regenerating elemental chlorine from hydrogen chloride.

8. (a) What weights of sodium and chlorine (in kilograms) would theoretically be expected from the consumption of 10,000 ampere hours of electricity in a Down's cell (nonaqueous variant) electrolyzing molten sodium chloride with 90% current efficiency? Neglect associated calcium.
(b) Give two reasons why the current efficiency for this electrochemical process is noticeably lower than the usual experience of chloralkali cells.

9. (a) What is the voltage efficiency of a Down's cell operating at 6 V, when the theoretical voltage required is 3.1 V?
(b) For current efficiency of 80% when operating at 6 V, what is the energy efficiency of this Down's cell?
(c) To push production higher, the Down's cell operation is to be raised to 7 V, which at the same time decreases the current efficiency to 78%. What is the energy efficiency of the cell under these conditions?
(d) For an electricity cost of 5¢/kWh what would be the respective power cost to produce 1 tonne of sodium (with associated chlorine) from each set of cell operating parameters?

10. The Salt process to chlorine using sodium chloride normally proceeds with about 90% conversion and 90% selectivity ("Industrial" yield) to chemically produce chlorine and sodium nitrate.
(a) What molar quantities of chlorine and sodium nitrate would be expected from the passage of 6 mol of sodium chloride once through this process?
(b) If the unreacted sodium chloride was recycled through the process until it was all converted (reacted) what total number of moles of chlorine and sodium nitrate would be obtained from the initial 6 mol of salt charged?

(c) What masses of sodium chloride and nitric acid would be theoretically required to produce 1 tonne of chlorine via the Salt process, and what mass of sodium nitrate would be coproduced?

FURTHER READING

L. Buffa, "Review of Environmental Control of Mercury in Japan," Rep. No. EPS-WP-76. Environment Canada, Ottawa, 1976.

M.O. Coulter, ed., "Modern Chlor-Alkali Technology." Society of Chemical Industry, Ellis Horwood, Chichester, UK, 1980.

R.W. Curry, ed., "Modern Chlor-Alkali Technology," Vol. 6, Special Publication No. 164. Royal Society of Chemistry, Cambridge, UK, 1995.

"Environmental Mercury and Man," Pollut. Pap. No. 10. Dept. of the Environment, H. M. Stationery Office, London, 1976.

F. Hine, M. Yasuda, and T. Yoshida, Studies on the Oxide-Coated Metal Anodes for Chloralkali Cells. *J. Electrochem. Soc.* **124**, 500 (1977).

L.C. Mitchell and M. M. Modan, Catalytic Purification of Diaphragm Cell Caustic. *Chem. Eng. (N.Y.)* **86**, 88 (1979).

M. Seko, Ion Exchange Membrane Chloralkali Process. *Ind. Eng. Chem., Prod. Res. Dev.* **15**, 286 (1976).

S. von Winbush and A. F. Sammells, Zinc Chloride Molten Salt Cell. *J. Electrochem. Soc.* **123**, 650 (1976).

REFERENCES

1. J. S. Sconce, ed., "Chlorine, Its Manufacture, Properties, and Uses," p. 85–89. Reinhold, New York, 1962.
2. W. J. Mead, ed., "Encyclopedia of Chemical Process Equipment," p. 81. Reinhold, New York, 1964.
3. E. A. LeSueur, Br. Pat. 5983 (1891); details cited by Warrington and Newbold [4].
4. C. J. Warrington and B. T. Newbold, "Chemical Canada," p. 38. Chemical Institute of Canada, Ottawa, 1970.
5. H. A. Sommers, *Chem. Eng. Prog.* **61**(3), 94 (1965).
6. C. J. S. Warrington and R. V. V. Nichols, "A History of Chemistry in Canada," p. 207. Chemical Institute of Canada/Pitman and Sons, Toronto, 1949.
7. R. C. Weast, ed., "Handbook of Chemistry and Physics," 56th ed, p. D-146. Chemical Rubber Publ. Co., Cleveland, OH, 1975.
8. E. J. Dickert, "Multistage Centrifugal Compressors," Reprint No. 116, Vol. 4, No. 1. Compressed Air and Gas Institute, Cleveland, OH, 1977.
9. "Kirk Othmer Encyclopedia of Chemical Technology," 3rd ed., Vol. 1, p. 709. Wiley, Toronto, 1978.
10. H. C. Twiehaus and N. J. Ehlers, *Chem. Ind. (N. Y.)* **63**, 230 (1948), *Chem. Abstr.* **45**, 4415 (1951).
11. R. N. Shreve and J. A. Brink, Jr., "Chemical Process Industries," 4th ed., p. 214. McGraw-Hill, Toronto.
12. F. A. Lowenheim and M. K. Moran, "Faith, Keyes, and Clark's Industrial Chemicals," 4th ed., p. 716. Wiley-Interscience, Toronto, 1975.
13. K. Viswanathan and B.K. Tilak, Chemical, Electrochemical, and Technological Aspects of Sodium Chlorate Manufacture. *J. Electrochem. Soc.* **131**, 1551–1559 (1984).
14. R. L. Dotson, *Chem. Eng. (N.Y.),* **85**(16, Pt.1), p. 106, July 17 (1978).
15. S. Puschaver, *Chem. Ind. (London)* p. 236, March 15 (1981).
16. R. Thompson, ed., "The Modern Inorganic Chemicals Industry," p. 106. Chemical Society, London, 1977.

17. S. Jensen and A. Jernelov, *Nature (London)* **223**, 753 (1969).

18. J. M. Wood, F. S. Kennedy, and C. G. Rosen, *Nature (London)* **220**, 173 (1968).

19. E. J. Laubusch, "Mercury Emissions Control in Scandinavia," Pamphlet No. R-105 (unpublished). Chlorine Institute, New York, 1970.

20. M. B. Hocking and J. F. Jaworski, eds., "Effects of Mercury in the Canadian Environment." National Research Council of Canada, Ottawa, 1979.

21. G. L. Bergeron and C. K. Bon, U. S. Pat. 2,860,952 (to the Dow Chemical Company) (1958).

22. D. M. Findlay and R. A. N. McLean, Can. Pat. 1,083,272 (to Domtar Inc., Montreal) (1979).

23. R. A. Perry, *Chem. Eng. Prog.* **70**(3) March (1974).

24. F. L. Flewelling, *Chem. Can.* **23**(5), 14 (1971).

25. M. J. Bumbaco, J. H. Shelton, and D. A. Williams, "Ambient Air Levels of Mercury in the Vicinity of Selected Chloralkali Plants," Rep. EPS 5-AP-73-12. Environment Canada, Ottawa, 1973.

26. A. Jernelov and T. Wallin, *Atmos. Environ.* **7**, 209–214 (1973).

27. A Chloralkali Plant is Converted for Weyerhauser, *Hydrocarbon. Process.* **54**(6), 107 (1975).

28. M. B. Hocking and M. Gellender, *Chem. Int.*, p. 7, April (1982).

29. D. R. Beaman and D. M. File, *Anal. Chem.* **48**(1), 101–n110 (1976).

30. Dow Bioclarification Test Looking Good, *Chem. Eng. News* **54**(32), 24, Aug. 2 (1976).

31. P. Kohl and K. Lohrberg, Material Problems in the Three Versions of Chloralkali Electrolysis. *J. Appl. Electrochem.* **19**, 589–595 (1989).

32. H. B. Johnson, cited in Brine Electrolysis: A More Efficient Cathode, *Chem. Eng. News* **58**(14), 38, April 7 (1980).

33. E. G. Grot, U. S. Pat. 3,718,627 (1973).

34. Chloralkali Membrane Data, *Chem. Eng. News* **58**(30), 23, July 28 (1980).

35. F. Hine and A. J. Acioli Maciel, Combination of the Amalgam Cell and the Membrane Cell Processes for Chloralkali Production. *J. Appl. Electrochem.* **22**, 699–704 (1992).

36. R. E. Means and T. R. Beck, A Techno/economic Review of Chloralkali Technology. *Chem Eng. (N.Y.)* **91**, 46–51, Oct. 29 (1984).

37. Y. Ito, S. Yoshizawa and S. Nakamatsu, *J. Appl. Electrochem.* **6**, 361 (1976).

38. I. Fukuura, Development of β-Alumina Ceramics for a Separator for Molten NaCl Electrolysis. *In* "Applications of Solid Electrolytes" (T. Takahashi and A. Kozawa, eds.). pp. 71–74, JEC Press, Cleveland, OH, 1980.

39. "United Nations Statistical Yearbook 1993," 40th ed. United Nations, New York, 1995, plus earlier editions.

40. Facts and Figures for the Chemical Industry, *Chem. Eng. News* **74**(26), 38–79, June 24 (1996), and same feature articles of June in earlier years.

41. F. S. Taylor, "A History of Industrial Chemistry," p. 192. Arno Press, New York, 1972.

42. Chlorine Recovery (Kel-Chlor Process) Pullman Kellogg, *Hydrocarbon. Process.* **56**, 139, Nov. (1979).

43. P. J. Thomas, *Chem. Ind. (London)*, p. 249, March 15, 1975.

44. Y. Ding and J. Winnick, Electrolytic Recovery of Chlorine from Hydrogen Chloride Gas with Fused Molten Salt Electrolyte. *J. Appl. Electrochem.* **26**, 143–146 (1996).

45. Chlorine Production is Higher, *Can. Chem. Process.* **59**(5), 8, May (1975).

46. More HCl Producers Turn to Chlorine Burning, *Chem. Eng. News* **57**(27), 9, July 2 (1979).

Detail of the towers used for hydrogen sulfide stripping, Sulfinol solvent recovery, etc.

9
■ SULFUR AND SULFURIC ACID

Vice and virtue are products like sulphuric acid and sugar.

—*H. A. Taine (1828–1893)*

9.1. COMMERCIAL PRODUCTION OF SULFUR

Sulfur is widely distributed in the earth's crust, but only to the extent of about 0.1% by weight. It is found here chiefly as the element, as sulfides, and as sulfates. The annual mass of global sulfur produced is about twice that of sodium hydroxide, another important commodity chemical, so sulfur is significant in the chemical marketplace. However, in keeping with the the varied geologic forms in which sulfur occurs, its nonuniform distribution in the crust and the differences in the degree of industrialization of a country, there is considerable variation in the level of production, by country (Table 9.1). The influence of these diverse factors is reflected in the per capita level of production. Poland and Canada, which have natural factors that make sulfur production advantageous, produce around 150 and 300 kg per capita per year. These annual per capita production figures are much higher than the 25 to 50 kg per capita per year of Japan, the U.S.A., and the U.S.S.R., other large-scale producers. Thus, the level of sulfur production of a country is markedly influenced by natural and market factors. Per capita sulfur consumption is a good indicator of the level of industrial activity of a country.

Diverse methods have been used to recover this element commercially to cope with the varied forms in which sulfur occurs (Table 9.2). For example, about 85% of Canada's sulfur production is from sulfides removed from natural gas to "sweeten" it. That is, it results from, or is an involuntary by-product of natural gas production and is not a product sought for its own sake. Poland, on the other hand, obtains about 85% of its annual sulfur by

▮ **TABLE 9.1 Major World Producers of Sulfur**[a]

	Thousands of metric tonnes				
	1960	1970	1980	1990	1994[b]
Canada	249	4,440	7,405	6,790	9,140
China	244	250	2,300	5,370	6,030
France	791	1,736	2,213	1,050	1,100
W. Germany	84	176	1,775	1,550	1,240
Japan	256	343	2,784	2,630	2,900
Mexico	1,349	1,381	2,252	2,410	2,920
Poland	25	2,711	5,535	4,900	2,380
Spain	42	1,278	1,236	1,150	702
S. Africa	n/a	16	618	683	524
U.S.A.	5,898	8,678	11,866	11,600	11,500
U.S.S.R.	863	1,600	11,000	9,030	—
Other	8,769	9,728	7,047	10,637	—
World	18,690	32,480	56,635	57,800	51,000

[a]Selected from data in Minerals Yearbooks [1].
[b]Estimates.

employing Frasch recovery of natural sulfur, a process which is more responsive to markets in its volume of production.

For the world's two largest producers of sulfur, the U.S.A. and the former U.S.S.R., the picture is much more diverse. The Frasch process still dominates the American sulfur industry, for reasons which will become evident later. The working of pyrites for sulfur recovery provides the largest contribution to the Russian sulfur industry, and is also of historical interest.

▮ **TABLE 9.2 Sources of Sulfur Production of Major Producing Countries by Method of Recovery, Given as Percentages of the Total**[a]

	Canada	Poland[b]	U.S.A.	U.S.S.R.[b]	World
Primary elemental:					
Frasch	—	84.3	53.8	8.2	25.3
Native	—	9.4	—	25.4	6.7
Pyrite	0.2	—	2.7	32.3	18.2
By-product:					
Coal	—	—	—	0.4	0.1
Metallurgy	12.2	5.4	8.5	21.0	13.7
Natural gas	81.0	—	14.8	10.9	21.5
Petroleum	2.6	0.5	19.5	1.8	8.8
Tar Sands	4.0	—	—	—	0.5
Miscellaneous	—	0.4[c]	0.7	—	6.2
Totals, percent	100.0	100.0	100.0	100.0	100.0
(Thousand tonnes)	(7,405)	(5,535)	(11,866)	(11,000)	(56,635)

[a]Table compiled from 1980 data selected from Minerals Yearbooks [1].
[b]Estimates throughout.
[c]Produced from gypsum.

The large-scale production of sulfur as a process secondary to natural gas production, such as has developed in Canada since 1951 when the first plant was built [2], also can dislocate any logical local pricing structure. It is evident from Fig. 9.1 that, within a transportation cost differential, Canadian ancillary sulfur production tended to hold American sulfur prices down. Rising energy costs have added significantly to the cost of Frasch sulfur, and the development of a more orderly marketing structure in Canada has served to bring North American sulfur prices to a more realistic level.

Traditionally almost 90% of all sulfur produced has been converted to sulfuric acid [3]. While this is a broadly based market, it is not a volume-flexible one. The only other significant uses of sulfur are by the pulp and paper industry, about 5%, and in the manufacture of carbon disulfide, 2.5%, neither significant enough to have a major impact on sulfur markets. Hence, a search for new viable uses for sulfur was stimulated by the low prices and the widespread surplus of sulfur during the late 1960s and early 1970s. The sulfur (or sulfur dioxide) production associated with other industries such as natural gas processing or the smelting of sulfide ores has recently become a particularly important contributor to the sulfur industry. This has occurred as sulfur gas containment, rather than discharge and dispersal, has been

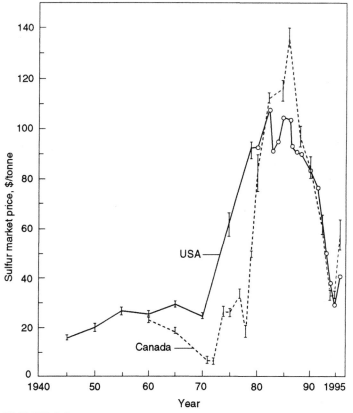

FIGURE 9.1 Influence of market and nonmarket factors on the U.S. dollar price of sulfur. Spot prices are given for the United States post-1978 to simplify the comparison.

increasingly imposed on this sector with the growing recognition of environmental damage from this source (Chap. 3).

Closed-foam sulfur slabs poured *in situ* from a melt have been successfully tested for highway subgrade insulation in arctic regions for prevention of frost-induced damage [4]. As little as 10 cm of foam could replace 120 cm of gravel in this application. Incorporation of up to about 40% sulfur by weight into asphalt formulations produces a finished paving with better cold flexibility and sufficient hardness to be useful under high summer temperatures [5]. A paving material is also possible in which the sulfur plus an "additive" entirely replaces petroleum-based asphalt as an aggregate binder [6]. Sulfur may also be used as a traffic paint, applied as a melt. This produces a pale yellow, durable highway marker which can accept traffic within a few seconds after its placement.

Construction applications for sulfur have also been tested. Interlocking bricks made predominantly of sulfur have been found to speed construction, but have an inherent fire hazard which restricts their use. Concrete blocks impregnated with molten sulfur have shown a more than tenfold increase in compressive strength, from 5.9 to 83.9 MPa (megapascals; 1 MPa $\equiv 10^6$ N/m^2), at the same time as an improvement in tensile strength from 1.3 MPa to 8.5 MPa for a 13–15% sulfur content [7]. It also improves its chemical resistance. Sulfur-coated bamboo has been found to be an economical yet effective concrete reinforcing agent as a replacement for steel, in areas where bamboo occurs naturally.

Adding sulfur to a sulfur-deficient soil can boost crop yields by 1000% or more, particularly for seed crops having a high sulfur content [8]. A deficiency of sulfur in the soil has traditionally been corrected by the application of sulfate-containing fertilizers such as ammonium sulfate and superphosphate (Section 10.5). Today, however, methods of sulfur application have been extended to include finely divided sulfur itself in weakly granulated form, or as a 10–12% solution of sulfur in anhydrous ammonia as ways of raising the sulfur analysis of the fertilizer formulation. Other variations along this theme include the coating of highly soluble nitrogenous fertilizers, such as urea, with sulfur to produce a slow release, and hence greater long-term effectiveness of the sulfur and other nutrients [9].

Some of these uses are evidently developing into viable markets for sulfur, although a leveling off of the increases in the sulfur production from recovery processes coupled with increased energy costs and increased sulfur demand for traditional uses has, to a significant extent, offset this pressing need for new markets.

9.2. PROPERTIES OF ELEMENTAL SULFUR

Sulfur is an intriguing element in the multitude of forms in which it can exist. Under ordinary conditions it occurs as a solid which exists in the well-known bright yellow form as either rhombic or monoclinic crystals, or as a dark, amorphous, moldable mass referred to as plastic sulfur (Fig. 9.2). The rhombic form is most stable at room temperature. The transitional equilibrium between the rhombic and the monoclinic forms occurs at 95.5°C. Just above this tem-

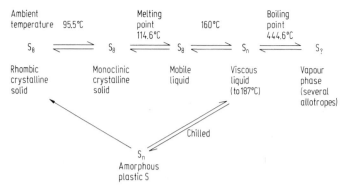

FIGURE 9.2 Allotropic forms and physical properties of sulfur.

perature, at 114.6°C, it melts to a transparent, pale yellow mobile liquid. In all the above forms sulfur occurs as molecules of eight-membered, crown-shaped rings.

At somewhat below 160°C the rings start to break up into chains. At about 160°C the ring-bonded form of sulfur is largely lost to a much darker, reddish-brown, viscous linear polymeric form, where n, the number of bonded sulfur atoms, is significantly more than eight. Above 187°C depolymerization of this linear form occurs, decreasing the viscosity again. Rapid chilling of the viscous liquid gives the dark plastic sulfur at room temperature, thought to be a linear molecular sulfur occurring in helices. Slower cooling of liquid sulfur allows the transitions through the eight-membered ring, and the monoclinic and rhombic crystal forms, and yields the more familiar yellow powder modification of sulfur.

9.3. SULFUR RECOVERY BY MINING AND RETORTING

Of both historical interest and as a general method still used to produce nearly a quarter of the world's sulfur (Table 9.2) is the application of conventional mining techniques to bring to the surface lumps of either a volcanic or one of the many pyritic forms of sulfur. Some of the pyritic forms are pyrite (FeS_2) itself, chalcocite (Cu_2S), and chalcopyrite ($CuFeS_2$). The sulfur content of the raw mineral is usually 25–35%, but may run as high as 50%. To obtain the sulfur in separated form, the original procedure was to pile the lumps of ore outside and seal these with clay or earth. Burning a part of the contained sulfur sealed into these mounds, with careful control of the air, generated sufficient heat to melt any elemental sulfur present and thermally decompose the pyrite (Eq. 9.1–9.3) [10].

$$3 \ FeS_2 + 8 \ O_2 \rightarrow Fe_3O_4 + 6 \ SO_2 + \text{heat} \qquad 9.1$$

$$S + O_2 \rightarrow SO_2 + \text{heat} \qquad 9.2$$

$$8 \ FeS_2 \rightarrow S_8 + 8 \ FeS \qquad 9.3$$

This primitive method gave sulfur recoveries of about 50% of that contained in the ore, the rest being burned to provide the heat requirement of the process. More advanced sulfur recovery techniques using dual chamber furnaces are capable of recovering as much as 80% of the contained sulfur.

Sulfur dioxide coproduced by this method was generally converted to sulfuric acid, at least in the early operations, by using the chamber process. Using this source for the sulfur dioxide for acid making put all of the volatilized arsenic vapor from the pyrites directly into the sulfuric acid product. But this aspect was not considered to be of serious concern at the time since most of the sulfuric acid was consumed in the Le Blanc process to produce sodium carbonate (Eq. 9.4, 9.5) causing arsenic volatilization and loss with the hydrogen chloride of the first step [11].

$$2 \ NaCl + H_2SO_4 \rightarrow 2 \ HCl + Na_2SO_4 \qquad\qquad 9.4$$

$$Na_2SO_4 + 2 \ C + CaCO_3 \rightarrow CaS + Na_2CO_3 + 2 \ CO_2 \qquad 9.5$$

If sulfuric acid production was the primary goal of the operation, the air supply to the piles of ore was not restricted and the whole of the contained sulfur was converted to sulfur dioxide (Eq. 9.1, 9.2). On the other hand, if the sulfur dioxide could not be used on site it was also possible to practice sulfur dioxide reduction back to elemental sulfur by using incandescent coke [11,12] (Eq. 9.6).

$$SO_{2(g)} + C_{(s)} \rightarrow CO_{2(g)} + S_{(g)} \qquad\qquad 9.6$$

9.4. FRASCH SULFUR

Native sulfur deposits were discovered in 1865 while prospecting for oil in a dome-like formation in Louisiana. These beds lay under a thick bed of unstable quicksand, eliminating conventional mining as an economic recovery method. But from initial successful experiments in 1891, H. Frasch developed a method using superheated water and air to produce sulfur from this deposit on a commercial scale by 1902. During the development period a skeptical detractor promised to "eat every barrel of sulfur produced." Since this first large-scale production by the Frasch method the procedure has been the chief source of the success of the United States as a dominant sulfur producer. The state of Texas alone supplied 70% of the world's sulfur by this means in 1935. Essentially the same method has also been used for most of the sulfur produced by Poland in recent years, although it is not referred to as the Frasch process there [13].

A primary geologic requirement for the success of the Frasch process lies in the rather unique salt dome structures which occur in the coast area of the Gulf of Mexico, in the states of Louisiana and Texas, and in parts of Mexico (Fig. 9.3). Only about 28 of the some 400 or so known dome formations in the area have produced commercial quantities of sulfur, because only a relatively few possess all the correct geological features to be exploited in this way [14]. Nevertheless, enormous reserves are accessible from a single dome with the right features. One dome in Texas, for example, has produced 12

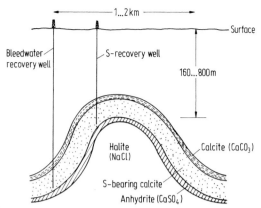

FIGURE 9.3 Vertical cross section of a typical salt dome formation occurring in the area around the Gulf of Mexico. Approximate placements of sulfur recovery and bleed water wells are shown.

million tonnes of sulfur during its useful life [15]. Offshore Frasch capability is now tapping into reserves of 67 million tonnes of sulfur along with oil and gas recovery off the Louisiana coast [16].

Frasch sulfur recovery today is still practiced much as originally developed and initially involves conventional oil well drilling equipment to reach the sulfur-bearing calcite zone. After placing a string of three concentric pipes into the well (Fig. 9.4), superheated water at 150 to 165°C is pumped down to the formation in the outermost, 15–20 cm diameter pipe to melt the native sulfur in place. After passage of hot water for some time, heated compressed air is forced down, inducing vigorous agitiation of the superheated water and molten sulfur mixture present. Slugs of water, sulfur, and air are forced to enter the intervening sulfur return pipe. The outermost pipe carrying superheated water serves to keep the sulfur (melting point 114.6°C) molten in the return pipe, which is jacketed by it. At the same

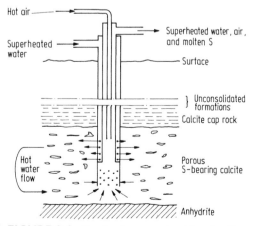

FIGURE 9.4 Details of a conventional Frasch sulfur string. Superheated water (140–160°C, 17 bar) enters outermost pipe, and air at 150–200°C and 34 bar enters the innermost pipe.

time, the heated compressed air both agitates the mixture in the formation, and by means of the air lift principle (Fig. 9.5) and some air incorporation into the sulfur, assists in bringing slugs of molten sulfur (density 2.0 g/cm^3) with water to the surface.

At the surface, tanks which are kept hot receive this mixture, still under pressure, and allow a lower liquid sulfur phase to separate from the water. Much of this product is simply filtered, while still molten, to remove any excess carbonaceous material, before it is shipped in this state via insulated tank trucks, rail tank cars, or barge. Traces of heavy petroleum which also occasionally occur with the sulfur from dome sources are removed by sublimation of the sulfur from the oil, if necessary.

For a solid sulfur product the sulfur layer is pumped to large outside vats where it is allowed to cool. The simplest containment measure for this method of storage consists of a sheet metal rim which is used to hold the molten sulfur until it solidifies. Once the product is solid, the metal rim is moved upward ready to receive the next layer of molten sulfur. The large, compact dense masses of product thus formed represent one of the least expensive and best environmentally controlled forms of storage of excess sulfur inventory. For delivery, the large mass is broken up, piecemeal, using small explosive charges or mechanical means.

In other sulfur recovery operations better control of product particle size is obtained. The molten sulfur is chilled on a steel belt to produce a roughly 0.5-in.-thick flake sulfur, or sprayed into water or tumbled in drums to obtain granulated sulfur. Pelletized sulfur, similar in particle size range to the granulated variety, is also obtained when a melt is sprayed into the top of a tower to form droplets of molten sulfur which harden into shot-sized beads as they fall through a current of air. In this form it is referred to as prills [17]. The narrow size range of particles of prilled sulfur, as well as the negligible dust content in this form, makes it more convenient to use. It normally commands a price premium.

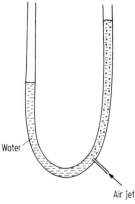

Water

Air jet

FIGURE 9.5   The air lift principle shown applied to a U-tube which illustrates that the tube leg with a mixture of bubbles and water will rise to a higher level to counterbalance the leg having water alone.

9.4.1. Environmental Aspects of Frasch Operations

A major cost item in Frasch processing is fuel for water heating, usually heavy oil or natural gas, since some 4 to 50 tonnes of superheated water are required for each tonne of sulfur recovered at the surface [18]. Some hot water is lost from the actual producing zone via undergound fractures and by slumping into the cavities from which sulfur (and some soluble salts) has been withdrawn. But much of this "leaked" water is recovered via "bleed wells" located downslope on the dome from the producing zone (Fig. 9.3). The residual heat content of bleed water and production waters which return sulfur to the surface may be salvaged by reheating the once-used water (with some withdrawal to allow a partial purge of impurities) to the required 140–165°C production temperatures. This is usually carried out indirectly using steam rather than by a direct firing boiler, to decrease the corrosion and fouling problems which can occur from the high total solids content of the recycle water. The scale of water heating capacities of these plants ranges from 4 to 40 million liters per day (1–10 million U.S. gallons/day).

Direct risk to boiler equipment and sulfur storage vats from overburden slumping is avoided by locating these surface facilities a kilometer or so away from producing wells. Risk to other parties not involved in the Frasch operations is minimized by purchase of the surface rights overlying the areal extent of the mined dome. Production upsets from slumping or subsidence are minimized by providing double sets of connecting pipe and much of the associated equipment to minimize downtime (lost production) in the event of a rupture. The extended period of formation heating required before the sulfur production rate becomes steady and the problems associated with pipe blockage by solidified sulfur make it important that the injection of superheated water be maintained as much as possible without interruption, once production from a particular well has been started.

Bleed water or process water in excess of injection requirements, which contains dissolved salts at similar concentrations to sea water, may be safely discarded to the sea after heat exchange energy recovery and treatment to remove toxic sulfides. Initial heat exchange to fresh process water not only conserves fuel but decreases calefaction (thermal pollution) of the receiving body of water. Prolonged aeration of these streams for both stripping and oxidation of sulfide has been used to decrease the sulfidic toxicity of this stream before discharge. When this is carried out in the presence of nickel salts, or by the addition of acid which also increases stripping through the loss of hydrogen sulfide, it has been found possible to reduce the sulfide content down to acceptable levels in 5 to 10 hr. Alternatively, hypochlorite (Eq. 9.7) or permanganate have both been found to be effective [19].

$$Na_2S + 4\ NaOCl \rightarrow Na_2SO_4 + 4\ NaCl \qquad 9.7$$

Chemical oxidants are used when only small volumes of sulfidic aqueous wastes are to be treated because of the higher reagent costs. Flue gas scrubbing of the waste water for sulfide removal has also been proposed, but has the disadvantage that this may only be used when it is acceptable to vent hydrogen sulfide with the spent flue gases at the site of operations.

9.5. SULFUR FROM SOUR NATURAL GAS

In the same way as Frasch sulfur recovery rapidly became the dominant source of the world's sulfur in the early 1900s, sulfur recovered from the hydrogen sulfide contained in sour natural gas as well as other petroleum streams began to exceed that produced by the Frasch process in 1970 [20]. This dominance of sulfur from petroleum sources has been given added impetus by the dramatically increased energy costs of the 1970s which has had a negative impact worldwide on Frasch production. It has also been stimulated by the increased demand for natural gas, which has given a boost to the secondary, sulfur-recovery aspects of natural gas processing. Canada, a major contributor to sulfur from petroleum sources (Table 9.2), started processing of sour natural gas in 1951, and Canadian production from this source probably peaked in 1976. However, as tar sands development begins to supply a larger fraction of Canadian petroleum, sulfur recovery from these bituminous sources may begin to dominate the Canadian petroleum-recovered sulfur supply. At Lacq, in France, the large amounts of sulfur recovered from natural gas are leading to the development of a sulfur-based chemicals industry at this site.

Why should sulfur recovery from natural gas be so important to the natural gas industry? First, hydrogen sulfide, in the presence of moisture is extremely corrosive to steel pipe lines. Hence, it is desirable to remove it as near as possible to the producing well head. Second, while some natural gas streams are virtually free of hydrogen sulfide as they emerge from the deposit, many gas wells yield a product containing 5–25% hydrogen sulfide by volume, and occasional wells as much as 85% H_2S [4]. The latter wells were usually capped because the expense of sulfur recovery was so great as to make the production of natural gas unprofitable. While natural gas has a reputation for being a clean and easily controlled fuel, if burned with even 5 to 25% H_2S present, as commonly obtained from the well head, it would generate flue gases with a very high sulfur dioxide content. Also the hydrogen sulfide itself and its sulfur dioxide combustion product are both very toxic so that there is a safety aspect to its removal as well. Hence, it is only after sour natural gas is treated to remove sulfur compounds to obtain a "sweetened" natural gas, a precombustion control measure, that this fuel can truly be regarded as clean.

9.5.1. Amine Absorption Process for Hydrogen Sulfide Removal

Sulfur recovery from sour natural gas is conducted in two stages. The first stage involves removal of the reduced sulfide gases from the natural gas stream. This is usually achieved by scrubbing the gases with an amine solution, for example by countercurrent contacting of the raw gases with a solution of monoethanolamine (MEA) in diethylene glycol (DEG) (about 1 MEA:2 DEG by volume). Monoethanolamine is a condensation product of ethylene oxide with ammonia (Eq. 9.8), and is a weakly basic organic liquid.

$$NH_3 + H_2C\overset{\displaystyle O}{\overset{\displaystyle /\,\backslash}{-}}CH_2 \rightarrow H_2N-CH_2CH_2-OH \qquad\qquad 9.8$$

$$\text{boiling point } 170°C$$

When contacted with hydrogen sulfide in natural gas under high pressures and at near ambient temperatures in a tray-type absorber (Fig. 9.6), this solvent mixture forms a monoethanolammonium salt which remains dissolved in the absorbing fluid to collect as a solution at the bottom of the absorbing unit (Eq. 9.9).

$$H_2N-CH_2CH_2-OH + H_2S \rightleftharpoons \overset{+}{H_3N}-CH_2CH_2-OH \atop HS^- \qquad 9.9$$

This scrubbing mixture also dries the incoming natural gas. The light hydrocarbon component of the gas stream is only very slightly soluble in the absorbing fluid and passes through the absorber unchanged.

If used on a one-pass basis, the volumes of monoethanolamine required would be prohibitively expensive. However, since the hydrogen sulfide is only weakly associated to the monoethanolamine, it is readily driven off and the MEA recovered by heating this solution indirectly with steam in a regenerator or stripper (Fig. 9.6; Eq. 9.10).

$$\overset{+}{H_3N}-CH_2CH_2-OH \atop HS^- \xrightarrow{100-140°C} H_2NCH_2CH_2OH + H_2S \qquad 9.10$$

In this way the volatile hydrogen sulfide (b.p. $-60.7°C$) is separated as an overhead gas stream from the monoethanolamine (b.p. $170°C$) and diethyleneglycol (b.p. $245°C$) which emerge as a regenerated solution, hot, from the bottom of the stripping column. As a heat conservation measure the hot, H_2S-lean monoethanolamine stream from the stripper is normally heat exchanged with the relatively cool H_2S-rich stream from the base of the absorber, sometimes with additional cooling with process water before it enters the absorber.

Most natural gas as obtained from the well contains almost no carbon dioxide although some wells yield gas containing up to 10% carbon dioxide and a few contain more than this [22]. Carbon dioxide in the presence of moisture or other proton source is also acidic in nature. If present in the gas it is also collected by the monoethanolamine with about the same efficiency as hydrogen sulfide, although it reacts more slowly (Eq. 9.11–9.14; Table 9.3).

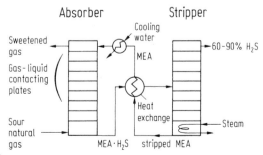

FIGURE 9.6 Flowsheet giving the schematic details of Girbotol and related sour gas sweetening processes. Twenty to 24 plates are used in the absorbers and strippers [21] for efficient contacting and desorption with monoethanolamine (MEA).

TABLE 9.3 Solubilities of Carbon Dioxide and Hydrogen Sulfide in 15.3% by Weight Monoethanolamine in Water[a]

Gas partial pressure (mm Hg)	Moles carbon dioxide per mole amine		Moles hydrogen sulfide per mole amine	
	40°C	100°C	40°C	100°C
1	0.383	0.096	0.128	0.029
10	0.471	0.194	0.374	0.091
50	0.542	0.299	0.683	0.203
100	0.576	0.347	0.802	0.279
500	0.672	0.458	0.959	0.536
1000	0.727	0.509	—	—

[a]Data selected from *Chemical Engineers' Handbook* [23].

Moist conditions:

$$CO_2 + H_2O \rightarrow H_2CO_3 \qquad 9.11$$

$$H_2CO_3 + H_2O \rightarrow H_3O^+ + HCO_3^- \qquad 9.12$$

$$H_2CO_3 + H_2N-CH_2CH_2-OH \rightarrow \begin{matrix} + \\ H_3N-CH_2CH_2OH \\ HCO_3^- \end{matrix} \qquad 9.13$$

Dry conditions:

$$CO_2 + 2\,H_2N-CH_2CH_2-OH \rightarrow \qquad + \qquad 9.14$$
$$HO-CH_2CH_2-NH-COO^- + HOCH_2CH_2-NH_3$$

While carbon dioxide removal does raise the fuel value of the natural gas, it also has the undesirable effect of raising the acid gas loading of the absorbing stream in the natural gas sweetening plant, decreasing the possible gas throughputs.

Tertiary amines, dry, do not react with carbon dioxide and yet still rapidly form a salt with hydrogen sulfide. This selectivity feature allows preferential absorption of hydrogen sulfide over carbon dioxide from a natural gas stream. Realization of this has led to the use of methyldiethanolamine ($CH_3N(CH_2CH_2OH)_2$) as an absorbing fluid in situations where both acid gases occur in the natural gas stream. Even for wet gas streams, where the formation of carbonic acid (Eq. 9.11) could cause interaction and absorption on the tertiary amine, it was found that the salt-forming reaction with carbon dioxide was relatively much slower than with hydrogen sulfide. Hence, a short natural gas contact time for the absorbing fluid with moist natural gas streams effectively minimizes carbon dioxide absorption and yet still efficiently remove hydrogen sulfide.

A further variable of potential concern in gas processing is the presence of carbon disulfide (CS_2) or carbonyl sulfide (COS) in the natural gas stream, whether from natural sources or, for the former component at least, occasionally arising from the use of carbon disulfide to remove sulfur blockages from sour gas wells or components of the Girbotol plant. These particular sulfur compounds react with monoethanolamine to yield complex thiazolidine and oxazolidine heterocycles and polymerization products of these, which are not dissociated in the regeneration step of the gas cleaning plant [24] (Eq. 9.15).

$$
\begin{array}{ccc}
\underset{\text{thiazolidine-2-thione}}{
\begin{array}{c}
\text{H}_2\text{C}\!-\!\text{CH}_2 \\
\text{S} \quad \text{N}\!-\!\text{H} \\
\text{C} \\
\Vert \\
\text{S}
\end{array}}
&
\underset{\text{oxazolidine-2-thione}}{
\begin{array}{c}
\text{H}_2\text{C}\!-\!\text{CH}_2 \\
\text{O} \quad \text{N}\!-\!\text{H} \\
\text{C} \\
\Vert \\
\text{S}
\end{array}}
&
\underset{\text{2-oxazolidine}}{
\begin{array}{c}
\text{H}_2\text{C}\!-\!\text{CH}_2 \\
\text{O} \quad \text{N}\!-\!\text{H} \\
\text{C} \\
\Vert \\
\text{O}
\end{array}}
\end{array}
\qquad 9.15
$$

Diethanolamine $(HN(CH_2CH_2OH)_2)$ and diisopropanolamine are far less susceptible to this irreversible reaction than monoethanolamine and hence have largely replaced MEA at locations where carbon disulfide and carbonyl sulfide occur to a significant extent in the natural gas.

A minor variant to the amine scrubbing process described above is the Sulfinol process, which still uses an alkanolamine base, diisopropanolamine (35%), but in a solvent consisting of a mixture of sulfolane [40%, tetramethylene sulfone, $(CH_2)_4SO_2$, which is a good hydrogen sulfide solvent] and water [25]. Other processes are based on hydrogen sulfide absorption in aqueous alkaline carbonate solutions, such as is used with the Catacarb and Benfield systems (Eq. 9.16, 9.17).

Absorption:
$$
Na_2CO_3 + H_2S \rightarrow NaHCO_3 + NaSH \qquad 9.16
$$

Regeneration:
$$
NaHCO_3 + NaSH \xrightarrow{\text{heat}} Na_2CO_3 + H_2S\uparrow \qquad 9.17
$$

Still further process variations use physical absorption of hydrogen sulfide by the solvent, rather than chemical reaction with it, to effect sweetening of natural gas.

9.5.2. Claus Process Conversion of Hydrogen Sulfide to Sulfur

Hydrogen sulfide separated from natural gas by amine scrubbing is a highly odorous, toxic, low boiling gas and as such is difficult to store or ship in large quantities. Hence, a Claus unit closely associated with each amine scrubber is used to convert the hydrogen sulfide gas primary product of the scrubber to elemental sulfur, a commodity which is much easier and safer to store and ship in large quantities.

The technology of Claus conversion of hydrogen sulfide to sulfur was worked out in Germany in about 1880, but it was not until 1940 that this process was commercially adopted in the United States. By 1967 annual sulfur production in the U.S.A. from this process had already reached 4.8 million tonnes, and by 1970 exceeded U.S. Frasch production of sulfur for the first time [10].

Two reactions are employed in Claus units, the first a simple combustion of one-third of the hydrogen sulfide stream in air, carried out in a waste heat boiler to capture the excess heat evolved as steam (Eq. 9.18).

$$H_2S_{(g)} + 3/2\ O_{2(g)} \rightarrow SO_{2(g)} + H_2O_{(g)} \quad \Delta H = -519\ kJ\ (-124\ kcal) \quad 9.18$$

Enough heat is retained in the boiler combustion gas stream for the catalytic stage of the process to proceed using this (Fig. 9.7). The sulfur dioxide/steam output is blended with the remaining two-thirds of the hydrogen sulfide. The resulting gas mixture is then fed to heated iron oxide catalyst beds where sulfur dioxide is reduced, and the hydrogen sulfide is oxidized, both to a sulfur product, with the further evolution of heat (Eq. 9.19).

$$SO_{2(g)} + 2\ H_2S_{(g)} \underset{300-320°C}{\rightleftharpoons} 3\ S_{(\ell)} + 2\ H_2O_{(\ell)} \quad \Delta H = -143\ kJ\ (-34\ kcal) \\ 9.19$$

A portion of the heat obtained from this reaction is also recovered as steam which, together with that produced by the sulfur-burning boiler, is frequently sufficient to heat the stripper of the associated Girbotol (amine scrubber) plant. The net overall Claus chemistry thus involves oxidation of hydrogen sulfide with air to yield sulfur and water (Eq. 9.20).

$$H_2S + 1/2\ O_2 \rightarrow S + H_2O \quad\quad\quad 9.20$$

In fact, in most Claus reactors this same overall reaction occurs in the first hydrogen sulfide combustion unit, so that a part of the final sulfur product is already present in the combustion gases of the first unit.

The hydrogen sulfide to sulfur conversion part of this process (Eq. 9.19) is a reaction in which the equilibrium lies very close to 100% on the sulfur side at temperatures below 125°C, but drops to barely 50% at 560°C [26]

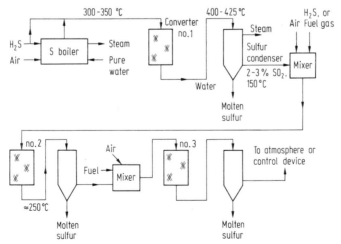

FIGURE 9.7 Schematic diagram of a three-stage Claus reactor unit showing steam generation both by direct combustion of one-third of the Girbotol output in a hydrogen sulfide fueled boiler, plus further steam recovery from the sulfur condensers. These are designed to remain hot enough to keep sulfur liquid.

(Fig. 9.8). Above this temperature, the equilibrium climbs again, to reach almost 75% at 1300°C. With suitable catalysts the rate can be made acceptably fast at temperatures of 300°C or so, sufficient to keep the sulfur in the vapor state and obtain about 90% conversion of hydrogen sulfide to sulfur (theoretical equilibrium conversion at 300°C would be about 93%). This would seem to give excellent sulfur recoveries. But when it is considered that many Claus plants, such as those operating at Lacq in France and at Waterton and Aquitaine in Canada, are producing sulfur on the scale of 2000–3000 tonnes per day, a 10% sulfur discharge would still represent hundreds of tons of sulfur dioxide loss per day. For the province of Alberta alone losses from this source could amount to up to 5000 tonnes per day. However, if the sulfur vapor from one Claus stage is condensed out, and the residual gases are blended with further hydrogen sulfide and passed over catalyst, the equilibrium of Eq. 9.19 is displaced further to the right to obtain up to 94–95% conversion to sulfur and process containment is greatly improved (Fig. 9.7). A third Claus reactor stage can similarly boost recoveries to the 96–98% range [28]. But even this is still not deemed to be adequate control in Japan and the U.S.A., or for large gas processing plants in Alberta where recoveries of about 99.5% are required. This can theoretically be achieved with a fourth Claus stage but the order of magnitude diminished sulfur return for the large increased investment required for the additional stage makes this alternative economically unattractive relative to other types of control devices.

9.5.3. New Developments and Emission Controls, Claus Technology

Primary Claus plant effluent gas from a single stage may still contain 2 to 3% sulfur dioxide, which would represent both a significant loss of feedstock and an emission problem if not further processed. Modeling experiments could help [29]. Both problems may be alleviated by addition of extra stages of Claus reactors but only to approach the limit imposed by the equilibrium relation of

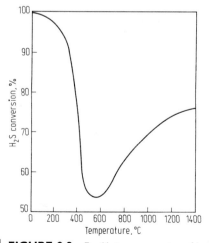

FIGURE 9.8 Equilibrium conversion of hydrogen sulfide to sulfur vapor based on thermodynamic calculations, one atmosphere total system pressure and with no sulfur removal. (Compiled from data from [26,27].)

Eq. 9.19 (Fig. 9.8) at Claus operating temperatures, or about 1% sulfur dioxide. The Sulfreen process takes advantage of Claus chemistry at temperatures of 125–150°C to use the higher equilibrium sulfur conversion accessible at these temperatures. Low reactor operating temperatures are combined with the practice of sulfur removal as a liquid from the gas phase as it forms, which further displaces the equilibrium to the right. This gives the process a theoretical 100% conversion efficiency of sulfur dioxide to sulfur [30]. As a result of the condensed sulfur, which gradually accumulates on the surface of the catalyst, the reactor has to be purged with a periodic high-temperature sulfur vaporization cycle. Exit gas sulfur dioxide concentrations from the process are decreased to the 2000–2500 ppm range, as compared to the 1% (10,000 ppm) or so experienced by the normal Claus technology. Actual sulfur containments of 99.9% have been claimed for the Sulfreen process [31].

A related procedure is used in the Westvaco process, except that sulfur dioxide is catalytically oxidized to sulfur trioxide using activated carbon at 75–150°C. The sulfur trioxide is then hydrated to sulfuric acid which is absorbed onto the active carbon [32]. Sulfur recovery from the sulfuric acid is as sulfur dioxide, which is formed in a regenerator by raising the temperature of the carbon and adding hydrogen sulfide.

Operation of the initial hydrogen sulfide burner in a mode to favor initiation of the high-temperature Claus reaction in this unit rather than to optimize on percent sulfur dioxide produced has also been found to raise sulfur recoveries [33]. So has the installation of a continuous ultraviolet spectrophotometer to monitor the relative ratios of sulfur dioxide and hydrogen sulfide in the exit gases from a Claus first-stage reactor. In this way a firmer, appropriate control basis is provided for oxygen or reducing gas addition prior to the subsequent stages.

Other integral measures to improve sulfur recoveries include the use of oxygen-enriched air for primary hydrogen sulfide combustion [34,35]. This would usefully serve to raise partial pressure of the sulfur gases throughout the process, for the same total system pressure. The integral inclusion of water removal as a part of the sulfur-forming reaction should also serve to raise conversion efficiencies by displacing the equilibrium of Eq. 9.19 further to the right [27]. Thus far, neither of these alternatives appears to have been tested under process conditions.

A further alternative to consider in this category is the suggestion that hydrogen sulfide be simply thermally decomposed to sulfur and hydrogen (Eq. 9.21) [36].

$$2 \ H_2S \rightarrow 2 \ H_2 + 2/x \ S_x \qquad \Delta H = +159 \text{ kJ } (+38 \text{ kcal}) \qquad 9.21$$

Thermodynamically it appears that burning only a part of the hydrogen product should be sufficient to provide the energy required for thermal decomposition. Another suggestion proposes the use of focused sunlight as the source of the heat [37].

Add-on tail-end emission control options include the use of polyethylene glycol scrubbing (the Clauspol process [33]), and the use of methane or hydrogen as a sulfur dioxide reducing agent in a follow-up stage, rather than hydrogen sulfide (the SCOT process; *Shell Claus off-gas treating* [31,33]).

Hydrogen sulfide formed in the SCOT process is scrubbed and recycled to the front end of the Claus sequence. Another control option is to react the dilute sulfur dioxide stream with oxygen and water vapor to form sulfuric acid (Eq. 9.22), and absorb this onto activated carbon at 75 to 150°C [38].

$$SO_2 + H_2O + 1/2 \ O_2 \rightarrow H_2SO_4 \qquad\qquad 9.22$$

A 90% sulfur dioxide stream, a high enough concentration for recycling, is then regained by a combination of heating in the absence of air and partial reduction with hydrogen sulfide, as already described in detail. There are also a number of other variations of these themes.

9.6. SULFURIC ACID

Sulfuric acid, though not often evident in the final product, plays some chemical or refining function in the preparation of a very large number of chemicals (Table 9.4). This wide utilization has occurred because it is a strong, inorganic acid that is also low in cost. Prices over the period 1954 to 1974 have ranged from $25 gradually rising to $36 per tonne in the U.S.A [41]. This relatively low cost has meant that its chemical properties of high acidity, dehydrating action, and ability to sulfonate (as oleum), oxidize, and react with unsaturated hydrocarbons have all been employed commercially on a large scale.

TABLE 9.4   Use Pattern for Sulfuric Acid in Industry as Percent of Total Consumption[a]

	U.S.A. 1975/78	United Kingdom 1976	W. Germany 1976
Phosphate fertilizers	38	32	15[c]
Ammonium sulfate	7		6
Detergents	—	11	[c]
Fibers and cellulose film	—	9	46
Petroleum refining	5–8	—	—
Alcohols	7	—	[d]
Titanium dioxide	5	15[b]	22
Iron and steel pickling	3	—	11
Explosives	2.5	—	—
Other chemicals	10	16	[d]
Battery acid	0.5	—	—
Miscellaneous	19–22	17	[d]
	100	100	100

[a]Prepared from data of *U.N. Statistical Yearbooks* [39] and *Chemical and Engineering News* [40]. Due to differences in the breakdown of the listings used, a dash does not necessarily indicate zero consumption of a category.
[b]Includes other paint and pigment uses as well as titanium dioxide.
[c]Detergent consumption included in phosphate fertilizer catgory.
[d]Plastics, petrochemicals, and miscellaneous uses included with fibers and cellulose film.

The large volume and wide application of sulfuric acid in the chemical and petroleum refining industries has for a long time meant that the per capita production of sulfuric acid by a country is one of the better chemical activity indicators of industrial development. Less circumstantial anomalies occur in a country-by-country listing of annual sulfuric acid production (Table 9.5), a chemical product, than for the per capita sulfur production, an extractive product. Thus average sulfuric acid production levels of the developed countries are from 50 up to 200 kg per capita per year as compared to values of generally less than 5 kg per capita for Third World countries.

World sulfuric acid production has grown by a factor of 1.4 to 1.6 every 10 years since 1930, with an accelerated rate of near doubling during the 1950 to 1960 period, a rate which has slowed again since. A growth in sulfuric acid production exceeding this rate is thus evidence that a country is in the process of development of a technological base, whereas those countries with a parallel or only slightly slower growth rate in their sulfuric acid production generally possessed well-developed technologies prior to the period considered (Table 9.5).

Two processes are used commercially to produce sulfuric acid, the contact process, where the important sulfur dioxide to sulfur trioxide oxidation step is accomplished heterogeneously over a solid catalyst with air, and the chamber process, where this transfer of oxygen from air to sulfur dioxide is accom-

▓▓▓▓ **TABLE 9.5   Major World Producers of Sulfuric Acid[a]**

	Thousands of metric tonnes					
	1950	1960	1970	1980	1990	1995
Australia	617	1,776	1,762	2,175	1,464	n/a
Belgium	880	1,403	1,794	—	1,906	1906[b]
Canada	686	1,517	2,475	4,295	3,560[c]	4,220
China	—	—	—	7,640	11,969	17,410
France	1,215	1,983	3,682	4,943	3,771	2,377
Germany	1,446[d]	3,170[d]	4,435[d]	4,108[d]	3,652[d]	3,725
India	104	354	1,189	—	3,272	n/a
Italy	1,276	2,299	3,327	2,822	2,038	2,178
Japan	2,030	4,452	6,925	6,777	6,887	6,888
Mexico	43	249	1,235	—	455	195[b]
Poland	285	685	1,901	3,019	1,721	1,244[b]
Spain	456	1,132	2,021	—	2,848	n/a
U. Kingdom	1,832	2,745	3,352	3,376	2,000	977
U.S.A.	11,820	16,223	26,784	40,071	40,222	43,267
U.S.S.R.	n/a	5,398	12,059	23,033	27,267	—
Other	2,310	4,834	13,239	—	47,482	—
World	25,000	48,220	86,180	102,259	135,918	126,799

[a]Derived from data in *U.N. Statistical Yearbooks* [39] and June issues of *Chemical and Engineering News* [40]. Those countries that have had an annual production in excess of two million metric tonnes are listed.
[b]Values for 1992.
[c]Value for 1989.
[d]Production of the former West Germany.

plished with a gaseous catalyst. Of the two, the contact process is used to produce more than 95% of the supply of sulfuric acid at present, both in Europe and in North America.

9.6.1. Contact Process Sulfuric Acid

The original conception of the contact process is credited to a patent issued to P. Phillips in 1831 [42], but the practice of the principal components taught by this patent took nearly 50 years to bring to commercial success. The key step, the reaction of sulfur dioxide and air over a yellow-hot platinum surface to obtain sulfur trioxide, took extensive development work to obtain reasonable conversions. Coupling this initial catalytic oxidation to the hydration of the sulfur trioxide product was eventually achieved on the scale of 17,000 tonnes per year by 1880, rising to 105,000 tonnes per year by 1890, by BASF (Badische Anilin-und-Soda-Fabrik), in Germany.

The chemical reactions involved in the operation of modern contact acid plants are those outlined in the patent which laid the basis for this industry. However, many improvements in practice, occasioned by the present more detailed knowledge of the catalytic gas phase kinetics involved in the sulfur dioxide oxidation step, as well as a better present-day appreciation of the gas–liquid equilibria associated with the hydration step, have now been instituted.

Initially sulfur is burned in air to produce sulfur dioxide and heat (Eq. 9.23).

$$S_{(g)} + O_{2(g)} \rightarrow SO_{2(g)} \qquad \Delta H = -297 \text{ kJ } (-70.9 \text{ kcal}) \qquad 9.23$$

Many operations also obtain sulfur dioxide for sulfuric acid production via the oxidation of pyrites or from the roasting of other sulfidic ores. But because of the additional capital cost involved in dust removal equipment to clean sulfur dioxide from these sources, they are of secondary importance. Roaster sources of sulfur dioxide are more often prompted by emission control incentives of metallurgical operations than by sulfuric acid production incentives.

The second reaction, oxidation of sulfur dioxide to sulfur trioxide with air, is a somewhat less exothermic, equilibrium reaction (Eq. 9.24).

$$SO_{2(g)} + 1/2 \; O_{2(g)} \stackrel{V_2O_5}{\rightleftharpoons} SO_{3(g)} \qquad \Delta H = -98.2 \text{ kJ } (-23.5 \text{ kcal}) \qquad 9.24$$

It is catalyzed in most modern plants by a by a 6–10% vanadium pentoxide coating on a support such as powdered pumice or kieselguhr. Supported platinum is a more active catalyst than vanadium pentoxide. It gives the same sulfur trioxide for the same contact time at lower initial gas temperatures, and also was used in many of the early contact acid plants. But because of its significantly higher cost and its susceptibility to poisoning, particularly by arsenic, vanadium pentoxide has virtually taken over this function [43]. A modern, potassium hydroxide promoted vanadium pentoxide catalyst bed is certainly quite efficient, being capable of virtually 100% sulfur dioxide conversion at 380–400°C with the correct initial gas proportions and sufficient contact time. It also has the durability (freedom from poisoning, etc.) to maintain nearly this level of activity over a 20-yr life. A fraction of the catalyst is usually replaced annually to maintain high activity.

A catalyst is a substance which speeds up a chemical reaction but is not itself consumed in the reaction. However, a catalyst cannot alter the equilibrium position of a chemical reaction, i.e., in this case the relative proportions of sulfur dioxide and sulfur trioxide present after the reaction. Thus, the reaction rate for an equilibrium reaction, such as that represented by Eq. 9.24, is the speed with which equilibrium is reached (not the speed to complete conversion).

The equilibrium for a gas phase reaction may be written in a parallel manner to the operation of the law of mass action in solution equilibria (Eq. 9.25), in this case the component concentrations being expressed in terms of partial pressures.

$$K_p = \frac{SO_3}{P_{SO_2} \times (P_{O_2})^{1/2}} \qquad\qquad 9.25$$

The value of the equilibrium constant for the correct ratio and concentrations of sulfur dioxide and oxygen (Eq. 9.25) has been determined experimentally at a number of temperatures (Table 9.6). The Le Chatelier principle states in essence that if a system in equilibrium is disturbed by a change in conditions, the position of the equilibrium will shift in the direction that will minimize the effect of the change on the system. Since this reaction is exothermic in the direction toward the sulfur trioxide product, from the Le Chatelier principle it would be expected that the equilibrium would lie more toward sulfur trioxide at low temperatures than at high temperatures.

Considering the equilibrium data, at 400°C, with a value of K_p of 397, the equilibrium of Eq. 9.24 is about 96% on the side of sulfur trioxide. But at this temperature the time required for sulfur dioxide to react with oxygen is relatively long, requiring a large reactor and large catalyst volume (and thus higher capital costs to use these conditions) to obtain significant sulfuric acid production rates.

At 500°C the rate of reaction is about 100 times as fast as at 400°C, requiring a much smaller reactor volume for the same sulfuric acid through-

**TABLE 9.6 Partial Pressure
Equilibrium Constants for the
Oxidation of Sulfur Dioxide[a]**

Temperature (°C)	K_p
400	397
500	48.1
600	9.53
700	2.63
800	0.915
900	0.384

[a]Conditions, one atmosphere total system pressure, partial pressures of components given in atmospheres. Data selected from Bodenstein and Pohl [44].

put, but the equilibrium constant, Kp, drops to about 50. Hence, at this temperature only about 85% of the sulfur is present as sulfur trioxide.

At 600°C, the rate of reaction is some 30 to 50 times faster again, requiring an even smaller reactor for the same throughput, but the rate of dissociation of sulfur trioxide to sulfur dioxide becomes appreciable. The value of K_p drops to about 10, giving only about 60–65% of the sulfur as sulfur trioxide at this temperature, and the remainder as sulfur dioxide. For process purposes there is no point in considering the sulfur equilibrium situation for any higher temperatures than this since with a promoted vanadium pentoxide catalyst bed at 600°C a 2–4 sec contact time is already sufficient to obtain essentially equilibrium concentrations at this temperature.

For process optimization therefore, advantage is taken of the very fast reaction rates at 550 to 600°C to operate at these temperatures to about 60 to 65% conversion. The gas mixture is then cooled to 400–450°C, generating further steam, to take advantage of the more favorable equilibrium at this temperature, before being passed over three (or more) additional catalyst beds to reach about 97 to 98% sulfur dioxide conversion to sulfur trioxide. Occasionally additional air is added at this stage to assist in displacing the equilibrium further to the right.

Since 1.5 moles (volumes) of gas on reaction are converted to 1 mole (volume) of gas in the oxidation of sulfur dioxide it would be expected that carrying out this process under pressure, again by the principle of Le Chatelier, would tend to drive the reaction more to the right. This has been confirmed in practice, but the improvement obtained has not been worth the additional capital costs required to operate the whole process under pressure [45]. The same sort of effect may also be achieved by raising the sulfur dioxide and oxygen concentrations, still in a total system pressure of 1.013×10^5 kP (equivalent to 1 atm; 1 kP = 1 N/m^2), as could be obtained by carrying out the sulfur combustion in oxygen-enriched air. But again, so far, economics have dictated against this option. New oxygen enrichment developments may change this.

These chemical principles are put into practice in a contact acid plant by first burning an atomized jet of filtered molten sulfur in a stream of dry air (Fig. 9.9). The air required is dried, prior to combustion, by upward countercurrent passage through a tower containing a stream of nearly concentrated sulfuric acid, which is trickled down an acid-resistant chemical stoneware packing. Preliminary air drying is necessary to avoid corrosion problems from the moist gases which would otherwise result after combustion, and to decrease the problems from sulfuric acid mist formation on eventual disposal of the spent waste gases. With a stoichiometric excess of air for this reaction a gas stream containing 9 to 12% sulfur dioxide in air at a temperature of nearly 1000°C is obtained.

A fire tube boiler is used to reduce gas temperatures to the 400 to 240°C range, simultaneously generating steam, before passage of these hot gases over the first catalyst bed of the four-pass converter. Sulfur dioxide is 60–65% converted to sulfur trioxide at this stage, simultaneously raising gas temperatures to about 600°C from the exotherm of the reaction and in so doing taking advantage of the very rapid reaction rates at these temperatures.

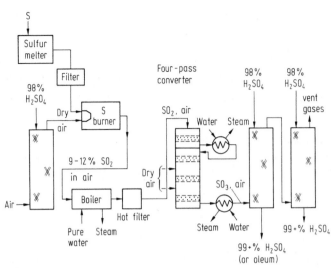

FIGURE 9.9   Contact process for making sulfuric acid and oleum from sulfur. Additional heat control measures are usually present for the last three stages of conversion.

Then, to take advantage of the higher equilibrium proportion of sulfur trioxide accessible at lower temperatures, the gases are again cooled to about 400°C, generating further steam, before passage over the second catalyst bed. Conversion to 80–85% sulfur trioxide occurs in the second catalyst bed accompanied by a smaller temperature rise. Cooling by gas-gas exchange and/ or by addition of small amounts of dry air at ambient temperatures is used to bring the reacting gases to the 400°C range for each of the third and fourth catalyst stages. Under ideal conditions these last two stages, in combination, give an overall 98% conversion of sulfur dioxide (or sulfur) to sulfur trioxide.

After cooling to near ambient temperatures the sulfur trioxide concentration is about 10% by volume. This product is absorbed in concentrated or nearly concentrated sulfuric acid, where both absorption and hydration occur via countercurrent contact in a chemical stoneware packed tower (Eq. 9.26).

$$SO_3 + H_2O \rightarrow H_2SO_4 \qquad \Delta H = -130 \text{ kJ } (-31.1 \text{ kcal}) \qquad 9.26$$

Using nearly concentrated sulfuric acid for hydration reduces the vigor (the exotherm) of this reaction. Because 98% sulfuric acid is also the concentration having the lowest vapor pressure (boiling point 338°C), using this concentration for hydration decreases the tendency to form sulfuric acid mists on absorption. The stream leaving the bottom of the absorber is then cooled, product acid is drawn off, and water is added to the remaining acid. Water addition to the process occurs from both the diluted acid produced by the air drier and by the metered addition of pure water, to produce the concentration required for absorption. Commercial grades of sulfuric acid give an analysis of about 96% H_2SO_4 (freezing point, about −7°C) to reduce risks of freezing during storage and transport, since pure 100% sulfuric acid freezes at 10.3°C. The contact acid plant product is also quite pure, normally containing less than 20 μg/g (ppm) sulfur dioxide and less than 10 μg/g iron, or nitrogen oxides.

If oleum, a solution of sulfur trioxide in sulfuric acid, is the desired product from the contact plant, this is normally obtained from the first sulfur trioxide absorber. Absorption here, where the highest sulfur trioxide concentration is available in the entering gases, yields up to 20% oleum directly. This is equivalent to 20 kg of sulfur trioxide dissolved in 80 kg of 100% H_2SO_4. At this concentration the oleum vapor pressure becomes so high that sulfur trioxide loss occurs almost as readily as absorption so that this is also the limit of oleum concentrations available by direct absorption. Higher concentration oleums may be obtained by distillation of the sulfur trioxide (boiling point 45°C) from 20% oleum, and condensing this now 100% sulfur trioxide vapor stream into 20% oleum until the desired concentration of oleum is obtained. Several commercial oleum grades are marketed, including 60% and 100%. At 60% and higher concentrations it is thought that a significant fraction of the mixture actually exists as pyrosulfuric acid, $H_2S_2O_7$, a different compound [46]. Pure sulfur trioxide, 100% oleum, is marketed with about 0.25% of stablizers such as sulfur, tellurium, carbon tetrachloride, or phosphorus oxychloride added to prevent crystallization or polymerization. Sulfur trioxide is easier to handle in the liquid state. Also, if either crystallization or polymerization were to occur during storage or shipping, these processes are sufficiently exothermic that they have the potential to raise the temperature of the bulk of stored sulfur trioxide to above its boiling point. If this should happen it risks sulfur trioxide release or vessel rupture from overpressure.

Operating a contact sulfuric acid plant essentially as described above thus allows straightforward production of concentrated sulfuric acid, oleums, and pure sulfur trioxide which may be used in sulfonations or to chemically rejuvenate process-diluted sulfuric acid by simple addition. Typical raw material and utility requirements are given in Table 9.7, from which it can be seen that the contact process is actually a net *producer* of high-pressure steam, sometimes a useful feature in a chemical complex.

Provision of the sulfur dioxide feed gas for a contact plant is dominated by direct sulfur combustion, but many other sources may also be tapped.

TABLE 9.7 Typical Raw Material Requirements and Utilities Consumed in the Production of One Tonne of 100% Sulfuric Acid[a]

	Contact plant, 180 tonne/day	Chamber process, 50 tonne/day
Sulfur, kg	337–344	337
Nitrogen oxides, kg (or ammonia burned)	—	2–2.5
Water, L (process, plus cooling)	16,700	10,000
Electricity, kWh	5–10	15–16.5
Steam, kg	1,000 (credit)	—
Air, m³	7,800	8,600

[a]Compiled from Lowenheim and Moran [41], Bodenstein and Pohl [44], and Shreve [47].

Pyrite, FeS_2, is burned (Eq. 9.27) and other sulfidic minerals are roasted (Eq. 9.28, 9.29), the latter primarily for their metal values rather than for the sulfur dioxide.

$$FeS_2 + 5/2\ O_2 \rightarrow FeO + 2\ SO_2 \qquad\qquad 9.27$$

$$CuS + 3/2\ O_2 \rightarrow CuO + SO_2 \qquad\qquad 9.28$$

$$ZnS + 3/2\ O_2 \rightarrow ZnO + SO_2 \qquad\qquad 9.29$$

Extensive dust removal facilities are required to clean up the sulfur dioxide stream from these sources, which adds significantly to the capital cost of the plant and serves to offset the raw material cost advantage conferred by these sulfur sources. Sometimes hydrogen sulfide is simply burned to produce sulfur dioxide when the source of the hydrogen sulfide is near the producing sulfuric acid plant. If this method is used then elemental sulfur production by the Claus process is bypassed.

If, however, none of these more straightforward sulfur sources is available and anhydrite ($CaSO_4$) is available on-site, or close by, then it is possible to practice thermal reduction of anhydrite with coke to obtain sulfur dioxide [48] (Eq. 9.30, 9.31).

$$CaSO_4 + 2\ C \xrightarrow[\text{heat}]{} CaS + 2\ CO_2 \qquad\qquad 9.30$$

$$CaS + 3\ CaSO_4 \rightarrow 4\ CaO + 4\ SO_2 \qquad\qquad 9.31$$

Again the sulfur dioxide gas stream requires efficient dust removal prior to conversion. By using the lime clinker produced from this process as raw material for cement production (about equal amounts of sulfuric acid and clinker are produced) the economics have been sufficiently favorable to have had roughly 20–25% of United Kingdom (ICI) sulfuric acid produced via this means at one time.

9.6.2. Chamber Process Sulfuric Acid

The chamber process for the production of sulfuric acid is by far the older of the two large-scale commercial processes. Records exist at least back to 1746, when Dr. Roebuck in Birmingham was burning sulfur and nitre (KNO_3) in the presence of steam in lead-lined rooms (chambers, hence the name given to the process) to produce sulfuric acid [11]. Less detailed references exist to a similar practice by B. Valentine as far back as the late 1400s. In addition there are even less well-documented reports of the use of processes of this type dating back to 1000 A.D.

The early chamber process facilities were relatively small-scale operations practiced from the experience that the desired product was obtained by carrying out certain steps, but with little appreciation of the detailed chemistry involved. The science of chemistry itself had not developed sufficiently to be able to determine the chemical details when this process was first practiced. Even if it had, the number and complexity of the reactions involved were sufficient to forestall complete understanding until about 1910. This was, however, the only existing process for sulfuric acid production until the late nineteenth century and until as late as 1945 was used to produce close to one-half of the sulfuric acid produced in the United Kingdom and the United States. Since then, however, the proportion of the total sulfuric acid produced via the chamber process has rapidly declined, in the U.S.A. to about 10% of the total in 1958, and further to only about 0.5% of the total in 1973, mainly because of the lack of flexibility in the concentrations of acid produced by this process.

In the Roebuck version, as in the present version of this process, nitrous acid is the active oxidant used to convert sulfurous acid to sulfuric acid (Eq. 9.32).

$$H_2SO_3 + 2\ HNO_2 \rightarrow H_2SO_4 + 2\ NO + H_2O \qquad 9.32$$
$$\text{nitric oxide}$$

At the normal operating temperature (about 110°C) the nitrous acid is in the vapor state, hence oxygen transfer is via a gas phase catalyst, unlike the heterogeneous solid phase catalyst employed in the contact process. Sulfurous acid is obtained from the steam hydration of sulfur dioxide produced by the burning of sulfur, and the nitrous acid via the hydration of nitrogen oxides generated by the heating of potassium nitrate (Eq. 9.33, 9.34).

$$2\ KNO_3 \rightarrow K_2O + NO_2 + NO + O_2 \qquad 9.33$$

$$NO + NO_2 + H_2O \rightarrow 2\ HNO_2 \qquad 9.34$$

The thermal decomposition of potassium nitrate is not well understood, and at best may be nonstoichiometric [49]. As operated in the 1750s, the 50–60% (50°Bé) sulfuric acid product was collected from the floor of the lead-lined chamber, and the associated gaseous nitric oxide was vented to the atmosphere.

In the early days of the process, it had been thought that the proportion of potassium nitrate heated to the amount of sulfur burned was crucial. The importance of the presence of air for reoxidation of nitric oxide to nitrogen dioxide (Eq. 9.35) was only discovered in 1806, and has been made use of by operators profitably since that date [11].

$$2\ NO + O_2 \rightarrow 2\ NO_2 \qquad 9.35$$

Since the potassium nitrate (nitre) was an expensive ingredient this discovery alone significantly improved the economics.

It was not until 1827 that Gay-Lussac developed the absorption tower that then made it possible to capture nitrogen oxides from the sulfuric acid chamber(s) to produce nitrosyl sulfuric acid, so-called "nitrous vitriol" (Eq. 9.36).

$$NO + NO_2 + 2\ H_2SO_4 \rightarrow 2\ ONOSO_3H + H_2O$$

nitric	nitrogen	nitrosyl	
oxide	dioxide	sulfuric	9.36
	(nitrogen	acid	
	IV oxide)		

This improved not only the economic but also the environmental aspects of this early process, except that it was not easy to return the active nitrogen oxides to the working chambers without diluting the solution of nitrosyl sulfuric acid in sulfuric acid with water. Most producers did not want to do this and, therefore, did not put it into practice because of the high cost of reconcentrating the diluted acid to commercial strength after the release of nitrogen oxides.

The development of the Glover tower, in 1860, allowed both nitrogen oxide release by water dilution (Eq. 9.37) and, in the same unit, reconcentration of the acid via the hot gases generated from sulfur combustion.

$$ONOSO_3H + H_2O \rightarrow H_2SO_4 + HNO_2 \qquad\qquad 9.37$$

This additional innovation now made the combination of the Glover tower front end unit, as a generator-concentrator, and the Gay-Lussac tower tail gas recovery unit more attractive to sulfuric acid producers, but still led to only slow adoption. Even by 1890 only about half of the sulfuric acid plants in the U.S.A. used the nitrogen oxide conserving towers as an integral part of their processes. But gradually the ability to recycle the nitrogen oxides plus the trend toward an increase in the scale of chamber operations caused a marked drop in price of acid, from £33–38 per tonne in 1800, £3.75 in 1820, down to £1.30 per tonne in 1885. Thus, with the lesser chemical background possessed by the early chamber process operators, it took some 110 years, from the early initial appreciation of the requirements for sulfuric acid formation in 1746, to the time that the knowledge of the nitrogen oxide recycle and sulfuric acid concentration steps of the Glover tower were worked out in 1859, for the full technology of this process to be developed. This is more than twice as long as the time required for the development of the more recent contact process, which occurred when the science of chemistry was further advanced.

The modern format of a chamber plant parallels these early developments quite closely (Fig. 9.10). Significant differences exist only in the Glover tower accessories for the makeup of any small amounts of lost nitrogen oxides which now occurs either via the heating of nitric acid (Eq. 9.38), or by the combustion of ammonia in air (Eq. 9.39, 9.40).

$$2\ HNO_3 + NO \rightarrow 3\ NO_2 + H_2O \qquad\qquad 9.38$$

$$2\ NH_3 + 5/2\ O_2 \rightarrow 2\ NO + 3\ H_2O \qquad\qquad 9.39$$

$$2\ NO + O_2 \rightarrow 2\ NO_2 \qquad\qquad 9.40$$

The nitrogen oxides produced by either method supplement the bulk of the nitrogen oxides which are supplied by recycle of nitrosyl sulfuric acid at the Glover tower by Eq. 9.37. In this way, the total amount present is adequate to provide sufficient nitrous acid (Eq. 9.34) for the oxidation of sulfurous to sulfuric acid. This oxidation occurs in both the Glover tower and the chambers. So the lead-lined, quartz-packed Glover tower performs multiple functions in the process. It hydrates the sulfur dioxide, which enters at the bottom, and the nitrogen oxides, and it concentrates the incoming 50°Bé (62%) sulfuric acid entering the top of the tower, to about 60°Bé (78%) sulfuric acid,

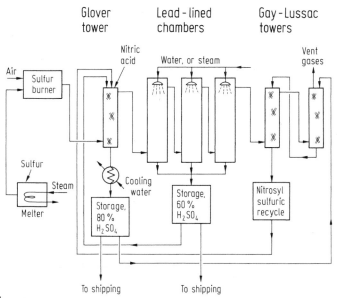

FIGURE 9.10 Simplified flowsheet of the chamber process for the production of sulfuric acid. The Glover tower, used for concentrating the sulfuric acid and recycling the nitrogen oxides recovered by the Gay-Lussac tower, actually comprises a volume of only 1–1.5% of the whole chamber set. The three to six chambers are made of wood or steel, lined with lead, size about 7–10 m roughly square cross section and 15–35 m long. Water addition is by steam or liquid water as necessary for temperature control.

which emerges from the bottom. Sulfuric acid of about 78% concentration is ordinarily the most concentrated product of the chamber process plant. At best, under normal operating conditions, only about half of the direct product of the chamber plant could be of 78% concentration and the remainder would be the product collected directly from the chambers which would be about 60% concentration.

The Baumé density scale for specifying sulfuric acid concentrations, which is still in use today, is a holdover from the early days of acid manufacture when there was no clear understanding of either acid composition or the exact correlation between density and concentration. For these reasons, it is an arbitrary density scale based on Eq. 9.41, which was originally devised for specifying sugar concentrations in water [50].

$$\text{Specific gravity} = \frac{m}{(m - d)} \qquad 9.41$$

where m = 145, and d = degrees Baumé reading
The relationship between some common process and trade acid concentrations and degrees Baumé and density readings is given in Table 9.8.

Most of the formation of sulfuric acid in the chamber process occurs in the liquid layer on the walls of the lead chambers and in the enclosed vapor space. While the sulfurous acid oxidation reaction, Eq. 9.32, is relatively rapid, at the moderate 110–120°C temperature of the chambers, sufficient

TABLE 9.8   **Relationship between Degrees Baumé and Density for Some Common Sulfuric Acid Concentrations**[a]

Concentration % H_2SO_4 (wt.)	Degrees Baumé	Density at 15.6°C	Remarks
0.0	0.0	1.00	pure water
10.9	10.0	1.075	"dilute" acid
26.0	23.0	1.190	
29.8	26.0	1.219	
33.5	29.0	1.250	conc. of battery acid
37.5	31.80	1.281	
62.5	50.0	1.526	"chamber" or "fertilizer" acid
73.1	57.0	1.647	$H_2SO_4 \cdot 2H_2O$
77.7	60.0	1.706	"Glover tower" acid
84.5	63.2	1.773	$H_2SO_4 \cdot H_2O$
90.0	65.1	1.8144	
93.2	66.0	1.8354	"Oil of vitriol"
95.0	[b]	1.8337	ordinary concentrated acid
100.0	[b]	1.8305	"monohydrate" sulfuric acid

[a]Data compiled with some calculation from Faith *et al.* [51] and Lange [52].

[b]Baumé readings become unreliable indicators of sulfuric acid concentration because of a decrease in density at these high concentrations.

volume has to be provided for virtually all of the sulfurous acid to be oxidized before the residual gases and vapors move to the Gay-Lussac tower. Thus these units must be large.

Spent chamber gases move into the bottom of the quartz-packed Gay-Lussac tower while 60°Bé (about 78%) sulfuric acid (part of the Glover tower product) is trickled in at the top. In this way the nitrogen oxides present are captured, forming nitrosyl sulfuric acid for recycle (Eq. 9.36), and the residual gases, mainly nitrogen and water vapor, are discharged.

Therefore, the two conveniently available concentrations of sulfuric acid from a chamber plant are about 60 and 80%. Higher concentrations require additional capital investment in a separate concentrator unit which uses distillation to remove water. This lack of flexibility for higher acid concentrations is probably the main reason for the small commercial operating capacity of this process today. Traces of oxides of arsenic, nitrogen, and selenium and of sulfates of iron, copper, mercury, zinc, and lead may also be present in the product acid depending on the presence of these components in the sulfur or sulfide minerals used to provide the sulfur dioxide feed gas. For applications such as the production of fertilizers these impurities can be tolerated, but for applications requiring high-purity acid they need to be removed.

9.7. EMISSION CONTAINMENT FOR SULFURIC ACID PLANTS

From the stoichiometry of the balanced combined equation representing the raw material requirements (Eq. 9.42), the theoretical sulfur, oxygen, and water requirement for any given quantity of sulfuric acid production may be calculated.

$$S + 3/2\ O_2 + H_2O \rightarrow H_2SO_4$$

molecular weights	32.06	32.00	18.02	98.08	
moles		$\times\ 1.5$			9.42
mass ratios	32.06	48.00	18.02	98.08	

The theoretical sulfur requirement works out to be 326.9 kg per tonne (Eq. 9.43), noticeably less than the 337 to 344 kg of sulfur per tonne experienced by many sulfuric acid producers (Table 9.7).

$$\frac{32.06\ \text{g/mol}}{98.08\ \text{g/mol}} \times 1\ \text{tonne} \times 1000\ \text{kg/tonne} \qquad 9.43$$

Material balance and conservation of matter considerations allow us to say that this 10 to 14 kg per tonne difference between the theoretical and actual sulfur consumption, less a proportion to account for any impurities which may be present in the feed sulfur, represents the sulfur loss from the process.

The experience reported above represents 95 to 97% sulfur to sulfuric acid conversion efficiencies. With sulfuric acid plants of 200 tonnes per day capacity becoming common and the occasional plant operating on the scale of 1800 tonnes per day [41], even these high conversion efficiencies are not enough to avoid local emission problems. Thus, the United Kingdom has a requirement of 99.5% containment of the sulfur burned as a feedstock, while the U.S.A. requirement is for 99.7% sulfur dioxide to sulfur trioxide conversion efficiency [53]. These amount to very similar regulations for the two countries. With modern process modifications and emission control devices in place these particular requirements are being met.

9.7.1. Contact Process Sulfuric Acid Emission Control

The main pollutant losses from a contact acid plant occur through the absorber exit gases, but the origins of these losses are a composite of several process variables. The gas composition of this exhaust stream is mainly nitrogen and oxygen. Prior to adoption of abatement measures it also contained 0.13 to 0.54% with an average of 0.26% by volume sulfur dioxide, and 39 mg/m^3 to 1730 mg/m^3 (1.1–48.8 mg/ft^3) with an average of 457 mg/m^3 (12.9 mg/ft^3) of sulfuric acid mists. Sulfuric acid mists are harmful to plant life and to human populations. Effects of the sulfur dioxide were considered in Chapter 2. Since there is contact of nitrogen and oxygen with heated metal surfaces in a contact acid plant there is also the potential for the formation and loss of nitrogen oxides (NO$_x$) in these exit gases. But the concentration of NO$_x$ would be expected to be less than that present in fossil-fueled power station stacks, and certainly less than obtained from an uncontrolled chamber acid plant.

The discharge of unconverted sulfur dioxide may be decreased by optimizing the sulfur dioxide to oxygen ratio entering the converter [54]. If, simultaneously, the thickness of the catalyst beds in the four-pass converter is increased, or the gas velocity through the converter is reduced, the increased gas to catalyst contact time obtained will also decrease the percent of unconverted sulfur dioxide. A regular maintenance schedule for cleanup and partial replacement of the catalyst of the first catalyst bed helps to maintain conversion efficiencies. A water or ammonia scrubber for tail gas cleanup from the last absorber (Fig. 9.11) coupled with the above steps also significantly improves the emission level. Sale of the recovered ammonium sulfate from an ammonia scrubber spent liquor can help to offset installation and operating costs of this unit.

But more than any other single measure, the interpass absorption (IPA) system has introduced a process integral measure which has the effect of reducing sulfur dioxide emissions to an order of magnitude below that obtained from a normal four-pass converter, that is, to the range of 0.01 to 0.03% of the sulfur feed 55. This system employs intermediate cooling and sulfur trioxide absorption after the gas mixture has gone through only three stages of conversion. By removing much of the sulfur trioxide product of the conversion <u>before</u> the last catalyst stage, the equilibrium of the conversion reaction of Eq. 9.24 is strongly shifted to the right. Any residual sulfur dioxide still present after absorption is then much more completely converted to sulfur trioxide as it passes through the last stage of the converter at a moderate 400–420°C (Fig. 9.12). The net effect of these modifications brings the overall sulfur conversion to acid to the 99.7% range. Combining IPA with operation under a pressure of about 5×10^5 kP (about 5 atm) for both conversion and absorption has been found to yield further improvement in conversion efficiencies to the 99.8 to 99.8% range, as practiced by the Ugine Kuhlmann process [56].

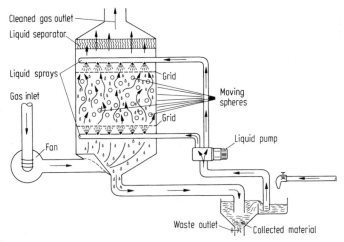

FIGURE 9.11   Floating bed scrubber, such as might be used to control sulfur dioxide and sulfuric acid mists. (Courtesy of European Plastic Machinery Mfg., A/S, Copenhagen.)

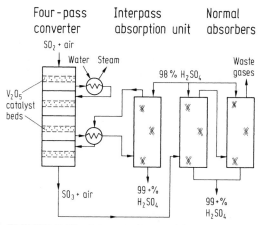

FIGURE 9.12 Schematic diagram of the converter section of a contact sulfuric acid plant employing an interpass absorption system for both better sulfur conversion and emission abatement. Product acid (or oleum) is cooled indirectly with process water prior to storage for sale.

Sulfur consumption for this system is 330 kg per tonne, very close to the theoretical 326.9 kg.

The acid mist emission problem arises from one or more factors. Water vapor in the air feed to the sulfur burner may cause mists because, as the water vapor plus sulfur trioxide stream drops below the dewpoint (condensation) temperature, aerosol formation occurs. If this cooling happens in the absorbers, mist losses can occur. Mist formation is minimized or eliminated by adequate drying of the air fed to the sulfur burner with concentrated sulfuric acid.

Water may also get into converter gases from the combustion of hydrocarbons present in the sulfur, a consideration with the dark grades of Frasch sulfur which may contain up to 0.3% bitumen. This produces water vapor at a point in the process beyond where the air drying tower can help. The best that can be accomplished in this situation is to keep the gas mixture above the dewpoint throughout, until it enters the absorbers, to minimize corrosion problems in the intervening process equipment, and to rely on ancillary devices to minimize mist discharge after absorption.

Occasionally mist problems occur during startup when the concentration of acid used for air drying is too low or while oleum is being produced in the first absorber, both of which promote aerosol formation [57]. These causes are normally intermittent and discharge prevention relies on mist control equipment.

The control device most frequently employed is the Brink mist eliminator, which consists of an impaction filter made of fiberglass packed between stainless steel screens. These units are useful for the exit gas stream from the last absorber, and for the intermediate dried air stream leaving the sulfuric acid drier. They may also be used for the sulfur dioxide air stream after an IPA unit, if present. Mist eliminator mass efficiencies are 97 to 98% which reduces sulfuric acid aerosol concentrations from the 850 to 1275 mg/m^3

[24 to 36 mg per standard cubic foot (scf)] range to 18 to 21 mg/m^3 (0.5 to 0.6 mg/scf).

Water, or a closed-cycle solvent, may be used with scrubbers, sequenced either following a mist eliminator or to replace a mist eliminator in its function at the last absorber. Mass efficiencies of 90 to 99% have been reported for water-wash scrubbers with acid mist loadings of 3.5 to 7 g/m^3, even for particle size distributions determined to be 60% less than 3 μm. Electrostatic precipitators have also been used for mist elimination [55].

9.7.2. Emission Control for Chamber Process Acid

The main point of loss of both sulfur and nitrogen oxides is via the exit gas stream of the Gay-Lussac towers. Concentrations of both gases are regularly 0.1–0.2 % by volume, with nitrogen dioxide comprising 50 to 60% of the total nitrogen oxides. There is also loss of sulfuric acid mist at 180–1000 mg/m^3 (5–30 mg/ft^3), but which usually comprises less than 0.1% of the acid produced [55].

To decrease the concentrations of these components in the exit gases requires low gas velocities, low temperatures, and high impaction efficiencies in the Gay-Lussac tower. Two Gay-Lussac towers operating in series also helps [58]. If the rate of oxidation of sulfurous acid in the chambers is increased by increasing the concentration of nitrogen oxides, exit gas sulfur dioxide concentrations will be reduced. This measure requires efficient control of the nitrogen oxides at the tail-end of the process. A final water scrubbing of the Gay-Lussac tower exit gases can decrease sulfur dioxide concentrations by 40% and nitrogen oxide concentrations by 25%. By routing the scrubber effluent as feedwater to the chambers, both of the captured feedstock components are recycled back into the process.

Sulfur dioxide and nitrogen oxides in power station flue gases have both been controlled by adsorption on alumina. This may also control chamber plant emissions. A high concentration of sulfur dioxide and nitrogen oxides produced on regeneration could beneficially be fed back to the chambers. A related procedure which uses a solution of sodium hydroxide and sodium sulfite in water could be workable, but probably at higher cost [59].

9.8. RECYCLING OF SULFURIC ACID

Motivated partly by decreasing supplies and increasing costs of elemental sulfur, but more from the high costs of environmentally acceptable disposal methods for spent sulfuric acid, an increasing number of acid consumers are recycling used acid. The costs of disposal options such as barge transport of waste acid 180 km out to sea have now become comparable to the costs of recycling options [60], which has stimulated the adoption of these alternatives by major acid users. For example, in West Germany by 1976 some 13% of the total sulfuric acid used by each of two largest use categories, organic chemicals and inorganic pigments (Table 9.4), was already being recycled. The percentage of acid recycle by these sectors is likely to increase further while at the same time adoption of acid recycle by smaller sulfuric acid users also grows. Even if the high cost of small-scale recycle options is unattractive on an individual basis, collective recycle arrangements may still prove to be practical.

The properties of the spent sulfuric acid to be dealt with vary widely. It might be acid which is simply diluted, but otherwise virtually uncontaminated, to acid which is still quite concentrated but contaminated with metallic or other inorganic or organic impurities, to acid which is both diluted and contaminated. Measures for recovery and recycle therefore vary in complexity from simple reconcentration to water removal accompanied by various chemical purification steps. Occasionally thermal destruction of impurities accompanied by dissociation of the acid and recycle of the sulfur content for acid remanufacture may be required. For these reasons the appropriate acid recycle method very much depends on the use to which the original acid was put, and the condition of the spent acid obtained. Specific examples follow.

Regeneration of high acid concentrations from sulfuric acid which has been diluted by water but is otherwise uncontaminated, such as is obtained from air or other gas drying functions, can be accomplished by simply boiling off the water in either a batch (pot) or continuous (heat exchanger) mode [61]. Temperatures of the order of 300°C are required for product acid concentrations of 95% or better at normal atmospheric pressure. If the pressure is reduced to 20 mm Hg, water removal may be accomplished at about 200°C. In both cases lead or lead-lined equipment is necessary for the dilute acid stages to avoid corrosion problems. Steel may be used for containment of concentrations above 95% (65°Bé).

Regeneration of high concentrations of sulfuric acid may also be achieved by oleum or sulfur trioxide addition to diluted acid. The inventory of acid in circulation is increased by a corresponding amount (Eq. 9.26). A third method of reconcentration, useful when the acid consumption occurs as a part of, or adjacent to a contact sulfuric acid plant, is to pass the diluted acid through the acid plant absorption tower. This option amounts to on-site addition of sulfur trioxide, and the increased acid inventory obtained can be sold to markets through the normal producer channels.

Large-scale nitrations, which use a mixture of high concentrations of nitric plus sulfuric acids, yield a spent acid which consists mainly of diluted sulfuric acid but also contains traces of residual nitric acid and organics. Recovery involves removal of the organics by extraction with a solvent such as toluene. The aqueous acid raffinate is then sparged with steam to remove most of the residual nitric acid. Any residual nitric acid is destroyed by treatment with hydrogen peroxide and hydroxylamine. Subsequent reconcentration regains a useful purity of high-strength sulfuric acid [62]. A simpler but less rigorous procedure for the recovery of sulfuric acid used in nitrations involves stirring of the spent acid with activated carbon (1–3%) and diatomaceous earth (2–5%) to yield directly a chamber grade acid suitable for reconcentration. Simple denitration of spent nitration acid by addition of ammonium ion or urea has also been tested [63].

Methods have also been developed for destruction of the organic content of spent acid using hypochlorite or hydrogen peroxide [64]. If the concentration of organics in the waste acid is not high even the residual nitric acid content may be adequate to destroy these, in the process consuming much of the residual nitric acid [61] (e.g., Eq. 9.44, 9.45).

$$5 \text{ C (in organic contaminants)} + 4 \text{ HNO}_3 \rightarrow 5 \text{ CO}_2 + 2 \text{ N}_2 + 2 \text{ H}_2\text{O}$$
$$9.44$$

$$\text{C (in organic contam.)} + 2 \text{ ONOSO}_3\text{H} \rightarrow \text{CO}_2 + 2 \text{ H}_2\text{SO}_4 + \text{N}_2 \quad 9.45$$
$$\text{(unbalanced)}$$

Metal ion impurities in spent acid impose different recovery problems. For instance, iron residues in the spent acid from titanium dioxide manufacture may be removed by crystallization of first $FeSO_4.7H_2O$ followed by crops of various iron(II) salts or by electrodialysis [65,66]. Incidentally electrodialysis has also been used for sulfuric acid recovery from wastewater, either directly or after preconcentration on cation exchange resin [67].

Removal of copper ion from plating bath liquors has been accomplished electrochemically, by plating it out, and also by extraction with tributyl phosphate [68]. The tributyl phosphate extract brings nickel, antimony, arsenic, and copper all into the organic phase, leaving a significantly cleaner acid raffinate. Phosphate extraction has also been used for iron removal from pickling waste liquors.

Sometimes, however, the spent acid is so severely contaminated with inorganics or intractable tars that high-temperature oxidation of the impurities plus thermal dissociation of the sulfuric acid for sulfur dioxide recovery becomes the only feasible option to recycle the acid [69] (Eq. 9.46).

$$2 \text{ H}_2\text{SO}_4 \rightarrow 2 \text{ SO}_2 + 2 \text{ H}_2\text{O} + \text{O}_2 \qquad 9.46$$

Temperatures of about 1100°C are necessary in the presence of added air to promote complete oxidation of any organic impurities and thermolysis of the acid. Oxygen-enriched air may be used with advantage [70]. Subsequently, the sulfur dioxide is recovered and passed through a conventional contact plant for regeneration to sulfur trioxide and thence via rehydration to concentrated sulfuric acid (Eq. 9.25, 9.26).

If disposal appears to be the only practical alternative despite this range of recycle options, then pH neutralization prior to discharge is preferred, and is required by many jurisdictions. One inexpensive method employs the waste product from dry lime-based sulfur dioxide emission control systems [71] (Eq. 9.47).

$$\text{CaO} + 1/2 \text{ O}_2 + \text{SO}_2 \rightarrow \text{CaSO}_4 \qquad 9.47$$

Only a fraction of the lime employed in such processes is consumed. The residual calcium oxide, comprising up to two-thirds of weight of the precipitator catch from these control systems, provides the inexpensive and convenient source of base for neutralization of waste but unrecoverable sulfuric acid. By using the residual chemical activity of both waste streams, the disposal problems and costs of dealing with them are decreased.

It is also possible to use ammonia to neutralize waste acid, which consumes a more costly base. However, the ammonium sulfate obtained can be sold as a valuable fertilizer constituent, to help offset the cost of the base.

REVIEW QUESTIONS

1. A natural gas stream at 50 atm absolute pressure and 40°C contains 600 mm Hg partial pressure of hydrogen sulfide, and enters a Girbotol scrubber at 100 m³ min⁻¹. Not more than 1 mm Hg partial pressure of hydrogen sulfide can remain after cleaning. Apply the data given below to a sketch of the Girbotol process for the removal of hydrogen sulfide from sour natural gas and, assuming ideal gas behavior, calculate the following:

   (a) How many moles of H_2S per minute must be removed from this sour natural gas stream to meet these requirements? Assume ideal gas behavior.

   (b) How many kilograms per minute of scrubber liquid comprising 15.3 wt % (2.5 M) monoethanolamine in ethylene glycol which would be required to remove the hydrogen sulfide from this gas stream? The H_2S partial pressure at the top of the stripper is 300 mm Hg, and at the amine exit, 10 mm Hg.

H_2S partial pressure (mmHg)	Moles hydrogen sulfide per mole amine	
	40°C	140°C
1	0.128	0.016
5	0.271	0.025
10	0.374	0.040
50	0.683	0.091
100	0.802	0.124
300	0.931	0.200
600	0.970	—

2. It has been proposed that hydrogen sulfide be thermally decomposed to recover both hydrogen and sulfur products. Why might this be a commercially attractive alternative to the Claus process? *Note:*

$2 H_2S_{(g)} \rightarrow 2 H_{2\ (g)} + 2/x\ S_{x\ (g)}$ $\Delta H = +159$ kJ $(+38$ kcal$)$
$H_{2\ (g)} + 1/2\ O_{2\ (g)} \rightarrow H_2O_{(g)}$ $\Delta H = -238$ kJ $(-56.9$ kcal$)$ at 100°C

3. (a) Iron pyrites (FeS_2) may be burned to form Fe_2O_3, and sulfur dioxide feed for a sulfuric acid plant. How many kilograms of pyrites would theoretically be required to produce one tonne of 100% H_2SO_4 in this way?

   (b) If pyrites and raw sulfur cost $5 and $25 per tonne, respectively, other factors being equal, which would be the more economical feed, and how much less would the raw material cost be per tonne of sulfuric acid produced (provide answers to the nearest cent)?

4. (a) How many kilojoules of energy is theoretically available on the combustion of 1000 kg of sulfur initially at 25°C producing sulfur dioxide also finally at 25°C?

(b) From a raw water supply at 20°C, what mass of 100°C steam could theoretically be produced from the energy available as calculated from part (a)?

5. (a) For a contact process sulfuric acid plant, using a stoichiometric ratio of dry air to sulfur in its sulfur burner, what percent by volume sulfur dioxide would be produced? (Dry air is 21% oxygen, 78% nitrogen, 1% argon by volume. Assume ideal gas behavior.)
   (b) If twice the stoichiometric requirement of air for the oxidation of sulfur dioxide to sulfur trioxide is now added, and a 97% conversion is obtained, what percent by volume concentrations of sulfur trioxide and sulfur dioxide would leave the four-pass converter?

6. A 200 tonne/day contact sulfuric acid plant burning elemental sulfur costs $1.5 million to build. Smelter by-product sulfur dioxide is available as an alternate feed at no raw material cost, but for the same capacity requires a total investment of $3.75 million to utilize this feed. Assuming continuous operation, what would the "breakeven" price for sulfur have to be to make the utilization of smelter sulfur dioxide attractive if the company has to pay 10% interest on capital borrowed to build either plant? (Use estimates based on the first year of operation only. Assume the same labor costs.)

FURTHER READING

D.J. Bourne, ed., "Symposium on Sulfur Utilization; a Progress Report." American Chemical Society, Washington, DC, 1978.

K.S. Gaur, Pollution Control with SO_2 Recovery; Wellman-Lord Sulfur Dioxide Recovery System. *Pollut. Eng.* **10**, 51 (1978).

A.G. Maadah and R.N. Maddox, Predict Claus Products. *Hydrocarbon Process.* **57**, 143, Aug. (1978).

W.H. Megonnell, Efficiency and Reliability of Sulfur Dioxide Scrubbers. *J. Air Pollut. Control Assoc.* **28**, 725 (1978).

J.B. Pfeiffer, ed., "Sulfur Removal and Recovery from Industrial Processes." American Chemical Society, Washington, DC, 1975.

M.E.D. Raymont, ed., "Sulphur: New Sources and Uses." ACS Symp. Ser. No. 183. American Chemical Society, Washington, DC, 1983.

J.R. Shafer, R.W. Grendel, and D.R. Pogue, Pollution Control Practices; Sulfuric Acid Plants for Handling H_2S Gases. *Chem. Eng. Prog.* **74**, 62 (1978).

REFERENCES

1. "Minerals Yearbook, Metals and Minerals," Vol. 1, 1994. U.S. Dept. of the Interior, Bureau of Mines, Washington, DC, 1996, and earlier editions.

2. A.H. Vroom, *Hydrocarbon Process.* **51**(7), 79, July (1972).

3. "Canadian Minerals Yearbook 1978," p. 437. Department of Energy, Mines and Resources, Ottawa, 1980.

4. Sulfur-based Foam Protects, *Chem. Eng. News* **55**(14), 24, April 4 (1977).

5. Sulfur-Asphalt Blend, *Chem. Eng. News* **57**(42), 8, Oct. 15 (1979).

6. Test Road Paved, *Chem. Eng. News* **58**(35), 25, Sept. 1 (1980).

7. M. Gellender, *Can. Chem. Process.* **62**(1), 14, Jan. (1978).

8. Sulfur Response, *Chem. Can.* **24**(7), 5, Summer (1972).

9. Sulphur-coating Urea in a Spouted Bed, *Sulphur,* **134**, 35, 40 (1978); cited by "Minerals Yearbook 1978–79," p. 897. U.S. Bureau of Mines, Washington, DC, 1980.

10. "Kirk-Othmer Encyclopedia of Chemical Technology," 2nd ed., Vol. 19, p. 337. Wiley, Toronto, 1969.

11. F.S. Taylor, "A History of Industrial Chemistry," Reprint ed., pp. 184, 189. Arno Press, New York, 1972.

12. Coal Converts Sulfur Dioxide to Sulfur, *Des. News,* **34**, 18, Nov. 20 (1978).

13. Polish Sulphur Industry Revisited, *Sulphur,* **144**, 17 (1979).

14. "Thorpe's Dictionary of Applied Chemistry," Vol. IX, p. 218. Longmans, Green, Toronto, 1954.

15. F.A. Lowenheim and M.K. Moran, "Faith, Keyes, and Clark's Industrial Chemicals," 4th ed., p. 786. Wiley-Interscience, New York, 1975.

16. J.M. Ackerman, Main Pass - Frasch Sulphur Mine Development. *Min. Eng. (Littleton, Colo.)* **44**, 222–226, Mar. (1992).

17. Sulfur Prilling Tower Starts Operating, *Chem. Eng. News* **58**(13), 37, March 31 (1981).

18. R.N. Shreve, "Chemical Process Industries," 3rd ed., p. 322. McGraw-Hill, Toronto, 1967.

19. H.F. Lund, ed., "Industrial Pollution Control Handbook," p. 14–9. McGraw-Hill, New York, 1971.

20. L.F. Hatch, *Hydrocarbon Process.* **51**(7), 75, July (1972).

21. R.F. Smith and A.H. Younger, *Hydrocarbon Process.* **51**(7), 98, July (1972).

22. "Acid Gas Content of Alberta Natural Gas," listing to April 30, 1974. Energy Resources Conservation Board, Calgary, Alberta, 1974.

23. J.H. Perry, ed., "Chemical Engineer's Handbook," 4th ed., p. 14-10. McGraw-Hill, New York, 1969.

24. J.B. Osenton and A.R. Knight, "Reaction of Carbon Disulphide with Alkanlolamines Used in the Sweetening of Natural Gas", Can. Nat. Gas Proc. Assoc. Meeting, Calgary, Alta., Nov. 2, 1970; and L.D. Polderman, C.P. Dillon, and A.B. Steele, *Oil and Gas J.* **54**(2), 180 (1955).

25. D.M. Considine, ed., "Chemical and Process Technology Encyclopedia," pp. 12–15. McGraw-Hill, New York, 1974.

26. Pollution Control in Claus Sulphur Recovery Plants, *Sulphur,* **109**, 36, Nov/Dec. (1973).

27. M.J. Pearson, *Hydrocarbon Process.* **52**(2), 81, Feb. (1973).

28. H. Krill and K. Storp, *Chem. Eng. (N.Y.)* **80**(17), 84, July 23 (1973).

29. W.D. Monnery, W.Y. Svrcek, and L.A. Behie, Modelling the Modified Claus Process Reaction Furnace and the Implications in Plant Design and Recovery. *Can. J. Chem. Eng.* **71**, 711–724 (1993).

30. P. Grancher, *Hydrocarbon Process.* **57**, 257, Sept. (1978).

31. Sulfur Recovery Routinely Hits 99+%, *Can. Chem. Process.* **62**(5), 21, May (1978).

32. F.G. Ball, G.N. Brown, J.E. Davis, A.J. Repik, and S.L. Torrence, *Hydrocarbon Process.* **51**(10), 125, Oct. (1972).

33. P. Grancher, *Hydrocarbon Process.* **57**, 155, July (1978).

34. B.M. Khudenko, G.M. Gitman, and T.E.P. Wechsler, Oxygen Based Claus Process for Recovery of Sulfur from H_2S Gases. *J. Environ. Engin.* **119**(6), 1233–1251 (1993).

35. J.B. Hyne, *Chem. Can.* **30**(5), 26, May (1978).

36. M.E.D. Raymont, *Hydrocarbon Process.* **54**(7), 139, July (1975).

37. O.A. Salman, A. Bishara, and A. Marafi, An Alternative to the Claus Process for Treating Hydrogen Sulfide. *Energy (Oxford)* **12**(12), 1227–1232 (1987).

38. F.J. Ball, G.N. Brown, J.E. Davis, A.J. Repik, and S.L. Torrence, *Hydrocarbon Process.* **51**(10), 125, Oct. (1972).

39. "United Nations Statistical Yearbook 1993," 40th ed. United Nations, New York, 1995, and earlier editions.

40. Facts and Figures for the Chemical Industry, *Chem. Eng. News* **74**(24), 38–79, June 24 (1996); and June issues of earlier years.

41. F.A. Lowenheim and M.K. Moran, "Faith, Keyes and Clark's Industrial Chemicals," 4th ed., p. 795. Wiley-Interscience, New York, 1975.

42. "Kirk-Othmer Encyclopedia of Chemical Technology," 2nd ed., Vol. 19, p. 441. Wiley, New York, 1969.

43. A. Phillips, in "The Modern Inorganic Chemicals Industry" (R. Thompson, ed.), p. 183. Chemical Society, London, 1977.

44. M. Bodenstein and W. Pohl, Z. Elektrochem. 11,373 (1905); cited by Shreve [18].

45. R.A. Bauer and B.P. Vidon, Chem. Eng. Prog. 74, 68, Sept. (1978).

46. F.A. Cotton and G. Wilkinson, "Advanced Inorganic Chemistry," 3rd ed., p. 180. Interscience, New York, 1972.

47. R.N. Shreve, "Chemical Process Industries," p. 353. McGraw-Hill, New York, 1945.

48. D.M. Samuel, "Industrial Chemistry-Inorganic," 2nd ed., p. 80. Royal Institute of Chemistry, London, 1970.

49. G.L. Billington, Chem. Brit. 16, 452 (1980).

50. R.C. Weast, ed., "Handbook of Chemistry and Physics," 51st ed., p. F-3.Chemical Rubber Publishing Co., Cleveland, OH, 1970.

51. W.L. Faith, D.B. Keyes, and R.I. Clark, "Industrial Chemicals," 3rd ed., pp. 744, 752. Wiley-Interscience, New York, 1965.

52. N.A. Lange, ed., "Lange's Handbook of Chemistry," 10th ed., p. 1156. Mcgraw-Hill, New York, 1969.

53. M.F. Tunnicliffe, Chem. Br. 14, 2 (1978).

54. J.R. Donovan, J.S. Palermo, and R.M. Smith, Chem. Eng. Prog. 74, 51, Sept. (1978).

55. E.F. Spencer, Jr., in "Industrial Pollution Control Handbook" (H.F. Lund, ed.), p. 14-3. McGraw-Hill, New York, 1971.

56. R.A. Bauer and B.P. Vidon, Chem. Eng. Prog. 74, 68, Sept. (1978).

57. D.R. Duros and E.D. Kennedy, Chem. Eng. Prog. 74, 70, Sept. (1978).

58. J.A. Kent, ed., "Riegel's Handbook of Industrial Chemistry," 7th ed., p. 66. Van Nostrand-Reinhold, New York, 1974.

59. H. Takeuche and Y. Yamanaka, Ind. Eng. Chem. Process Des. Dev. 17, 389 (1978).

60. R.F. Vaccaro, G.D. Grice, G.T. Rowe, and P.H. Wiebe, Water Res. 6, 231 (1972).

61. H.R. Kueng and P. Reimann, Chem. Eng. (N.Y.) 89, 72, April 19 (1982).

62. K. Blanck, B. Leutner, and W.D. Back, Ger. Offen. 2,831,941, (1980), (to BASF, A.-G.); Chem. Abstr. 93, P10344z (1980).

63. Japan Kokai Tokkyo Koho 79 46,198, (1979), (to Bayer A.-G.); Chem. Abstr. 91, P59581x (1979).

64. E.I. Elbert, N.E. Tsveklinskaya, R.A. Bovkun, V.N. Stepanov, N.V. Grazhdan, and A.S. Zlobina, U.S.S.R. Pat. 601,222, (1978); Chem. Abstr. 89, P8414w (1978).

65. I. Smith, G.M. Cameron, and H.C. Peterson, Acid Recovery Cuts Waste Output. Chem. Eng. (N.Y.) 93(3), 44–45, Feb. 3 (1986).

66. J.J. Barney and J.L. Hendrix, Ind. Eng. Chem. Process Res. Dev. 17, 148 (1978).

67. B.R. Nott, Ind. Eng. Chem. Process Res. Dev. 20, 170 (1981).

68. A. De Schepper and A. Van Peteghem, Can. Pat. 1,044,825, (1978); Chem. Abstr. 90, P139717r (1979).

69. U. Sander and G. Daradimos, Chem. Eng. Prog. 74, 57, Sept. (1978).

70. Anonymous, ICI's Sulphuric Acid Recovery Project Cuts Out Dumping. Process Eng. 73, 21, Feb. (1992).

71. P.-W. Lin, Environ. Sci. Techol. 12(9), 1081 (1981).

10

PHOSPHORUS AND PHOSPHORIC ACID

*. . . Marine Acid combined with Phlogiston [which]
results in a kind of Sulphur . . . that . . . takes fire
of itself upon being exposed to open air.*
—*Andrew Reid, 1758*

Without phosphorus there would be no thought.
—*Lüdwig Büchner (1824–1899)*

10.1. PHOSPHATE ROCK DEPOSITS AND BENEFICIATION

Phosphorus occurrence in the lithosphere is predominantly as phosphates, PO_4^{3-}, although a rare iron-nickel phosphide, schreibersite $((Fe,Ni)_3P)_8$ is also known in nature [1]. For this reason, phosphates are the primary source of elemental phosphorus for chemical process requirements. Only 0.20 to 0.27% phosphate (0.15 to 0.20% as P_2O_5; 0.07 to 0.09% as P) is present in ordinary crustal rocks, which is the form in which the bulk of the phosphorus is present in the lithosphere.

Fortunately, however, a significant fraction of phosphate deposits have a much higher P_2O_5 equivalent concentration than the average crustal concentrations. About 80% of current world production is derived from the sedimentary phosphorites of marine origin, containing phosphate concentrations of 29 to 30% (as P_2O_5 equivalent) in unweathered deposits and 32 to 35% in leached deposits [2]. These frequently occur as a variety of apatite in beds of loosely consolidated granules. Second in commercial importance are the igneous apatites [general formula: $Ca_5(Cl, F, OH)(PO_4)_3$] which average about 27% P_2O_5 and comprise some 15% of current phosphate production mainly from deposits in the former U.S.S.R., South Africa, Uganda, and Brazil. Guano sources, mostly resulting from deposition from sea fowl, can range from 10 to 32% P_2O_5 depending on whether the deposits are relatively modern or leached. The source of most of the material produced from Nauru, for example, in the Western Pacific, and from Christmas Island in the Indian Ocean (Table 10.1), is from large reserves of this origin.

■ TABLE 10.1   Major World Producers of Phosphate Rock[a]

Producer	Thousands of metric tonnes				
	1960	**1970**	**1980**	**1990**	**1994**
China	300	1,700	10,726	21,552	24,000
Christmas Island	580	989	1,713	—	—
Israel	224	880	2,307	2,428	3,600
Jordan	362	913	3,911	5,925	3,500
Morocco	7,492	11,424	19,341	21,396	18,000
Nauru	1,248	2,200	2,087	926	613
Senegal	198	998	1,408	2,147	1,600
South Africa	268	1,685	3,185	3,165	2,880
Togo	—	1,508	2,933	2,314	1,800
Tunisia	2,101	2,969	4,502	6,259	5,500
U.S.S.R.	7,000	17,800	30,300	33,500	—
U.S.A.	17,797	35,143	54,415	46,343	41,100
Vietnam	543	455	90	274	—
Others	1,332	2,956	22,068	—	—
World[b]	39,445	81,620	158,986	157,226	124,000

[a]Producers of over one million tonnes of phosphate rock in 1977. Average phosphate content ranged from 29 to 36% (as P_2O_5). Data assembled from *U.N. Statistical Yearbooks* [3], *Energy Mines and Resources Canada* [4], and Leyshon and Schneider [5].
[b]World totals include unlisted producers.

The United States and Russia are dominant producers (Table 10.1) and users of phosphate, and the state of Florida is by far the largest single producing area. From phosphorite deposits about l0 m thick covering an area of about 5000 km², Florida alone supplied one-third of the world total in 1979. Phosphorite deposits may require beneficiation by froth flotation for removal of slimes (clays and other finely divided material) to raise the phosphate content of the ore for market. Igneous apatite deposits require only simple ore selection and crushing to market size requirements. Guano deposits merely require loading to ocean freighters for transport. Thus, phosphate rock marketing is not very dependent on mineral technology; instead it is more dependent on solids material handling and shipping economies for competitive export. Prices in 1995 ranged from about US$25 per tonne (run of mine) in Tampa, Florida to about US$50 per tonne in Casablanca (Morocco) at 77% TPL (triphosphate of lime)

10.1.1. End Use Areas for Phosphate Rock

About 87% or about 32 million tonnes of the phosphate rock consumed in the U.S.A. in 1978 went to agricultural uses, the remainder going as feedstock for elemental phosphorus production [6]. Phosphate rock of fluorapatite stoichiometry ($CaF_2 \cdot 3Ca_3(PO_4)_2$) which is destined as a mineral supplement for animal feeds or as a bulking agent in fertilizers is normally defluorinated. This is accomplished by heating in a rotary kiln with silica and steam (Eq. 10.1).

$$3 \ CaF_2{\cdot}3Ca_3(PO_4)_2 \ + \ SiO_2 \ + \ H_2O \xrightarrow[\text{ca. } 1500°C]{}$$
$$\underset{\text{fluorapatite}}{}$$
$$SiF_4{\uparrow} \ + \ 2 \ HF \ + \ 3 \ CaO \ + \ 9 \ Ca_3(PO_4)_2$$

10.1

The volatile fluorides formed by this process are captured by water scrubbers (Section 10.4.3). However, about 58% of agricultural rock, or 50% of the total, ultimately goes into wet process phosphoric acid production (Section 10.4). The 13% of the total destined for nonagricultural uses is a feedstock for elemental phosphorus production, from which both high-purity phosphoric acid and other phosphorus derivatives are made (Table 10.2). Nearly half of this, about 6% of the total, goes into the manufacture of phosphate builders for detergents, about a tenth into food and beverage additives, and the remainder into a multitude of small-scale applications.

10.1.2. Environmental Impacts of Phosphate Rock Processing

The principal environmental concerns of phosphorite surface mining operations relate to water consumption, storage of large volumes of waste slimes from beneficiation, and reclamation of the mined-out area. Some concern has also been expressed about exposure to elevated radiation levels on reclaimed land from the exposed slightly elevated uranium concentrations. Natural dewatering of the waste slimes produced is a slow process; it can take 2 yr before the solids content reaches 25 to 30% [7]. This effectively decreases water recycle capability and increases the volume of the stored slimes, which consequently can exceed the volume of the mined-out areas. Mining of igneous apatites, however, does not pose any significant environmental problems particular to the ore recovery. Guano sources, which typically involve surface mining and no beneficiation, only have surface reclamation to look after as necessary.

TABLE 10.2   Relationship of P_2O_5 Content and P_4 Content to Phosphate Rock Containing Different Concentrations of Calcium Phosphate, $Ca_3(PO_4)_2{}^a$

Calcium phosphate content (%) (equivalent to % BPL, or % TPL[b])	Percent P_2O_5 equivalent	Percent P_4 equivalent
—	100.00	43.64
100	45.76	19.97
80	36.61	15.98
60	27.47	11.98
40	18.30	7.99

[a]Pure fluorapatite $CaF_2{\cdot}3Ca_3(PO_4)_2$ analyzes 92.26% $Ca_3(PO_4)_2$, 42.26% P_2O_5, and 42.26% P_2O_5, and 18.44% P_4 equivalent.

[b]BPL and TPL are still the commonly used trade designations for bone phosphate of lime and triphosphate of lime, respectively, which are used synonymously with calcium phosphate $(Ca_3(PO_4)_2)$.

10.2. ELEMENTAL PHOSPHORUS

Phosphorus was first isolated in the seventeenth century, by a procedure which gave a recovery of about one ounce of phosphorus from a hogshead of urine [8]. In 1769, a more convenient method that also made larger quantities accessible was developed by Scheele. This involved treating crushed bones with sulfuric acid, extraction of the phosphoric acid from the gypsum also produced with small amounts of water, followed by evaporation of the water, mixing of the residue with charcoal, and vigorous heating (Eq. 10.2, 10.3).

$$Ca_3(PO_4)_2 + 3\ H_2SO_4 + 6\ H_2O \rightarrow 3\ CaSO_4 \cdot 2H_2O\downarrow + 2\ H_3PO_4$$
from crushed   gypsum   10.2
bone   (insoluble)

$$4\ H_3PO_4 + 10\ C \xrightarrow[\text{heat}]{\text{white}} P_4\uparrow + 6\ H_2O + 10\ CO \qquad 10.3$$
white
(or yellow)
phosphorus

The phosphorus vapor which formed was captured by condensation with water, since yellow phosphorus spontaneously catches fire in air. Produced in this manner, phosphorus was a small-scale article of commerce in France and Britain in the eighteenth century, mainly for match production.

Elemental phosphorus, as obtained by this procedure, is a soft solid at ordinary temperatures. It forms tetrahedral molecules of four phosphorus atoms each, with three bonds to each atom. The six bonds of a white phosphorus molecule are at 60° to each other placing them under great strain. This is thought to be the origin of the high reactivity (e.g., the spontaneous flammability in air). In the dark it is seen to luminesce, the property which originally gave this element its name. The other properties important to both its original isolation and its commercial recovery are its boiling point of 280°C, low enough to encourage easy volatilization as it is formed in the heated reduction mixture and yet high enough to permit condensation of the vapor by a shower of cold water. By being insoluble in and denser than water, direct contact ("spray," rather than heat exchange) condensation is the most efficient and least expensive option for vapor capture and gives the product an im-

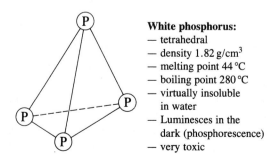

White phosphorus:
— tetrahedral
— density 1.82 g/cm³
— melting point 44 °C
— boiling point 280 °C
— virtually insoluble
  in water
— Luminesces in the
  dark (phosphorescence)
— very toxic

mediate protective covering of water to avoid its contact with air. The moderate melting point permits easy transfer of phosphorus in the liquid state merely by keeping the protective water layer warm.

10.2.1. Electric Furnace Phosphorus

The forerunners of the modern electric furnace for phosphorus production were first operated in England by J. B. Readman (1888) and Albright and Wilson (1893) the latter unit with a production capacity of 180 tonnes per year [1]. By allowing control of the heating level independently of the atmosphere above the heated charge, unlike earlier methods which required a combustion source for the white heat necessary, this development revolutionized commercial production of phosphorus.

Today, some 13 to 15% of American phosphate rock produced is consumed for elemental phosphorus production. Prior to charging to the phosphorus furnace, the phosphate rock of about 65 to 70 BPL level [65 to 70% bone phosphate of lime, $Ca_3(PO_4)_2$, from the early origins] is sintered (1200–1250°C) in a rotary nodulizing kiln to marble to golf ball-sized porous lumps. This preliminary step increases charge porosity in the furnace, facilitating release of phosphorus vapor and also decreases the amount of entrained dust that is carried out of the furnace with the product. The nodulized rock is fed in a carefully metered ratio with silica, sand, and coke to the furnace of steel lined with monolithic carbon and fitted with three adjustable carbon electrodes. Three-phase AC power at 200–300 V is fed to the electrodes which heat the charge to 1260–1480°C, sufficient to fuse it and initiate the fluxing and reducing reaction leading to phosphorus formation. This may be a composite of phosphate partial reduction by silica followed by reduction of the phosphorus pentoxide (P_4O_{10}) on contact with the incandescent coke (Eq. 10.4, 10.5).

$$2\ Ca_3(PO_4)_2 + 6\ SiO_2 \rightarrow \underset{\substack{\text{sublimes} \\ \text{at } 300°C}}{P_4O_{10}} + 6\ CaSiO_3 \qquad 10.4$$

$$P_4O_{10} + 10\ C \rightarrow \underset{\text{vapor}}{P_4\uparrow} + 10\ CO \qquad 10.5$$

Or it may be that silica merely serves as a fluxing agent or solvent for the reduction, in which case the overall process is best represented by Eq. 10.6.

$$\underset{\text{BPL}}{2\ Ca_3(PO_4)_2} + \underset{\text{flux}}{6\ SiO_2} + 10\ C \xrightarrow{1300–1400°C} \underset{\text{vapor}}{P_4\uparrow} + \underset{\text{slag}}{6\ CaSiO_3} + 10\ CO \qquad 10.6$$

$$\Delta H = +2840\ kJ\ (+680\ kcal)$$

At this temperature electrical energy provides the heat requirement of the endotherm of this reaction, an indirect electrical contribution to the chemical change, which is therefore called an electrothermal, rather than an electrochemical process.

The carbon monoxide, phosphorus vapor mixture is first fed to an electrostatic precipitator where any entrained dust is removed while the gases are still hot (Fig. 10.1). In this way, nonvolatile contaminants are excluded from the condenser. The water spray condenser is used to capture most of the liquid white phosphorus and separate it from the noncondensible carbon monoxide. The direct contact condenser may also be followed by an indirect, tubular-type condenser to further raise phosphorus containment efficiency [9]. The product is stored in the liquid state in sumps, under a layer of hot water which maintains the phosphorus in this form for easy transfer and pumping via hot water-jacketed pipes. Typical specifications of this phosphorus are P_4, >99.85%; Pb, <2 ppm; As_2O_3, <250 ppm; and F, <3 ppm. The cleaned carbon monoxide gas stream, which comprised about 93% of the gases leaving the furnace, is used to fuel the phosphate rock nodulizing kiln (Fig. 10.1).

The white or yellow phosphorus allotrope, the direct product, is only occasionally used on site. This is because the important products particularly phosphoric acid, involve the shipping of considerably more mass than the element. Thus, molten white phosphorus is shipped by road or rail, in insulated tank cars of some 50 tonnes capacity, and by sea in mild steel, beehive-shaped tanks of 1100 tonne capacity, under both water and inert gas.

Small amounts of red and black phosphorus, two more stable phosphorus allotropes, are also produced for special purposes from the white phosphorus product of the electric arc furnace. Red phosphorus is obtained by heating white phosphorus at 400°C for several hours, which yields a complex polymeric material considerably more stable than the white variety. Red phosphorus is not only stable in air, but far less toxic than white phosphorus. Black phosphorus, of a different, more complex structure, is obtained by heating the white variety at 220 to 370°C for 8 days, and in addition requires either pressures exceeding 10^4 kg/cm^2 or a seed crystal of black phosphorus for its formation. This product has a density of 2.70 g/cm^3, much higher than either of the other allotropes, has a structure resembling graphite, is a good electrical conductor, and can be lit with a match only with difficulty [10] (Table 10.3).

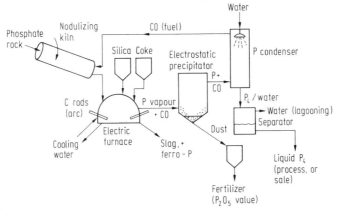

FIGURE 10.1 Flowsheet for the electric furnace process for the production of elemental phosphorus. Phossy water from the phosphorus phase separator (sump) is normally maintained in a closed-circuit holding pond system which also provides the feedwater for the phosphorus direct contact condenser. Some facilities have a further indirect, heat exchange condenser for the carbon monoxide stream, after the direct contact condenser.

TABLE 10.3   Properties of the Allotropic Forms of Phosphorus

	Phosphorus allotrope		
	White	**Red**	**Black**
Density, g/cm^3	1.82	2.34	2.70
Melting point, °C	44	ca. 400	—
Flammability in air	spontaneous	when heated to 200°C	when heated to >550°C
Solubility	soluble in CS_2 and many organic solvents; about 3 mg/L[a] in water at 15°C	insoluble	insoluble
Toxicity[b] acute (human, approx.)	< 0.2 g	relatively harmless	relatively harmless
chronic	1 mg/day	—	—

[a]Value from the *Handbook of Chemistry and Physics* [11]. More recent data gives 0.56 mg/L at 20°C.
[b]Details given in Idler *et al.* [12].

A single present-day phosphorus furnace produces from 60 to 160 tonnes of phosphorus per day, almost up to the *annual* production figures of the early arc furnaces, and requires a power supply of some 90,000 kW for a single furnace at the upper end of this production range. Power consumption per tonne of phosphorus produced varies with the percent calcium phosphate in the rock (% BPL level) and furnace size among other factors but ranges around 12,000 to 14,000 kWh/tonne (Table 10.4). Hence, power costs are a major component of electric furnace phosphorus production. This has

TABLE 10.4   Operating Experience of Various Capacities of Electric Furnaces for Phosphorus Production, Values per Tonne of Phosphorus Produced[a]

Furnace size (kW)	15,000	25,000	Large	Large
Raw materials:				
Phosphate rock, kg	10,000	9,340	7,250	6–10,000
Silica, kg	1,500	1,560	2,660	0.5–3,500
Coke, kg	1,500	1,550	1,430	1,200–1,800
Power, kWh	14,320	11,830	11,850	11,600–16,000
Electrode carbon, kg	—	—	15	20–60
Water:				
Spray condenser, L	13,500	400	13,000	—
Products:				
Phosphorus, kg	1,000[b]	1,000	1,000	1,000[b]
Ferrophosphorus, kg	300	360	90	90–380
Slag, kg	8,900	8,100	7,100	7,100–8,900
Gases, as CO, kg	2,800	2,820	3,300	—

[a]Data compiled from Kent [7], Ellwood [9], Kirk-Othmer [10], Thompson [13], and *Canadian Chemical Processing* [14].
[b]For 87–90% elemental phosphorus recovery from phosphate rock.

prompted a reexamination of fossil-fueled (petroleum coke-based) sources of heat for rotary kiln combustion to provide the energy of the endotherm of the reaction [15].

Raw material addition and phosphorus vapor removal from the furnace occur on a more or less continuous basis. Removal of the molten calcium silicate slag, more than 7 tonnes of which is formed for each tonne of phosphorus, and ferrophosphorus, which is formed from iron present in the charge, takes place periodically through appropriately placed tap holes on the lower furnace wall. The slag may be made into a rock wool insulating material or may be crushed for use in glass making or liming of soils but is more frequenty employed as an aggregate for highway or railway roadbed construction. Ferrophosphorus is normally sold for steel making, providing a convenient source for phosphorus addition to alloy steels.

10.2.2. Uses of Elemental Phosphorus

A significant, early, small-scale use of white (or yellow) phosphorus used to be for the manufacture of matches and fireworks. But the debilitating occupational illness of phossy jaw among workers, and the occasional hot pockets and singed purses among the public, led to its discontinuance for this application by the early l900s [12]. Today either phosphorus or phosphorus sesquisulfide (P_4S_3; Eq. 10.7), which are both much safer than yellow phosphorus, are used for matches, flares and other kinds of incendiary devices.

$$P_4 + 3\,S \xrightarrow[\text{heat}]{CO_2} P_4S_3 \qquad\qquad 10.7$$

Both of these more recent ignition components possess a lower vapor pressure, are much less toxic, and are much less likely to ignite inadvertently.

The phosphorus halides continue to be an important class of chemical intermediates and are prepared by combination of the elements (Eq. 10.8–10.10).

$$P_4 + 6\,Cl_2 \rightarrow 4\,PCl_3 \qquad\qquad 10.8$$

$$2\,PCl_3 + O_2 \rightarrow 2\,POCl_3 \qquad\qquad 10.9$$

$$P_4 + 10\,Cl_2 \rightarrow 4\,PCl_5 \qquad\qquad 10.10$$

The products phosphorus trichloride, phosphorus oxychloride, and phosphorus pentachloride are each useful as intermediates for the direct preparation of phosphite and phosphate esters used as extractants and plasticizers, for the manufacture of gasoline and oil additives and pesticides. However, most of the phosphorus produced, some 90%, is used for the preparation of high-purity grades of phosphoric acid (Section 10.3) and various phosphate salts for detergent formulation from this [16]. The decline in the phosphate content of detergents as a result of concern about phosphate pollution in water courses has been the main factor responsible for the steady decline in the U.S. production of phosphorus in recent years, from about 540 thousand tonnes in 1970 to about 390 thousand tonnes in 1976 (Table 10.5). Production since

TABLE 10.5 **Annual Production of Yellow Phosphorus by Selected Countries**

	Thousands of metric tonnes[a]			
	1965	1976	1980	1992
Canada	(27)	(85)	(85)	(89)[b]
China	(ca. 18)	(35)		(150)
E. Germany	(18)	(15)		
France	(14)	17		(15)
Italy	(13)			(15)
Japan	(30)	19		
U.K.	(32)			
U.S.A.[c]	(555)	397	392	323[b]
U.S.S.R.	(ca. 100)	(250)		(425)
W. Germany	(73)	(80)		

[a]Countries having a 1965 annual production capacity of more than 10,000 tonnes are listed. Sources of information include Kirk-Othmer [10], Thompson [13], and *Chemical and Engineering News* [17]. When only capacity data was available, this is given in parentheses.
[b]Data for 1990.
[c]U.S. production in 1970 and 1994 was 542,000 and 231,000 tonnes, respectively.

that time has only slightly further declined now with detergent consumption of phosphorus stabilized at about 45% of the total. Prices range around $1.20 per kilogram, depending on purity.

10.2.3. Environmental Aspects of Phosphorus Production

Electric furnace phosphorus production is relatively tolerant of the impurities present in the poorer grades of phosphate rock which contain lower concentrations of calcium phosphate. A high silica content in the phosphate rock can be accommodated by an appropriate decrease in the proportion of silica fed to the phosphorus furnace, and the presence of carbonaceous matter may be allowed for similarly. The presence of more or less iron oxide affects the yield of ferrophosphorus by-product, again without seriously affecting operation. In fact, occasionally lumps of iron will be deliberately added to the furnace charge to boost the yield of ferrophosphorus at times when the market for this alloying material is strong.

A primary area of environmental concern in the operation of a phosphorus furnace is with potential losses of dissolved and suspended elemental phosphorus in water since even low concentrations of the element present a potent toxicity hazard to freshwater and marine animal life. Care is also taken to avoid losses of fluoride to air or freshwater, which might occur particularly during process upsets, since fluorides may cause damage to adjacent plant life or freshwater organisms [18].

The nodulizing of phosphate rock before it is charged to the furnace is a measure designed both for good gas release from the heated furnace charge and to decrease dust entrainment in the large volume of gaseous products produced by the furnace. If fluorapatite (theoretical 3.7% F) comprises a significant fraction of the phosphate rock feed used, the rock will contain about 2.8% fluoride. In the rotary nodulizing kiln, dusts are entrained in the hot gases and about 8% of the fluoride present is volatilized to hydrogen fluoride and silicon tetrafluoride (Eq. 10.1). The dust is collected from the kiln exhaust gas stream and the fluorides are subsequently removed by scrubbing with water [19]. Hydrogen fluoride is extremely soluble in water and is effectively trapped in this form, while silicon tetrafluoride reacts with water to form soluble fluosilicic acid and colloidal silica (Eq. 10.11).

$$3 \ SiF_4 + 2 \ H_2O \rightarrow 2 \ H_2SiF_6 + SiO_2 \qquad\qquad 10.11$$

A safeguard additional to nodulization of the charge to prevent dust entry to the phosphorus condenser from the furnace gas stream is provided by two-stage electrostatic precipitation of any involatiles from the primarily carbon monoxide and phosphorus vapor mixture leaving the furnace. This measure not only raises the purity of the condensed phosphorus recovered from sumps, but also assists in decreasing the amount of phosphorus condensed in water in colloidal form rather than as easily settled droplets, at the next stage.

Condensation of phosphorus vapor by direct contact with a water shower in a mild steel tube is both the most economical method of obtaining the efficient heat transfer necessary and immediately provides the protective water covering required for the condensed phosphorus to prevent inflammation in air. At the same time, it places dissolved (solubility about 3 μg/mL [20]) and colloidal, highly toxic elemental phosphorus into the water layer. For this reason phosphorus condensers are normally operated on a closed-circuit, water recycle basis from a lagoon or holding pond reservoir.

During electrothermal phosphate reduction, 80 to 90% of the fluoride contained in the rock remains with the slag as its calcium salt [19] (Eq. 10.12).

$$CaF_2 \cdot 3Ca_3(PO_4)_2 + 9 \ SiO_2 + 15 \ C \rightarrow 3/2 \ P_4\uparrow + CaF_2 + 15 \ CO\uparrow + \atop 9 \ CaSiO_3 \qquad 10.12$$

Since calcium fluoride is insoluble in water, fluoride departure from the process in this form is relatively innocuous. This fluoride present in the slag, however, may be mobilized in the water phase by procedures which contact the hot slag with water to cause fracture to produce a broken slag appropriate for fill uses. The fluoride not captured in the slag is converted to the volatile substances silicon tetrafluoride and hydrogen fluoride, both of which leave the furnace with the phosphorus vapor and carbon monoxide but are captured in the water stream of the phosphorus condensers (Eq. 10.11). Thus, the water from the phosphorus condenser contains not only dissolved and colloidal phosphorus, but also dissolved fluoride and the complex fluosilicate anion.

Thus, the water outflow from the sumps fed by the phosphorus condensers quite ordinarily contains 1700 ppm P_4, which is joined by further phossy water streams such as that displaced by incoming phosphorus in the water-blanketed storage tanks. This stream also contains scrubbed fluorides [20]. Gradual buildup of these contaminants occurs in the recycle pond water to the point where it may contain phosphates and phosphorus at 1.7 g/L (specified as P_2O_5), and fluoride and ammonium ion each at 10 g/L (when ammonia is used for pH control). Either from the gradual accumulation of excess dissolved solids, which start to affect process requirements and aggravate corrosion problems, or from an excess of precipitation over evaporation rates in the holding ponds of the area in which the plant operates, a portion of the pond water content must be continuously or periodically purged from the closed system. Percolation of the discharged water through a pile of granulated slag has been found to decrease the fluoride concentration from about 1% (10,000 ppm) to 30 ppm by reaction of the fluoride with residual calcium slags in the slag-forming insoluble calcium fluoride [19]. The eluate from this procedure may be further treated with lime in a second lagoon, if necessary, to decrease fluoride concentrations to lower than this (Eq. 10.13).

$$2\ NH_4F\ +\ Ca(OH)_2 \rightarrow CaF_2\downarrow\ +\ 2\ NH_4OH \qquad 10.13$$

Alternatively the wastewaters may be treated with aluminum slags and/or sodium hydroxide in a variety of ways to generate normal, or ammonium cryolite of marketable value for aluminum production (Eq. 10.14, 10.15).

$$6\ NH_4F\ +\ NaAlO_2\ +\ 2\ NaOH \rightarrow Na_3AlF_6\ +\ 6\ NH_3\ +\ 4\ H_2O \qquad 10.14$$

$$12\ NH_4F\ +\ Al_2(SO_4)_3 \rightarrow 2\ (NH_4)_3AlF_6\ +\ 3\ (NH_4)_2SO_4$$
$$\text{ammonium} \qquad 10.15$$
$$\text{cryolite}$$

Alternatively, magnesium fluosilicate, used for preservation of Portland cement surfaces, or sodium fluosilicate for the fluoridation of water supplies may be prepared and marketed (Eq. 10.16, 10.17).

$$MgO\ +\ H_2SiF_6 \rightarrow MgSiF_6\ +\ H_2O \qquad 10.16$$

$$2\ NaOH\ +\ H_2SiF_6 \rightarrow Na_2SiF_6\ +\ 2\ H_2O \qquad 10.17$$

The phosphorus content of the recycle pond water can vary widely depending on recycle rates and process upsets but seldom goes below about 23 ppm P_4 at the inlet and averages 0.3 to 0.5 ppm occasionally rising to 1.0 to 2.0 ppm at the exit. These concentrations are much lower than those directly from the phosphorus condensers, but are still not low enough for safe discharge since phosphorus concentrations of a few parts per billion are still toxic to fish [20].

Since phosphate (PO_4^{3-}) is relatively innocuous at these concentrations it might be expected that aeration of phossy water should rapidly render it safe, but several tests have shown oxidation by aeration to be both relatively slow and ineffective [12,20]. Ozone or potassium permanganate have been found to be effective oxidants, but are too expensive for commercial use. Chlorine, too, is an effective treatment but metering control relative to phosphorus content of the wastewater has been found to be difficult, and an excess of chlorine was found to be not much better than the pollutant it was intended to neutralize. The most effective combination for phossy water treatment appears to be addition of lime and one or more chemical coagulants, coupled with settling and occasionally centrifugation leaving treated wastewater containing 50 parts per billion phosphorus (as P_4) or less. If these measures are still inadequate for a particular situation then 24 hr or longer holding of the treated wastewater in a large settling lagoon provides a further safeguard and an additional decrease in phosphorus concentrations to 2–3 μg/L [21].

10.3. PHOSPHORIC ACID VIA PHOSPHORUS COMBUSTION

Three main routes are employed for the commercial production of phosphoric acid. The "dry process," also called the "combustion process" and yielding a "furnace acid," obtains phosphoric acid by hydration of phosphorus pentoxide obtained via the combustion of yellow phosphorus. The other two processes, referred to as "wet processes," produce phosphoric acid by acidulation of phosphate rock with sulfuric acid, the Dorr process, or with hydrochloric acid, the Haifa process. The wet processes are discussed separately later.

For combustion, liquid phosphorus is displaced from a closed, steam-heated storage tank by metering in hot water (Fig. 10.2), to a burner which uses compressed air to finely atomize the phosphorus. Additional combustion air then accomplishes complete oxidation producing phosphorus pentoxide and a great deal of heat (Eq. 10.18).

$$4\,P + 5\,O_2 \rightarrow P_4O_{10} \qquad \Delta H = -3010\ kJ\ (-720\ kcal) \qquad 10.18$$
$$\text{melts } 580\text{--}585°C,$$
$$\text{sublimes } 300°C$$

The combustion chamber may be operated as a separate unit and the phosphorus pentoxide (named from P_2O_5, originally thought to be the molecular formula of the P^{5+} oxide) vapor ducted via an intermediate cooler to the base of the hydration tower. Or the combustion chamber itself may form the base of the hydration tower. In either case, cooling by water addition for hydration and indirect cooling with water jackets are employed to absorb the heat produced by the reaction.

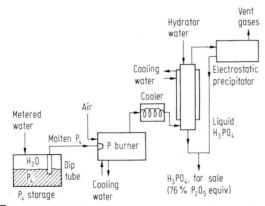

FIGURE 10.2 Diagram of process paths for the production of phosphoric acid by the combustion of elemental phosphorus. Graphite tubes are frequently employed to provide the initial cooling for the extremely corrosive combustion products.

Hydration is accomplished in a packed, acid-proof steel tower by passing in phosphorus pentoxide vapor at the base, and trickling in phosphoric acid and water at the top, thereby accomplishing both cooling and hydration (Eq. 10.19).

$$P_4O_{10} + 6\ H_2O \rightarrow 4\ H_3PO_4 \qquad \Delta H = -188\ kJ\ (-45\ kcal) \qquad 10.19$$

Hydration, too, releases some additional heat. The nature of phosphorus pentoxide and the absorption process inevitably leaves as much as 25% of the oxide plus a phosphoric acid mist in the exit gases from the absorption tower which are captured on passage through an electrostatic precipitator. By variations of the process details and equipment grades of phosphoric acid from 75–105% (ortho, or superphosphoric acid) H_3PO_4 may be made in this way [22].

Phosphoric acid production by phosphorus combustion is usually accomplished in a stepwise manner as outlined, with intermediate isolation of the phosphorus. But it may also be conducted by direct contact of phosphorus vapor from the furnace of a phosphorus plant with an airstream and the phosphorus pentoxide produced passed directly into a hydrator without collection of the intermediate phosphorus as a liquid. Direct conversion to phosphoric acid in this way reflects the fact that the bulk of white phosphorus produced proceeds directly to phosphoric acid production. Furnace phosphoric acid is pure enough for most uses as obtained directly from the process. But for food-grade applications traces of arsenious oxide must be removed (Table 10.6). Arsenic, present in the feed phosphorus to the extent of 50–180 ppm (as As_2O_3 equivalent) mainly because of the similarity of its chemical properties to those of phosphorus (same group in the periodic table), ends up

TABLE 10.6 Partial Analysis of a Typical Food-Grade Phosphoric Acid Produced by Phosphorus Combustion (75.0% H_3PO_4; 54.3% P_2O_5 equiv.)[a]

Component	Concentration ppm	Component	Concentration ppm
Na_2O	200	H_2S	0.1
O_2 consumed[b]	20	As_2O_3	0.05
Fe	2	Al, SO_4 each	0.0
Cl	2	CaO, K_2O each	0.0
F	0.4	SiO_2	0.0
Cr	0	color	10[c]
Pb	0.2	turbidity	1[c]
Cu	0.1	odor	none

[a]Data from Beveridge and Hill [29].
[b]A measure of the phosphorus present in oxidation states lower than phosphate. Determined in a manner similar to that used for the measurement of BOD.
[c]In American Public Health Association (APHA) units.

in the product acid on oxidation and hydration [1]. For a food-grade product this may be removed by the addition of the requisite amount of sodium sulfide or hydrogen sulfide which precipitates it as the arsenic III and V sulfides (e.g., Eq. 10.20).

$$As_2O_3 + 3\ H_2S \rightarrow As_2S_3\downarrow + 3\ H_2O \qquad\qquad 10.20$$

Any excess hydrogen sulfide is subsequently removed from the acid by air stripping, after which the acid is clarified by passage through a sand filter. By this means arsenic (as As_2O_3) will be brought to below 0.1 μg/g. The only other impurities present at levels above 1 μg/g will be Na_2O, 0.02%, and iron and chloride at about 2 μg/g each [10].

10.3.1. Environmental Factors of Furnace Phosphoric Acid

Losses of phosphoric acid mists or phosphorus pentoxide fumes, either from corrosion-induced rupture of ductwork or a vessel, or from a process-related parameter, represent the most commonly experienced emission control problems with furnace acid plants. When this occurs, a white plume, or "ghost," remains visible downwind of the point where the steam component has dissipated. Corrosion problems are now better understood so that more components are constructed of stainless steel, or if high temperatures are also involved the equipment is lined with graphite [22]. Heat exchanger tubes for the cooler situated between the phosphorus burner and the hydrator towers are also made of graphite, and the gas path in the electrostatic precipitator consists of carbon tubes with cathode elements of stainless steel wire to ensure reliable operation.

Process parameters of importance for mist and fume control include measures to minimize the moisture content of the combustion air fed to the phosphorus burners. A high moisture content in this gas stream tends to aggravate mist formation. Solid phosphorus pentoxide has been found to be very difficult to dissolve in either water or phosphoric acid, leading to the suggestion that temperatures in the hydrator be maintained high enough that absorption takes place from the vapor phase. Further studies have shown that uptake of phosphorus pentoxide vapor in 70% phosphorus acid is only about 60% and climbs with increasing acid concentrations up to about 88% phosphoric acid. More recent studies have shown that two fiber beds operated in series, the first bed as an agglomerator, and the second as a collector, may also be used to efficiently control mists and fumes from furnace acid plants [24]. Mass containment efficiencies of better than 99.9% were reported for a median aerosol particle diameter of 1.1 to 1.6 μm.

10.4. PHOSPHORIC ACID USING SULFURIC ACID ACIDULATION

The oldest process route to phosphoric acid (see Section 10.2) and still the lowest cost route to this product, at least for fertilizer grades, is via the addition of quite high concentrations of sulfuric acid to finely ground phosphate rock. This acidulation releases phosphoric acid from the calcium phosphate salts present and produces insoluble gypsum (calcium sulfate dihydrate) which can be removed by filtration (Eq. 10.21).

$$CaF_2 \cdot 3Ca_3(PO_4)_2 + 10\ H_2SO_4 + 20\ H_2O \rightarrow$$
$$6\ H_3PO_4 + 10\ CaSO_4 \cdot 2H_2O + 2\ HF \qquad 10.21$$

The fluoride ion present in the phosphate rock of the majority of sources is of no commercial value, and in fact is more of a nuisance to phosphoric acid production by this method since strong acid acidulation releases virtually all of the fluoride originally present in the rock. Considering just the phosphoric acid-forming part of the overall acidulation reaction, this is seen to be a moderately exothermic process [1] (Eq. 10.22).

$$Ca_3(PO_4)_{2(s)} + 3\ H_2SO_{4(\ell)} + 6\ H_2O_{(\ell)} \rightarrow 2\ H_3PO_{4(\ell)} + 3\ CaSO_4 \cdot 2H_2O_{(s)}$$
BPL of 70–75% $\Delta H = -321\ kJ\ (-76.4\ kcal)$ 10.22

Overall plant design and operational features, then, emphasize the need to obtain as high a concentration of phosphoric acid as possible and the arrangement of acidulation and vapor control equipment to permit safe containment of the incidental fluorides. This is accomplished with adequate cooling measures in place to permit operation of the whole process at low to not more than moderate acidulation temperatures.

Operation of the process at not more than moderate temperatures is necessary to avoid the precipitation of anhydrite (anhydrous calcium sulfate) or calcium sulfate hemihydrate ($CaSO_4 \cdot 0.5H_2O$) which tend to form at high operating temperatures. Both of these tend to precipitate in a finely divided form which is difficult to filter [1]. The less hydrated forms of calcium sulfate, if formed, may also revert to gypsum at cooler downstream portions of the process and cause deposition and plugging of equipment. Gypsum, the coarser dihydrate of calcium sulfate which is easily filtered, while it may form at higher temperatures, is only stable at temperatures below about 40°C and then only for phosphoric acid concentrations below 35% (by weight). Hence the desirability of keeping acidulation temperatures low.

Severe corrosion conditions which exist throughout this process have led to the use of wood, lead-lined wood, polyester-fiberglass, and rubber-lined steel for construction of much of the equipment, with dense carbon or a nickel alloy being favored for heat exchangers.

10.4.1. Operation of the Acidulation Process

Phosphate rock of 70 to 75% BPL, or as high as is reasonably obtainable, is finely ground in a ball mill and then mixed with cooled recycled phosphoric acid-gypsum slurry in a digestion tank (Fig. 10.3, 10.4). At this stage the only reaction which occurs is between acid and any carbonates present in the rock, and between phosphates and low concentrations of sulfuric acid which may be present, in this way minimizing foaming (Eq. 10.23) and heat evolution on initial acid contact.

$$3\ CaCO_3 + 2\ H_3PO_4 \rightarrow Ca_3(PO_4)_2 + 3\ CO_2 + 3\ H_2O \qquad 10.23$$

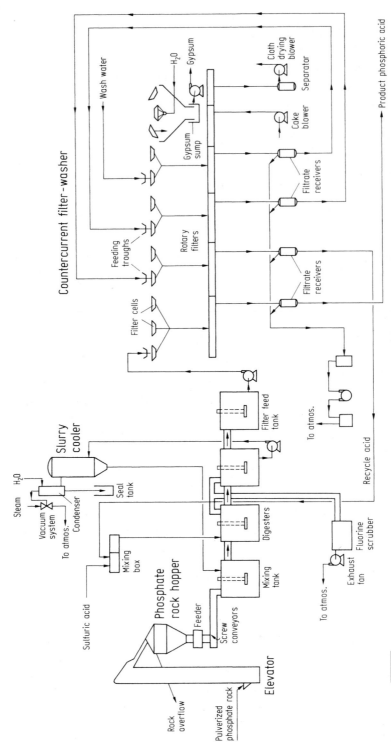

FIGURE 10.3 Process outline for phosphoric acid production by sulfuric acid acidulation of phosphate rock. Fume control and slurry evaporative cooling systems are shown, together with a countercurrent filtration and washing system which serves to maximize product acid concentrations accessible without evaporation. (From Shreve and Brink [23] by permission, McGraw-Hill.)

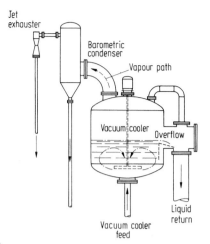

FIGURE 10.4 Details of operation of a slurry vacuum cooler, in which a small water jet exhauster maintains the vacuum, largely produced by direct steam condensation and the barometric leg of the barometic condenser. (Reprinted from Slack [47], p. 275, by courtesy of Marcel Dekker, Inc.)

The newly formed mixed slurry is then moved into the next digestion tank where fresh sulfuric acid mixed with relatively dilute recycle phosphoric acid (about 50 to 55% sulfuric acid concentration) is added and vigorous mixing continued. Acidulation continues through two or more additional digestion tanks combined with several passes of the partially reacted slurry through a slurry cooler. In the cooler direct cooling occurs by evaporation of a part of the water in the slurry under reduced pressure as one of the measures to avoid anhydrite formation. This also assists in keeping the final phosphoric acid concentration high. Digestion requires a period of 4 to 8 hr while keeping temperatures to below 75 to 80°C, after which time the slurry proceeds to the product acid recovery system [16]. Under these acidulation conditions gypsum, the desired hydrate, is initially precipitated. Provided the slurry is cooled or that filtration is accomplished after a reasonably short time at these temperatures, the gypsum does not revert to troublesome anhydrite or the hemihydrate.

Recovery of phosphoric acid from the fully digested slurry is by continuous countercurrent filtration and washing in moving belt (e.g., Fig. 10.5), moving pan, or horizontal rotary filters [1]. Monofilament polypropylene cloth is a durable filter element in this service. All filtration methods initially filter the cooled, fully digested gypsum phosphoric acid slurry to yield a filtrate of 40 to 45% (29–33% P_2O_5 equivalent) phosphoric acid. This is the maximum concentration directly obtainable by this process. The gypsum filter cake, still damp with 40 to 45% phosphoric acid, is washed with dilute phosphoric acid collected from a later wash stage and this filtrate is used to dilute the fresh incoming sulfuric acid before it contacts the powdered rock. This wash sequence is repeated from once, to three or four times, each time using the filtrate from the subsequent wash stage as the wash liquor, until the last stage where pure water is used. In this way the gypsum is virtually freed of adhering

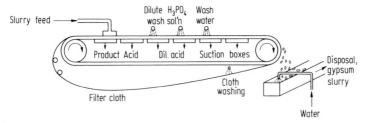

FIGURE 10.5 Diagram illustrating the mode of operation of a moving belt filter to obtain phosphoric acid recovery and permitting countercurrent washing of the gypsum filter cake.

residual acid and in the process the countercurrent reuse of the wash water has also minimized the amount of dilute phosphoric acid to be incorporated into the product, in this way maintaining the product acid concentration as high as is feasible.

The crude acid obtained from this process is not very concentrated, nor very pure. Concentration by boiling off water also reduces the solubility of many of the dissolved impurities present, such as calcium, fluoride ions, aluminum, and sulfate ions, causing precipitation and scaling of these materials on heat transfer surfaces. As a result, rather specialized heat transfer equipment is used to achieve this without necessitating frequent equipment cleaning and overhaul. Submerged combustion units having no intervening heat transfer surfaces or the direct application of hot air or flue gases through carbon nozzles discharging into a shallow stream of acid in an open trough both successfully achieve this [10, 25]. Multiple effect evaporation has more recently been shown to also be feasible, as long as it is accompanied by forced circulation [26].

Concentration alone serves to remove most of the impurities through precipitation, and about 40% of the dissolved fluoride by volatilization, and produces about 70% phosphoric acid (50 + % P_2O_5) suitable for many uses. But for solution fertilizers, where residual impurities may cause precipitation on long-term storage or for the manufacture of a food-grade acid, further steps are necessary [27]. Fluoride is stripped to less than 50 μg/g as volatile hydrogen fluoride, by sparging the crude concentrate with superheated steam. Or it may be removed by the addition of powdered silica, which forms silicon tetrafluoride which is removed by sparging with air or steam [10] (Eq. 10.24).

$$SiO_2 + 4\ HF \rightarrow SiF_4 + 2\ H_2O$$
$$\text{b.p. } -86°C$$

10.24

Arsenic, which is also extracted from the phosphate rock by acid, is removed by precipitation as the sulfide after addition of sodium sulfide [28] (Eq. 10.19). Addition of a slight excess of sulfuric acid serves to precipitate the bulk of any dissolved calcium (Eq. 10.25) after which filtration of the treated acid through a sand bed produces a clarified, substantially pure product.

$$Ca(H_2PO_4)_2 + H_2SO_4 + 2\ H_2O \rightarrow CaSO_4 \cdot 2H_2O\downarrow + 2\ H_3PO_4$$

10.25

A few additional purification refinements may be required depending on the origin of the phosphate rock used and particular impurities common to that deposit, to produce a food-grade phosphoric acid (Table 10.5). Overall phosphate recoveries from the rock are about 90%.

10.4.2. New Developments in Sulfuric Acid Acidulation

Accessible process choices have been surveyed and the initial separation of calcium sulfate from phosphoric acid as the hemihydrate has been exploited to a commercial-scale operation [29,30]. In this process, the initial hemihydrate separation is followed by a second stage where the hemihydrate, with the help of further added sulfuric acid, is recrystallized as conventional gypsum. It is reported that this process yields both capital and operating economies while the recovery of phosphate from rock is increased to better than 98% (conventionally about 90%).

In another variant of the hemihydrate process, the foam layer formed by reaction with rock carbonates (Eq. 10.26) is used to distribute a more concentrated sulfuric acid acidulant [31].

$$2\ CaCO_3 + 2\ H_2SO_4 + H_2O \rightarrow 2\ CaSO_4 \cdot 0.5H_2O + 2\ CO_2\uparrow + 2\ H_2O$$

$$10.26$$

The high temperatures employed, 105–120°C, favor rapid reaction and hemihydrate formation and directly yield a more concentrated phosphoric acid, saving on energy costs for phosphoric acid concentration.

A single, isothermal reactor design is said to more reliably yield gypsum crystals than multiple-reactor designs with poorer temperature control, which has also been claimed for a two-vessel loop system [32,33]. Higher reactor circulation rates and increased agitation coupled with other refinements have also been found to improve acid recovery [34].

Incorporation of methanol at the rate of 2 kg/kg phosphorus pentoxide in the dried acidulate has been found to improve significantly the purity of the phosphoric acid product in a two-stage process. This modification is especially advantageous if a low-grade phosphate rock provides the rock feed [35]. Acetone has also been found of value in an experimental variant which enables much less expensive sulfurous acid, rather than sulfuric acid, to be used as the acidulant [36]. In this case the solvent serves to form an (α-hydroxysulfonic acid with the sulfurous acid (Eq. 10.27) which is sufficiently strong to attack phosphate rock at a reasonable rate.

$$\begin{array}{c} CH_3 \\ \diagdown \\ C=O \\ \diagup \\ CH_3 \end{array} + SO_2 + 2H_2O \rightarrow \begin{array}{c} H_3C \quad OH \\ \diagdown \diagup \\ C \\ \diagup \diagdown \\ H_3C \quad SO_3^- \end{array} + H_3O^+ \qquad 10.27$$

Post acidulation applications of solvent either as an extractant [37] or as a precipitant [26] have been found to be valuable for acid purification, most recently using methyl isobutyl ketone [38].

Methods of decreasing the fluoride concentrations of phosphoric acid have recently been surveyed [39]. Apart from endorsement of the addition of a reactive form of silica prior (Eq. 10.22) to stripping, the procedure of acid dilution prior to reevaporation has also been recommended, the additional water vapor removal apparently assisting purification by entraining hydrogen fluoride and silicon tetrafluoride.

Uranium recovery from the wet process phosphoric acid filtrate has been of interest for many years [40]. Several U.S. operations recover a total of 200,000 kg/yr of U_3O_8 (yellow cake) from this source [41]. Prior cleanup of dissolved organics in the raw phosphoric acid with activated carbon improves the ease of uranium extraction by assisting clean phase separation. Vanadium recovery from wet process acid has also been of interest, particularly from Idaho-Montana-Wyoming ores [1].

10.4.3. Emission Control Measures for Wet Process Acid

Contact of the fluoride present in fluorapatite phosphate rocks with a strong mineral acid mobilizes much of the fluoride to hydrofluoric acid (HF: boiling point, 19°C; Eq. 10.21) and silicon tetrafluoride (SiF_4; boiling point, -86°C; Eq. 10.24). Fluorides have an acute oral toxic dose as sodium fluoride of about 200 mg/kg body weight in rabbits, and can also show severe skeletal effects on from chronic exposure [42]. For these reasons fluorides are easily the most important components requiring control in a wet process phosphoric acid plant. Industrial hygiene requirements for workers have been set at 2.5 mg/m^3 for dusts and 3 ppm for hydrogen fluoride.

From an initial approximately 3.8% fluoride which may be present in the phosphate rock (e.g., for fluorapatite) the fluoride is distributed to virtually all the product streams of the acid, but the proportions found in each stream vary widely depending on the origin of the rock and the process details. One study found that, of the original fluoride present in the rock, about 29% remained in the gypsum filter cake (much less than that captured in the slag of a phosphorus plant), 15% was evolved to air, and about 55% ended up in the crude phosphoric acid. It is unclear whether the gypsum-bound fluoride is mobile or not since studies of the emissions to air from gypsum settling ponds demonstrated a rate of fluoride emission of 0.43 kg/hectare. The pond water pH in this study was 1.6 to 1.8 at the same time as the fluoride concentration stood at 2800–5100 mg/L. This information would suggest that separation of the fluoride disposal from the gypsum disposal functions or liming of the ponds could have significantly decreased these discharge levels. Certainly more recent studies suggest that it may be possible to bind up to 85% of total fluoride and immobilize it in the gypsum [43].

Fluoride losses to air occur in dusts raised in phosphate rock grinding operations, which have been controlled by means of a baghouse. Small amounts of further dust loss occur from the first digester, as well as fluoride loss mainly as silicon tetrafluoride vapor from the whole digestion train (1200–3500 mg F/m^3) as well as the filtration and vacuum areas. A system of covers for the digestion vessels, and hoods over the filtration area, all exhausted to a venturi scrubber system using low water flow rates on a recycle basis now appear to best meet these control requirements [44]. Equilibrium pond water fluoride concentrations of 0.2% can result from this contribution from scrubber fluoride loading. The safety of disposal of a part of excess pond water volumes may be raised by prior fluoride removal via smaller separate basins for single or double liming the effluent before discharge (Eq. 10.13). It has been proposed that this procedure be used for production of calcium fluoride from pond water.

Crude 28 to 41% phosphoric acid (20 to 30% P_2O_5 equivalent) may contain as much as 1 to 2% fluoride, or about 55% of the fluoride originally present in the rock, mostly as fluosilicic acid (H_2SiF_6) or convertible to fluosilicic acid [45]. Normal evaporative concentration serves to volatilize about two-fifths of this which can be increased to a larger proportion by variations of this technique [1] (see Section 10.4.1). The vapor control procedures just outlined effectively capture this volatilized fluoride (Fig. 10.6). If, however, concentration of the crude acid is not required it may be purified by the addition of sodium carbonate. Sodium fluosilicate precipitates, and may be filtered, leaving a phosphoric acid containing only about 0.1% fluoride (Eq. 10.28).

$$H_2SiF_6 + Na_2CO_3 \rightarrow Na_2SiF_6\downarrow + H_2O + CO_2 \qquad 10.28$$

Sodium fluoride is soluble. The sodium fluosilicate may be purified and marketed directly for purposes such as fluoridation of water supplies, or it may be used to prepare other salts, for example, synthetic cryolite (Na_3AlF_6), valuable for aluminum smelting.

Gypsum, while a relatively inert by-product of wet process acid production, may pose an environmental problem to wet process acid operations merely because of the large mass obtained. Some 5 to 7 tonnes are produced for each tonne of phosphoric acid (100% basis), depending on the BPL level of the phosphate rock used. Early disposal by some facilities was simply into the nearest watercourse via a slurry pipeline, but with the realization that even relatively noninteractive suspended solids in freshwater streams can have severe effects on both water quality and diversity of stream life from a smothering and siltation effect, this practice has been largely discontinued. Use of the gypsum for wallboard manufacture has been studied

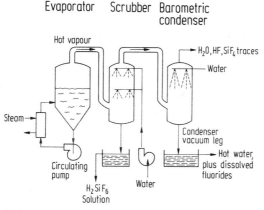

FIGURE 10.6 Emission control devices used with the vapor from a phosphoric acid evaporator to decrease fluoride emissions. The water used to condense steam by direct contact with cold water also serves to control fluoride vapor losses quite efficiently.

with some success [46], and where supplies of natural gypsum are scarce, such as in the United States and the former West Germany, this source is used [47]. Application of phosphogypsum as a source of sulfur dioxide for sulfuric acid production (Section 9.6.1) has not attracted much interest. Thus, inland facilities have largely settled on gypsum dispersal into extensive holding ponds, or lagoons laid out with landfill objectives. Many coastal facilities in Europe dump waste gypsum into the sea. One operation has found a 90-m outfall satisfactory for the daily disposal of 400 to 500 tonnes of gypsum, the dispersal of which is assisted by the codischarge of some 550 m^3/hr of seawater used for cooling.

10.5. PHOSPHORIC ACID USING HYDROCHLORIC ACID ACIDULATION

Other acids than sulfuric have been used to obtain phosphoric acid from phosphate rock, among them nitric acid and hydrochloric acid. Process difficulties in obtaining a pure phosphoric acid using nitric acid have resulted in this chemistry only being utilized to prepare granular fertilizers [27] (Eq. 10.29, 10.30).

Acidulation:
$$CaF_2 \cdot 3Ca_3(PO_4)_2 + 20\ HNO_3 \rightarrow 6\ H_3PO_4 + 10\ Ca(NO_3)_2 + 2\ HF \quad 10.29$$

Ammoniation (full):
$$6\ H_3PO_4 + 10\ Ca(NO_3)_2 + 18\ NH_3 \rightarrow 6\ (NH_4)_3PO_4 + 10\ Ca(NO_3)_2$$
$$10.30$$

The hydrochloric acid acidulation process, which was originally developed by Israel Mining Industries at Haifa as a production-scale process to phosphoric acid [48], has been adopted to such an extent that the licensed capacity stood at 50,000 tonnes of P_2O_5 equivalent per day in 1965. Trade hydrogen chloride or hydrochloric acid is normally two to three times the price of sulfuric acid. Hence this process, which is commonly referred to as the Haifa or IMI process, is primarily of interest to integrated chemical operations having a surplus of hydrogen chloride. This might arise, for example, from the operation of a chlorinated solvents plant or from the manufacture of toluene diisocyanate (Eq. 10.31–10.33), since the markets for anhydrous hydrogen chloride have traditionally been limited.

$$2\ Cl_2 + CH_4 \rightarrow CH_2Cl_2 + 2\ HCl \qquad 10.31$$

$$CH_3C_6H_3(NH_2 \cdot HCl)NHCOCl$$
$$+ COCl_2 \xrightarrow{\text{heat}} CH_3C_6H_3(NHCOCl)_2 + 2\ HCl$$
$$10.32$$

$$CH_3C_6H_3(NHCOCl)_2 \xrightarrow{\text{heat}} CH_3C_6H_3(NCO)_2 + 2\ HCl \qquad 10.33$$

Acidulation of fluorapatite with hydrochloric acid produces phosphoric acid in a very similar manner to sulfuric acid acidulation, except that in this case the coproduct obtained, calcium chloride, is very water soluble (Eq. 10.34).

$$CaF_2 \cdot 3Ca_3(PO_4)_2 + 20\ HCl \rightarrow 6\ H_3PO_4 + 10\ CaCl_2 + 2\ HF \qquad 10.34$$

Phosphoric acid recovery from HCl acidulation of phosphate rock, therefore, is not a relatively simple solid-liquid separation as it is from sulfuric acid acidulation; instead it requires a solvent extraction step to obtain the product free of dissolved calcium chloride. Suitable solvents must be polar to permit good solution of phosphoric acid, while having relatively low water solubility to minimize solvent losses. Israel Mining Industries prefers C_4 to C_5 alcohols, though some producers prefer these to be mixed with a water-immiscible solvent such as kerosene. Others prefer trialkylphosphates such as tri-n-butylphosphate or tri-2-ethylhexylphosphate for this function, but all have similar physical properties [49]. It is appropriate to mention here that solvent extraction has also been teamed up to sulfuric acid acidulation plants by using solvent discrimination to obtain a purified phosphoric acid extract and leave behind most of the water-soluble, solvent insoluble impurities [50].

10.5.1. Product Recovery by Solvent Extraction

To establish extraction efficiencies for a substance A one first needs to know, or must experimentally determine, the partition coefficient, K_p, which relates the concentration of solute A in the extraction solvent to the concentration of the solute in the extracted phase, at equilibrium (Eq. 10.35).

$$K_p = \frac{[A]_{solvent}}{[A]_{water}} \qquad\qquad 10.35$$

This holds true at any given temperature as long as the distributing species, solute A, exists in the same form in both phases. Otherwise a more complex "distribution ratio" must be employed. For increased refinement of extraction efficiency determination it may also be necessary to employ activities, rather than concentrations in some cases [51].

To illustrate the partition coefficient aspect of extraction with a simple example, for the fluid system carbon tetrachloride/water and using ammonia or iodine as the solute, respectively, the partition coefficient K_p at 25°C is 0.0042 and 55 [52]. Thus, with ammonia, this solute, being polar, would primarily move to the water phase on extraction, whereas with nonpolar iodine, the higher concentration would be in the nonpolar solvent phase. Obviously, using carbon tetrachloride as the extractant, iodine would be much more readily extracted from water than ammonia.

The other features of importance in industrial extractions are whether one or more than one extraction stage is to be used, which will depend in part on the partition coefficient of the system of interest, and on what flow pattern of solvent and extracted phase (the raffinate) is to be used, if more than one extraction stage is employed. Single-stage extraction requires a mixer, to bring about intimate contact between the two phases, followed by a settler which allows phase separation and a means for independent removal of the two phases (Fig. 10.7).

For a solute with a relatively poor partition coefficient one of the two arrangements of multistage extraction, crosscurrent or countercurrent may be selected (Fig. 10.8). To decide which would be the advantageous choice, assume we have a 1000 L/hr aqueous waste stream containing 0.02 M benzoic

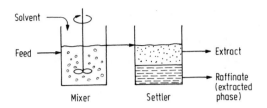

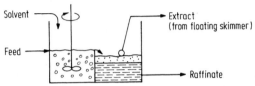

FIGURE 10.7 Arrangement for the physical components of a mixer settler, such as that used for solvent extraction processes. Units may be separated or combined, in either case yielding discrete solvent and raffinate (extracted) phases with equilibrium, or near equilibrium concentrations of solute.

acid which is to be extracted with 300 L/hr of benzene for benzoic acid recovery. To accomplish this extraction in a single-stage system, start by assuming complete immiscibiity of benzene and water, as a simplification. Using the partition coefficient, K_p, for this system, which is 4.3 at 20°C, one can set up an equation to calculate the concentration of benzoic acid in each of the streams produced. To do this, the value for the concentration of benzoic acid in the first extract $[BzOH]_E$ is obtained from the known equilibrium which must exist between the two streams of Eq. 10.36. This value is substituted into Eq. 10.37, which defines the balance of concentrations of benzoic acid which must exist between the incoming and outgoing streams.

$$[BzOH]_E = K_p[BzOH]_R, \text{ from } K_p = \frac{[BzOH]_E}{[BzOH]_R} \qquad 10.36$$

$$\begin{array}{cc} \textit{Feed Streams} & \textit{Product Streams} \\ S[BzOH]_S + F[BzOH]_F = E_1[BzOH]_{E_1} + R_1[BzOH]_{R_1} \end{array} \qquad 10.37$$

where S, F, E_1, and R_1 represent *volumes* of solvent, aqueous benzoic acid feed stream, the first extract and the first raffinate, respectively. The [BzOH] term

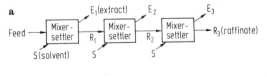

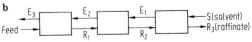

FIGURE 10.8 Possible arrangements of mixer settlers and process streams for multistage extraction processes. (a) Crosscurrent arrangement. (b) Countercurrent arrangement.

with its various subscripts represents the *concentration* of benzoic acid present, in moles per liter, in the subscripted stream.

Knowing this, one can derive Eq. 10.38, which allows calculation of the concentration of benzoic acid in the first raffinate, $[BzOH]_{R_1}$.

$$[BzOH]_{R_1} = \frac{[BzOH]_F + \left(\frac{S}{F}\right)[BzOH]_S}{1 + \left(\frac{S}{F}\right)K_P} \qquad 10.38$$

$$\text{For single-stage extraction, } \frac{S}{F} = \frac{E_1}{R_1}$$

By substitution of known values into this expression the concentration is found to be 9×10^{-3} M. Thus, one extraction stage as described has dropped the benzoic acid concentration in the effluent to just under one-half the original value. Similarly, by substituting this concentration value and K_P into Eq. 10.36, the concentration of benzoic acid obtained in the benzene extract is found to be 0.04 M, a ratio corresponding to the benzoic acid partition factor between these two liquids. The same equation may also be used to calculate the single-stage extraction parameters for any other solute of interest, provided that K_P for the system is known or may first be determined experimentally.

In a similar manner one can derive an equation to calculate the benzoic acid in the raffinate for a three-stage, crosscurrent system (Eq. 10.39; cf. ref. [53]).

$$[BzOH]_{R_3} = \frac{[BzOH]_F}{\left[1 + \left(\frac{S}{F}\right)K_P\right]^n} \qquad 10.39$$

where, for three stages, $n = 3$. If fresh extracting solvent volumes for each stage differ, the denominator requires expansion to individual terms for each solvent volume (Eq. 10.40).

$$\left[1 + \left(\frac{S_1}{F}\right)K_P\right]\left[1 + \left(\frac{S_2}{F}\right)K_P\right]\left[1 + \left(\frac{S_3}{F}\right)K_P\right]\cdots\left[1 + \left(\frac{S_i}{F}\right)K_P\right] \qquad 10.40$$

Assuming that 100 L volumes of benzene are used for each of the three extraction stages in a system of this kind, the benzoic acid concentration in the third-stage raffinate is found to be 0.007 M (Eq. 10.41), or about one-third that of the incoming stream.

$$[BzOH]_R = \frac{0.02 \text{ mol/L}}{\left[1 + \frac{100}{1000}(4.3)\right]^3} = 0.007 \text{ mol/L} \qquad 10.41$$

Carrying this approach to three-stage, countercurrent extraction and solving the pairs of equations for each of the three stages of the process one can derive

Eq. 10.42 which allows the determination of the benzoic acid concentration in raffinate 3, the aqueous effluent stream.

$$[BzOH]_R = \frac{[BzOH]_F}{\left[1 + \left(\frac{S}{F} \cdot K_p\right)\right]\left[1 + \left(\frac{S}{F} \cdot K_p\right)^2\right]}$$

10.42

Substituting into this expression, one obtains a value of 0.003 mol/L for the benzoic acid concentration in the final raffinate, one-seventh that of the incoming stream. Comparing the calculated results for the three modes of extraction demonstrates the clear superiority of the extraction efficiency countercurrent contacting method and provides the rationale for the frequent use of this arrangement for industrial processes using extraction. The additional advantage gained by any of the extraction modes as the total solvent volume employed is increased and with one additional stage of countercurrent contacting is shown by the calculated data of Table 10.7. These data clearly show the choice to be made between increasing the volume of extracting solvent or the mode and number of extracting stages to achieve the level of solute recovery desired.

10.5.2. Haifa (or IMI) Phosphoric Acid Process Details

A slight excess of the theoretical requirement of hydrochloric acid, based on the BPL level [$Ca_3(PO_4)_2$ concentration] in the rock, is used for acidulation to avoid the formation of monocalcium phosphate ($CaHPO_4$) which is insoluble and hence would result in phosphate losses. The acidulation reaction (Eq. 10.34) is more rapid with hydrochloric acid than the equivalent reaction with sulfuric acid, and yields soluble products from both the calcium and phosphate. Thus, there is neither a need to heat the mixture to speed up the reaction, nor any particular crystal form problems requiring temperature stabilization.

About 10% of the raw rock, comprising silica and the like, does not dissolve but this material is readily removed by settling the dilute slurry in a thickener (Fig. 6.6). Occasionally coagulants may be used to facilitate this

TABLE 10.7 Calculated Molar Benzoic Acid Concentrations in Water after Extraction with Benzene[a]

Volume of benzene (L)	Single-stage extraction	Three-stage crosscurrent	Countercurrent	
			Three stage	Four stage
100	0.0014	0.013	0.012	0.010
300	0.009	0.007	0.003	0.001
600	0.006	0.003	0.0007	4×10^{-5}
1200	0.003	0.001	0.0001	8×10^{-7}

[a]Calculated for a 1000 L aqueous wastewater stream after extraction with benzene in various modes at 20°C. Initial benzoic acid concentration in water is 0.02 mol/L, and $K_p = 4.3$.

separation of insolubles. The clarified solution from the thickeners is then extracted with a C_4, C_5 alcohol mixture or trialkylphosphate in a series of three or more mixer settlers, producing a solvent phase rich in phosphoric acid, and a raffinate of calcium chloride brine freed of phosphate. Presence of the calcium chloride salt in the aqueous phase undoubtedly assists in driving the phosphoric acid transfer to the organic phase.

To recover the phosphoric acid, the solvent phase is back-extracted with pure water in a further mixer settler (Fig. 10.9). In the absence of dissolved calcium chloride the aqueous phase now accepts the bulk of the phosphoric acid from the solvent, mixed with some residual hydrochloric acid and solvent. Distillation of this stream, usually in three separate evaporators, enables simultaneous recovery of solvent plus unused hydrochloric acid, which are both recycled, and phosphoric acid of appropriate concentrations for sale [55].

The product acid from this process is often purer than that obtained from sulfuric acid acidulation, and in addition this process is claimed to recover 98 to 99% of the available P_2O_5 in the rock, as opposed to the 94 to 95% experience of the most efficient sulfuric acid processes. Refined and food grades of phosphoric acid may be obtained after chemical treatment, stripping, and if necessary anion and cation exchange.

An important economic feature of Haifa process operation is the need to maintain good solvent containment, which extends to solvent recovery from the brine raffinate as well as from the washed waste solids and the extracted

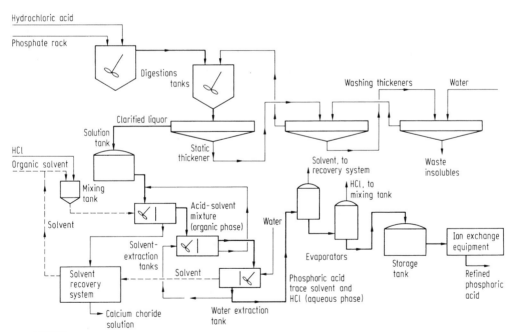

FIGURE 10.9 Flowsheet for the Haifa (Israel Mining Industries) hydrochloric acid acidulation route to phosphoric acid. Digestion for a short time is followed by solvent extraction in polyvinyl-chloride mixer settlers, which tolerate the corrosive conditions. (Reprinted by special permission from *Chemical Engineering* (1962), Ref. [54]. Copyright © 1962, by McGraw-Hill, Inc., New York, NY 10020.)

acid. By these means it has been found possible to maintain solvent losses at not more than about 5 kg/tonne P_2O_5 equivalent produced [56].

10.5.3. Haifa Process By-Products and Waste Disposal

Fluoride containment still has to be considered in the operation of this process, as with sulfuric acid acidulation, since the acid treatment similarly mobilizes chemically bound fluoride. Problems from this source are minimized by prior calcination of the feed phosphate rock, as has been practiced for the early Japanese experimental operations using this process. In any case, fluoride tends to follow the aqueous phase throughout to end up as a component of the calcium chloride raffinate solution. From this stream, fluoride may either be recovered for sale or removed from the calcium chloride stream and discarded by methods which have been discussed (Section 10.4.3).

The washed insoluble slurry from acid digestion is generally discarded to landfill, or as a slurry to the sea if the plant operates on ocean frontage.

The calcium chloride brine, contains about 2 tonnes of the salt for each tonne of phosphoric acid (100% basis) produced and may be sold for such uses as highway deicing, as a low-temperature heat transfer fluid, or as a heat sink as has been outlined in more detail with the Solvay process (Section 7.4). If not sold, in the absence of ocean frontage the waste brine may pose a disposal problem. Deep well disposal into a brine aquifer may be used for inland locations, but geological and water table factors do not always make this possible locally. However, the smaller waste load compared to the Dorr process and the easier disposal of the calcium chloride brine to the sea keeps the Haifa process of interest to phosphoric acid producers [57].

10.6. MAJOR PRODUCERS AND USERS OF PHOSPHORIC ACID

Available production figures for many of the large world producers of phosphoric acid are given in Table 10.8. This product does show a more generally increasing trend with time since fertilizer demands on phosphoric acid output have generally climbed.

End uses of phosphoric acid are highly dependent on the process used to produce it. About 875,000 tonnes (as P_2O_5) of furnace phosphoric acid was produced in the United States in 1984, about 7% of the total U.S. production [58]. Nearly half of this furnace acid is still dedicated to the production of sodium, potassium, and calcium phosphates for use in detergents and commands an 80% or so price premium over fertilizer grade, wet process acid. A further 5% is destined as a food-grade product which is mostly used as salts in foods, bakery goods, and soft drinks. Some of this grade is also used as a supplement in animal feeds.

These are relatively small-scale uses compared to the consumption of about 85% of the total phosphoric acid production in the United States by fertilizer manufacturers, mostly for ammonium phosphates and triple superphosphate (Chap. 11). This consumption picture may be slightly distorted since the United States is a major exporter of phosphoric acid (about 10% of production).

█ **TABLE 10.8 Phosphoric Acid Production by Selected Countries**[a]

	Thousands of metric tonnes of P_2O_5 equivalent[b]				
	1976	1978	1980	1990	1995
Belgium	120	118			
Canada	520	611	808	402	
Finland	104	126			
France	—	338	203		
Italy	426[c]	—		247	
Japan	—	198			
U.K.	482	500			
U.S.A.	7210	8750	9840	10,294	11,884

[a]Sources of data include *U.N. Statistical Yearbooks* [3], and *Chemical and Engineering News* [17]. Includes both furnace acid and wet process products.
[b]To obtain production as 100% H_3PO_4, entry shown should be multiplied by 1.38 (i.e., 196 ÷ 142).
[c]Specified as 82% H_3PO_4.

REVIEW QUESTIONS

1. (a) Calculate the theoretical quantity of phosphate rock (fluorapatite), silica, and coke required by the electric furnace process to produce 1 tonne of elemental phosphorus.
   (b) Give reasonable explanations for the discrepancies between the theoretical requirements, as calculated in part (a), and the amounts of each of these raw materials used in practice.

2. What quantity of elemental phosphorus and water would theoretically be required using the furnace acid process to produce 1 tonne of 100% equivalent phosphoric acid?

3. A company is considering construction of a phosphoric acid plant employing Haifa process technology which is to use its own by-product hydrochloric acid for acidulation. Present-day cost of sulfuric acid is $110 per tonne.
   (a) Assuming quantitative reactions on fluorapatite in both cases and no solvent losses, what "breakeven" value would the company have to assign to its by-product hydrochloric acid (100% basis) to compete with Dorr process operation using sulfuric acid?
   (b) If the company were able to apply a by-product credit of $40 per tonne for Haifa coproduced calcium chloride, and was not able to sell the Dorr coproduced gypsum, what hydrochloric acid "breakeven" value should be assigned? (That is, what HCl price would now still be competitive with Dorr acid under these circumstances?)

4. A manufacturer has to choose between by-product anhydrous hydrochloric acid and 100% sulfuric acid available at $50 and $40 per tonne, respectively, for acidulation of fluorapatite ($CaF_2 \cdot 3Ca_3(PO_4)_2$) to produce phosphoric acid.

(a) What would be the respective acid costs for each of these alternatives per tonne of 100% equivalent phosphoric acid, assuming 100% conversions and yields?

(b) If there was a market for by-product calcium chloride at $20 per tonne, and no market for gypsum, by what dollar value per tonne of 100% phosphoric would the costs of the hydrochloric acid-based process be decreased by the sale of the by-product?

5. A company has to make the decision between using Haifa technology and Dorr technology to build a new phosphoric acid plant. It has access to an extensive source of by-product hydrochloric acid at $50 per tonne (100% basis).

(a) Assuming quantitative reactions in both cases and no solvent losses, at what price would sulfuric acid have to be available for the company to be in a "breakeven" raw material cost situation?

(b) If the company were able to obtain a price of $30 per tonne for the by-product Haifa calcium chloride and was not able to sell the gypsum if it operated the Dorr process, what would be the "breakeven" price of sulfuric acid in this case? In other words, what would the sulfuric acid price have to be in this case to give an equivalent cost phosphoric acid product?

6. (a) Calculate the masses of gypsum ($CaSO_4 \cdot 2H_2O$) and calcium chloride which would theoretically be obtained from fluorapatite using the Dorr and the Haifa processes, respectively, per tonne of phosphoric acid produced (100% basis).

(b) How would the nature of the by-product produced in each case affect the environmentally acceptable disposal options that a phosphoric acid producer has, when disposal is land based and when disposal can take place to sea?

FURTHER READING

Y. C. Athanassiaclis, "Air Pollution Aspects of Phosphorus and its Compounds," NTIS PB-188-073. Washington, DC, 1970.

G. L. Bridger, C. B. Drees, and A. H. Roy, New Process for Production of Purified Phosphoric Acid and/or Fertilizer Grade Dicalcium Phosphate from Various Grades of Phosphatic Materials. *Ind. Eng. Chem. Process Des. Dev.* **20**, 416 (1981).

M. S. Silverstein, G. F. Nordblum, C. W. Dittrich, and J. J. Jakabein, Stable Red Phosphorus, *Ind. Eng. Chem.* **40**, 301 (1948).

S. Skolnik, G. Tarbutton, and W. E. Bergman, Conversion of Liquid White Phosphorus to Red Phosphorus. *J. Am. Chem. Soc.* **68**, 2310 (1946).

L. K. Thompson, S. S. Sidhu, and B. A. Roberts, Fluoride Accumulations in the Soil and Vegetation in the Vicinity of a Phosphorus Plant. *Environ. Pollut.* **18**(3), 221 (1979).

E. N. Walsh, ed. Phosphorus Chemistry: Developments in American Science, American Chemical Society, Washington, DC, 1992.

REFERENCES

1. J. R. Van Wazer, ed., "Phosphorus and its Compounds," Vol. 2, pp. 957, 1025, 1158. Interscience, New York, 1961.

2. E. J. Griffith, A. Beeton, J. M. Spencer, and D. T. Mitchell, eds., "Environmental Phosphorus Handbook," p. 24. Wiley, New York, 1973.

3. "United Nations Statistical Yearbook 1993," 40th ed., United Nations, New York, 1995, and earlier editions.

4. "Phosphate," Miner. Bull. MR 160. Energy Mines and Resources Canada, Ottawa, 1976.

5. D. W. Leyshon and R. T. Schneider, "The Evaluation of Phosphate Rock," 2nd Chemical Congress of the North American Continent, Las Vegas, Nevada, Abstr., FERT 26 (1980).

6. "Mineral Facts and Problems, 1980 ed.," Bull. No. 671, p. 663. U. S. Bureau of Mines, Washington, DC, 1980.

7. J. A. Kent, ed., "Riegel's Industrial Chemistry," p. 652. Reinhold, New York, 1962.

8. F. S. Taylor, "A History of Industrial Chemistry," Reprint ed., p. 203. Arno Press, New York, 1972.

9. P. Ellwood, *Chem. Eng.* **72**, 54 (1965).

10. "Kirk-Othmer Encyclopedia of Chemical Technology," 2nd ed., Vol. 15, pp. 257–276. Interscience, New York, 1968.

11. R. C. Weast, ed., "Handbook of Chemistry and Physics," 51st ed., p. B-117. Chem. Rubber Publ. Co., Cleveland, OH, 1970.

12. D. R. Idler, G. L. Fletcher, and D. F. Addison, "Effects of Yellow Phosphorus in the Canadian Environment." National Research Council of Canada, Ottawa, 1981.

13. R. Thompson, ed., "Industrial Inorganic Chemicals: Production and Uses," pp. 373–390. The Royal Society of Chemistry, Cambridge, UK, 1995.

14. Super Problems Belabor Super Plant, *Can. Chem. Process.* **55**(10), 64, Oct. (1971).

15. F. Leder, W. C. Park, P.-W Chang, J. D. Ellis, J. A. Megy, R. A. Hard, H. E. Kyle, J. Mu, and B. W. Shaw, New Process for Technical Grade Phosphoric Acid. *Ind. Eng. Chem. Process Des. Dev.* **24**, 688–697 (1985).

16. F. A. Lowenheim and M. K. Moran, "Faith, Keyes and Clark's Industrial Chemicals," 4th ed., p. 641. Wiley-Interscience, New York, 1975.

17. Facts and Figures for the Chemical Industry, *Chem. Eng. News* **74**(26), 38–79, June 24 (1996), and June issues of earlier years.

18. D. Rose and J. R. Marier, "Environmental Fluoride." National Research Council of Canada, Ottawa, 1977.

19. J. C. Barber and T. D. Farr, *Chem. Eng. Prog.* **66**(11), 56, Nov. (1970).

20. J. C. Barber, *Chem. Eng. Prog.* **65**(6), 70, June (1969).

21. D. R. Idler, *Chem. Can.* **21**(11), 16, Dec. (1969).

22. J. Q. Hardesty and L. B. Hein, *in* "Reigel's Handbook of Industrial Chemistry" (J. A. Kent, ed.), 7th ed., p. 537. Van Nostrand-Reinhold, Toronto, 1974.

23. R. N. Shreve and J. A. Brink, Jr., "Chemical Process Industries," 4th ed., pp. 253, 256. McGraw-Hill, New York, 1977.

24. J. F. Coykendall, E. F. Spencer, and O. H. York, *J. Air. Pollut. Control Assoc.* **18**(5), 315 (1968).

25. Acid's Value Enhanced, *Chem. Eng. (N.Y.)* **71**(1), 26, Jan. 6 (1964).

26. G. Kleinman, *Chem. Eng. Prog.* **74**(11), 37, Nov. (1978).

27. J. F. McCullough, *Chem. Eng. (N.Y.)* **83**, 101, Dec. 6 (1976).

28. G. S. G. Beveridge and R. G. Hill, *Chem. Process. Eng. (N.Y.)* **49**(8), 63, Aug. (1968).

29. G. S. G. Beveridge and R. G. Hill, *Chem. Process. Eng. (N.Y.)* **49**(7), 61, July (1968).

30. W. E. Blumrich, H. J. Koening, and E. W. Schwer, *Chem. Eng. Prog.* **74**(11), 58, Nov. (1978).

31. TVA Displays Energy Saving, *Chem. Eng. News* **56**(46), 32, Nov. 13 (1978).

32. P. L. Olivier, Jr., *Chem. Eng. Prog.* **74**(11), 55, Nov. (1978).

33. L. E. Bostwick, *Chem. Eng.* **77**(8), 100, April 20 (1970).

34. C. Earl, A. Davister, and F. Thirion, High Strength Phosphoric Acid Process. *Chem. Eng. Prog.* **82**, 34–37, Oct. (1986).

35. G. L. Bridger and A. H. Roy, *Chem. Eng. Prog.* **74**(11), 62, Nov. (1978).

36. *Chem. Eng. News* **54**, 32, Sept. 6 (1976).

37. J. Bergdorf and R. Fischer, *Chem. Eng. Prog.* **75**(11), 41, Nov. (1978).

38. M. Feki, M. Fourati, M. M. Chaabouni, and H. F. Ayedi, Purification of Wet Process Phosphoric Acid by Solvent Extraction. *Can. J. Chem. Eng.* **72**, 939–944 (1994).

39. W. E. Rushton, *Chem. Eng. Prog.* **74**(11), 52, Nov. (1978).

40. A. P. Kouloheris, *Chem. Eng. (N.Y.)* **87**, 82, Aug. 11 (1980).

41. *Chem. Eng. News* 55(42), 17, Oct. 17 (1977); 57(42), 10, Oct. 15 (1979).

42. F. A. Patty, ed., "Industrial Hygiene and Toxicology," 2nd ed., Vol. 2, p. 832. Interscience, New York, 1963.

43. A. W. Frazier, J. R. Lehr, and E. F. Dillard, *Environ. Sci, Technol.* 11(10), 1007 (1977).

44. C. Djololian and D. Billaud, *Chem. Eng. Prog.* 74(11), 117, Nov. (1969).

45. P. S. O'Neill, *Ind. Eng. Chem., Prod. Res. Dev.* 19, 250 (1980).

46. D. Kitchen and W. J. Skinner, *J. Appl. Chem. Biotechnol.* 21(2), 53, 56, Feb. (1971).

47. A. V. Slack, ed., "Phosphoric Acid," Vol. 1, Part II, p. 503. Dekker, New York, 1968.

48. R. Blumberg, *in* "Solvent Extraction Reviews," (Y. Marcus, ed.), Vol. 1, p. 93. Dekker, New York, 1971; R. Blumberg, D. Gonen, and D. Meyer, *in* "Recent Advances in Liquid-Liquid Extraction" (C. Hanson, ed.), p. 93. Pergamon, Oxford, 1971.

49. Dow to Test Phosphoric Acid Process, *Chem. Eng. News* 41(41), 35, Oct. 14 (1963).

50. I. Raz, *Chem. Eng. (N.Y.)* 83, 52, June 10 (1974).

51. R. M. Diamond and D. G. Tuck, in Progress in Inorganic Chemistry, Vol. 2, F. A. Cotton, ed., p. 109, Interscience, New York, 1960.

52. R. H. Perry, ed., "Chemical Engineers Handbook," pp. 14–45. McGraw-Hill, New York, 1969.

53. D. F. Rudd, G. J. Powers, and J. J. Siirola, "Process Synthesis," p. 141. Prentice-Hall, Englewood Cliffs, NJ, 1973.

54. Hydrochloric-Based Route to Pure Phosphoric Acid, *Chem. Eng. (N.Y.)* 69(52), 34, Dec. 24 (1962).

55. A. Baniel, R. Blumberg, A. Alon, M. El-Roy, and D. Goniadski, *Chem. Eng. Prog.* 58(11), 100, Nov. (1962).

56. Israelis Pioneer New Route, *Chem. Eng. (N.Y.)* 69(12), 88, June 11 (1962).

57. Anonymous, New Life Seen for Phosphoric Acid Process. *Chem. Eng. Prog.* 89, 17, Jan. (1993).

58. Recovery Slows for Large Volume Acids, *Chem. Engin. News* 62(24), 10–12, June 11 (1984).

11
AMMONIA, NITRIC ACID, AND THEIR DERIVATIVES

. . . whoever could make two ears of corn, or two blades of grass, to grow upon a spot of ground where only one grew before, would . . . do more essential service . . . than the whole race of politicians put together.
—*Jonathan Swift (1667–1745)*

It has long been known . . . as . . . volatile alkali, hartshorn, spirit of sal ammoniac . . . at present generally by the name of ammonia.
—*John Dalton, 1810*

11.1. AMMONIA

11.1.1. Background

Early samples of ammonia were obtained by means of the bacterial action on the urea present in urine (Eq. 11.1), or by the dry distillation of protein-containing substances such as bone, horns, and hides.

$$H_2N-CO-NH_2 + H_2O \xrightarrow{\text{bacteria}} CO_2 + 2\ NH_3 \qquad 11.1$$

Small amounts were also obtained as a by-product from the manufacture of coke or coal gas from coals. Ammonia is still recovered at the rate of 134,000 tonnes per year as an incidental part of U.S. coal-based operations, but this amounts to less than 1% of current American ammonia production [1] (Table 11.1). By far the bulk of ammonia produced today is by the direct combination of the elements. Other synthetic processes have been tested and found to be either impractical for commercial exploitation or have gradually been supplanted by direct elemental combination for economic reasons.

There are several examples of what turned out to be commercially impractical methods to produce ammonia. The French Serpek process depended on the formation of aluminum nitride from the heating of bauxite in nitrogen [5] (Eq. 11.2).

$$Al_2O_3 + N_2 + 3\ C \xrightarrow{1600°C} 2\ AlN + 3\ CO \qquad 11.2$$

311

TABLE 11.1   Production of Synthetic Ammonia by Selected Areas

	Thousands of tonnes[a]			
	1970	1980	1990	1995
Canada	1,220	2,556	3,602	4,655
China	—	12,150	—	—
France	1,618	2,085	1,478	1,478
W. Germany	—	2,040	1,616	2,992[b]
Italy	1,277	1,714	1,482	n/a
Japan	3,261	2,573	1,831	1,831
Mexico	—	1,883		
United States	12,541	17,262	15,390	16,152
U.S.S.R.	7,638	16,732	n/a	n/a
World	(62,500)			

[a]Production *capacities* given in parentheses. Data compiled from Kirk-Othmer [2], *Chemical and Engineering News* [3], *Canadian Chemical Processing* [4], and data supplied by the Verband der Chemischen Industrie, e.V. (VCI).
[b]After reunification.

The aluminum nitride was then hydrolyzed with water to yield ammonia and aluminum hydroxide (Eq. 11.3).

$$AlN + 3\ H_2O \rightarrow NH_3 + Al(OH)_3 \qquad 11.3$$

The aluminum hydroxide could subsequently be converted to alumina in pure form for aluminum production. Again high temperatures were used to fix elemental nitrogen in the cyanamide route, developed by Frank and Caro in Germany in 1898. In this process an electric arc furnace provides the very high temperatures required to obtain calcium cyanamide from calcium carbide in a manner similar to that used for phosphorus production (Eq. 11.4). Hydrolysis of the calcium cyanamide then produces ammonia and calcium carbonate (Eq. 11.5). Cyanamide production for many years predominantly went into ammonia synthesis.

$$CaC_2 + N_2 \xrightarrow{\ 900\text{--}1000°C\ } CaNCN + C \qquad 11.4$$

$$CaCN_2 + 3\ H_2O \rightarrow 2\ NH_3 + CaCO_3 \qquad 11.5$$

The direct synthesis of ammonia has now displaced cyanamide hydrolysis for this purpose, though cyanamide is still produced on a very large scale for other purposes.

Ammonia was probably first synthesized directly from its elements in the laboratory by Dobereiner in 1823, as a secondary result of experiments designed to study the direct combination of gaseous hydrogen and oxygen [6]. However, it was not until the early 1900s, as Nernst developed the theory by which equilibrium concentrations could be approximated from thermochemical calculations and Haber proved the direct reaction of nitrogen and hydrogen on a laboratory scale, that any promise of adoption of this process on an industrial scale developed. An early bench scale reactor produced about 80 g of ammonia per hour. Later, in collaboration with Badische Anilin-und Soda-Fabrik (BASF), this was rapidly scaled up to plants operating in Germany which produced 27 tonnes per day by 1913, and 180,000 tonnes per year for the country by 1918. Haber in 1918, and Bosch in 1931, earned Nobel prizes for their respective contributions to this area of chemistry, and have jointly lent their names to the process which originated as a result of their work. Today there are many single-train ammonia plants worldwide operating at capacities of 900 tonnes per day, and some as large as 1650 tonnes per day [2].

11.1.2. Principles of Ammonia Synthesis: The Haber or Haber-Bosch Process

Equation 11.6, which represents the ammonia synthesis reaction, establishes the stoichiometry of 2 mol (volumes) of ammonia produced for each 4 mol (volumes) of initial gases reacting.

$$N_2 {}_{(g)} + 3 H_2 {}_{(g)} \rightleftharpoons 2 NH_3 {}_{(g)} \qquad \Delta H = -92.0 \text{ kJ } (-22.0 \text{ kcal}) \qquad 11.6$$

The principle of Le Chatelier states that when a change is placed on a system in equilibrium that equilibrium will shift so as to reduce the effect of the change (quantities, pressure, temperature etc.). From this principle it would be expected that increasing the pressure on this system would increase the equilibrium ammonia concentration. It does (Table 11.2).

The ammonia synthesis reaction is also faster, that is, equilibrium conditions are reached more rapidly, at higher temperatures. At the same time the reaction is exothermic to the extent of 46 kJ mol^{-1} (about 54 kJ mol^{-1} at normal conversion temperatures), so that as the synthesis temperature is raised the ammonia equilibrium is shifted to the left, again in accordance with Le Chatelier's principle (Table 11.2). Thus, while raising the synthesis temperature does give a higher reaction rate allowing a greater volume of production to be obtained in the same size reactor, it also displaces the equilibrium of the synthesis (Eq. 11.6), to the left giving a smaller potential nitrogen

TABLE 11.2 Equilibrium Percent Concentrations of Ammonia at Various Temperatures and Pressures[a]

Temperature, °C	Absolute pressure (atm)						
	1	10	50	100	300	600	1000
200	15.3	50.7	74.4	81.5	89.9	95.4	98.3
300	2.2	14.7	39.4	52.0	71.0	84.2	92.6
400	0.4	3.9	15.3	25.1	47.0	65.2	79.8
500	—	1.2	5.6	10.6	26.4	42.2	57.5
600	—	0.5	2.3	4.5	13.8	23.1	31.4
700	—	—	1.1	2.2	7.3	11.5	12.9

[a]From reacting a gas mixture of 1 mol nitrogen to 3 mol hydrogen under the indicated conditions. Compiled from the data of Haber et al. [7], and Larson [8]. Nielson gives more recent and more detailed data for conditions of commercial interest [9].

conversion to ammonia. Fortunately, a substantial increase in the pressure on the system significantly improves the equilibrium conversion to ammonia to make the whole exercise practical. For these reasons the majority of synthetic ammonia processes tend to cluster around 100–300 atm pressure and 400–500°C operating temperatures in an attempt to maximize conversions and conversion rates while moderating compression costs. The Claude and the Casale ammonia synthesis processes, operating in France at 900 atm and in Italy at 750 atm, respectively, are among the highest pressure industrial processes. The Claude process relies on a nitrogen to hydrogen feed gas system connected to multiple, small inside diameter (10-cm) bored nickel chrome ingots to contain the enormous pressures and in return gains conversions to ammonia of 40% or better. This compares to conventional converter dimensions of 7 to 10 m high by 3 m in outside diameter which requires 20-cm-thick alloy steel walls for pressure containment.

Important to the successful operation of ammonia converters was the discovery of a suitable catalyst which would promote a sufficiently rapid reaction at these operating temperatures to utilize the moderately favorable equilibrium under these conditions. Otherwise higher temperatures would be required to obtain appropriately rapid rates, and the less favorable equilibrium under these conditions would necessitate higher pressures as well, in order to maintain a sufficiently favorable percent conversion. The original synthesis experiments were conducted with an osmium catalyst. Haber later discovered that reduced magnetic iron oxide (Fe_3O_4) was highly effective, and that its activity was further enhanced by the presence of the promoters alumina (Al_2O_3; 3%) and potassium oxide (K_2O; 1%), probably from the introduction of iron lattice defects. Iron with various proprietary variations still forms the basis of all ammonia catalyst systems today.

Space velocity in the broad sense refers to the volume fed to a reactor per hour, divided by the volume of the reactor. For liquid reaction streams this relationship is straightforward. For gases, however, the space velocity is defined as being the volume of gases corrected to 0°C and 760 mm Hg (1 atm) passing through the reactor (or catalyst) volume per hour, a measure of the gas-catalyst contact time for heterogeneous reactions (Eq. 11.7).

$$\text{space velocity} = \frac{\text{volumetric feed rate}}{\text{volume of the reactor}} \qquad 11.7$$

or, for gases, (gas volume corrected to 0°C, 760 mm Hg):

$$\text{space velocity} = \frac{m^3 \text{ gas(es) hr}^{-1}}{m^3 \text{ catalyst volume}}$$

For an initial doubly promoted magnetic iron oxide catalyst, working on a nitrogen:hydrogen mole ratio of 1:3 at a pressure of 100 atm and 450°C, and a space velocity of 5000 (hr^{-1}), approximately 13–15% ammonia could be expected [10]. Much lower conversions are obtained from singly promoted or nonpromoted iron.

A space velocity of 5000 hr^{-1} (corrected to 0°C, 760 mm) corresponds to a gas change rate in the catalyst space of 132.4 m^3 of gas per cubic meter of catalyst space per hour at operating conditions (450°C, 100 atm; Eq. 11.8), or an actual gas-catalyst contact time of about 27 sec. Increasing the space velocity decreases the gas-catalyst contact time and in so doing decreases the percentage of nitrogen and hydrogen converted to ammonia.

Space Velocity Relationships:

$$\text{space velocity} \propto \frac{1}{\text{contact time}} \qquad\qquad 11.8$$

Thus, 5000 hr^{-1} space velocity is equivalent to a contact time of 1/(5000 hr^{-1}), or 2×10^{-4} hr, or 0.72 (2×10^{-4} hr $\times$ 3600 sec/hr) sec contact time, if gases were at 0^{0}C (273 K) and 1 atm. At constant space velocity then, contact time $\propto$ pressure. Thus changing only the pressure to 100 atm raises the contact time to 100 $\times$ 0.72 sec, or 72 sec.

$$\text{contact time also} \propto \frac{1}{\text{absolute temperature}}$$

Thus, simultaneously raising the temperature has the effect of decreasing the actual contact time to

$$72 \text{ sec} \times \frac{1}{(723 \text{ K}/273 \text{ K})} = 22.19 \text{ sec,}$$

or 27.2 sec contact time, under converter operating conditions

Gas changes per hour under converter operating conditions, therefore, will be: (3600 sec/hr)/(27.19 sec/change) = 132.4 changes per hour.

Doubling the space velocity of a converter may well drop the ammonia conversion percentage to, say, 10%, which may appear to be a change in the wrong direction. But 10% of twice the original volume of reacting gases passing through the converter still means that the volume of ammonia recovered is roughly one-third more or 1000 m^3/hr NH$_3$ (10% of 10,000 m^3) rather than 750 m^3/hr NH$_3$ as obtained at 5000 hr^{-1} space velocity. Hence, high space velocities of 10,000–20,000 hr^{-1} or more are common for commercial ammonia converters.

For this combination of reasons, most large ammonia plants have to recycle a large proportion of the converter exit gases in the form of unreacted nitrogen and hydrogen. Thus, a significant component of the engineering of a modern ammonia synthesis unit is concerned with pressure and heat conservation to enable this recycle to be conducted as economically as possible.

11.1.3. Feedstocks for Ammonia Synthesis, Air Distillation

Nitrogen for ammonia synthesis may be either separated from air by liquid air distillation to obtain the nitrogen component, or it may be obtained by consumption of the oxygen of air by the burning of a fuel in some air-restricted oxidative process to leave a nitrogen residue.

Cryogenic, low-temperature technology is called on to separate nitrogen from liquid air. Initial steps involve dust removal, usually by filtration, and removal of the normal carbon dioxide present in air (about 350 ppm), either by scrubbing with an organic base such as monoethanolamine (MEA) or with aqueous alkali (Eq. 11.9–11.13).

$$CO_2 + H_2O \rightarrow H_2CO_3 \qquad\qquad 11.9$$

MEA absorption:

$$H_2CO_3 + H_2NCH_2CH_2OH \xrightarrow[\text{ambient temp}]{\text{high pressure}} HCO_3\text{-} + H_3\overset{+}{N}CH_2CH_2OH$$

$$11.10$$

Regeneration (separate unit):

$$HCO_3^- + H_3\overset{+}{N}CH_2CH_2OH \xrightarrow[\text{heat}]{\text{low pressure}} H_2NCH_2CH_2OH + H_2O + CO_2$$

$$11.11$$

Alkali absorption:

$$H_2CO_3 + NaOH \rightarrow NaHCO_3 + H_2O \qquad\qquad 11.12$$

$$H_2CO_3 + 2\ NaOH \rightarrow Na_2CO_3 + 2\ H_2O \qquad\qquad 11.13$$

Purification is an important preliminary step, particularly in operating areas near a refinery or petrochemical complex, since accumulation of combustible dusts or condensed vapors such as methane or acetylene in cryogenic lines containing liquid oxygen introduces serious operating risks. And, of course, the temperatures in the region of −200°C required to liquefy air are sufficiently low to cause carbon dioxide to solidify, which could plug operating components. After purification the air is normally dried to as low a dewpoint as feasible using towers packed with an adsorbent such as activated alumina, a molecular sieve (e.g., Union Carbide type 13X), or silica gel. This helps to avoid problems from water vapor. Two or more drying towers operated on swing cycles provide continuous water adsorption capacity from one tower, while the other is regenerated by heating the adsorbent under reduced pressure.

Liquefaction of the purified air is accomplished using the Joule-Thompson effect, which is the cooling effect obtained from a compressed gas when it is allowed to expand. By using this expansion-cooling effect repetitively, and by employing the chilled expanded gas to prechill the compressed gas before expansion, air may be liquefied by employing compression pressures of only about 10 atm (about 150 psig, Fig. 11.1). To accomplish liquefaction of air by direct compression alone is not possible at temperatures above the critical temperatures of the component gases, 126.1 K (−147.1°C) for nitrogen and 154.4 K (−118.8°C) for oxygen, and even at these temperatures, pressures of 33.5 and 49.7 atm, respectively, would be required for condensation. But by using the technique of repeated compression, then external cooling of the compressed gases to well below their critical temperatures makes expansion-cooling in a well-insulated system sufficient to liquefy a part of both gases as they approach their atmospheric pressure boiling points: nitrogen, 74.8 K (−195.8°C), and oxygen, 90.2 K (−183.0°C).

On fractionation (fractional distillation) of liquid air, nitrogen, the lower boiling constituent, distills first. The oxygen fraction is frequently vented un-

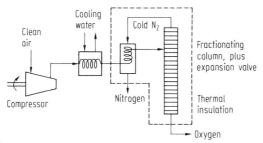

FIGURE 11.1 Simplified flowsheet for the separation of nitrogen from liquid air.

less there is a local petrochemical or metallurgical application which can use it. Liquid air distillations are commonly conducted at a pressure of about 5 atm, which both raises the column operating temperatures and increases the difference in boiling points for the component gases to be separated, improving the separation efficiency. This is particularly useful if argon (0.93% in air) recovery is practiced. With a boiling point of 87.5 K ($-185.7°C$) argon would be lost with the oxygen stream without these precautions. Incidentally, the close boiling points of argon and oxygen represent physical evidence of the monoatomic versus diatomic existence of these gases, argon with an atomic and molecular weight of 39.95, and oxygen with an atomic weight of 16.00 and molecular weight 32.00.

When liquid air distillation is used as the source of nitrogen, the hydrogen also required for ammonia synthesis is obtained from a variety of sources. Some is obtained as the coproduct from the electrolytic production of chlorine and caustic soda (Chap. 8), some from refinery sources as a by-product of cracking processes or olefin synthesis, some from the water gas reaction, and some is produced specifically for the purpose by the electrolysis of alkaline water (e.g., by Cominco, Trail, Eq. 11.14).

$$2 \; H_2O \; + \; \underset{\substack{\text{ca. 5 molar} \\ (26\%)}}{KOH} \; \xrightarrow{\text{electrol.}} 2 \; H_2 + O_2 + KOH \qquad 11.14$$

Despite the apparently straightforward nature of these procedures to the feedstocks for ammonia synthesis and the "free" source of nitrogen from the air, this combination of approaches has become feasible only for relatively small ammonia plants of around 100 tonnes per day, or for special process situations where abundant hydrogen is available. When it is realized that several ammonia plants of capacities of 900 tonnes per day are now operating and a few as large as 1650 tonnes per day it puts this process sequence into proper perspective.

11.1.4. Ammonia Feedstocks, Reforming and Secondary Reforming

Hydrogen on a sufficient scale for ammonia synthesis is accessible from coal (indirectly), petroleum, natural gas, and water, and all of these sources are used to some extent. In the original Haber-Bosch process, for example, coke derived from coal was used as the source of hydrogen via the water gas reaction. To do this, initially the coke is burned in air until it reaches red heat (Table 11.3), and then the air feed stream is replaced by steam (Eq. 11.15, 11.16).

Combustion phase:

$$C + air \rightarrow CO_2 + 4 N_2 + heat \quad (air\ is\ approx.\ O_2:4\ N_2) \qquad 11.15$$

Water gas reaction:

$$C + H_2O \xrightarrow[ca.\ 1000°C]{} CO + H_2\ (positive\ \Delta H) \qquad 11.16$$

The hot coke in an endothermic reaction produces mainly carbon monoxide and hydrogen, and cools down. The process is periodically switched to combustion air to regain the high coke temperatures necessary for the water splitting reaction. Since carbon monoxide is relatively difficult to separate from hydrogen, and also because it is possible to obtain a further mole of hydrogen from each mole of carbon monoxide, the whole water gas stream is normally put through a shift conversion step (Eq. 11.17).

Shift conversion:

$$CO_{(g)} + H_2O_{(g)} \xrightarrow[\sim 500°C]{FeO + Cr_2O_3} CO_{2(g)} + H_{2(g)} \qquad 11.17$$

$$\Delta H = -41.2\ kJ\ (-9.8\ kcal)$$

Here, the carbon monoxide is converted to carbon dioxide at more moderate temperatures in the presence of a chromium-promoted iron catalyst. Carbon dioxide is readily separated from the hydrogen by absorption in water under pressure (Eq. 11.18).

TABLE 11.3   Relationship between the Thermally Induced Color of a Substance and Its Temperature[a]

Color of emission	Temperature (°C)	Applications
Dark red	585–870	Al melts at 660°, soft glass at ca. 700°, potters decorating kilns, reformer tubes
Cherry red	880–980	Ag melts at 962°, fired red clay products
Orange red	990–1090	Au melts at 1064°, Cu at 1083°
Yellow	1100–1220	Lime kiln, cement kiln operation
White	1230–1280	Lime kiln, cement kiln operation
Brilliant white	1290–1520	Ni melts at 1453°, firing of porcelain, spark plug and power line insulators
Dazzling white	1540–1800	Fe melts at 1535°, firing temp. of silica bricks and fire brick

[a]Temperature relations selected from Lange [11].

$$CO_{2(g)} + H_2O_{(\ell)} \xrightarrow{25 \text{ atm}} H_2CO_{3(\ell)} \qquad \Delta H = -38.5 \text{ kJ } (-9.2 \text{ kcal})$$

$$11.18$$

This leaves a hydrogen stream ready for ammonia synthesis, after a final cleanup. The nitrogen required is normally taken from a part of the gas stream obtained from the heating, coke combustion phase of the water gas reaction sequence, which is depleted in oxygen. Carbon dioxide removal from the nitrogen feed component is effected in the same manner as from the hydrogen.

Where natural gas supplies are abundant and available at reasonable cost, however, an alternative reforming and secondary reforming sequence is generally favored over the water gas route for ammonia feedstock production, for economic reasons [12]. Coal-based ammonia requires nearly twice as much capital investment as a plant based on natural gas, and environmental problems are often greater with the former (see also Section 11.1.7 on new developments). As an overview, reforming and secondary reforming of natural gas amounts to a burning of methane in the presence of steam and a deficiency of air to give a mixture of hydrogen, nitrogen, and carbon dioxide.

Reforming can use any kind of petroleum feedstock, not just natural gas, although the latter is favored because it has the highest hydrogen content and is the easiest to clean prior to use. If the gas has been thoroughly desulfurized at the well head the only additional pretreatment required is passage through a bed of activated alumina or bauxite for an absorptive purification. Use of either a gas oil type of distilled petroleum feedstock or a residual fuel oil for synthesis gas generation requires much more stringent precautions to remove sulfur since this is one of a series of elements which can permanently decrease the activity of (poison) the catalyst system employed for ammonia synthesis [13]. Copper, arsenic, and phosphorus are also serious catalyst poisons.

Regardless of the petroleum source employed, the chemistry to obtain ammonia feedstocks is similar. Since methane is of dominant importance this will be used as an example to describe the steps required. Initially methane is mixed with steam and passed into heat resistant nickel-chromium-iron alloy tubes containing a supported nickel catalyst. The tubes are heated externally by a further portion of methane consumed as fuel (Fig. 11.2, Eq. 11.19).

Primary steam reforming:

$$CH_{4\ (g)} + H_2O_{(g)} \xrightarrow[\text{400 psi, 800°C}]{\text{Ni catalyst}} CO_{(g)} + 3 H_{2(g)}$$

$$11.19$$

$$\Delta H = +208 \text{ kJ } (+49.7 \text{ kcal})$$

This reaction requires continued external combustion of methane to maintain the red heat (Table 11.3) of the reformer tubes, since it is an endothermic reaction. It produces most of the hydrogen required for ammonia production.

Air plus methane are then added to the exit gas stream from the steam reformer in the appropriate proportions to provide all of the eventual nitrogen requirement for ammonia synthesis, plus some further hydrogen (Eq. 11.20).

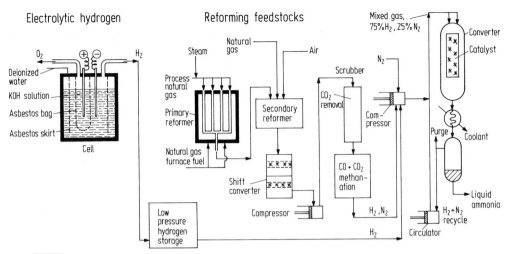

FIGURE 11.2 Outline of the main components of an ammonia synthesis plant using reforming and secondary reforming as the principal sources of hydrogen. Electrolysis of water is used to supplement this.

Secondary air reforming:

$$CH_{4(g)} + air_{(g)} \xrightarrow[950°C]{Ni\ catalyst} CO_{(g)} + 2\ H_{2(g)} + 2\ N_{2(g)} \qquad 11.20$$

Air is approx. $1/2\ O_2 : 2\ N_2$. $\quad \Delta H = +35.6$ kJ (+8.5 kcal)

This reaction is also endothermic and requires external fuel combustion to maintain the high reformer tube temperatures. About one-third of the methane consumed at this stage is used to heat the primary steam and secondary air reformers.

As with the water gas product of the Haber-Bosch route to hydrogen, the carbon monoxide produced by the two reforming reactions may be converted to a further mole of hydrogen via shift conversion. The coproduced carbon dioxide may also be more readily removed than carbon monoxide from the product gases, hydrogen and nitrogen (Eq. 11.17, 11.18). To accomplish this, water is sprayed into the very hot gas mixture emerging from the air reformer to bring the temperature down to the 400°C range for shift conversion, which also provides the steam necessary for this step. Since this step is exothermic it is usually carried out in two separate units to permit better temperature control of the process and also, once started, it does not require external heat input. At a space velocity of about 100, carbon monoxide conversion to carbon dioxide is achieved to a residual monoxide concentration of less than 1% [14].

The proportions of the two reforming reactions and shift conversion are so controlled that the gas mixture obtained contains nitrogen and hydrogen in the mole ratio (volume ratio) of 1:3. But this mixture still contains 20 to 30% carbon dioxide, resulting from the shift conversion reaction and traces of unconverted carbon monoxide. Carbon dioxide can yield carbonates and carbamates in the ammonia synthesis cycle, which are undesirable because they would deposit in the piping. Also oxygen, or any of its compounds such as carbon monoxide, water, etc., is also an ammonia catalyst poison [13]. Consequently they must be removed.

Carbon dioxide can be removed by scrubbing the pressurized gas mixture with water (Eq. 11.19). In an ordinary scrubber 21.6 volumes of gas are absorbed per volume of water at 25 atm, 12°C. However, energy costs for moving the large volumes of water are significant. Also, even if a low-energy contactor is used, such as a U-tube in a deep well, it has been found that at sufficiently high pressures for efficient carbon dioxide removal 3 to 5% hydrogen loss occurs. To avoid these hydrogen losses gas liquid equilibria are employed which are more favorable than the one for carbon dioxide-water. Several systems are based on employment of monoethanolamine (e.g., the Girbotol process) which is capable of decreasing the carbon dioxide content to less than 50 ppm [15] (Eq. 11.9–11.12). The potassium carbonate-bicarbonate cycle also permits the absorbing base to be recycled, but can only decrease the carbon dioxide content to 500–1000 ppm (Eq. 11.21, 11.22).

Absorption:

$$CO_2 + H_2O + K_2CO_3 \rightarrow 2\ KHCO_3 \qquad\qquad 11.21$$

Regeneration:

$$2\ KHCO_3 \xrightarrow[120°C]{} K_2CO_3 + CO_2 + H_2O \qquad\qquad 11.22$$

The efficiency may be improved somewhat if a two-stage contactor with potassium carbonate solutions is used, particularly if the second stage is operated at high pressure. Sodium carbonate is a less costly absorbent, but also less efficient. Other acid gas removal techniques use physical absorption of carbon dioxide in organic solvents, such as the dimethyl ether of polyethylene glycol, or methanol (the Selexol or Rectisol processes), but for absorption to be effective these require high gas pressures. Regeneration of physical absorption solvents is by pressure letdown plus some air stripping (sparging of air through the solvent).

The traces of residual carbon dioxide and any remaining carbon monoxide are then treated catalytically with hydrogen in a methanation step, which removes the catalyst poisoning effect of these oxygen-containing components (Eq. 11.23, 11.24).

Methanation:

$$CO + 3\ H_2 \xrightarrow[315°C]{Ni\ catalyst} CH_4 + H_2O \qquad\qquad 11.23$$

$$CO_2 + 4\ H_2 \rightarrow CH_4 + 2\ H_2O \qquad\qquad 11.24$$

If, however, the ammonia plant is obtaining nitrogen from an air separation plant, carbon monoxide and carbon dioxide (boiling points -205°C; and -78.5°C, sublimes) may be condensed and removed by scrubbing with liquid nitrogen.

11.1.5. Ammonia Synthesis

Whether feedstocks are obtained from air separation plus supplementary hydrogen sources or from a reforming and secondary reforming sequence, the feedstock requirements are the same. Nitrogen and hydrogen gas are required in a mole ratio (at ordinary pressures close to a volume ratio) of 1:3, and raised to very high pressures in the 100–900 atm (10,130–91,200 kPa; 1500–13,500 psi) range, depending on the synthesis system employed [16]. Choice of synthesis pressure used depends on the balance to be struck between the lower compression energy costs but lower conversions (more recycle cost) at low synthesis pressures, and higher compression costs but higher conversions at high synthesis pressures.

Early in the development of ammonia synthesis technology it became apparent that iron and high carbon steels became brittle and underwent intergranular corrosion on exposure to high pressures of hydrogen or ammonia, particularly at high temperatures. This caused a modification of the structure of the steel by a little understood process. This structural alteration weakened the steel, and made vessels so attacked unsafe to use at high pressures after a short time. This process, termed "hydrogen embrittlement," is now understood to involve adsorption of hydrogen onto the steel, followed by ionization (Eq. 11.25).

$$2\ H_{2(g)} \xrightarrow{\text{adsorption}} 2\ H_{2(ads)} \xrightarrow{\text{ionization}} \underset{(+4e^-)}{4\ H^+} \xrightarrow{C_{(Fe)}} CH_{4(Fe)} \rightarrow CH_{4(g)}$$

$$11.25$$

The hydrogen ions formed diffuse into intergranular carbon and react with it, forming methane bubbles. These internally stress the steel, making it much less ductile and causing intergranular cracks.

Since we have gained a better understanding of hydrogen embrittlement phenomena, the design of the pressure shell of all ammonia converters has incorporated features to avoid weakening by this kind of attack [17]. All use the incoming nitrogen/hydrogen stream to keep the inner wall of the pressure shell cool by passing this entering stream immediately inside the pressure shell, thus eliminating one contributing factor to embrittlement, the heat (e.g., Fig. 11.3). By using alloy steels incorporating titanium, vanadium, tungsten, chromium, or molybdenum, methanation of any adsorbed/absorbed hydrogen does not occur, since these metals form very stable carbides. Instrumentation has also been developed to monitor continuously hydrogen penetration rates, which adds a useful safeguard. Wrapped converter designs are used where only the inner layer of about a 1.2-cm-thick plate is made leak tight, and the outer multilayers of tightly wrapped 0.6-cm plate are deliberately perforated with small holes to release any harmless diffused hydrogen to provide another safeguard.

Ammonia conversion takes place at 400–500°C. Most of the input gas heating is provided by indirect heat exchange with already converted exit gases via an exchanger inside the pressure shell of the converter (Fig. 11.3). During the fraction of a second contact time the feed gas stream forms a gas mixture containing 10–20% ammonia in residual nitrogen and hydrogen, and

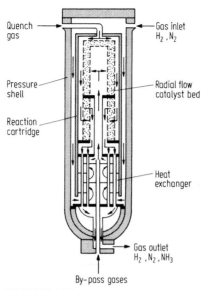

Quench gas

Gas inlet
H_2, N_2

Pressure shell

Radial flow catalyst bed

Reaction cartridge

Heat exchanger

Gas outlet
H_2, N_2, NH_3

By-pass gases

FIGURE 11.3 Diagram showing design features of a large-scale ammonia synthesis converter. The quench and bypass inlets for feed gas inputs, by foregoing heat exchange before entering the catalyst beds, provide one means of temperature control on reacting gases. The fraction of "inerts" in the feed gas stream is another (see text).

the pressure drops slightly in the process from both flow resistance through the catalyst beds, and the slight gas volume decrease incurred by the reaction itself. Catalyst life of up to 15 yr can be achieved with careful feed and recycle gas treatment.

Ammonia is recovered from this gas mixture by condensation (on cooling the gases) at converter exit pressures. This practice has been adopted for several reasons. First, it conserves compressor capacity and recompression energy requirements once the liquefied ammonia has been separated, by maintaining the unreacted nitrogen and hydrogen at nearly the inlet pressure required for recycle to the converter again. Second, ammonia (boiling point -33°C at normal atmospheric pressure) is far easier to condense out of the gas mixture under high pressures than at atmospheric pressure. For processes operating at pressures of 400 atm or above, ordinary process cooling water temperatures (10 to 20°C) are sufficient to cause liquefaction of the ammonia. It is only at pressures below this that refrigeration becomes necessary (Table 11.4). The crude liquid ammonia product is then fed to a high-pressure separator, and following this to a low-pressure separator from which it proceeds to product storage spheres or shipping units (Fig. 11.4).

Ammonia may also be recovered from converter exit gases by absorption into water when ammonia solutions are to be marketed. Over 600 volumes of ammonia are absorbed per volume of water at 25°C, 1 atm. Solutions containing 28 to 30% NH_3 by weight are obtained, which is concentrated ammonium hydroxide (Eq. 11.26).

$$NH_3 + H_2O \rightleftharpoons NH_4OH \qquad\qquad 11.26$$

TABLE 11.4   Equilibrium Ammonia Vapor Concentrations in a 1:3 Nitrogen:Hydrogen Gas Mixture above Liquid Ammonia at 15°C[a]

Pressure (atm)	Approximate equilibrium ammonia concentrations (%)
50	18
100	9
300	6
600	4
1000	3

[a]Data from Kirk-Othmer [18].

Before recycle of the uncondensed and unreacted gases from ammonia recovery, the inerts, mainly argon and methane, have to be bled from the system. Otherwise their concentrations build up, which dilutes the active synthesis gases in the system even to the point of blowout. "Blowout" is obtained when the heat of reaction of low residual concentrations of active gases is inadequate to sustain converter catalyst bed temperatures and the synthesis reaction fails. This dilution is intentionally practiced to control converter temperatures, which may be necessary particularly when fresh, highly active catalyst has been placed. The purge rate used ultimately determines the concentration of inerts in the system. For instance, if the purge rate corresponds to 1% of the rate of addition of make-up gas, and the make-up gas contains 0.1% inerts, then the steady-state concentration of inerts in the converter feed would be 100 × 0.1%, or 10%. At this purge rate 0.9 mol of nitrogen and hydrogen would be discarded for every 100 mol of nitrogen and hydrogen added. At a purge rate of 10% of the make-up rate the steady-state concentration of inerts in the converter drops to 1% but 9.9 moles of nitrogen and hydrogen are discarded for each 100 moles added. The hydrogen purged is not a complete loss since it may be recovered, or the hydrogen and methane

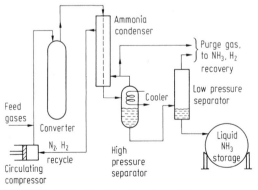

FIGURE 11.4   Simplified flowsheet of the gas recycle and product recovery systems for ammonia synthesis, from the gas mixture leaving the converter.

in the purge gas are at least used to fuel the reformers. But burned hydrogen is not as productive as converted hydrogen so that this conflict of objectives between raising the concentration of inerts fed to the converter versus raising the loss rate of the feed gases as reformer fuel represents another case for process optimization.

Occasionally an ammonia plant practices argon recovery by distillation of liquefied purge gases, particularly if air separation is used to provide at least a part of the nitrogen [19]. But when ammonia synthesis is associated with air separation, very often most of the argon and methane are removed in the liquid nitrogen scrub for carbon dioxide cleanup, giving very low concentrations of inerts in the make-up gas. Recovered argon is used as an illuminating gas for decorative lighting, as an inert filler for filament light bulbs, and as an inert blanketing gas for the shielded welding of reactive metals.

A very slight deviation from a 1:3 nitrogen to hydrogen mole ratio in the make-up gases fed to a converter can gradually build up to a significant difference from this ratio in the converter feed, because of the multiple recycle. The desired ratio is directly stabilized to some extent by the purge rate, and may also be compensated for at regular intervals by adjusting the reformer gas feed rates to affect the proportion of nitrogen to hydrogen in the make-up gas.

Liquid ammonia from the low-pressure separator will give an analysis 99.5+% ammonia. This product is led to refrigerated, insulated storage tanks. Ammonia storage capacity needs to be considerable because of the seasonal nature of its major market, fertilizers. For shipment, it is fed from storage to insulated tank trucks, tank cars, or barges for delivery. Markets and production facilities in the southwestern United States are so well developed that in this area an extensive pipeline network over 4000 km in extent has been constructed capable of anhydrous ammonia deliveries of up to 4500 tonnes per day. This ensures continuity and efficiency of supply. Smaller quantities are also shipped as ammonia solutions (ammonium hydroxide), which contain 28 to 30% ammonia by weight, or in pressurized cylinders.

11.1.6. Major Uses of Ammonia

The main use for ammonia is for fertilizer related applications, which represented some 63% of the U.S. market share in 1968 and climbed to about 80% of the total by 1973, where it has since remained (Table 11.5). The major proportion of this dramatic increase is for direct application of ammonia to the soil, either as the gas flowing under pressure to tractor-drawn knives which discharge just under the soil surface or, for smaller agricultural operations, as the 28–30% solution. The incentives for this large increase are the high nitrogen content, ca. 82%, and the strong adsorption by organic matter and clays in the soil to promote efficient retention. Ammonium nitrate, another popular nitrogen component for fertilizers, consumes ammonia both to make the nitric acid required (Section 11.2) and to form the salt from this. Fertilizer nitrogen from ammonium nitrate, however, is losing favor to urea, both in the U.S.A. and in Canada. Ammonium phosphate, ammonium sulfate, and other variations are also popular nitrogen components of fertilizers.

████ **TABLE 11.5 Trends in the Use Pattern for Ammonia in the United States since 1968**[a]

Use	Percent of total		
	1968	1973	1983
Fertilizer:			
Ammonia, aqua ammonia	6.2	27.1	25
Ammonium nitrate	21.6[b]	18.9	
Nitric acid	25.0	—	20
Urea	10.1	13.9	20
Other	—	20.5	15[d]
Industrial:			
Plastics, fiber monomers[c]	18.9	8.7	10
Explosives	[b]	4.0	5
Miscellaneous	17.8	6.9	5[e]
Total	99.6	100.0	100

[a]Compiled from data of Kirk-Othmer [2], Matasa and Tonca [13], and issues of *Chemical and Engineering News* and the *Chemical Marketing Reporter*.
[b]Includes fertilizer and explosive applications.
[c]Includes amines produced but not destined for polymer applications.
[d]As ammonium phosphate.
[e]Includes refrigerants, 2%, and water purification, froth flotation, and other important applications for small amounts of ammonia.

Smaller scale uses of ammonia are mostly indirect, such as the manufacture of explosives via nitric acid or using the new ammonium nitrate/fuel oil or metallized ammonium nitrate slurry explosive systems. Some is consumed in the preparation of monoamines and diamines as monomers for nylons, in the manufacture of acrylonitrile for Acrilan, and melamine (via urea) for the production of melamine-formaldehyde resins (Chap. 21, 23). The low cost of ammonia and its favorable compression-liquefaction properties make it ideal as the working refrigerant in large-scale freezing systems such as for frozen food processing and storage, artificial ice rinks, and the like. But its significant toxicity and flammability make it of only limited use in household refrigerators in favor of the much more costly but less toxic and nonflammable chlorofluorocarbons and derivatives. Ammonia is also employed as the cation in some sulfite-based wood pulping processes, and as a means of stabilizing residual chlorine in municipal treatment for potable water, among myriads of other small-scale applications.

11.1.7. New Developments in Ammonia Synthesis

Even with the large recent price increases for petroleum-based feedstocks it is still held that natural gas is likely to remain a significant, if not the dominant source of the hydrogen required for ammonia synthesis for the foreseeable future [20]. Production facilities that have selected and are committed to this hydrogen source have centered their attention on innovations to boost energy efficiency or on hydrogen production and recycle efficiencies to obtain operating economies.

Modifications using waste heat recycle and the more extensive use of heat exchange to simultaneously cool product streams and warm feed streams together with other improvements are anticipated to reduce the energy cost for ammonia production from 35.2 million Btu per tonne to 28.7 million Btu per tonne, a saving of some 18% [21]. A new type of ammonia converter which uses intermediate indirect cooling of the reacting gas stream between a succession of three catalyst beds, rather than quench gas cooling, (Fig. 11.3), gives a higher concentration of ammonia in the converter exit gas. This promising development has parallels with contact sulfuric acid technology. Other process refinements include adoption of two-stage centrifugal compressors driven by steam turbines for primary compression, in place of reciprocating compressors [22]. An analysis of computer control of ammonia plant operation has demonstrated that economies can be achieved through better regulation of the nitrogen:hydrogen synthesis gas ratio sufficient to write off the installation cost of the control system in as little as 1 yr [23]. Faster startups on commissioning of new ammonia plants and planned maintenance procedures can also contribute to higher actual operating capacities. Shutdowns have been surveyed as to cause and the results interpreted to minimize this source of inefficiency [24]. Recommendations have also been made to minimize turnaround time (time to get back in production again) in the event of a shutdown [25].

The efficiency of hydrogen production is being boosted by reducing the steam to carbon ratios in the steam reforming step from 4.5 to about 3, and by improving the carbon dioxide removal system [26]. Skillful control of the concentration of inerts in the recycle gas decreases both feed gas (nitrogen and hydrogen) loss as well as the loss of residual ammonia present in the purge gas when this is burned as reformer fuel. And if efficient control of purge ratios is coupled to recovery of the ammonia and hydrogen from the purge gas prior to combustion, hydrogen utilization efficiencies can be raised to about 99.5% from the 92 to 95% normally experienced [27].

It has been proposed that water be used to a greater extent than at present to provide the hydrogen requirement [28] (Eq. 11.27–11.29).

$$2\ CaO + 2\ I_2 \xrightarrow[\text{ca. }100°C]{} 2\ CaI_2 + O_2 \qquad 11.27$$

$$CaI_2 + H_2O \xrightarrow[500°C]{} CaO + 2\ HI \qquad 11.28$$

$$2\ HI \xrightarrow[300-700°C]{} H_2 + I_2 \qquad 11.29$$

This indirect route to cracking water requires 20% less heat to produce hydrogen over existing carbon-based cracking processes. Or water may be electrolyzed to provide the dominant source of hydrogen using electrolyzers built to operate at 400 atm pressure. If liquid nitrogen was used as the nitrogen source, this option to ammonia could be practised without compressors. As long as low-cost power is available for electrolytic hydrogen production this provides an interesting alternative process sequence to ammonia.

The original Haber-Bosch process relied on coal or coke to produce the hydrogen requirement but has since been largely replaced by natural gas or petroleum-based technologies. However, as these sources have become more expensive there has been a resurgence of interest in the coal-based systems 29. The dominant sequences which are emerging use coal gasification as the first stage, followed by shift conversion and gas cleanup to obtain ammonia feedstocks (e.g., Eq. 11.30).

$$
\begin{array}{l}
\text{Coal} \\
\text{gasification to} \rightarrow \\
\text{approximately}
\end{array}
\left\{
\begin{array}{ll}
CO & 31\% \\
H_2O_{(g)} & 31\% \\
H_2 & 24\% \\
CO_2 & 13\% \\
H_2 & 1\%
\end{array}
\right.
\begin{array}{l}
\text{1. shift conversion} \\
\text{2. clean up} \\
\longrightarrow \\
\text{3. N}_2 \text{ wash, addition}
\end{array}
\quad
\begin{array}{l}
\text{feedstocks} \quad 11.30 \\
N_2 : 3\ H_2
\end{array}
$$

The methanol-based Rectisol process is favored for gas cleanup because it depends on physical absorption at high pressures, and permits the recovery of separated hydrogen sulfide and carbon dioxide streams on solvent regeneration. These recovery considerations are more important with coal feedstocks because of the higher sulfur gas content in the gasifier product. Mass balance predictions certainly appear to make this alternative worth considering (Eq. 11.31), although the increased technological complexity over well-tested natural gas reforming sequences and greater emission control problems tend to be important negative factors.

$$
\left\{
\begin{array}{l}
\text{Coal} \quad\quad\quad 2250 \text{ t/day} \\
2600 \text{ tonnes/day} \longrightarrow \\
4.2\%\ S
\end{array}
\right.
\left\{
\begin{array}{l}
\text{gasified,} \\
\text{cleaned,} \longrightarrow \\
N_2 \text{ added}
\end{array}
\right.
\left\{
\begin{array}{l}
NH_3,\ 1500 \text{ t/day} \\
S,\ 107 \text{ t/day} \quad 11.31 \\
\text{wet ash, } 550 \text{ t/day}
\end{array}
\right.
$$

(+ 350 t/day burned

for steam, flue gas

scrubbed for SO_2 removal)

Solar ammonia production has been demonstrated to occur naturally in desert areas via exposure of nitrogen plus water vapor to catalytic sand grain surfaces and has been confirmed by model experiments in the laboratory [30]. The details of this photochemical water splitting reaction are being further investigated with a view to potential industrial applications (Eq. 11.32).

$$
2\ N_2 + 6\ H_2O \xrightarrow{h\nu} 4\ NH_3 + 3\ O_2 \qquad 11.32
$$

Indirect solar energy utilization via the surface to depth temperature differential existing in warm oceans has also received consideration. This would be exploited using large-scale floating facilities. All of these solar ammonia projects are experimental prospects at present.

11.1.8. Environmental Concerns of Ammonia Production

A recent analysis of the emissions and their discharge rates allows gaseous pollutant emission factors per tonne of ammonia produced to be estimated for relatively old and newer production facilities (Table 11.6).

Sulfur dioxide emissions are low when natural gas feedstocks are used in old or new facilities since sulfur compounds present are removed at the well head (see Chap. 9). Traces of sulfur dioxide present in the reformer product, however, must be removed since sulfur is one of the permanent catalyst poisons. Sulfur dioxide is normally removed by adsorption, and the sulfur dioxide discharge occurs when the adsorptive bed is heated for regeneration. The mass rate of emission with natural gas feedstocks is small, of the order of 60 kg SO_2 per day for a 1000 tonne per day ammonia plant. Hence, there has not been a need to change this purging practice for new plants. Use of fuel oil or coal for hydrogen generation will produce more associated sulfur dioxide on reforming and therefore will require integral sulfur dioxide containment [31].

Nitrogen dioxide discharge results from the oxidation of atmospheric nitrogen on combustion of reformer fuel. This occurs to some extent whenever hot nitrogen and oxygen contact heated metal surfaces, with a contribution

TABLE 11.6   Gas Emission Rates from U.S. Ammonia Production Facilities[a]

Source of emission	Emission rate, kg/tonne of ammonia produced			
	SO_2	NO_2	CO	NH_3
Old Plants[b]				
Natural gas cleaning	0.05–0.7	—	—	—
Reformer	0.03–0.3	0.6	—	—
Carbon dioxide removal	—	—	0.03	—
Copper liquor scrubbing	—	—	91.5	3.2
Ammonia synthesis	—	—	—	1.6
Ammonia loading	—	—	—	0.5
Totals	0.08–1.0	0.6	91.5	5.3
Modern Plants[c]				
Natural gas cleaning	0.05–0.7	—	—	—
Reformer	0.03–0.3	0.5	—	—
Carbon dioxide removal	—	—	0.03	—
Copper liquor scrubbing	—	—	—	—
Ammonia synthesis	—	—	—	1.6
Ammonia loading	—	—	—	0.2
Totals	0.08–1.0	0.5	0.03	1.8
Weighted average emission rates, old and modern facilities (kg/tonne NH_3 produced)	0.4	0.6	6.0	1.3

[a]Selected from summary tables of the National Research Council [1].
[b]Plants using ammoniacal cuprous solutions for carbon monoxide removal.
[c]Plants using methanation for residual carbon dioxide and carbon monoxide removal; ammonia-synthesis purge gas is burned as fuel.

from the combustion of traces of residual ammonia still present in purge gases when burned. Nitrogen oxidation also occurs, and on a much larger scale from the operation of fossil-fueled power stations. The improvement observed for modern ammonia plants results from increased reformer energy efficiency which requires less fuel to be burned per unit of hydrogen and better control of the ammonia content of purge gases. These measures plus a decrease in the excess air used for reformer combustion improve energy efficiency and decrease NO_x emissions.

Ammonia plants which use aqueous ammoniacal cuprous chloride or formate ion for final stage scrubbing to remove traces of carbon monoxide have experienced significant carbon monoxide losses when the aqueous scrubber liquor is regenerated. This can amount to about 60 tonnes of carbon monoxide per day for a 600 tonne per day ammonia plant. Only relatively slight losses occurred on regeneration of the scrubbing liquor of the carbon dioxide scrubbing circuit because of the very high selectivity for carbon dioxide of the solutions used for this step. More recently built ammonia plants which employ methanation (Eq. 11.24, 11.25) to remove the last traces of both carbon monoxide and carbon dioxide avoid the major carbon monoxide discharge source altogether. They only experience losses of the order of 30 kg of carbon monoxide per day for a 1000 tonne per day facility from traces that are captured in the carbon dioxide scrubber circuit.

Ammonia losses from the synthesis stage primarily arise from purge gas combustion and leakage. Little change is seen in the loss rate from this area with newly adopted technology. But as more facilities adopt purge gas ammonia and hydrogen recovery systems the loss rate from this operating area should decrease [22]. Improved loading techniques, including provision of ammonia line vapor recompression and continuous control of storage tank vapors have resulted in the decreased losses. Realistic guidelines of daily average ammonia discharge rates of 1 kg per tonne (design) and 1.5 kg per tonne (operating) have recently been set for one jurisdiction compatible with the loss rates outlined above [32].

Water impacts from ammonia producing facilities can occur through thermal loading, from discharge of large volumes of cooling water, or from the discharge of dilute ammonia solutions which may also contain organics from recovered process condensates. Lagooning, forced-air cooling, and increased use of heat exchange of hot exit gases to preheat entering raw materials are among the remedies for thermal impacts (Chap. 3). Steam stripping may be used to recover ammonia from dilute aqueous waste streams. However, this is a relatively expensive, energy-intensive procedure for the small quantities likely to be recovered. A better expedient might be to use these streams, which typically contain 400–2000 ppm ammonia, to simultaneously irrigate and fertilize local croplands. Permissible and desirable criteria for ammonia in public water supplies have been set at 0.5 and 0.01 mg L^{-1}, respectively [33]. These standards limit the acceptable volumes of such waste streams which can be discharged into surface waters.

Noise problems in large ammonia plants from the movement of large gas volumes, the operation of compressors and the like, mainly affect operating employees but have nevertheless received attention to meet recently placed noise criteria. Procedures to locate and localize or dampen (attenuate) noise sources which produce sound levels in excess of 90 dBA have been described [34].

11.2. NITRIC ACID

11.2.1. Background

Nitric acid (*aqua fortis*) was known and its chemistry practiced in the Middle Ages. It was obtained by heating hydrated copper sulfate or sulfuric acid with sodium nitrate (saltpeter or niter) and cooling the vapors generated to obtain a solution of nitric acid [5] (Eq. 11.33).

$$NaNO_3 + H_2SO_4 \xrightarrow{\text{ca. } 200°C} NaHSO_4 + HNO_3 \qquad 11.33$$
$$\text{"niter"} \qquad\qquad\qquad \text{"niter cake"}$$

While it is theoretically possible to obtain 2 mol of nitric acid for each mole of sulfuric acid used, in practice the 900°C temperature required to achieve this makes this stoichiometry impractical. However, it was possible to use the residual acid value of the niter cake to make hydrochloric acid from salt (Eq. 11.34).

$$NaHSO_4 + NaCl \rightarrow Na_2SO_4 + HCl \qquad 11.34$$
$$\text{"niter cake"} \qquad\quad \text{"salt cake"}$$

For nitrogen fertilizers and for the nitric acid requirement for the manufacture of explosives in the nineteenth century, natural sodium (or potassium) nitrate provided the only source. India produced some 30,000 tonnes per year by 1861 and Chile, in 1870, was exporting some 90,000 tonnes annually which gradually climbed to about 1.4 million tonnes per year, in both cases as sodium nitrate.

Present-day nitric acid production is almost entirely via the oxidation of ammonia and absorption of the oxidation products in water. The chemistry of this process was proven experimentally by Kuhlmann in 1839, but had to wait for the development of an economical route to ammonia before it could become commercially significant [35]. Ostwald, working in Germany in about 1900, reexamined and extended Kuhlmann's data and established the proper conditions required for the ammonia oxidation step. Very shortly after this, operating processes based on these principles were assembled both in Germany and in the United States and since then production levels have gradually risen, so that in the United States since 1980 7 to 8 million tonnes (100% basis) of nitric acid have been produced annually (Table 11.7).

11.2.2. Nitric Acid by Ammonia Oxidation, Chemistry and Theory

Modern nitric acid production amounts to a catalytic oxidation of ammonia in air followed by absorption of the secondary oxidation products in water to yield nitric acid. Unlike the reaction sequence for ammonia feedstock preparation, all of the reaction steps in this sequence are exothermic.

TABLE 11.7   Production of Nitric Acid (100% Basis) by Selected Countries[a]

| | Thousands of metric tonnes | | | | | |
	1965	1970	1975	1980	1990	1995
Australia	25	126	174	180		
Belgium	616	718	794	1167		
Canada	344	503	365	713	965	991
Finland	5	261	398	422		
France	2319	2575	3287	—		
W. Germany	2599	3254	3035	3173	1880	2354[b]
Italy	946	1038	967	1011	1040	578
Japan	246	474	553	577		
Sweden	190	270	337	346		
United Kingdom	—	—	2550[c]	2825		
United States	4444	6897	6418	8102	7033	7822
World	15,260	20,681	24,852			

[a]Compiled from Kirk-Othmer [2], *Chemical and Engineering News* [3], and *Statistics Canada* [36], plus data supplied by the Verband der Chemischen Indusrie e.V. (VCI). U.S. production in thousands of tonnes for earlier years was 1960, 3007; 1965, 4445; 1970, 6899.
[b]After reunification.
[c]Estimated, from 1976 data.

Initial ammonia oxidation is conducted using a mixture of ammonia gas (9–11%) in air which is passed through multiple layers of fine platinum-rhodium alloy gauze (Eq. 11.35).

$$4\ NH_{3(g)} + 5\ O_{2(g)} \xrightarrow[\text{ca. } 900°C]{Pt/Rh} \underset{\text{nitric oxide}}{4\ NO_{(g)}} + 6\ H_2O_{(g)} \qquad 11.35$$

$$\Delta H = -907\ kJ\ (-217\ kcal)$$

While this equation represents a very close approximation of the stoichiometry of the process under these conditions, the detailed chemistry is complex and poorly understood. The exotherm of the oxidation is sufficient to cause a temperature rise of about 70°C for each 1% of ammonia in the mix, so that with prewarming of one or both component streams it can keep the alloy gauze at close to the optimum 900°C [37]. Operating at a gauze temperature of 500°C produces mostly the relatively unreactive nitrous oxide as the ammonia oxidation product (Eq. 11.36) which would be lost to acid production.

$$2\ NH_{3(g)} + 2\ O_{2(g)} \xrightarrow[\text{slow}]{500°C} \underset{\text{nitrous oxide}}{N_2O + 3\ H_2O} \qquad \Delta H = -276\ kJ\ (-66\ kcal) \qquad 11.36$$

This is the motivation to provide conditions which maintain gauze temperatures of about 900°C.

It may seem to be a backward step to have prepared ammonia from nitrogen and hydrogen, only to turn around and burn the ammonia to obtain nitric oxide. However, if one considers the endothermic nature of the direct combination reaction between nitrogen and oxygen (Eq. 11.37) it becomes apparent why nitric oxide yields, even at carbon arc temperatures of around 3000°C, are not more than a few volume percent.

$$N_{2(g)} + O_{2(g)} \rightarrow 2\ NO_{(g)} \qquad \Delta H = +90.3\ kJ\ (+21.6\ kcal) \qquad 11.37$$

Using ammonia as the starting material for nitric oxide preparation bypasses the enormous energy requirement for elemental nitrogen bond dissociation, and in so doing achieves the product of interest via a net exothermic (thermodynamically favorable) process.

Nitric oxide obtained from ammonia combustion is then further oxidized to nitrogen dioxide in another, less exothermic step (Eq. 11.38).

$$2\ NO_{(g)} + O_{2(g)} \xrightarrow[\text{slow}]{} 2\ NO_{2(g)} \qquad \Delta H = -113\ kJ\ (-27\ kcal) \qquad 11.38$$

The nitrogen dioxide obtained immediately participates in a relatively rapid equilibrium to dinitrogen tetroxide (Eq. 11.39).

$$2\ NO_{2(g)} \rightleftharpoons N_2O_{4(g)} \qquad \Delta H = -57.4\ kJ\ (-13.7\ kcal)$$
$$\text{brown} \qquad \text{colorless}$$
$$\text{m.p.}\ -11.2°C$$
$$\text{b.p.}\ \ 21.2°C$$

11.39

At 100°C the equilibrium composition lies at approximately 90% NO_2, 10% N_2O_4 whereas at 21°C only about 0.1% NO_2 is present [38]. This equilibrium may be important for acid production since there is evidence that dinitrogen tetroxide is the molecular species that reacts with water for acid formation.

The final step in acid formation is absorption of nitrogen dioxide in water the net result of which is given by Eq. 11.40.

$$3\ NO_{2(g)} + H_2O_{(\ell)} \rightarrow 2\ HNO_{3(aq)} + NO_{(g)} \qquad \Delta H = -139\ kJ\ (-33.1\ kcal)$$
$$11.40$$

This is not as straightforward a transformation as it may at first appear since it involves both chemical combination with water and a net redox reaction for the nitrogen (4+) in nitrogen dioxide to nitrogen (5+) in nitric acid. One nitrogen dioxide thus takes up two electrons (is reduced) to become nitrogen (2+) in nitric oxide to balance this oxidation. The actual chemistry involved is probably better represented as, first of all, a formation of both nitric and nitrous acids (Eq. 11.41).

$$2\ NO_2 + 3\ H_2O \rightarrow HNO_3 + HNO_2 \qquad 11.41$$

The nitrous acid, being unstable in the presence of any strong mineral acid, disproportionates to yield further nitric acid and nitric oxide (Eq. 11.42).

$$3\ HNO_2 \rightarrow HNO_3 + 2\ NO + H_2O \qquad 11.42$$

Not all of the nitrogen in the ammonia feed ends up as nitric acid. Some of the ammonia reacts with oxygen of the air to yield elemental nitrogen from ammonia nitrogen, and some reacts with nitric oxide causing loss of potential product from both ammonia and oxidant consumption (Eq. 11.43 and 11.44).

$$4 \, NH_{3(g)} + 3 \, O_{2(g)} \rightarrow 2 \, N_{2(g)} + 6 \, H_2O_{(g)} \quad \Delta H = -1267 \, kJ \, (-303 \, kcal)$$

$$11.43$$

$$4 \, NH_{3(g)} + 6 \, NO_{(g)} \rightarrow 5 \, N_{2(g)} + 6 \, H_2O_{(g)} \quad \Delta H = -1806 \, kJ \, (-432 \, kcal)$$

$$11.44$$

Both reaction 11.43 and 11.44 are highly favored thermodynamically, even relative to the desired reaction, Eq. 11.37. The platinum gauze catalytic surface serves to boost the proportion of nitric oxide obtained and decrease nitrogen formation by these pathways, probably by accelerating the rate of the desired reaction without affecting the rates of the others. Use of multiple platinum-rhodium or platinum-palladium alloy gauzes further raises the ammonia conversion to 99% (for 8.3% in air; gauze temperature 930°C), over the 97.5% accessible with platinum alone. Use of alloy gauzes also decreases the metal erosion rate.

The important aspect of the ammonia oxidation reaction, Eq. 11.35, from a process design standpoint is that since 9 mol react to produce 10 mol of product, there is little volume change from this step. Therefore, the equilibrium of this reaction will be little affected by a change in operating pressure. But an increase in pressure pushes more material through the same sized equipment in the same time, lowering the capital cost of the plant per unit of product produced. If the pressure increase is moderate, say, 5 to 10 atm, the metal wall of the vessels used does not have to be significantly thicker to contain it, and this modification provides a further process benefit by slightly increasing the concentration of the product nitric acid from the 50–55% obtained when operating at atmospheric pressure, to 57–65% when under pressure.

Increasing the pressure of the feed gases to the ammonia converter also effectively increases the space velocity through this unit, which slightly decreases the conversion efficiency for ammonia oxidation. But this reaction is so rapid, being essentially complete in 3×10^{-4} sec at 750°C, that additional layers of catalytic gauze *almost* compensate for this stage of the process.

The nitric oxide oxidation and water absorption reactions, Eq. 11.38 and 11.40, are both much slower than the ammonia oxidation reaction, Eq. 11.35, and involve a significant volume decrease on reaction, 3 mol (volumes) to 2, and 3 mol to 1, respectively. Thus the absorbers, where the bulk of these reactions occur, must be large to provide sufficient residence time, and cooled to favor the equilibria in the desired direction. In the absorbers, because of this volume decrease on reaction, raising the pressure achieves a significant improvement in performance, in accord with Le Chatelier's principle. For an increase to 8 atm from 1, the rate of the very slow nitric oxide reoxidation reaction in particular (Eq. 11.45) is accelerated by a factor of the cube of this pressure increase, or 512 times [39].

$$\text{rate} = k_1(P_{NO})^2(P_{O_2}) - k_2(P_{NO_2})^2 \qquad 11.45$$

For an operating process absorber, normally not very close to equilibrium, the rate of the reverse reaction (the second term in this rate equation) can be ignored. The pressure increase also boosts the rate of nitrogen dioxide absorption, once formed, in water. In practice this pressure increase means that the absorber volume required per daily tonne of nitric acid production capacity can be decreased from about 35 m³ to about 0.5 m³ [40].

11.2.3. Process Description

To put these principles into practice, liquid ammonia is first vaporized by indirect heating with steam, and then filtered to reduce risk of catalyst contamination. This produces an ammonia gas stream at about 8 atm pressure without requiring mechanical compression. An air stream is separately compressed to about the same pressure, preheated to 200–300°C, and filtered prior to mixing with the ammonia (about 10%) gas stream immediately before conversion. This mixture is passed through the red hot platinum-rhodium gauze to produce a hot gas mixture of nitric oxide and water vapor plus the unreacted nitrogen and oxygen components of air (Fig. 11.5), with a yield efficiency (selectivity) under these conditions of about 95%.

Hot converter gas products are cooled in a waste heat boiler prior to absorption, to give simultaneous production of steam and a product gas stream at more moderate temperatures ready for catalyst recovery. The high-pressure nitric acid process, as described here, experiences a gauze metal loss rate of 250–500 mg/tonne of acid produced. This can be kept to the lower end of the range by efficient filtration. If operated at atmospheric pressure, catalyst loss rates amount to about 50 mg/tonne. After filtration the gas mixture is quickly chilled by heat exchange with process water to condense some dilute nitric acid from the product gases. The dilute nitric acid condensate is also filtered for further catalyst recovery and then trickled into the top of the absorber to raise the acid concentration. The gas stream still contains nitric

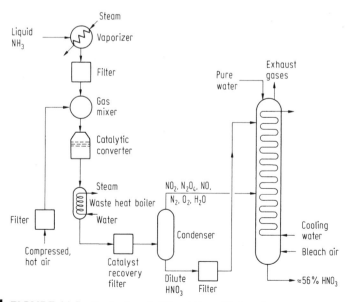

FIGURE 11.5 Production of nitric acid by oxidation of ammonia.

oxide, but with the elapse of time and having been cooled now also contains nitrogen dioxide and dinitrogen tetroxide (Eq. 11.38, 11.39) plus the residual air. This stream is fed into the absorber at an intermediate level.

In the absorbers nitric oxide is oxidized to nitrogen dioxide at the same time as nitrogen dioxide is absorbed in water to give nitric acid plus further nitric oxide. This is not lost but is being continuously reoxidized to nitrogen dioxide like the primary nitric oxide. Cooling water keeps the temperatures down to maintain favorable oxidation and solution equilibria. As the concentration of acid builds up to equilibrium levels in the downward flow of absorption water some dissolved (but unreacted) nitrogen dioxide may also be present, which gives the acid a brown color. This nitrogen dioxide is stripped from the product acid by purging the last few plates of the absorber with "bleach air," to give a colorless to pale yellow product acid containing 60–62% by weight HNO_3 as the bottom product of the absorber. The bleach air, at the same time as stripping excess dissolved nitrogen dioxide from the product, also assists in maintaining an oxygen excess in the middle and upper reaches of the absorber which favors the formation of nitrogen dioxide from nitric oxide. Residual gases are vented from the top of the absorber via an expander turbine which recovers energy from the vented gas sufficient to provide about 40% of the power required for driving the inlet air compressor mounted on a common shaft.

11.2.4. Nitric Acid Concentrations and Markets

The ammonia oxidation product is sold in several technical and commercial market grades ranging from about 50% (by weight) HNO_3, up to what is referred to by chemists as ordinary concentrated nitric acid (68%) which is the maximum boiling point or azeotropic composition (Table 11.8). These are the concentrations obtained either directly from the various types of nitric acid plant or obtained by raising the concentration of the acid plant product by simple distillation. The concentrated product ranges in color from water

TABLE 11.8 Industrial Market Grades of Nitric Acid

Degrees Baumé[a]	Density (g/mL at 20°C)	Nitric acid concentration (%)
36	1.330	52.3
38	1.3475	56.5
40	1.3743	61.4
42	1.4014	67.2, ordinary conc. nitric
—	1.4521	80.0
—	1.4826	90.0, or higher, is fuming nitric
—	1.4932	95.0
—	1.5492	100.0, anhydrous nitric, unstable[b]

[a]°Bé, an arbitrary specific gravity scale base on density, where specific gravity = $145/(145 − °Bé \text{ reading})$.

[b]Decomposes to NO_2, NO, and water above the freezing point. Hence, fuming grades are more useful at somewhat lower concentrations.

white, through yellow to brown, as the dissolved nitrogen dioxide content gradually rises in the product with time from the operation of the slow equilibrium reaction, Eq. 11.46.

$$4 \text{ HNO}_3 \rightleftharpoons 2 \text{ H}_2\text{O} + 4 \text{ NO}_2 + \text{O}_2 \qquad\qquad 11.46$$

These grades or concentrations are used in the production of ammonium nitrate, which at present consumes some 65–75% of the total nitric acid made [41], as well as for many of the other more minor uses. Some of these are, for example, adipic acid production (5–7%), military and industrial explosives (3–5%), isocyanates for polyurethane manufacture (1–2%), nitrobenzene (1–2%), and potassium nitrate preparation from the chloride (about 1%). A miscellany of even smaller scale uses consumes the remaining 10 to 15% of the total acid produced.

For forcing nitrations and other purposes which require concentrations of nitric acid above 68% by weight, special techniques have to be used since this is the maximum concentration which may be achieved by simple distillation. Nitric acid concentrations above 86% comprise ordinary concentrated nitric acid plus dissolved nitrogen dioxide. These concentrations are dark brown in color, and tend to lose the dissolved nitrogen dioxide relatively rapidly if open to the air. For this reason these are collectively referred to as "fuming nitric acid" and require considerable caution in their use because of their significantly greater nitration and oxidation reactivities. A standard commercial fuming nitric is sold containing 94.5–95.5% HNO_3.

One method used commercially to obtain nitric acid concentrations above 68%, particularly by larger facilities, is to dehydrate the azeotropic composition or slightly less than this, using concentrated sulfuric acid. Countercurrent passage of hot nitric acid vapor against the dehydration acid in a tower packed with chemical stoneware achieves direct dehydration to 90+% HNO_3 and produces a diluted sulfuric acid stream (Fig. 11.6). If this concentration process is only practiced on a small scale, the sulfuric acid may be reconcentrated by addition of oleum, and a portion of the buildup of sulfuric acid in this circuit may be used to formulate a commercial nitrating mixture with some of the fuming nitric acid made. Operation of the concentrator on a larger scale, however, requires the use of a sulfuric acid boiler to reconcentrate the dehydration acid. Thus, the disadvantages of sulfuric acid dehydration are the

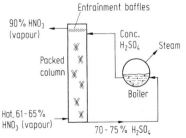

FIGURE 11.6 Dehydration of the nitric acid from an ammonia oxidation plant to fuming grades using concentrated sulfuric acid.

large heat input required to reconcentrate the high boiling point sulfuric acid, and the noticeable sulfate contamination of the nitric acid product from the direct contact dehydration.

Another procedure also used for production of fuming nitric which employs 72% aqueous magnesium nitrate ($Mg(NO_3)_2 \cdot 4H_2O$) as a dehydrating agent gets around some of these difficulties [42]. As in sulfuric acid dehydration, hot nitric acid vapor is fed countercurrently to hot, molten 72% $MgNO_3$, this time in a multiple-plate column of stainless steel, to achieve similar results. But here, with the common anion of the dehydrating agent and nitric acid, there is no sulfate contamination and the diluted magnesium nitrate product (approx. $Mg(NO_2)_2 \cdot 6H_2O$; m.p. 89°C) from the bottom of the column requires less heat for reconcentration than sulfuric acid.

11.2.5. Nitric Acid Process Variants and New Developments

Absorption trains of early ammonia oxidation processes to nitric acid were constructed of chemical stoneware or acid-proof brick, which restricted acid production to near normal atmospheric pressure because of the low strength of the structural materials. The discovery that Duriron (silicon-iron) or high chrome stainless steels could tolerate these corrosive conditions well allowed the adoption of pressure absorption. This measure markedly decreased the size of the absorbers required and reduced nitrogen oxide stack losses over what was possible at atmospheric pressure. Pressure operation was easiest to achieve by compression of the feed gases at the front end of the process. In this way acid production is obtained at very comparable capital costs per unit of product as obtained by operation at atmospheric pressure.

However, an attendant cost of pressure operation throughout the process is nitric oxide yields of only about 90% from the ammonia combustion stage, by recent improvements raised to about 95%, as compared to yields of 97 to 98% possible at atmospheric pressure. For this reason atmospheric pressure processes are still viable and are still operating in North America and elsewhere. For the greater ammonia oxidation efficiencies possible at atmospheric or even slightly negative pressures and to retain the much more efficient absorption at elevated pressures many European producers use split-pressure processes [43]. This combination, which makes more efficient use of ammonia than of either of the other technologies discussed, is of particular value when ammonia costs are high. But, in general, because of the need to construct the compressor for raising the pressure of oxidized converter gases out of stainless steel or other more exotic metals, the capital costs of such split-pressure plants are 1.5 to 2 times the cost of the other two alternatives. Nevertheless, the worldwide escalation in ammonia costs has led to the adoption of the split-pressure process for some new facilities being constructed in Canada.

A relatively recent development in catalyst support systems, either replacing half of the normal 5–10% rhodium-platinum alloy gauzes by non-noble metal supports or by moving to completely non-noble metal catalysts, is said to result in economies in catalyst cost without adversely affecting operating efficiency [44].

Technologies for concentrating nitric acid have been developed to circumvent the azeotropic restrictions to distillative preparation by obtaining a much higher concentration nitric oxide, nitrogen dioxide gas stream from the acid condenser, of the order of an NO$_x$ vapor pressure of 2–3 atm [41]. Absorption of this gas in cold concentrated (68%) nitric acid yields a superazeotropic product containing 80% or more nitric acid. Distillation of this now gives an overhead stream of 96–99% nitric acid, requiring only six theoretical plates for separation from a column bottom stream of the azeotropic composition (Fig. 11.7). Brief operating details of this Espindesa process are available [45].

11.2.6. Emission Control Features

The chief environmental problem of nitric acid plant operation is discharge of residual nitric oxide and nitrogen dioxide from the vent stack of the absorber. For a high-pressure process of the 1960s without the emission controls described here, the total concentration of nitrogen oxides discharged typically amounted to about 0.3% by volume [33]. A more detailed breakdown is given in Table 11.9. Nitric acid plants are not the only source of nitrogen oxides (NO$_x$) but they do correspond to a relatively large point source of emissions, unlike the more diffuse discharge resulting from automobiles. The need to regulate NO$_x$ discharges arises from the implication of nitrogen oxides in photochemical air pollution problems (Chap. 2). The toxic effects of nitrogen dioxide itself may be felt during photochemical smog episodes.

Nitrogen dioxide is a brown irritating gas which exists in equilibrium with dinitrogen tetroxide at ordinary ambient conditions (Eq. 11.39). It is dangerous to humans at concentrations above 50 ppm [48]. An industrial hygiene standard of 5 ppm for an 8-hr workday has been set. Colorless nitric oxide is not in itself an irritant, but at high concentrations (1000–2000 ppm) in air it can cause loss of consciousness and convulsions. Even though nitric oxide *itself* is not frequently a problem, and its discharge is not immediately visible,

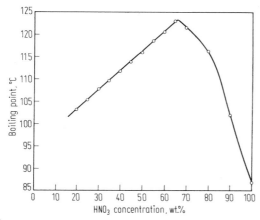

FIGURE 11.7 Plot of the boiling point of nitric acid versus concentration.

TABLE 11.9   Typical Composition of Nitric Acid Plant Vent Gases, Prior to Expansion (At about 30°C, 7.3 Atm)[a]

	Percent, without controls	Percent with methane catalytic reduction
Nitric oxide	0.05–0.20	0.01–0.20
Nitrogen dioxide	0.05–0.20	
Oxygen	3.0	0.5–2.0
Water	0.6–0.7	2.5–5.0
Nitrogen	96+	92–93.5
Nitric acid	<0.0	<0.0
CO_2	0.0	0–2.0

[a]Compiled from the data of Lund [33], Gillespie *et al.* [46], and Boyars [47].

it does indirectly contribute to atmospheric nitrogen dioxide loadings because the nitric oxide in the vent gases is relatively rapidly oxidized to nitrogen dioxide on exposure to the additional oxygen present in air (Eq. 11.38). And nitrogen dioxide, if present in vent gases above about 100 ppm (by volume), gives a visible brown discharge plume from the vent stack.

First-generation emission control measures centered on catalytic reduction of nitrogen oxides using methane (natural gas) or hydrogen (Fig. 11.8). For methane reduction, tail gases from the absorber vent are preheated to about 400°C, and then blended with the appropriate proportion of methane before passage over platinum or palladium catalytic surfaces for reduction [33]. The concentration of nitrogen oxides is decreased by about 90% (Table 11.9), from about 0.3% to 0.01–0.2% (by volume) depending on conditions, from the reactions of Eq. 11.47 to 11.49.

$$\underset{\text{generate heat}}{} \quad CH_4 + 2\ O_2 \xrightarrow{\text{cat.}} CO_2 + 2\ H_2O \qquad\qquad 11.47$$

$$CH_4 + 4\ NO_2 \rightarrow 4\ NO + CO_2 + 2\ H_2O \qquad\qquad 11.48$$
$$\text{brown} \qquad \text{colorless}$$

$$CH_4 + 4\ NO \rightarrow 2\ N_2 + CO_2 + 2\ H_2O \quad \text{slow} \qquad 11.49$$

Methane is first rapidly oxidized by the excess oxygen normally present in the vent gases (for reoxidation of nitric oxide to nitrogen dioxide in the absorbers) which generates heat but does nothing to alleviate emissions. Nitrogen dioxide is also rapidly reduced to nitric oxide which at least serves to remove the brown color from the vent stack discharge, a "decolorization" level of control. And if the reduction unit is large enough to provide adequately low space velocities for the much slower nitric oxide reduction reaction (Eq. 11.49), and the added methane is also sufficient to do this then true "abatement" is achieved, at additional increments of capital and fuel cost [46] (Fig. 11.8). Fortunately, some of the additional fuel cost of either level of control may be recovered from steam generation and from the energy boost provided to the tail gas expander by the heated, rather than near ambient, temperature exhaust gas discharge.

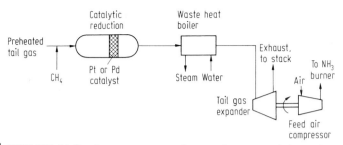

FIGURE 11.8 One emission control system for nitric acid plant tail gas cleanup via catalytic reduction.

The present regulatory requirements for nitric acid plant discharges in the United States now stand at 1.5 kg of NO_x (as NO_2 equivalent) per tonne of acid equating to about 230 ppm NO_x, and the discharge must be colorless (opacity of 10% or less) for new plants. Older plants are likely to be allowed a maximum of 400 ppm (by volume) NO_x [49]. For European jurisdictions the requirements vary but generally lie near the proposed requirements for older U.S. plants, whereas the U.S.S.R., probably because of either very large single-train facilities or because of a concentration of several fertilizer works in proximity, has set a much tighter requirement of 0.55 kg NO_x (calculated as NO_2) discharge rate per tonne of acid.

With careful operation the methane catalytic reduction route is capable of meeting the more stringent regulatory requirements. Capital costs dictate the maximum economic size of absorbing towers so that early designs were only sufficient to achieve 0.1 to 0.3% NO_x in the tail gases. New plants with larger absorbers and employing refrigeration chilled water systems particularly on the upper trays of the absorbers are able to reach tail gas NO_x concentrations of <200 ppm directly. Use of booster compressors to permit absorption at high pressures has also been found to be an economic measure to improve absorption efficiencies, at least for very large plants.

From a raw material utilization standpoint as well as for fuel conservation (methane or hydrogen) reasons, discharge controls integral with the process absorbers have obvious advantages for new plants. Older facilities which have to meet postconstruction regulations have a choice of add-on units. Nitric oxide and nitrogen dioxide may be reduced by ammonia using alumina-supported platinum catalysts without requiring sufficient ammonia to consume the residual oxygen [50] (Eq. 11.50, 11.51).

$$6\ NO + 4\ NH_3 \rightarrow 5\ N_2 + 6\ H_2O \qquad\qquad 11.50$$

$$6\ NO_2 + 8\ NH_3 \rightarrow 7\ N_2 + 12\ H_2O \qquad\qquad 11.51$$

This has been shown to be a feasible option to confer fuel and capital cost savings while meeting stringent regulatory requirements. The processing train used is very similar to that given for methane reduction (Fig. 11.8).

Physical absorption in scrubbers which employ water or nitric acid has also been employed, as has chemical absorption by alkaline solutions, or solutions of urea in water. All can reduce tail gas NO_x concentrations to below 200 ppm [51]. Physical adsorption on molecular sieves is also a feasible control method. Activated carbon cannot be used because of oxidation hazards. Molecular sieve and liquid physical adsorption was improved if both nitric oxide and nitrogen dioxide were present in the tail gas rather than just nitric oxide, perhaps from the formation of dinitrogen trioxide (N_2O_3). An advantage of the use of any of these physical adsorption systems is that the NO_x collected is recoverable on regeneration of the adsorbent.

Comparative summaries of abatement measures have been published [49, 50].

11.3. COMMERCIAL AMMONIUM NITRATE

11.3.1. Background

Production of ammonium nitrate is the largest single end use for nitric acid and has ranked 10th to 12th in volume of all inorganic chemicals produced in both Canada and the Unites States over the last 10 years (Table 11.10). This high volume has been stimulated by the demand for ammonium nitrate as a fertilizer component and as an ingredient in explosives, combined with its simplicity of production and low cost, about $170 per tonne in the United States in 1995. Fertilizer value is derived both from the high total nitrogen content (~33%) split between fast acting nitrate nitrogen and the slower acting ammonium nitrogen, plus the ease of marketing a solid product.

Explosives applications have been varied from blends with nitroglycerin to make this safer for use in mines, to admixtures with smokeless powders and nitro compounds such as trinitrotoluene, to the more recent developments of blends of low-density ammonium nitrate prills with fuel oil (so-called ANFO explosives), or in water slurries with metal powder boosters producing low-cost "heaving" power [54]. All explosives applications rely on the large

TABLE 11.10   Production of Ammonium Nitrate and Urea in Selected Countries[a]

	Thousands of tonnes				
	1970	1975	1980	1990	1995
Canada: NH_4NO_3		875	873	1031	1055
Urea		329	1274	2490	2550
Japan: NH_4NO_3	37	115	123	—	—
Urea	2463	3162	1770	776	755
U.S.A.: NH_4NO_3	5938	6430	7792	6425	7255
Urea	2829	3446	7103	7368	7074

[a]Compiled from FAO UN [52], Japan economic yearbooks [53], data supplied by Verband der Chemischen Industrie e.V. (VCI), and issues of *Chemical and Engineering News*.

volume of gas released on vigorous detonation of the ammonium nitrate component (Eq. 11.52).

$$2\ NH_4NO_{3(s)} \rightarrow 2\ N_{2(g)} + 4\ H_2O_{(g)} + O_{2(g)} \qquad 11.52$$

The oxygen is consumed by the fuel oil or a readily oxidizable metal such as aluminum or magnesium present in the explosive formulation to contribute further heat and power to the detonation process. Means to improve the normal handling of ammonium nitrate to decrease the explosion risk under ordinary shipping and storage conditions have been proposed [55].

Under milder conditions a small-scale but important application is the preparation of inhalation-grade nitrous oxide for use as an anaesthetic by heating pure ammonium nitrate from about 200 to 250°C (Eq. 11.53).

$$NH_4NO_3 \rightarrow N_2O + 2\ H_2O \qquad 11.53$$

Using this method the anaesthetic "laughing gas" is obtained, virtually free of nitric oxide or nitrogen dioxide.

11.3.2. Production of Ammonium Nitrate

Early methods employed simple neutralization of concentrated aqueous ammonium hydroxide by economic concentrations of nitric acid in the correct proportions, followed by evaporation of much of the water and crystallization of the product. However, because of high capital, labor, and energy costs of this process relative to newer procedures, this batch process is being superseded by the Fauser process or variations. The Fauser process contacts ammonia gas with concentrated nitric acid and uses the heat of reaction to evaporate a part of the original water [14]. The Stengel process uses sufficiently high preheat temperatures for ammonia and nitric acid such that complete evaporation of the residual water occurs when the heat of reaction is added to this [16].

In the Stengel process ammonia gas preheated to 145°C is blended with about 60% nitric acid preheated to 165°C under pressure in a packed stainless steel reactor (Fig. 11.9). This is now a routine procedure, yet the mixing of a

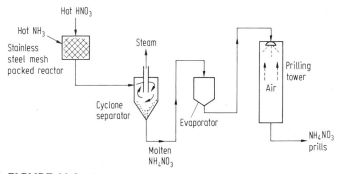

FIGURE 11.9 Ammonium nitrate by the Stengel process. Intermediate evaporation before prilling or granulation would be employed for the production of a high-density product.

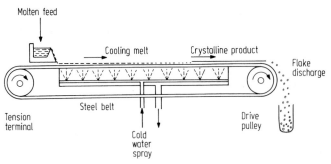

FIGURE 11.10   Steel belt cooler for the preparation of a flaked ammonium nitrate (or other) product.

hot strong oxidant with a hot oxidizable material is a process that is fundamentally unnerving to a chemist. Doing this provides sufficient additional heat from the neutralization reaction to nearly completely evaporate the residual water [56] (Eq. 11.54).

$$NH_{3(g)} + HNO_{3(aq)} \rightarrow NH_4NO_{3(aq)}$$
$$\Delta H = -86.1 \text{ kJ } (-20.6 \text{ kcal})$$

11.54

The steam produced is removed from the ammonium nitrate (m.p. 170°C) via the vortex finder of a cyclone separator, and the 99+% molten salt proceeds to either a cooled stainless steel belt to produce a flaked product (Fig. 11.10), or to a 30-m prilling tower where passage of the droplets of melt through a countercurrent dry air stream produces shot-sized prills (beads) of ammonium nitrate. Concentrated hot solutions of ammonium nitrate are explosively sensitized by the presence of traces of acid so that care is taken in these latter stages of working to add sufficient ammonia to the wet melt to keep the pH above about 5.

The product is hygroscopic so that it is often coated with an anticaking compound such as clay or diatomaceous earth before being placed in bulk storage in large moisture-tight silos. Bulk shipment is by covered gondola rail cars and in smaller quantities in vapor tight bags of polyethylene or waxed paper.

The process itself and environmental aspects of this have been analyzed for efficiency of operation and air and water pollution control measures [57].

11.4. PRODUCTION OF UREA

The first synthesis of urea was conducted by Wohler in 1828 via the heating of ammonium cyanate (Eq. 11.55).

$$NH_4OCN \rightarrow H_2NCONH_2$$

11.55

Not only did this preparation mark the first synthesis of urea, but it provided a landmark by being the first time that a compound produced by living things was synthesized in the laboratory. As such this event served to open up the study of organic compounds, which was formerly almost sacrosanct, to emerge today with a vast area of products and services which are based on this area of chemistry.

Production of urea on a commercial scale did not assume much significance until about 1930. Early processes were based on calcium cyanamide hydrolysis in the presence of a base (Eq. 11.56, 11.57).

$$CaNCN + 2\ H_2O \rightarrow Ca(OH)_2 + H_2NCN \qquad 11.56$$

$$H_2NCN + H_2O \xrightarrow{OH^-} H_2NCONH_2 \qquad 11.57$$

Since about 1950 most urea producing units have been based on ammonia-carbon dioxide feedstocks put through high-pressure equipment. Over the last 10 yr in North America urea production volumes have grown faster than ammonium nitrate for the supply of fertilizer nitrogen (Table 11.10).

Urea from ammonia and carbon dioxide in the various process sequences operated requires, first of all, reaction of these feedstocks under a pressure of 100 to 200 atm to form ammonium carbamate (Eq. 11.58).

$$NH_{3(g)} + CO_{2(g)} \rightarrow \underset{\substack{\text{carbamic} \\ \text{acid}}}{H_2NCOOH} \xrightarrow{NH_3} \underset{\substack{\text{ammonium} \\ \text{carbamate}}}{H_2NCOONH_{4(s)}} \qquad 11.58$$

$$\Delta H_{25°} = -159\ kJ\ (-38\ kcal)$$

This thermodynamically favorable step of the process is followed by an endothermic thermal decomposition of ammonium carbamate, in a concentrated solution, to give a 50–60% conversion to urea (Eq. 11.59).

$$H_2NCOONH_4 \rightarrow H_2NCONH_2 + H_2O$$
$$\Delta H = +31.4\ kJ\ (+7.5\ kcal) \qquad 11.59$$

A "once-through" modification of a urea process is shown in Fig. 11.11. Recycle of unconverted ammonium carbamate and of ammonia and carbon

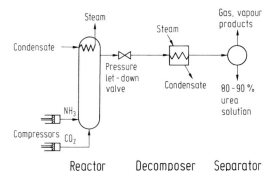

FIGURE 11.11 Production of urea by a simple, one-pass reactor-separator system.

dioxide formed in part from ammonium carbamate decomposition (Eq. 11.60) is practiced in other variations on this theme [58].

$$H_2NCOONH_4 + H_2O \rightarrow (NH_4)_2CO_3 \rightarrow 2\ NH_3 + CO_2 + H_2O \qquad 11.60$$

This assists in minimizing the environmental impacts of operation of urea plants. However, despite these recycle measures there is still a small ammonia loss rate, estimated to average 0.6 kg per tonne of urea produced [1]. This loss rate is, however, well below one published guideline for urea plants of 2.7 kg NH_3 per tonne [32].

The principal end use of urea is to provide nonionic combined nitrogen for solid fertilizer formulations. Some urea is also converted to biuret ($H_2NCONHCONH_2$, also called carbamoyl urea) and to a sulfur derivative which are sold as Kedlor [59] and Urasil [60] cattle feed supplements, respectively. Urea is also used to a lesser extent in the manufacture of plastics components, such as melamine, and as a component of urea-formaldehyde resins used as adhesives. It is also used as a component of foam-in-place formulations as a rigid insulating material [61].

11.5. SYNTHETIC FERTILIZERS

Six of the 12 largest volume inorganic chemicals produced in the United States in 1980 directly or indirectly form part of all the various chemical fertilizer formulations produced. Such is the importance and significance of the fertilizer market in terms of the volume of activity of the chemical process industries.

The value of the chemical fertilizer product, of course, lies in the significantly enhanced agricultural productivity possible with their use. This advantage is more easily afforded by the large highly mechanized farming operations of the Western world, so that the per capita consumption in these countries runs from 20 to about 100 kg per capita per year as compared to usage rates in the neighborhood of 1 kg per capita per year for countries such as Bangladesh, mainland China, India, and Pakistan. While the per capita fertilizer consumption for these latter countries is small, the effectiveness of chemical utilization to foodstuffs and other agricultural production, kilogram for kilogram, is probably greater. The importance of chemical fertilizer applications in the stabilization of these major agricultural economies should not be underestimated.

11.5.1. Fertilizer Composition

Plant growth and development requires a total of nine major nutrients or macronutrients to proceed normally [62]. Three of these, carbon, hydrogen, and oxygen, are obtained from air (as carbon dioxide) and soil (as water). Another three, calcium, magnesium, and sulfur, are present to a sufficient extent in most ordinary soils. And another three, fixed nitrogen, phosphorus, and potassium, are present to a small extent in most soils but are rapidly depleted from the soil when the bulk of an agricultural plant is removed from the soil on harvesting. Returning the unused portions of plants to the soil can serve to return some of these elements. But removal from whole plant cropping and from soil bacterial processes which causes loss of nitrogen in the form of ammonia or elemental nitrogen, requires regular addition to the soil to maintain fertility. Compost, manure, and specially treated sewage sludges comprise "organic" methods to return these elements in low analysis forms. However, supplementation by the use of chemical fertilizers can achieve significantly improved yield benefits.

The major constituents or macronutrients present in most commercial fertilizers are nitrogen, which contributes to early plant development and greening; phosphorus, which assists with early growth and seed or fruit development; and potassium, which is used for the production of cellulose and starches. For the fertilizer analysis the three principal ingredients are specified in order as follows: nitrogen, as % N; phosphorus, as % P_2O_5 (equivalent), and potassium, as % K_2O (equivalent) as three numbers in sequence separated only by hyphens. Thus, 15-15-15, one of the popular high analysis fertilizers, would contain 15% N, 15% P_2O_5, and 15% K_2O. High analysis fertilizers offer simplicity and low cost shipping and handling. With the development of agricultural machinery for efficient distribution of these materials, more concentrated fertilizers occupy an increasingly important place in the range of fertilizer products marketed (Table 11.11). Some of the other common commercial designations are 0-20-20, 2-12-16, 3-12-12, etc.

In addition to the conventional listing of major constituents, a commercial blend may be formulated to contain one or more of the secondary constituents, calcium, magnesium, or sulfur, which are used to correct local soil deficiencies. These are normally specified in a separate listing. The importance of these too cannot be overlooked as, for example, fertilization of a sulfur deficient soil with soluble sulfate has given over 1100% rapeseed (canola) crop improvement [63].

There are also the trace elements boron, copper, iron, manganese, zinc, molybdenum, and occasionally chlorine which are important for some plant metabolic pathways and which may be deficient in the soils of some areas [64]. These are added as required to special fertilizer formulations made for the purpose, and are also specified separately from the major constituents.

TABLE 11.11 Commercial Mixed Fertilizers Shipped in Canada in 1969[a]

Analysis	Thousand tonnes	Analysis	Thousand tonnes
0–20–20	29.0	5–10–15	22.0
2–12–10	4.2	5–20–10	43.8
2–12–12	11.6	5–20–20	102.6
2–16–6	1.4	6–12–8	5.7
3–15–6	3.3	6–12–12	86.1
4–12–10	2.2	10–10–10	129.4

[a]Data from *Canadian Chemical Processing* [4].

In addition to the chemically important constituents many fertilizers have fillers and/or conditioners added. Fillers can include sand, gypsum, or ground phosphate rock (only very slightly water soluble, hence *not* part of P_2O_5 specified), and assist in even distribution of the fertilizer over small areas, such as in home gardens. Sand may also serve a conditioning function by assisting in the breakup of clays, but to be effective requires higher application rates than are normally used for fertilizers. Other soil conditioners, such as dried peat moss, specially processed wood waste, some types of shredded and expanded synthetic polymers (e.g., Krylium), and expanded mica (vermiculite), are used to improve soil structure. Since the conditioners have a much lower density than the active chemical constituents or the fillers, they are packaged and used separately from these.

11.5.2. Formulation of Major Active Constituents

Appropriate ingredients for the nitrogen component include ammonia gas itself which contains 82.5% N, the highest feasible nitrogen content available in a commercially useful compound. But this source is restricted to specialized application methods. Ammonia solutions, which are essentially concentrated ammonium hydroxide of about 25% N, still require specialized equipment and can add only nitrogen. Urea (46.6% N) is also used, either as the neat solid or as an aqueous solution.

For applications of a combination of major nutrients, solid forms are normally preferred for lower cost of shipment and convenient broadcasting in croplands. Thus urea and ammonium nitrate (35% N) serve as nitrogen-containing constituents in solid fertilizer blends. In the form of ammonium sulfate (21.2% N) nitrogen addition also adds soluble sulfate as a separately specified secondary nutrient. Of course, nitrogen may also be added in forms chemically combined with other major constituents such as in diammonium phosphate ($(NH_4)_2HPO_4$) or potassium nitrate, for example. A few plants, such as peas, clover, and alfalfa, have nodules containing nitrogen-fixing bacteria attached to their roots and are able to utilize elemental nitrogen from the air to provide their nitrogen requirements. Genetic engineering experiments may increase the number of species of plants able to do this, and in this way decrease the agricultural demand for synthetic fertilizer nitrogen [65].

Finely ground phosphate rock ($Ca_3(PO_4)_2$) is occasionally added to fertilizer formulations as a diluent or filler. However, because phosphate rock has only a very limited water solubility its action as a phosphate nutrient is small and very slow. Hence, this ingredient is not allowed to be included in the % P_2O_5 analysis appearing on fertilizer packaging. Before use as a filler phosphate rock is normally defluorinated by heating with silica and steam to decrease the risk of soil contamination by fluoride (Chap. 10).

A more soluble form of phosphate ($Ca(H_2PO_4)_2$), variously called "superphosphate" (of lime), monocalcium phosphate, or calcium dihydrogen phosphate, is made by the acidulation of phosphate rock with about 65% sulfuric acid (Eq. 11.61).

$$\text{CaF}_2\cdot3\text{Ca}_3(\text{PO}_4)_2 + 7\ \text{H}_2\text{SO}_4 + 14(\text{or more})\ \text{H}_2\text{O} \rightarrow$$
$$3\ \text{Ca}(\text{H}_2\text{PO}_4)_2 + 7\ \text{CaSO}_4\cdot2\text{H}_2\text{O} + 2\ \text{HF}\uparrow \qquad 11.61$$

calcium	gypsum	collected
dihydrogen		in scrubbers
phosphate		

The thick slurry of calcium dihydrogen phosphate and gypsum obtained is piled outside for 8–10 weeks to complete the reaction (to "cure") after which it is milled to produce a granular product, and then bagged for market. Thus the product, superphosphate, contains the inert gypsum component as well as the fertilizer active calcium dihydrogen phosphate, which is some 1000 times more soluble than calcium phosphate. The theoretical content of active material for the composition resulting from the reaction of Eq. 11.57 is 21.7% P_2O_5 equivalence, just a shade more than the actual 18–21% P_2O_5 range of the commercial product. The presence of silica or iron oxide in the original phosphate rock, plus the formation of the monohydrate of calcium dihydrogen phosphate $(\text{Ca}(\text{H}_2\text{PO}_4)_2.\text{H}_2\text{O})$ or some dicalcium hydrogen phosphate (CaHPO_4), all serve to cause these slightly lower than theoretical fertilizer values.

As with the nitrogen component, various other alternatives are available for the phosphate rock acidulation step. For instance, if acidulation is conducted with nitric, instead of sulfuric acid, calcium nitrate is formed instead of gypsum (Eq. 11.62).

$$\text{CaF}_2\cdot3\text{Ca}_3(\text{PO}_4)_2 + 14\ \text{HNO}_3 + 3\ \text{H}_2\text{O} \rightarrow 3\ \text{Ca}(\text{H}_2\text{PO}_4)_2\cdot\text{H}_2\text{O} +$$
$$7\ \text{Ca}(\text{NO}_3)_2 + 2\ \text{HF}\uparrow \qquad 11.62$$

This produces both P_2O_5 and nitrogen value from the single step. The mixture may be dried and used as obtained, or it may be partially or fully ammoniated to raise the nitrogen analysis still further. Full ammoniation would correspond to the stoichiometry of Eq. 11.63.

$$3\ \text{Ca}(\text{H}_2\text{PO}_4)_2\cdot\text{H}_2\text{O} + 7\ \text{Ca}(\text{NO}_3)_2 + 12\ \text{NH}_4\text{OH} \rightarrow$$
$$3\ \text{Ca}(\text{NH}_4)_4(\text{PO}_4)_2 + 7\ \text{Ca}(\text{NO}_3)_2 + 12\ \text{H}_2\text{O} \qquad 11.63$$

Of course the ammoniation may also be carried out independently of the use of nitric acid if only the slower acting ammonia nitrogen rather than a mixed nitrogen contribution in the formulation is wanted.

Another option to contribute phosphate to a fertilizer formulation is to use phosphoric instead of sulfuric acid for initial acidulation of phosphate rock. In this way the only solid product of acidulation is the soluble calcium dihydrogen phosphate free of gypsum diluent product with about three times the "available P_2O_5" of superphosphate (Eq. 11.64).

$$\text{CaF}_2\cdot3\text{Ca}_3(\text{PO}_4)_2 + 14\ \text{H}_3\text{PO}_4 + 10\ \text{H}_2\text{O} \rightarrow$$
$$10\ \text{Ca}(\text{H}_2\text{PO}_4)_2\cdot\text{H}_2\text{O} + 2\ \text{HF}\uparrow \qquad 11.64$$

This is the reason for the commercial reference to this material as "triple superphosphate." This product too may be ammoniated, or mixed with other nitrogen-containing constituents such as urea or ammonium nitrate nitrogen to provide both nitrogen and phosphorus for plant growth.

Elemental phosphorus itself has been tested as a fertilizer in its more stable red allotropic form but was not found to be commercially useful, presumably because of low availability of phosphorous [66].

Potassium chloride, "muriate of potash," is the principal component used as such to supply the potassium macronutrient in fertilizer formulations. Some crops with a low chloride tolerance, such as tobacco, may require the chloride to be replaced by nitrate, carbonate, or sulfate anion to be useful (e.g., Eq. 11.65).

$$KCl + HNO_3 \rightarrow KNO_3 + HCl\uparrow \qquad\qquad 11.65$$

11.5.3. Environmental Aspects of Fertilizer Production and Use

Emissions during the production phase of most fertilizer constituents have been discussed with the details of the processes used to produce the chemical. However, the production of superphosphate and triple superphosphate involves procedures with potential emission problems which differ from those of phosphoric acid production. There is the potential loss of fluoride, primarily as calcium fluoride in the dusts from the primary grinding of phosphate rock, or from milling of the cured product. Hydrogen fluoride, silicon tetrafluoride, and fluosilicic acid vapors are also formed when acid is added to the phosphate rock. Dusts may be controlled in cyclones, for which 80–95% mass collection efficiencies for a particle size range of 40% <10 μm has been found [33]. Vapors may be controlled by dry methods involving adsorption on chalk or limestone followed by capture in a cyclone, which gives 95% control [67]. Or methods employing water scrubbers may be used, the most efficient of which achieve 98% fluoride vapor containment [68]. Of course, while scrubbing with water or aqueous solutions can effectively clean waste gases of contaminants, it also generates an aqueous effluent high in fluoride which must be treated prior to discharge (see Chap. 10).

Use of chemical fertilizers on agricultural land introduces the risk of contamination of surface waters from the leaching of soluble constituents from the soil after application. Important precautions to minimize this are the use of good farming practices, application of fertilizer at no more than the recommended rate, and maintaining soil structure through measures such as plowing in a fallow crop. Use of chemical fertilizers at no more than the recommended rate gives the added advantage of providing the best crop return for the fertilizer investment. Maintenance of soil structure helps by providing an insoluble matrix on which complexation assists to hold the soluble fertilizer constituents in place, and minimizes erosion losses of the soil itself.

Long term concerns have been expressed that soil applications of fixed nitrogen in various forms provides more substrate for ammonia and nitrous oxide losses to the air via bacterial processes [69]. These sources of fixed nitrogen, which contribute about one-third of the total atmospheric loading (the remainder arising from natural sources), are in turn thought to contribute to depletion of the ozone layer through photochemical reactions.

REVIEW QUESTIONS

1. The Farmers Cooperative ammonia plant has two sources of hydrogen available, 25 tonnes per day via a chlor-alkali facility and a further 60 tonnes per day from operation of its own reforming and shift conversion processes.
(a) Assuming there are no losses, calculate the daily production of ammonia in tonnes.
(b) A part of the ammonia produced is sold directly. One hundred tonnes of ammonia per day is consumed by their nitric acid plant, which operates with 95% yield (selectivity) on ammonia. Calculate the tonnes of ammonium nitrate which would be available if the total nitric acid production was used in its manufacture.

2. (a) An ammonia plant uses steam and air reforming of methane together with shift conversion to provide the nitrogen and hydrogen feed for ammonia conversion. Assuming that air comprises oxygen and nitrogen in the mole ratio 1:4 and ideal gas relations, calculate the correct multiples of the three reforming reactions which would give rise to a gross nitrogen to hydrogen mole ratio of 1:3 with most efficient use of methane (i.e., no wasted carbon monoxide).
(b) If the plant in part (a) produces 600 tonnes of ammonia per day (minimum economic size for new ammonia plants), how many tonnes of carbon dioxide would be discharged daily?

3. The inert component (Ar, CH_4) of ammonia plant make-up gas is 1.5% by volume. What purge rate of the recycle gases is necessary, if there is a 10% conversion to ammonia, for the concentration of inerts in the feed gases to the converter to be kept to 15% of the total volume?

4. Assuming ideal gas behavior what would be the gas-catalyst contact time (seconds) for a space velocity of 5000 hr-1 and a gas mixture at:
(a) 0°C and 1 atm pressure?
(b) 0°C and 200 atm pressure?
(c) 450°C and 200 atm pressure?

5. The tail gas nitrogen dioxide concentration from the stack of a nitric acid plant measures 0.20% by volume at 25°C and 1.013×10^5 N m^{-2} (newtons per square meter, i.e., a pressure equivalent to 1 atm).
(a) To what concentration, in mg/m^3 at the same temperature, would this correspond?
(b) In the absorbers, where the pressure is 7.091×10^5 N m^{-2} at an operating exit temperature of 25°C, what would the nitrogen dioxide concentration be, expressed in mg/m^3?

6. (a) For a 225 tonne per day nitric acid (100% basis) process using ammonia as the feedstock, assume that all of the nitric oxide oxidation to nitrogen dioxide and the nitrogen dioxide hydration reactions occur in the absorbers. If the combustion gas stream and water for absorption enter the absorber at 25°C and the product

acid also leaves at 25°C, how many liters of absorber cooling water would be required per day to keep the temperature rise of this cooling stream to 10°C?

(b) What approximate proportion of the total nitric acid plant cooling water requirements (i.e., fraction of the total heat removal required) does your answer to part (a) represent?

7. Calculate and give the fertilizer analyses as they would be commercially cited for the following:
   (a) Pure potassium nitrate
   (b) Fully ammoniated triple superphosphate $[Ca\{(NH_4)_2PO_4\}_2]$
   (c) Pure potassium sulfate
   (d) A 30:70 mixture by weight of potassium chloride and fully ammoniated triple superphosphate, respectively

8. (a) What percentage of sand by weight would have to be blended with ammonium nitrate to quote a fertilizer analysis of 15-0-0 on the package?
   (b) What would be the fertilizer analysis that should be quoted on the package for a mixture of 40% by weight ammonium nitrate, 30% potassium chloride, and 30% triple superphosphate $(Ca(H_2PO_4)_2)$?

9. (a) What is the relative cost per tonne of fertilizer nitrogen when added via ammonia at $210 per tonne, ammonium nitrate at $150 per tonne, or urea, H_2NCONH_2, at $200.00 per tonne (provide answers to the nearest cent)?
   (b) Briefly comment on other factors which might be important, apart from lowest cost per tonne of added nitrogen, in the selection of these three sources of N in the formulation of a blended fertilizer.

10. Calculate and give the fertilizer analyses as they would be commercially cited for the following:
   (a) Superphosphate made by nitric acid acidulation of phosphate rock $(CaF_2 \cdot 3Ca_3(PO_4)_2)$
   (b) Ingredient of part (a), blended with 25% by weight potassium chloride
   (c) Pure ammonium phosphate
   (d) A 75:25 mixture of ammonium phosphate and potassium chloride

FURTHER READING

R. M. Harrison and H. A. McCartney, Some Measurements of Ambient Air Pollution Arising from the Manufacture of Nitric Acid and Ammonium Nitrate Fertilizer. *Atmos. Envir.* **13**, 1105 (1979).

K. V. Reddy and A. Husain, Vapor-Liquid Equilibrium Relationship for Ammonia in Presence of Other Gases. *Ind. Eng. Chem. Process Res. Dev.* **19**, 580 (1980).

REFERENCES

1. "Ammonia." Subcommittee on Ammonia, Committee on Medical and Biologic Effects of Environmental Pollutants, National Research Council, University Park Press, Baltimore, 1979.
2. "Kirk-Othmer Encyclopedia of Chemical Technology," 3rd ed., Vol. 2, p. 470, Wiley, New York, 1978.
3. *Chem. Eng. News* 74(26), 38–79, June 24 (1996), and earlier issues.
4. *Can. Chem. Process.* 54(9), 55, Sept. (1970).
5. T.H. Chilton, "Strong Water; Nitric Acid: Sources, Methods of Manufacture, and Uses." MIT Press, Cambridge, MA, (l968).
6. J.W. Dobereiner, *Ann. Chim. (Paris)* 24(2), 91 (1823); cited by Kirk-Othmer, 2nd ed., Vol.2, p. 268, Wiley, New York, 1968.
7. F. Haber, S. Tamaru, and C. Ponnaz, Z. *Elektrochem.* 21, 89, 128, and 191, (1915).
8. A.T. Larson, *J. Am. Chem. Soc.* 46, 367 (1924), and earlier refs.
9. A. Nielson, "An Investigation on Promoted Iron Catalyst for the Synthesis of Ammonia," 3rd ed., J. Gjellerups Forlag, Denmark, 1968; cited by Kirk-Othmer [2].
10. P.H. Emmett and J.T. Kummer, *Ind. Eng. Chem.* 35, 677 (1943).
11. N.A. Lange, ed., "Handbook of Chemistry," 10th ed., p. 915. McGraw-Hill, Toronto, 1969.
12. Methane Reforming to Stay, *Chem. Eng. News* 59(35), 39, Aug.31 (1981).
13. C. Matasa and E. Tonca, "Basic Nitrogen Compounds, Chemistry, Technology, Applications." Chem. Publ. Co., New York, 1973.
14. F.A. Lowenheim and M.K. Moran, "Faith, Keyes and Clark's Industrial Chemicals," 4th ed., Wiley-Interscience, New York, 1975.
15. R. Wood, *Process. Eng. (London)*, p. 7, Sept. (1976).
16. R.N. Shreve and J.A. Brink, Jr., "Chemical Process Industries," 4th ed. McGraw-Hill, New York, 1977.
17. M.R. Louthan, Jr., *Process Ind. Corros.*, p. 126 (1975).
18. R.E. Kirk and D.F. Othmer, eds.,"Encyclopedia of Chemical Technology." Interscience, New York, 1941.
19. Argon Recovery Plant, *Chem. Eng. News,* 58(4), 32, Jan. 28 (1980).
20. Methane Reforming to Stay, *Chem. Eng. News* 59(35), 39, Aug. 31 (1981).
21. Ammonia Plants More Efficient, *Can. Chem. Process.* 64(10), 10, Oct. (1980).
22. M. Lauzon, *Can. Chem. Process.* 65(3), 42, May (1981).
23. L.C. Daigre, III and G.R. Nieman, *Chem. Eng. Prog.* 70(2), 50, Feb. (1974).
24. G.P. Williams and J.G. Sawyer, *Chem. Eng. Prog.* 70(2), 45, Feb. (1974).
25. J.G. Sawyer and G.P. Williams, *Chem. Eng. Prog.* 70(2), 62, Feb. (1974).
26. Ammonia Unit, *Chem. Eng. News* 58(33), 24, Aug.18 (1980).
27. Monsanto Sells Hydrogen-Recovery, *Chem. Eng. News* 58(18), 8, May 5 (1980).
28. Japanese Technologists, *Can. Chem. Process.* 59(5), 8, May (1975).
29. L.J. Buividas, *Chem. Eng. Prog.* 77, 44, May (1981).
30. Desert Sands Catalyze Ammonia Formation, *Chem. Eng. News* 56(46), 7, Nov. 13 (1978).
31. TVA Ammonia-from-coal Project, *Chem. Eng. News* 57(33), 27, June 4 (1979).
32. "Guidelines for Limiting Contaminant Emissions to the Atmosphere from Fertilizer Plants and Related Industries in Alberta." Standards and Approvals Division, Alberta Dept. of the Environment, Edmonton, 1976.
33. H.F. Lund, ed., "Industrial Pollution Control Handbook." McGraw-Hill, Toronto, 1971.
34. T. Dear, *Chem. Eng. Prog.* 70(2), 65, Feb. (1974).
35. C.F. Kuhlmann, *Justus Liebigs Ann. Chem.* 29, 272 (1839), and reference cited therein.
36. "Statistics Canada," Catalog 46-002. Supply and Services Canada, Ottawa, 1982.
37. R. Thompson, ed., "The Modern Inorganic Chemicals Industry," p. 221. Chemical Society, London, 1977.
38. F.A. Cotton and G. Wilkinson, "Advanced Inorganic Chemistry; A Comprehensive Text," 3rd ed., p. 357. Interscience, New York, 1972.
39. M. Bodenstein, Z. *Electrochem.* 34, 183 (1918).
40. R.M. Stephenson, "Introduction to the Chemical Process Industries," p. 136. Reinhold, New York, 1966.

41. M.C. Manderson, *Chem. Eng. Prog.* **68**(4), 57, April (1972).

42. D.J. Newman and L.A. Klein, *Chem. Eng. Prog.* **68**(4), 62, April (1972).

43. New Generation of Nitric Acid Plants, *Chem. Eng. News* **54**(52), 33, Dec. 20 (1976).

44. Technology Newsletter, *Chem. Week* **108**(13), April 1 (1970).

45. L.M. Marzo and J.M. Marzo, *Chem. Eng. (N.Y.)* **87**, 54, Nov. 3 (1980).

46. G.R. Gillespie, A.A. Boyum, and M.F. Collins, *Chem. Eng. Prog.* **68**(4), 72, April (1972).

47. C. Boyars, *Ind. Eng. Chem., Prod. Res. Dev.* **15**(4), 308 (1976).

48. F.A. Patty, ed., "Industrial Hygiene and Toxicology," 2nd ed., Vol. 2, p. 918. Interscience, New York, 1963.

49. W. Frietag and M.W. Packbier, *Ammonia Plant Saf.* **20**, 11 (1978).

50. M. Yamaguchi, K. Matsushita, and K. Takami, *Hydrocarbon Process.* **55**, 101, Aug. (1976).

51. C.G. Swanson, Jr., J.V. Prusa, T.M. Hellman, and D.E. Elliott, *Pollut. Eng.,* **10**(10), 52, Oct. (1978).

52. "1974 Annual Fertilizer Review." Food and Agricultural Organization, United Nations, Rome, 1975.

53. "Japan Economic Yearbook 1981/82," p. 69. The Oriental Economist, Tokyo, 1981, and earlier issues.

54. I. Dunstan, *Chem. Br.* **7**(2), 62, Feb. (1971).

55. C. Boyars, *Ind. Eng. Chem., Prod. Res. Dev.* **15**(4), 308 (1976).

56. Ammonium Nitrate, *Hydrocarbon Process.* **58**(11), 135, Nov. (1979).

57. M. Bryson, Experimental Design Boosts Production Yields, *Chem. Eng. (N.Y.)* **102**, 155 (1995).

58. E. Otsuka, S. Inoue, and T. Jojima, *Hydrocarbon Process.* **55**(11), 160, Nov. (1976).

59. Dow Feed-Grade Buiret, *Chem. Eng. News* **49**(25), 23, June 21 (1871).

60. A New Liquid Cattle Feed Supplement, *Can. Chem. Process.* **56**(10), 4, Oct. (1972).

61. Product Profile Urea, *Can. Chem. Process.* **61**(10), 50, Oct. (1977).

62. C.J. Pratt, *Sci. Am.* **212**(6), 62, June (1965).

63. Sulphur Response, *Chem. Can.* **24**(7), 5, Summer (1972).

64. R.J.P. Williams, *Chem. Br.* **15**(10), 506 (1979).

65. Nitrogen Fixation Research Advances, *Chem. Eng. News* **58**(49), 29, Dec. 8 (1980).

66. H.P. Rotbaum and W. Kitt, *N. Z. J. Sci.* **7**, 67 (1964).

67. K. Karbe, *Chem. Eng. (London)* **221**, 268 (1968).

68. N.L. Nemerow, "Industrial Water Pollution, Origins, Characteristics, and Treatment," p. 613. Addison-Wesley, Don Mills, Ontario, 1978.

69. Fertilizer May Deplete Ozone Layer, *Chem. Eng. News* **56**(40), 6, Oct. 2 (1978).

12
■ ALUMINUM AND COMPOUNDS

Science got the beautiful metal aluminum out of the clay which ignorance trod underfoot.
—John Cunningham Geikie, 1976

12.1. HISTORICAL BACKGROUND

From its relatively small-scale utilization of the order of 1/100th that of copper, lead, and zinc prior to 1900, to its present scale of production of two to four times that of these more traditional nonferrous metals (Table 12.1), aluminum is a metal that has come of age in the twentieth century. Oersted, in Denmark, was the first to produce aluminum in 1825, obtained in impure form by the reduction of aluminum chloride with potassium amalgam. Wohler, two years later, obtained higher purity metal and more fully described its properties. Henri St.-C. Deville put aluminum production into commercial practice in France by 1845 using sodium fusion to reduce aluminum chloride (Eq. 12.1).

$$AlCl_3 + 3\ Na \rightarrow Al + 3\ NaCl \text{ (conducted as a melt)} \qquad 12.1$$

By 1852 the metal sold commercially for about \$545 per pound. An elaborately enameled fan (now housed at the Smithsonian, Washington, DC) was made from aluminum for the Paris Exposition of 1867 while it was still classed as a precious metal, and it is said that Napoleon III had a tea service made from it. Process improvements such as cryolite fluxing of the melt and lower sodium costs brought the price of the metal down to US\$8 per pound by 1886, which removed its status appeal.

In the early part of 1886, Paul Heroult in France and Charles Hall in the U.S.A. independently devised electrolytic methods to reduce combined aluminum to the metal, and first made this possible on a large scale. In their

TABLE 12.1 Relative Growth Rates of World Production of Some Nonferrous Metals Relative to Their Scale of Production in 1900[a]

Year	Aluminum		Copper		Lead		Zinc	
	Thous. tonnes	Multiple of 1900	Thous. tonnes	Multiple of 1900	Thous. tonnes	Multiple of 1900	Thous. tonnes	Multiple of 1900
1900	5.7	1	499	1	877	1	479	1
1960	4,670	819	4,400	8.82	2,630	3.00	3,070	6.41
1970	9,666	1,696	6,227	12.45	3,299	3.76	4,905	10.24
1980	15,368	2,696	6,194	12.4	2,957	3.37	4,354	9.09
1990	21,846	3,833[b]	9,680	19.4[b]	5,950	6.78[b]	7,180	15.0[b]

[a]Mass data in thousands of metric tonnes compiled from Kirk-Othmer [1], and Minerals Yearbooks [2].
[b]1990 production as a multiple of 1960 values was aluminum, 4.68; copper, 2.20; lead, 2.26; zinc 2.34.

TABLE 12.2 Annual Production of Aluminum by the World's Major Producers, in Thousands of Metric Tonnes and Percent of World[a]

	1945 Mass	1945 %	1965 Mass	1965 %	1970 Mass	1980 Mass.	1980 %	1990 Mass	1994 Mass	1994 %
Canada	196	22.1	762	11.3	964	1,068	6.9	1,567	2,250	11.8
France	37.2	4.2	340	5.1	380	432	2.8	325	400	2.1
Japan	5.4	6.1	323	4.8	733	1,091	7.1	51	17	0.1
Norway	7.0	0.8	277	4.1	530	651	4.2	867	857	4.5
U.S.A.	449	50.6	2,498	37.1	3,607	4,654	30.3	4,121	3,300	17.3
U.S.S.R.	86.3	9.7	1,279	19.0	1,098	1,787	11.6	—	—	
W. Germany	20.0	2.3	234[b]	3.5	308[b]	731[b]	4.8	720[b]	—	
Other	37.3	4.2	1,014	15.1	2,046	4,954	32.3	—	—	
World	887.0	100.0	6,727	100.0	69,666	15,368	100.0	21,846	19,100	100.0

[a]List includes all countries whose annual primary aluminum production exceeded 400,000 tonnes in 1980. Compiled and calculated from data of *Mineals Yearbooks* [2].
[b]Totals for Germany after partition.

procedure, the essence of which is still used today, alumina (Al_2O_3) was dissolved in a bath of molten cryolite. Passage of a direct current through this bath, via an electrolytic process (not electrothermal, as used for phosphorus), produced a layer of molten aluminum on the bottom of the cell. This procedure both raised the purity of the metal obtained and at the same time further lowered the price. The advantages in terms of cost and ease of availability that the Hall-Heroult process contributed coupled with the lightness and corrosion resistance of this metal led to a rapid increase in consumption (Table 12.1).

Continued growth in the production of aluminum is unlikely to be hampered by a shortage of mineral since it is estimated that the earth's crust consists of about 8% aluminum, chiefly as aluminosilicates. Even though aluminum is the most abundant metallic crustal element, bauxite ores suitable for aluminum recovery only occur in more limited areas where natural processes such as leaching have served to concentrate the aluminum-containing minerals relative to the average crustal content. Since the free metal is quite chemically reactive it is never found in nature in this form.

With the rapid acceptance of the Hall-Heroult electrolytic method of aluminum production, facilities using this process have tended to be contructed in areas with abundant, low-cost electric power (Table 12.2). In addition, to minimize shipping costs it is usual practice to process bauxite, the crude ore, at the mine site to produce purified alumina for shipment to smelters. This means that a further factor for the siting of aluminum smelters is that they provide easy shipping access to permit economical alumina delivery from the mine. Thus, the production of aluminum from bauxite logically can be considered in two steps: the first, production of high-purity alumina from the working of natural bauxite deposits, and the second, electrolytic reduction of alumina to the metal.

12.2. ALUMINA FROM BAUXITE: THE BAYER PROCESS

Bauxite, the principal ore used for aluminum smelting, is named after Les Baux, Provence, the village where the first deposits were discovered. Australia is now the world's largest producer of bauxite. Bauxite contains hydrated alumina equivalent to as much as 40–60% Al_2O_3 with much of the other siliceous materials leached out over time. However it still contains 10–30% iron oxide, silica, and other impurities, making it unsuitable for direct electrolysis. The first commercial-scale recovery of alumina from bauxite was practiced by Henri Deville, but by 1900 this was largely replaced by the more economical process devised by Bayer in Austria, based on caustic extraction.

Alumina recovery from bauxite by extraction with sodium hydroxide, now frequently referred to as the Bayer process, relies on the amphoterism of aluminum for its success (Fig. 12.1). Details of the particular alumina extraction procedure required depend on the particular form of hydrated alumina which occurs in the bauxite being processed.

Preliminary to any chemical processing the coarse ore is mechanically reduced to a finely divided form and stirred with the requisite concentration of

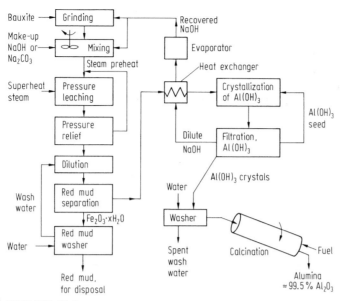

FIGURE 12.1 Production of alumina from bauxite via the Bayer process.

aqueous base prior to pressure leaching. The alumina from trihydrate bauxite ($Al_2O_3\cdot3H_2O$; natural bayerite or gibbsite) is relatively easy to dissolve out using 15–20% aqueous sodium hydroxide, under pressure, at temperatures of 120–140°C (Eq. 12.2).

$$Al(OH)_3 + NaOH \rightarrow \underset{\text{soluble}}{AlO\cdot ONa} + 2\,H_2O \qquad 12.2$$

If there is any significant proportion of monohydrate alumina ($Al_2O_2\cdot H_2O$; e.g. diaspore) present in the ore, both more concentrated sodium hydroxide (20–30%) and higher temperatures and pressures, 200–250°C and up to 35 atm, are required to effectively leach out the alumina content (Eq. 12.3).

$$AlO\cdot OH + NaOH \rightarrow AlO\cdot ONa + H_2O \qquad 12.3$$

The more severe conditions will also extract trihydrate alumina from bauxite so that the deciding factor in the digestion conditions required is the presence of a significant concentration of monohydrate alumina. Fortunately trihydrate alumina is the dominant aluminum species present in the major world deposits of Africa, Australia, the Caribbean, Central and South America, and the United States. European bauxites are mainly the monohydrate.

Iron oxide, clays, and most other impurities do not dissolve under alumina digestion conditions, and are quite finely divided (1–10 μm). By adding wash water to decrease the viscosity somewhat, it is possible to decant the still hot sodium aluminate solution from the slowly precipitating red muds. These muds, which are red from the high iron content, are then washed with water to minimize losses of alumina and base on ultimate disposal of the muds by lagooning. By using this wash water for dilution of the next digester output, alumina and base are not lost from the Bayer circuit. Some more recently constructed facilities use pressure filtration for both removal and washing of red muds.

Aluminum hydroxide crystals are obtained from the supernatant sodium aluminate solution after a final polishing filtration by ensuring a $Na_2O:Al_2O_3$ ratio of 1.5–1.8:1, controlling the solution temperature to about 60°C, and seeding with crystals from an earlier crystallization. Even with seeding the crystallization process is slow, requiring 2–3 days, but sufficiently large crystals are obtained by the reverse of the solution process (Eq. 12.4) to enable recovery of the product by filtration.

$$AlO \cdot ONa + 2 H_2O \rightarrow Al(OH)_3 + NaOH \qquad 12.4$$

The aluminum hydroxide product, while still on the filters, is washed with water and then dried. Evaporation of most of the water from the filtrates allows return of the sodium hydroxide solution to the initial grinding and pressure leaching circuits for reuse at appropriate concentrations for these steps.

Calcination of the aluminum hydroxide at about 1200°C in either fluidized bed or rotary calciners finally yields alumina of 99.5% purity (Eq. 12.5).

$$2 Al(OH)_3 \rightarrow Al_2O_3 + 3 H_2O \qquad 12.5$$

The chief contaminant is 0.3–0.5% sodium oxide, which fortunately does not affect electrolysis, with <0.05% calcium oxide, <0.025% of silica or iron oxide, and <0.02% of any other metallic oxide [3]. Apart from metal production, some of this alumina produced at 1200°C goes into the manufacture of synthetic abrasives and refractory materials. Activated alumina, destined for adsorptive uses, is produced in the same way except that more moderate calcining temperatures of about 500°C are employed, producing a highly porous product with excellent surface activity. The volumes of alumina from the world's major producers is given for several recent years in Table 12.3.

Gallium is a trace contaminant of chemical and commercial interest which is present to the extent of 60–200 µg/g in Bayer process liquors. Situated in the same group as aluminum in the periodic table its very similar chemical properties cause it to be carried through the alumina purification and electrolytic steps along with the aluminum. Production is on the scale of 50,000 kg/yr, mostly in Australia, Germany, and Russia [2]. It is used in such fascinating applications as electro-optical devices (LEDs and the like), in the GaAs used for the microchannel plates that guide and amplify photons in night vision goggles and sights, and for the integrated circuits in one of the fastest digital signal processors made. Improvement of recovery methods has been examined [5].

It is also possible to obtain alumina from clays or bauxite via leaching with sulfuric acid, a technique which is especially useful with high silica or high iron oxide bauxites (Eq. 12.6).

$$clays + H_2SO_4 \rightarrow Al_2(SO_4)_3 + water\text{-}insoluble\ residue \qquad 12.6$$
$$soluble$$

■ **TABLE 12.3 Major World Producers of Alumina**[a]

	Thousands of metric tonnes				
	1972	**1980**	**1985**	**1990**	**1994**
Australia	3,068	7,246	8,792	11,200	12,900
Canada	1,149	1,202	1,020	1,090	1,170
France	1,274	1,173	734	606	438
Jamaica	2,087	2,456	1,513	2,870	3,220
Japan	1,644	1,936	978	481	300
Surinam	1,378	1,316	1,000[b]	1,530	1,500
U.S.A.	6,114	6,810	3,456	5,230	4,860
U.S.S.R.	2,300	2,700	3,500	5,900	—
W. Germany	916	1,608	1,657	922	—
Yugoslavia	135	1,635	1,138	1,090	—
Others	3,535	3,709	7,820	11,681	—
World total	23,600	33,426	31,617	42,600	42,000

[a]Includes all those countries producing more than a million tonnes of alumina annually by 1980. Compiled from data in *Minerals Yearbooks* [2] and de Magalhaes and Tubino [5].
[b]Estimate.

Crystals of pure aluminum sulfate hydrate ($Al_2(SO_4)_3 \cdot 18H_2O$) are obtained from the leach solution, which may then be calcined to yield alumina (Eq. 12.7).

$$Al_2(SO_4)_3 \cdot 18H_2O \rightarrow Al_2O_3 + 3\ SO_3 + 18\ H_2O \qquad 12.7$$

Cryolite ($AlF_3 \cdot 3NaF$), the major, though largely reused component of the electrolytic bath is a rare natural mineral originally found only in Gothaab, Greenland. Most present aluminum smelters function using synthetic cryolite prepared from alumina and hydrogen fluoride in the presence of sodium hydroxide (Eq. 12.8, 12.9).

$$\begin{array}{c} CaF_2 + H_2SO_4 \rightarrow 2\ HF + CaSO_4 \\ \text{fluorspar} \end{array} \qquad 12.8$$

$$\begin{array}{c} 6\ HF + Al(OH)_3 + 3\ NaOH \rightarrow Na_3AlF_6 + 6\ H_2O \\ \text{as} \\ \text{briquets} \end{array} \qquad 12.9$$

In this manner a cell electrolyte having characteristics identical to the natural mineral is readily available for use by smelters.

12.3. ALUMINUM BY THE ELECTROLYSIS OF ALUMINA

Reduction of alumina to aluminum is an electrolytic process like the dominant method for chlorine and caustic production from sodium chloride, and not an electrothermal process (e.g., phosphorus). The required ionic mobility could theoretically be provided by melting alumina, of melting point 2050°C.

TABLE 12.4 Typical Melt Composition for Electrolytic Reduction of Alumina[a]

Component	Range (%)
Cryolite, Na_3AlF_6	80–85
Fluorspar, CaF_2	5–7
Aluminum fluoride	5–7
Alumina	2–8

[a]Data from Kirk-Othmer [1]. Normally alumina reduction pots are operated on the acid side, i.e., a net AlF_3:NaF mole ratio of 1.2 to 1.5:1.

But the temperatures required are so high (above the useful range of most refractory lining materials) that they make this technically unfeasible. Mobility can be provided by dissolving the alumina in an ionizing solution, although not an aqueous one since alumina is insoluble in water. Alumina is, however, sufficiently soluble in molten cryolite (Na_3AlF_6, melting point 1006°C) to enable electrolysis to be carried out. In practice, addition of a few percent of calcium fluoride and/or aluminum fluoride depresses the melting point of the electrolyte some 4–5°C for each 1% of additive [1], which enables electrolysis at ca. 950°C, temperatures technically a little easier to achieve (Table 12.4). While the aluminum reduction process is primarily electrolytic much of the power required is consumed through cell resistance to keep the reduction pot temperatures sufficiently high to maintain the electrolyte fluid state.

Virtually all alumina electrolysis cells consist of a rectangular steel shell lined with a 25–35 cm layer of baked and rammed dense carbon, which pro-

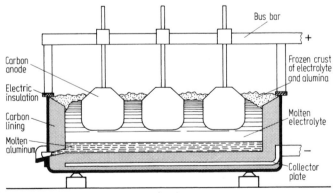

FIGURE 12.2 Cross-sectional diagram of a typical aluminum reduction pot using prebaked carbon anodes. The refractory brick lining and the frozen crust beside and on top of the molten electrolyte provide thermal insulation which raises overall energy efficiencies. (Courtesy of Aluminum Co. of America.)

vides both chemical resistance and the cathode contact with the electrolyte via steel bus bars imbedded in the carbon. Normal lining life is 4 to 6 yr, after which it is replaced as large preformed slabs. Once a reduction pot has been started the bulk of the cathodic current to the carbon lining is via the pool of newly formed molten aluminum in the bottom of the cell (Fig. 12.2).

Anode elements are commonly prebaked low ash carbon blocks, low ash since any ash residue ends up in the electrolyte. These are electrically connected to copper or aluminum bus bars (heavy electrical conductors) suspended over the cell, which also provide mechanical support and a means for vertical adjustment of the anode elements. Another anode variant, the Soderberg paste option, uses a good grade of petroleum coke formed into a paste with hard pitch, to which electrical contact is maintained and mechanical adjustment provided by using specially shaped steel pins (Fig. 12.3). As the baked portion of this anode is gradually consumed, the paste approaches the molten electrolyte and the volatile components in the paste vaporize leaving a hard baked working anode element. Either type of anode element is consumed at the rate of 1–2 cm/day during normal operation, requiring periodic vertical adjustment to maintain an anode-aluminum metal pool spacing of about 5 cm.

To start operation of a pot, anodes are lowered to the lining, the solid electrolyte and alumina components are placed in it, and then the power is turned on. Initial melting of the pot charge is by resistance heating. A typical cell consumes from 50,000 to 200,000 A, at an operating voltage of about 4.5 to 5.5. As the charge melts the anode elements are gradually raised to the normal about 5-cm spacing, and electrolysis begins.

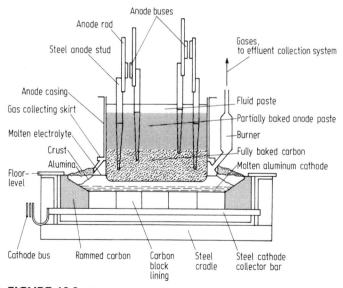

FIGURE 12.3 Diagram of an aluminum reduction pot showing the arrangement of the principal components of a vertical stud Soderberg paste anode [6]. (Reprinted courtesy of Alcan Smelters and Chemicals Ltd.)

At the cathode, aluminum ions pick up electrons to form elemental aluminum (Eq. 12.10), which, being more dense than the molten cryolite, gradually accumulates as a pool beneath the electrolyte melt layer and rests on the carbon lining.

$$Al^{3+} + 3\ e^{-} \rightarrow Al^{\circ} \qquad\qquad 12.10$$

For a 125,000 A cell the metal reduction capacity is 810–910 kg/day. Normally, alumina additions are made to each cell at about four hourly intervals. However, if the alumina consumption in any one cell for any reason exceeds this rate of routine addition and the alumina content of the electrolyte drops to about 2% alumina or less, there is a sudden rise in the cell operating voltage to about 50 V, caused by polarization of the anode from the near stationary film of carbon monoxide and carbon dioxide on its surface [1]. This situation, called the "anode effect," is detected via a 50–60 V light wired in parallel to each cell in a multicell potline. Normal cell operating voltages are too low to cause bulb incandescence, but an alumina-deficient situation initiates fluorine as well as oxygen deposition at the anodes.

During the occurrence of an anode effect the higher anode-cathode voltage differential may also cause arcing to occur across the anode-aluminum pool gap. These severe conditions cause the abnormally present fluorine, like oxygen, to also react with the hot carbon-forming carbon tetrafluoride and smaller amount of hexafluoroethane (Eq. 12.11, 12.12).

$$2\ F_2 + C\ (hot) \rightarrow CF_4 \qquad\qquad 12.11$$

$$3\ F_2 + 2\ C\ (hot) \rightarrow C_2F_6 \qquad\qquad 12.12$$

These fluorocarbons and carbon monoxide effectively combine to stabilize the polarization of the anode elements by formation of a gas envelope around them [7]. That it is these occurrences which cause the rapid rise of the operating voltage of a cell is confirmed by this effect being temporarily alleviated by stirring the electrolyte, which momentarily clears the gas envelope from the anode surface. To correct this situation effectively, fresh powdered alumina, already prewarmed on the electrolyte crust of the cell, is broken into the molten electrolyte with a remotely controlled pneumatic hammer. The anode polarizing gas used to be dispelled by poking wood poles through the crust to flush out the polarizing gases by both the stirring action and the rapid flow of combustion gases from burning of the pole. Now, however, air lances are more often used to accomplish this task.

At about daily intervals, as sufficient aluminum accumulates in each, a stout pipe attached to an evacuable vessel is broken through the crust on the cell and dipped into the pool of molten aluminum. The pressure in the receiving vessel is decreased, which pulls the melt into the container, a process which is repeated for each cell down the potline, to collect the molten aluminum in a central holding furnace ready for fabrication.

At the same time as aluminum is being reduced, oxygen ions migrate to the suspended anode elements and in turn are oxidized to atomic and possibly some molecular oxygen (Eq. 12.13).

$$O^{2-} - 2\ e^{-} \rightarrow 2\ [O];\ (2\ [O] \rightarrow O_2) \qquad\qquad 12.13$$

Because both forms of oxygen are generated in close contact with a hot carbon element, virtually all of it immediately reacts to form carbon monoxide and carbon dioxide (Eq. 12.14–12.16).

$$C \text{ (hot)} + [O] \rightarrow CO \qquad\qquad 12.14$$

$$CO \text{ (hot)} + [O] \rightarrow CO_2 \qquad\qquad 12.15$$

$$C \text{ (hot)} + O_2 \rightarrow CO_2 \qquad\qquad 12.16$$

However, as this carbon dioxide moves from the anode region it contacts hot aluminum vapors, and possibly sodium vapors and fine liquid metal droplets (fog), which immediately oxidize to form the corresponding metal oxide and carbon monoxide (Equations 12.17, 12.18).

$$3 CO_2 + 2 Al \rightarrow 3 CO + Al_2O_3 \qquad\qquad 12.17$$

$$CO_2 + 2 Na \rightarrow CO + Na_2O \qquad\qquad 12.18$$

Thus, the gas mixture collected from the skirted anode area consists of carbon monoxide (from both incomplete initial oxidation, plus metal reduction of carbon dioxide) and carbon dioxide in a weight ratio which varies between 1:2 and 1:3. The toxic carbon monoxide in these hot gas mixtures normally burns spontaneously as it escapes the confining crust. For the vertical Soderburg pots, where the gases are captured by the anode skirt, a small burner is used to convert the carbon monoxide content to carbon dioxide at each cell gas exit before the effluent gas mixture proceeds to treatment systems.

The overall electrochemical process being considered is closely represented by Eq. 12.19.

$$Al_2O_3 + 3/2\ C \rightarrow 2\ Al + 3/2\ CO_2 \qquad \Delta H = +1096\ kJ\ (+262\ kcal)$$
$$12.19$$

This highly endothermic energy balance in the direction given is supplied electrochemically during the electrolysis. Using this endotherm in the Gibbs-Helmholtz equation allows determination of the theoretical equilibrium voltage required for deposition of aluminum, which is 1.7 V. It is generally thought that there is no electrochemical assistance from carbon oxidation by the anode product, hence this electrochemical potential is the same as the standard reduction potential for aluminum, which is -1.706 V [8]. The actual operating potential required is significantly higher than this, 4.5 to 5.5 V, which is needed to overcome electrolyte resistivity (Table 12.5) over the 5-cm electrode gap and other factors, as well as to move the process from an equilibrium to a producing situation. Thus, voltage efficiencies range from 24 to about 38% with the bulk of production taking place at the upper end of this range, much lower than any of the common chloralkali electrolysis cells. As mentioned previously this electrochemical work required for alumina electrolysis is the result of overall cell resistance effects, and appears in the form of heat which keeps the cell contents molten. The smaller surface to volume ratio for the larger electrolysis cells (up to 260 kA) confers some voltage efficiency contribution which improves overall energy efficiencies slightly over the values obtained with smaller cells.

TABLE 12.5 Electrochemical Parameters of Importance to Hall-Heroult Electrolysis of Alumina[a]

Reduction potentials, V:	
Theory	1.706
Normal operating	4.5–5.5 (to 7 occasionally)
Anode effect, at ca. 2% Al_2O_3	30–50
Operating capacities, kA	50–260
Anode characteristics:	
For prebaked: Resistivity	5–6 ohm/cm
Current density	6.5–13 kA/m^2
For Soderberg: Resistivity	6.5–7.8 ohm/cm
Current density	6.5–9 kA/m^2
Electrode spacing	ca. 5 cm
Electrolyte bath resistivity, at 950°C	0.50 ohm cm
Current efficiency, 100-kA cells	ca. 90%
Normal energy efficiency	34–38%
Typical anode gas composition[b]	85% CO_2, 15% CO

[a]Calculated and compiled from Kirk-Othmer [1], *World Mineral Statistics* [4], and *The Chemistry of Aluminum* [9].
[b]Gas composition is given as experienced for 90% current efficiencies. At 80% current efficiency the gas composition will be approximately 75% CO_2, 25% CO, i.e., the CO content is inversely proportional to current efficiencies.

Like all electrochemical processes the quantity of product obtained for a given current flow/time period is related through Faraday's law, i.e., that in an electrochemical reaction 1 g equivalent weight of substance is deposited on the passage of 96,494 coulombs (one Faraday) of electricity. Since each aluminum ion, to be deposited, requires three electrons (Eq. 12.10), a gram equivalent weight of aluminum is one-third of its atomic weight (26.9815 g/mol ÷ 3 electrons/mol) or 8.994 g. Thus, 8.994 g of aluminum would be expected to be deposited in the cell, for each Faraday of electrons. But in practice only about 90% of this amount is collected, the lower experience being the result of such processes as the reaction of anode carbon dioxide with already reduced aluminum or sodium, which returns these cell "products" on which power has already been expended to the electrolyte as oxides. It has been found that the addition of aluminum fluoride to the electrolyte tends to decrease the concentration of reduced species in the electrolyte, and in this way raises current efficiencies.

Combining the voltage and current efficiency factors, the overall energy efficiency of the cell is usually about 34% (0.90 × 0.38) although during periods of high voltage operation this can drop as low as 22%. In production experience terms, this normally means a power consumption of 17,600 kWh/tonne of aluminum for the electrolytic step. Power and other quantitative aspects of aluminum production such as anode consumption rates are given in Table 12.6. For quantitative calculations which reflect quite closely the actual stoichiometry experienced, the molar relationships given in Eq. 12.20 have been used.

███ **TABLE 12.6 Materials and Energy Required to Produce One Tonne of Commercial Aluminum**[a]

Alumina consumption:	
Theory	1889.5 kg
Practice	1900–1901 kg
Carbon anode material:	
Carbon, theory[b]	445 kg
Carbon, practice	450–550 kg
Pitch, Soderberg	100–200 kg
Electrolyte:	
Cryolite	30–70 kg
Aluminum, fluoride	ca. 40 kg
Power	15,000–17,600 kWh
Labor	13–31 man-hours

[a]Calculated from data of Kirk-Othmer [1], *World Mineral Statistics* [4], *The Chemistry of Aluminum* [9], Patterson [10], and OECD [11].

[b]Assuming that this is pure carbon, as is nearly true with prebaked anodes, and the stoichiometry of Eq. 12.19

$$Al_2O_3 + 2\,C \rightarrow 2\,Al + CO + CO_2 \qquad\qquad 12.20$$

Purity of the direct Hall-Heroult product is very good, about 99.5 to 99.7% aluminum, and adequate for most uses requiring the pure metal, and for alloys. The chief impurities present are iron, 0.06 to 0.10%, and silicon, 0.04 to 0.30%, and with a total for all other impurities seldom exceeding 0.10% [1]. Analysis of both the aluminum and the electrolyte of each cell is normally conducted on at least a weekly basis. Among other things this practice provides a warning when the carbon lining of a pot has worn through to the steel shell ahead of schedule, which is seen as an increase in the iron content of the aluminum. This particular cell may then be taken out of production and repaired before aluminum and electrolyte break through the outer steel shell.

Electrolytic or fractional crystallization methods are employed to produce 99.99% pure aluminum. Electrolytic methods employ a melted aluminum-copper alloy (about 75:25) as the bottom layer, overlain by a high content of a barium salt (either chloride or fluoride) to produce a high-density electrolyte, topped by a layer of the electrolytically refined aluminum as it forms. By passage of a direct current through the cell floor as the anode and suspended carbon cathodes there is a gradual accumulation of refined aluminum in the top layer. Voltages, and hence voltage efficiencies, are similar to those required for primary aluminum production. Zone refining is resorted to when 99.999% or higher aluminum purities are required. Recycling of scrap aluminum at present accounts for some 20% of aluminum produced. This requires less than 5% of the energy required to produce the metal from alumina [12].

12.4. NEW DEVELOPMENTS IN ALUMINUM PRODUCTION

Despite the world's abundant reserves of bauxite the deposits are not, in many cases, in the major aluminum producing areas. This factor has stimulated the extensive testing of alternative sources of aluminum [10]. High alumina clays, shales, and other alumina-containing minerals have all been used as alumina sources, and some of these alternatives, such as nepheline syenite $((Na,K)(Al,Si)_2O_4)$ in the U.S.S.R., are actually used. Tests in the United States are being performed on extraction of alumina from kaolin clays (30–35% Al_2O_3) with hydrochloric or nitric acids. Alumina recovery from the clays removed from coal (ca. 28% Al_2O_3) during washing, and from coal ash, if found to be economically feasible, have the dual advantage of providing a resource from large-scale waste materials. French and Canadian ventures have studied alumina extraction from clays and shales using sulfuric and hydrochloric acids [13]. Recent developments in alumina processing technology have been reviewed [14].

Sumitomo Chemical technology consists of a number of small changes to conventional Hall-Heroult electrolysis cells which combine to cut power consumption to some 14,000 kWh/tonne from the normal experience of 16,000 to 18,000 kWh/tonne [15]. In addition to reduced power consumption, better emission control, extended cell life, and decreased labor requirements are also achieved by these changes. The improvements are obtained through better external cell insulation, changes in what is basically a Soderberg anode design, and better fume containment by cell skirt construction modifications.

Advances in alumina reduction technology have been surveyed [16]. The most innovative developments involve changes in the smelting technology employed. Aluminum can be obtained by metallic, carbon, or electrolytic reduction of appropriate compounds. The earliest processes successfully used metallic reduction via potassium amalgam or sodium fusion to produce aluminum from aluminum chloride (Equations 12.21, 12.22).

$$AlCl_3 + 3\ K_{(Hg)} \rightarrow Al + 3\ KCl + Hg \qquad 12.21$$

$$AlCl_3 + 3\ Na \rightarrow Al + 3\ NaCl \qquad 12.22$$

As Hall-Heroult technology became available these processes were displaced because of higher costs.

Reduction of alumina with carbon is fraught with problems, not the least of which is the very high temperatures required (about 2000°C) and the formation of quite stable aluminum carbide (Eq. 12.23).

$$2\ Al_2O_3 + 9\ C \rightarrow Al_4C_3 + 6\ CO \qquad 12.23$$

Metallic aluminum is a sufficiently active metal that it reduces carbon dioxide to carbon monoxide (Eq. 12.17) so that the favorable aspects of a solid-gas, alumina-carbon monoxide reduction cannot be utilized. However, good yields have been obtained experimentally by carrying out the reduction in two steps. The first step requires formation of the carbide (Eq. 12.23), following which the temperature of the system is raised another 100–200°C to obtain aluminum by the reaction of aluminum carbide with alumina (Eq. 12.24).

$$Al_4C_3 + Al_2O_3 \rightarrow 6\ Al + 3\ CO \qquad 12.24$$

At around 2100°C, the temperature required for the second step, aluminum (melting point 660°, boiling point 2467°C) itself has sufficient volatility to cause metal loss and migration problems.

Carbothermic reduction in the presence of an alloying element such as copper, iron, or silicon to decrease aluminum vapor pressures decreases volatility problems but requires a second stage to recover aluminum from the alloy product. It may be selectively dissolved from the alloy with a more volatile metal, such as mercury, lead, or zinc, and then the aluminum recovered by distillation. Or, the tendency for aluminum halides to form more volatile monohalides at high temperatures which revert to the trihalides at lower temperatures (Eq. 12.25) may also be employed.

$$AlCl_3 + 2\ Al\ \underset{\substack{\text{In alloy}}}{\overset{1300°C}{\underset{800°C}{\rightleftharpoons}}} 3\ AlCl \qquad\qquad 12.25$$

In this procedure, extensively tested by Alcan at one time, dissolution of aluminum from the alloy using aluminum trichloride at about 1300°C produces the volatile aluminum subhalide [17]. Cooling the subhalide moderately returns a pool of molten product aluminum, and aluminum trichloride (boiling point ca. 183°C), which serves as a working fluid and can be recycled. Despite problems with the aluminum recovery methods there remains an interest in nonelectrolytic aluminum production such as that offered by carbothermic processes [18].

New technology for electrolytic aluminum production employing aluminum is also stimulating fresh interest, because of the about 30% power savings possible [19]. Since aluminum chloride melts at much lower temperatures and forms a much more fluid melt than the standard Hall-Heroult electrolyte matrix, much higher voltage efficiencies are possible, but sublimation and control problems limit the utility of direct, one-component electrolytic methods. The basis of this idea is employed in the process developed by C. Toth of Alcoa which has the additional advantage of enabling clay sources of alumina to be tapped [20] (Fig. 12.4). While technical details of this process have not been

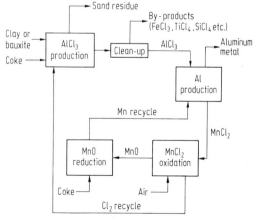

FIGURE 12.4 Simplified flowsheet of new Alcoa aluminum chloride-based route to aluminum. (Information from *Canadian Chemical Processing* [19] and *Chemical and Engineering News* [21].)

released, production probably involves the sequence of reactions given in Eqs. 12.26–12.29.

$$2 \ Al_2O_3 + 6 \ Cl_2 + 3 \ C \rightarrow 4 \ AlCl_3 + 3 \ CO_2 \qquad 12.26$$

$$2 \ AlCl_3 + 3 \ Mn \xrightarrow[\text{anodes}]{\text{electrolysis}} 2 \ Al + 3 \ MnCl_2 \qquad 12.27$$

$$2 \ MnCl_2 + O_2 \ (\text{from air}) \rightarrow 2 \ MnO + 2 \ Cl_2 \ (\text{Deacon related}) \qquad 12.28$$

$$2 \ MnO + C \rightarrow 2 \ Mn + CO_2 \qquad 12.29$$

Prototype results with this process were sufficiently attractive for Alcoa to have started production in 1976 using this technology on a scale of about 13,500 tonnes/yr [15].

12.5. EMISSION CONTROL PRACTICES

12.5.1. Environmental Impacts of Bayer Alumina from Bauxite

The mining of bauxite for an alumina product has the same kind and degree of environmental impact as imposed by any type of strip mining operation. These effects may be minimized by spoils replacement and aesthetic contouring, as well as by the replanting of hardy grasses and shrubs, where appropriate.

Processing of bauxite to produce alumina also produces volumes of "red muds" which contain principally water, iron oxide, silica, and the oxides of titanium, chromium, vanadium, and aluminum. The solids in this mixture do eventually settle to a relatively high solids content sludge, so that a moderately sized holding pond may be used for many years.

However, as environmental concerns, land costs, and raw material costs rise, alumina production facilities are seeking ways to decrease or eliminate the volume of this red mud waste. Alcoa has devised the "Alcoa-Bayer" process to enable recovery of much of the formerly discarded alumina fraction, as has been lost in the red muds, as combined alumina precipitated with silica and sodium hydroxide [22]. By adding limestone and sodium carbonate and calcining the mixture to a clinker, it is possible to extract the alumina values out with water and return this extract to the main Bayer circuit. Decreasing the alumina content of the red muds in this way not only directly decreases red mud volumes by removing a former significant constituent, but it also indirectly decreases the volume of waste from the decreased volume of bauxite that has to be processed for a given number of tonnes of alumina product.

Other tests have shown that the residual Al^{3+} and Fe^{3+} ions present in red muds can be effective and inexpensive flocculants for sewage treatments which are particularly efficient at removing phosphate. When processed to a finely divided dry powder, this might prove to be a generally attractive option to yield a salable commodity from a waste material. More recently dried, pelletized red mud, which is essentially iron oxide, has been reduced to steel via an electric prereduction and smelting process [23]. This technology could be particularly useful in bauxite processing areas where conventional iron ores are low grade, scarce, or nonexistent. Alumina calcination and grinding particulate discharge regulations prior to 1971 were set at 100 and 150 mg/m^3 air (standard conditions), respectively, by the VDI (German Engineering Society) [24].

12.5.2. Aluminum Smelter Emission Control

Emission problems of conventional aluminum smelters center on fluoride losses, which preabatement (before ca. 1972) was at the rate of some 21 kg/tonne of aluminum produced (Table 12.7). The bulk of this fluoride loss occurred from the operating electrolytic cells, and two-thirds or more of this was gaseous fluoride. The chief constituents of the fluoride discharge are known to be cryolite (Na_3AlF_6), aluminum fluoride, calcium fluoride, chiolite ($Na_5Al_3F_{14}$), silicon tetrafluoride, and hydrogen fluoride. A rough guide as to mass discharge rates may be obtained from consumption rates of these commodities by the industry, in relation to the volume of aluminum produced (Table 12.7). Most of these substances are lost as fumes or vapors, except the hydrogen fluoride gas evolution which results from the reaction of traces of moisture present in alumina added to the cell (Eq. 12.30).

$$3\ H_2O + 2\ AlF_3 \rightarrow Al_2O_3 + 6\ HF \qquad 12.30$$

This mode of fluoride loss is the reason that alumina shipment and transfers are conducted with minimum exposure to air, when moisture adsorption would occur. In addition to losses of these substances the extremely stable fluorocarbons, carbon tetrafluoride, and hexafluoroethane, which are relatively nontoxic, are known to form during an anode effect [6] (Eq. 12.11, 12.12). Practical abatement measures have been proposed [28].

TABLE 12.7 Preabatement and Postabatement Atmospheric Fluoride Emission Rates in Primary Aluminum Smelters Operating in the U.S.A. and Canada[a]

	Kilograms of fluoride emitted per tonne aluminum		
	Uncontrolled potlines[b]	Partial abatement	Efficient abatement
Prebaked anodes:			
Gaseous	14	13	—
Particulate	9	8	—
Total	23	21	1.7
Range	12–33	5–21	0.6–2.3
Vertical stud Soderburg anodes:			
Gaseous	18	13	3.4
Particulate	2–9	2–4	ca.1
Total	20	21	4.4
Range	18–27	16–22.5	1.1–4.3

[a]Compiled from Brewer *et al.* [6], Iverson [25], Rose and Marier [26], and OECD [11,27].
[b]In 1973 represented only 3% of primary aluminum producing facilities in the U.S.A. [22].

While fluoride emissions have undoubtedly been the impact area of greatest concern, a number of other gases, some arising from the carbon anode baking plant, are also significant. These include carbon monoxide, sulfur dioxide, hydrogen sulfide, nitrogen oxides (NO and NO_2), carbonyl sulfide (COS), and carbon disulfide. In a smelter employing Soderberg anodes, in which the pitch component of the carbon paste is largely vaporized during the *in situ* baking, there will also be measurable generation of hydrocarbon vapors and smokes, including polynuclear aromatic hydrocarbons. The concentration of hydrocarbon vapors collected by the hooding can run to about 3%, but this is decreased to about 0.1% on passage through the carbon monoxide burner [29]. Loss estimate ranges in kilograms per tonne of aluminum produced are sulfur dioxide, 1–30; particulates (including alumina, soot, etc.) 5–10; and hydrocarbon vapors (Soderberg type cells only), from 0.25–2 [22].

There are also problems dealing with liquid wastes, such as fluoride-contaminated discharge water from wet scrubbers, or drainage of precipitation from areas where spent pot linings are discarded, since during use the carbon becomes impregnated with the fluoride constituents of the electrolyte. Water used for cooling metal castings or transformers is not contaminated. Disposal of spent pot linings or discarded prebaked anodes requires care to minimize problems of the type mentioned above. These are solved by some facilities by fluoride recovery followed by recycle of the carbon content.

Of all the pollutants outlined, fluorides represent the aluminum smelter discharge component of greatest hazard to plant life, and also indirectly to grazing and predator animals [30]. For example, while the injury threshold to plants for sulfur dioxide is about 0.1 ppm (100 ppb) the equivalent value for fluoride to fluoride-sensitive plants is <1 ppb [23]. Thus, since fluoride emission has generally represented the largest mass loss to the atmosphere and it may disperse widely by a variety of ways it has the potential to produce a significant biotic impact [31]. This, therefore, is the area that has received the most significant control attention [32].

12.5.3. Smelter Emission Control Strategies

Aluminum smelting control systems may be considered in two groupings. The primary control system includes the hoods and ductwork for each cell, and the common duct which receives the vent gases from a bank of cells. While the primary system is able to collect and deal with the bulk of the fume and vapor discharge from a cell, particularly when the electrolyte crust betwen the collection skirt around the anode(s) and the rim of the cell is intact, many normal cell operations require this crust to be broken. Scheduled additions of alumina (several times a day), removal of aluminum (about daily), and periodic electrolyte analysis, as well as dealing with the occasional "anode effect" all require breaking the electrolyte crust. For the period of the operation, plus some additional time for a new crust to form, the loss rate to the cell working area for most types of cells is significantly raised. It is this type of loss that the secondary collection system is designed to handle [21]. This system consists of a series of forced circulation or convective roof exhaust vents together with floor level inlet vents in the sidewalls of the building. These measures help to minimize the occupational particulate exposure of cell room operating staff [33].

The primary collection system, where the bulk of the cell fluoride loss is captured, may simply discharge to a tall stack [22]. This aids in dispersal, which may be sufficient for a small smelter. But it does nothing to decrease mass discharge rates, it does not allow recovery of fluoride for reuse, and it has been found to cost more to operate in some cases than several true abatement methods which employ fluoride capture.

Wet scrubbers are the fluoride containment system that has been used the longest for cleanup of primary system gases before discharge, partly because of the high affinity of hydrogen fluoride for water. The best of these systems, which use a wet scrubber followed by a wet electrostatic precipitator, have been found to be 98–99% efficient for collection of both gaseous and particulate fluorides [25]. Most wet scrubber configurations have no difficulty achieving 98+% mass containment for gaseous fluorides although some have difficulty trapping as much as 90% of the particulate mass and achieve efficiencies of little better than 10% on the particle fraction less than 5 μm in diameter [22]. The other consideration which has stimulated the development of alternative primary control systems is the need to have a method of dealing with spent scrubber liquor, preferably to regenerate fluorides of value for electrolyte make-up (Eq. 12.14, 12.15). This can be relatively straightforward if a sieve plate column is used for scrubbing, since a spent scrubber liquor containing 4% hydrogen fluoride by weight may be obtained directly [25] (Fig. 12.5).

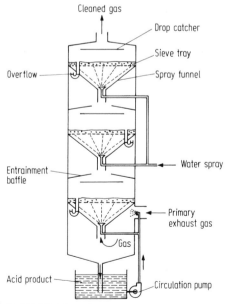

FIGURE 12.5 Hydrogen fluoride removal from smelter primary gas collection system via a sieve tray gas scrubber (only a section shown). Note the recycle of acid product for initial hydrogen fluoride absorption from the exhaust gases to maximize the concentration of recovered acid. (Adapted from Ball and Dawson [29], with permission.)

Various dry scrubbing configurations have been applied to primary control systems for fluoride including passage of the gases through coated filters or fluidized beds, or injection of alumina into the waste gas stream followed by recovery of the alumina plus adsorbed fluorides. The dry filter is more efficient than most wet scrubber arrangements for capture of fluoride particulates but the early versions were only about 90% efficient at controlling gaseous fluoride [21]. However, both the fluidized alumina bed (Fig. 12.6) and the alumina injection dry scrubbing systems (Fig. 12.7) make good use of the high adsorptive power of γ-alumina to control both gaseous and particulate fluoride as efficiently as good wet scrubbers, 98+% [34, 35]. This form of alumina is made at a lower degree of calcination (lower aluminum hydroxide calcination temperatures), possesses a mosaic crystal structure, and has a high specific surface area of 60+ m^2/g, highly suitable for adsorption [36]. Ordinary α-alumina, which forms a part of most commercial aluminas, is obtained from calcination temperatures above 1000°C as hard particles which are resistant to hydration or adsorption and is consequently less useful in this function. These two dry methods have the further advantage that the spent alumina merely has to be heated to remove adsorbed hydrocarbons which are burned; at the same time aluminum fluoride is formed (Eq. 12.31).

$$6\ Al_2O_3 \cdot HF \rightarrow 2\ AlF_3 + 5\ Al_2O_3 + 3\ H_2O \qquad\qquad 12.31$$

This allows the captured fluoride to be returned to the cells at the time alumina is added. These methods are used by several large North American aluminum smelters [37].

The philosophy of improvement of secondary control systems, that is, control of the gaseous and particulate discharges to the air of the buildings in which pot lines are operating, has been twofold. An efficient primary control system removes the need for secondary control. But where this is not the

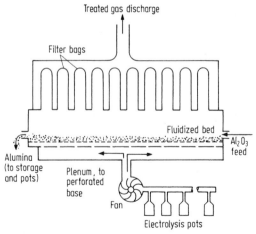

FIGURE 12.6 Typical arrangement of an alumina fluidized bed for primary emission control in electrolytic smelting of aluminum.

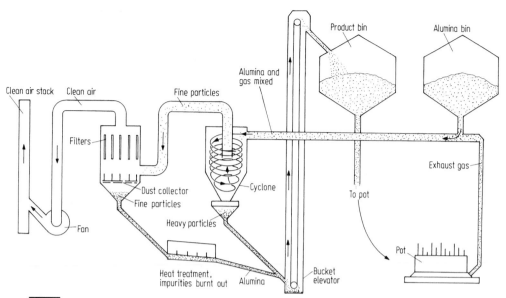

FIGURE 12.7 Typical arrangement of dry scrubber components, as used in aluminum smelter primary emission control. Bucket elevator is used to raise spent, fluoridated alumina to bin used to feed alumina to operating pots. Captured fluoride is directly recycled. (From Brewer *et al.* [6], reprinted courtesy of Alcan Smelters and Chemicals Ltd.)

case, such as with older facilities where gas collection efficiencies for primary control systems have been 40–60%, a relatively heavy loading of emissions is left to the wall ventilators and roof exhaust fans of the secondary control systems to disperse [22]. Low primary collection efficiencies also meant that the mass discharge rate of pollutants from the secondary control system was large enough to also require treatment, as abatement became a smelter requirement, to obtain any reduction in gross smelter discharge rates. At the secondary treatment stage the concentration of pollutants in air is much lower, and the volume of air to be treated is very much larger than that obtained from the primary collection system, because of the restricted air flow into the hooded cell enclosures of the latter. The large size of treating equipment required to handle the large volumes of air moved through the secondary collection system makes secondary treatment costs very high relative to the overall degree of improvement achieved [25]. This last factor is what has provided a strong stimulus for improvement of the collection efficiency of the primary control systems of the electrolytic cells, even by retrofitting older facilities where this is feasible.

By improvements in the proportion of the cell superstructure which is hooded, as well as by such procedural changes as enclosing the crust breaking/ alumina addition operations and increasing the level of sophistication used in timing these have combined to obtain an increase in primary collection efficiencies to about 95%, at least for new facilities [29]. Cell operations, such as electrolyte analyses and periodic aluminum removal, require the opening of hooded access doors, which makes 95% containment about the practical limit for primary control. Some of these improvements comprise a part of the technology package being promoted by Sumitomo [38].

The major advantage to secondary control systems contributed by primary emission control of 90+% collection efficiency is that secondary air emission abatement procedures are less critical. This means that the pollutant mass discharge rate from the roof vent secondary control system now sometimes exceeds that from the cleaned gas discharge from the primary control system. As the containment efficiencies of primary control systems improve, the need for secondary control is minimized. However, there will always be a ventilating requirement, if for no other reason than to shed the enormous heat load from the operating pots.

Liquid effluents from the operation of wet scrubbers and collected precipitation runoff from the operating and waste disposal sites require treatment prior to discharge. Fluoride removal is straightforward using lime addition to a one- or two-stage effluent lagoon system (Chap. 10). But hydrocarbons and phenols, mainly contributed by effluents from scrubbers operating on the gases from Soderberg cells, have proved to be more of a problem [22]. Traces of fluorides and other salts present in treated waste water can be safely discharged into seawater with a high salt concentration [39], but would require prior treatment for hydrocarbon removal. Discharge of fresh wastewater into the marine environment should be to the bottom of the receiving waters. Bottom discharge to the denser saltwater promotes active vertical mixing, particularly when into zones with active tidal exchange. Without these measures stratification of the fluoride-containing waters into top layers will tend to occur and dispersal could be slow [26].

Pot linings which are discarded after 3–5 years of continuous use still contain 10–15% of absorbed fluorides, which could be an environmental hazard. However, increased costs for electrolyte fluoride replacement have stimulated more smelters to crush used pot linings and recover the fluoride content by extraction with aqueous base. The filtered extract is then used for the preparation of synthetic cryolite. If spent pot linings are discarded to a landfill, any drainage from the site has to be treated to remove fluoride. This method of disposal also has a risk of occasional fires from spontaneous ignition, which have to be dealt with, and could restrict subsequent uses.

12.6. PROPERTIES AND USES OF ALUMINUM

Special properties of the metal itself have led to the rapid rise in the applications and in the volume of production of aluminum. Its natural corrosion resistance, due to a firmly bound chemically inert oxide film, contributes to its utility for many uses. The construction industry uses aluminum for window frames, exterior doors and their fittings, decorative and insulating panels, siding, roofing, etc. Anodizing, an anodic electrical passivation of the alumina surface in chromic or sulfuric acid, is used to further enhance corrosion resistance for some construction applications as well as for some smaller articles [40]. Anodizing can also be used to permanently impart colors other than the natural silvery-gray of aluminum to the surface of the article. Some highly polluted atmospheres will cause pitting or formation of a poorly bound powdery white coating on aluminum articles exposed to these conditions, but otherwise no coating finish is required. Mercury too, should not be allowed to come into contact with an aluminum surface since amalgamation destroys the protective ability of the surface-oxidized aluminum. This is the reason for the strict ban on air freight shipment of mercury.

Aluminum, with a density of 2.70 g/cm^3, is the lightest of all metals which are permanent (stable) in air, except for magnesium (1.74 g/cm^3) and beryllium (1.85 g/cm^3). This property, plus its relatively low cost adds to its advantages in construction-related uses, and also makes it a valued component in the transportation industry. Many components of aircraft, the load carrying components (rather than running gear) of unit rail cars and rapid transit systems, truck cabs and bodies, and some engine components are made of aluminum or alloys. In the automotive industry the applications are more extensive, including not only some of the above, but also wheels, trim, decorative components, and transmission housings. Light weight is also an important rationale for the choice of Al/Mg or Al/Mg/Si alloys in many types of small boats for use in fresh- or saltwater.

Aluminum is also an excellent electrical and thermal conductor, the latter being a contributing property to the choice of this metal in some of the transportation uses outlined above. The intrinsic electrical resistance (i.e., for a test piece of the same dimensions) is 2.655 $\mu\Omega/cm$ which gives it 63% of the conductivity of copper and 60% of that of silver. But, for an equivalent *weight* of conductor per unit length, aluminum is slightly more than twice as good a conductor as copper, and even better than this relative to silver, the reason for its use in long-distance high-tension transmission lines. Often, the cables used will be a composite of a high tensile strength steel core surrounded by a larger number of aluminum conductors. Great care has to be taken in making proper electrical connections to aluminum because of the nonconductivity of the thin aluminum oxide surface coating always present, complicated by the plasticity and high coefficient of thermal expansion of the metal itself. These properties have led to problems in some aluminum house wiring installations.

Aluminum's good thermal conductivity (and nontoxicity) has led to extensive use in domestic cookware as well as in industrial heat exchangers, where its very good chemical resistance (e.g., to hydrogen peroxide, nitric acid) is also valued. For cryogenic applications such as components of liquid air distillation plants, the conductivity, low-temperature toughness, and high reflectivity are also important properties. Related to cryogenic uses, aluminum also has a very low diffusivity (ease of gas transmission through intact metal), less than half that of steel, a property valuable in the construction of double-walled evacuated vessels. Some steels also become very brittle at low temperatures, another reason for its unsuitability relative to aluminum for these types of applications.

Compatibility with most foodstuffs plus its light weight and relatively high strength have led to extensive use of aluminum in the food packaging industry. Aluminum cans are used for soft drinks and beer, an active area for metal recycling, as well as for aerosol, meat, prepared pudding, and fish cans. Improved metal properties and packaging technologies have reduced the weight of aluminum per can to one-third of that used 20 years ago. Heavy foil is used for frozen food packaging, which doubles as a preparation container as well. Lighter foil is used for all types of packaging either on its own or in laminates for foodstuffs, tobacco and confectionery products, and for other materials where its light weight, conformability, and low gas transmission characteristics prove useful.

12.7. PREPARATION AND USES OF ALUMINUM COMPOUNDS

Alumina itself, the product of the Bayer process, is important for many industrial applications other than for the production of aluminum. The red mud wastes, which comprise the impurities removed from bauxite during alumina preparation, are employed in steel making, the manufacture of red pigments, and for the production of chemicals used for water and sewage treatment [41].

By heating the washed aluminum hydroxide (alumina hydrate) to relatively moderate temperatures of about 500°C, most of the water is removed without significantly altering the ionic arrangement, leaving a highly porous, surface active product. Activated alumina, obtained in this way, is an invaluable industrial dehydrating agent and is also extensively employed as a catalyst support in the petrochemical industry. The large surface area, commonly 40–65 m^2/g, provides for good distribution of intrinsic activity of the catalyst, or contributes to the effectiveness of other catalytically active agents previously placed on this surface. Alumina-silica composite systems are also of catalytic value.

Calcination of washed aluminum hydroxide at 1100–1300°C causes shrinkage of the particles as they dehydrate and yields a denser, harder product, an (α-alumina, useful as a constituent of abrasives. The low porosity and high temperature stability (up to ca. 1800°C) of this product also make it appropriate for the production of refractories, either as bricks using a little fireclay as a binder, or as a loose fill. Ceramics, with an alumina content of 85–95% and fired for up to 2 days at 1450°C, possess superior strength, hardness, and electrical and thermal insulating properties. Typical uses of these materials include as spark plug insulators and textile guides as well as demanding applications in the chemical industry. Composites of α-alumina fiber with a metal are showing great promise in achieving from four to six times improved stiffness, and two to four times increased fatigue resistance over the unreinforced metal [42].

Aluminum sulfate ($Al_2(SO_4)_3$) or alum is an aluminum salt valuable for water treatment (Chap. 5) and also extensively used in papermaking and to a smaller extent in the leather, drug and cosmetic, and dye industries [43]. Strictly speaking an "alum" is a double salt of a monovalent metal sulfate and a trivalent metal sulfate crystallized with 24 waters, i.e., $(M^+)_2SO_4 \cdot (M^{3+})_2(SO_4)_3 \cdot 24H_2O$. The monovalent cation is usually sodium, potassium, or ammonium, and the trivalent cation, aluminum. Sometimes chromium or iron is substituted for the aluminum. However, since most traditional applications for alums use the value of the trivalent ion, aluminum sulfate hydrate itself has gradually become known in industry as alum, or "filter alum." It is made by stirring bauxite with hot sulfuric acid (Eq. 12.32), settling, and then treating with a reducing agent such as sodium sulfide to convert any ferric salts to the less colored ferrous equivalents.

$$Al_2O_3 \; (+ \; Fe_2O_3, \; SiO_2, \; etc.) + H_2SO_4 \rightarrow Al_2(SO_4)_3 + residue \qquad 12.32$$
bauxite

The decanted solution is then concentrated hot, allowed to cool, and crystals of the octadecahydrate, $Al_2(SO_4)_3 \cdot 18H_2O$, are filtered off. Partial dehydration of the crystals during drying increases the aluminum analysis of the commercial product to approximately an $Al_2(SO_4)_3 \cdot 14H_2O$ stoichiometry. One variant of this route, used for applications such as dyeing which demand a low iron content, is the use of purified alumina rather than bauxite as the source of the aluminum ion. Alum production by either method provides a potentially profitable outlet for large quantities of by-product sulfuric acid, particularly if there is a local market for a solution of the alum, since solutions may be economically shipped short distances without evaporation.

Aluminum chloride is another commercially important aluminum compound, valued as a Lewis acid catalyst in many types of petrochemical reactions and for other process functions. The usual procedure is to melt clean factory scrap (offcuts from door and window manufacturing operations and the like) and then pass in dry chlorine gas, beneath the surface of the melt (Eq. 12.33).

$$2 \; Al \quad + 3 \; Cl_2 \rightarrow \quad AlCl_3$$
$$\text{m.p. } 660°C \qquad\qquad \text{m.p. } 190°C \qquad\qquad 12.33$$
$$\text{(at 2.5 atm)}$$

The melt is hot enough to sublime the aluminum chloride out of the melt as it forms (sublimes at $>178°C$), and the product is continuously collected on a water-cooled, scraped surface heat exchanger. The largest single use for this anhydrous product is for commercial alkylation reactions, and in particular the alkylation of benzene to ethylbenzene enroute to styrene (Chap. 19). Aluminum chloride made by this route is only slightly more expensive (at $1.80/kg) than aluminum because of the large molar quantity of relatively inexpensive chlorine (23 cents/kg) added to the relatively expensive aluminum (ca. $1.60/kg) [44]. Liquid (solution in water; specific gravity 1.28) aluminum chloride is a by-product of the manufacture of the anhydrous material. The solution may also be produced by treating alumina with hydrochloric acid (Eq. 12.34).

$$Al_2O_4 + 6 \; HCl \rightarrow 2 \; AlCl_3 + 3 \; H_2O \qquad 12.34$$

This product is employed in the preparation of antiperspirants and some types of pigments.

Virtually equivalent purity anhydrous aluminum chloride (ca. 99.8%) is also available as a side stream from Alcoa's new aluminum chloride route to the metal [45]. This process is said to require 30% less energy than either of the procedures described above, and gives a fine-grain, free flowing product that is achieved from a fluid bed reactor.

REVIEW QUESTIONS

1. (a) What mass of sodium would be required theoretically to produce 1.00 tonne of aluminum metal from aluminum chloride?
   (b) How would the theoretical and actual number of Faradays compare for electrochemical reduction of sodium, then use of sodium for Al^{3+} reduction to 1 tonne of aluminum versus direct electrochemical reduction of Al^{3+}?
   (c) How would the theoretical and actual power (kWh/tonne) requirements compare for the two options of part (b)?
2. What is the cause of the "anode effect" as it relates to an aluminum electrolytic cell, and how is this problem corrected?
3. Outline the options that an alumina production facility has to convert the iron-rich "red muds" (a potential waste disposal problem) to salable products. Give equations for their production and describe their applications.
4. What is a "dry scrubber" and how is this device used to help control fluoride loss during the electrolytic production of aluminum?
5. Describe, with equations, the primary means by which aqueous fluoride loss is controlled for aluminum and for phosphorus production facilities.

FURTHER READING

Aluminum Co. of Canada Ltd., "Handbook of Aluminum," 3rd ed. Montreal, 1970.

American Society for Metals, "Aluminum: Properties, Physical Metallurgy and Phase Diagrams," Vol. I. Metals Park, OH, 1967.

National Research Council, "Fluorides." Committee on Biologic Effects of Atmospheric Pollutants, Washington, DC, 1971.

Singmaster and Breyer, "Air Pollution Control in the Primary Aluminum Industry." Singmaster and Breyer Consultants, New York, 1973.

Surinam: 62 Years of Bauxite Development, 12 Years of Alumina, Aluminum Production, *Eng. and Mining J.* **178**, 75 (1977).

REFERENCES

1. "Kirk-Othmer Encyclopedia of Chemical Technology," 2nd ed., Vol. 1, p. 929. Wiley, New York, 1963.
2. "1994 Minerals Yearbook," Vol. 1. U.S. Dept. of the Interior, Bureau of Mines, Washington, DC, 1996, and earlier years.
3. "Kirk-Othmer Encyclopedia of Chemical Technology," 3rd ed., Vol. 2, p. 129. Wiley, New York, 1978.
4. "World Mineral Statistics 1975–79." Institute of Geological Sciences, H. M. Stationery Office, London, 1981, and 1972–1976 issue.
5. M.E. Afonso de Magalhaes and M. Tubino, Recovering Gallium from Residual Bayer Process Liquor, *JOM,* **43**, 37–39, June (1991).
6. R.C. Brewer, J.B. Brodie, L.D. Kornder, W. Papenbrock, and W.G. Wallace, "Environmental Effects of Emissions from the Alcan Smelter at Kitimat, B.C." Ministry of Environment, Province of British Columbia, 1979.
7. "McGraw-Hill Encyclopedia of Science and Technology," Vol. 1, p. 326. McGraw-Hill, New York, 1977.

8. R.C. Weast, ed., "Handbook of Chemistry and Physics," 51st ed., p. D-111. Chem. Rubber Publ. Co., Cleveland, OH, 1970.

9. "The Chemistry of Aluminum." Aluminum Co. of Can. Ltd., Montreal, 1970.

10. S.H. Patterson, *Amer. Sci.* **65**, 345 (1977).

11. Organization for Economic Cooperation and Development, "Pollution Control Costs in the Primary Aluminum Industry." OECD, Paris, 1977.

12. R.D. Peterson, Issues in the Melting and Reclamation of Aluminum Scrap, *JOM* **47**, 27–29, Feb. (1995).

13. Nonbauxite Aluminum Technology, *Chem. Eng. News* **52**, 19, Dec.16 (1974).

14. J.V. Thompson, Alumina: Simple Chemistry—Complex Plants. *Eng. Min. J.* **196**, 42–49, Feb. (1995).

15. M.K. McAbee, *Chem. Eng. News* **53**(31), 19, Aug. 4 (1975).

16. M.V. Chaubal, J.L. Anjier, B.J. Welch, R.D. Peterson, M.A. Smith, J.H. van Linden, and G.J. Kipouras, Light Metals 1994: Advances in Aluminum Production, *JOM* **46**, 14–23, Aug. (1994).

17. Recovery of Aluminum, *Can. Chem. Process.* **51**(3), 75 (1967).

18. New Aluminum Smelting Process, *J. Met.* **29**, 6, Nov. (1977).

19. Alcan's Work, *Can. Chem. Proc.* **57**(3), 52, Feb. (1973).

20. Aluminum Process, *Chem. Eng. News* **57**, 23, April 30 (1979).

21. New Processes Promise Lower Cost Aluminum, *Chem. Eng. News* **51**, 11, Feb. 26 (1973).

22. N.L. Nemerow, "Industrial Water Pollution, Origins, Characteristics and Treatment," p. 502. Addison-Wesley, New York, 1977.

23. Red Mud Converted to Steel, *Chem. Eng. News* **57**(9) 26, Feb. 26 (1979).

24. H.F. Lund, ed., "Industrial Pollution Control," p. 4–20. McGraw-Hill, New York, 1971.

25. R.E. Iverson, *J. Met.* **25**, 19, Jan. (1973).

26. D. Rose and J.R. Marier, "Environmental Fluoride." National Research Council of Canada, Ottawa, 1977.

27. Organization for Economic Cooperation and Development, "Air Pollution by Fluorine Compounds from Primary Aluminum Smelting," OECD, Paris, 1973.

28. M.J. Gibbs and C. Jacobs, Reducing PFC Emissions from Primary Aluminum Production in the United States. *Light Met. Age* **54**, 26–34, Feb. (1996).

29. D.F. Ball and P.R. Dawson, *Chem. Process. Eng. (N.Y.)* **52**(6), 49, June (1971).

30. C.E. Carlson, W.E. Bousfield, and M.D. McGregor, *Fluoride* **10**(1), 14, Jan. (1977).

31. M.B. Hocking, D. Hocking, and T.A. Smyth, *Water, Air, Soil Pollut.* **14**, 133 (1980).

32. B. Bohlen, *Chem. Eng. (London)* **221**, CE 266 (1968).

33. F. Akbar-Khanzadeh, Exposure to Particulates and Fluorides and Respiratory Health of Workers in an Aluminum Production Potroom with Limited Control Measures. *Am. Ind. Hyg. Assoc. J.* **56**, 1008–1015 (1995).

34. A.S. Reid, G.J. Gurnon, W.D. Lamb, and K.F. Denning, Effect of Design and Operating Variables on Scrubbing Efficiencies in a Dry Scrubbing System for V.S. Pots. *109th Am. Inst. Met. Eng. Annu. Meet.*, New York, p. 759–781 (1980).

35. D. Rush, J.C. Russell, and R.E. Iversen, *J. Air Pollut. Control Assoc.* **23**(2), 98 (1973).

36. J. Miller and D.F. Nasmith, Operation of Alcan/ASV Dry Scrubbing Units with Various Aluminas. *102nd Am. Inst. Met. Eng. Annu. Meet.*, A.V. Clack, ed., New York, pp. 191–208 (1973).

37. C.C. Cook, G.R. Swany, and J.W. Colpitts, *J. Air Pollut. Control Assoc.* **21**(8), 479 (1971).

38. Aluminum Cell Modifications Cut Energy Use, *Chem. Eng. News* **53**(31) 19, Aug. 4 (1975).

39. R.M. Harbo, F.T. McComas, and J.A.J. Thompson, *J. Fish. Res. Board Can.* **31**, 1151 (1974).

40. A. Jenny and W. Lewis, "The Anodic Oxidation of Aluminum and its Alloys." C. Griffin, London, 1950.

41. D.M. Samuel, "Industrial Chemistry—Inorganic," 2nd ed. Royal Institute of Chemistry, London, 1970.

42. Alumina Fibers Used to Strengthen Metals, *Chem. Eng. News* **58**(26), 24, June 30 (1980).

43. Aluminum Sulfate, *Can. Chem. Process.* **60**(9), 130, Sept. (1976).

44. "Chemical Marketing Reporter," **246**(24), p. 26, June 30. Schnell Publishing Company, New York, 1995.

45. R. Wood, *Process. Eng. (London)* **57**(9), 9, Sept. (1976).

Gyratory crusher used for initial breakdown of ore from lumps of up to a meter to less than 12 centimeters in diameter. Sign in foreground warns personnel of the risk of flying rocks while steel and rock walls protect other areas of the site from this hazard (see Fig. 13.1).

13

ORE ENRICHMENT AND SMELTING OF COPPER

Chawke may not beare the price of Cheese, nor cop-
per be currant to goe for paiment.
                    —*Thomas Nashe, 1590*

The vessel, though her masts be firm,
Beneath her copper bears a worm.
                    —*Henry D. Thoreau (1817–1862)*

13.1. EARLY DEVELOPMENT

Copper rivals gold as one of the oldest metals employed by humans. Its first use about 10,000 years ago was stimulated by the natural occurrence of the metal in lumps or leaves in exposed rock formations, so called "native copper." These exposures enabled the fashioning of simple tools directly by hammering and heat working of these fragments, and in so doing signaled the end of the Stone Age. The natural occurrence of elemental copper is a feature of its relatively low standard reduction potential of +0.158 V, significantly below hydrogen in the electromotive series and hence relatively easily accumulated in elemental form as a consequence of normal geologic processes. This contrasts with the much more easily oxidized aluminum, which lies above hydrogen in the electromotive series with a reduction potential of −1.71 V, and hence is never found in elemental form in nature.

More purposeful recovery of the metal by mining is known to have occurred since about 3800 B.C. from early workings discovered in the Sinai. Mines operating in Cyprus around 3000 B.C. were later taken over by the Roman Empire, and the metal product was called *cyprium*, later simplified to *cuprum*, the origin of the Latin name still used for the metal. The incentive for the development of these early metal mines probably came from the experiments of early man who first relied upon simple hammering to shape native copper lumps. Subsequently this was elaborated to heat working and hammering, eventually leading to a need to melt and cast the metal to allow more fabrication scope [1]. It must have first been discovered that coal or

383

charcoal firing was necessary to obtain sufficiently high temperatures to melt the native metal, which is apparently not possible with even a large, hot wood fire. At some time after this must have come the revelation that some highly favorable ores of copper, which did not originally look like copper, could be reduced to the metal under these conditions [2]. The series of steps required for early experimenters to proceed from a knowledge of the requirements to melt metal to the intentional practice of ore reduction by heat in the presence of carbon is still open to speculation. The reddish sheen of the polished metal undoubtedly provided impetus to the early practitioners to develop fabrication methods for decorative and artwork, weapons, tools, and other practical needs.

From these early beginnings world copper production reached about 8100 tonnes per year by 1750, 19,000 by 1860, and 50,000 by 1880 [3]. Production of the eighteenth and early nineteenth century was dominated by smelters in Mansfeld, Germany, and Swansea, Wales, the latter center accounting for some three-quarters of known world production in 1860. However,the importance of both centers for the smelting of indigenous ores today is slight to nonexistent (Table 13.1). The Rio Tinto Co. in Spain was dominant by 1880 producing in the neighborhood of 25,000 tonnes of the metal per year, and

TABLE 13.1   **Major World Producers of Copper Concentrates and Primary Smelted Copper**[a]

	1960		1970		1980		1990	1994[b]
	Thousands of metric tonnes, copper content							
	Ore	**Metal**	**Ore**	**Metal**	**Ore**	**Metal**	**Metal**	**Metal**
Australia	111	72	142	111	217	171	192	315
Canada	399	361	610	453	710	463	476	515
Chile	536	502	711	658	1068	953	1210	1280
China	n/a	n/a	100	100	200	200	359	480
Japan	89	188	120	513	53	861	893	1030
Peru	182	164	218	177	365	350	196	244
Poland	11	22	83	69	346	364	331	360
South Africa	46	46	148	149	215	181	176	166
U.S.A.	980	1119	1560	1489	1168	1008	1160	1310
U.S.S.R.	n/a	n/a	925	925	900	905	1370	—
West Germany	2	62	1	80	1	160	180	210
Yugoslavia	33	36	91	108	134	110	174	—
Zaire	302	302	387	386	459	426	346	60
Zambia	550	568	836	683	596	617	384	381
Other	347	220	718	364	945	−37[d]		
World totals	3640[c]	3670[c]	6500	6320	7890	6828	9680	9750

[a]Data for 1975 and earlier compiled from *U.N. Statistical Yearbooks* [4], and more recent data from *Minerals Yearbooks*, [5]. The interfacing years may not entirely correspond.
[b]Estimates.
[c]Excludes China, Czechoslovakia, Hungary, North Korea, and the U.S.S.R.
[d]Negative value due to rounding and the inclusion of undifferentiated production, as to whether primary or secondary (recycled), e.g., for Poland.

this country still has a significant output, though production levels have fluctuated widely in recent years. By 1900, the Anaconda Company, Montana, was producing around 50,000 tonnes per year. All of the production development described above relied on relatively rich (for copper) grades of ore containing from 2–5% copper [6].

Much larger scale mining and smelting of formerly substandard ores (ores of too low a copper analysis for practical smelting) became feasible and were practiced following the development around 1920 of froth flotation methods of ore enrichment. World production, which had reached levels of 1 million metric tonnes by 1912, climbed further to 1.5 million metric tonnes by 1930 with further growth to a volume of 2.3 million tonnes for both 1940 and 1950, stimulated jointly by growing demand and the accessibility to lower grade reserves that the new enrichment methods provided. Froth flotation enrichment of ores has also stimulated a decreased dependence of smelter location on ore location since ore concentrates of 25 to 30% of a high value metal such as copper selling at 2.00–2.70 US$/kg during the 1990 to 1994 period, could justify shipping significant distances for metal recovery. For this reason some countries are large-scale ore producers only, e.g., Papua New Guinea and the Philippines; some produce both ore and metal but more of the former, e.g., Australia and Canada, many countries produce roughly equivalent amounts of ore and metal; and a few, e.g., Japan and West Germany, produce metal from mostly imported ore concentrates (Table 13.1). Another feature characteristic of the value of the metal is that copper is extensively reused. About one-third of the current annual consumption is provided by recycled scrap copper, and it is estimated that nearly 60% of the virgin metal produced is ultimately reused.

13.2. MINING OF ORES AND BENEFICIATION

Copper, being a less active metal, occasionally occurs naturally in elemental forms (e.g., native copper deposits of the "thumb" of Michigan near Lake Superior, and at Afton Mine, near Kamloops, British Columbia). However, the bulk of the commercial metal is produced from sulfide or oxide minerals. Common sulfide minerals are chalcopyrite ($Cu_2S \cdot Fe_2S_3$; 34.5% Cu), chalcocite (Cu_2S; 79.8% Cu), covellite (CuS; 66.4% Cu), and hornite ($Cu_2S \cdot CuS \cdot FeS$; 55.6% Cu), with many variations possible, both among these minerals and in admixtures with other sulfides. Sulfides of arsenic, antimony, gold, mercury, silver, and zinc, in particular, are found with copper ores because of the similarity of some of their chemical properties. The oxide minerals, which are less common than the sulfides, occur as cuprite (red copper oxide, Cu_2O; 88.8% Cu), tenorite (black copper oxide CuO; 79.8% Cu), malachite (basic, or green copper carbonate, $CuCO_3 \cdot Cu(OH)_2$; 57.3% Cu), azurite ($2CuCO_3 \cdot Cu(OH)_2$; 55.3% Cu), and other variations.

In the Americas, the minerals containing copper occur chiefly in porphyry form, i.e., they are igneous in origin and are widely dispersed in the rock. In Africa and Russia, however, porphyry ores account for less than 20% of the total. In all areas copper *ores*, or deposits of copper minerals which contain a sufficiently high copper concentration to permit economic metal recovery, have a relatively low copper analysis, in the range of 0.3 to 4% Cu. Many mines operate today on ores of 1% or lower copper content. Thus, a significant consideration for economical metal production is low-cost removal of igneous rock and recovery of the copper bearing mineral from this.

To minimize ore recovery costs, open pit mining is used wherever feasible. For instance, in the U.S.A. in 1970, open pit mines accounted for 84% of mine output (e.g., by the Kennecott Copper Corporation, Salt Lake City) and only 16% was recovered from underground mining operations. Additionally, by the use of low-cost explosives such as ammonium nitrate/fuel oil, or aluminum powder supplemented or slurry variants of this, and large-scale loading and haulage units, mining costs to withdraw the ore can generally be kept to less than 25% of the cost of the refined metal. Overburden containing little or no metal value is removed to a spoils disposal area, and the ore with some preselection proceeds to the concentrating circuit.

Concentrating, because of the wide dispersal and hence initial inaccessibility of the valued mineral in the rock, first requires size reduction. Crushing, in two stages in gyratory cone (Fig. 13.1) or jaw crushers, decreases the initial, randomly sized masses to first 15–22 cm average diameters, and then in a second unit to 2–5 cm (walnut-sized) fragments. The walnut size range is then suitable as feed to rod or ball mills (Fig. 13.2) for final wet grinding to 65 to 200 mesh. A "100-mesh" sieve has a nominal 100 openings per lineal inch, equivalent to 10,000 openings per square inch. Thus, a particle passing through a 100-mesh sieve will have an average diameter of 0.01 in., less the sieve wire thickness of 0.0043 in. or about 0.0057 in. (ca. 0.149 μm). For this particular example then, the 65- to 200-mesh range equates to particle

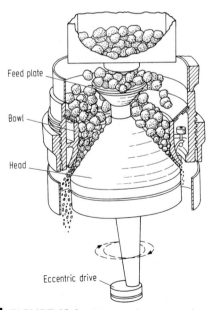

FIGURE 13.1   Diagram of a cone crusher. (Courtesy of Inco Ltd., Toronto.)

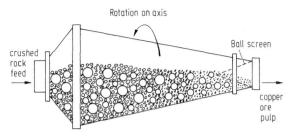

Rotation on axis

crushed
rock
feed

Ball screen

copper
ore
pulp

■■■ **FIGURE 13.2** Cross section of a ball mill used for the wet grinding of copper ore preparatory to froth flotation.

diameters in the 230 to 74 μm (0.0090–0.0029 in.) range, or averaging 150 μm (0.006 in.). Attainment of a larger particle size range than this decreases grinding costs, but also decreases mineral recovery from the rock by not exposing all copper minerals on the surface of a particle, which is required for flotation. Aiming at a smaller size range than this ensures mineral liberation from the associated rock, thus increasing the selectivity of separation in flotation, but raises grinding costs and may permit losses of nonfloated mineral-containing fines from the flotation circuit [7]. For these reasons the ground ore in water produced by the ball or rod mill is sorted in a cyclone (or similar) classifier into a fines stream, which is fed to the flotation circuit, and a coarse stream, which is returned to the grinding circuit for further size reduction.

13.2.1. Ore Beneficiation by Froth Flotation

Froth flotation is a procedure used to raise the low initial concentrations of minerals in ores to values that can be economically treated by smelting or hydrometallurgical extraction processes. For copper, the 25–30% range is suitable for economical smelting. The froth flotation technique was originally developed in about 1910 to raise the copper concentrations of the strip mined ores of Bingham Canyon, near Salt Lake City [8], and was further perfected for the differential separation of lead, zinc, and iron sulfides at Trail, B.C., at about the same time [9]. Now flotation technologies are widely used for such diverse tasks as the beneficiation of lower grade Florida phosphate ores from 30–40% to 60–70% concentrations of calcium phosphate (BPL), the separation of about 98% potassium chloride from sylvinite, a natural mineral mixture of potassium and sodium chlorides, and for several other mineral separations. It is also an important technique for bitumen separation from tar sand, removal of slate from coal, and removal of ink from repulped paper stock preparatory to the manufacture of recycled paper stock.

In general, flotation separations rely on the attachment of properly prepared mineral surfaces to air bubbles in water. Surfaces which are easily wetted by water, such as ordinary ground rock, are not attracted to bubbles and thus sink. Elemental sulfur, graphite (and other forms of elemental carbon), and talc, minerals with layer lattice type structures, have hydrophobic surfaces which are attracted to the surface of an air bubble and, if small enough, may be lifted to the surface of the water and separated from the water as a froth layer.

This distinction in behavior in aerated water is primarily dependant on the surface activity, or more precisely wettability, natural or induced, of the material under study. Three types of behavior are observed (Fig. 13.3). If a solid surface is hydrophilic, when a bubble is pushed against it the contact angle θ, is zero (nonexistent) or very small and the surface is relatively easily water wetted. If the bubble is released from a holder it will not stick to the surface, nor will small particles of the same material stick to a bubble. Finely ground material of this type, such as rock gangue, will sink.

If the test surface is neither readily wetted nor hydrophobic, i.e., it is indifferent to water, the contact angle is zero or very small and poor differentiation of the unprepared surface can be expected. However, the behavior of wetted and indifferent surfaces can often be readily modified by treatment with surface active agents to obtain either a sink or a float behavior, as desired.

If the solid surface is hydrophobic it will tend to develop contact with the air of a bubble pressed against it to give distinct contact angles. For example, the contact angle for paraffin wax is 104°. In this situation, there is a significant force component perpendicular to the solid surface to γLA, the liquid-air interface of the force vector diagram (Fig. 13.3), and hence there exists a significant force retaining the bubble to the solid surface. Small particles of this type of material will readily cling to small air bubbles and may be raised to a froth in a flotation cell.

The specific reduction in energy, WSA, from particle adhesion to, rather than release from a bubble by interaction of the three relevant interfaces is given by Eq. 13.1.

$$\text{WSA} = \gamma\text{SL} + \gamma\text{LA} - \gamma\text{SA} \qquad 13.1$$

This equation is credited to Dupré and S, L, and A represent the solid, liquid, and air surfaces, respectively [10] (Fig. 13.3). Combining this equation with the Young equation, 13.2,

$$\gamma\text{SA} = \gamma\text{SL} + \gamma\text{LA} \cos\theta \qquad 13.2$$

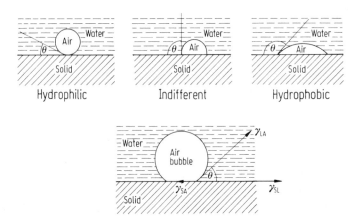

FIGURE 13.3 Behavior at the air-water interface of three types of solid surfaces and the relationship of the three interfacial tensions involved.

which relates the equality of the components of the three interfacial tensions involved, allows simplification of the Dupré equation to the form of Eq. 13.3.

$$WSA = \gamma LA(1 - \cos \theta) \qquad 13.3$$

Since γLA for water (ca. 70 dynes/cm^2) is readily determined and for a solution of a surfactant in water it is lower than for pure water and relatively constant, the contact angle θ is thus also theoretically shown to be a significant variable factor in the work required to obtain particle adhesion. However, it is known that several other factors influence particle separations such as particle and bubble size ranges, the nature and age since grinding of the particle surface, the pH of the aqueous phase, and its influence on particle-bubble interaction via neutralization of electrostatic charges on particles. Hydrodynamic effects which occur in agitated mineral suspensions are also of importance [11].

Fine particles of many sulfide minerals including those of copper are, like sulfur, naturally hydrophobic. If they are subjected to flotation soon after grinding they will tend to stick to bubbles and to be raised to a froth. Silica and gangue are wetted by water and will tend to sink.

The natural differentiation of valuable mineral from gangue in this way is good, but the "sticking" of sulfide minerals to air bubbles can be improved by the addition of a collector to the water phase. Typical collectors for sulfide minerals are xanthates ($ROC(S)S^-Na^+$), e.g., potassium ethyl xanthate, $CH_3CH_2OC(S)S^-\ K^+$, or dithiophosphates (($RO)_2P(S)S^-Na^+$), e.g., sodium diethyldithiophosphate, $(CH_3CH_2O)_2P(S)S^-Na^+$, which all have in common an oil-attractive (or hydrophobic) group and a sulfide-attractive group. When added to the pulp of ground-up ore in water at the rate of 10 to 45 g per tonne of ore processed just before flotation, these collectors confer greater than the natural water repellency and air avidity to the particles of the desired mineral. In this way the proportion of the total copper sulfide minerals recovered from the gangue and the rejection of gangue in the floated material are both improved.

If a bubble carrying air-avid mineral particles reaches the surface of the flotation unit and then bursts, the raised mineral particles will simply sink again. Bubble stability is maximized by addition of a foam stabilizer or frother which assists in generating a sufficiently stable foam layer on the flotation unit to enable the foam plus associated mineral to be skimmed from the water phase. The frother also puts an oily phase on the surface of each bubble as it forms, which helps the mineral-gangue differentiation function of the collector. Typical frothers, used at the rate of 20 to 45 g per tonne of ore, are oily materials of no more than slight water solubility such as pine oil (a mixture of terpenes) or a long chain (C5 or higher) alcohol such as 1-pentanol.

In addition to the two primary flotation agents there are also a number of modifying and conditioning reagents which are important to control a variety of ore and flotation circuit variables [11]. For instance, lime is a depressant for pyrite (FeS_2) in the flotation separation of copper sulfides. Lime would normally be added to the grinding circuit of a copper concentrator some time before flotation is carried out for oxidation of the pyrite and the pH regulating effect to take hold. Sodium carbonate and sodium hydroxide are also common pH regulating agents. Molybdenite (MoS$_2$), a valuable by-product of copper ore processing, is first floated with copper sulfides then depressed, to the underflow of a separate flotation cell, by the addition of colloidal dextrin to the pulp feed. There are also additives, such as sodium cyanide, which are used occasionally to cause depression of some minerals, as well as dispersants or deflocculants to avoid nonselective aggregation of particles of mineral value with particles of gangue [12].

Separation of copper sulfide minerals from the gangue is normally carried on in a series of large rectangular steel tanks of capacities from 2 to 28 m^3 equipped with stirrers and compressed air inlets (Fig. 13.4). These provide vigorous agitation and air blowing through the properly conditioned ground ore pulp. The froth collected from a series of rougher cells is then passed to a set of cleaner cells to further raise the copper sulfide concentrations (Fig. 13.5). The gangue-rich underflow from the rougher cells is passed through a series of scavenger cells to collect any residual copper sulfide values not picked up by the rougher cells. Dried flotation concentrates contain from 20 to 30% copper, and are frequently in the 25–28% range. In the concentration process, 80 to 90% of the metal content of the ore feed is recovered. The discarded gangue may still contain ca. 0.05% copper, which may or may not be processed further to recover additional copper values (Section 13.5.2).

Copper (or metal) recovery percentages of the flotation circuits of a copper mine/mill are extremely important. For example, it is estimated that for a plant processing 50,000 tonnes of ore containing 1% copper per day, increasing the copper recovery from 86 to 87% of that contained in the ore could increase revenues by approximately $4 million annually. This revenue increase, estimated from a value of $1.10/kg allowed for the copper content of the concentrate when the metal was selling at $2.25/kg, would be obtained

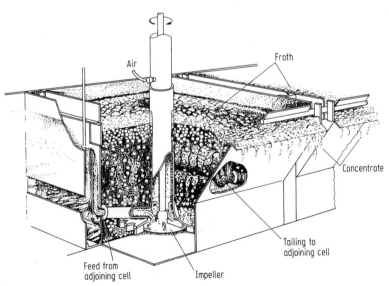

FIGURE 13.4 A cutaway drawing of a froth flotation unit for the concentration of copper ores with a detail photomicrograph showing mineral-loaded bubbles. (Courtesy of Inco Ltd., Toronto.)

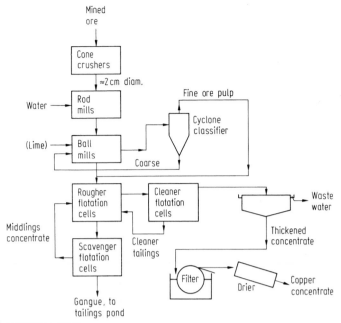

FIGURE 13.5 Flowsheet of copper sulfide ore pathway, mined ore to dried concentrate ready for smelting. In some operations there may be additional cells and the thickener, such as a molybdenite (MoS$_2$) recovery circuit or the like.

for the same overhead costs; hence the importance of seemingly minor flotation variables. Flotation reagents in the parts per million range are placed in the wet pulp prior to flotation at a cost of 10–35 cents per tonne of ore. Recent developments have been reviewed [13].

13.3. SMELTING AND REFINING OF CONCENTRATES

Much of the copper smelted today still uses procedures very similar to those developed by the early Welsh copper smelting industry [2]. The chemistry involved, which may be divided into four more or less discrete steps, is better understood today.

Roasting involves heating of the concentrate in the presence of air to temperatures of 750–800°C in a multiple hearth, or fluidized bed roaster (Fig. 13.6). The charge remains more or less solid throughout. This completes the drying step and removes a part of the sulfur from both the copper and iron present, producing oxides from some of the iron (Eq. 13.4–13.7).

$$2 \; CuS \rightarrow Cu_2S + S \qquad\qquad 13.4$$

$$FeS_2 \rightarrow FeS + S \; \text{(pyrite pyrolysis)} \qquad\qquad 13.5$$

$$S + O_2 \rightarrow SO_2 \qquad\qquad 13.6$$

$$2 \; FeS + 3 \; O_2 \rightarrow 2 \; FeO + 2 \; SO_2 \qquad\qquad 13.7$$

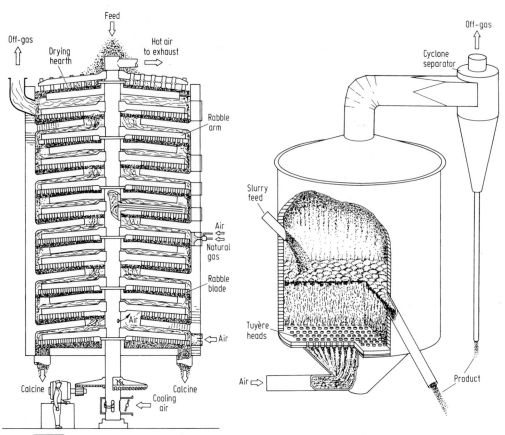

FIGURE 13.6   Diagrams of multiple hearth type and fluidized bed roasters, such as are used to decrease the sulfur content of copper sulfide ores. (Courtesy of Inco Ltd., Toronto.)

Some external fuel, usually oil or gas, is required initially to raise the roaster temperature to the desired range, and smaller amounts subsequently to maintain these temperatures since the sulfur combustion is highly exothermic and provides some of the heat. If the concentrate analyzes more than about 25% sulfur and is dry, sufficient heat is evolved from the sulfur burned to maintain the roaster temperature once started. This favorable situation is referred to as "autogenous roasting."

Some sulfur trioxide also forms in the roaster, even though this is not the direct objective of the operation. This forms small amounts of ferrous sulfate in the roaster output (Eq. 13.8).

$$SO_3 + FeO \rightarrow FeSO_4 \qquad\qquad 13.8$$

Many modern roasting operations now collect and use the sulfur dioxide present in roaster exit gases to produce sulfuric acid. Fluidized bed roasters, in particular, are favored for this purpose since the sulfur dioxide concentrations obtained may be as high as 15%, as opposed to the 3–6% range normally expected from the early traveling grate or multiple hearth roaster designs.

Roasting is followed by reverberatory smelting, carried out on the still solid, hot roaster output. Silica, as a fluxing agent, and slag from a previous converter stage are added to the roasted concentrate and then external heat is applied by the combustion of powdered coal, oil, or natural gas. The flames and hot gases are reflected down from the ceiling of the furnace onto the mixed charge (hence, "reverberatory" furnace). This stage of smelting, where the charge temperature is gradually raised from about 600 to 1000°C, accomplishes removal of more of the sulfur, by burning. Much of the iron is also removed as it combines with the fluxing agent to form a lighter ferrous silicate slag which floats on top of the copper melt, as it forms (Eq. 13.9–13.13).

$$Cu_2O + FeS \rightarrow Cu_2S + FeO \qquad\qquad 13.9$$

$$2\ Cu + FeS \rightarrow Cu_2S + Fe \qquad\qquad 13.10$$

$$2\ Fe + O_2 \rightarrow 2\ FeO \qquad\qquad 13.11$$

$$Cu_2S + 2\ CuO \rightarrow 4\ Cu + SO_2 \qquad\qquad 13.12$$

$$FeO + SiO_2 \rightarrow FeSiO_3\ (slag) \qquad\qquad 13.13$$

The transfer of a part of the iron-bound sulfur to copper, which is an objective of this stage, is the result of the change in the relative heats of formation of cuprous and ferrous sulfides at these high temperatures (e.g., at 1300°C, $\Delta H = -174$ and 34.4 kJ, respectively) relative to those at 25°C ($\Delta H = -79.5$ and 95.1 kJ, respectively).

The product of the reverberatory smelting stage is matte copper, containing primarily copper metal and cuprous sulfide. The analysis will range from 15% (particularly with nonconcentrated ores) to 50% copper [14]. Ideally, the copper concentration in matte copper is kept at 40–45% Cu to maintain the matte as an effective collector of precious metals, and to keep the copper loss in the slag to a minimum. Insufficient sulfur (as Cu_2S, melting point 1100°C) in the matte also makes it difficult to keep the charge molten (melting point of Cu, 1083°C; which is eutectic depressed in the matte copper).

In converting, the third stage of copper smelting, air is blown through the mass of molten matte copper to complete the oxidation of sulfides, since by this stage the bulk of the iron has been removed (Fig. 13.7) (Eq. 13.14, 13.15).

$$2\ Cu_2S + 3\ O_2 \rightarrow 2\ Cu_2O + 2\ SO_2 \qquad\qquad 13.14$$

$$Cu_2S + 2\ Cu_2O \rightarrow 6\ Cu + SO_2 \qquad\qquad 13.15$$

Any residual iron present is oxidized and forms a slag layer with an additional small portion of added silica (Eq. 13.13). This slag layer also captures any traces of oxidized arsenic and antimony, if present, that have not already been volatilized by the blowing. The less volatile metal oxides remaining are lithophilic ("rock-loving"), and thus tend to be absorbed by the slag layer. For the same reason, some oxidized copper also enters this slag layer, but is recaptured when the slag is used as a part of the charge for subsequent reverberatory smelting. Any precious metals present, silver, platinum, and gold are relatively stable toward oxidation and remain with the copper. The product of converting is termed "blister copper," from the pock marks on the surface

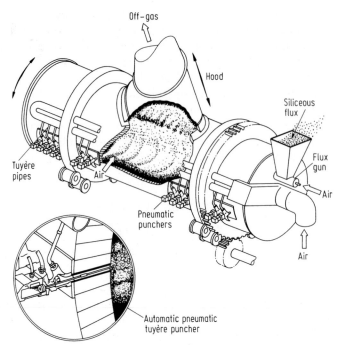

FIGURE 13.7   Diagram of a side blown converter used to separate an iron-rich ferrosilicate slag from a sulfur-rich matte copper. (Courtesy of Inco Ltd., Toronto.)

formed by the escape of sulfur dioxide bubbles from the metal as it cools and solidifies. It is virtually free of sulfur, but contains from 0.6 to 0.9% oxygen (ca. 5.4–8.1% Cu_2O) at this stage from the vigorous oxidation.

Fire refining, the final stage of copper smelting, is used to reduce the small percentage of cuprous oxide present in the blister copper back to elemental copper. It is often carried out in the converter itself, once the air flow has been cut off and while the charge is still molten. Originally fire refining was accomplished by the addition of coke and green wood poles, which were firmly pushed under the surface of the melt, to bring about reduction of cuprous oxide by the carbon of both additives. Vaporization of the moisture and partial combustion of the green wood also vigorously agitated the melt via the subsurface release of steam, carbon dioxide, carbon monoxide, and hydrogen. The hydrogen contributed to the reducing action of the carbon (Eq. 13.16–3.18).

$$2\ Cu_2O + C \rightarrow 4\ Cu + CO_2 \qquad 13.16$$

$$Cu_2O + CO \rightarrow 2\ Cu + CO_2 \qquad 13.17$$

$$Cu_2O + H_2 \rightarrow 2Cu + H_2O \qquad 13.18$$

"Poling" is now usually carried out by passing natural gas or ammonia into the melt by water (or gas) cooled, hollow steel lances. Fire-refined copper, the product of this smelting stage, gives a copper analysis of 99% or better. Fire-refined copper can be used directly for about 15% of copper uses, including some alloys whose properties are not seriously affected by small amounts of impurities. Metal properties are seriously impaired for some uses, however, by the presence of traces of impurities, which makes further refining necessary. Phosphorus, arsenic, aluminum, iron, or antimony in particular cause relatively sharp reductions in electrical conductivity, i.e., some 6% loss is experienced with the presence of as little as 50 ppm phosphorus [6]. For these reasons, as well as from an interest in the recovery of the precious metal values which are often found in the fire-refined copper, the bulk of fire-refined copper is further purified by electrorefining.

13.3.1. Electrorefining of Smelted Copper

Electrolytic purification of smelted copper removes contaminants which adversely affect electrical conductivity, malleability, and other properties of the metal. It also permits recovery of the precious metal content of the fire-refined product. Anodes for electrorefining must first be cast from the smelted metal, with a uniform thickness across each anode as well as from one anode to the next. Anodes which are too thick give too much scrap anode material for recasting when the spent anodes are changed. Anodes which are too thin or irregular in thickness can cause pieces of metal to drop off in the electrolytic cells, resulting in electrical shorts and decreased current efficiencies.

Anode slabs are connected in parallel to heavy bus bars (high current conductors) and placed in the electrolytic cells. These are long rectangular vessels of wood or cement lined with lead or plastic (Fig. 13.8). Interleaved between the thick anode castings of fire-refined metal to be purified are thin starting sheets of elecrolytically pure copper which are connected to the cathode (negative) bus bar. The electrolyte consists of about 20% sulfuric acid (ca. 2 molar) which also contains 20 to 50 g/L (0.3 to 0.8 molar) cupric ion. Traces, up to a combined total of 20 to 30 g/L arsenic, antimony, bismuth, and nickel gradually accumulate in the electrolyte as a result of the dissolution of these electrolytically more active metals (Table 13.2) from the anodes. A low concentration of chloride (ca. 0.02 g/L) supplied by sodium chloride or

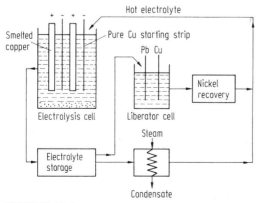

FIGURE 13.8 Diagram of a section of an electrolytic cell for the purification of copper.

TABLE 13.2 Standard Reduction Potentials for a Series of Elements of Importance to Electrolytic Purification of Copper[a]

Ion reduced	Standard reduction potential (V)
Ca^{2+}	−2.76
Mg^{2+}	−2.38
Al^{3+}	−1.71
Fe^{2+}	−0.41
Ni^{2+}	−0.23
Pb^{2+}	−0.126
H^+	0.0
Sb^{3+}	+0.145
As^{3+}	+0.234
Bi^{3+}	+0.320
Cu^{2+}	+0.158
Te^{4+}	+0.63
Ag^+	+0.80
Pd^{2+}	+0.83
Pt^{2+}	ca. +1.2
Au^{3+}	+1.42
Au^+	+1.68

[a]Arranged in order from the most, to the least active elements.

hydrochloric acid is also maintained in the electrolyte to ensure that silver is precipitated on the bottom of the cell as finely divided silver chloride, rather than being deposited with the copper on the cathodes. As electrolysis proceeds copper is dissolved from the anodes, and is simultaneously deposited from the electrolyte onto the cathodes causing a movement of metal between the electrodes (Eq. 13.19, 13.20).

$$\text{Anode reaction: } Cu^0 - 2\ e^- \rightarrow Cu^{2+}_{(aq)} \qquad 13.19$$

$$\text{Cathode reaction: } Cu^{2+}_{(aq)} + 2\ e^- \rightarrow Cu^0 \qquad 13.20$$

So there is no overall electrochemical reaction, hence, the theoretical voltage required is zero. The applied voltage of about 0.2 V is required for the process solely to drive the electrolytic purification in the desired direction. Indirect heating of the electrolyte to about 60°C decreases the solution viscosity. This helps to maintain a high production rate at these low operating voltages. Current densities of about 240 A/m^2 are normal, although higher densities of up to 350 A/m^2 are successfully employed with periodic current reversals (PCR) [10]. Current efficiencies for copper deposition are of the order of 90–95%, tending toward the lower end of this range with PCR, so that the overall power requirement for copper purification by electrolysis is relatively low. Power consumption of several cell types in commercial electrorefining service is in the range of 176 to 220 kWh per tonne of copper cathode deposited.

During the electrolysis the metals present in the anode which are of lower electrochemical activity than copper, chiefly silver, tellurium, gold, and traces of platinum, are precipitated as slimes on the floor of the electrolysis cells as they are released from the anodes (Table 13.3). Any mercury, which is frequently present in the ore, is mostly volatilized in the hot smelting stages prior to electrorefining. Lead, which is dissolved from the anode as Pb^{2+}, is immediately precipitated as the insoluble sulfate. Nickel and many other metals more electrochemically active than copper, the sulfates of which are soluble, accumulate in the electrolyte. The accumulated slime sludges are periodically pumped from the floor of the cells and dried prior to recovery. By a complex series of steps which include sulfuric acid acidification, roasting, copper electrowinning, silver cementation, fusions, and further electrowinning operations a precious metal alloy called Doré metal analyzing 8–9% gold and 85–90% silver is recovered from the residual copper, selenium, tellurium, and lead. The value of the precious metal content recovered from fire-refined copper by electrorefining is sufficient to pay for much of the cost of the electrolytic step.

During normal cell operation the anode contaminants that are more electrochemically active than copper (and less than hydrogen), mainly arsenic, antimony, and bismuth, accumulate gradually in the electrolyte. These are kept from depositing on the refined copper cathodes partly by maintaining a much higher concentration of cupric ion in the electrolyte. Fortunately for the electrorefining process, cuprous oxide is usually the main impurity of fire-refined copper. Since each mole of the cuprous oxide present in the anodes and exposed to the electrolyte puts a mole of cupric ion into solution by an auto-redox reaction (Eq. 13.21), without electrochemically removing a like amount onto the cathodes, the concentration of copper ion in the electrolyte rises more rapidly than the concentration of the other contaminants.

$$Cu_2O + 2\ H^+ \rightarrow Cu^{2+} + CuO + H_2O \qquad\qquad 13.21$$

TABLE 13.3 Analytical Ranges of Various Products of the Copper Refining Industry, U.S.A. and Japan[a]

	Anode (fire-refined) copper	Electrorefined copper	Dry, unprocessed anode slime
Copper, %	99.0–99.6	99.95	20–40
Oxygen, %	0.1–0.3	0.02–0.04	—
Gold, g/tonne	3–140	0.07–0.3	1,700–10,000
Silver, g/tonne	70–3,400	2–20	34,000–300,000
S, As, Sb, or Bi, %	0.003–0.2	0.00001–0.0002	0.5–5.0
Lead, %	0.01–0.15	0.0002–0.001	2–15
Nickel, %	0.01–0.10	0.0001–0.002	0.1–2
Selenium, %	0.01–0.06	0.0003–0.001	1–20
Tellurium, %	0.001–0.02	0.0001–0.001	0.5–8

[a]Compiled from data of Hofman [6] and Kirk-Othmer [14].

Incidentally, the elemental copper formed by Eq. 13.21 is the chief source of the copper present in anode slimes. Electrolyte is continually recirculated through the electrolysis cells to minimize the buildup of impurities. When the copper content of the electrolyte reaches 40–50 g/L (from the initial 12–18 g/L), it must be purified before reuse. The concentrations of copper, arsenic, antimony, bismuth, and nickel must all be substantially decreased. A small portion (ca. 2–3% of the total volume, per day) of the electrolyte stream is continuously bled from the main electrolyte stream for this purpose.

Most of the excess copper present in spent electrolyte is electrolytically removed from solution in liberator cells, which are very similar to the electrolysis cells used for electrowinning. These have a pure copper cathode, on which copper is deposited (Eq. 13.20), and a lead anode, rather than the copper anode used in electrorefining. The lead anode reforms sulfuric acid from the hydrogen ions released from water, instead of contributing dissolved lead to the electrolyte (Eq. 13.22).

$$H_2O - 2\ e^- \rightarrow 2\ H^+ + \frac{1}{2}\ O_2 \qquad\qquad 13.22$$

Antimony and bismuth tend to precipitate on the bottom of the cell toward the later stages of this electrolysis, and arsenic reacts with the acid present under electrolytic conditions to form highly toxic arsine gas (Eq. 13.23).

$$As + 3\ H^+ + 3\ e^- \rightarrow AsH_3\ (-0.54\ V)$$
$$\text{arsine, b.p. } -55°C \qquad\qquad 13.23$$

For this reason, liberator cells are normally hooded and well vented to avoid problems from the evolution of these gases. The "liberated" electrolyte is then further treated to remove nickel, either by crystallization as nickel sulfate or by dialysis, before it is returned to the electrolysis circuit for reuse.

Vigorous air agitation has been found to boost cathode purities when electrorefining at very high (ca. 226 A/m^2) current densities and suggestions have been made to improve electrorefining and electrowinning operations generally [15,16].

13.3.2. Fabrication and End Uses

For uses of pure copper or critical alloys the 99.95% copper cathodes obtained from electrolytic refining have to be washed, dried, and remelted before being cast into stock shapes useful for fabrication. The cast ingots can then be rolled into sheets (or foil), or hot or cold drawn into wire or pipe. Working the metal in these ways hardens and strengthens it, that is, it raises the tensile yield point from 30,000–40,000 lb/in.^{2} to 60,000–70,000 lb/in.^{2} [6]. With prolonged work this process can result in cracks or fractures in the metal. The metal may be made malleable again, or annealed, by heating to a temperature of 500–700°C followed by quenching (rapid chilling). The annealed stock may then be worked again and sold either in a work-hardened or annealed state, depending on the requirements of the application.

The high electrical conductivity of copper, 0.586×10^6 mho cm at 20°C, or about 95% of that of the most conducting metal, silver, has led to the large-scale use of copper in wiring and electrical machinery of all types. More than half of the metal produced in Japan, the U.S.A., and Western Europe is used in electrical or electronic applications [17]. Even though aluminum possesses only some 62% of the conductivity of copper, because of its light weight and relatively lower cost it has become competitive with copper for some electrical applications.

The relative thermal conductivities of the metals quite closely parallel electrical conductivities, so that again copper, with a conductivity of 0.934 cal/cm^2/cm/sec/C° at 20°C (3.98 W/cm/K at 300 K) emerges as the best low-cost metal for efficient heat transfer. This property has led to its employment in industrial heat exchangers as well as in food processing and domestic cookware. Cookware is usually lined with stainless steel to prevent the possibility of copper uptake by food. Copper is an essential trace element in human nutrition and the average American intake has been established to be 2 to 5 mg daily [18]. However, it has been known to cause mild toxic symptoms from ingestion of citrus juices that have been in long-term contact with, or cooked in, copper vessels [19]. Copper and its salts have a relatively low mammalian toxicity, although instances of acute copper poisoning have been described (from 1–12 g of copper sulfate). Chronic toxicity is known to lead to liver cirrhosis and nerve, brain, and kidney damage (Wilson's disease).

Excellent resistance to corrosion both promotes the use of copper in some of the applications mentioned above and favors its use as an expensive, though extremely long-lived roofing material. This property has also led to the use of copper in high-quality piping and for gutters and drains.

Its malleability and corrosion resistance are also factors in the extensive use of copper as a coinage metal, although undoubtedly here its beauty is an important contributing factor. The attractive red color is also the stimulus for its use in many types of decorative interior fittings.

About one-quarter of the copper produced goes directly into copper alloys, mostly as fire refined, plus some electrolytically refined. The two principal alloy groups are the brasses, which comprise copper alloyed primarily with zinc, and the bronzes which are mainly copper/tin alloys. Both of these alloy groups are stronger than copper, one motivation for their use, but there are other reasons which favor their choice. The brasses are less expensive than pure copper, hence their employment in such uses as cartridge cases for ammunition. While the brasses are less corrosion resistant than copper this property is still sufficiently good for brasses to be used in automobile radiators where its good heat transfer capabilities and higher strength as well as lower cost favor this choice. The zinc content of brasses ranges all the way from 5%, in gilding metal, through 30% and 40% in cartridge brass and Muntz metal, respectively, which are the brass alloys containing the highest proportion of zinc. Some brasses also contain 1–3% lead which improves machinability.

The tin content of bronzes is lower, ranging from 1.25 to 8% in the phosphorbronzes which also contain 0.25% phosphorus, and confer not only higher strength but also better corrosion resistance in the alloy than in pure annealed copper. These alloys are therefore ideally suited to marine exposure in such applications as ships' propellers and other marine exterior fittings as well as for water lines carrying saltwater and the like. A low coefficient of friction between bronze and steel and the superior strength of bronze over copper have led to applications of bronze as a bearing metal and as the casting metal for gears [20].

Two special copper alloys, beryllium copper and nickel silver, although only used on a small scale, contribute valuable metal properties for special uses. Two percent beryllium added to copper gives greater fatigue resistance to the metal and confers a nonsparking (on impact) quality to tools made of this alloy. This property is important for work in flammable or explosive atmospheres. Nickel silver, a copper/nickel/zinc alloy with an appearance very like silver, is important as the strong base metal for silver-plated tableware.

A small amount of copper is consumed for copper salts such as copper sulfate, which are of some importance as agricultural fungicides. The salt is employed as a spray or a dust. Care should be taken to ensure that inhalation contact is minimized. Copper salts are also used to correct very occasional local soil deficiencies in this element. Copper sulfate itself is also an excellent algicide in swimming pools and fish rearing ponds.

13.4. EMISSION CONTROL PRACTICES

13.4.1. Mining and Concentration

Problems at the mining stage, whether by open pit or underground methods, center around overburden and waste rock disposal and the potential contamination of surface water courses from disposal of mine drainage waters. Waste rock disposal in mined-out areas is being practiced where feasible but because the crushed material does not pack as densely as the bedrock from which it was derived some supplementary surface disposal is required. By skillful placement and profiling followed by plantings to encourage reestablishment of flora, dry land surface disposal can be accomplished in an aesthetically acceptable manner.

Percolate from such dry land disposal sites, as well as mine drainage waters, often contains traces of dissolved copper and other metals such as iron (Table 13.4). It also usually has a low pH. These impurities all arise from bacterial action on the sulfidic minerals present (e.g., Eq. 13.24).

$$2\ CuS + 4\ O_2 \xrightarrow[\text{bacteria}]{} 2\ Cu^{2+} + 2\ SO_4^{2-} \qquad 13.24$$

Copper may be recovered from such waste streams by cementation, or by cementation plus liming. These processes can recover metal from such waste streams and decrease the toxicity risks on stream discharge (Eq. 13.25–13.28).

Cementation:
$$Cu^{2+} + Fe \rightarrow Fe^{2+} + Cu\ (\text{``cement copper''}) \qquad 13.25$$

Liming:
$$Cu^{2+} + Ca(OH)_2 \rightarrow Cu(OH)_2\downarrow + Ca^{2+} \qquad 13.26$$

$$Fe^{2+} + Ca(OH)_2 \rightarrow Fe(OH)_2\downarrow + Ca^{2+} \qquad 13.27$$

$$2\ Fe^{2+} + 1/2\ O_2 + 2\ H_2O + Ca(OH)_2 \rightarrow 2\ Fe(OH)_3\downarrow + Ca^{2+} \qquad 13.28$$

TABLE 13.4 Partial Analysis of Typical Copper Mine and Smelter Waters[a]

	Ion concentrations (mg/L)					
	Cu^{2+}	Fe^{2+}	Fe, total	Zn^{2+}	H_2SO_4	pH
Mine waters:						
Typical	970	8,350	10,500	4,000	—	1.4
S. American	120	2,000	4,000	1,000	—	1.4
Canadian	156	—	176	328	<10	3.8
Typical wash water:	100	>5	10	24	100	2.8
Electrolytic	240	—	3	<1	760	2.0
Copper refinery	446	—	4.5	43	1,810	1.5

[a]Typical data from Barthel [21].

Cementation involves the passage of the solution containing copper ions over a large surface area of scrap iron followed by settling. Finely divided copper is precipitated. Subsequent liming removes most of the iron and raises the pH to more normal values before discharge. More recently, four-stage solvent extraction in the presence of proprietary complexing agents has been tested and found to be capable of decreasing aqueous Cu^{2+} concentrations from about 400 to about 6 mg/L [21].

It is possible to recirculate a large fraction of the water used for froth flotation by decantation of a clear supernatant liquor from a settling lagoon or thickener overflow. However, froth flotation is sensitive to low concentrations of surface active agents. This means that the froth flotation plant cannot operate as a completely closed circuit since buildup of impurities in the water could markedly affect flotation efficiencies. Traces of metals in the flotation waters effluent, some suspended and some dissolved, are captured prior to discharge by liming and settling as outlined for mine drainage waters or tailings percolate.

Other types of water effluent problems can arise from froth flotation operations. For example, with multimineral flotation when separate concentrates are required of each of say copper, lead, and zinc, copper would normally be floated first, then lead, and finally zinc sulfide. This separation sequence requires addition of sodium cyanide as a depressant. The waste water stream from this process can contain 50 mg/L or more of dissolved cyanide which has to be detoxified before discharge. Calcium hypochlorite has been found to be effective to neutralize the cyanide by oxidation to cyanate (Eq. 13.29).

$$2\ NaCN + Ca(ClO)_2 \rightarrow 2\ NaCNO + CaCl_2 \qquad 13.29$$

When treatment is undertaken on a large scale, however, chlorine addition has been found to be both more convenient and more economical (Eq. 13.30, 13.31).

$$Cl_2 + H_2O \rightarrow HCl + HClO \qquad 13.30$$

$$2\ Ca(OH)_2 + 2\ HCl + 2\ HClO \rightarrow CaCl_2 + Ca(ClO)_2 + 2\ H_2O \qquad 13.31$$

If the chlorine addition is undertaken prior to pH reduction with lime the reaction sequence is thought to involve intermediate formation of cyanogen chloride which is decomposed to cyanate on addition of base (Eq. 13.32, 13.33).

$$NaCN + Cl_2 \rightarrow CNCl + NaCl \qquad\qquad 13.32$$

$$2\ CNCl + 2\ Ca(OH)_2 \rightarrow Ca(CNO)_2 + CaCl_2 + 2\ H_2O \qquad 13.33$$

This waste treatment sequence can reduce feed values of 68 mg/L cyanide and 42 mg/L copper to concentrations of about 0.1 mg/L for both contaminants [22]. This result required 3.4 kg chlorine/kg cyanide in the effluent, and a 15–30 min residence time for destruction of the cyanide only. Three to four times this residence time was required for neutralization of copper cyanide. Ideally, the oxidant added should be sufficient to convert cyanate to carbon dioxide and nitrogen to avoid the possibility of reversion of cyanate back to cyanide after discharge. Ozonation has also been found to be effective for the neutralization of cyanide in waste waters [23].

Disposal of the separated, finely pulverized gangue may be to underground or strip mined areas. However, excessive disposal volumes are a problem. Dry land disposal may be used, followed by aesthetic contouring and reestablishment of plant life. This method requires monitoring and control measures for any dissolved metals in percolate from the tailings. Drainage directly to a small lake or a water course is not permitted by pollution control agencies (e.g., [36]). Direct tailings disposal to large lakes and ocean inlets has been practiced, but is rarely acceptable now [24].

Solution mining can generate complex recovery problems, particularly with in-place leaching methods. When mining has been completed the fractured deposit requires extensive flushing to remove leaching solution residues. Deep well injection or various solution concentrating procedures are used for ultimate disposal of waste solutions from these flushing operations.

13.4.2. Smelter Operation

Copper smelting is associated with dust production from materials handling and grinding, and sulfur dioxide from sulfides. Upwards of a tonne of sulfur dioxide is produced for every tonne of copper obtained from sulfidic minerals. In the early days of the industry this sulfur dioxide was simply allowed to dissipate in the air around a smelter. The severe effect on local air quality prompted mixing of the smelter gases with large volumes of air before discharge to obtain decreased ground level sulfur dioxide concentrations. Stack heights gradually increased over the years to exploit dilution effects for larger volumes of sulfur dioxide, culminating in the construction of the world's tallest stack, 380 m high, by the International Nickel Company (Inco) at Sudbury for this purpose, in 1970. At the same time, some smelters adopted sulfur dioxide containment and utilization for sulfuric acid production as a means to decrease discharge problems. The sulfuric acid by-product was sold to obtain a process credit. Containment and sulfur dioxide utilization have now been adopted by Inco, as well as more generally by the industry, as the limited atmospheric capacity for dispersion of enormous quantities of sulfur dioxide worldwide has become recognized [25]. This desirable approach is now enforced by the Environmental Protection Agency in the U.S.A. with a requirement that 95% (or better) of the sulfur in the ores being processed by a smelter must be contained during smelting.

The potential quantity of dust discharged per tonne of ore processed varies with the particular stage of processing. Crushing and dry grinding stages as well as final rotary drying are likely to contribute the most to dust losses. Ore grinding wet avoids dust problems and improves control of the final particle size. For all processing stages the proportion of copper present, about 10% in roaster dusts, 25% in reverberatory furnace dusts, and 45–55% in converter dusts, is sufficiently high to provide material recovery, as well as emission control incentives to capture these.

Simple gravity settling of dust has been used for many years with some 84% mass containment success for copper blast furnace discharges [6]. Sometimes the dust settling and gas carrying functions are combined in a single large cross-sectional area (slow gas flow) balloon flue (Fig. 13.9). Tighter emission control requirements as well as an interest in cleaner acid plant operation, however, have required employment of electrostatic precipitation and/or scrubbers for smelter flue gas treatment [12]. Electrostatic precipitation, with a control efficiency of about 98% by mass on 0.1–100 μm diameter particles, allows control of the fume component as well as the larger particles [26]. Fumes form from sublimation, or from condensation and solidification of such components as oxides of antimony, arsenic, lead, zinc, and sulfates. Water, sulfuric acid, and mercury may also have been condensed and adsorbed onto the fume particles. Dusts containing these elements are treated for recovery of the components of interest, e.g., mercury by sulfatization [27,28]. Baghouses are also efficient for dust control. The bag fabric has to be compatible with acid gas contact and flue gases temperatures have to be kept within the working range of the fabric. Additional blower energy is required to force gases through a filter fabric. An electrostatic precipitator also has an energy cost to maintain the electrical charge on the active elements.

The problem with sulfur dioxide containment is that many smelter units produce sulfur dioxide concentrations of 1–2% whereas the minimum economic concentration for sulfur dioxide conversion processes (e.g., for sulfuric

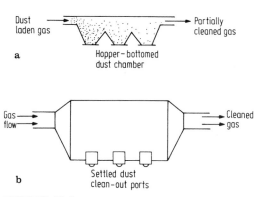

FIGURE 13.9 Operating details of (a) a gravity settling chamber and (b) a section of a balloon-flue for removal of the coarser dusts from copper smelter operation.

acid or sulfur production) is 3.5–4%. Thus, the economic solution to containment requires either modification of smelter processes to obtain higher sulfur dioxide concentrations directly, or capture of sulfur dioxide at relatively low concentrations followed by regeneration of a high concentration sulfur dioxide gas stream (Chap. 3). Throw-away approaches to sulfur dioxide containment, lime or limestone wet scrubbing, are worthy of consideration by small-scale smelters (Eq. 13.34–13.36). The gypsum-calcium sulfite product of this control system is only suitable for landfill.

$$Ca(OH)_2 + H_2O + SO_2 \rightarrow CaSO_3 + 2\ H_2O \qquad 13.34$$

$$CaSO_3 + 1/2\ O_2 + 2\ H_2O \rightarrow CaSO_4 \cdot 2H_2O \qquad 13.35$$

$$CaCO_3 + H_2O + SO_2 \rightarrow CaSO_3 + CO_2 + H_2O \qquad 13.36$$

This choice adds an operating cost to the smelter without contributing any by-product credit and hence is adopted only when required.

A switch from chain grate or multiple hearth type roasters to fluidized bed types with improved air to hot sulfide ore particle contact and less entrained air can raise the sulfur dioxide from 1% to 12% in the roaster "off-gases." Fluid bed, autogenous (self-heating) roasting of this type solves the sulfur dioxide containment problem for concentrates of high sulfur ores. However, ore concentrates of lower sulfur content require external fuel combustion to provide sufficient heat for adequate roasting. Addition of external combustion gases to the roaster gases dilutes the sulfur dioxide to less than economic concentrations. This disadvantage has led to the abandonment of roasting as a separate smelter operation in locations which have a low sulfur to copper ratio in the concentrate [14].

Continuous smelting processes or flash smelting processes using oxygen-enriched air are being employed to combine two or three smelting stages, roasting, reverberatory smelting, and/or converting, into one [29]. In this way sulfur removal is obtained with good control, and high sulfur dioxide concentrations are obtained in the exit gases produced. Outokumpu furnace designs using air feed for sulfur combustion produce exhaust gases containing 14% sulfur dioxide. Use of a top blown rotary converter (TBRC) also promises to streamline copper production by merging smelting stages and by allowing the use of oxygen-enriched air for blowing [30]. Use of oxygen-enriched air is not possible in converters because of heat damage to tuyeres (air nozzles). If oxygen-enriched air is used for sulfur combustion it is possible to obtain flue gases containing 75 to 80% sulfur dioxide, sufficiently high to practice direct sulfur dioxide recovery, if desired [31]. These joint developments have been adopted by several new smelters [32,33].

Enriched sulfur dioxide streams combined with lower sulfur dioxide flue gases can yield a blended gas stream of 4 to 6% sulfur dioxide suitable for acid production. Smelter acid is not as pure as the acid produced from sulfur combustion, so it fetches a lower price. Nevertheless, the smelter product is quite suitable for many uses, e.g., fertilizer phosphate production. Thus a reasonable by-product credit is possible by this sulfur dioxide conversion choice. It has been estimated that smelters which exercise this choice produce 4 tonnes of sulfuric acid for each tonne of copper [31]. Smelter sources contributed more than 6% of the sulfuric acid produced in the United States in 1965 and more than 60% of the Canadian total for 1976 [34].

Sulfur dioxide containment and subsequent reduction to a sulfur product has also been tested. The American Smelting and Refining Company's process involves combustion of the cooled and cleaned smelter gases with methane (Fig. 13.10; Eq. 13.37).

$$4\ SO_2 + 2\ CH_4 + air \xrightarrow{1250°} S + H_2S + COS + SO_2 +$$
$$\text{(unbalanced)} \qquad\qquad 2\ CO_2 + 3\ H_2O + 4\ N_2 \qquad 13.37$$

Carbonyl sulfide and some of the residual sulfur dioxide are then reacted at 450°C in the presence of a bauxite catalyst to raise the yield of sulfur (Eq. 13.38).

$$2\ COS + SO_2 \rightarrow 2\ CO_2 + 3\ S \qquad\qquad 13.38$$

After electrostatic precipitation of the products of these two steps, the remaining 2:1 ratio of hydrogen sulfide and sulfur dioxide are reacted in the presence of a Claus-type catalyst to add to the sulfur yield (Chap. 3; Eq. 13.39).

$$2\ H_2S + SO_2 \rightarrow 3\ S + 2\ H_2O \qquad\qquad 13.39$$

Actual sulfur recoveries were about 90% with a possibility of raising this to 95%.

The U.S. Bureau of Mines has tested a pilot-scale sulfur dioxide to sulfur conversion process which involves initial absorption of sulfur dioxide in an aqueous solution of citric acid, $HOC(CH_2CO_2H)_2CO_2H$, and sodium citrate

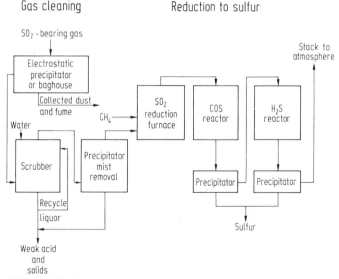

FIGURE 13.10 Flowsheet of the American Smelting and Mining Company's process for conversion of sulfur dioxide to sulfur. (Adapted from U.S. Bureau of Mines [35].)

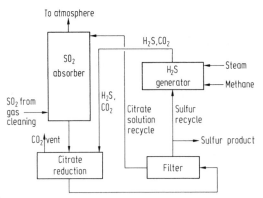

FIGURE 13.11   The citrate process for sulfur dioxide capture and conversion to elemental sulfur developed by the U.S. Bureau of Mines. Details available from U.S. Bureau of Mines [35].

(Fig. 13.11). This solution is then contacted with hydrogen sulfide to reduce the sulfur dioxide to elemental sulfur, which is then readily filtered from the resulting slurry (Eq. 13.40–13.42).

$$SO_2 + H_2O \rightarrow H_2SO_3 \qquad\qquad 13.40$$

$$H_2SO_3 + H_3Cit \rightarrow H_2SO_3 \cdot H_3Cit \qquad\qquad 13.41$$

$$H_2SO_3 \cdot H_3Cit + 2\ H_2S \rightarrow 3\ S + H_3Cit + 3\ H_2O \qquad\qquad 13.42$$

The filtrate is returned to the absorber for reuse. A part of the product sulfur is reacted with methane and steam to produce the hydrogen sulfide required for reduction of the citrate complex (Eq. 13.43).

$$4\ S + CH_4 + 2\ H_2O \rightarrow 4\ H_2S + CO_2 \qquad\qquad 13.43$$

This is over 90% efficient for reverberatory furnace flue gases containing 1.5% sulfur dioxide.

13.5. HYDROMETALLURGICAL COPPER RECOVERY

13.5.1. Basic Principles

Hydrometallurgy, or the application of aqueous solutions for metal recovery from ores, has been practiced for copper recovery for many years. The original impetus for solution methods for copper extraction before the development of froth flotation technology was the existence of large ore bodies of low copper grade which were uneconomic to work using conventional smelting methods. Hydrometallurgy also holds a complementary position to concentrator technologies for its ability to scavenge copper from concentrator tailings, since froth flotation is poor at floating copper oxides which are readily extracted by appropriate aqueous solutions [36]. Most recently, hydrometallurgical copper recovery methods have been developed to handle concentrates of more complex ores with the objective of eliminating the sulfur dioxide discharge problems of conventional smelting. Thus, while all hydrometallurgical methods essentially have a low air pollution potential, some recent methods developed use chemistry which yields elemental sulfur directly from the sulfur minerals present in the ores, to give a cleaner process.

Hydrometallurgical copper recovery can be conveniently considered in two stages: the extraction stage, where the various forms of copper in the ore are placed into an aqueous solution, usually as cupric ion, and the recovery stage, where dissolved copper is regained as solid, pure, or nearly pure copper metal ready for fabrication or final smelting stages.

The most common solvent is 5 to 10% sulfuric acid in water. Diluted smelter acid can provide this at low cost. This can dissolve a number of copper minerals, not just the oxide (Eq. 13.44–13.47).

$$CuO + H_2SO_4 \rightarrow CuSO_4 + H_2O \qquad\qquad 13.44$$

$$Cu_2O + 2\ H_2SO_4 + 1/2\ O_2 \rightarrow 2\ CuSO_4 + 2\ H_2O \qquad\qquad 13.45$$

$$CuSiO_3 \cdot 2H_2O + H_2SO_4 \rightarrow CuSO_4 + SiO_2 + 3\ H_2O \qquad\qquad 13.46$$

$$Cu_2(OH)_2CO_3 + 2\ H_2SO_4 \rightarrow 2\ CuSO_4 + CO_2 + 3\ H_2O \qquad 13.47$$

These reactions, in particular carbonate solvation, deplete the sulfuric acid content of the extraction solution. Later metal recovery from the solutions, however, returns an equivalent acid strength to the water. However, in the presence of air *Thiobacillus ferro-oxidans* converts sulfide minerals to sulfates, which contributes to the acid content of the solvent when metal is recovered from the solution.

Metal may be extracted by leaching in place, by heap or percolation leaching, or by leaching with mechanical agitation in separate tanks. Leaching in place requires prior fracturing of the ore body by explosives to provide cracks or passages for percolation of the leach solutions. Copper-containing percolate is recovered for recirculation and eventual metal recovery by pumping this solution from lower galleries of the in-place leaching zone. This method may be used with low-grade zones of an ore body loosened in place and with ores which have been passed through crushing stages and then returned to the mine for disposal.

Heap and percolation leaching are used with low-grade ores which are crushed to an average 1-cm particle diameter for this purpose. Heap percolation is practiced on large-scale, 100,000-tonne by 35-m-high flattened top ore heaps constructed so as to allow even distribution of liquor and air throughout the heap. A series of vats containing the crushed ore is employed for percolation leaching, which relies on either alternate filling and emptying cycles or continuous slow flow through the series of vats to gradually put the metal into solution. Percolation leaching ensures thorough contact of the percolating solution with the whole of the crushed ore and has been practiced for many years at Chuquicamata, Chile. It provides better control of extraction conditions than heap leaching, which is important to the Rio-Tinto Company, Spain, but it does so at a higher capital cost.

Agitation leaching is used with very finely divided ores which may have already been partially processed and would not be practical for percolation methods. This material is sufficiently fine that it is fairly readily kept in aqueous suspension. Thus, for agitation leaching the finely ground pulp is placed with a sufficient quantity of leach solution to obtain a very fluid mass, in vats equipped with air-lift (a pachuca) or mechanical agitation to maintain the whole mass in suspension [12]. Extraction is orders of magnitude faster than either in-place or heap/percolation leaching methods. The liquor containing the dissolved metal values is collected as the overflow from a thickener. The slurry underflow from the thickener containing the gangue is filtered to collect residual copper-containing liquor and washed before it is discarded.

Once the copper and any other metals of interest have been obtained in solution, the difficult and usually time-consuming part of the process is over. Copper is recovered from these solutions by a variety of methods.

Cementation involves precipitation of the copper by passage of the solution over scrap iron. This requires about 1 hr of contact time and produces a finely divided form of the metal analyzing ca. 90% Cu (dry) by replacement of the dissolved copper by iron. In theory this should only require 0.879 kg of iron for each kilogram of copper obtained (Eq. 13.25). Normal experience, however, is an iron consumption of 1.3 to 3 kg/kg copper [37]. In the United States in recent years some 10% of U.S. production in 1973 has been from cement copper, mostly by the Kennecott Copper Corp. and Anaconda Co.

Copper may also be recovered from leach solutions electrolytically. This process, called electrowinning, requires use of an insoluble anode such as hard lead, comparable to the liberator cell used for liquor purification in copper electrorefining. Consequently, there *are* net electrochemical reactions involved in electrowinning (Eq. 13.20, 13.22), as opposed to the situation with electrorefining, so that about 1.7 V are required for this step. This results in an electrical power consumption for electrowinning of about 2.8 kWh/kg copper, much higher than the approximately 0.2 kWh/kg required for electrorefining.

13.5.2. New Hydrometallurgy Developments

More selective copper recovery from leach solutions may be obtained by solvent extraction using a dissolved, copper-selective complexer such as an hydroxyoxime (Fig. 13.12) in the organic phase. For example, complexation of 2 mol of this selective additive to cupric ion confers a marked preference of the copper (II) species for the organic, rather than the aqueous phase, which allows it to be efficiently captured from a dilute aqueous solution [39]. The metal complex is stabilized by the formation of two six-membered rings which include the metal ion and stabilizes the complexed, over the uncomplexed, state. After separation of the metal-rich organic phase from the original aqueous scavenging solution the copper may be back-extracted to a fresh aqueous acid phase in preparation for electrowinning or other methods of recovery of the metal values [40] (Fig. 13.13). Final copper recovery from the aqueous solution is greatly facilitated by the higher concentration and purer state of the cupric ion, which compensates well for the cost of the extraction step. Metal values from the primary leach solutions may be concentrated prior to

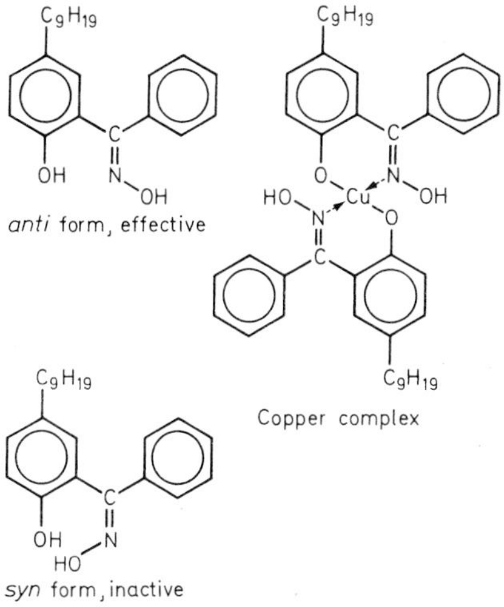

FIGURE 13.12 Structure of the two forms of the copper-selective complexer, LIX65N, and the kerosene-soluble copper complex formed on contact of the anti form with cupric ion. (From Bailes *et al.* [38]. Excerpted by special permission from CHEMICAL ENGINEERING [Aug. 1976]. Copyright © 1976, by McGraw-Hill Inc., New York, NY 10020.)

metal recovery in other ways, among them ion exchange techniques and adsorption/desorption from activated carbon [41].

Anaconda's development of a sulfuric acid based leaching process employs a number of innovations [42]. Concentrates are leached with about 90% sulfuric acid (much more concentrated than usual) to yield a solution of dissolved metal sulfates. Sulfide sulfur is released in elemental form, in the process reducing a portion of the sulfuric acid to sulfur dioxide and water. Purification of the solution, followed by treatment with hydrogen cyanide and sulfur dioxide then yields a solid phase of high-purity cuprous cyanide, all accomplished at atmospheric pressure. High-purity copper, 99.99% Cu, is obtained on reduction of the separated cuprous cyanide with hydrogen (Eq. 13.48).

$$CuCN + 1/2\ H_2 \rightarrow Cu + HCN \qquad\qquad 13.48$$

The recent Sherritt-Cominco copper process also introduces some hydrometallurgical refinements into sulfuric acid leaching procedures to make the method applicable to chalcopyrite ($CuFeS_2$) ores [43]. After thermal rearrange-

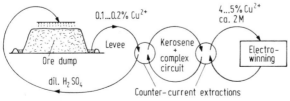

FIGURE 13.13 Countercurrent solvent extraction, using a copper-selective complexer in the organic phase used to raise the concentration and purity of copper ion solutions fed to electrowinning.

ment and reduction of the concentrate with hydrogen it is leached first with sulfuric acid to dissolve out the iron [44] (Eq. 13.49).

$$FeS + H_2SO_4 \rightarrow FeSO_4 + H_2S \qquad\qquad 13.49$$

The hydrogen sulfide is recovered for processing to sulfur and the iron-containing solution is separated for iron precipitation. The copper, as Cu_5FeS_4, remains undissolved in the concentrate up to this stage. It is separately dissolved out of the concentrate with sulfuric acid under oxidizing conditions (Eq. 13.50).

$$Cu_5FeS_4 + 6 H_2SO_4 + 3 O_2 \rightarrow 5 CuSO_4 + FeSO_4 + 4 S + 6 H_2O \qquad 13.50$$

After purification of this copper-containing solution, metallic copper is obtained either by electrowinning or hydrogen reduction. Claims of at least 98% copper recovery, gold and silver capture at least equivalent to conventional smelting, recovery of 60 to 90% of the sulfur in elemental form, and recovery of other metal values in salable form have been substantiated [45].

Another hydrometallurgical variant appropriate for chalcopyritic ores also adaptable to porphyry copper and oxides is the Cyprus Metallurgical Processes Corporation's ("Cymet") chloride leaching process and adaptations. Copper concentrates are first leached with an aqueous ferric chloride solution. Cupric (Cu^{2+}) ion in the ore reacts with the chloride leach medium to form cuprous ion and dissolves as cuprous chloride (CuCl). After a series of several more steps cuprous chloride is crystallized from the solution in pure form [46]. Final metal recovery is obtained by hot reduction of the cuprous chloride with hydrogen (Eq. 13.51).

$$CuCl + 1/2 H_2 \rightarrow Cu + HCl \qquad\qquad 13.51$$

This sequence of reactions has been successfully tested with 680,000 kg of concentrate analyzing about 28% copper, 30% iron, and 30% sulfur, including elemental sulfur recovery.

Ammonia complexation has also been employed, primarily for nickel-copper separations but also for copper recovery, when this is from native copper or oxide ores [12]. The dissolving process is thought to involve formation of soluble cuprous ammonium carbonate [47] (Eq. 13.52, 13.53).

$$Cu + Cu(NH_3)_4CO_3 \rightarrow Cu_2(NH_3)_4CO_3 \qquad\qquad 13.52$$

$$2 Cu_2(NH_3)_4CO_3 + 4 (NH_4)_2CO_3 + O_2 \rightarrow \\ 4 Cu(NH_3)_4CO_3 + 4 H_2O + 2 CO_2 \qquad\qquad 13.53$$

Most interesting developments relate a combination of electrolytic metal recovery from heap leached solutions with the actual copper extraction process [48]. In laboratory-scale tests, where the finely ground (2–3 μm) copper ore was both leached and electrolytically deposited in a single electrochemical cell, an "electroslurried" product of 99.9% copper was obtained [49]. While this is not electrical conductor purity, it is very respectable for a single-stage electrohydrometallurgical process capable of recovering 98% of the copper present in the ore feed. The innovations and evolution of these technologies have been evaluated [50,51]

REVIEW QUESTIONS

1. (a) Explain the meaning of the term "contact angle" and its relevance to the beneficiation of low-grade copper ores.
   (b) Why are good recoveries obtained with froth flotation of sulfide ores of copper, while the oxides are poorly reclaimed?
   (c) What influence(s) does the particle size of the ore have on the efficiency of copper recovery from froth flotation, and how does time since grinding affect this?
2. (a) What is the actual power required (kWh/tonne) for the electrolytic production of 1.00 tonne of each of the following: chlorine, sodium metal, aluminum, and electrorefined copper?
   (b) Calculate the theoretical power required for 1.00 tonne of electrorefined copper and for 1.00 tonne of copper recovered by electrowinning. Why is there a difference?
3. (a) Why does the pH of the copper-containing electrolyte drop during electrowinning?
   (b) How does this pH change help the hydrometallurgical component of the process?
4. (a) What mass of iron is required theoretically to recover 1.00 tonne of "cement copper"?
   (b) What factors contribute to the higher than theoretical amount of iron required for cementation in practice, and how do these affect the process?
5. Compare and contrast the niches served by froth flotation, hot smelting, hydrometallurgical extraction, electrowinning, and cementation for copper recovery and production. Include consideration of environmental factors.

FURTHER READING

C.R. Bayliss, Modern Techniques in Electrolytic Refining of Copper. *Electron Power* **22**, 773 (1976).

J. Leja, "Surface Chemistry of Froth Flotation." Plenum, New York, 1981.

J.F. Lonergan and A.D. Robson, eds. "International Symposium on Copper in Soils and Plants, Murdoch University, 1981." Academic Press, Sydney, 1981.

J.O. Nriagu, ed. "Copper in the Environment." 2 vols. Wiley, New York, 1979.

M.W. Ranney, ed. "Flotation Agents and Processes: Technology and Applications." Noyes Data Corp., Park Ridge, NJ, 1980.

P.A. Spear and R.C. Pierce, "Copper in the Aquatic Environment: Chemistry, Distribution, and Toxicology." National Research Council, Ottawa, 1979.

REFERENCES

1. H.H. Coghlan, "Notes on the Prehistoric Metallurgy of Copper and Bronze in the Old World." University Press, Oxford, 1951.
2. F.S. Taylor, "A History of Industrial Chemistry," p. 37. Arno Press, New York, 1972.
3. R. Chadwick, *Chem. Br.* **17**(8), 369 (1981).
4. "United Nations Statistical Yearbook 1979/80," United Nations, New York, 1981, and earlier issues.
5. "Minerals Yearbook, Metals and Minerals," Vol. 1, 1994. U.S. Department of the Interior, Bureau of Mines, Washington, DC, 1996, and earlier issues.
6. H.O. Hofman, "Metallurgy of Copper." McGraw-Hill, New York, 1914.
7. J. Leja, *J. Chem. Educ.* **49**(3), 157 (1972).
8. D.W. Fuerstenau, ed., "Froth Flotation," 50th Aniversary Volume. American Institute of Mining, Metallurgical, and Petroleum Engineers, New York, 1962.
9. C.J.S. Warrington and R.V.V. Nichols, "A History of Chemistry in Canada," p. 15. Pitman and Sons, Toronto, 1949.
10. "Kirk-Othmer Encyclopedia of Chemical Technology," 3rd ed., Vol. 6, p. 819. Wiley, New York, 1979.
11. V.I. Klassen and V.A. Mokrousov, "An Introduction to the Theory of Flotation" (translated by J. Leja and G.W. Poling), 2nd ed. Butterworth, London, 1963.
12. J.R. Boldt, Jr. and P. Queneau, "The Winning of Nickel." Longmans of Canada, Ltd., Toronto, 1967.
13. R.R. Klimpel, Technological Trends in Flotation Chemistry. *Min. Eng. (Littleton, Colo.)* **47**, 933–942 (1995).
14. "Kirk-Othmer Encyclopedia of Chemical Technology," 2nd ed., Vol. 6, p. 131. Wiley, New York, 1965.
15. W.W. Harvey, M.R. Randlett, and K.I. Bangerskis, *J. Met.* **30**, 32, July (1978).
16. A.S. Gendron, R.R. Matthews, and W.C. Wilson, *Can. Min. Metall. Bull.* **70B**, 166, Aug. (1977).
17. W. Gluschke, J. Shaw, and B. Varon, "Copper: The Next Fifteen Years," p. 30. D. Reidel (The United Nations), Dordrecht, The Netherlands, 1979.
18. G.D. Clayton and F.E. Clayton, eds., "Patty's Industrial Hygiene and Toxicology," 3rd ed., Vol. 2A, p. 1620. Wiley, New York, 1981.
19. "McGraw-Hill Encyclopedia of Science and Technology," Vol. 3, p. 641. McGraw-Hill, New York, 1982.
20. R.C. Weast, ed., "Handbook of Chemistry and Physics," 51st ed., p. F-16. Chem. Rubber Publ. Co., Cleveland, OH, 1970.
21. G. Barthel, *J. Met.* **30**, 7, July (1978).
22. A.G. Eccles, *Can. Mining Metall. Bull.* **70**(785), 141, Sept. (1977).
23. G.I. Mathieu, "Ozonation for Destruction of Cyanide in Canadian Gold Mill Effluents: A Preliminary Evaluation," Canmet Rep. 77-11. Dept. of Supply and Services, Ottawa, 1977.
24. M.J.R. Clark and T.O. Morrison, "Impact of the Westmin Resources Ltd. Mining Operation on Buttle Lake and the Campbell River Watershed." Waste Management Branch, Ministry of the Environment, Victoria, B.C., 1982.
25. Inco Entering Acid Production, *Can. Chem. Process.* **65**(2), 10, March 27 (1981).
26. P.L. Magill, F.R. Holden, and C. Ackley, eds., "Air Pollution Handbook." McGraw-Hill, New York, 1956.
27. J. Rastas, E. Nyholm, and J. Kangas, *Eng. Min. J.* **172**, 123 (1971).
28. J. Kangas, E. Nyholm, and J. Rastas, *Chem. Eng. (N.Y.)* **78E**, 55 (1971).
29. Copper Smelting Is Going Continuous, *Can. Chem. Process.* **60**(4), 21, April (1976).
30. Copper Refining Via Rotary Smelting, *Can. Chem. Process.* **60**(1), 6, Jan. (1976).
31. T.J. Chessell, *Can. Chem. Process.* **63**, 21, March 21 (1979).
32. New Copper Recipe, *Dravo Rev.*, Fall, p. 2 (1978).
33. Afton, New Canadian Copper Mine, *World Min.* **31**(4), 42, April (1978).
34. P.D. Nolan, *Can. Chem. Process.* **61**(4), 35, April (1977).
35. "Control of Sulfur Oxide Emissions in Copper, Lead, and Zinc Smelting," Inf. Circ. 8527. U.S. Bureau of Mines, Washington, DC, 1971.

36. W.A. Triggs and A.M. Laird, *Min. Mag.* **124**(6), 438 (1971).

37. Copper Industry Uses Much Scrap Iron, *Environ. Sci. Technol.* 7(2), 100 (1973).

38. P.J. Bailes, C. Hanson, and M.A. Hughes, *Chem. Eng. (N.Y.)* **83C**, 86, Aug. 30 (1976).

39. D.S. Flett, *Chem. and Ind. (London)*, p. 706, Sept. 3 (1977).

40. M.C. Kuhn, *Min. Eng. (Littleton, Colo.)* **29**, 79 (1977).

41. F. Habashi, *Chem. Eng. News* **60**(6), 46, Feb. 8 (1982).

42. Chemical Route to Copper, *Chem. Eng. (N.Y.)* 77(8), 64, April 20 (1970).

43. C-S Process Offers Unique Steps, *Can. Chem. Process.* **63**, 25, March 21 (1979).

44. D.E.G. Maschmeyer, P. Kawalka, E.F.G. Milner, and G.M. Swinkels, *J. Met.* **30**, 27, July (1978).

45. D.E.G. Maschmeyer, E.F.G. Milner, and B.M. Parekh, *Can. Min. Metall. Bull.* **71**, 131, Feb. (1978).

46. Cyprus Reveals Details, *Eng. Min. J.* **178**, 27, Nov. (1977).

47. B.D. Pandey, V. Kumar, D. Bagehi, and D.D. Akerkar, Extraction of Nickel and Copper from the Ammoniacal Leach Solution of Sea Nodules by LIX 64N, *Ind. Eng. Chem. Res.* **28**, 1664–1669, Nov. (1989).

48. C.C. Simpson, Jr., *J. Met.* **29**, 6, July (1977).

49. Lower Energy Process for Copper Refining, *Chem. Eng. News* **58**(23), 7, June 9 (1980).

50. J.B. Hiskey, Technical Innovations Spur Resurgence of Copper Solution Mining. *Min. Eng. (Littleton, Colo.)* **38**, 1036–1039 (1986).

51. N. Arbiter and A.W. Fletcher, Copper Hydrometallurgy-Evolution and Milestones. *Min. Eng. (Littleton, Colo.)* **46**, 118–123 (1994).

Tapping an electric furnace for molten alloy steel.

14

■ PRODUCTION OF IRON AND STEEL

You should hammer your iron when it is hot.
                    —Publilius Syrus, ca. 42 B.C.

Strike the iron whilst it is hot.
                    —Francois Rabelais (1494–1553)

Strike now, or else the iron cools.
                    —William Shakespeare (1564–1616)

14.1. EARLY HISTORY AND DEVELOPMENT

Iron, at 5% of the earth's crust, is the most abundant metal after aluminum and the fourth most abundant element present in the surface rocks of the earth. Iron is also the preeminently useful metal of our society for the manufacture of machinery of all types and for constructional purposes. Iron itself is a widely used, versatile metal. This factor and an increasingly wide variety of steels (iron alloys) combine to make the tonnage of iron produced each year easily exceed the combined annual production of all the other metals used by our society.

Iron, like aluminum, is relatively easily oxidized and so is seldom found in nature in metallic form, except in recent meteorites. In fact, sideritic (iron-containing) meteorites were probably the origin of the first iron worked by early peoples. It is also possible that lumps of iron-rich ores placed around early, large, very hot campfires were reduced to a spongy type of iron mixed with slag. Slag could then have been mostly beaten out from these lumps and simple tools fashioned in the same way from this quite malleable form of iron. Since it is unlikely that meteoritic iron would be as common as the lumps of native copper found in many areas, historians believe that the iron age followed the copper and bronze ages in prehistoric stages of metallurgical development, following the Stone Age [1]. Some parts of the world without any significant surface exposures of copper mineral, however, could have experienced direct development from the Stone to the Iron Ages. In either case the use of iron was certainly known by the Egyptians dating back to 3 to 4000 B.C., both for tools and simple jewelery [2].

The methods used by the early small-scale smelters of the ore were probably simpler versions of the methods used in Europe before about 1750. At this time a furnace was built of a fire-resisting clay and fired by burning charcoal which was raised to smelting temperatures by a pair of continuously operated foot bellows. Selected ore, of 40% or more iron, was then added in lumps a few at a time and the charcoal fired heating continued for several hours. The furnace was then allowed to cool, the front was broken open, and the lumps of spongy iron mixed with slag removed. Hammering was used to drive out most of the slag and to shape the residual iron. Direct reduction of iron ore as accomplished by these methods, in the absence of liquefaction, produced a metallic iron which was low in carbon and hence ductile enough to be fashioned by beating.

These small-scale operations permitted the production of only a few kilograms of iron from each firing. The development of the forerunners of the modern blast furnace in rudimentary form is believed to have occurred in Belgium in about 1400. These larger units, although still fired by charcoal, could produce several hundred kilograms every 24 hours but also required mechanization to provide a sufficient air blast, usually provided by water power.

The greater cost and organization required for blast furnace construction and operation caused large blast furnaces to grow alongside the earlier processes and only very gradually displaced the early smaller scale operations.

To provide the charcoal required for both firing and the reduction process took enormous quantities of wood and caused rapid depletion of forest stands near smelters, both in Europe and North America [3]. Depletion of charcoal resources, and the increasing availability of coal led to many attempts to fire blast furnaces directly with coal, with uniformly poor results. The sulfur content of the coal contributed to a high sulfur content in the iron which made it brittle when hot, an undesirable property. Gradually it was learned that prior coking of the coal could be used to obtain equivalent quality of iron from coke of coal origin as could be obtained using wood charcoal. Only subsequently was it learned that sulfur had been the cause of the early problems with iron quality, and that coking got rid of this problem by getting rid of the sulfur before it contacted the ore. As coke came into more general use and became available in quantity the size of blast furnaces being built also became greater, developments which coincided with the invention of the steam engine which was ideally suited to provide the increasing demand of power for the air blast.

Impurities such as sand, clay, etc., present in iron ore were combined with lime produced by the addition of either limestone or oyster shells into the furnace charge [4]. The slag so formed melted at the blast temperatures, and being lighter than the melt of iron, floated on top. Slag and iron were withdrawn separately, from upper and lower tapholes, at appropriate intervals.

From these early and longer term developments the iron and steel industry has grown to the point where world production of iron ore, based on iron content, exceeded half a billion tonnes in 1978 (Table 14.1) [5]. While the

TABLE 14.1 Production of Pig Iron by the Major World Producers, in Thousands of Metric Tonnes[a]

	1950	1960	1970	1980	1990
Australia	1,116	2,922	5,769	7,407	6,188
Brazil	729	1,750	4,296	13,116	21,141
Canada	2,266	4,025	8,424	11,182	7,344
China	2,392[b]	27,500	22,000	30,604	62,380
India	1,708	4,260	7,118	8,718	12,000
Japan	2,299	12,341	69,714	89,130	80,229
Sweden	837	1,627	2,842	2,596	2,851
Turkey	—	—	1,034	2,072	32,829
U.K.	—	—	17,677	6,316	12,320
U.S.S.R.	25,000[c]	46,800	85,933	107,283	110,166
U.S.A.	60,217	62,202	85,141	63,748	49,668
West Germany	9,511	25,739	33,897	34,055	29,585
Other	6,825	69,434	114,666	134,869	126,355
World totals	112,900	258,600	439,800	520,096	553,056

[a]Countries producing more than 20 million tonnes per year of iron ore (Fe content basis) or 29 million tonnes per year of pig iron in 1978 included. Data from *U.N. Statistical Yearbooks* [5].
[b]Value for 1944, published data closest to 1950.
[c]Value for 1952, published data closest to 1950.

major producers of ore are generally also major producers of pig iron, this is by no means universally the case. Sweden, Australia, Brazil, Canada, and India each mine and process iron ore corresponding to multiples of 8, 7, 4, 2, and 2 times the mass of pig iron they produce, respectively. Japan and West Germany are currently importing nearly all (99.5% and >98%, respectively) of their iron ore requirements, an almost complete dependence on imported ores. The United States depends on imported ores for about one-third of requirements, a proportion which has been fairly steady, at least for the last 10 years.

Steel, derived both from new pig iron produced from ores as well as from recycled steel obtained from structural shapes, old automobiles, ships, and the like, is a better defined material of construction for most uses. World steel production currently hovers around 1.33 times world pig iron production, a reflection of the recycled metal content. Again, we have some countries, like China and the U.S.A., whose national steel consumption exceeds even their very large scale of production by about one-third and about one-fifth, respectively (Table 14.2). France and the United Kingdom each produce slightly more steel than they consume, whereas Italy and West Germany each produce close to 1.33 times their annual consumption. And Japan, even with a very large per capita consumption, produces some 1.67 times its annual requirements. This is undoubtedly a reflection of the growth in export significance of shipbuilding, automobile manufacture, and heavy engineering generally to the Japanese economy, as compared to those of the other large-scale steel producers. Even though some 55 nations were reporting steel production on

TABLE 14.2 Production and Consumption of Steel by the Major Steel Producing Countries, in Thousands of Metric Tonnes[a]

| | 1950 | 1960 | 1970 | 1978 | 1978 Consumption | | 1980 | 1990 |
	Production	Production	Production	Production	Total	kg/capita	Production	Production
China	540	18,460	18,000	31,780	42,706	46	39,996	41,936
France	8,652	17,284	23,773	22,841	19,199	360	23,176	19,304
Italy	2,362	8,231	17,277	24,283	18,822	332	26,501	25,647
Japan	4,848	22,144	93,322	102,105	61,507	535	111,395	110,339
U.K.	16,555	24,680	28,316	20,311	19,429	348	11,277	17,841
U.S.S.R.	27,000	65,318	115,889	151,453	155,000[b]	591[b]	147,941	162,326
U.S.A.	87,848	90,092	119,309	124,314	146,445	670	101,456	89,726
West Germany	12,121	34,110	45,040	41,253	32,209	525	43,838	38,433
Other	26,074	65,597	132,974	185,860	209,883	1 to 756[c]	196,404	267,389
World	186,000	345,916	593,900	704,200	705,200	166	701,984	772,941

[a]Countries producing more than 20 million tonnes of steel per year, from *U.N. Statistical Yearbooks* [5].
[b]Estimated by extrapolation, 1970–1976 period. 1978 not available.
[c]Range of kilograms per capita consumption in 1978; e.g., Afghanistan, Burma, and Ethiopia with 1 kg per captia, Czechoslovakia with 756 kg per capita. See text for explanations.

some scale in 1970, the top 8 countries produced more than three-quarters (77.6%) of the world's steel in that year. Large-scale steel production is becoming less exclusive since by 1978 the same 8 countries produced only about 70% of the world's steel.

14.2. REDUCTION OF IRON ORES

14.2.1. Direct Reduction

A relatively large number of naturally occurring minerals contain iron but only a small number are rich enough in iron and occur in extensive enough deposits to be classed as ores. The important ores in North America are the oxides magnetite (Fe_3O_4, 72.4% iron), hematite (Fe_2O_3, 70.0% iron), and limonite, a hydrated iron oxide ($2Fe_2O_3 \cdot 3H_2O$, 59.8% iron). In Europe, including the U.K., an iron carbonate siderite ($FeCO_3$, 48.2% iron) is also an important ore for pig iron production, though this ore is of less significance in North America. The sulfide minerals pyrite (FeS_2) and chalcopyrite ($CuFeS_2$) are more important for their sulfur content than for their iron content. While the iron content of deposits of these minerals is never quite as high as the calculated percentages given for the pure mineral, many commercial ores are mined with an iron content in the 50–60% range and some deposits are worked with iron compositions of as high as 66–68% [6].

If the ore is contained in a hard rock matrix it is mined from underground or surface deposits by drilling and blasting followed by recovery with power shovels. Many deposits are sufficiently loosely consolidated that direct recovery with power shovels is possible, by either open cut or open pit operation. Both types of ore are also frequently beneficiated (cleaned of extraneous material) at the mine site. Beneficiation may involve any of rough sorting, crushing, screening, magnetic separation (magnetites only), washing (for clay or sand removal), and drying, all designed to improve the grade of the ore prior to shipping or smelting.

Direct reduction processes of various kinds, similar in nature to prehistoric reduction methods in that at no time during the process is the iron content entirely fused (melted), are still used on a small scale to produce spongy and powdered iron products. In practice all direct reduction methods involve contacting of a solid or gaseous reductant, at temperatures of 1000 to 1200°C, with a high analysis magnetite or hematite ore to effect reduction without melting [7]. Both solid reductants such as powdered coal and coke, and gaseous reductants such as natural gas and hydrogen, are being used. The advantage of direct reduction methods for iron production is that the iron product obtained is low in carbon and can therefore be used for steel making directly, without requiring carbon removal as a preliminary step. For this reason the direct reduction product is occasionally referred to as "synthetic scrap," since it is possible to use this type of iron to replace part or all of the scrap component of steel making. Catalyst applications also provide a small but important market for both the spongy and the iron powder forms of directly reduced iron. The very high porosities and specific surface areas make these forms of iron far more valuable than iron filings or turnings for catalyst applications.

Various feed composition and processing constraints required by direct reduction methods have led to blast furnace reduction of iron ores as the dominant method used to obtain metallic iron from its ores. Direct reduction methods remain in favor in parts of the world where there is abundant natural gas but negligible coal reserves, such as in Iran, Qatar, and Venezuela. They are also of value for small-scale iron making operations which feed their product directly to an electric furnace for steel making.

14.2.2. Blast Furnace Reduction of Iron Ores

Reduction of iron ore to the metal in a blast furnace requires the reduced metal to be melted in the process, the key physical difference between direct reduction and blast furnace reduction. Two to three tonnes of iron ore are required for each tonne of iron produced. The main impurity in iron ore is generally silica. To avoid contaminating the iron with silica, limestone or dolomite ($CaCO_3$ + $MgCO_3$) is also added to combine with the silica and to provide a lithophilic phase to dissolve up any other acidic oxides which do not react with lime (Eq. 14.1–14.3).

$$CaCO_3 \rightarrow CaO + CO_2 \qquad\qquad 14.1$$

$$CaO + SiO_2 \rightarrow CaSiO_3 \text{ (slag)} \qquad\qquad 14.2$$

$$CaO + Al_2O_3 \rightarrow Ca(AlO_2)_2 \qquad\qquad 14.3$$

In this way a separate molten slag phase is formed from the ore impurities, which can be easily separated from the molten iron. In addition, some 550–1000 kg of coke is used per tonne of iron produced, to provide both much of the heat and the reducing agents necessary for the formation of pig iron.

The whole reduction process is carried out in a blast furnace consisting of large, almost cylindrical stack up to 10 m in diameter and 70 m high, which is tapered to a larger diameter at the bottom to avoid bridging (jamming) of ore as it moves down the stack (Fig. 14.1). The outer shell is constructed of heavy (ca. 2-cm-thick) riveted or welded steel plates, which is then lined with silica-containing (upper region) or carbon (hearth area) refractory bricks to make the structure resistant to the intense heat. This unit functions as a heterogeneous, countercurrent reactor in which the solid reactants are charged in at the top, and the firing air is blown in at the bottom. The ore, limestone, and coke, in lumps to permit movement of gases through the charges, are added in alternating layers to the top of the furnace using a skip cart on rails. A double bell arrangement at the top of the furnace prevents gas loss from the furnace during the charging process by only opening one bell at a time during the addition steps. The remaining raw material, the firing air, is preheated to about 850°C in regenerative stoves and then blown through water-cooled steel tuyeres (air nozzles) at the hearth area of the furnace.

Initial warming of a newly lined blast furnace is conducted gradually, over several days, to avoid any unnecessary thermal stresses. Fueled first by wood

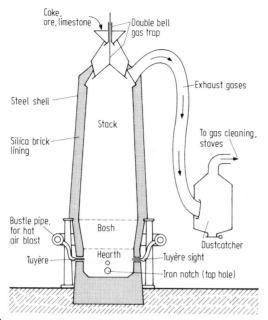

Coke,
ore, limestone
Double bell
gas trap

Exhaust gases

Steel shell

Stack

To gas cleaning,
stoves

Silica brick
lining

Bustle pipe,
for hot
air blast
Bosh

Dustcatcher

Tuyère
Hearth
Tuyère sight

Iron notch (tap hole)

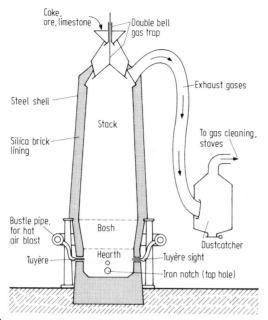

FIGURE 14.1 Diagram of the operating details of a blast furnace for the production of pig iron.

or oil and then by coke, the approach to normal operating temperature by the furnace is gauged from peepholes through the tuyeres. As these temperatures are reached ore and limestone are blended with the coke additions through the double bell arrangement at the top of the furnace. Heat, provided by primary combustion in the oxidizing zone near the hot air blast of the tuyeres is sufficient to cause carbon dioxide and water vapor (from the air) reduction (Eq. 14.4–14.6).

Primary combustion and reductant formation:

$$C + O_2 \rightarrow CO_2 \qquad\qquad 14.4$$

$$CO_2 + C \text{ (white hot)} \rightarrow 2\ CO \qquad\qquad 14.5$$

$$H_2O + C \text{ (white hot)} \rightarrow CO + H_2 \qquad\qquad 14.6$$

Carbon monoxide is the primary agent responsible for reduction of the iron oxides in the ore, assisted by carbon and to a small extent by hydrogen (Eq. 14.7–14.12).

Primary reducing reactions:

$$\text{ca. } 500°C\ 3\ Fe_2O_3 + CO \rightarrow 2\ Fe_3O_4 + CO_2 \qquad\qquad 14.7$$

$$\text{ca. } 850°\ Fe_3O_4 + CO \rightarrow 3\ FeO + CO_2 \qquad\qquad 14.8$$

$$\text{ca. } 900°\ FeO + CO \rightarrow Fe + CO_2 \qquad\qquad 14.9$$

$$\text{ca. } 900°\ Fe_2O_3 + 3\ H_2 \rightarrow 2\ Fe + 3\ H_2O \qquad\qquad 14.10$$

The solid-solid or liquid-solid reactions between ferric and ferrous oxides and carbon also occur at temperatures above about 500°C [6].

$$Fe_2O_3 + C \rightarrow 2\ FeO + CO \qquad\qquad 14.11$$

$$FeO + C \rightarrow Fe + CO \qquad\qquad 14.12$$

The iron, with its melting point depressed to about 1100°C by dissolved carbon (melting point of pure iron is 1535°C), drips through the charge and collects on the hearth. Being the denser fluid product of the blast furnace, it forms a lower phase as it accumulates on the hearth. Molten pig iron is periodically withdrawn from this accumulation by drilling or knocking out a clay plug placed in a lower tap hole in the side of the hearth.

The slagging reactions (Eq. 14.1–14.3) start to occur as the charge moves down the furnace and the limestone or dolomite used as a flux reaches temperatures of about 900°C and higher. As carbon dioxide loss occurs, the lime (and magnesia, if present) start to react with and dissolve other oxide impurities of the charge into the slag layer. It is also possible to reduce silica and phosphate by the same kinds of reactions used to reduce iron which, in the process, contaminates the pig iron (Eq. 14.13, 14.14).

$$SiO_2 + 2\ C \rightarrow Si + 2\ CO \qquad\qquad 14.13$$

$$P_2O_5 + 5\ C \rightarrow 2\ P + 5\ CO \qquad\qquad 14.14$$

The presence of the basic slagging ingredients minimizes, but does not completely prevent these reductions from occurring. The molten slag, which is less dense than iron, floats on top of the iron layer and is periodically withdrawn in this form through an upper tap hole in the hearth of the blast furnace. About 0.5–1.5 tonnes of slag are produced for each tonne of pig iron, the low end of the range being obtained with high-grade ores while using a coke with a low ash content for reduction. Thus utilization of this by-product of pig iron production is a significant part of normal blast furnace operation. Slag is used for road aggregate (ca. 50–60% of the total), in crushed form as a stone component of asphalt, and in granulated form (by running the melt into water) for use as an ingredient in the manufacture of cement (ca. 10–20% of the total). It is also used to a small extent (ca. 5%) for the preparation of mineral wool ("rock wool") insulation materials. The approximate chemical composition of blast furnace slag is given in Table 14.3.

The carbon monoxide produced in the furnace is only partially consumed on its passage through the granular charge (Table 14.3). But capture of this important fuel gas for energy recovery, by the double bell closure at the top of the furnace, is only a relatively recent (ca. 1840) innovation in the long history of iron making. After gas collection from the upper part of the furnace the exit gas is passed to a dust catcher for removal of most of the entrained dust and is then used, partly to fuel three or four regenerative stoves and partly for other purposes. These regenerative stoves, which are designed to have a large thermal mass from an extensive internal firebrick checker-work of narrow passages, are first heated by blast furnace gas. Then, when the temperature of the brick checker-work of one stove has been raised sufficiently, the blast furnace gas is switched to fuel the burner of another stove.

TABLE 14.3 Approximate Composition Ranges of the Products of Blast Furnace Operation[a]

Iron		Slag		Exit gases	
Component	% by wt.	Component	% by wt.	Component	% by wt.
Iron	90–94	Lime	30–48	Nitrogen	58–60
Carbon	3–4.5	Silica	28–42	Carbon monoxide	27–30
Silicon	1–2.5	Alumina	10–20	Carbon dioxide	8–14
Phosphorus	0.04–2.5	Magnesia (MgO)	2–14	Hydrogen	0.4–3
Manganese	0.04–0.7	Sulfur	1–3	Water vapor	0.4–1.0
Sulfur	0.02–0.2	Iron oxide (FeO)	0.3–2		
		Manganese oxide	0.2–1.5 (MnO)		

[a]Compiled from data of Dearden [2] and Kirk-Othmer [6].

At this point the preheated stove is used to heat the air blast being fed to the tuyeres in the hearth area of the blast furnace. By a fuel gas and air switching arrangement like this, blast air is continuously preheated using the fuel value of the furnace exit gases, and in so doing decreases the overall coke consumption to significantly below that required before gas collection was practiced.

Originally the molten iron product from the blast furnace, as it was drained from the lower notch on the hearth, was led to a series of shallow trenches shaped in sand. As the molten iron moved along each long trench, it filled a series of short smaller trenches terminated by roughly rectangular hollows each of which, in cooling, eventually formed a 20–25 lb ingot of iron. This product became known as "pig iron" by the appearance parallels between this manual method of forming ingots to a sow suckling her young. Today pig iron ingots are made on a continuous pig-casting machine, which consists of an endless chain with metal molds into which the molten iron is ladled. A water spray, as the molds with molten metal move through it, quickly chills the melt allowing pig production at the rate of two tonnes per minute or more. A lime or similar wash, sprayed into the insides of the emptied molds as they return to the ladle for refilling, prevents the pig iron from sticking to the molds.

Blast furnace production of iron allows the hot, newly reduced product to trickle through the bed of heated coke to the hearth. Since carbon is somewhat soluble in molten iron, pig iron usually contains from 3–4.5% carbon. It also contains smaller percentages of other reduced elements such as silicon, phosphorus, manganese, etc., generated by the same reducing processes that yielded the iron (Table 14.3). Primarily from the effect of the high carbon content on the iron crystal structure, the blast furnace product is brittle, hard, and possesses relatively low tensile strength. Hence the crude pig iron product of the blast furnace is little used directly in this form.

After remelting in a cupola furnace and some minor composition adjustments pig iron is used to make various types of cast irons with improved properties. Molten gray cast iron, for instance, is poured into molds to make automotive engine blocks. The hardness and graphite content of this material contribute to its desirable wear resistance in this application. The variety of cast irons comprise an important field of iron metallurgy. Properties may be adjusted by microstructural constituents and crystal structures cover a wide range of possibilities. The corrosion resistance of some cast irons makes these useful for the fabrication of certain types of chemical reactors. The low cost of the raw material means that the wall thickness of a reactor may be made sufficiently thick that some corrosion can be tolerated without perforation. Construction of low-speed flywheels from cast iron and its use as a ballast in ships depend on low cost and high density. However, most pig iron moves directly to one or another of the various steel making processes, either still in molten state or in ingot form.

A number of advantages are obtained from pig iron production in very large furnaces, one of which is the lowered coke consumption from the lower furnace heat losses which results from the smaller surface to volume ratio. This factor, together with more extensive use of richer, beneficiated ores which also may be partially prefluxed, use of all charges in a more uniform size range (by prior agglomeration), plus additions of oxygen and steam to the blast air, have combined to nearly halve the coke consumption per tonne of pig iron produced in American practice since 1950 [8]. The addition of steam to the blast has been a growing practice, which refines the control of the blast temperature as well as providing additional reducing gas at the bosh zone via Eq. 14.6 [7]. Occasionally fuel oil or natural gas is used to supplement combustion and reducing capacity of a blast furnace. Recent energy balance data incorporating many of these variations are given by Kirk-Othmer [9].

An idea of the scale of operation of a typical modern intermediate size blast furnace incorporating some of these innovations may be gained from consulting Table 14.4. The largest blast furnaces now operating in the U.S.A.

▬▬ TABLE 14.4 Approximate Material Balance Data for a Modern Blast Furnace Producing 600 Tonnes of Pig Iron per Day[a]

Raw materials		Product and by-products	
Beneficiated iron ore (50–60% Fe)	1000–1200 tonnes	Pig iron	600 tonnes
Hard coke	300–600 tonnes	Slag	300–400 tonnes
Limestone (or dolomite)	200–300 tonnes	Dust	60–80 tonnes
Blast air, at >800°C and 550 kPa (80 psi)	2000–2400 tonnes 1.5–$1.9 \times 10^6 m^{3b}$	Exhaust gases	2800–3400 tonnes, 2.2–2.6×10^6 m^{3b}
Water, as steam	20–40 tonnes		
Water, cooling	14,000–18,000 tonnes		
Electricity	2400–28,600 kWh		

[a]Data compiled from Dearden [2], McGannon [7], *Encyclopedia of Science and Technology* [8 and Kirk-Othmer [9].
[b]Volumes at standard conditions, i.e., 15°C and 1 atm.

and Japan are capable of producing 10,000 tonnes of pig iron per day, more than 16 times this capacity. The total U.S. iron production in 1979 of some 80 million tonnes was produced by 168 large and small furnaces, mostly operating in the Pittsburgh and Chicago areas.

14.3. THE MAKING OF MILD AND CARBON STEELS

Pig irons contain 6 to 8% of total impurities which give the product brittle, low-strength characteristics. Broadly speaking pig irons may all be classified into two groups, those that are low in phosphorus and those that are high (Table 14.5) [10,11]. Mild steels and high-carbon steels are made by reducing the concentrations of the impurities present in pig irons and controlling the final concentration of carbon. Steels are considerably stronger materials and in addition may have other specific desirable properties contributed to them by control of composition and by heat, or other types of subsequent treatments. As a result of this improvement in properties more than 90% of all pig iron produced, as ingots, hot metal, or molten metal proceeds directly to various steel making procedures. This is particularly true at integrated operations where the steel making is closely associated with blast furnace operation. Over 90% of all steels made, the "common steels" of Table 14.5, are carbon steels. Thus, the bulk of the following discussion is centered on this important group of products. Mild steels, the product used for the vast majority of structural and engineering purposes, contain 0.15 to 0.25% carbon. High-carbon steels, defined as steels containing 0.6 to 0.7% carbon, possess important heat treating qualities which can permit them to hold a sharp "edge." Mild steels are not usually heat treated.

The processes used in making common steels may be classified on a chemical basis using the composition of the refractory linings used for the vessels in which they are made and the primary composition of the slag. The processes used may be further subdivided according to whether they are based on pneumatic, open hearth, or electric furnace technologies.

▬▬ **TABLE 14.5 Percentage of Impurities in Representative Pig Irons from the Two Main Types of Ores, Compared to the Compositions of Common Steels**[a]

| | Pig iron | | |
Element	No. 1 (%)	No. 2 (%)	Common steels (%)
Carbon	3.5–4.0	3.0–3.6	0.1–0.74
Silicon	2.0–2.5	0.6–1.0	0.01–0.21
Sulfur	<0.040	0.08–0.10	0.028–0.055
Phosphorus	<0.040	1.8–2.5	0.005–0.054
Manganese	0.75	1.0–2.5	0.35–0.84

[a]Compositions in percent by weight compiled from Allen [10] and Bashworth [11].

Acidic steel making employs silica (SiO_2) refractory linings in the equipment used, and generates silica-rich slags via additions of sand for removal of impurities. Silica, being a nonmetallic oxide, allows removal of carbon, manganese, and silicon during the metallurgical steps but cannot achieve removal of sulfur or phosphorus. Thus, pig irons rich in sulfur phosphorus or both elements cannot be satisfactorily processed under these conditions. Basic steel making procedures use magnesia (MgO) refractory linings and limestone or dolomite added to the charge as a slag former. Equipment with a basic lining is not only capable of removing carbon, manganese, and silicon, but is also capable of removing phosphorus and sulfur by capture of their oxides in the basic slag. Addition of basic slag formers to an acid-lined vessel or vice versa is never practiced because doing so drastically shortens lining life as a result of their direct interaction.

The technological classification is based on the method used to effect removal of pig iron impurities. All pneumatic methods rely on the heat content of the molten charge plus the heat of the refining reactions themselves to provide the heat required by the process. Purification is effected by blowing air or oxygen through or onto the molten charge. Open hearth methods all rely on the combustion of an external fuel, usually natural gas or oil, to provide the heat required for the process. Hence, stoves and their regenerative heating capabilities (as with blast furnace gas) are important for open hearth operation. Electric furnace methods derive heat requirements from electric arc or resistance sources, and as such permit control of the atmosphere above the melt independently of the heating process. This capability has important consequences, particularly for the production of special alloy steels.

14.3.1. Pneumatic Steel Making: The Bessemer Process

The earliest of the pneumatic methods, the Bessemer process, was devised and patented in England in 1855 by Henry Bessemer [12]. Originally prompted by an interest in improving the quality and strength of cast iron gun barrels to obtain greater range from lighter guns, this process was developed from the perception by Bessemer that the carbon in molten iron could be burned out by air blowing. It was anticipated that this method could raise the speed and decrease the cost of the existing crucible method of steel making.

Bessemer steel is made in a Bessemer converter, a pear-shaped vessel with a double bottom lined with silica, and capable of being tilted on its axis. A typical converter can hold 15 to 25 tonnes of metal at one time (Fig. 14.2). It is charged with molten pig iron while the converter is lying on its side. The air pressure to the perforated inside base of the unit is then turned on and the vessel gradually raised to the upright position so that the air is sparged through the molten pig iron via the inner perforated floor of the converter. Blowing is continued for a period of 15 to 20 minutes until the operator, from observing the color and luminosity change of the flame at the mouth of the converter, judges the refining process to be complete. Carbon, as well as silicon, manganese, and some iron are removed, largely via reactions 14.15 to 14.19.

$$2 \; Fe + O_2 \rightarrow 2 \; FeO \qquad\qquad 14.15$$

$$2 \; Mn + O_2 \rightarrow 2 \; MnO \qquad\qquad 14.16$$

$$Si + O_2 \rightarrow SiO_2 \qquad\qquad 14.17$$

$$2 \; C + O_2 \rightarrow 2 \; CO \qquad\qquad 14.18$$

$$C + O_2 \rightarrow CO_2 \qquad\qquad 14.19$$

The involatile products of these reactions form an acidic slag ($SiO_2 \cdot FeO$; $SiO_2 \cdot MnO$ etc.,) which forms a separate lighter phase on top of the molten iron. These exothermic reactions occurring during the blow are sufficient to raise the temperature of the converted metal from about 1350°C to around 1600° or higher, adequate to keep the residual iron molten. This temperature rise is necessary since the melting point of the iron phase rises as the concentration of impurities is decreased by the conversion process.

The Bessemer process is fast; one converter is capable of producing nearly a tonne of steel per minute. This compares to 10 to 14 days for smaller quantities per unit via the crucible methods current at the time the Bessemer process was developed. The product is of acceptable quality for some uses, at least when the pig iron is derived from low phosphorus ores and is smelted in blast furnaces using low sulfur cokes. However, with high phosphorus pig irons the decarbonized product is "cold short" (brittle at ambient temperatures) which makes it next to useless for the vast majority of applications of the steel [12]. Another difficulty is that if the ore or the smelting process produces a pig iron which is high in sulfur (e.g., from sulfur-containing cokes) the Bessemer product is "hot short," or brittle when hot. This made hot shaping or working of the product impossible, further adding to steel making difficulties with this product.

The speed of Bessemer steel making is both an advantage and a disadvantage, in that there is no time during the process for product sampling,

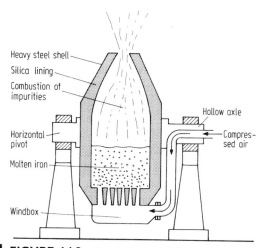

FIGURE 14.2 Cut-away view of the main operating details of a Bessemer converter during a "blow." It is filled and emptied by tilting onto its side, while the air flow is off.

analysis, and composition adjustment. Thus, much of the product quality depends on the skill and judgment of the operator. Overblowing, which results in total carbon removal plus some iron oxide and iron nitride formation, may be partially corrected by blending the blown product with a small amount of hot melt pig iron and mixing well [7]. Other minor adjustments may be made in the ladle receiving the Bessemer product. But removal of the nitrogen is never easy, even with strong deoxidizers such as aluminum, and yet this is essential to obtain good stamping quality steels (e.g., steels destined for shaping into auto or appliance body panels). The combination of these stringent material requirements and the short composition adjustment time thus has restricted production of acid Bessemer steels (produced in converters with acidic linings) to the low carbon, mild steels. However, steel making using these technological elements is still being practiced on a small scale in the U.K. and other European countries as a method for producing low-cost steels.

14.3.2. The Basic Bessemer or Thomas Process

The problem of phosphorus or sulfur removal from pig irons high in these elements was solved by the use of a basic refractory lining, such as magnesia (MgO) or lime (CaO), in the converter. When this was accompanied by the addition of limestone or dolomite with the molten charge of pig iron, phosphorus and sulfur constituents were removed in the basic slag layer that was formed. These improvements were developed in London in 1878 by the cousins S.G. and P.G. Thomas, and were adopted by the U.K. steel industry for the processing of moderate phosphorus content ores. The steel makers of Lorraine, Germany, however, enthusiastically adopted the Thomas process to achieve phosphorus removal from their high-phosphorus ores and obtain improved steel quality [10].

Phosphorus removal in this process is thought to occur primarily as the oxide, whereas elemental sulfur is thought to form a slag constituent by direct replacement of the oxygen of calcium oxide (Eq. 14.1, 14.20–14.22).

$$4\,P + 5\,O_2 \rightarrow 2\,P_2O_5 \qquad\qquad 14.20$$

$$4\,P_2O_5 + 4\,CaO \rightarrow 4\,CaO{\cdot}P_2O_5\ \text{(the slag)} \qquad\qquad 14.21$$

$$2\,S_{(Fe)} + 2\,CaO_{(slag)} \rightarrow 2\,CaS_{(slag)} + O_{2(Fe)} \qquad\qquad 14.22$$

An important attraction of the acidic Bessemer process and of the basic Bessemer or Thomas process is the rapid steel making capability, about a tonne a minute per converter, that these methods provide. However, both processes tend to cause some nitriding, which is the absorption of nitrogen from the blowing air by the hot metal (Table 14.6). Nitrides in a finished steel increase its susceptibility to strain aging, or embrittlement under stress, a feature which makes these products unsuitable for any cold metal shaping or finishing operations, such as wire drawing. This disadvantage and the inability to handle large quantities of scrap metal or solid iron charges have decreased the relative importance of both of these pneumatic processes in modern steel making to 10% or less of the total [10]. There has been limited adoption of variants of these methods such as oxygen-enriched air, oxygen and steam, or

■ **TABLE 14.6 Nitrogen Content of Blast Furnace Metal (Pig Iron) and Various Steels[a]**

Iron product	Range of nitrogen content (%)
Pig iron	0.002–0.006
Bessemer steels	0.010–0.020
Duplex steels[b]	0.005–0.008
Open hearth steels	0.004–0.007

[a]Compiled from McGannon [7] and Allen [10]. [b]See text for details.

oxygen and carbon dioxide to decrease the nitriding tendency. Pure oxygen introduces a too severe thermal stress on a converter. Use of a pneumatic process as a preliminary to an open hearth furnace (duplex steel making) or as a preliminary to both open hearth and electric furnaces (triplex steel making) decreases the nitrogen content of the product and improves steel making flexibility.

14.3.3. The Open Hearth Process

Improved furnace combustion efficiency was the primary objective of C.W. Siemens, from Germany, who conducted most of his combustion trials in Britain. He applied regenerative methods to recover waste heat from a combustion process by passing the combustion gases through a chamber containing an open, firebrick checkerwork. When the checkerwork reached nearly the combustion gas temperature, the hot gas flow was diverted to another chamber of firebrick checkerwork to heat this. At the same time the air flow to the combustion process was directed through the preheated chamber of hot checkerwork in order to preheat the combustion air. This air preheat improved the efficiency of the combustion process itself and enabled much higher ultimate temperatures to be reached. The two sets of firebrick checkerwork were operated on swing cycles such that as the air preheat checkerwork cooled down the air stream was redirected to the other, already hot checkerwork, and the hot combustion gas stream was returned to reheat the initial, now cool checkerwork. In this way a continuous heat recovery from the spent combustion gases was obtained. By 1857 these developments alone gave fuel savings of some 70 to 80% for industries such as glass making [12].

These combustion developments were first tested for steel making in 1863 in France, under license from Siemens, by a father and son team, E. and P. Martin. Siemens himself tested the furnace for this purpose in Wales, three years later, and both operations were pronounced a success. These were the origins of the open hearth process, now often referred to as the Siemens-Martin process after the joint contributions of these inventors.

The open hearth furnace uses the Siemens regenerative principle to raise maximum combustion temperatures to about 1650°C and possesses a heating capability independent of the heat content or impurity burning reactions of the iron charge. Because of this, the charge placed in the open hearth furnace may be any ratio of scrap to hot metal up to 100% scrap, since the furnace has the capability to melt this. Usual practice, however, is to use a roughly 50:50 mix of scrap to melt [13].

The steel making components of an open hearth furnace consist of a rectangular heating zone with a floor containing a long shallow depression of 200 to 500 tonnes capacity into which the charge is placed (Fig. 14.3). Gas or oil burners at each end are provided with preheated combustion air via hot checkerwork, and the hot gases are deflected down onto the charge by a convex-down, reverberatory roof. So a combination of radiant heating and hot gas contact with the charge transfers sufficient heat to melt it. To protect the structural components of the depression holding the charge from the intensely high temperatures and also to aid in removal of phosphorus and sulfur impurities from the charge the depression is lined with a magnesia (MgO) brick refractory.

Steel making begins by placing the solid components of the charge, the limestone, any ore requirement, and rusty scrap steel into the depression, cold. The scrap is preferably rusty to provide oxygen which will assist in impurity removal. Iron ore (e.g., Fe_2O_3) may also be added to provide oxygen. The fuel, usually natural gas, coal gas, or oil, is then ignited at the burners and the initial charge is heated until it just starts to melt. During this process some oxidation also inevitably occurs as a result of the oxidizing atmosphere of the furnace. At these temperatures carbon dioxide is chemically oxidizing toward iron, and causes some additional formation of ferrous oxides, apart from that present initially. Once the initial solid components are melted, the blast furnace melt, at 1300–1350°C, is added causing initiation of the main carbon removing reaction (Eq. 14.23).

$$2 \, C_{(Fe)} + O_{2(Fe)} \rightarrow 2 \, CO_{(gas)} \qquad 14.23$$

Oxygen, from the scrap, or the ore, or dissolved in the molten iron from air, reacts with the carbon present in the blast furnace melt to form carbon mon-

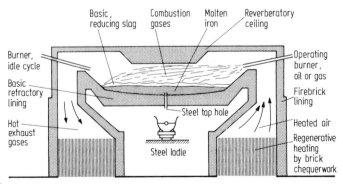

FIGURE 14.3 Operating details of an open hearth (Siemens-Martin) furnace for steel production.

oxide. As carbon monoxide bubbles up through the charge it serves to stir the fluid mass and aids in the rise of silica, manganese oxide, phosphorus oxides, etc., to form a slag layer. While these reactions are going on, limestone decarbonization also occurs and the carbon dioxide formed further contributes to the carbon monoxide formed by its conversion with dissolved carbon (Eq. 14.1, 14.24).

$$CO_{2(Fe)} + C_{(Fe)} \rightarrow 2\ CO_{(gas)} \qquad 14.24$$

Thus, the slag formed on the open hearth charge is quite frothy. It is pushed off the pool of molten steel beneath it with a type of hoe, once or twice during the process. Blast furnace metal melts at about 1130°C, whereas pure iron, more closely representative of steels, melts at 1535°C. For this reason, as carbon removal from the blast furnace melt proceeds the temperature of the charge has to be raised to avoid solidification of the charge. Since the bulk of heat transfer in the open hearth process occurs via direct contact of hot combustion gases with the surface of the charge, it is important that the frothy slag layer not be allowed to get too thick or heating may become ineffective.

When a satisfactory composition of the charge is obtained, the product is tapped from a hole in the base of the depression and cast into large ingots ready for final shaping. Open hearth processing is quite slow, 6 to 12 hours per batch; about half of this time is for charging. But this does provide time for sampling, analysis, and adjustment of the carbon content by additions of further scrap or hot blast furnace metal. Some alloys of up to 10% of another metal are also possible in an open hearth process. Thus greater product flexibility is possible than with the pneumatic steel making processes. Material balances of the various steel making processes are given in Table 14.7.

Open hearth processes employing silica refractory linings are also known [7]. However, since this type of lining does not provide phosphorus or sulfur removal capability these are usually reserved for open hearth operation on solely scrap iron charges, or on the occasional high-grade iron ores free of these elements. The independent heating capability of an open hearth furnace frees it from necessarily being associated with a blast furnace operation and allows steels to be made locally entirely from scrap.

14.3.4. Electric Furnace Steel

Early development of practical furnaces heated by an electric arc at around 1876 again owed much of their development to the contributions of C.W. Siemens. Again it was others who first utilized these principles in steel processing. Stassano, in Italy, succeeded in melting steel scrap in 1899 using this furnace design, with an arc struck between closely spaced carbon electrodes, although he was unable to reduce iron ores by this method. The designs of Paul Heroult, working in France at about the same time, were more successful and became the forerunners of electric furnace steel making processes used to the present day [2].

Heroult used experience from his electrolytic method for reduction of alumina and abandoned the indirect arc electric furnace designs, in which the arc was struck between opposed carbon electrodes. Instead he used direct arc

■ **TABLE 14.7   Raw Materials Required to Make One Tonne of Ingot by the Various Steel Making Processes[a,b]**

| Raw material | Basic Bessemer process | Hot metal, fixed open hearth furnace | Electric arc process | | Basic (LD) oxygen furnace |
			Common steels	Special alloy steels	
Iron, kg:					
Molten furnace	1050–1150	550	nil	nil	850–1025
Cold pig	nil	slight	some[c]	0–75	nil
Scrap iron (or steel), kg:	0–80	525	1075	1000–1100	100–300
Oxidizing ore or millscale[d], kg:	low to 60	100	12.5	2–18	7–75
Fluxes, as limestone etc., kg:	110–180	200	55	25–75	60–80
Finishing alloys:	ca. 10	14	10	various	6–7
Oxygen, m^3 (15°C, 1 atm):	nil	varies	150 ft^3	varies	1700–2000
Primary fuel, kJ (Btu):					
As natural or coal gas	nil	4.2–5.3 × 10^6 (4–5 × 10^6)	nil	nil	nil
Electricity, kWh	nil	nil	500–550	650–750	nil

[a]Adapted from data in Allen [10].
[b]Amounts actually consumed in the process are given in kilograms, unless otherwise stated.
[c]Some used occasionally.
[d]Millscale is the iron oxide (mostly Fe_3O_4) "crust" removed from steel billets during rolling operations.

designs which still employed the electrodes to carry power to the heating zone of the furnace. But the heating effect was obtained by two arcs, one struck between one electrode and the surface of the metal and the other between the other electrode and the metal. The metal itself served as the power conductor between the electrodes, thus producing two arcs in series from the same current flow (Fig. 14.4). The double or multiple arcs in direct contact with the metal provided by the direct arc design gave much improved heat transfer efficiency and contributed significantly to improved electrical efficiency per

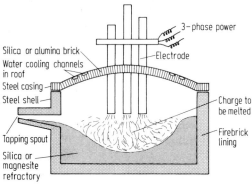

■ **FIGURE 14.4**   Schematic design of a direct arc electric furnace for steel making. The whole unit may be tilted to the side for emptying the finished steel melt.

mass of metal melted, over the earlier indirect arc designs. This heating principle also forms the basis of the design of all modern steel making arc furnaces [14].

Even though electric power is more expensive than natural gas, oil, or coal gas as a source of energy, its higher efficiency of utilization in the arc furnace makes up for this. It is estimated that an open hearth furnace, even with regenerators, is only able to utilize 15 to 20% of the heat energy in the fuel supplied to it for steel making whereas the parallel value for the electric furnace is of the order of 55 to 60% [2].

Electric furnace steel making, like the open hearth process, has a means of charge heating independent of a hot metal charge or the combustion of impurities. Since 1900 it has been primarily used for the remelting of clean scrap, particularly of valuable alloys. But since about 1950 electric furnace steel making has been much more widely adopted because it permits control of both the heat input *and* the atmosphere above the charge. It does not need an oxidizing atmosphere as is required for both the pneumatic and open hearth processes. An appropriate atmosphere can be provided for the type of steel being made, a particularly important factor for some of the alloy steels. Special versions of arc furnaces are also used to permit melting under vacuum for products destined for particularly demanding applications where small amounts of dissolved gases in the steel could adversely affect properties.

For an electric furnace with basic refractory, while the power is off, a small amount of limestone is charged initially to provide the basic slag-forming components. This is followed by the remainder of the charge consisting entirely of scrap steel for simple refabrication, or scrap steel plus alloying metal(s) if special alloys are being produced. If carbon steels are being made, a small amount of ore plus scrap to provide oxygen and a significant component of solid or molten pig iron will be added. For furnaces producing specialized alloys the total charge volumes may be only 1 to 5 tonnes although electric furnaces accepting charges as large as 150 tonnes are in common use. On completion of the charging step the arcs are struck and heating begins. An equilibrium reducing atmosphere of carbon monoxide and carbon dioxide is soon produced in the vapor space above the charge. A period of 2 to 3 hours is required for the usual charge to melt completely, after which samples can be taken to check for composition.

If the carbon content is still too high for carbon steel or mild steel preparation, further oxygen can be added using a water-cooled lance to discharge the gas close to the surface of the melt (about 1 to 1.5 m above) while the arc is off. The initial slag containing silica, MnO, FeO, P_2O_5, and calcium sulfide in a lime base, called an "oxidizing slag," is skimmed off. Then just before pouring the steel it is "killed" (deoxidized) by addition of one or more star-shaped pieces of aluminum to the melt, depending on the oxygen content determined by analysis. The easily oxidized aluminum combines with the oxygen of any residual iron oxides forming further slag and decreasing the incidence of imperfections in castings made from the melt (Eq. 14.25).

$$2 \text{ Al} + 3 \text{ FeO} \rightarrow \text{Al}_2\text{O}_3 + 3 \text{ Fe} \qquad 14.25$$

For the preparation of alloy steels such as chromium, vanadium, tungsten, or manganese, most of the alloying metal is in place at the start of the melt. To prevent oxidation of the alloying metal a reducing slag of burned lime (CaO) and powdered carbon is used. This provides both the protection of the reducing atmosphere of carbon monoxide and carbon dioxide in the vapor space above the metal, as well as the protective layer on the metal. On completion of melting a sample is taken and analyzed to determine whether any final adjustment of alloying metal concentration is necessary before poring.

The stainless steels require an electric furnace for their preparation. The corrosion resistance of this class of alloy, conferred by the presence of about 18% chromium plus 8% nickel in the iron, is a valuable feature of its use in industrial vessels, the food industry, and for interior and exterior fittings.

The stainless steels are only one example of the present day versatility of the electric furnace in steel making. The electric furnace has a high flexibility to accept solid, molten, or a mix of these types of charges, to operate with acidic or basic linings depending on need, and to provide an oxidizing, reducing, or neutral (inert) atmosphere. The combined appeal of these attributes, plus its ability to operate under clean conditions and under a partial vacuum for steels destined for critical service have led to the rapid adoption of electric furnace methods as a component of steel making practice worldwide [15].

Electric furnaces which employ induction heating have also been used sporadically for steel making. In this procedure high-frequency electric power of 1000 to 2000 Hz is passed through a water-cooled copper coil (conductors are actually heavy walled copper pipe) surrounding a nonmagnetic, nonconducting crucible made entirely of refractory. In this way, the charge placed in the crucible serves as both the "core" and the single turn secondary "winding" of a step-down transformer. The power flow and eddy currents generated in the charge as a result of this placement brings about rapid melting and permits any composition adjustments to be made. Electrical consumption is about the same as for an electric arc furnace, 660–880 kWh per tonne of steel, but the capital cost of an induction furnace averages two to three times that of an arc furnace, mainly for the motor-alternator and condensers required to provide the high-frequency AC current. For steel requiring clean conditions and for specialty steels induction furnaces fill an important need [10].

14.3.5. Oxygen Steel Making: The LD or BOS Process

The use of essentially pure oxygen for the oxidation of iron impurities became a commercial possibility as oxygen became available on a tonnage scale in the 1930s. Early development of this process, originally called the "Linzer Dusenverfahren," took place in Linz and Donawitz in Austria during the 1930s and 1940s [12]. Either the original process name or the places of development led to the identification of this development as the LD process. The first two oxygen converter plants using this principle became operational in Austria in 1952 and 1953.

The LD process, also referred to as basic oxygen steel making (BOS, also refers to the product, basic oxygen steel), is carried out in a basic oxygen furnace (BOF), all terms used synonymously to refer to these developments. The vessel used for these processes is similar to the Bessemer converter, but is lined with a basic refractory. From a water-cooled nozzle located in the neck of the vessel a stream of oxygen is directed onto the surface of the mostly molten blast furnace metal charge. Impurities in the hot metal combine rapidly with the high-oxygen content atmosphere. The resulting oxides then combine with lime, or limestone which is also added to the charge, to form a slag layer to permit removal of the impurities.

Oxygen steel making is rapid, like the older pneumatic processes. Tap to tap times are normally in the range of 18 to 20 minutes and the product is not subject to nitriding since nitrogen contact is avoided. Also, because the heat production from impurity oxidation is roughly equivalent to Bessemer expectations and there is no heat loss to nitrogen of air, it is possible to incorporate up to 30% by weight of scrap steel or high-grade iron ores into the 70% or so of molten blast furnace iron which is the usual charge in the LD process [13]. This ability to accept a proportion of solid charge with the blast furnace melt contributes flexibility to the process. To do all this requires some 65 m^3 of oxygen per tonne of steel (about 2100 ft^3/ton, at 16°C, 1 atm), a net raw material cost to the process. But in return for this raw material cost, the time from charging to tapping the steel product for the larger LD furnaces of 200 and 300 tonnes capacity is still only about twice as long as for the smaller LD units, about 40 minutes [16]. In comparison to the 10- to 12-hour turnaround time for a single charge to a 300-tonne capacity open hearth furnace the cycle time for an oxygen steel making unit relates very favorably. Several other versions of oxygen steel making using variants of the equipment described here are also practiced [3]. In total, oxygen steel making methods are used to produce about one-half of the steel of the U.K. and the U.S.A., about 70% of the steel in West Germany, and about 80% of the total in Japan, and these proportions of the total are likely to grow still further.

14.4. PROPERTIES AND USES OF IRON AND STEELS

The wrought iron of historical importance was produced by reducing the iron ore without melting the iron component and then beating out most of the ferrous silicate slag. This product was virtually pure iron containing traces of carbon (0.02–0.04%) and strings of residual slag (1.0–2.0% $FeSiO_3$). The low carbon content was a result of reduction without fusion and contributed the good malleability, ductility, and the associated good cold working properties that are synonymous with wrought iron. Nineteenth century iron works could form sheets and bars by cold rolling of blooms (balls of reduced iron from which much of the slag has been removed by mechanical hammering) of wrought iron. Cold rolling hardens and generally strengthens the iron, a process called work-hardening, which is a desirable procedure for some end uses (Tables 14.8 and 14.9) [17, 18]. If, however, the iron shapes are required to be ductile for subsequent shaping, etc., the cold rolled shapes must first be annealed, or heated to a high temperature and then allowed to cool slowly to resoften the metal for continued cold forming.

■ **TABLE 14.8 Percentage Composition of Various Forms of Iron and Carbon Steels**[a]

Element	A Ingot iron	B Wrought iron	C Gray cast iron	D Carbon steels, SAE 1020	E Ultra-high-strength steel, 300-M
Iron	99.9+	97.5[b]	94.3	99.1	94.0
Carbon			3.4	0.2	0.43
Silicon			1.8	0.25	1.6
Manganese			0.5	0.45	0.80
Nickel					1.85
Chromium					0.85
Molybdenum					0.38
Vanadium					0.08

[a]Compiled and recalculated from data of Perry [17] and Lange [18].
[b]Residue consists of slag.

Wrought iron is still produced on a small scale by pouring molten Bessemer iron into slag that is molten but below the melting point of the iron, and by other similar methods. The spongy mass which forms is then mechanically treated to form a wrought iron very similar in properties to the earlier hand-worked material. Residual slag stringers left in the wrought iron contribute to the toughness of the iron in this form, and also make a small con-

■ **TABLE 14.9 Selected Properties of Various Forms of Iron and Carbon Steels**[a]

Material type[b] and condition	Yield[c] strength (MPa)	Tensile[c] strength (MPa)	Elongation[d] in 5 cm (%)	Hardness[e] Brinell	Density (g/cm³)	Melting point (°C)
A Annealed	130	260	45	67	7.86	1530
Hot rolled	200	310	26	90		
B Hot rolled	210	330	30	100	7.70	1510
C As cast	—	170	0.5	180	7.20	1230
D Annealed	260	450	30	130	7.86	1515
Hot rolled	290	470	32	135		
Hardened[f]	430	620	25	179		
E Hardened[g]	1650	2000	10	535	7.84	1504

[a]Compiled and recalculated from data of Perry [17] and Lange [18].
[b]Labeled compositions are those given in Table 14.8.
[c]Megapascals convertible to pounds per square inch by multiplying by 145. Yield strengths correspond to the stresses causing permanent deformation of the metal (i.e., beyond the elastic limit). Tensile strengths correspond to stresses at the point of ultimate metal failure.
[d]A measure of ductility.
[e]The Brinell hardness number is based on the indentation of the surface under test by a ball of specific diameter, usually 10 mm, when pressed into the test surface. It is usually expressed in kilograms per square millimeter, thus small numbers represent low hardness.
[f]Hardened by water quench followed by 540°C temper.
[g]Hardened by oil quench followed by 320°C temper.

tribution to corrosion resistance. The ductility, weldability, and toughness of wrought iron continue to make it valuable for many of the same uses as the original hand-produced material, for example, pipe, boiler tubes, rivets, heavy chains and hooks, and ornamental iron work.

14.4.1. Iron-Carbon Alloys: Cast Iron and Steels

One of the most important characteristics of iron products which enables a wide difference in spectrum of properties to be achieved is the concentration of carbon present. Ordinary steels are iron-carbon alloys, which are simply referred to as steel, and are so important that this alloy comprises more than 98% of all iron alloys produced. In most iron-carbon alloys the carbon is present as iron carbide, Fe_3C, also called cementite. Since the carbon content of cementite is only 6.69%, a small change in the carbon content of an iron causes a large change in the concentration of the cementite present in the iron. Cementite is soluble in molten iron, one of the reasons why carbon is accumulated in the product of the blast furnace process for reduction of iron ores. This is an advantage, since the melting point of the iron-cementite mixture is depressed to about 1100°C, permitting accumulation of the product in the base of the furnace at temperatures considerably below the 1535° melting point of pure iron. At the same time it is a disadvantage since, in the process, rather more cementite is dissolved in the molten iron than is wanted for most uses.

If the percentage of carbon in molten iron is 1.7% or less, which amounts to 25.4% or less cementite, it remains dissolved in the iron when solidified by quick cooling (e.g., when quenched in water) to form only one type of crystal. The crystals of this solid solution have cementite evenly distributed in the crystals of γ-iron and yields an extremely hard, brittle product called martensite. If the iron containing from about 0.82–1.7% carbon is allowed to cool more slowly, cementite can separate from iron and the result is an intercrystalline structure of cementite and pearlite. The formation of pearlite, which consists of alternate layers of ferrite (pure iron) and cementite, contributes significantly softer, more malleable, and tougher properties to the iron than that obtained from a quenched product. As the amount of carbon in the iron is brought down below about 0.82%, slow cooling allows ferrite plus pearlite (interlayered ferrite and cementite) phases to separate, again giving a relatively soft, tough product. More detail concerning these phase changes and their effect on properties of carbon steels is available from the more extensive accounts [7].

Iron-carbon alloys are commonly referred to as carbon steels or simply steel, and are classified in accordance with their carbon content. Thus mild steels normally contain about 0.25% carbon, medium steels about 0.45% carbon, and high carbon steels about 0.7 to 1.7% carbon. Both medium and high carbon steels are commonly tempered (heat treated) to optimize their properties although any of the carbon steels are amenable to property modification by heat treating [19]. In fact, that is the general definition of a steel: an iron-carbon alloy that contains less than 1.7% carbon and is amenable to hardening by quenching (Tables 14.8 and 14.9). As a rule, the strength and hardenability of a steel is raised at the same time as the toughness and ductility are decreased, with an increase in the carbon content of a steel. A recent user classification system has been proposed [20].

To obtain optimum properties from heat treating the finished steel article is first hardened by heating to red heat followed by quenching by dipping in water, brine, or oil any of which may also be cold or warm depending on the degree of hardness desired (Table 14.10) [21]. The martensite structure obtained in the iron by this treatment is very hard and a tool at this stage will hold a cutting edge well, but it will also be rather brittle and in this condition can fracture quite easily. It will also be under severe internal stress because of the rapid cooling rate, which for a small article will tend to increase its susceptibility to fracture under shock. A large part may actually cause it to crack as a result of the quenching alone. For this reason large parts will be oil or air quenched to reduce this risk. To gain the desired degree of toughness, the quenched article is tempered (or "drawn down") by reheating to a particular controlled temperature according to the end use desired in an air or oil bath, and then allowed to cool slowly. This combined treatment gives a cutting product or spring the desired combination of hardness (sharp edge retention) and toughness (resistance to shattering under sudden physical shock).

If a carbon steel is required in a ductile state either for intermediate cold forming (e.g., auto body panels) or machining, or in its final use form, then it is annealed. To accomplish this the finished part is heated to above the iron phase transition temperature, about 650°C, and then allowed to cool very slowly. Annealing not only accomplishes a softening and ductility improving effect on the steel (Table 14.9) but also achieves stress relief. Large cold-formed or welded industrial vessels or ships' components will have stresses induced in the structure from the fabrication methods used, which may be removed by an annealing procedure such as the one outlined. In this way the service life and reliability of the unit may both be greatly improved.

When the carbon content of the iron rises to above 1.7% (>25.4% cementite) it is classed as a cast iron. At this concentration the proportion of cementite becomes so high that it is no longer completely soluble even in solid γ-iron and forms a separate cementite phase on hardening from the melt. Cast iron may be made by mixing molten pig iron with molten steel scrap, or by melting steel scrap in direct contact with coke in a cupola furnace (virtually

TABLE 14.10   Tempering (Draw Down) Temperatures for Some Common Uses of Carbon Steels[a]

End use	Reheat temperature (°C)	Color of oxide layer
Razor blades, twist drills	232	pale straw yellow
Axes, taps, and dies	255	brown-yellow
Cutlery	277	purple
Watch spring	288	bright blue
Chisels	300	dark blue

[a]Compiled from Dearden [2], Oberg *et al.* [21], and other sources.

a small version of a blast furnace). It will ordinarily contain 2.0–4.0% carbon, only a little less than pig iron, together with 0.5–3.0% silicon, 0.5–1.0% manganese, and traces of phosphorus and sulfur. The hard brittle nature of many cast irons, hot or cold, makes these impossible to shape by rolling or forging (hammering while hot) the solidified metal, so that casting followed by machining if required are the usual methods of fabrication. Fortunately, as in the original blast furnace reduction of iron from its ores, the presence of the cementite and other impurities depresses the melting point of the iron to the 1100–1200°C range making both the melting process and the casting step easier than it would be for pure iron.

It is possible to produce a moderately tough cast iron by special alloying techniques, and by control of the microstructure through the selection of the appropriate cooling cycle. And some versions of cast iron may be converted to malleable iron by annealing for a period of several days. The long anneal causes dissociation of the cementite (Fe_3C), the cause of the hardness, into ferrite (iron) and spherical zones of graphite making the product relatively soft and significantly tougher than the material as originally cast. With these special irons primary fabrication by casting can be followed by other shaping techniques.

14.4.2. Alloy Steels

Mild steels and high carbon steels may be fabricated and heat treated in ways to achieve most combinations of properties which may be required. But plain carbon steels suffer several serious drawbacks, primarily in the areas of strength, toughness, and hardness for some applications, and in corrosion resistance. For example, heat treating of very large machine parts made of carbon steels can only achieve very shallow surface hardening. Also, heat treating of large parts introduces severe internal stresses and strains which cause distortion of the part from the originally machined shape. In small parts this distortion may be acceptable or compensated for, but it may be sufficiently great with large parts to introduce fitting problems and a tendency to crack. With a situation such as this, a drastic quench cannot be used and therefore optimum hardness of the carbon steel is not achievable. Alloying with elements other than carbon can enable achievement of hardenability, a term reserved to mean hardness at depth, with a less severe (such as an air) quench. Addition of molybdenum, for example, not only achieves hardenability but at the same time contributes other desirable properties (Table 14.11).

Examples of the kinds of properties achievable with steels using several different alloying elements are given in Table 14.11. Many hundreds of alloy steels are available, including other elements than those listed, to tailor-make a steel to suit a particular application. For example, nickel-chromium combinations confer stainlessness, as outlined in the table, more than 20 different commercial variations being available to achieve specific objectives. The stainless steels, from their general utility, probably represent the largest single class of alloy steels. Aluminum may be added in trace amounts to tie up oxygen and avoid the deleterious properties induced by the presence of small amounts

TABLE 14.11 Compositions and Applications of Some Common Examples of Alloy Steels[a]

Alloy steels	Percentages of alloying elements	Properties achieved	Typical uses
Chromium	Cr 0.6–1.2, C 0.2–0.5	Improved strength, toughness, heat resistance	Improved castings, stainless steels
	Cr 16–25, Ni 3.5–20	Combination confers stainlessness, heat resistance, strength	Stainless steels: chemical, medical, and food processing equipment
	Cr > 12%, Ni > 8%	As above, plus hardenability	Stainless cutlery
Copper	Cu 0.10, C ca. 0.2	Improved corrosion resistance	Copper-bearing steels, bridges, exterior steelwork
Nickel	Ni 6–9, C 0.2–0.5	Low-temperature toughness	Snowmobile parts, icebreakers, armor plate (as 6.5% Ni)
	Ni 30–45%	Low coefficient of thermal expansion, i.e., 0.8×10^{-6} as compared to 10 to 20×10^{-6} for most other steels	Invar: graded seals to glass, length standards (surveying, etc.), compensated pendulum and balance wheels
	Ni 60, Cr 16	Heating and electrical resistance units	Nichrome heating elements
Manganese	Mn 10–15, C ca. 1	Tough, acquires hardness on repeated impact	Rock crusher jaws, power shovel bucket teeth, burglar-proof safes
Molybdenum	Mo 5–9	Hardenability, high-temperature strength	High-speed tools, steam boilers and pipes, jet engine turbines
Silicon	Si 3–5	Low hysteresis and eddy current losses, increased magnetic permeability	Electric power transformer plates, motor parts
	Si 13–15, C 0.8	Superior hardness, acid resistance	For example Duriron (14.5% Si), industrial valves piping
Tungsten	W 5, C 0.5	Assists in retaining hardness while hot	Tungsten steels, twist drill bits
	W 17, Cr 10, V 0.3, C 0.7	Same as above, remains hard and strong even when hot	High-speed steel, high-speed cutting, and machine tools
Vanadium	V 0.2–0.5, + Cr ca. 1	High strength, toughness, heat treatability	Gears, driveshafts, axles, ball bearings, mechanics tools

[a]Compiled from data of McGannon [7], Perry [17], and Lange [18].

of ferrous oxide. A small percentage of manganese is useful to tie up any residual traces of sulfur as well as for the other functions of this alloying metal given in the table. While ferrous sulfide introduces brittleness in the steel making hot working impossible, manganese sulfide does not interfere in this way which thus permits normal forging operations to be carried out without introducing defects. Boron, too, is a relatively recently discovered alloying element, a few hundredths of a percent contributing low-cost hardenability. Coating developments with hard carbides such as titanium promise to increase the life of cutting tools made with the new high-speed tool steels [22].

14.5. EMISSION CONTROL IN IRON AND STEEL PRODUCTION

For an industry which involves large-scale handling of finely divided solid materials, the operation of large-scale combustion processes for coking, etc., to emission control has to be a priority. Containment is necessary for maintenance of public health in associated built-up areas, for industrial hygiene requirements of plant operators, for better product yields from the raw materials entering the processes used, and for aesthetic reasons.

Many technological changes which were introduced to the iron and steel industry as a means of improving operating efficiency also contributed significantly reduced levels of emission per tonne of iron or steel produced. For instance, blast furnace top gas containment measures were introduced to conserve the carbon monoxide fueling capacity of these gases for preheating of blast air. At the same time this measure served to decrease both particulate and gaseous pollutant discharges from blast furnace operations. Until very recently blast furnace operating procedures allowed both some intentional gas bleeding (pressure relief), and accidental gas losses containing up to 35 g/m^3 (15 grains/ft^3) particulate, as well as gaseous pollutants [23]. These relief and accidental discharges are now regulated by passage through at least two levels of control devices before discharge. In these ways the industry has progressed from no containment to 90–98+% containment to its and society's mutual benefit. A similar progression has occurred with the BOP (basic oxygen process) for steel making. Since the air and water emission control have many overlapping aspects, these aspects will be treated by the emission control area rather than process area.

14.5.1. Air Pollution Control

Discharges to the air from facilities for producing iron and steel may be conveniently considered in sequence, from the raw material preparation area to the fabrication of finished products. Thus, coke plants and sintering plants used to prepare the blast furnace fuel and ore components come first, followed by the various steel making procedures.

In the preparation of raw materials as blast furnace feed, the coke ovens used to prepare a dense, desulfurized metallurgical coke from coking grade coals can produce a significant mass of particulate discharge in the absence of control measures. A battery of coke ovens used to process 4500 tonnes of coking coal per day and to yield about 3200 tonnes of coke is estimated to generate a solid discharge of about 0.1% of the mass of coal processed, or about 4500 kg of particulate per day [24]. This particulate emission occurs at all stages of coke preparation, including charging, pushing, and quenching stages of the operation as well as from leaks in the ovens themselves, thus making control measures difficult. No estimates of current feasible particulate emission control efficiencies are available.

While the coking process does produce a blast furnace coke feed substantially free of sulfur, the gaseous product, coke oven gas, has a sulfur gas content of 900–1100 g/m^3 (at 15°C, 1 atm) [23]. The principal sulfur-containing constituent of coke oven gas is hydrogen sulfide, which may be removed either by the vacuum carbonate or Stretford processes. The sulfur gas removal efficiency of the Koppers Company's vacuum carbonate process is about 90%, whereas the Stretford process can maintain 99% containment. The choice of desulfurization process depends on the sulfur gas product desired: Sulfuric acid is the vacuum carbonate product and elemental sulfur for the Stretford (Chap. 3, 9). Condensible hydrocarbons such as benzene (and other aromatics) and phenols have always been recovered [25].

Sintering machines are used to prepare large nodules from beneficiated iron ore fines and iron oxide containing dusts recycled from particulate emission control equipment. Uncontrolled operation, it is estimated, would discharge particulate at the rate of about 0.3% of the mass of sinter produced or about 2700 kg from a machine producing 900 tonnes of sinter per day. Control measures rely on cyclones, which are estimated to decrease the particulate emission to about one-quarter of the uncontrolled levels. It is also possible to use the sintering machine as a roaster to enable sulfur removal from sulfur-containing iron ores. This produces an amenable ore feed, but it also produces sulfur dioxide in the waste gas stream. No concentration or mass emission rate data for sulfur dioxide in sinter plant exhaust gas is available, since this has not normally been recovered. But a mathematical model which enables estimation has been described [13].

Particle loading of the top gases leaving a blast furnace is very dependent on the particle size distribution of the raw materials charged to the furnace. With no particular care taken to eliminate fines from the ore or flux, dust loss rates of about 200 kg per tonne of pig iron are produced. This corresponds to a particulate loading of about 2 g/m^3. Screening of the charge to remove fines roughly halves this loss rate and at the same time improves blast furnace operation by providing a freer flow of gases through the charge (cf. Table 14.4). The greatest reduction in dust carryover in the top gases, to the range of 15–20 kg/tonne of pig iron produced, is obtained by prior briquetting, pelletizing, or nodulizing of the charge before placement [26]. When carried out with the flux and coke also incorporated into the lumps charged, blast furnace gas flow is further improved.

All the measures described above amount to precombustion or prereduction methodologies. In addition to these dust *avoidance* techniques, modern blast furnaces also usually require three stages of gas cleaning for the top gases before these are employed as a fuel for blast stove heating and other energy requirements. A dust catcher situated close to the furnace captures the coarse particulate (see Fig. 14.1) and then by either low and high energy scrubbers, or by a low energy scrubber followed by electrostatic precipitation the residual midsize range and aerosol particles are removed [24]. By these measures the actual particulate loss rate amounts to 1–18 kg per tonne of pig iron produced. Magnetic filtration has been proposed as a safer method than electrostatic precipitation for iron particulate removal from combustible blast furnace gases, to avoid explosion risks [27].

All steel making procedures produce exhaust gases having a moderate particle loading before treatment, but none as severe as from blast furnace operation. Dust loadings of 0.5–1.4 g/m^3 are experienced during the charging phase (loading ore, flux, scrap, etc.) of an open hearth furnace [25]. Higher dust losses, 4.6–7.0 g/m^3, are experienced when an oxygen lancing (of basic oxygen process steel making utility) stage is employed in open hearth operation. Originally, open hearth furnaces were operated without emission control devices. However, electrostatic precipitation, for which 90–97% particulate control efficiencies are reported, or a wet scrubber is now used for particulate control [13, 25]. The wastewater stream from a wet scrubber requires treatment. Particles are removed by chemical coagulation and settling followed by pH neutralization of the acidic supernatant liquor [28]. When a scrubber is employed as part of the system used to clean up blast furnace top gases, the waste liquor is normally alkaline from absorbed limestone. Hence when this is the control system used, the two scrubber waste liquors may be used to neutralize each other.

For Bessemer, electric arc, and basic oxygen furnaces pollutant containment is practiced via a hood with connecting ductwork placed over the opening of the vessel during active steel making phases [29, 30]. The collected gases are cleaned by electrostatic precipitation, scrubbers, or with a baghouse [23]. Particulate collection efficiencies of 80–90% are cited for the former two methods. Particulate mass control efficiencies of 98 to 99% have been reported for baghouses.

Ingot reheating prior to hot rolling or forging of the final products use normal combustion processes to provide reheat energies. Hence there is little control concern for this area of operations.

14.5.2. Water Pollution Control

Preparation of blast furnace coke involves the indirect heating of metallurgical coal to 1000–1100°C in the absence of air in a battery of refractory brick-lined coke ovens. These procedures are collectively referred to as a "by-product coke plant" from the association of by-product recovery with the coke formation process. The coal charge is heated until pyrolysis is complete and all of the volatile matter has been vaporized, a process which takes 16 to 24 hours. Then the residual lumps of coke, still hot, are pushed out of the oven through a quenching shower of water and into a rail car for final shipment. About 700 kg of coke plus a number of volatile products are recovered from each tonne of metallurgical coal heated. More details on the coking process itself are available [31].

About 42 m^3/tonne coke (10,000 U.S. gal/ton) of water is used in a coking operation, mostly for indirect cooling of the volatiles stream to collect coal tar, water, and other condensibles. This warmed water stream only requires thermal loading considerations on discharge. The aqueous condensate stream also contains ammonia, sulfides, hydrocarbons, phenols, etc., all condensed from the coal volatiles (Tables 14.12 and 14.13) [32, 33]. This stream requires treatment before discharge, both from a chemical recovery standpoint and for emission control.

About 2 m^3 of water per tonne of coke (ca. 500 U.S. gal/ton) is used for quenching purposes, about a third of which is vaporized in the coke cooling process. The vaporized fraction is discharged as steam but the residual water, which will contain suspended coke fines ("coke breeze") plus low concentrations of cyanides and phenols, also requires treatment before discharge. Coke fines are removed by settling and when recovered are blended into a mixture of ore and limestone for briquetting as blast furnace charge, or are used as fuel for sintering. Recycling much of the supernatant water minimizes the volume that requires treatment.

The aqueous condensate stream is the most complex to deal with. Phase separation produces a water phase and an oily tarry phase, each of which is then treated separately. Phenol is kept in the aqueous phase during ammonia recovery by converting it to the less volatile sodium salt (Eq. 14.26).

$$PhOH + NaOH \rightarrow NaOPh + H_2O \qquad 14.26$$

Ammonia is then removed from the aqueous phase by distillation and either recovered by scrubbing with sulfuric acid to produce ammonium sulfate, which is sold as a fertilizer constituent, or it is discarded by discharge to air. Economical reacidification of the aqueous ammonia recovery residue with flue

TABLE 14.12 Typical Characteristics of the Wastewater Stream from the Ammonia Still of a Battery of By-Product Coke Ovens[a]

Parameter	Significance (ppm by weight)
5 day BOD, 20°C	3974
Suspended solids:	
Volatile	153
Total	356
Nitrogen:	
As ammonia, NH$_3$	187
Organic and NH$_3$	281
Phenol	2057
Cyanide	110
pH	8.9

[a]Data selected from the American Iron and Steel Institute [32].

■■■ **TABLE 14.13 Approximate Coke and Volatile Product Yields from One Tonne of Metallurgical Coal**[a]

Solids and condensed liquids (kg)		Gaseous products (kg)	
Coke, with coke breeze	750	Ammonia	3.5
Light oil	0.5	Hydrogen sulfide	3
Heavy creosote	4.5	Carbon dioxide	9
Pitches	24	Nitrogen	4
Benzene	7.5	Hydrogen	16
Toluene	3.0	Carbon monoxide	45
Xylene	0.8	Methane	65
Naphthalene oil	4.0	Ethane	5.5
Anthracene oil	7.0	Ethylene	10
Other organics	1.3	Other gases	5
Aqueous liquor	60	Total gases	143.5[b]

[a]Data selected and calculated from that of Kent [33].
[b]About 305 m^3 at 15°C, 1 atm.

gas and subsequent distillation allows recovery of both phenol and sodium carbonate (Eq. 14.27, 14.28).

$$CO_2 + H_2O \rightarrow H_2CO_3 \qquad 14.27$$

$$2\ NaOPh + H_2CO_3 \rightarrow Na_2CO_3 + 2\ PhOH \qquad 14.28$$

Cyanide may be detoxified to cyanate by oxidation with chlorine or hypochlorite (Chap. 5). Or the oxidation may be conducted by air in the presence of sulfur dioxide and copper ion at pH 9 or 10 [34]. Finally BOD reduction to 80 to 90% of the original values will be achieved via either biological waste treatment or alkaline chlorination methods before discharge of this stream.

The separated organic phase is commonly used as fuel. Otherwise it will be put through a sequence of washes and stills for chemical recovery, or it may be segregated into portions destined for each appropriate end use area at some coking locations [33]. Details of some of the products recoverable are given in Table 14.13. About one-third of the coke oven gas is burned to provide the heat for coal carbonization and the rest is usually used on site for other energy needs.

Blast furnace water requirements are mainly for water jacket cooling of the critical tuyere and bosh areas of the furnace. The primary concern on discharge of this water is to minimize thermal loadings. Water is also frequently used for scrubbing of blast furnace top gases, and less often for fracturing (breaking up) of hot slag. These water requirements require settling and appropriate chemical neutralization before discharge.

Steel making techniques all have a cooling water requirement of some 12–15 m^3/tonne (ca. 3000–3700 U.S. gal/ton) as well as a much smaller requirement of about 80–100 L/tonne (ca. 20–25 U.S. gal/ton) for scrubbing of exhaust gases from the various processes [28]. Cooling water disposal requi-

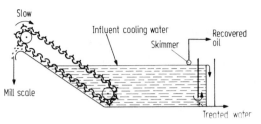

FIGURE 14.5 Schematic diagram of a scale pit showing how simultaneous oil and scale removal is achieved from steel rolling mill waste waters.

rements are similar to those from any other source. Treatment of scrubber effluent will be similar to the procedures used with blast furnace scrubber effluent, except that steel making waste streams are usually acidic whereas blast furnace effluent is normally alkaline. After particulate removal, pH neutralization can be by stream blending, particularly if the steel mill is operated in close proximity to supporting blast furnaces.

Water effluent problems of hot and cold steel rolling mills differ from those of other areas of the complex. Even though the primary water requirement in a hot rolling mill is for cooling of the heavy iron or steel rolls used for shaping the steel, the cooling function here is by direct spray onto the rollers. Thus, the water picks up small particles of iron, scale (mostly iron oxide), and lubricating oil in its passage across the face of the rollers, which are retained in suspension as the water reaches mill collection channels. Some additional water, used in high-pressure jets to flush scale from a steel sheet as it forms, is also collected in the same wastewater system. A large tank, called a "scale pit," serves the combined functions of scale settling and oil creaming to simultaneously remove the bulk of both of these contaminants before discharge of the water stream (Fig. 14.5). A system of paddles on endless chains is used to move settled scale out of the scale pit, up an inclined plane to drain off most of the fluid, and thence to a collection for incorporation into blast furnace feed [25]. Creamed oil collected on the water surface of the scale pit is removed by suction pumping through a skimmer floating on the fluid surface. If the cleaned exit water from the scale pit still does not meet local effluent guidelines it can be treated further by chemical or biological methods before discharge.

14.5.3. Recycle Precautions

Scrap steel is extensively recycled without incident, as mentioned in the introductory comments. However, before it is accepted into a steel remelting yard it should be checked for any stray radioactive material. Of the 21 known accidents of steel contamination with cobalt 60, cesium 137, or other radioactive materials, 8 occurred in the 1992–1993 period [35]. Checking is simple and inexpensive compared with the potentially serious outcome and possibly costly recovery from accidental radioactivity put into steel.

REVIEW QUESTIONS

1. (a) What are the usual benefits of large scale contributed by very large blast furnaces as compared to smaller units?
   (b) What additional benefit is also obtained in this case which reduces the raw material requirements per unit of pig iron product, and how does this happen?
2. What would be the theoretical coke (take as 100% C) required for only the reduction reactions of 1000 tonnes of iron ore if the ore is:
   (a) 100% by weight Fe_2O_3?
   (b) 50% Fe_2O_3, 30% SiO_2, 20% $Ca_3(PO_4)_2$?
   (c) How do your answers compare to the actual requirements?
3. An 8-tonne charge of molten iron in an electric furnace contains 0.20% Fe.
   (a) What mass of aluminum should be added to be just sufficient to deoxidize this melt (reduce the ferrous oxide to iron)?
   (b) How would the aluminum oxide produced be removed from the iron?
4. Compare the types of impurities which may be best removed during steel processing when using acidic (SiO_2-based) vessel linings and when using basic (MgO-based) linings.
5. What would be the theoretical oxygen demand of a wastewater stream from an integrated steel mill which contains:
   (a) 200 ppm (by wt.) ammonia (pH about 4)?
   (b) 900 ppm phenol (C_6H_5OH)?
   (c) Give an alternative method to effectively treat each of the above waste streams that does not involve long-term biopond aeration.

FURTHER READING

Air and Water Quality Control at Stelco's Hilton Works, *Iron Steel Eng.* **53**, 75, Nov. (1976).

H.B.H. Cooper and W.J. Green, Energy Consumption Requirements for Air Pollution Control at an Iron Foundry. *J. Air Pollut. Control Assoc.* **28**, 545 (1978).

D. Marchand, Possible Improvement to Dust Collection in Electric Steel Plants and Summary of all Planned and Existing Systems in the Federal Republic of Germany. *Ironmaking Steelmaking* 3(4), 221 (Discuss. p. 230) (1976).

N.T. Stephens, J.M. Hughes, B.W. Owen, and W.O. Warwick, Emissions Control and Ambient Air Quality at a Secondary Steel Production Facility. *J. Air Pollut. Control Assoc.* **27**, 61 Jan. (1977).

REFERENCES

1. J. Newton Friend, "Iron in Antiquity," p. 27. Griffin, London, 1926.
2. J. Dearden, "Iron and Steel Today," 2nd ed., p. 19. Oxford University Press, Oxford, 1956.
3. W.K.V. Gale, "Iron and Steel." Longmans, Green, London, 1969.
4. J.B. Pearse, "A Concise History of the Iron Manufacture of the American Colonies up to the Revolution and of Pennsylvania until the Present Time." Burt Franklin, New York, 1970, reprint of the 1876 edition.

5. "United Nations Statistical Yearbook 1993," 40th ed., United Nations, New York, 1995, and earlier editions.

6. "Kirk-Othmer Encyclopedia of Chemical Technology," Vol. 8. Interscience, New York, 1952.

7. H.E. McGannon, ed., "The Making, Shaping and Treating of Steel," 9th ed. United States Steel, Pittsburgh, 1971.

8. "McGraw-Hill Encyclopedia of Science and Technology," Vol. 13, p. 98. McGraw-Hill, New York, 1971.

9. "Kirk-Othmer Encyclopedia of Chemical Technology," 3rd ed., Vol. 13, p. 743. Wiley-Interscience, New York, 1981.

10. J.A. Allen, "Studies in Innovation in the Steel and Chemical Industries." A.M. Kelley, New York, 1968.

11. G.R. Bashworth, "The Manufacture of Iron and Steel," 3rd ed., Chapman & Hall, London, 1964, cited by Allen [10].

12. W.K.V. Gale, "The British Iron and Steel Industry." David and Charles, Newton Abbot, 1967.

13. C.S. Russell and W.J. Vaughan, "Steel Production: Processes and Residuals." Johns Hopkins University Press, Baltimore, 1976.

14. "Kirk-Othmer Encyclopedia of Chemical Technology," 2nd ed., Vol. 18, p. 715. Interscience, New York, 1969.

15. G.E. Wittur, "Primary Iron and Steel in Canada." Department of Energy, Mines and Resources, Ottawa, 1968.

16. M. Finniston, *Chem. Ind. (London)*, June 19, 501, (1976).

17. R.H. Perry, ed., "Chemical Engineers Handbook," 4th ed. McGraw-Hill, New York, 1969.

18. N.A. Lange, ed.,"Lange's Handbook of Chemistry," 10th ed. McGraw-Hill, New York, 1969.

19. Potentials for More High-strength, Low-alloy Steels in Autos, *Met. Prog.* **109**(2), 26 Feb. (1976).

20. J.T. Sponzilli, C.H. Sperry, and J.L. Lytell, Jr., *Met. Prog.* **109**(2), 32 Feb. (1976).

21. E. Oberg, F.D. Jones, and H.L. Horton, "Machinery's Handbook," 20th ed. Industrial Press, New York, 1975.

22. R.F. Bunshah and A.H. Shabaik, *Res./Dev.* **26**(6), 46 June (1975).

23. J. Szekely, ed., "The Steel Industry and the Environment." Dekker, New York, 1973.

24. "Handbook of Environmental Control," Vol. 1., CRC Press, Boca Raton, FL, 1972.

25. H.F. Lund, ed., "Industrial Pollution Control Handbook," McGraw-Hill, New York, 1971.

26. Waste Iron-bearing Fumes, *Chem. Eng. News* **55**(16), 17, April 18 (1977).

27. Magnetic Filter May Cut Pollution, *ChemEcology,* p. 11, July (1980).

28. N.L. Nemerow, "Industrial Water Pollution, Origins, Characteristics, Treatment." Addison-Wesley, Reading, MA, 1978.

29. T. Cesta and L.M. Wrona., Optimization of BOF Air Emission Control Systems. *Iron Steel Eng.* **72**, 23–31, July (1995).

30. J.A.T. Jones, Interactions Between Electric Arc Furnace Operation and Environmental Control. *Iron Steel Eng.* **72**, 37–46, Dec. (1995).

31. R.N. Shreve and J.A. Brink, Jr., "Chemical Process Industries," 4th ed., McGraw-Hill, New York, 1977.

32. "Annual Statistical Report." American Iron and Steel Institute, New York, 1949, cited by Nemerow [28].

33. J.A. Kent, ed., "Riegel's Handbook of Industrial Chemistry," 7th ed. Van Nostrand-Reinhold, New York, 1974.

34. Process Destroys Cyanide, *Can. Chem. Process.* **66**(6), 8, Sept. 10 (1982).

35. A. LaMastra, The Changing Face of Radioactivity in Steel. *Iron Steel Eng.* **72**, 44–45, July (1995).

15

■ PRODUCTION OF PULP AND PAPER

...thou has caused printing to be used and, contrary to the king his crown, and dignity, thou hast built a paper mill.
—*William Shakespeare, 1591*

15.1. BACKGROUND AND DISTRIBUTION OF THE INDUSTRY

The production of wood pulp, and paper from this, is a primary industry in the sense of its utilization of wood as its chief raw material. Pulp and paper is also a secondary industry in the sense that it consumes large quantities of bulk inorganic chemicals such as chlorine, sodium hydroxide, pigments, etc., produced by the primary (commodity) chemical industry. This business area consumes close to 10% of the inorganic chemicals produced in the countries in which it is a dominant industry. Thus it not only contributes directly to the economy of these areas, but also indirectly by its purchase from the primary areas of chemicals production.

Development of a large-scale pulp and paper industry is favored by the availability of an extensive domestic wood supply, e.g., Brazil, Canada, Finland, Sweden, U.S.A., and the U.S.S.R. (Table 15.1) [1,2]. Japan is alone among the very large scale pulp producers in relying on imported pulp logs and pulp for her pulp and paper industry. Norway on the other hand, with a large forest potential, does not appear to have exploited this to the same extent as other nations with this resource. Brazil, a relatively new pulp producer, is one of the area's most committed to the "farming" of wood on a continuous harvest basis using fast growing varieties of trees. The extent of exploitation of the available wood supply by all the major pulp producing countries has aroused concern about the inability of natural processes to continue to provide wood at the rate that it is being cut. In most cases this has resulted in better organized care and harvesting of wood, and increased

TABLE 15.1 Major World Producers of Wood Pulp, in Thousands of Metric Tonnes[a]

	1950	1960	1970	Chemical 1980	Mechanical 1980	1990
Brazil	139	453	811	3,256	105	4,844
Canada	7,686	10,397	16,609	4,765	8,625	16,466
China			1,220	6,450	375	17,057
Finland	1,913	3,700	6,471	4,350	1,569	8,777
France	500	1,139	1,787	4,891	261	7,049
Japan	748	3,524	8,801	15,414	2,674	28,088
Norway	1,073	1,645	2,182	784	589	1,819
Sweden	3,448	5,611	8,142	4,648	1,534	8,419
U.S.A.	13,471	22,966	39,304	52,601	4,238	71,965
U.S.S.R.	2,046	3,213	6,679	7,379	1,354	10,718
West Germany	914	1,452	1,732	6,974	606	11,873
Other	2,682	5,480	11,034			
World	34,620	59,580	104,772	144,805	26,143	239,050

[a]Includes totals of both chemical and mechanical pulp where undifferentiated. Pulp weight specified on an air dry basis, i.e., 10% moisture. Data compiled from Reference Tables [1] and *U.N. Statistical Yearbooks* [2].

encouragement of renewal by reforestation practices which are more effective than the natural processes.

Financially, an active pulp and paper industry provides a significant fraction of the gross national product of the countries where this is a major business area. In Canada this amounts to some $8 billion annually. Of 20 manufacturing categories in Canada, the annual value of pulp and paper production is only exceeded by the food and beverage sector, and the transportation equipment sector at $25.4 billion and $19.9 billion, respectively, and about equaled by the value of petroleum production. What is perhaps of equivalent importance to a national economy, because of its ability to earn international exchange, is the value of exports in this commodity area. This factor is particularly significant to countries which produce pulp and paper on a very large scale and have a relatively small population and hence consumption, for example, Canada, Finland, Norway, and Sweden (Table 15.2). Brazil is likely, very shortly, to join this group.

In paper consumption the largest producers, U.S.A. and Canada, are also the largest consumers, at least on a per capita basis (Table 15.3), the United States alone consuming one-half of the total world production. They, and Japan, the runner-up after the U.S.A. on a consumption rate by country basis, all still show a trend of increasing annual newsprint consumption of 5 to 30% over the 9-year period. Denmark, the Netherlands, Sweden, and the United Kingdom, all nations known to be great newspaper subscribers, all lie in the mid range for per capity newsprint consumption, the last two showing a 36 and 14% decrease in per capita consumption over the same period. Among the smaller scale newsprint users, per capita consumption trends ranged from

▮ Table 15.2 Growth in Exports of Wood Pulp, by Countries Exporting More Than 500,000 Tonnes in 1979, in Thousands of Metric Tonnes[a]

	1950	1960	1970	1979
Brazil	0	—	40	582
Canada	1,675	2,360	5,063	7,090
Finland	1,056	1,587	2,057	1,865
Norway	546	802	981	573
South Africa	0	74	278	529
Sweden	2,091	2,931	3,762	3,519
U.S.A.	87	1,036	2,808	2,662
U.S.S.R.	n/a	257	448	680
Other	243	632	1,410	2,673
Total	5,698	9,679	16,847	29,173

[a]Data from Reference Tables [1].

a 100% increase for China, to a stabilized consumption for India, Iraq, and Ethiopia, to decreases of 58% and more for Malawi and Uganda.

We may have originally learned to make paper by careful observation of wasp nest construction. Early papermaking was accomplished one sheet at a time, by hand, by methods which originated in China at about 105 A.D. From that time till about 1760 papermaking remained a batch operation, single

▮ TABLE 15.3 Annual Consumption of Newsprint by a Selection of Countries Both by Country, in Thousands of Metric Tonnes (Mg), and Per Capita[a]

	1970		1980		1991	
	Total (Mg)	Per capita (kg)	Total (Mg)	Per capita (kg)	Total (Mg)	Per capita (kg)
U.S.A.	8,924	43.6	10,673	46.9	12,274	48.6
Canada	657	30.8	918	38.1	416	15.4
Australia	449	35.9	557	37.9	561	32.4
Denmark	149	30.2	153	29.8	212	41.3
Netherlands	379	29.1	458	32.4	465	30.9
Sweden	343	42.7	296	35.6	414	48.1
U. K.	1,544	27.8	1,381	24.5	1,851	32.2
Japan	1,973	18.9	2,703	23.1	3,821	30.8
West Germany	1,077	17.7	1,394	22.6	—	—
France	606	11.9	629	11.7	729	12.8
U.S.S.R.	928	3.8	1,604	4.1	1,324	4.7
China	571	0.74	528	0.53	768	0.66
India	182	0.33	310	0.45	450	0.52
Afghanistan	—	—	0.1	0.006	—	—
Ethiopia	0.90	0.036	2	0.05	2.3	0.045
Other	3,773.1	—	—	—	—	—
Total, world mean	21,563	5.86	—	—	—	—

[a]Data from U.N. Statistical Yearbooks [2].

sheets being prepared by dipping a mold (or screen) into a vat holding a suspension of fiber in water. After draining and pressing of the excess water the sheets were dried in the open air. N.L. Robert, in 1761, took the first steps toward developing a machine for papermaking on a continuous basis [3]. It was not until 1822 that a Fourdrinier paper machine, initially operated in France, was first operated as a continuous papermaking machine. Thus, papermaking by anything like the methods used today is a relatively recent activity.

A suspension of pulp in water is the common raw material for virtually all types of papermaking. The pulp is mostly cellulose and is derived from wood fibers separated from whole wood. Other sources such as cotton and linen rags (as offcuts from clothing manufacture) which are virtually 100% cellulose are also used to produce small quantities of the very highest quality papers. Where wood may be scarce, straw, cornstalks, hemp, jute, or bagasse (sugar cane fiber) may also be employed as raw materials. Papermaking from bagasse has to cope with carbon particles left on the stems when cane field foliage is burned just prior to sugar harvest, and with the high proportion of pith. Esparto grass, a tough dryland cover, and bamboo are also used. The very rapid growth of bamboo allows "cropping" for papermaking and with prior crushing between steel rollers the tough internodes may be used. Modern chemical pulping processes are also capable of pulping annual plants, which permits the use of a range of easily renewable raw materials. But by far the dominant raw material used for papermaking worldwide is wood, so the detailed discussion which follows will center on the methods used with this source of pulp.

15.2. WOOD COMPOSITION AND MORPHOLOGY

The woody part of a tree provides both mechanical support to hold the tree upright for optimum exposure of the leaves to sunlight and air, and serves as a conduit to carry water and trace nutrients from the roots to the leaves and photosynthetic products to where needed. Hollow, interconnected fibers composed mostly of cellulose and oriented along the axis of the tree provide both of these functions (Fig. 15.1) [4].

Cellulose, the main component of the conduit fibers, is a polymer of β-D-glucose (Fig. 15.2). The initial step in its formation is the photosynthetic reaction, which produces glucose from water and carbon dioxide (Eq. 15.1).

$$6 \text{ H}_2\text{O} + 6 \text{ CO}_2 \xrightarrow[\text{sunlight}]{\text{chlorophyll}} \underset{\text{glucose}}{\text{C}_6\text{H}_{12}\text{O}_6} + 6 \text{ O}_2 \qquad 15.1$$

Glucose in solution is carried in the sap to provide energy for plant metabolism, by the reverse of the photosynthetic reaction, and building blocks to the cambium layer of the wood, where the new growth occurs. The cambium layer comprises the interface between the wood and the bark (the "sapwood"). This is where new fibers form by processes which deposit insoluble cellulose from the soluble glucose and cellobiose, a disaccharide of glucose. Rings of

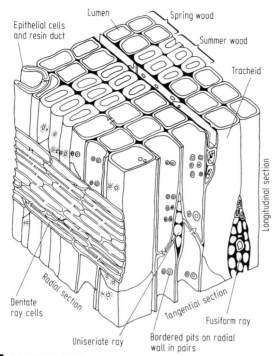

Lumen
Spring wood
Epithelial cells and resin duct
Summer wood
Tracheid
Longitudinal section
Radial section
Dentate ray cells
Tangential section
Fusiform ray
Uniseriate ray
Bordered pits on radial wall in pairs

FIGURE 15.1 Enlarged cubic section of a typical softwood. (From Kirk-Othmer [4], reprinted courtesy of John Wiley and Sons, Inc.)

fibers which form in the spring during a period of rapid growth are relatively thin walled and have a large lumen (bore, or duct). These are collectively referred to as springwood. In summer, when growth is slowed, fibers are formed with thicker walls and smaller lumens, which are collectively referred to as summerwood. It is these differential growth features which give rise to the characteristic annual rings of wood, and which also allow longer term historical climatic variations to be estimated from cross sections of old specimens. The core of the trunk or limb inside the cambium growth layer is dead wood and is termed heartwood.

Cellulose is formed by elimination of water from adjacent β-anomers of the two cyclic forms of glucose (Eq. 15.2), hence it is a "dehydro" polymer.

$$n\ C_6H_{12}O_6 \rightarrow (C_6H_{10}O_5)_n + n\ H_2O \qquad n = \text{about } 10,000 \qquad 15.2$$

Formation of cellulose from the α-anomer is what gives the polymer its characteristic linear form which favors strong intermolecular hydrogen bonding between adjacent polymer chains. With some 10,000 glucose units (molecular weight about 1.5 million) per chain, individual molecules can be up to 0.012 mm in length [5]. This high molecular weight and the close intermolecular association makes fibers composed of cellulose both strong and insoluble in water. Both factors also confer cellulose with significant resistance to hydrolysis because of the relative inaccessibility of the acetal links between adjacent glucose units. The hydrolytic stability, strength, and insolubility are necessary

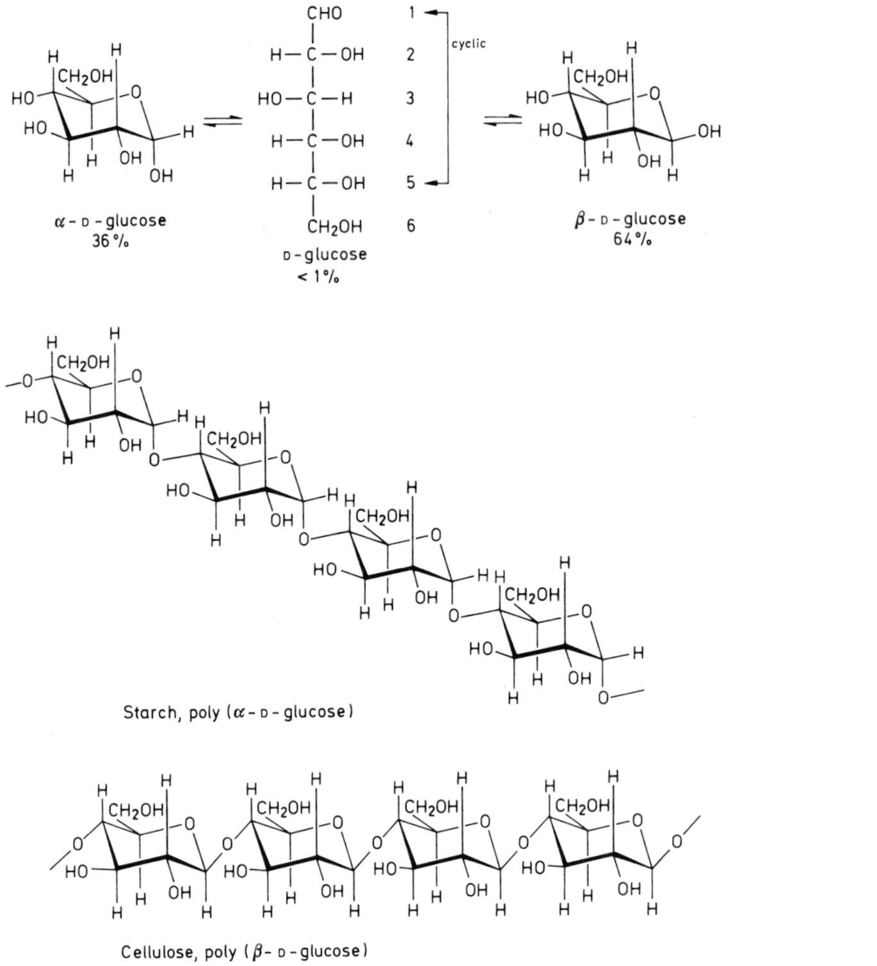

FIGURE 15.2 Linear and stereochemical representations of the cyclic forms of glucose, and their relationship to the two regular dehydropolymers of glucose.

features of the structural role of cellulose in the plant as well as being important for papermaking and paper properties.

Chemically cellulose is a two-dimensional high molecular weight polymer with no conjugation, hence it is white (colorless) in color. It has no taste or odor and is insoluble and relatively stable in water or aqueous alkali. It is, however, soluble in high concentrations of aqueous mineral acids, and may be regenerated to lower molecular weight fragments of the solid on marked dilution with water. Cellulose does not melt on heating and is relatively stable to heat, not decomposing until it reaches 260–270°C. It is also relatively stable toward oxidation. This set of properties is clearly highly suitable for a papermaking raw material. Chemically speaking, cellulose is the same from one species to the next, although there are morphological differences between the cellulosic fibers from different species.

Lignin, which functions primarily as an interfiber bonding agent, is also an important component of wood. It is a three-dimensional, crosslinked polymer of heterogeneous structure composed of predominantly n-propylbenzene units joined by a variety of ether $(C-O-C)$ and carbon-carbon $(C-C)$ links (Fig. 15.3). Lignin is also biosynthetically derived from glucose, like cellulose although a little less directly [6]. Because of the high aromatic-phenolic content in the polymer, lignins are all more or less brown in color, and are relatively readily oxidized. The molecular weight of natural lignins, *in situ,* is not well defined. Even in the natural state it probably consists of polymers comprising a range of molecular weights. Molecular weights determined on isolated lignins mostly range from 1000 to 20,000, depending to a significant extent on the origin of the lignin and the method used to determine the molecular weight.

Lignin is relatively stable to most aqueous mineral acids but is soluble in hot aqueous base and hot aqueous bisulfite ion (HSO_3^-). It has low softening and melting points, e.g., native spruce lignin softens at 80–90°C (wet), 120°C (dry), and melts at 140–150°C. Higher temperatures than this cause carbonization. The lignins from different species of tree, and even from the same species growing in different locations are significantly chemically different from one another, particularly when comparing softwoods and hardwoods.

Less dominant components of wood are the hemicelluloses, so called because at first it was thought that this component consisted of lower molecular weight biosynthetic intermediates en route to cellulose. However, the hemicelluloses are soluble in aqueous alkali, which was one of the early means for differentiating them from cellulose. This property is also a clue to the amorphous nature of hemicelluloses, which are branched polysaccharides of a variety of sugars, all of lower molecular weight than cellulose. None have simple trivial names. The name of the particular hemicellulose is derived from the sugar monomer from which the polymer is composed, for example, xylose forms xylans, galactose forms galactans, mannose forms mannans, etc. As with lignin, the chemical structure and composition of the hemicellulosic fraction of wood also differs from species to species in woods, grasses, and annual plants. For example, with hardwoods the xylans predominate and with softwoods the glucomannans are usually the main hemicellulose component. The exact hemicellulosic content of any particular wood is heterogeneous. There is also evidence that the xylan from a softwood is not identical with the xylan from a hardwood. To add to these complications, isolation of a particular hemicellulosic fraction from wood inevitably modifies it somewhat, so that it is difficult to determine what the original composition of the hemicellulose was. Thus the structure of hemicelluloses as they occur in the natural state remains rather poorly defined, much like lignins.

The amorphous nature and lower molecular weight of hemicellulose makes it an important component of papermaking, probably by its ability to improve interfiber bonding. But there is inevitably some hemicellulose loss from the pulp, particularly with some chemical pulping processes, because hemicellulose is less stable to acid and alkali than cellulose. The sum of the cellulose and hemicellulose components in wood or pulp is collectively

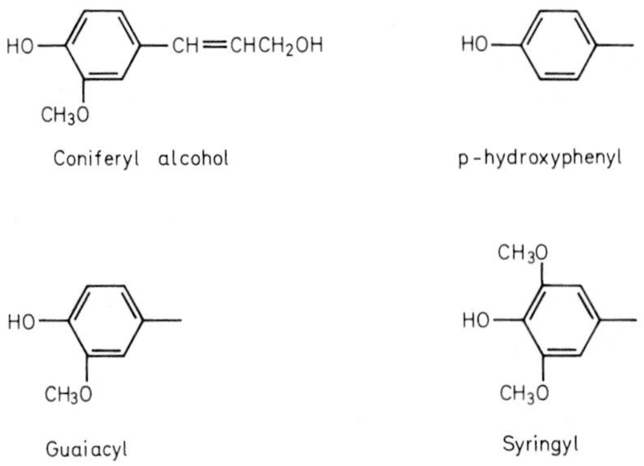

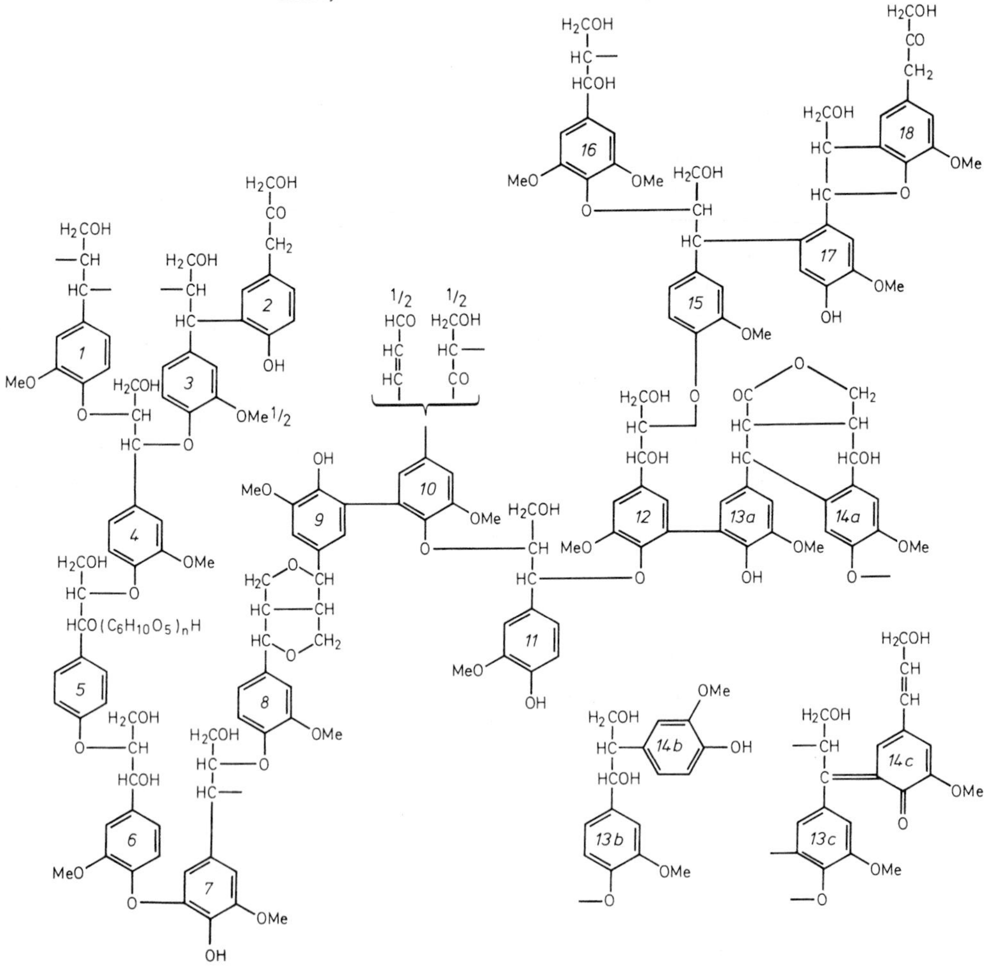

FIGURE 15.3 Structure and identity of decomposition fragments and an approximation of the structure of a softwood (spruce) lignin. (Reprinted from Bolker [5, p. 579], courtesy of Marcel Dekker, Inc.)

referred to as holocellulose, because of the high value placed on the hemicellulose component. Some types of chemical pulps (the so-called "dissolving pulps") are destined for end uses requiring pure cellulose such as for the manufacture of cellophane or cellulose acetate. For these applications the presence of hemicelluloses causes problems with filtration, etc. Thus, when pulping to produce a dissolving pulp, the process is modified to remove as much of the hemicellulose as possible.

In addition to these dominant components of dry wood, there is a group of miscellaneous, nonpolymeric, or very low molecular weight polymeric materials which are soluble in cold water and are relatively easily separated from the wood. These are collectively referred to as extractives and comprise from 3–8% by weight of dry wood. Despite their lower importance in paper the extractives can cause processing problems with some species of wood. As an example of this, the presence or formation of fatty acids from some species of wood during pulping can cause foaming problems, either during pulping or on waste stream disposal, or both. Fatty acids and the resin acids formed from oxidized terpenes can also cause corrosion problems.

The proportions of these wood components vary from species to species, and with specimens of the same species grown under different conditions. However, in general hardwoods (from deciduous trees) contain more cellulose and less lignin than softwoods (coniferous trees; Table 15.4). Also keep in mind that the chemical nature of the lignin, hemicelluloses, and miscellaneous components will also differ significantly from one species to another. These factors will affect the details of the optimum pulping method to be used for any particular species. The objective of all pulping processes is to produce a fiber suspension in water suitable for papermaking from a wood raw material. Fiber separation may be obtained by largely physical work, or by chemical treatment, or by a combination of these methods.

TABLE 15.4 Typical Approximate Percentage Compositions of Dry Wood and Cotton, on an Extractive-Free Basis[a]

Constituent	Softwood (e.g., fir)	Hardwood (e.g., aspen)	Cotton
Holocellulose			
Cellulose	42	42–48	99
Hemicelluloses	27	27	1
Total holocellulose	69	75	100
Lignin	29	15–20	0
Pectin, starch, ash, etc.	2	4	0
Total	100	100	100

[a]Analysis after thorough extraction with water and nonpolar organic solvent and drying. Compiled from data in Kirk-Othmer [4] and Bolker [5]

15.3. PREPARATION OF WOOD FOR PULPING

After wood has been harvested, today employing extensive mechanization, and brought in to the pulp mill by water, rail, or road it has to be prepared for pulping. Bark removal, or debarking, is a preliminary step common to all pulping processes. Leaving the bark in place as the wood is pulped could slightly increase the fiber yield, at least for some woods and some pulping processes. However, for most species of wood it would contribute an undesirable highly colored, nonfibrous constituent to the pulp with most species of wood, the reason for its removal.

With smaller logs a drum debarker is commonly used. This consists of a steel cylinder 3–4 m in diameter and 15–25 m long, placed on a horizontal or near-horizontal axis and capable of slow rotation on its axis. It is fitted with welded, log-tumbling baffles inside and has longitudinal slots cut into the periphery to allow bark fragments to fall out or be flushed out with water. Slow rotation of the drum containing a load of logs removes the bark by a combination of abrasion and impact attrition, after which the debarked logs are removed from the outlet gate. Drum debarkers may be operated dry, producing a dry bark by-product of high fuel value to the pulp mill, but sometimes slowing the debarking step as a result. Or they may be operated wet which gives faster debarking, particularly with hot water and cold or frozen logs. But this results in a bark of lower fuel value that requires a drying step before combustion. Wet debarking also produces a wastewater stream which requires treatment.

The bark of very large trees is usually removed in an in-line debarker. A horizontal single log at a time is moved slowly, in an axial direction, while at the same time being rotated slowly on its axis. At the same time rapidly rotating slack chains (chain flails) beat the bark off, or a "hydraulic debarker," a water jet operating at 1000 psi (lb/in.^{2}) or higher and directed to the wood/bark interface, literally blasts the bark off. Another version of in-line debarker for large logs moves the log axially, without rotation. At the same time a large stout ring fitted with spring-loaded knives pointing radially inward is revolved around the log in a lathe-like fashion, removing the bark as the log moves through.

When the bark is removed from a cord (128 ft^3; 3.625 m^3) of wood, 70 to 80 ft^3 (1.98–2.27 m^3) of debarked solid wood is normally obtained, a yield of 55–63%. Larger logs can also give larger yields of wood. The volume measure used by the pulp and paper industry is the cunit (c unit), meaning 100 ft^3 (equivalent to 2.83 m^3) of bark-free wood, i.e., the raw material suitable for papermaking.

Once the bark is removed, small logs which are to be reduced to a pulp suspension in water by grinding (one form of mechanical pulping) proceed directly to the stone groundwood operation. Very large logs which cannot be accommodated by the stone groundwood equipment directly will be sawn to manageable sizes of block before grinding. Water channels are frequently used to float both types of wood to grinding operations. While stone grinding of wood uses short lengths of whole logs, or sawn rectangular blocks of wood for processing (see later), all of the other methods of pulp production use a chipped wood feed.

Chipping, the process of reducing a log to chips of about $2.5 \times 2.5 \times 0.5$ cm thick, is a common preliminary to all other pulping methods except stone groundwood. Wood chips are more convenient and uniform for solids transport within the mill complex by conveyor belt or pneumatic delivery systems than are whole logs. Also a chip medium is more amenable to direct physical conversion to fibers because of the much easier penetration of heat and moisture through the thin sections of wood. A chip format for chemical pulping provides faster, and more uniform penetration of chemicals into the wood when digesting under pressure.

A chipper consists of a 2–4 m diameter heavy steel disk fitted with 5 or 6 radially mounted steel knives, each of which protrude about 2.5 cm (the chip length desired) from one face of the disk [7]. A spout, aimed at about 30° from the axis of rotation, directs the log to be chipped toward the rotating disk, end on and at the ideal angle to minimize the rotational energy required to drive the chipper. In the chipping process, the angular feed also produces a splayed angular edge cut to the chip which optimizes the ease of liquor penetration when pulped. As the chips are cut, slots through the steel disk ahead of each of the knives allow the chips to pass straight through the disk, aided by blowing vanes attached to the opposite face of the chipper from the knives.

A chipper of the type described, driven by a 500-hp (ca. 800-kW) electric motor is capable of reducing a 0.6 m diameter $\times$ 8 m log to chips in less than 30 seconds and could produce something like 50 tonnes of chips per hour. More recent models obtain much greater production rates by using 2000–2500 hp (1600–2000-kW) drive motors.

It has been proposed that chipping be conducted in two stages to obtain chips with less compression damage. Whatever method of chipping is used uniformity of size is important to the yield and the strength of the pulp produced. Any fines have to be screened out and cooked under different conditions than regular sized chips. Any oversize material has to be further processed for use. Blade sharpness, speed of disk rotation, weight of log(s) in the spout, and whether the butt or the top of the log is fed in first, all affect chip consistency and quality.

15.4. MECHANICAL PULPING

The wood pulps produced which rely solely on physical work on the log to reduce the wood to a fiber suspension in water are collectively referred to as mechanical pulps. By the nature of the method used to produce it, the pulp consists of a mixture of all of the insoluble constituents present in the original wood, less a small amount that is degraded during the pulping step which is lost in soluble form in the water phase. Thus the yield, or the weight of pulp obtained from the weight of wood pulped is normally high, 90% or better. Low-cost papers are made from this pulp. With care while pulping light colored woods of favorable species of trees such as white spruce, the pulp obtained is light in color. The whiteness and strength of the low-cost papers made from mechanical pulps may both be improved by blending some fully bleached chemical pulp into this before papermaking.

Relatively coarse mechanical pulps, with only 10–15% incorporation of bleached chemical pulp to provide sufficient strength, are used in the preparation of hanging stock (wallpapers). Newsprint, which is used for newspapers, inexpensive magazines, paperback books, and other low-cost publishing requirements employs a blend of 15–35% chemical pulp stocks into the mechanical pulp, to obtain an economically thin sheet of sufficient strength to avoid breakage in modern high-speed printing presses. And about 10% or so of mechanical pulps, particularly of stone groundwood types, may be blended into the pulps used for some fine papers as a means of improving the smoothness and opacity.

15.4.1. Stone Groundwood

Whole, debarked logs or sawn wood blocks with their axes parallel to the axis of the stone (Fig. 15.4) are pressed against a large grindstone rotating in a pit of water, to produce a suspension of 3–4% groundwood pulp in water. The axis of the log is kept parallel to the axis of the stone to obtain as long a fiber from the wood as possible and at the same time to minimize the grinding power requirement. Friction between the log and the stone produces heat which softens the lignin and allows cellulose fibers to be torn out with minimum fiber damage. Grinding a log end first into the stone or under very cold conditions produces a wood flour rather than a fibrous pulp, which is of little papermaking utility and would expend more power. To assist the thermal softening of lignin a system of showers with flow control and a pit water level below the base of the stone are used to obtain better regulation of stone and wood temperatures [8].

Stone dimensions range from 1.8 to 2 m in diameter by 1 to 1.5 m wide. They are made of synthetic grits such as aluminum oxide or silicon carbide, and are embedded in a ceramic matrix for a long working life under the high centrifugal forces involved under hot, wet conditions. To drive a working stone of this size at the required 250–300 rpm (peripheral speeds of 25–30 m/sec), requires 4000 to 5000 hp (ca. 3200–4000 kW) is required, usually provided by large electric motors. Grinding pressure, or the force with which the logs are gravity fed or hydraulically pressed to the stone, is the means used to regulate both stone speeds and the motor power requirement. Mechanical power, in the range of 1600 to 1750 hp hours per tonne of pulp for newsprint grades, is the chief production cost for groundwood pulping. For hanging stock, the coarsest pulps, the power requirement is slightly less than this, and for book papers slightly more.

The last step in stone groundwood production is a screening of the fiber suspension in water to remove shives (fiber bundles, or small splinters) from the stock. If paper is made from stock containing shives it will have low strength properties in the area around each shive because of poor interfiber cohesion. This makes the sheet susceptible to tearing either on the papermak-

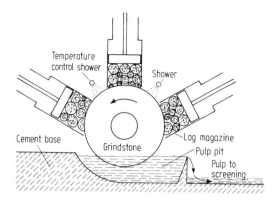

Three-pocket hydraulic grinder

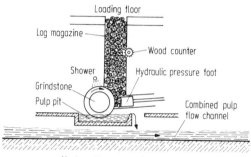

Hydraulic magazine grinder

FIGURE 15.4 Grindstone and log feed configurations used for the production of stone ground-wood pulps from whole logs or blocks.

ing machine itself or during subsequent printing. The separated shives are either discarded (burned when dry) or fed in a water suspension to a refiner for defiberization.

15.4.2. Chip Refiner Groundwood

It became possible to produce a pulp from chips, as well as from whole logs, when the process of refiner mechanical pulping was developed to accomplish this during the 1950s. This convenient method of producing mechanical pulp was rapidly adopted as a complement to stone grinding by many newsprint mills, but did not, however, completely displace stone grinding as a mechanical pulping procedure because of the differing qualities of the pulp produced by the two methods.

The machine used to convert chips to a fiber suspension in water, called a refiner, consists of two stout steel disks of about 1 m diameter fitted with hardened steel bars and depressions on one face of each (Fig. 15.5). The barred faces of the two disks are placed together, and the spacing between the bars of the facing disks is adjusted to control the fineness of the fiber suspension produced. One disk may be stationary and the other rotated, or

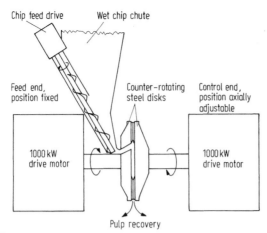

FIGURE 15.5   Operating details of a refiner used to reduce a wood chip feed to pulp.

they may be counter-rotating (rotated in opposite directions) in either case by powerful electric drives of up to 10,000 hp (ca. 8000 kW).

Wet chips, together with some additional water, are fed to the center of the space between the two refiner disks while they are rotating. The adjacent faces of the two disks are each slightly concave so that the chips are able to drop into the wider space existing between the disks, near their centers. Friction between the chips and the disks produces heat and steam which softens the lignin and starts the defiberization process. As the chip fragments get smaller they are carried by the continuing flow of water, fresh chips, and steam toward the periphery of the disks, where the spacing between the disks is less, and the bar/depression frequency is higher (Fig. 15.5). In this way the chips are rapidly reduced to a separated, individual fiber suspension in water, as the fiber fragments reach the periphery of the disks. Peripheral speeds of refiner disks may be as high as 160 m/sec, significantly greater than the usual peripheral speeds of grinders [4].

To obtain a well-processed pulp for newsprint manufacture, passage of the chips and the partially refined pulp through two or three refiners operating in series is usually required. Paper produced from this type of pulp is generally stronger, from the longer average fiber lengths produced, than paper produced from stone grinding of the same wood. At the same time, and for the same reason, it also generally has a higher surface roughness and lower opacity than stone groundwood newsprint, although these factors are affected by refiner adjustments which increase or decrease the power requirement to obtain the pulp (Table 15.5). Chip refining, however, does have a somewhat higher power cost for pulp production, say, about 80 to 100 hp days/air dry tonne for spruce, as opposed to about 80 to 90 when stone grinding. These are all reasons for the continuance of stone grinding as a component of the mechanical pulping operations of many newsprint mills. Many newsprint mills use blends of refiner and stone groundwood pulps with a proportion of chemical

■ **TABLE 15.5 Comparison of Properties of Papers Produced from Spruce Stone or Refiner Groundwood Pulps at Similar and Dissimilar Work Inputs and Related Terminology**[a,b]

| | Stone groundwood | Refiner groundwoods | |
		Moderate work	Higher work
Power, hp days/air dry ton	76	75.3	89
Can. standard freeness,[b] mL	115	164	101
Burst factor	12.1	12.6	18.2
Tear factor	56.7	87.8	86.0
Breaking length, m	2,920	2,680	3,200
Brightness, GE or Elrepho	61.0	59.0	60.5
Opacity, %	95.0	92.5	95.0

[a]Properties selected from Britt [8].

[b]Terminology:

Canadian standard freeness (CSF) is a measure of the ease of water flow through a standard perforated plate, specified in mL, from a one-liter sample of pulp at 20°C and 0.3% consistency (weight fraction of pulp in the suspension). Drained water from the sample passes into a dual outlet funnel, one outlet higher than the other. The amount of water collected from the upper "overflow" outlet, in mL, is recorded as the CSF. The rate of water removal on the paper machine of a "free" pulp will be faster than the rate of water removal from a "slow" (slow draining) pulp.

Burst factor. The pressure developed by a fluid, in pounds per square inch, required to split open a sample of the paper when it covers a circular hole 1.20 inches in diameter and is protected by a rubber diaphragm, is defined as burst (lb/in.^{2} or g/cm^2). When burst is divided by the basis weight, e.g., in g/m^2, it is called burst factor.

Tear corresponds to the tensile force in grams required to continue a tear that has already been started in a single sheet of paper. Corresponds to the "internal tear resistance" as opposed to "edge tear resistance," which is frequently a much higher value. Tear, divided by basis weight, is the tear factor.

Breaking length is a hypothetical value calculated from tensile strength and material density measurements, to give a relative measure of the length of a strip of paper which is just self-supporting, when it is held vertically. Generally speaking, the higher this value, the higher the tensile strength of the sample.

Brightness is a measure of overall reflectivity of a paper to white light, or whiteness. It is based on a scale of 100 for pure magnesium oxide and 8 for carbon black, when the reflectance of a beam of light, of wavelength 457 nm, is compared at angle of 45° from the axis of the incident light.

Opacity is a measure of sheet "see-through." It is obtained by dividing the reflectance of the sheet when backed by a black body by the reflectance of the same sheet when backed by a white body having an effective reflectance of 89% (absolute), and expressed as percent.

pulp to raise the strength. However, it is possible to produce a good newsprint solely from refiner groundwood pulps because of the longer average fiber lengths. One further advantage of the refiner pulping method is its capability to produce a useful pulp from planer shavings or sawdust, options not possible with stone grinding. While it is not possible to make a useful paper from sawdust pulp alone, it can be blended with more conventional pulps to produce a paper sheet to meet a printer's specifications.

15.4.3. Thermomechanical Pulp

As more experience was gained, it was realized that stone or refiner mechanical pulping at higher temperatures decreased the power input required per tonne of pulp produced. Carrying this realization a step further, the idea of subjecting chips to a short period of preheating before refining was developed with the expectation that this might both decrease refining power requirements and produce a longer fibered pulp potentially able to yield a stronger paper than the product from conventional refining. Combinations of these features were the benefits achieved by the first commercial installations which began operation in the early 1970s.

Chips, shavings, or sawdust are first washed to remove any grit or other potentially damaging contaminants, and then digested with steam at 120 to 130°C, 170–315 kPa (25 to 45 psi) for 2–3 minutes [9]. While still under pressure, and with only a 60–75% water content (much less than required for conventional chip refining), the fragmented wood is refined and the pressure then released, producing a fluffed up, bulky pulp. After addition of a small amount of water and a further stage of refining at atmospheric pressure a pulp is obtained which has longer average fiber lengths, higher strength (from less damaged fiber), and a lower fines content than either stone groundwood or conventional chip refiner groundwood pulps.

The favorable strength properties of thermomechanical pulp (TMP) allow it to be used to replace part or all of the chemical pulp component in a newsprint furnish, the replacement rate ranging from 2 to 4 tonnes for every 1 tonne reduction in chemical pulp content. Since the yield from wood for TMP, about 90%, is nearly twice that of chemical pulps, chemical pulp replacement by TMP in newsprint yields a significant saving in wood requirements for newsprint production. The saving in wood far surpasses the additional refining energy requirement for thermomechanical pulping. For coniferous wood TMP requires about 2000–2500 kWh/tonne, about 1000 more than stone grinding at about 1200–1650 kWh per air dry tonne. A further advantage of TMP over the chemical pulps it can partially replace is more straightforward effluent treatment.

The exposure of the pulp to heat and air in TMP processing causes some darkening of thermomechanical pulps. The use of an aqueous sulfite pretreatment before TMP processing helps to reduce the loss of brightness on refining. This variation of TMP processing is still referred to as a mechanical pulping process since refining (mechanical work) is still the primary process used to pulp the chips. It is referred to as chemithermomechanical pulp(ing) (CTMP) to distinguish it from ordinary TMP [10].

15.5. CHEMICAL PULPING PROCESSES

Processes which involve the use of chemicals for either softening or removal of much or all of the lignin, to assist the process of fiber separation from the wood, are called chemical pulping processes. The simpler processes of pulping use a chemical presoak of whole (small) logs or chips prior to the use of conventional mechanical pulping methods of defiberization to produce a fiber suspension in water. These presoak methods are usually reserved for the mechanical pulping of hardwoods and are referred to as chemimechanical pulping, not as true chemical pulping processes. At the other extreme the more rigorous chemical pulping methods remove all or nearly all of the lignin from wood chips leaving cellulose fiber as intact as possible. In this case a fiber suspension in water is obtained directly, with little or no mechanical work required for fiber separation. Despite the selectivity aim of chemical pulping methods to soften and dissolve lignin alone without affecting the cellulose, there is inevitably some cellulose chain shortening and some end group conversion to carboxyl that occurs at the same time. Nevertheless, since about 1930 the range of chemical pulping techniques that have been developed is adequate to produce satisfactory pulps for papermaking from virtually any species of tree as well as from grasses and annual plants. Thus, a much wider range of raw materials can be handled by chemical pulping methods than is possible by straight mechanical pulping procedures.

15.5.1. Chemimechanical Pulping

Direct grinding or refining of hardwoods (from deciduous trees) tends to give pulps with a short average fiber length which would produce a relatively low strength sheet on the paper machine. This is particularly the case on mechanical pulping of *hard* woods from deciduous trees since the fibers in place are shorter and finer and since there is a greater tendency for fibers to be fragmented, instead of whole, as they are torn out from the denser wood structures.

However, if the wood is given a presoak in a hot solution of sodium hydroxide it is possible to obtain a more useful pulp with less fiber damage. Temperatures may be near ambient up to 80°C, depending on the species to be pulped and treatment time. Small amounts of sulfur dioxide may be added to the liquor to decrease the tendency of the wood to darken under the aerated alkaline conditions. Whole logs or blocks require a 4- to 10-hour presoak whereas chips only need a 10- to 15-minute contact time. In this way, pulp yields of 85–90% are obtained with properties parallel those expected for groundwood pulps generally, and from species of tree not otherwise amenable to mechanical pulping. Chemical pretreatment also significantly decreases the power input required to reduce the wood to fibers. Even mixed blends of soft, and hardwood chips have been found amenable to chemical pretreatment before mechanical pulping to still yield useful pulps. This approach thus simplifies the newsprint product for mills using a mixture of coniferous and deciduous woods.

15.5.2. Semichemical Pulping

The essence of semichemical pulping methods, as the name implies, is that chips are subjected to a more intensive chemical treatment than a simple presoak but not as severe as for chemical pulping, to cause a partial chemical delignification. Intermediate intensity chemical treatment is achieved by a number of means, among them decreased ratios of chemicals to wood, decreased cooking times and/or temperatures, and the use of chemicals which pulp at near neutral pH rather than under highly acidic or highly alkaline conditions. Following the cook, the chips require much lower power inputs per tonne of pulp for refining than required for straight mechanical pulping, because of the chemical softening and partial solubilization of the lignin. Partial lignin solubilization during the pulping process lowers the yields of pulp to 65 to 80%, less than that given by mechanical or chemimechanical methods.

At least half a dozen variants of semichemical pulping have been practiced. Some of these, namely, neutral sulfite semichemical (NSSC) pulping, softwood bisulfite high yield pulping, and softwood sulfate pulping for linerboard production, are still common [4]. Among these, NSSC pulping has a number of differentiating features of interest. Hence this procedure, which is primarily of value in the pulping of hardwoods, will be discussed here as an example of a semichemical pulping procedure.

The digestion liquor used for NSSC pulping essentially consists of a solution of sodium sulfite and sodium carbonate in water, but at the proportions and the pH used, this produces a mixture of sodium sulfite and sodium bicarbonate. The make-up sodium carbonate is either purchased or made from aqueous solutions of low-grade sodium hydroxide by contacting this with the carbon dioxide of flue gas (Eq. 15.3, 15.4).

$$CO_2 + H_2O \rightleftharpoons H_2CO_3 \qquad\qquad 15.3$$

$$H_2CO_3 + 2\,NaOH \rightarrow Na_2CO_3 + 2\,H_2O \qquad\qquad 15.4$$

Elemental sulfur is burned for the first stage of sodium sulfite preparation. Repeated contact of the sulfur dioxide formed with the aqueous sodium carbonate in a gas-liquid absorption unit (sulfiting tower, Fig. 15.6) produces the approximate proportion of sodium sulfite and sodium bicarbonate (about 1 mol:2 mol) required (Eq. 15.5, 15.6).

$$SO_2 + H_2O \rightleftharpoons H_2SO_3 \qquad\qquad 15.5$$

$$H_2SO_3 + 2\,Na_2CO_3 \rightarrow Na_2SO_3 + 2\,NaHCO_3 \qquad\qquad 15.6$$

Liquor recycle to the sulfiting tower is continued until the pH of the underflow reaches 8.5, at which time a portion of this stream may be continually bled off for use as pulping digestion liquor. In this situation the actual proportions of salts present in solution are about 82% Na_2SO_3, 4% $NaHSO_3$, and 14% $NaHCO_3$. For some species of wood or for different pulping objectives liquor of about pH 8 may be used. This is made by more prolonged contact between the sulfur dioxide and sodium carbonate solutions (Eq. 15.7).

$$H_2SO_3 + 3\,Na_2CO_3 + SO_2 \rightarrow 2\,Na_2SO_3 + 2\,NaHCO_3 + CO_2 \quad 15.7$$

Care is taken to ensure that the digestion liquor does not become too acid, otherwise cellulose degradation during the pulping step becomes more significant.

For NSSC pulp production the effective chemical concentrations are about 120 g/L of sodium sulfite and about 20 g/L of sodium carbonate. Chips to be

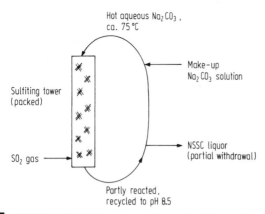

FIGURE 15.6 Preparation of neutral sulfite semichemical pulping liquor via countercurrent contacting of a sulfur dioxide gas stream with aqueous sodium carbonate.

pulped are cooked in this two buffer system at temperatures of 140–170°C for 1–4 hours depending on the type of wood being pulped and the pulping objective. High-grade pulps destined for bleaching will be pulped at ratios of sodium sulfite to wood of about 15–20:100 by weight, significantly more than is used for coarser pulps. They also demand higher temperatures and/or longer times, and give correspondingly lower yields of finished pulp because of the greater degree of delignification achieved. Coarser pulps destined for board production (cardboard, fluting medium, or box liners) will be subjected to milder conditions and shorter times. This produces higher yields and retains individual fiber stiffness, a property important for products such as corrugating medium. The dual buffer system exerts its pulping action by sulfonation of lignin present in the chips and probably by reduction as well [8]. Both processes give soluble lower molecular weight lignin fragments from the original insoluble high molecular weight lignin.

The cooked pulp is filtered from the pulping liquor, washed, and then refined to complete the defiberization. A high yield of the clean, refined pulp is obtained, which still contains some lignin. The lignin content and the dark color of high-yield NSSC pulps makes this product generally unsuitable for high-grade papers. The light color of lower yield grades minimizes bleaching costs for incorporation of this product into intermediate or coarser paper grades. There is also no significant smell problem associated with the pulping step, as there is with some of the other chemically based pulping processes.

A problem with NSSC pulping is the difficulty of recycle or disposal of the waste (spent) pulping liquor. It can be concentrated and burned but this produces sodium sulfate, not sulfite suitable for reuse. If there is a kraft pulping operation near the semichemical pulping operation, the sodium sulfate may provide both sodium ion and sulfate make-up chemical for the kraft mill. If the kraft mill is three or more times the scale of the NSSC mill rate, the kraft process is capable of profitably using all of the spent liquor of the latter. It is possible to obtain sulfite from sulfate by a complex series of steps but these procedures add significantly to the costs and thus makes recycle less attractive.

Alternatively the spent NSSC liquor may simply be discharged. But untreated, this liquor has a very high BOD and several toxic constituents so that disposal to freshwater has been discontinued for many years. Disposal in the open ocean, at depth, in areas of vigorous currents could probably be accepted. However, mills fronting on the sea are usually on sheltered inlets or fjords so they have the same restrictions as mills operating on freshwater. Details of other possibilities are discussed with emission control aspects of sulfite pulping.

15.5.3. Chemical Pulping

Wood chips supply the fiber raw material for chemical pulping operations, as for most chemimechanical and semichemical methods. In chemical pulping, the treatment with chemicals is severe enough to dissolve all or much of the lignin originally present in the wood. The chemicals employed operate at low or high pH (not near neutral), higher ratios of chemicals to wood are used, and the cooking conditions are usually more severe than for the processes discussed thus far. Thus, the pulps obtained require little or no mechanical work to obtain a fiber suspension in water from the original wood.

Photomicrographs of chemical pulps show that there is very little fiber damage. Consequently papers made from chemical pulps, particularly from kraft chemical pulps, are strong. Since there is not only lignin loss, but for some chemical pulping processes (for instance, the acid sulfite process) hemicellulose as well, the yields of pulp from straight chemical processes are in the 40 to 60% range, much lower than that achieved by the straight mechanical procedures. Chemical pulps are more or less readily bleached to high brightness (whiteness) levels because they consist of almost pure cellulose. This factor also means that the brightness and the strength properties are both much more stable than for mechanical (newsprint-type) pulps. Paper produced from bleached chemical pulps is more pleasing in appearance and more durable than papers produced from other types of pulps. But these papers are also more expensive because of the increased wood and chemical cost required for their manufacture. The various chemical pulping processes in use may be conveniently classified according to the pH of the digestion liquor used to carry out the pulping.

15.5.4. Acid Sulfite Pulping

After the discovery of the softening action of sulfurous acid on wood by Benjamin Tilghman in the U.S.A. in about 1857, the idea was developed experimentally as a chemical method for wood pulping by 1867 [8]. It was adopted on a commercial scale in the next few years, both in the U.S.A. and elsewhere, and rapidly became the dominant chemical pulping method until about 1937, when the volume of kraft pulp produced first exceeded the volume of sulfite pulp.

Digestion liquor for acid sulfite pulping consists of a solution of sulfur dioxide in water containing lime. Sulfur dioxide is obtained by sulfur combustion or by the roasting of pyrite. The lime component is added to the liquor by passage of the cold sulfur dioxide solution down one or more Jenssen towers packed with limestone, in a manner very similar to the sulfiting towers used to prepare NSSC pulping liquor (Eq. 15.8, 15.9).

$$SO_2 + H_2O \rightleftharpoons H_2SO_3 \qquad 15.8$$

$$2\ H_2SO_3 + CaCO_3 \rightarrow Ca(HSO_3)_2 + H_2O + CO_2 \qquad 15.9$$

Since calcium sulfite is not very soluble in water, the last stage(s) of liquor preparation involve absorption of sulfur dioxide into the cold liquor under pressure. By this means an eventual ratio of about 1 mole of calcium hydrogen sulfite (calcium bisulfite) to a total of 4 to 5 moles of (sulfur dioxide + sulfurous acid) is obtained. At this stage the pulping liquor is an aqueous solution containing a net 6.5 to 9% dissolved sulfur dioxide in all forms.

The terminology used describes the total dissolved (sulfur dioxide + sulfurous acid) as "free SO_2," and the sulfur dioxide tied up in calcium hydrogen sulfite as "combined SO_2" (Eq. 15.10).

$$\begin{array}{ccc} SO_2,\ H_2SO_3 & Ca(HSO_3)_2 & \\ \text{"Free } SO_2\text{"} & \text{"Combined } SO_2\text{"} & 15.10 \end{array}$$
$$\text{Mole ratio:} \quad 4\text{--}5 \quad : \quad 1$$

However, the effective sulfur dioxide accessible for pulping is a composite of these two forms of sulfur dioxide on the basis that the "available SO_2" = "free SO_2" + 1/2 "combined SO_2." The reason for this definition is that, as sulfur dioxide is consumed or vented from the digestion vessel during pulping, the equilibrium represented by Eq. 15.11 is displaced to the right.

$$Ca(HSO_3)_2 \rightarrow CaSO_3 + H_2O + SO_2 \qquad 15.11$$

As this occurs one-half of the combined SO_2 is released to the solution to become available for pulping.

Conditions required for sulfite pulping vary with the species of wood being pulped, the degree of delignification desired, and the chemical to wood ratio being used. Normal range is 8–12 hours at 110–145°C, frequently following a 2-hour liquor penetration period. Typical pulping liquor to wood ratios on the West Coast of North America are 182,000 L (50,000 U.S. gal) of liquor containing 8.80% total SO_2 and 1.10% combined SO_2 at 70°C, to 36 tonnes of bone dry wood. Of course it is not necessary that the chips be dry for digestion. But the moisture content must be known to have a consistent ratio of active chemicals to the dry wood content of the chips.

The heat and low pH conditions bring about hydrolytic fragmentation and sulfonation of the lignins in the chips. These changes solubilize a large fraction of the lignin. The sulfuric acid concentration in the liquor is minimized by burning the sulfur or pyrite in a deficiency of air and then rapidly chilling the gases produced. Without these precautions the lower pH could cause cellulose hydrolysis and reduce the yield of pulp. Cellulose is less susceptible to sulfonic acids in the presence of the weaker, buffering acids also present.

Acid sulfite pulping produces a light-colored, easily bleached pulp. After washing and screening and even without bleaching it has a brightness (whiteness) compatible with direct blending into mechanical pulps for newsprint production. With two or three bleaching stages high brightness bond papers may be produced from these pulps. However, because of the less than optimum paper strength properties obtained and because chemical recovery procedures are expensive (see the Further Reading section), acid sulfite pulping has become much less important in recent years. It is still of value for the manufacture of specialty pulps for uses such as tissues and printing papers [11], and for the production of dissolving pulp for conversion to cellulose derivatives employed as viscose rayon and thermoplastic materials. Yield advantages and lower bleaching requirements than necessary for kraft pulps promise a continued place for sulfite-based pulping processes for these applications [12].

15.5.5. Other Sulfite-Based Pulping Processes

Acid sulfite pulping, originally developed around low-cost calcium ion, suffers from a number of disadvantages relative to alternatives. Among these, calcium sulfite is virtually insoluble, which introduces difficulties during the preparation of digestion liquor; chemical recovery is complex making it uneconomic; and disposal of the waste liquor is difficult (Section 15.5.2). Because of these disadvantages, most modern sulfite mills have switched from calcium-based to magnesium-, sodium-, or ammonium-based systems which circumvent many of these difficulties.

With magnesium-, sodium-, or ammonium-based sytems the bisulfite and sulfite salts are all soluble at all proportions in the presence of sulfurous acid. Even magnesium sulfite, with a solubility of about 1.25 g/100 mL, cold, is about 160 times as soluble as calcium sulfite at the same temperature and its solubility <u>increases</u> with temperature. So liquor preparation with these sulfite salts is easier, whether for acid sulfite, bisulfite, or neutral (NSSC) pulping conditions. And, in fact, sulfite pulping has also been experimentally tested under alkaline conditions. For ammonium-based systems, ammonium hydroxide is contacted with a sulfur dioxide gas stream for liquor preparation. Magnesium-based systems use a magnesium hydroxide slurry to contact the sulfur dioxide gas stream. Sodium-based systems normally employ sodium carbonate lumps in a sulfiting tower, in a method similar to that used for NSSC liquor preparation, although sodium hydroxide may also be used for this purpose if it is available at low cost.

Ammonium, magnesium, and sodium-based sulfite pulping systems offer a versatility of operating pH not possible with calcium. However, sodium ion is the only one useful throughout the whole pH range even to monosulfite and alkaline sulfite pulping liquor preparation (Table 15.6) [13]. With sodium carbonate as the base it is possible to use carbonate to sulfurous acid ratios ranging from 1:3 where the active pulping species are bisulfite (HSO_3^-) and sulfurous acid (acid sulfite pulping), to 1:2 where the principal pulping species is bisulfite, such as is used in bisulfite pulping. At the 1:1 to 2:1 range of ratios of carbonate to sulfurous acid, sulfite and bicarbonate are the active pulping agents, as in NSSC pulping. And when one reaches the 3:1 to 4:1 range of ratios, sulfite and carbonate are the dominant anions present, important in the experimental alkaline sulfite pulping method (Table 15.7). Analogous results may be achieved with ammonium and magnesium ions, at least with pHs up to the bisulfite range.

Acid sulfite and bisulfite pulping processes yield light colored pulps, suitable for direct blending with groundwood pulps for newsprint production. These are also pulps which may be bleached easily for fine paper production, and require low input of beating energy to obtain good sheet formation (even distribution of fiber) and strength. They may also be readily purified with solutions of sodium hydroxide for the preparation of dissolving pulps. But to offset these advantages, long cooking times are required, and either disposal or recovery of the chemicals from the spent liquor present problems. It is, however, possible to profitably produce the valuable chemical intermediate and flavor constituent vanillin from the lignin fragments present in sulfite (or kraft) waste liquor [14,15]. At one time the Ontario Paper Company, operating at Thorold, Ontario, produced 60% of the world supply of vanillin from this source.

TABLE 15.6   Chemical Pulping Processes Classified According to Operating pH, and Pulp End Uses

Pulping process	Liquor pH at 20°C	Base options	Appropriate end uses
Acid sulfite	1–2	Ca^{2+}, Mg^{2+}	1. Chemical conversion to derivatives of cellulose, e.g., viscous rayon, cellulose acetate.
		Na^+, NH_4^+	2. Filler pulps for paper requiring little strength, e.g., writing papers. Provides bulk, cushioning properties.
Bisulfite	3–5	Mg^{2+}, Na^+,	1. Unbleached—in newsprint (25–30%, rest groundwood)
		NH_4^+	2. Bleached—fine papers (up to 100%)
Neutral sulfite semichemical[a]	6–8	Na^+, NH_4^+	1. Fluting board—provides stiffness, crush resistance to corrugated cardboard
			2. Coarse wrapping papers
Monosulfite	8–10	Na^+, NH_4^+	Same as for neutral sulfite
Alkaline sulfite (experimental)	11–13[b]	Na^+	1. Unbleached—box liner-board other packaging materials
			2. Bleached—high-quality papers
Soda	11–13	Na^+	Outmoded for chemical pulping; see text.
Kraft	11–13	Na^+ (+ Na_2S)	1. Unbleached—packaging materials: boxboard, sack and bag papers
			2. Semibleached—newsprint (25–35%)
			3. Bleached—high-quality papers of all types
Prehydrolysis kraft	11–13	Na^+, (Na_2S)	Chemical conversion

[a]Not generally used as a full chemical pulping process, but may also be used for the production of fine papers. Details of this process are included here to demonstrate the pulping versatility of sulfite ion over the full range of pH conditions.

[b]Only at the laboratory stage at present, but see *Canadian Chemical Processing* [13].

███ **TABLE 15.7   Ratios of Sodium Carbonate to Sulfurous Acid^a Required for Sulfite Pulping at Various pH's**

Process	Required mole ratio of		Active pulping entities present
	Na$_2$CO$_3$:	H$_2$SO$_3$	
Acid sulfite,	1	3	H$_2$SO$_3$, HSO$_3^-$
pH ca. 1–2		(and excess)	
Bisulfite,	1	2	HSO$_3^-$
pH ca. 3–5			
NSSC^b,	1	1	SO$_3^{2-}$
pH ca. 6–8	3	2	SO$_3^{2-}$, HCO$_3^-$
NSSC, and monosulfite	2	1	SO$_3^{2-}$, 2 HCO$_3^-$
pH ca. 8–10			
Alkaline sulfite^c,	3	1	SO$_3^{2-}$, 2 CO$_3^{2-}$
pH ca. 11–13	4	1	SO$_3^{2-}$, 3 CO$_3^{2-}$

^{a}The existence of sulfurous acid (H$_2$SO$_3$) is somewhat hypothetical, but is a convenient concept for the considerations required here. Sulfite salts are common.
^{b}Neutral sulfite semichemical pulping.
^{c}Experimental in nature at present, see *Canadian Chemical Processing* [13].

15.5.6. Alkaline Pulping: The Soda Process

The older soda process, which involves the pulping of chips by cooking with aqueous sodium hydroxide at 160–170°C, was originally devised by Watt and Burgess in England in 1851. It requires long cooking times and gives poor yields of a dark pulp which still has a significant lignin content, even with hardwoods, which give the most favorable results. Also the strength properties of soda pulps are second rate. Any "soda" mills still operating generally employ a small amount of sulfide in the pulping liquor, which is not strictly soda pulping, but should be classed as a modified process bearing more than a superficial resemblance to kraft pulping. Thus, at present virtually all alkaline pulping, worldwide, is conducted by the kraft process.

15.5.7. Alkaline Pulping: Background of the Kraft Process

The kraft, or sulfate, process is today the preeminent chemical pulping procedure. In Canada alone, 80% of all the chemical pulp produced is by this process, and worldwide some 85% of the total is via this route. The current prominence of this pulping procedure warrants a rather complete discussion of the details of this process than devoted to the other chemical pulping methods.

This process, in which wood chips are cooked in a pulping liquor consisting of a solution of sodium hydroxide and sodium sulfide in water (so-called "white liquor"), was originally developed by Dahl, in Germany, in 1879 [16]. The presence of the sodium sulfide accelerates the pulping rate so that kraft pulping requires a shorter cooking time than soda pulping. The sulfide also serves to stabilize the cellulose somewhat, so that the pulps produced suffer minimal fiber damage and hence make a strong paper sheet. This high-strength pulp product is the origin of the name given to the process, "kraft" being German for "strength." Factors which contribute to the greater strength of pulps made by this process include not only the presence of the sodium sulfide and the shorter cooking time (and consequently less cellulose degradation) but also the lower ratio of chemicals to wood required to obtain effective pulping. A further advantage of kraft pulping lies in its ability to handle hardwoods or softwoods, even of resinous or pitchy species, since the liquor composition is such that the resins are dissolved. Resinous species are more difficult to handle by any of the acidic sulfite pulping processes.

A feature of the kraft process vital to its continued success is the integral, well-tested chemical recovery system. The digestion liquor for each batch of chips to be pulped is mainly obtained from the chemicals recovered from the spent liquor of previous digestions, and has approximately the composition given in Table 15.8. For a pulping context it is usual to specify all of the components present in the digestion liquor on a "Na_2O equivalent" basis, which has the effect of putting the active constituents on the same sodium ion content basis. Thus, the actual concentration of sodium hydroxide present for a 73 g/L, Na_2O equivalent is given by Eq. 15.12.

$$Na_2O + H_2O \rightarrow 2\ NaOH$$
73 g/L (Na_2O equiv.) $\div$ 62 g/mol (Na_2O) $\times$ 40 g/mol (NaOH) $\times$ 2 mol/mol
Thus, NaOH from Na_2O = 94.2 g/L NaOH   15.12

By a similar calculation using the stoichiometry of Eq. 15.13 the actual concentration of sodium sulfide present works out to 39.5 g/L.

$$Na_2O + H_2S \rightarrow Na_2S + H_2O \qquad 15.13$$

For pulping purposes, the "effective alkali" present is somewhat higher than the actual sodium hydroxide concentration present in the white liquor because of the existence of a hydrolytic equilibrium between sodium sulfide and sodium hydrogen sulfide (Eq. 15.14).

$$Na_2S + H_2O \rightarrow NaOH + NaHS \qquad 15.14$$

TABLE 15.8   Typical Analysis of White Liquor Used for Kraft Pulping[a]

| Component | Concentrations, as | | Function |
	Na$_2$O equivalent (g/L)	Molarity (mol/L)	
NaOH	73.0	2.34	Primary pulping agent.
Na$_2$S	31.4	0.51	Accelerates pulping, raises yield of cellulose and hemicellulose by protective action.
Na$_2$CO$_3$	18.2	0.29	Negligible. Presence incidental as a result of chemical recovery system.
Na$_2$SO$_4$, Na$_2$SO$_3$, Na$_2$S$_2$O$_3$	ca. 1.0	—	None. Traces present incidentally.

[a]Data from Kirk-Othmer [4] and Britt [8]. Liquor with the composition given here will normally be diluted to a net sodium hydroxide concentration of about 50 g/L Na_2O.

Thus, the effective alkali comprises the concentration of sodium hydroxide present, plus one-half of the concentration of sodium sulfide present, when both are expressed as Na_2O, or for the typical analysis of Table 15.8, about 88–89 g/L Na_2O. Specified in molar terms, since the molar concentration of sodium hydroxide derived from sodium sulfide is equivalent to the initial molar concentration of sodium sulfide (Eq. 15.14), the effective alkali concentration of the white liquor of Table 15.8 would be 2.85 M.

"Sulfidity" is also an important parameter relating to the liquor composition for kraft pulping. It refers to the concentration of sodium sulfide divided by the concentrations of sodium sulfide plus sodium hydroxide—the "active alkali" as opposed to "effective alkali," all expressed in terms of their Na_2O equivalents. Eq. 15.15 gives the method of determining sulfidity, expressed as a percentage.

$$\% \text{ Sulfidity} = (Na_2S)/(Na_2S + NaOH) \times 100 \qquad 15.15$$

The desirable range for sulfidity is 15 to 35%. Values lower than this tend to give incomplete pulping and weaker pulps, akin to the properties of pulp produced by the soda process. Higher values tend to give sluggish delignification, and also tend to raise potential sulfur losses in both the pulping and chemical recovery segments of kraft pulping.

15.5.8. Alkaline Pulping: Details of the Kraft Process

The cook is carried out in a digester, which is a steel or stainless steel cylindrical pressure vessel about 3 m in diameter and 16 m high about the height of a five-story building, usually fitted with external heat exchangers. For a digester this size, 50 to 60 tonnes of chips per batch (containing about 50% moisture) are loaded from the top, while steam is blown through the opening of the digester. Steam blowing serves to both displace much of the air from the digester during the loading process and on white liquor addition. It also aids in liquor penetration into the chip from the combined air displacement, warming, and wetting actions on the wood. Any knots or uncooked chips screened from a previous digestion will also be added at this stage. At the time of white liquor addition, sufficient "weak black liquor" is also added to bring the content of $NaOH + Na_2S$ present in the digestion liquor to the concentration range of about 50 g/L Na_2O equivalent, and to a liquor to wood ratio of about 16 kg of effective alkali for each 100 kg of dry wood. The heavy steel lid is bolted down and heating begun by pumping liquor from the digester, through the external steam-heated heat exchangers, and back to the digester. A common heating schedule is to take 1.5 hours to bring the temperature up to about 170°C and 620 kPa (kilopascals, i.e. 6 atm pressure, 90 psig), after which it is held at this temperature for a further 1.5 hours to complete the digestion step. Chemical to wood ratios and digestion times and temperatures will be varied somewhat depending on the species of wood being pulped, and the degree of delignification desired for the product.

After the digestion is completed a valve at the base of the digester is opened while it is still at operating temperature and pressure, causing a very rapid discharge of the contents of the digester into a blow tank. The fast pressure drop experienced by the pulp on discharge assists in blowing the fiber bundles apart, aiding in the production of a pulp from the chips.

Cooked pulp plus weak black liquor, spent white liquor which has picked up lignin and some hydrolyzed cellulose and hemicellulose, etc., proceeds to screens which serve to remove any uncooked fragments and knots from the pulp. The bulk of the weak black liquor is separated from the cooked pulp in a decker, basically a filter that captures the pulp on a screen and removes the bulk of the liquor by suction. The pulp is then rinsed in washers, units very like drum filters, equipped with two or three freshwater fed shower tubes which run parallel to the axis of the drum and which rinse off any residual weak black liquor from the pulp. At this stage the separated kraft process product is a fairly clean pulp, though still dark brown in color from residual lignin, etc. (i.e., the raw material for "kraft" paper bags, etc.). The weak black liquor filtrate contains all of the spent chemical constituents of the original white liquor plus the components of the wood which were dissolved during the course of the pulping process.

The stages just described, from the point of loading chips into a cylindrical digestion vessel, comprise *batch* pulping of wood. To maintain a continuous stream of cooked pulp for subsequent stages requires, say, six to ten of these digesters, operating in parallel in a staggered loading sequence, with each discharging five to seven loads of 12 to 15 tonnes (about one-quarter of the weight of wet wood charged to each digester) of cooked pulp from each load, per 24 hours. All of the stages described up to this point may also be conducted continuously in a single, more complex digestion unit performing all these functions in turn on a mass of chips which moves steadily through the unit. One such commercially available version is the Kamyr continuous digester (Fig. 15.7). Many pulp mills use a combination of batch and continuous digesters to obtain maximum flexibility in terms of the types of wood being pulped, and to obtain the desired degree of pulp delignification proportioned as required for the end use objectives of the pulp.

15.5.9. Kraft Cyclic Chemical Recovery

The objectives of chemical recovery from the weak black liquor separated from the pulp after kraft pulping are threefold. Recovery and regeneration of most of the chemicals employed in the original digestion liquor are primary objectives. A further objective is to generate a significant fraction of the total steam requirements of the pulp mill from combustion of the dissolved lignin residues and other organic constituents (comprising nearly half the original dry weight of the wood) contained in the weak black liquor. Burning the organic constituents present in the spent liquor also helps to accomplish the third objective, avoidance of the significant polluting potential on untreated discharge of spent pulping liquor into a stream or lake. An environmentally acceptable alternative to this, treatment of the spent liquor by artificial lagoon aeration, would be a net cost to the process rather than a net chemicals and energy credit.

Initially the weak black liquor is contacted with air, either by sparging air through a tank of the liquor, or by causing thin films of black liquor to

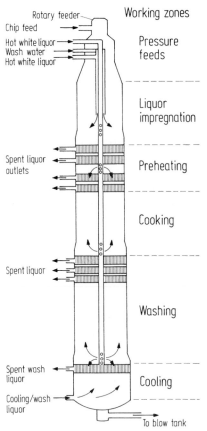

FIGURE 15.7 Schematic of the operating details of a Kamyr continuous digester.

fall through an air chamber. The objective of this step is to decrease the volatility and losses of foul-smelling volatile sulfur compounds which would otherwise occur during subsequent evaporation or combustion steps (Eq. 15.16–15.18).

Volatile sulfide loss:

$$H_2O + CO_2 \text{ (from combustion)} \rightarrow H_2CO_3 \qquad\qquad 15.16$$

$$Na_2S + H_2CO_3 \rightarrow Na_2CO_3 + H_2S\uparrow \qquad\qquad 15.17$$

$$2\,NaSCH_3 + H_2CO_3 \rightarrow Na_2CO_3 + CH_3SH$$
$$\text{methyl} \qquad\qquad 15.18$$
$$\text{mercaptan}$$

By oxidizing sodium sulfide to sodium thiosulfate and methyl mercaptan (methane thiol) to dimethyldisulfide, the first component is made virtually involatile, and the volatility of the second is greatly decreased (Eq. 15.19, 15.20).

Black liquor oxidation:

$$2\ Na_2S + 2\ O_2 + H_2O \rightarrow Na_2S_2O_3 + 2\ NaOH \qquad 15.19$$
$$\Delta H = -900\ kJ\ (-215\ kcal)$$

$$2\ CH_3SNa + 1/2\ O_2 + H_2O \rightarrow CH_3SSCH_3 + 2\ NaOH \qquad 15.20$$

After oxidation of the weak black liquor, of about 15–18% total solids content by weight, it is concentrated. Concentration before combustion is necessary to increase the energy recovery possible on combustion of the organic constituents. Otherwise a large proportion of the energy derived from ultimate combustion is consumed to evaporate the high water content of the liquor. Evaporation to 50–55% solids is usually carried out in three to six stages of multiple-effect evaporators. By operating under reduced pressure, and by using the steam obtained from the first and each of the subsequent stages to heat the liquor of the following stage, 5 to 5.5 kg of water removal can be achieved for each kilogram of steam used for heating the first stage in a sextuple-effect evaporator.

The last stage of black liquor evaporation, from 50–55% solids to about 65–70% solids, is conducted by direct contact of the black liquor with hot flue gases from the recovery boiler (Fig. 15.8). Increased surface area for evaporation and increased evaporation rates are obtained by slowly rotating parallel disks which are half immersed in the liquor and half exposed to the hot flue gases. In this manner direct contact evaporation also accomplishes a moderate scrubbing effect on the flue gases, at the same time consuming the bulk of the residual sodium hydroxide (Eq. 15.21, 15.22).

$$2\ NaOH + CO_2 \rightarrow Na_2CO_3 + H_2O \qquad 15.21$$

$$2\ NaOH + SO_2 \rightarrow Na_2SO_3 + H_2O \qquad 15.22$$

This process also transfers heat from the flue gases to the black liquor so that it leaves the last stage of evaporation hot. At this point it is referred to as strong black liquor. Addition of make-up chemicals as either sodium sulfate ("salt cake") or sulfur as necessary is carried out at this stage to replace any losses of sodium and sulfur occurring elsewhere in the process. In fact, "sulfate process" is synonymous with "kraft process" because the primary make-up chemical requirement of the kraft process is sodium sulfate, although sodium sulfate is not actually an active constituent of kraft pulping liquor.

The objectives of black liquor combustion in the furnace of the recovery boiler are complex. These include combustion of the organic components present in the strong black liquor to generate heat and in turn to produce steam, and to recover the inorganic chemicals originally present in the black liquor, in <u>reduced</u> form, that is, as sodium sulfide rather than sodium sulfate. Preheated (about 120°C) strong black liquor is first sprayed into the hot, drying zone of the recovery furnace (Fig. 15.9). The residual water in the black liquor practically instantaneously flashes off, producing a char, part of which adheres to the walls of the furnace. As the thickness of the layers of carbonized char on the walls builds up, pieces fall off and land on the hearth. At the hearth conditions are chemically reducing, maintained by providing inadequate air

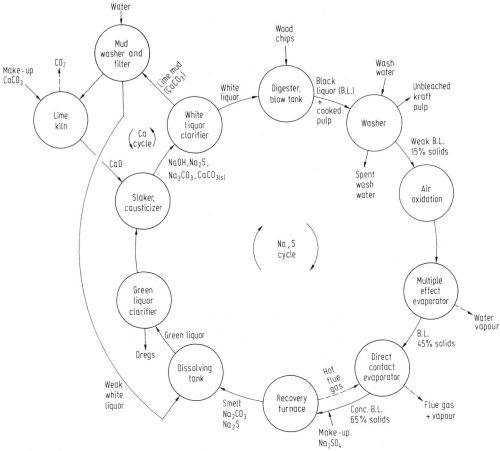

FIGURE 15.8 Cyclic process for recovery of chemicals and energy from the spent pulping liquor of the kraft process.

for total combustion in this zone, and temperatures of ca. 1000°C prevail, sufficiently high to provide the energy required for the endothermic nature of many of the reducing reactions. Here, sodium sulfate and sodium thiosulfate, produced at the black liquor oxidation stage, are reduced to sodium sulfide, and any remaining sodium hydroxide is converted to sodium carbonate (Eq. 15.23–15.28).

Examples of recovery furnace reducing zone reactions:

$$2\,C + O_2 \rightarrow 2\,CO \qquad\qquad 15.23$$

$$C + H_2O \rightarrow CO + H_2 \qquad\qquad 15.24$$

$$Na_2SO_4 + 4\,CO \rightarrow Na_2S + 4\,CO_2 \qquad\qquad 15.25$$

$$Na_2SO_4 + 2\,C \rightarrow Na_2S + 2\,CO_2 \qquad\qquad 15.26$$

$$Na_2S_2O_3 + 4\,CO + 2NaOH \rightarrow 2\,Na_2S + 4\,CO_2 + H_2O \qquad\qquad 15.27$$

$$Na_2CO_3 \rightarrow Na_2O + CO_2 \text{ (negligible, Chap. 7)} \qquad\qquad 15.28$$

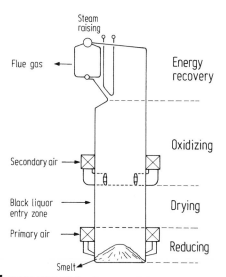

FIGURE 15.9 Diagram of the main components of a kraft recovery boiler, for strong black liquor combustion. (Reprinted from *Canadian Chemical Processing*, November 1970, courtesy of Southam Business Publications Ltd.)

The hearth is sloping so that as the sodium sulfide and sodium carbonate (melting points 950 and 851°C) accumulate in the char on the hearth the mixture, called smelt, trickles out. The smelt is immediately dissolved in lime mud wash water for the first stage in fresh liquor preparation. Dissolving of the molten smelt in water is a noisy procedure, not only from the large temperature differential between the smelt and the water but also from the exothermic nature of the solution process as well.

At the upper parts of the recovery furnace secondary air is admitted to obtain an oxidizing atmosphere, and the temperatures are still sufficiently high that any residual reduced compounds present in the gases are oxidized (Eq. 15.29–15.32).

Recovery furnace oxidizing zone reactions:

$$2\,CO + O_2 \rightarrow 2\,CO_2 \qquad\qquad 15.29$$

$$2\,H_2 + O_2 \rightarrow 2\,H_2O \text{ (minor)} \qquad\qquad 15.30$$

$$2\,H_2S + 3\,O_2 \rightarrow 2\,SO_2 + 2\,H_2O \text{ (traces)} \qquad\qquad 15.31$$

$$CH_3SH + 3\,O_2 \rightarrow SO_2 + CO_2 + 2\,H_2O \text{ (traces)} \qquad\qquad 15.32$$

As a result, gas temperatures rise to the neighborhood of 1250°C in this region of the furnace, from which heat recovery is practiced via steam superheating tubes and water tubes to generate high- and low-pressure steam. After heat recovery the flue gases are first passed through the direct contact evaporator, for additional heat utilization in the last stage of concentrating black liquor, and through electrostatic precipitators, for particulate removal, before delivery to the stack for discharge. The electrostatic precipitator catch is usually returned to the black liquor stream for recycling.

When smelt is dissolved in lime mud wash water the resulting solution is referred to as "green liquor," from the color imparted to it from suspended inorganic insoluble matter, or "dregs," present. The dregs, which consist mostly of carbon plus other inorganic insolubles such as iron, are settled out to leave a supernatant clarified green liquor which is essentially a solution of sodium carbonate and sodium sulfide in water.

For white liquor preparation, clarified green liquor is fed to a causticizer, which forms sodium carbonate on the addition of lime, leaving the sodium sulfide unchanged (Eq. 15.33).

$$Na_2CO_3 + Ca(OH)_2 + Na_2S \rightarrow 2\ NaOH + CaCO_3\downarrow + Na_2S \quad 15.33$$

The insoluble calcium carbonate (lime mud) is settled out from this mixture, filtered, and rinsed with water which is then subsequently used for dissolving smelt. The supernatant liquor is now a solution of sodium hydroxide, sodium sulfide, and a small amount of unreacted sodium carbonate of nearly the correct composition required for a new batch of white liquor. Calcium carbonate in the lime mud is recycled by first calcining to decarbonate, followed by slaking to regenerate the calcium hydroxide required for a subsequent causticization (Eq. 15.34, 15.35).

$$CaCO_3 \quad \underset{1200°C}{\rightarrow} \quad CaO + CO_2 \quad\quad 15.34$$

$$CaO + H_2O \rightarrow Ca(OH)_2 \quad\quad 15.35$$

Several important by-products may be derived and are recovered from many kraft pulping operations, among which are kraft turpentine, tall oil, and dimethyl sulfoxide. Kraft turpentine is condensed from digester relief gas while pulping coniferous species. Pine, for example, yields about 10 kg/tonne of pulp. Kraft turpentine contains a number of diterpenes such as α- and β-pinene, camphene, dipentene, etc., similar to the composition of the turpentine obtained from the dry distillation of pine wood [17] (e.g., naval stores industry). Crude kraft turpentine contains water, volatile acids, and reduced sulfides such as methyl mercaptan and hydrogen sulfide in addition to the terpenes. By careful washing and fractional distillation of this crude product several useful terpenes are recovered.

If tall oil is to be recovered, black liquor from softwood pulping is taken out at an intermediate stage of the multiple-effect evaporation when it contains about 30% total solids, and is allowed to stand [18]. Recovery is generally uneconomic from spruce or hardwoods. In 30% solids liquor the soaps (sodium salts of acids present) are insoluble, cream to the top of the vessel, and are skimmed off. The residual black liquor is returned to the evaporators. The soap yield, which can range from 10 to 200 kg/tonne of pulp (or even higher for pine), is then acidified and the free fatty acids and resin acids obtained are separated by distillation. The fatty acids recovered consist mainly of oleic and linoleic acids and are employed in soap manufacture and as the drying oil components of paints and varnishes [19] (Chap. 19). Resin acids consist of terpene acids such as abietic acid and its positional and reductive variants, and are mainly employed in paper sizing.

Dimethyl sulfoxide ((CH$_3$)$_2$SO; b.p. 189°C) is made from dimethyl sulfide recovered by condensation from kraft digester relief gases. This useful, polar, and water-miscible industrial solvent is produced by air oxidation of the sulfide in the presence of catalytic concentrations of nitrogen oxides.

15.6. BLEACHING OF WOOD PULPS

Mechanical and chemical pulps are normally more or less brown as initially obtained. The color is due in part to residual lignin or chemically modified lignin and miscellaneous color bodies present in the pulp. The objective of bleaching is to obtain the maximum increase in brightness possible consistent with the pulp end use, with a minimum yield loss and bleaching cost. Brightness is a term applied to the whiteness of a pulp, usually determined from a small paper sample or "handsheet" prepared manually from a pulp sample. Brightness is measured in a spectrophotometer at an angle of 45° to the incident beam of blue light of 457 nm. It is based on an arbitrary scale of 8 for the brightness of carbon black, and 100 for the brightness of pure magnesium oxide. On this basis, brightnesses of about 55 to 65 are common for unbleached mechanical pulps, which may be raised to the 65–75 range by bleaching. Unbleached sulfite pulps will have a brightness of 60–70, and kraft pulps of 40–45. Both may be bleached to brightnesses in the 80–85 range, the brightness of ordinary bond papers, with the appropriate number of bleaching stages. Very high brightnesses of 85 to 90 are achievable with chemical pulps using one or two additional bleaching stages. Because the bleaching procedures for the high lignin content mechanical and chemimechanical pulps differs significantly from the procedures used with chemical pulps which have a low to negligible lignin content, the two groups will be discussed separately. Neutral sulfite semichemical pulps are usually not bleached, since their uses do not require this.

15.6.1. Bleaching of Mechanical and Chemimechanical Pulps

Mechanical pulps, since they comprise the whole wood put into a pulp suspension in water, have approximately the same lignin content as the wood from which they were derived and an as-pulped brightness usually in the 50 to 65 range, depending on species pulped and process. Either a chemically oxidizing or a chemically reducing system, or sometimes both, is used with these pulps to gain a 10–15 point improvement in brightness. The bleaching methods used are selected to minimize attack of lignin, since the objective of these is to decolorize lignin without significant solubilization in order to maximize the yield of bleached pulp obtained. Brightening is the terminology used to reflect these objectives.

Oxidizing systems are based on peroxide, and may use solutions of either sodium peroxide or hydrogen peroxide in water, usually with added disodium ethylenediaminetetra-acetic acid (Na$_2$ EDTA) to suppress metal ion-catalyzed spontaneous decomposition of the peroxide. Capital P is the symbol used to designate a peroxide stage in bleaching. The active decolorizing agent is probably hydroperoxide anion, HOO^-, formed by an equilibrium between hydrogen peroxide and sodium hydroxide (Eq. 15.36, 15.37). It is also possible that peracetate is involved [20].

$$H_2O_2 + NaOH \rightarrow HOO^- + Na^+ + H_2O \qquad\qquad 15.36$$

$$H_2O + Na_2O_2 \rightarrow HOO^- + 2\,Na^+ + OH^- \qquad\qquad 15.37$$

Peroxide-induced lignin decolorization is achieved by hydroperoxide ion attack of the carbonyl carbon of the quinones and ketones which form a part of long conjugated chains (which contribute to the color) in the lignin. In this way the conjugation is interrupted, which reduces the intensity and wavelength of the contributing color [21] (e.g., Eq. 15.38, 15.39).

$$15.38$$

$$15.39$$

By this means peripheral, color-inducing functionalities are oxidized and decolorized over a bleaching period of 2–3 hours while the bulk of the lignin macromolecule remains intact (insoluble). Thus a brightness improvement of 8 to 10 points is possible and yet the yield loss is kept low. Unfortunately peroxide-based bleaching systems tend to be expensive relative to other alternatives for groundwoods, and hence this method is focused on pulps destined for tissue papers, paper napkins, and some specialty papers, rather than for newsprint applications.

Sodium bisulfite ($NaHSO_3$) was once used to some extent for reductive bleaching of mechanical pulps, but its action is slow and ineffective so that its use has been discontinued. Present reductive bleach systems for groundwoods employ either zinc or sodium hydrosulfite (or dithionites: ZnS_2O_4; or $Na_2S_2O_4$) dissolved in water. Zinc hydrosulfite for the former system, developed in the mid-1950s, was inexpensively made at the mill site by reacting powdered zinc with common pulp mill chemicals (Eq. 15.40).

$$Zn + 2\,H_2SO_3 \rightarrow ZnS_2O_4 + 2\,H_2O \qquad\qquad 15.40$$

This powerful reducing agent could produce brightness improvements of 10 to 12 points particularly if care is taken to exclude oxygen. Any oxygen present wastes a part of the active reducing capacity from the very rapid (probably only diffusion limited) reaction of zinc hydrosulfite with oxygen (Eq. 15.41).

$$2\ ZnS_2O_4 + 2\ O_2 + 2\ H_2O \rightarrow Zn(HSO_4)_2 + Zn(HSO_3)_2 \qquad 15.41$$

However, the disposal of spent groundwood bleach liquor from zinc hydrosulfite has posed zinc toxicity problems for shellfish [22], so that mills have now switched to sodium hydrosulfite. This is equally effective (although more expensive) and avoids this problem. Sodium hydrosulfite is made at the mill from purchased sodium borohydride (Eq. 15.42).

$$NaBH_4 + 8\ NaHSO_3 \rightarrow NaBO_2 + 4\ Na_2S_2O_4 + 6\ H_2O \qquad 15.42$$

If the two groundwood bleaching methods are to be used in sequence, a sulfur dioxide solution will be use as a stop bath for the pulp, after the peroxide stage, to ensure that any residual oxidants are kept out of the hydrosulfite stage. Combination bleaching will usually be conducted in this order, not the reverse, to obtain optimum results. Brightness improvements of 16 to 18 points are possible, better than with either bleach system alone but not quite an additive effect. The residual lignin present in these pulps makes the brightness unstable. This brightness instability may be observed by the significant darkening of a newspaper if it is left exposed to sunlight for an afternoon.

15.6.2. Bleaching of Chemical Pulps

High brightness papers are produced from chemical pulps which are cooked under conditions chosen to leave a low residual lignin content in the pulp. Bleaching of these pulps involves selective solubilization and washing out of lignin in the early stages, with some yield loss. This is followed by decolorization of any lignin residues present in the cellulose fiber in later stages, which results in a smaller yield loss. Thus, bleaching of chemical pulps results in a yield loss proportional to the number of bleaching stages.

The ease of bleaching of a chemical pulp is related to its initial brightness (color) and its lignin content. To establish this, a permanganate number (also kappa, or K-number) test, specified as the number of millileters of 0.10 N potassium permanganate which is decolorized by 1 g of dry pulp at 25°C in 5 minutes, is run on the pulp to determine its bleachability. Easy bleaching pulps give permanganate numbers of 6 to 10 or less, whereas pulps which give permanganate numbers of 20 or more are generally classified as unbleachable.

The chlorination, or *C stage* of bleaching is conducted at a pH of about 2 by a solution of chlorine in water. This relatively rapidly chlorinates and demethoxylates lignin via the action of both dissolved chlorine and of undissociated hypochlorous acid [23] (Eq. 15.43).

$$Cl_2 + H_2O \rightleftharpoons HCl + HOCl \qquad 15.43$$

A slower oxidation of lignin also occurs, serving in combination with the first two processes to yield more soluble, lower molecular weight fragments of lignin which may later be washed out of the pulp. Chlorination thus removes much of the lignin, though the pulp does not appear to be substantially brighter at this stage. Cellulose is not chlorinated under these conditions, although undissociated hypochlorous acid causes some oxidative degradation.

A caustic extraction, E, stage uses a solution of about 0.5–0.7% sodium hydroxide in water at 0–60°C to remove solubilized lignin produced by either the C or H (hypochlorite) stage. The high-pH conditions convert any phenolic hydroxyls to their sodium salts, raising their solubility, and convert any residual fatty acids and rosin acids present to salts, aiding in their removal by washing. Caustic extraction also removes some hemicellulose even though this component can be used for papermaking. Hemicellulose losses are minimized by keeping the caustic solutions dilute. Use of 1 to 5 M (4 to 20%) cold aqueous sodium hydroxide can maximize hemicellulose removal when required, such as for the preparation of dissolving pulps.

Hypochlorite (H stage) bleaching employs a solution of either calcium or sodium hypochlorite in water, made at the pulp mill by adding chlorine to a suspension of lime or sodium hydroxide in water (Eq. 15.44, 15.45).

$$2 \ Ca(OH)_2 + 2 \ Cl_2 \rightarrow Ca(OCl)_2 + CaCl_2 + 2 \ H_2O \qquad 15.44$$

$$2 \ NaOH + Cl_2 \rightarrow NaOCl + NaCl + H_2O \qquad 15.45$$

Excess base is used to ensure that the pH is kept at, or below 12, both to ensure that hypochlorite, OCl^- (and not hypochlorous acid, HOCl), is the bleach active species (see Fig. 15.10) and to stabilize the hypochlorite ion for this purpose. Both the C and H stages are designed to operate at pH's well clear of neutral to minimize the concentration of undissociated hypochlorous acid and attendant cellulose degradation. Hypochlorite ion is a less powerful oxidant than chlorine and hence more suitable for later bleach stages. Hypochlorite oxidizes and dissolves the chlorinated residues from the chlorination stage, the so-called "refractory material" remaining after the E stage. Some cellulose degradation is also experienced here.

A solution of chlorine dioxide, ClO_2, in water is the active agent in a D bleaching stage. Good bleaching action is obtained with less loss in pulp vis-

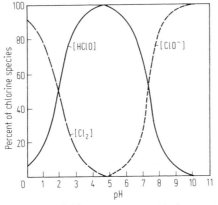

FIGURE 15.10   Variation with pH of proportions of chlorine, hypochlorous acid, and hypochlorite ion in 0.05 M chlorine water at 25°C. (Data from Nevell [23].)

cosity (i.e., less cellulose degradation) than the other procedures just described. Even solutions of chlorine dioxide are unstable so that this is usually made as needed, right at the mill site. It is obtained by treating an aqueous solution of sodium chlorate with sulfur dioxide (Eq. 15.46).

$$2 \text{ NaClO}_3 + \text{SO}_2 \rightarrow 2 \text{ ClO}_2 + \text{Na}_2\text{SO}_4 \qquad 15.46$$

The by-product sodium sulfate can provide chemical make-up for a kraft pulp mill. Chlorine dioxide is a milder oxidizing agent than hypochlorite and may be employed either on its own, as a separate bleach stage, or in combination with chlorine to replace a part of the stronger acting chlorine [24]. Chlorine dioxide may be economically produced at the same time as chlorine by employing sodium chloride or hydrochloric acid in the generators (Eq. 15.47, 15.48).

$$2 \text{ H}_2\text{SO}_4 + 2 \text{ NaClO}_3 + 2 \text{ NaCl} \rightarrow 2 \text{ ClO}_2$$
$$+ \text{ Cl}_2 + 2 \text{ Na}_2\text{SO}_4 + 2 \text{ H}_2\text{O} \qquad 15.47$$

$$4 \text{ HCl} + 2 \text{ NaClO}_3 \rightarrow 2 \text{ ClO}_2 + \text{Cl}_2$$
$$+ 2 \text{ NaCl (recycled)} + 2 \text{ H}_2\text{O} \qquad 15.48$$

Combination usage both gives pulps of higher strength than obtained using chlorine alone in the first stage, and produces much less chlorinated material in and greatly facilitates recovery of bleach plant effluents [25].

A peroxide stage, *(P stage)* is also occasionally used as a polishing bleach for chemical pulps, using the conditions discussed. Peroxide has also been tested for initial bleaching of softwood kraft pulps after an acid pretreatment. A significant reduction in waste loads is obtained but at high cost [26].

Oxygen bleaching, an *O stage*, has also become a part of the bleach sequence of many pulp mills. Delignification and decolorization are effected by treating the chemical pulp at medium to high consistencies (10–15%, up to about 30% pulp by weight) in a dilute solution of sodium hydroxide, with up to 5 atm pressure of oxygen [27]. Addition of a small amount of a magnesium or manganese salt serves to stabilize the cellulose against severe degradation. One of the principal motivations to use for an oxygen bleaching step are that it can replace a C E or C E H combination, and produces a decreased waste load of lower toxicity. It does, however, require rather different, moderate pressure bleach plant equipment to put into practice.

Easy bleaching (low permanganate number) sulfite pulps can be bleached to a brightness of 80 or so with a single H stage, and to 85 or higher with a C E H sequence. Three to five bleaching stages, e.g., C E H, C E H D, C E H E D sequences, are required for the sulfite pulps that are harder to bleach.

To contrast with sulfite pulps, dark kraft pulps can only be bleached to the 40 to 50 brightness range with a single hypochlorite stage. For semibleached grades of brightness, the 70 to 75 range, at least three or four bleach stages, e.g., C E D, C E H D, are required. A typical progression of pulp, chemicals, and water for a semibleached kraft pulp is given in Fig. 15.11. To obtain high or "super" brightness kraft pulps of 85–90 brightness levels, five to seven stages of bleaching are required. Sequences such as C E D E D, C E H D E D, C E H E D P, and C E H E D E D P are commonly used, which

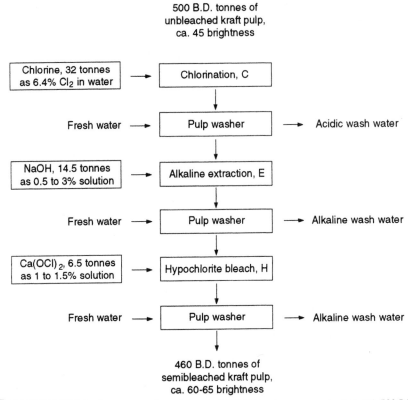

500 B.D. tonnes of
unbleached kraft pulp,
ca. 45 brightness

| Chlorine, 32 tonnes as 6.4% Cl_2 in water | → | Chlorination, C |

Fresh water → | Pulp washer | → Acidic wash water

| NaOH, 14.5 tonnes as 0.5 to 3% solution | → | Alkaline extraction, E |

Fresh water → | Pulp washer | → Alkaline wash water

| $Ca(OCl)_2$, 6.5 tonnes as 1 to 1.5% solution | → | Hypochlorite bleach, H |

Fresh water → | Pulp washer | → Alkaline wash water

460 B.D. tonnes of
semibleached kraft pulp,
ca. 60-65 brightness

FIGURE 15.11   Progression through a typical C E H bleach sequence starting with 500 B.D. (bone dry) tonnes of unbleached kraft pulp.

requires a period of 18 to 24 hours for passage of the pulp through the bleach plant.

15.7. MARKET PULP AND PAPERMAKING

Bleached or unbleached chemical pulps may progress along either of two paths. They may be fed as a dilute water suspension onto a fairly coarse wire mesh screen to form a thick pulp mat, which is then dried to about 10% moisture (roughly in equilibrium with atmospheric moisture). The mat is then cut into sheets of about 0.6 × 1.2 m, wrapped, and baled for shipment as 232 kg (500 lb) units of market pulp. Or they may be fed in diluted form directly onto the Fourdrinier wire of a paper machine, with the intention of producing directly one of the wide variety of trade papers. The sheets produced for market pulp are deliberately made thick and only loosely consolidated to facilitate repulping by stirring with water, at the destination point, for eventual papermaking. Mechanical pulps are much less color stable and would not as readily take to repulping without a loss in sheet properties. For these reasons they are nearly always converted to the finished product, newsprint or paper "boards" of various types, at the pulping site.

With any pulp source, eventual papermaking requires the appropriate single or a blend of pulps to be prepared into a furnish, which is a dilute pulp suspension (about 2%) in water ready to be fed to a paper machine. Before papermaking, chemical pulps may be beaten (mechanically worked in units similar to refiners) to a varying extent depending on the properties desired in the finished sheet. A typical furnish composition for a newsprint might be a blend of 60–70% groundwood (or mechanical) pulp, plus 30–40% sulfite, semibleached, or bleached kraft pulps to provide strength. Bag papers and board, such as linerboard for corrugated board manufacture would normally use 100% unbleached kraft pulp, hence the term "kraft paper." Board destined for use as corrugating medium would normally use a higher yield stiff pulp fiber such as that produced by the NSSC process. Bond or book papers would use fully bleached sulfite, rag, or kraft pulps, or blends of these.

The furnish is fed to the headbox of a paper machine (Fig. 15.12), in length about a city block. Sufficient additional water (white water) is added to bring the pulp concentration in the headbox down to about 0.5%. White water, the water removed from the fibers as the paper sheet is formed, is continuously recycled. This low concentration of pulp in water is necessary to obtain smooth formation or evenness of fiber distribution as sheet formation occurs. The diluted furnish is fed through an adjustable horizontal slot (the slice) in the headbox onto an endless wire screen (the Fourdrinier wire) which moves rapidly away from the headbox. Direct drainage plus suction boxes draw off more than 90% of the water while the sheet is still on the wire, enough that the delicate 20% solids sheet may be passed on to the press

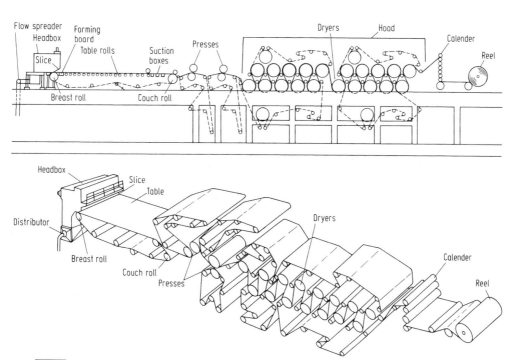

FIGURE 15.12 Principal components of a Fourdrinier paper machine. (From Kirk-Othmer [4], courtesy of Beloit Corporation.)

section of the paper machine, where the sheet is now supported on felts. By pressing the damp sheet into the drying felts the pulp content is nearly doubled, giving the sheet enough strength to carry it through the drying section. Here, 30 to 50 steam heated rolls rapidly evaporate the remaining water to give a sheet containing only 5 to 8% moisture as it emerges from the dryer. The last function of the paper machine carried out on the still warm paper sheet is a smoothing one, performed as the sheet passes through the calendar stack. Here, heavy smooth steel rollers serve to decrease surface roughness and eliminate as much as possible any residual "sidedness" of the sheet that occurred as it formed on the Fourdrinier wire.

Any adjustment of the final paper roll width, from the 7 to 10 m width produced by the typical paper machine, is carried out as a separate step on a slitter-rewinder, using the 20 tonne rolls of paper produced by the paper machine. The final step is a heavy paper wrapping which is applied to protect the contents from physical damage during shipment and to control the uniformity of the moisture content of the product.

Paper is an extraordinarily interesting commodity because of the wide variations possible in all its properties. While superficially it appears to be uniform, it is actually significantly anisotropic. Even the best papers are usually sided, that is they have a top side which is smoother and a wire side which is rougher. A newsprint will have a higher proportion of chemical fibers on the wire side than the top side of the sheet. The sheet will have a machine direction and cross direction to it, an anisotropy feature which is under some control by the paper machine operator. More of the fiber will be aligned close to the machine direction than the cross direction. For this reason tensile and folding strengths are greater in the machine direction than in the cross direction. This property is also important in printing, for instance, since books lie better with the pages having the machine direction up and down the page rather than across the page. There are many other properties under the control of the papermaker such as density, thickness, porosity, smoothness, stiffness, strength, etc., as well as specialized measurements which are used to quantify each of these properties which are beyond the scope of the present discussion.

15.8. PULPING EMISSION CONTROL MEASURES

Each of the pulping methods—mechanical, chemimechanical (presoak), semichemical, and the kraft and sulfite chemical processes—produce different types of wastes, so that emission control measures for each should be considered separately. Since the mechanical and kraft chemically based pulping methods dominate the world markets (Table 15.1) the discussion will focus on these two processes. Resource use and pollution control problems of the pulp and paper industry have been surveyed [28], and the details compared with those of competitive materials [29,30].

Wastes generated from wood preparation for pulping do not generally pose a major problem. Hydraulic debarker effluent amounting to some 14 m³/ tonne (3700 U.S. gal/tonne) is relatively clean, containing only about 10 ppm suspended solids as wood fines and grit and about 5 mg/L BOD (biochemical oxygen demand). Primary clarification is required to decrease the suspended solids, while the BOD level is regarded as inconsequential. Toxic rosin acids flushed from the bark/wood interface of some species are handled by dilution with other mill effluents. Many mills use drum, or in-line dry debarking systems, particularly when there is a limited water supply. The bark itself is normally burned in a bark-fueled (hog fuel) boiler with energy recovery as steam.

15.8.1. Effluents of Mechanical Pulping

The nature of mechanical pulping processes means that they do not directly present any air emission problems. Problems are, however, possible at the point of electric power generation, if this is from combustion sources. Pulp mills often use on-site thermal power generation. Control of these potential emission sources is discussed later.

There is, however, a high water requirement, estimated to be about 21 m^3 per tonne (5500 U.S. gal/tonne) of newsprint. This is used for log or block transport, wet grinding operations, and sheet preparation on the paper machine itself. The types and approximate levels of impurities present in the waste water stream are given in Table 15.9.

The philosophy of pulp mill emission control, particularly with the valuable fiber constituents which are sometimes present in waste streams, is to recover as much fiber as possible before employing other treatment measures. Recovery is obtained by sedimentation and/or flotation methods. Sedimentation may be conducted in a pond, which is periodically drained and the settled material removed. Or a clarifier (which operates in a similar manner to a thickener) may be used. The sediment collected contains 90–95% water plus fiber and grit solids. For disposal the waste is dewatered to 50–60% solids in a filter or centrifuge and then burned [31], which accomplishes both disposal and energy recovery.

Flotation methods allow recovery of fibers from water which are raised to the surface on bubbles as a thick mat which forms on the top of an aeration cell. This mat, which is virtually free of grit, is recycled to the pulp working

TABLE 15.9 Typical Characteristics of the Wastewater Produced by a Groundwood Pulping Operation[a]

Component	Value (ppm)
Total solids	1160
Suspended solids	600
Ash of above	60
Dissolved solids	560
Ash of above	240
BOD, 5 day	250

[a]Data selected from Nemerow [28].

stream. Slime growth in the aeration cell is discouraged by strong alternation of conditions such as pH, temperature, or chlorination level. Organomercury slime control agents were once used for this purpose.

After grit removal and fiber recovery, wastewater streams from mechanical pulping are treated in aeration lagoons to decrease the biochemical oxygen demand (BOD). A 40–75% decrease of the BOD in 4 days and a 90% or better decrease in 7 days can be achieved using a surface to volume ratio of about 1 hectare (1–2 acres) per million gallons per day of wastewater flow [32]. Floating mechanical aerators, rather than activated sludge treatment, are the optimum solution for BOD reduction. The large volume and good mixing provided by this system are tolerant of the shock loadings, and phosphate and nitrogen nutrient requirements are less than alternatives. Activated sludge treatment has been examined for thermomechanical pulping effluents [33].

15.8.2. Kraft Pulping Air Pollution Control

Air pollution problems center around control of the formation and discharge of reduced sulfur compounds, which cause severe odor problems (Table 15.10), and loss of particulates.

Reduced sulfur compounds, such as methyl mercaptan and dimethylsulfide, arise from the action of hydrosulfide and methyl sulfide anions on the methoxyl groups present in the lignin of wood (Eq. 15.49–15.51).

$$HS^- + lignin\text{–}O\text{–}CH_3 \rightarrow lignin\text{–}O^- + CH_3SH \qquad 15.49$$

$$CH_3SH + NaOH \rightarrow CH_3S^- + Na^+ + H_2O \qquad 15.50$$

$$CH_3S^- + lignin\text{–}O\text{–}CH_3 \rightarrow lignin\text{–}O^- + CH_3SCH_3 \qquad 15.51$$

In this way, about 3% of the total sulfide loading to the digester of some 80–100 kg per air dry tonne of pulp is converted to methyl mercaptan and dimethylsulfide. High sulfidities, high pulping temperatures, long cooks, the time, or pulping of hardwoods are all factors which tend to produce proportionately more of these sulfides [34]. The much higher methoxyl content of

---

■ **TABLE 15.10   Reduced Sulfides Contributing to the Odor Control Problems in Kraft Pulping[a]**

Compound	Boiling point (°C)	Henry's law constant at 38°C[b]	Odor threshold (ppb)	Industrial hygiene requirement (ppm)
Hydrogen sulfide, H_2S	−62	700	5	10
Methyl mercaptan, CH_3SH (methane thiol)	6	180	5	10
Dimethylsulfide, CH_3SCH_3	38	50	ca. 10	—
Dimethyldisulfide, CH_3SSCH_3	116	30	ca. 100	—

[a]Compiled from Sarkanen et al. [34], Karnovski [35], and Cederlov et al. [36].
[b]Equilibrium mole fraction in the gas phase divided by the mole fraction in a water phase at 38°C. If base is present the equilibria of the first two compounds will affect this ratio. H_2S, $k_1 = 2.1 \times 10^{-7}$, $k_2 = <10^{-14}$; CH_3SH, $k = 4.3 \times 10^{-11}$ for aqueous solutions at 100°C.

hardwood lignins nearly doubles the amount of volatile sulfides generated relative to that expected from softwoods. About 6 to 8% of the residual sulfide in the liquor becomes bound to lignin, mostly in soluble but involatile forms. The residue is converted to various forms of oxidized, dissolved sulfur.

Escape of the volatile sulfides can occur when excess digester pressure is vented, which occurs at intervals during the digestion, at the blow, when the contents of the digester are released to the blow tank, and during the last stage of black liquor evaporation in the direct contact evaporator, when there is rapid passage of a gas stream past the hot concentrated liquor. Details are given in Table 15.11. Control measures focus on sulfide containment from the major loss points, the direct contact evaporator and digester relief and blow gases. Black liquor oxidation (Fig. 15.8), decreases the volatile sulfide losses on direct contact evaporation. For optimum benefit the time interval between oxidation and evaporation should be kept short [37]. Otherwise sugar-based reductants still present in the liquor cause reformation of volatile sulfides on standing. Indirect black liquor evaporation for all stages has been advocated by some Swedish mills.

For control of digester relief and blow gases, they are chilled, which removes a "foul condensate" and the uncondensed gas fraction is moved on to a gas accumulator, which smooths out the variable gas flow rates. The accumulator is vented at a steady rate through a shower of weak black liquor which has some reserve alkalinity. This captures at least hydrogen sulfide and methyl mercaptan. Any residual dimethylsulfide and dimethyldisulfide which are not captured are destroyed by venting these into the combustion air stream of the lime kiln. Novel scrubbing techniques have been described [38].

Sulfide emissions may of course also be bypassed altogether by a change of the pulping process to avoid the use of sulfide [39]. Both the old soda process, and nitric acid pulping suffer from the inferior strength properties of the pulps produced. Use of anthraquinone with basically a soda pulping procedure could put this alternative on a more competitive basis [40].

TABLE 15.11 Total Reduced Sulfide Emissions from the Chemical Recovery Operations of a Kraft Pulp Mill, in kg S per Air Dry Tonne of Pulp Produced[a]

	Total reduced sulfide emissions (kg S/tonne)	
Source	No controls	With controls
Direct contact evaporator	7 −10	0.05–1
Recovery furnace	0.1 − 1	0.05–1
Digester and evaporators, noncondensible gases	1.0 − 1.5	0.0 −0.3
Black liquor oxidation	0.05− 0.2	0.05–0.2
Pulp washer hoods	0.05− 0.2	0.05–0.1
Dissolving tank vents	0.05− 0.2	0.05–0.1
Lime kilns	0.05− 1	0.05–0.1
Total	8.3 −14.1	0.03–2.8

[a]Compiled from data of Sarkanen *et al.* [34].

■ **TABLE 15.12 Particulate Emission Loads from the Chemical Recovery Operations of a Kraft Pulp Mill, per Air Dry Tonne of Pulp Produced[a]**

Potential source	Particulate composition	Particulate discharge (kg/tonne pulp)	
		No controls	With controls
Recovery furnace	Na_2SO_4, Na_2CO_3, $NaCl$,[b] etc.	100–200	1–10
Lime kiln	CaO dust	10–20	0.4–2
Dissolving tank vent	Na_2S, Na_2CO_3, Na_2SO_4, $NaCl$[b]	3–4	0.4–1
Totals, kg/tonne pulp		113–224	1.8–13

[a]Air dry tonne is defined as pulp with 10% moisture content, which is below the normal equilibrium moisture content of pulp exposed to ambient air.
[b]Salt build-up will affect coastal mills which use tidewater transport and storage of logs, particularly when good particulate emission control is in place (see text).

Particulate emissions can amount to as much as 0.25 tonne per tonne of pulp produced, particularly for an older pulp mill operating without controls (Table 15.12). Electrostatic precipitators, wet scrubbers, or occasionally both can achieve particulate containment of 95% or better [39]. The precipitator catch is returned to the black liquor stream for chemical recovery from the captured inorganic chemicals. Fumes lost from the dissolving tank vent are being captured by demister pads or small low-energy scrubbers which return the collected material to the green liquor circuit. Wet scrubbers are used for lime kiln dust containment. The waste liquor is used for slaking or other recycle functions in the lime circuit (Chap. 7).

15.8.3. Kraft Pulping Water Pollution Control

The principal wastewater producing sectors of a large fully bleached kraft mill are summarized in Table 15.13 [41] which gives ranges of values to cover normal operating variations as well as to cover a range of ages of pulp mills. Hot wastewater streams are rarely a problem because of blending with other wastewater streams. The total volumes of water consumed are so large, 20,000 to 30,000 U.S. gallons per tonne of unbleached kraft pulp and 40,000 to 55,000 U.S. gallons per tonne for fully bleached kraft pulp, that the blended wastewater stream is negligibly higher in temperature than the source water.

Water quality parameters which may be affected by kraft process effluents include wastewater streams which are high in suspended and dissolved solids, highly acidic or alkaline, and may have a high oxygen demand, or BOD. Other parameters include intensely colored streams which have a foaming tendency and streams that are toxic to fish, all effects from dissolved constituents. Useful reviews of wastewater treatment for the pulp and paper industry have been published [42]. Keep in mind that the chemical recovery part of kraft pulping has already served to decrease potential effluent loadings of about 1000 kg of BOD and 300 kg of dissolved inorganic chemicals per tonne of pulp which

TABLE 15.13 Aqueous Effluent Characteristics and Volumes per Tonne of Pulp (TP) from a Fully Bleached Kraft Pulp Mill Producing Market Pulp[a]

Water quality parameter	Waste stream source						
	Debarking	Pulping screening	Chemical recovery	Causticizing	Bleaching stages		Paper machine
					1st E	All other	
Volume,[b] m^3/TP	7.5–15.0	22.7–45.4	3.8–13.2	1.9–7.5	19.0–38.0	56.8–94.6	1.9–5.7
Suspended solids, kg/TP	10–15	5–15	0.5–1	1–10	1–2	1–3	0.5–1
Composition	fiber, grit	fiber		lime	fiber		
pH	6–8	6–9	6–9	8–10	9–11	2–3	5–6
BOD_5, kg/TP	3–4	6–15	1–2	1–2	5–10	10–12	0.5
Color, APHA units	300–700	1,000–1,500	200–400	400–600	20,000–30,000	1,000–1,500	low
Foam[c]	occasional	occasional	occasional	none[d]	yes, low	yes, high	none
Toxicity, acute bioassay	some	2nd most	none[d]	none[d]	most toxic	4th most	none

[a]Compiled from personal contacts and data from Nemerow [28], Bruley [32], and Lund [41].
[b]To convert to square meters per tonne of pulp, multiply these entries by 3.785×10^{-3}.
[c]Low and high refer to the relative ease of a foam being generated with these particular effluents.
[d]Effluent streams marked as having no toxicity are only nontoxic if no condensates are added to these streams. Condensate addition can contribute several toxic aromatic compounds to the stream.

would otherwise have to be treated. This is far greater than the total pollutant loadings of Table 15.13.

The need for treatment depends on, first, the degree to which the water quality parameters of the effluent stream differ from those of the receiving body of water and, second, on the gross effluent volume of any stream which does show a significant difference of one or more parameters. A low-volume stream which shows a significant deviation from normal in one or more water quality parameters may be adequately diluted on blending with 50,000 gallons of process water to no longer pose a problem. However, it is frequently easier to devise a treatment measure to improve the quality of differentiated streams than of a blended one.

Using the criteria outlined above the main problem streams from kraft pulping are those from the hydraulic debarker (if used), the pulping and screening area, the causticization section, and both bleach plant streams. Chemical recovery process wastewaters are relatively benign, as long as relief gas or evaporator foul condensates are kept separated from them [43]. Paper or pulp machine waters are quite clean, as long as specialty papers involving use of other chemicals, dyes, pigments, or fillers are not being produced.

A clarifier can be used to remove 90–95% of settleable material from waste streams containing suspended solids. The sludges obtained are dewatered by filtration or centrifugation to 40–50% solids or higher and then burned. In combustible sludges from the lime circuit clarifiers can be filtered and recycled to the lime kiln for burning.

The pH deviations from normal can be corrected by blending streams of opposing pH's as far as possible, *after* other required treatment steps. Final adjustment may be necessary through reagent addition, using 90+% sulfuric acid for correction of excess alkalinity, and lime or 50% sodium hydroxide for correction of excess acidity. An automated operating system has been described [41].

Reduction of the biochemical oxygen demand (BOD) is desirable, particularly for waste streams where the loading is 3 to 4 kg/tonne of pulp, or more. In concentration terms, BODs of 75 to 300 ppm (mg O_2/L) are common in kraft pulping effluents [28]. Biochemical waste treatment in artificially aerated waste lagoons, after adjustment of the water conditions to a pH in the 6.5–9 range, the temperature to 12–32°C, and addition of ammonium phosphate to provide nitrogen and phosphorus nutrients, accomplishes accelerated BOD reduction, as outlined in more detail earlier. Use of oxygen, rather than air, for treatment has also been suggested (e.g., Unox) [44].

The intense brown color of the effluent from the first caustic extraction or E stage, may be dealt with in a number of ways [45]. Addition of lime $(Ca(OH)_2)$ can decrease the intensity of the color by precipitating a lime/organics complex [46]. However, the mixed organic-inorganic sludge poses disposal problems. Activated carbon has also been used for color adsorption, but carbon regeneration is difficult and incomplete. It is, however, possible to operate using mill generated carbon [47]. Soil infiltration has been found to be effective [48], but its applicability depends on the soil type and land available at the mill site. Reverse osmosis and ozone treatment have also been suggested [49,50].

The toxicity of effluent may be from sulfides originating from digester relief gas or black liquor evaporation, from unsaturated fatty acids released from wood during pulping [51], or from chlorinated compounds produced in bleaching steps [52]. The sulfides from foul condensates may be detoxified by air or steam stripping of the condensate, plus capture of the volatiles released by capture in a black liquor scrubber, or combustion in the recovery boiler. Normal biodegradation is sufficient to deal with the toxicants released from wood, but chlorinated phenols are more resistant to this process [53]. The foaming tendencies of these effluents may be used to generate a head of foam in which the toxicants are concentrated, analogous to froth flotation for mineral separations. Separation of the foam removes these toxic compounds [54]. Use of lime addition for color removal also decreases toxicities via adsorption of these compounds onto the sludges as they separate. Toxic constituents may also be removed by adsorption onto activated carbon.

Another approach is to change the pulping technology. Use of oxygen, air pulping [55], and/or bleaching could avoid many of the problems outlined [56]. It has been found possible to recycle bleach plant effluents as well as spent liquor from pulping, in the process regaining the chemicals required for pulping and bleaching [25].

Waste streams with a tendency to produce foam can be controlled by adding a commercial antifoaming agent but many of these are also toxic, which could upset the operation of a biobasin or contribute a toxicant to the effluent. Foams can be generated in a treatment plant, then skimmed and burned simultaneously destroying foam and toxicants (Fig. 15.13). Mills operating in cold continental climates find that a 0.5- 1-m head of foam on a biopond can actually be an aid to maintaining optimum pond temperatures for continued BOD reduction under winter operating conditions. Without this assistance, floating plastic foam mats may be used to keep pond temperatures high enough to maintain biochemical and biological activity.

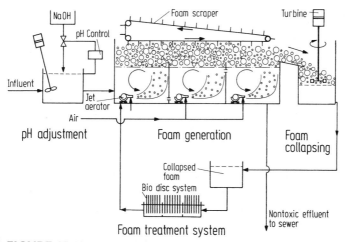

FIGURE 15.13 Proposed scheme for a foam production and separation device for the removal of both toxicants and foamable constituents from kraft process wastewater streams. (From Leach et al. [53], reprinted courtesy of Wheatlands Journals.)

15.9. ENVIRONMENTAL ASPECTS OF PAPERMAKING AND PAPER RECYCLING

Wastewater streams from paper machines while operating entirely on wood pulp do not pose serious waste loads on discharge. But when filled, coated, sized, or colored papers are being produced additives accumulated in the white water have to be substantially removed before water discharge. The volume of wastewater from this source that is required to be treated is minimized by operating the paper machine using 90% or higher water recycle. This measure also achieves a substantial direct saving of fillers, dyes, or the expensive titanium dioxide-based coatings by capture of these constituents into the paper sheet as it is formed. Treatment measures for the rejected fraction of recycled water, necessary to avoid buildup of resinous components, dissolved solids, or the establishment of microbiological growths, involve settling for waste streams containing primarily suspended solids, as already described. Color may be removed by activated carbon adsorption or one of the other color removal techniques already described under wastewater emission control.

Fiber reuse by repulping of printers offcuts, computer and office paper wastes, as well as the paper component of municipal garbage is also an excellent way to decrease the overall impact of papermaking operations. Wood requirements are greatly reduced, though not eliminated, since it is usual to blend virgin fiber with recycle stock. Also energy and chemical utilization are greatly reduced as are the waste loadings to air or water [57]. However, recycling does not eliminate external impacts since significant waste loads to water are generated from the repulping, de-inking, washing, and simple bleach stages required [58].

Waste paper stock destined for recycling is normally segregated into grades according to whether the original pulp was produced by chemical or mechanical methods. The separate stocks are then repulped (dispersed) in a hot dilute solution of sodium hydroxide, sodium phosphate, or sodium silicate in water, and cleaned via a series of raggers, cyclones, and screens. Standard printer's inks are removed by thorough dispersal and washing [59]. For the recently introduced thermoplastic inks used by laser printers, and some fax machines and photocopiers that use the same printing system, the polymeric pigments are present in the repulped stock as larger, hydrophobic particles, so are not removed by conventional wash de-inking. Froth flotation techniques are used to remove of the ink from these pulps via a pigment-rich froth.

After a further screening and wash, the stock is bleached with one or two stages of sodium peroxide, hypochlorite, or a chlorine dioxide-rich mix of chlorine and chlorine dioxide. On completion of these steps there is perhaps a 15–20% mass loss in reprocessing of chemical pulps and a 20–25% loss from groundwood pulps [60]. The product pulp may be incorporated into pulps prepared from new fiber, at varying proportions, to produce a sheet with little difference in final properties from a sheet produced entirely from new fiber [61]. Fine papers and newsprint may be made from the cleaner grades of recycle fiber; liner boards, corrugating medium, and asphalt shingle stock are prepared from the less clean grades.

REVIEW QUESTIONS

1. Biosynthesis of cellulose involves photosynthetic conversion of carbon dioxide to glucose, followed by dehydropolymerization of β-D-glucose to cellulose (polymerization accompanied by loss of water). What mass of carbon dioxide would theoretically be utilized by a tree to produce 1 tonne (1000 kg) of cellulose?

2. How do starch and cellulose differ stereochemically and functionally in the plant from one another, and why are these distinctions important in the utilization of cellulose rather than starch for papermaking?

3. (a) What weight of sodium hydroxide would be required to produce 10,000 liters of solution containing 53 grams of sodium carbonate per liter, after contacting with flue gas (for CO_2 content)?
   (b) How much sulfur would have to be burned to provide the sulfur dioxide for absorption into 10,000 liters of aqueous sodium carbonate (53 g/L) to produce a solution containing equal molar concentrations of sodium bicarbonate ($NaHCO_3$) and sodium sulfite (Na_2SO_3)? Assume no volume change on absorption.

4. A typical analysis for kraft white liquor for pulping is given below in terms of the sodium oxide equivalence. This puts all the dissolved components on the same sodium ion content basis. Using this information, calculate the actual molarities present of the three dissolved components in this sample of kraft white liquor.

	Na_2O equivalent content (g/L)
NaOH	73
Na_2S	31
Na_2CO_3	18

5. Clarified green liquor from the chemical recovery circuit of a kraft pulp mill has a density of 1.21 g/mL at 20°C and contains 20% sodium carbonate and 5% sodium sulfide, both specified by weight. On causticization with a 10% excess of lime ($Ca(OH)2$) 90% of the sodium carbonate is converted to sodium hydroxide.
   (a) What would be the percent <u>and</u> the molar concentrations of sodium hydroxide, sodium sulfide, and sodium carbonate in the clarified white liquor obtained? Assume no volume or solution density changes.
   (b) What would be the "effective alkali" concentration, as g/L of Na_2O equivalent, in this white liquor?
   (c) What would be the sulfidity of this solution?
   (d) Does this white liquor require any adjustment (make-up) in terms of concentration and/or chemical composition to be suitable for chip digesting in the Kamyr kraft method of pulping? If so, specify details.

6. The objective for emission of bivalent sulfur compounds (H_2S, CH_3SH, $(CH_3)_2S$, etc.) from an existing kraft pulp mill in British Columbia has been set on the basis of 3.74 lb of sulfur equivalent per ton of air dry pulp produced per day.

(a) If this limit was reached solely from methyl mercaptan (CH_3SH) for a mill producing 600 tons of kraft pulp per day, what mass of CH_3SH could be expected?

(b) What volume of air, in cubic meters, would be required to dilute the mass discharged to the point where it would be undetectable by the nose (0°C, 760 mm Hg)? The olfactory threshold of methyl mercaptan is 40 ppb by volume.

(c) Briefly outline any three control measures that could be taken to minimize emission of reduced sulfides from a kraft pulping operation.

7. (a) About 6.4% by weight chlorine in water is used in the first stage of bleaching of kraft pulps. What initial percent by weight and molarity of chlorine and hypochlorous acid would result if half the added chlorine reacted with water and half remained as dissolved chlorine?

(b) What masses (kg) of calcium hydroxide and chlorine would theoretically be required to make up 10,000 L of solution, 1.30% by weight in calcium hypochlorite, suitable for hypochlorite bleaching of pulp?

(c) What would be the molarity of 1.30% by weight calcium hypochlorite?

8. Assume complete biodegradation of a tonne of waste paper in a landfill according to the following equation:

$$(C_6H_{10}O_5)_n \xrightarrow[\text{H}_2\text{O, hydrolases}]{\text{Hydrolysis,}} n\ C_6H_{12}O_6 \xrightarrow[\text{methanogenic bacteria}]{\text{Anaerobic decomposition}} 3n\ CH_4 + 3n\ CO_2$$

What would be the masses of glucose, methane, and carbon dioxide produced if 100% hydrolysis occurred, followed by 90% anaerobic decomposition?

FURTHER READING

C.J. Biermann, ed., "Handbook of Pulping and Papermaking," 2nd ed., Academic Press, San Diego, 1996.

J.P. Casey, ed., "Pulp and Paper: Chemistry and Chemical Technology," 3rd ed., Vols. 1–4. Wiley, New York, 1979–1981.

W. Flaig, Slow Releasing Nitrogen Fertilizer from the Waste Product, Lignin Sulphonates. *Chem. Ind. (London),* June 16, p. 553 (1973).

J.J. Garceau, S.N. Lo, and L. Marchildon, A Cost Evaluation of Alternative Sulphite Spent Liquor Strategies. *Pulp Pap. Can.* 77(10), T174 (1976).

"Kirk-Othmer Encyclopedia of Chemical Technology," 3rd ed., Vol. 19. Wiley, New York, 1982.

W.F. Sinclair, "Controlling Pollution from Canadian Pulp and Paper Manufacturers: A Federal Perspective." Environment Canada, Ottawa, 1990.

REFERENCES

1. "Reference Tables," 35th ed., Canadian Pulp and Paper Association, Montreal, 1981.
2. "United Nations Statistical Yearbook 1993," 40th ed. United Nations, New York, 1995, and earlier editions.
3. R.H. Clapperton, "The Paper-making Machine." Pergamon, Toronto, 1967.
4. "Kirk-Othmer Encyclopedia of Chemical Technology," 2nd ed., Vol. 16, p. 680. Wiley, New York, 1968.
5. H.I. Bolker, "Natural and Synthetic Polymers: An Introduction." Dekker, New York, 1974.
6. K.V. Sarkanen and C.H. Ludwig, eds., "Lignins, Occurrence, Formation, Structure, Reactions." Wiley-Interscience, Toronto, 1971.
7. D.J. MacLaurin and A.M. Van Allen, *Pap. Trade J.* Nov. 25, p. 1 (1948).
8. K.W. Britt, ed., "Handbook of Pulp and Paper Technology," 2nd ed. Van Nostrand-Reinhold, New York, 1970.
9. J.C. Gauss and D.J. Wachowiak, *TAPPI J.* **62**(9), 27, Sept. (1979).
10. J.A. Wright, M.J. Sabourin, and W.S. Dvorak, Laboratory Results of TMP and CTMP Trials., *TAPPI J.* **78**(1), 91–96, Jan. (1995).
11. T. Fossum, *TAPPI J.* **65**(10), 69, Oct. (1982).
12. A. Wong, *TAPPI J.* **65**(10), 11, Oct. (1982).
13. Alkaline Sulfite May Challenge Kraft, *Can. Chem. Process.* **54**(11), 56, Oct. (1970).
14. M.B. Hocking, Vanillin: Synthetic Flavouring from Spent Sulfite Liquor. *J. Chem. Educ.* **74**(9), 1055–1059, Sept. (1997).
15. D. Craig and C.D. Logan, Method of Producing Vanillin and Other Useful Products from Lignosulfonic Acid Compounds. Can. Pats. 615,552 and 615,553, (to the Ontario Paper Company) (1961).
16. J.N. McGovern and R.R. Fuller, 1984: Centennial of Kraft Pulping Patent. *TAPPI J.* **67**(11), 48–49, Nov. (1984).
17. J. Drew and G.D. Pylant, Jr., *TAPPI J.* **49**(10), 430, Oct. (1966).
18. "Crude Tall Oil Recovery," Tech. Sect. Proc., 47th Annu. Meet. Can. Pulp and Paper Assoc., Montreal, 1961.
19. Tall Oil Fatty Acids, *Chem. Eng. News* **57**(40), 10, Oct. 10 (1979).
20. M.B. Hocking and J.P. Crow, On the Mechanism of Alkaline Hydrogen Peroxide Oxidation of the Lignin Model p-Hydroxyacetophenone. *Can. J. Chem* **72**, 1137–1142 (1994).
21. K.Kratzl, P.K. Claus, A. Hruschka, and F.W. Vierhapper, *Cellul. Chem. Technol.* **12**, 445 (1978).
22. D.V. Ellis, P. Gee, and S. Cross, *Water Pollut. Res. J. Can.* **15**(4), 303 (1981).
23. T.P. Nevell, *Chem. Ind. (London),* March 15, p. 253 (1975).
24. A. Teder and D. Tormund, *TAPPI J.* **61**(12), 59, Dec. (1978).
25. H. Rapson, C.B. Anderson, and D. Reeve, *Pulp Pap. Mag. Can.* **78**(6), T137, June (1977).
26. M. Ruhanen and H.S. Dugal, *TAPPI J.* **65**(9), 107, Sept. (1982).
27. L. Nasman and G. Annergren, *TAPPI J.* **63**(4), 105, April (1980).
28. N.L. Nemerow, "Industrial Water Pollution," p. 439. Addison-Wesley, Reading, MA, 1978.
29. M.B. Hocking, Paper versus Polystyrene, A Complex Choice. *Science* **251**, 504–505 (1991).
30. M.B. Hocking, Relative Merits of Polystyrene Foam and Paper in Hot Drink Cups: Implications for Packaging. *Environ. Manage.* **15**(6), 731–747 (1991).
31. T.R. Aspitarte, A.S. Rosenfield, B.C. Smale, and H.R. Amberg, "Methods for Pulp and Paper Mill Sludge Utilization and Disposal." U.S. Environmental Protection Agency, Washington, DC, 1973.
32. A.J. Bruley, "The Basic Technology of the Pulp and Paper Industry and its Waste Reduction Practices." EPS 6-WP-74-3. Environment Canada, Ottawa, 1974.
33. S.N. Lo, H.C. Lavallee, R.S. Rowbottom, M.M. Meunier, and R. Zaloum, Activated Sludge Treatment of TMP Mill Effluents. *TAPPI J.* **77**(11), 167–178, Nov. (1994).
34. K.V. Sarkanen, B.F. Hrutfiord, L.N. Johanson, and H.S. Gardner, *TAPPI J.* **53**(5), 766, May (1970).
35. M.A. Karnovski, *J. Chem. Educ.* **52**, 490 (1975).
36. R. Cederlov, M.L. Edfors, L. Friberg, and T. Lindval, *TAPPI J.* **48**(7), 405, July (1965).
37. J.M. Bentvelzen, W.T. McKean, and J.S. Gratzl, *TAPPI J.* **59**(1), 130, Jan. (1976).

38. A.J. Teller and L.L. Bebchick, *Pulp Pap. Mag. Can.* **81**(12), T358, Dec. (1980).
39. V.R. Parthasarathy, R.C. Grygotis, K.W. Wahoske, and D.M. Bryer, A Sulfur-free, Chlorine-free Alternative to Kraft Pulping. *TAPPI J.* **79**(6), 189–198, June (1996).
40. M.B. Hocking, H.I. Bolker, and B.I. Fleming, An Investigation of Anthraquinone-catalysed Alkaline Pulping via Component Modelling and Electron Spin Resonance Measurements. *Can. J. Chem.* **58**, 1983–1992 (1980).
41. H.F. Lund, ed., "Industrial Pollution Control Handbook," p. 18-1. McGraw-Hill, New York, 1971.
42. "Proceedings of Seminars on Water Pollution Abatement Technology in the Pulp and Paper Industry," Rep. EPS 3-WP-76-4. Environment Canada and the Canadian Pulp and Paper Association, Ottawa, 1976.
43. B.R. Blackwell, W.B. Mackay, F.E. Murray, and W.K. Oldham, *TAPPI J.* **62**(10), 33, Oct. (1979).
44. M.V. Nelson, *TAPPI J.* **63**(3), 61, March (1980).
45. R.J. Rush and E.E. Shannon, "Review of Colour Removal Technology in the Pulp and Paper Industry," Rep. EPS 3-WP-76-5. Environment Canada, Ottawa, 1976.
46. D.J. Bennett, C.W. Dence, F.L. Kung, P. Luner, and M. Ota, *TAPPI J.* **54**, 2019 (1971).
47. New Scrubbing and Effluent Treatment Processes use Activated Carbon, *Chem. Can.* **28**(11), 19, Dec. (1976).
48. J.C. Mueller, "B.C. Guidelines." B.C. Research, Vancouver, Feb. (1981).
49. H. Lundahl and I. Mansson, *TAPPI J.* **63**(4), 97, April (1980).
50. Oxidation with Ozone, *Can. Chem. Process.* **61**(5), 4, May (1977).
51. J.M. Leach and A.N. Thakore, *Prog. Water Technol.* **9**, 787 (1977).
52. A.N. Thakore and A.C. Oehlschlager, *Can. J. Chem.* **55**, 3298 (1977).
53. J.M. Leach, J.C. Mueller, and C.C. Walden, *Process Biochem.* **10**(1), 7 (1976).
54. K.S. Ng, J.C. Mueller, and C.C. Walden, *Can. J. Chem. Eng.* **55**, 439 (1977).
55. C. de Choudens and P. Monzie, *Pulp Pap. Mag. Can.* **78**(6), T118, June (1977).
56. J. Luo and P.K. Christensen, Oxygen Delignification with Magnesium as Base: A Good Solution for Sulfite Pulp. *TAPPI J.* **75**(6), 183–187, June (1992).
57. J.R. Fooks and C.H. Langford, *Environ. Lett.* **6**(3), 205 (1974).
58. G. Kaufman, Activated Sludge Treatment of De-Inking Waste Water at Fraser Inc., Thorold. *Pulp Pap. Can.* **86**(2), T36-T41, Feb. (1985).
59. L.D. Ferguson, Deinking Chemistry: Part 1. *TAPPI J.* **57**(7), 75–83, July (1992); Part 2. *Ibid.* (8), 49–58, Aug. (1992).
60. D.W. Duncan, *TAPPI J.* **62**(7), 31, July (1979).
61. R.C. Howard and W. Bichard, The Basic Effects of Recycling on Pulp Properties. *J. Pulp Pap. Sci.* **18**(4), J151–J159, July (1992).

16
FERMENTATION AND OTHER MICROBIOLOGICAL PROCESSES

Ale, man, ale's the stuff to drink
For fellows whom it hurts to think.
　　　　　　　—*Alfred E. Housman, 1896*

16.1. GENERAL MICROBIOLOGICAL PRINCIPLES

Strictly speaking, fermentation is the process of anaerobic breakdown or fragmentation of organic compounds by the metabolic processes of micro-organisms. However, fermentation processes can be considered generally to relate to the chemical changes of a substrate accomplished by selected micro-organisms or extracts of micro-organisms, to yield a useful product. This less specific definition includes microbiological processes carried out under anaerobic (fermentative, or in the absence of air), aerobic (respiratory), and enzymatic (via extracts) conditions.

Historically, microbes of one kind or another have been used for centuries for the preparation of a variety of foods and beverages. Each fermentation product probably arose from what was originally an accidental discovery, for example, grapes or grape juice fermenting on storage or mold growth on curds (milk solids). These were recognized as changes which increased the menu variety and the storage stability of the foodstuff, hence the appeal of the products of these processes. It is easy to speculate, then, that each accidental discovery led to experiments aimed at being able to obtain the same change in the character of the food intentionally and at will. From probable early beginnings of this kind, yeasts have been used in the arts of wine, beer, and spirits production, as well as for the *in situ* production of the carbon dioxide used for leavening of bread and other bakery goods. Bacteria have been employed for the preparation of a variety of cheeses, sauerkraut, and yogurt. Molds too, still have a significant utility in the preparation of certain cheeses.

However, it is only since about 1900 that fermentation processes have been employed on any major scale for the production of industrial products, in addition to foods.

The impetus for industrial utilization of micro-organisms is based on several factors which favor their use. Most importantly, a fermentation process is very often the only feasible way to obtain a complex, or a highly specific chemical transformation of the raw material. Also a chemical change conducted enzymatically (via micro-organisms or extracts) has a far lower energy barrier incurred by the process than a chemical synthesis route to the same change. This means that the process may very often be carried out enzymatically at ambient or near-ambient conditions. Under these conditions the enzymatic processes can be 10^9 to 10^{12} times as rapid as the corresponding chemical route to the same product [1]. Another feature of importance for the production of some complex biological products is that many enzymes are totally specific and stereoselective for the product of interest.

Under appropriate conditions the rate of reproduction of yeasts and bacteria is extremely rapid, with doubling rates measured in minutes, so that there is little delay in building up the number of individuals to maximize the conversion rate of raw material (Table 16.1) [2–4]. Their small size, and thus large surface to volume ratio, is also favorable for the rapid diffusion of substrate into the cell, and metabolized product out. But utilization of micro-organisms for chemical change does introduce a difference in the profile of a

TABLE 16.1   Key Details of the Four Principal Groups of Micro-Organisms Useful in Industrial Fermentations[a]

Class, complexity, and size	Reproduction	Occurrence (common uses)
Bacteria		
Simple, single chromosome (prokaryote), usually single-celled organisms, 1–6 μm diameter, spherical (coccus), or rod-like (bacillus) in shape	Asexually, by binary fission (simple cell division)	Air, water, and soil (sour milk, cheeses, yogurt, sauerkraut)
Yeasts		
Two or more chromosomes (eukaryote), single-celled near spherical, 5–12 μm diameter	Usually asexual by "budding," under some conditions sexual (via spore cells)	Air, water, and soil (beer, wine, spirits, bread)
Actinomycetes		
Multicellular organisms intermediate in complexity between bacteria and molds, filamentous bacteria whose mycelium is readily fragmented	See bacteria and molds/ fungi	Normally inhabit soil (antibiotics)
Molds, fungi		
Two or more chromosomes (eukaryote), readily visible multicellular filaments	Varies, generally asexually via wind-transmitted spores, also sexually via wind ascospore formation	Air, water, and soil (cheeses, antibiotics)

[a]Compiled from Kent [2], Shreve and Brink [3], and Phaff [4].

rate plot for the appearance of product. With a conventional batch reaction using only chemical starting materials, if all of the reactants are present in the reactor at the start of the process, an initial maximum rate is observed, which gradually tapers to lower rates. In contrast to this, chemical conversions using microbes usually start relatively slowly while the number of organisms builds up during a reproductive phase from the small number present on initial inoculation (Fig. 16.1). Following this lag phase or induction period, the maximum conversion rate of substrate to product is observed during the interval B–C. The interval C–D corresponds to the period when there is no further increase in numbers of organisms and the concentration of available substrate has decreased sufficiently to limit its availability to organisms, which then slows its rate of metabolic conversion to product.

For large-scale fermentations the proportion of time during which a fermenter is producing under high rate conditions should be arranged to be large relative to the total fermenter operating time to minimize operating costs. That is, the time period B–C, during which maximum production is observed, should be large relative to the total of the remaining intervals that the fermenter is occupied, as given by Eq. 16.1.

$$\text{If B–C} \gg \text{A–B} + \text{C–D} + \text{D–A},$$
$$\text{then batch operation is generally favored.} \quad 16.1$$

In this expression, interval D–A corresponds to the turnaround time, or time required to empty the fermenter, clean and sterilize it, refill it with fresh sterile substrate, and introduce a "starter" of the appropriate organism. If this situation does in fact hold, then batch operation of the fermentation process is commercially attractive.

If, however, B–C, the interval for maximum production rate, is short relative to the total fermenter operating time, as given by Eq. 16.2, then continuous operation is generally favored.

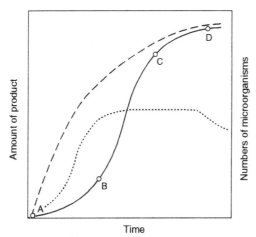

Time

FIGURE 16.1 A qualitive comparison of production curves via a conventional chemical process (-----) and a fermentation process (——) correlated to the propagation and metabolic rates of the micro-organism used (. . . .).

If B–C < A–B + C–D + D–A,
then continuous operation is generally favored.   16.2

With this situation there is a strong incentive to develop a continuous, rather than batch mode of operation for the process. A continuous mode of operation would be particularly attractive under these conditions for very large scale fermentations. Continuous fermentation is usually accomplished by staging the fermentation in several vessels rather than in a single one. Under these conditions it is possible to obtain a high fermentation rate continuously in one or more of the vessels to give the most efficient overall rate of conversion of substrate to product.

There are problems attached to operation of fermentations on a continuous basis, some of which are the potential for genetic variation with time of the micro-organism itself and the difficulty of maintaining ideal conditions for growth. Nevertheless several fermentation products are produced on a commercial scale using a continuous mode of operation. Baker's yeast, yeast from petroleum or petroleum products, and vinegar are all presently in large-scale production using continuous culture methods. Even in the highly market-sensitive brewing and winemaking industries, which have traditionally operated exclusively by batch fermentation and are still dominated by this mode of operation, continuous mode fermentations have now been proved to be feasible. Of course wastewater treatment plants also operate in a continuous mode. Wastewater treatment is also an example of a fermentative process, but one which employs a wide variety of micro-organisms to accomplish the collective changes desired, a nonspecific conversion (Chap. 5).

The requirements for producing a specific product via an industrial process using fermentation are, first and foremost, the need for a culture of a specific micro-organism that produces the desired end product. This is usually the dominant product, such as is seen for the anaerobic production of ethanol (ca. 95% selective) from sugar by yeasts. In this situation, the process is termed homofermentative. On the other hand there are also fermentations which produce several products, such as the simultaneous anaerobic production of butanol (60–70%), acetone (20–30%), and ethanol (about 10%) from sugars by *Clostridium acetobutylicum* [5,6]. This type of process, called heterofermentative, may also be useful if all the products are recoverable and salable.

Other fermentation requirements are that the microbe must be relatively easy to culture and to preserve in a dormant state for reference and stock renewal purposes. It should also be able to be easily maintained in an active state which has consistent genetic characteristics. Raw materials appropriate for conversion must be economic for the end product desired, i.e., sugars and starches for beverage alcohol, low-cost sugar waste streams, starches or cellulosic substrates for fuel alcohol, etc. Acceptable to good yields should be obtained, based on the relative prices of substrate and end product. The fermentation should proceed rapidly under conditions (pH, temperature, pressure, nutrients, etc.) that can be relatively easily provided. Finally, it should be possible to recover the product of the fermentation in a straightforward manner.

Many of the requirements summarized above are quite general for any viable industrial process. The key difference here is the participation of the micro-organism in the process, the locating, selection, and improvement of which is primarily the province of microbiologists. Cultures of the desired strains are maintained in a dormant state and are supplemented under scrupulous conditions. Portions of these cultures can then be taken and provided with ideal propagating conditions to build up the numbers of organisms available to provide a "starter" for inoculation to the medium of a full-scale fermenter. The medium, or growth environment, will consist of a solution of the primary substrate to be converted, usually in water. It will also have small amounts of nutrients, such as ammonium phosphate, to provide a source of nitrogen and phosphorus, and traces of other minerals such as may be required for favorable microbial growth and reproduction. After pH adjustment, and sterilization of the medium to avoid interference from any competing organisms, the temperature will be brought to the appropriate range and the starter added.

One remaining consideration which affects the mode of operation of industrial fermentations is whether it is to be operated as an aerobic or anaerobic process. Some micro-organisms can only function under aerobic conditions (in a respiratory mode) and are referred to as obligate aerobes. Others, such as the obligate anaerobes, can only function in the absence of air, i.e., in a strict fermentative sense. Many species of obligate anaerobes are so sensitive to air that they are killed on exposure to air. Yet a third group, referred to as facultative micro-organisms, is capable of functioning under either anaerobic or aerobic conditions. The metabolic processes and end products of facultative organisms may be significantly altered depending on whether the conditions provided are anaerobic or aerobic. Thus, both the range of substrates which may be utilized and the product(s) obtained may be changed depending on the metabolic mode forced onto the organism employed by the conditions provided. Yeasts in the presence of air, for example, ferment simple sugars to carbon dioxide and water. Under anaerobic conditions they produce carbon dioxide and ethanol instead.

The type of fermenter to be used will be determined by whether the fermentation is to be conducted aerobically or anaerobically, as well as other factors. However, even anaerobic fermentations such as those used for alcohol production, can be conducted in open-topped fermenters, when these are located inside a building. In this event the top of the fermenting medium is protected from contact with air by the blanket of coproduced carbon dioxide, which is heavier than air. Closed top fermenters are usually preferred for aerobic or anaerobic fermentation to reduce the risk of contamination of the process by unwanted organisms.

16.2. BREWING OF BEER

Brewing is a term used to refer to the large- or small-scale manufacture of beer. Beer production has been practiced for more than 8000 years. Certainly beers, the beverages produced from a fermented brew of cereal(s) and hops, have been produced at least since the dawn of recorded history and probably well before that [7]. Brewing is the sector of the fermentation industry with the highest volume of product and the highest value of annual sales, and is dominated by nations with a temperate climate (Table 16.2) [8,9]. In keeping with this, beer production also occupies a larger volume of total fermenter capacity than any other specific fermentation product, for example, a gross of over 128,000 m^3 for the U.K. Beer production on a global basis thus represents the most important sector of beverage alcohol production in terms of both volume and value of product. In the U.K. the only fermentative sector which exceeds brewing is the 2.8 million cubic meters vessel volume used for wastewater treatment.

Beer is a perishable product and is largely consumed in the country of origin. On this basis Australia, Czechoslovakia, and West Germany show the highest per capita production, producing over 140 L per person per year (Table 16.2). In contrast to this, the countries with a low level of production average out at less than 5 L per capita per year.

The manufacture of beer is an anaerobic fermentation process which uses varieties of yeast to produce a beverage of low alcohol content from a solution containing simple sugars in water. To avoid problems of corrosion, or product contamination from metals, all brewery processing is conducted in glass-lined or epoxy resin-lined steel, or stainless steel. The exception is the cereal cooker where copper is often used to obtain more efficient heat transfer. Three types of beer are recognized, based on their respective alcohol content which is specified by volume. Light beers are those containing less than about 4% ethanol. Regular beers run around 5% alcohol, and the malt liquors have an alcohol content in the 5.5 to 6.5% range.

TABLE 16.2   Annual Production of Beer by the Major World Producers[a]

	1968	1978		1980	1990
	Total (million L)	Total (million L)	Liters per capita[b]	Total (million L)	Total (million L)
Australia	1,408	2,001	148	2,023	1,939
Canada	1,506	2,129	93	2,266	—
Czechoslovakia	2,010	2,206	154	2,393	2,197
France	1,996	2,278	43	2,129	—
Japan	2,428	4,421	40	4,559	5,053
Mexico	1,252	2,257	47	2,688	3,873
Spain	1,026	1,735[c]	51	2,061	2,794
U. K.	5,143	6,642	120	6,484	7,080
U.S.S.R.	3,830	6,414	24	6,133	6,251
U.S.A.	13,790	18,000[c]	89	22,777	23,667
West Germany	7,373	8,792	145	8,932	9,138
Other	10,125	16,431	4.9	—	—
World	56,682	80,480	18.5	91,686	109,351

[a]All countries producing more than 10 billion (10^{10}) liters annually in 1978. From Lom [8], U.N. *Statistical Yearbooks* [9], and Kirk-Othmer [10].
[b]Calculated from 1978 production divided by 1979 populations.
[c]Estimated.

In outline, the aqueous substrate for beer production is made by brewing cereals with hops, which provide the characteristic bitter flavor component. Barley is the chief cereal ingredient which is first malted, both to give enzymes required to break down starches to simple sugars and also to contribute an important component of the flavor. Barley is a fairly expensive source of starch, so it is usually supplemented by the addition of less expensive starch adjuncts such as corn (maize), oats, millet, wheat, or rice. The adjunct makes little direct contribution to the flavor. Various strains of yeasts, selected for their efficient conversion of glucose to ethanol, are further developed by individual brewers to achieve particular product qualities.

16.2.1. Malting, Mashing, and Fermentation

The barley component of the brew is first malted. This begins with cleaned, graded whole barley grains, which are soaked in cold water until the desired moisture content of 44 to 48% has been reached. Then the grains are spread out to germinate. During the germination step, which requires a period of 5 to 8 days, the enzymes diastase and maltase are formed by the sprouting seeds for the primary purpose of mobilizing the starch reserves in the seed for growth utilization. Diastase promotes the hydrolytic conversion of starch, a polysaccharide, into the disaccharide, maltose. This hydrolysis is accomplished relatively easily since starch is a polymer of about 2000 units of the alpha anomer of glucose, in which the glucosidic links are relatively accessible to water. Maltase in turn rapidly hydrolyzes maltose, a disaccharide of two glucose units, to glucose itself. Once these enzymes have formed in the germinating barley it is heat-killed and dried in a current of hot air. This product, "light malt," is ready for incorporation into a brew. Or light malt may be cooked or roasted in a kiln to form a dark or "black malt," to contribute additional color and stronger flavor elements to particular types of beer. The enzymes of a dark malt may be inactivated. Careful control of temperature, humidity, and air flow is required through all these malting stages to produce an acceptable product.

The considerable time and space required for operating the malting step and the degree of industry standardization of this process has led to the virtual takeover of this step today by large, centrally located malting companies. Dried malt of various grades is shipped from the central processor on order, as required by individual brewers.

In mashing, the next step in the process, the dried malt and the starch adjunct (malt surrogate, 25–30% of the total starch) are ground up to coarse particles. Some of the ground malt and the whole of the adjunct are then mixed with water and boiled for half an hour in a cereal cooker to convert insoluble to soluble starch (Eq. 16.3, 16.4).

$$\begin{array}{ccc} \text{insoluble starch} & \rightarrow & \text{soluble starch} \\ \text{(intact cell membranes)} & & \text{(ruptured cells)} \end{array} \qquad 16.3$$

$$\begin{array}{ccc} (C_6H_{10}O_5)_n & \rightarrow & (C_6H_{10}O_5)_n \\ \text{intracellular} & & \text{extracellular} \end{array} \qquad 16.4$$

Meanwhile, in the mash tun (tub, or mixer), the remainder of the malt is suspended in water and warmed to about 50°C. The soluble starch in water from the cereal cooker, while still hot, is then added to the mash tun containing the warmed malt in water to produce a resultant temperature of 70–75°C (Fig. 16.2). At this temperature (*not* higher or enzymes are destroyed) about half an hour is provided for hydrolysis of the soluble starches to maltose and glucose (Eq. 16.5, 16.6).

$$\underset{\substack{\text{soluble starches} \\ \text{(extracellular)}}}{} \xrightarrow[\substack{\text{amylases,} \\ \text{e.g., diastase}}]{\text{hydrol.}} \text{disaccharide} \xrightarrow[\text{maltase}]{\text{hydrol.}} \underset{\text{(simple sugar)}}{\text{monosaccharide}} \qquad 16.5$$

$$\underset{\text{starch}}{(C_6H_{10}O_5)_n} \xrightarrow{n/2 \ H_2O} \underset{\text{maltose}}{n/2 \ (C_{12}H_{22}O_{11})} \xrightarrow{n/2 \ H_2O} \underset{\text{glucose}}{n \ C_6H_{12}O_6} \qquad 16.6$$

Under these conditions the mashing step effectively converts 75 to 80% of the starches present to simple sugars [8]. Higher mashing temperatures tend to give lower proportions of sugars to dextrins, and lower mashing temperatures tend to give higher proportions of sugars to dextrins. This step is sometimes referred to as saccharification, literally "sugar formation."

The product of the mashing stage, called a wort, consists of varying proportions of maltose and glucose, flavor constituents from the malt, plus dextrins (incompletely hydrolyzed starches of 30–100 glucose units), and undissolved fragments of grain, malt, and husks (Table 16.3) [10]. The hot wort is first lautered, a coarse filtration step, by passing it into a vessel with a perforated false bottom (lauter tun) which retains any undissolved grain, malt, and husk fragments. Filtered wort is then boiled for a period of 2 to 3 hours, hops being added after the first hour, usually in several portions. Hops, which consist of the bracteole or cone of a vine *Humulus lupulus*, provide the characteristic bitter flavor component of a beer contributed by the several humulones and related compounds [11] (Eq. 16.7) which are extracted from the hops during this boiling step.

TABLE 16.3 Breakdown of the Principal Carbohydrate Constituents of a Normal Wort Used for Making Beer[a]

Component	Concentration (g/L)
Maltose	57.7
Tri- to nonasaccharides	22.8
Higher carbohydrates	10.8
Total carbohydrates (91.0% of extract)	101.1
Fermentable carbohydrates (69.7% of extract)	77.4

[a]Data selected from the more detailed listing of Kirk-Othmer [10].

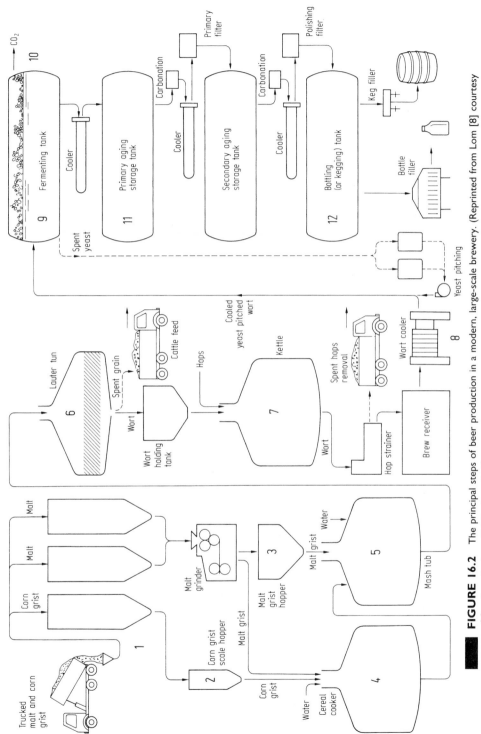

FIGURE 16.2 The principal steps of beer production in a modern, large-scale brewery. (Reprinted from Lom [8] courtesy of Southam Business Publications, Ltd.)

humulone   lupulone   16.7

These extracts are also bacteriostatic, one of the original reasons for using hops. This boiling period not only extracts the hoppy flavor constituents, but also serves to destroy residual enzymes, to coagulate undesirable proteins, and on completion to provide a sterile medium for subsequent fermentation. After boiling, the hot wort is allowed to settle to remove the coagulated proteins while it is kept covered to avoid contamination by any unwanted organisms. Subsequently it is cooled through a plate cooler (heat exchange to water) to 8–11°C. The product at this stage is referred to as hopped wort.

Before, or immediately after, the yeast is added for fermentation the hopped wort is oxygenated briefly. This helps to start the fermentation process more rapidly once the brew is placed under anaerobic conditions. A period of 6 to 7 days is required to complete the fermentation once the yeast is added. During the active phase this is accomplished by a yeast count in the fermenting wort of about 12 million/mL, originally added as a starter of yeast cells in a propagating suspension of about 1% of the volume of hopped wort to be fermented.

For each mole of glucose fermented about 80 kJ of heat is evolved (Eq. 16.8).

$$C_6H_{12}O_6 \xrightarrow[\text{(in yeast)}]{\text{zymase}} 2\ CH_3CH_2OH + 2\ CO_2$$
$$\Delta H = -74.5 \text{ to } -93.3 \text{ kJ } (-17.8 \text{ to } -22.3 \text{ kcal})$$
16.8

The heat evolution on maltose fermentation can be approximated by multiplying the figure taken for glucose by two and adding the further 15.9 kJ/mole required for maltose hydrolysis (Eq. 16.9). Fermentation temperatures of 15 to 18°C are desirable for beer production.

$$C_{12}H_{22}O_{11} + H_2O \xrightarrow{\text{maltase}} 2\ C_6H_{12}O_6 \quad \Delta H = -15.9 \text{ kJ } (-3.8 \text{ kcal})$$
16.9

Above this range some flavor elements are lost with the carbon dioxide bubbles leaving the wort and head (froth) formation of the finished beer is poor. Therefore, brewers normally use refrigeration to maintain fermentation temperatures in this optimum range.

Yeasts used in brewing (and for winemaking and spirits production) are homofermentative, that is, under anaerobic conditions they are about 95% selective for ethanol as the end product of glucose metabolism [12]. Thus Eq. 16.8 and 16.9 closely approximate the overall chemical change accomplished by fermentation, even though they do not do justice to the sequence of biochemical steps required. Using these equations it is possible to calculate the approximate alcohol concentration to be expected on the fermentation of any given initial concentration of fermentable sugars. During the fermentation step virtually all of the fermentable sugars present, which comprise about two-thirds of the original carbohydrate mashed, are converted to ethanol and carbon dioxide (Table 16.3). Tri- and higher polysaccharides which are not fermentable by yeast remain as a dissolved carbohydrate fraction in the beer. Most breweries collect the carbon dioxide formed during fermentation and clean it after compression by passage through two or more stages of active carbon adsorption, followed by a water wash. The 99.99% pure carbon dioxide is then liquefied for storage for later use as needed in packaging or inert gas blanketing. In continuous mode brewing systems part of the carbon dioxide produced is used to keep the yeast and the wort agitated and moving during the fermentation step.

After fermentation the brew is separated from the yeast by centrifuging, or by decantation through an outlet raised slightly above the bottom of the fermenter, which leaves nearly all of the yeast behind as a "heel" in the fermenter. Each fermentation produces enough yeast for the pitching (starting) of five further brews of equivalent size. After careful checking for strain purity and freedom from mutations the middle portions of this yeast will be reused, sometimes for as long as 3 months or 8 to 12 fermentations, before it is arbitrarily discarded. A starter freshly prepared from pure culture stock is then used to reinitiate the whole process. The decanted, fermented brew is then filtered. The opalescent product at this stage is referred to as a "green" beer.

Green beer requires aging at temperatures as near 0°C as possible for a period of 2 to 6 weeks, usually with at least one intermediate decantation plus filtration. During this period some further protein settling occurs, which helps to clarify the product, and a certain amount of flavor mellowing (maturation) is achieved to produce a more palatable product. The aged product, with an analysis as given in Table 16.4 [10,13], is now ready for final packaging and marketing.

Finished beer is marketed in two forms, bottled (or canned) and draft. The proportions of each sold varies widely depending on market area. In Canada over 95% bottled beer is sold in Quebec and only about 80% bottled in Alberta. Prior to packaging of either form of beer in North America, the mature product is usually given a final polishing filtration to yield the product of high brilliance (clarity) demanded by the market. In Europe some opalescence (haze) is generally accepted by customers, so that the final product is not usually filtered before packaging. The product is carbonated to the extent of about 0.4% carbon dioxide by weight (2.6 to 2.9% by volume), using a part of the purified carbon dioxide produced by fermentation [14]. It is then loaded into bottles previously flushed out with carbon dioxide if possible with the bottling equipment used. Following bottling, the shelf life of the bottled product is improved to about 3–4 months by pasteurizing for 6 to 10 minutes at 60°C. After pasteurization a simple residual "head" check determines the quality of the seal obtained on capping, and any imperfect seals are discarded. Optimum product shelf life is obtained under cool, dark storage conditions.

TABLE 16.4   Composition by Weight and Other Properties of a Typical Finished Beer[a]

	Composition[b]	Per 341 mL bottle,[c] approx.
Alcohol content, by weight	3.8–3.9%	11.3 g
by volume	ca. 5.2%	—
Degree fermentation	66%	—
Density	1.003–1.007 g/mL	—
Total unfermented carbohydrates	4.0–4.6%	12.6 g
Fermentable sugars (as maltose)	0.90%	2.6 g
Water	92%	270 g
Proteins (as N, × 6.25)	0.30–0.50%	1.3 g
Humulones	ppm range	—
Minerals, ppm		
Calcium (as Ca)	40	15 mg
Phosphorus (as P)	150–250	93 mg
Vitamins, ppb		
Thiamin (vitamin B_1)	20–90	19 µg
Riboflavin (vitamin B_2)	300–1300	270 µg
Pantothenic acid (vitamin B_5)	1000	340 µg
Pyridoxin (vitamin B_6)	400–800	205 µg
Nicotinamide	10,000	3400 µg
Inositol	29,000	9900 µg
Biotin (vitamin H)	10	3 µg
Other properties		
pH	4.1–4.4	—
Food energy (as alcohol, carbohydrates, proteins)	1780–1970 kJ/L (425–470 kcal/L)	607–672 kJ (145–160 kcal)

[a]Data compiled from Lon [8], Kirk-Othmer [10], and Donaldson and Lampert [13].
[b]By weight, except where otherwise stated.
[c]Equivalent to a 12-ounce bottle.

Draft beer is produced from the same brew, but has a lower level of carbonation, 2.2 to 2.4% carbon dioxide by volume. In Canada it is packaged in aluminum or stainless steel "half barrels" of 12.5 Imperial gallons capacity, and is not pasteurized. Thus, draft beer has a much shorter shelf life than bottled beer, about 21 days with refrigerated storage. For this reason a brewer apportions his newly brewed product appropriately to ensure that all that is packaged as draft is consumed within this period. The incentive for the brewer to use the draft package lies in the very much lower packaging cost in this form. In return, consumers enjoy a flavor unaffected by pasteurization and less affected by the higher carbonation levels required for bottled beer. Even for a large-scale brewery-bottling plant, the cost of bottling very often exceeds all the other costs of brewing.

16.2.2. Product Variety and Quality

Many factors other than packaging influence the quality and type of product emerging from the brewing process. To start with, the water quality and dissolved ion content are important, both for the chemistry of the starch conversion step and for the flavor of the finished beer. Very soft waters may require addition of small amounts of calcium chloride and/or sulfuric acid, or sometimes gypsum ($CaSO_4 \cdot 2H_2O$) to provide sufficient trace nutrients for the yeast and a sufficiently low pH for a good product.

A breakdown of the main ingredients used in brewing is given in Table 16.5. Lager beers, which would closely follow the tabulated proportions, have two distinctions. First they are produced by a bottom-fermenting strain of yeast, *Saccharomyces carlsbergensis,* named after its discovery and isolation by E.C. Hansen at the Carlsberg Institute, Copenhagen, in about 1870, or a variety of *Saccharomyces uvarum.* Bottom fermentating yeasts settle out or sink in the green beer when fermentation is complete. Two or more subsequent settlings or rests of the decanted lager beer from the fermentation tank, while chilled, is sufficient to remove much of the yeast as well as other undesirable protein constituents. Hence the name lager, since "lagern" is German for "to rest."

Lagers comprise more than 95% of the brew sold in the U.S.A., but just over 40% of the beer sales in Canada, and less than 5% of the beer sold in the U.K. More than 50% of the beer sold in Canada and nearly 90% of that sold in the U.K. is an ale. The basic difference between these brews is that an ale is produced by a top-fermenting yeast, *Saccharomyces cerevisiae,* or a related strain. Not only is the yeast skimmed from the top of the green beer on completion of fermentation, but fermentation is carried out at a slightly higher temperature, 15 to 20°C, which requires less time. Ales are also hopped at a higher rate than lagers, which gives the product a more bitter, hoppy flavor component.

Other brew variations are pilsner, a light beer of 3.4 to 3.8% alcohol content and a little higher hop content, originally produced in Pilsen, Bohemia, and bock beer, a heavy, darker, sweet product of lower hop content produced as a specialty product in the spring by German breweries. Porters and stouts are dark, heavy, strongly flavored English beers produced using

TABLE 16.5 Major Ingredients and Appropriate Current Costs of Brewing, Per Canadian Barrel and Per Hectoliter of Beer[a,b]

	Per Canadian barrel	Per hectoliter (100 L)
Brewer's malt	34–37 lb	11–12 kg
Malt adjunct(s)	12–14 lb	4–5 kg
Hops	0.27–1.5 lb	0.090–0.50 kg
Yeast	ca. 1 lb	ca. 0.3 lb
Production costs:		
Labor	$4.50	North American cost estimate
Power, steam	$4.00	
Bottling	$6.00	

[a]Canadian barrel of beer = 25 Imperial gallons (or 30.023 U.S. gallons); U.S. beer barrel = 31 U.S. gallons; English beer barrel = 36 Imperial gallons; One dozen small bottles = 0.90 Imperial gallons; one hectoliter (European marketing unit) = 100 L = 22.00 Imperial gallons.
[b]Ranges calculated from data of Shreve and Brink [3] and Kirk-Othmer [10].

more hops and using malts which have been baked (roasted or black malt), and little or no malt adjuncts. Porters and stouts, produced by top fermentation, are also sold in Canada but together comprise less than 1% of the Canadian brewing market.

The perishability of beer is attributable to its low alcohol content, which is inadequate to prevent attack by micro-organisms, and to the unstable nature of many of its constituents. Thus, bottled and draft beers are at their best immediately after packaging. Optimum keeping qualities are obtained by ensuring minimum oxygen content and aseptic conditions during aging and packaging, by keeping the product under cool conditions to slow any chemical changes, and by storage of the product in the dark or subdued light to avoid the generation of "light struck flavors" [15].

16.2.3. Brewing Emissions and Controls

Air emission problems of breweries are minor since mass emission rates are low and discharges are largely nontoxic. Emission control measures focus on containment of process odors. Water scrubbers are used on brew house stacks, to eliminate this potential problem area [8]. If spent yeasts and spent grains are dried on site, the resulting odors are controlled by afterburners on the dryer vent stacks [16]. The high temperatures in the presence of excess oxygen destroy the odors pyrolytically.

Wastewater streams with a high BOD and high suspended solids are the main aqueous emission problem areas. It is estimated that the equivalent of about 9 kg of BOD loading results from each 1000 L of beer (90 lb BOD/ 1000 Imp. gal.) produced in a commercial brewery, with a breakdown as to origin roughly as given in Table 16.6 [17]. Beer itself has a BOD in the 50,000 to 90,000 range and a COD some 20 to 30% higher than this. This problem has to be dealt with in the event of any spillage.

The heaviest aqueous BOD loading comes from the spent yeast, any beer spillage, and trub (accumulated clarification precipitates) streams. Where the local sewerage regulations permit, the brewery can simply pay a treatment surcharge to the local authority and discharge the brewery liquid wastes to

TABLE 16.6   Characteristics of Typical Aqueous Brewery Waste Streams[a]

Origin	Biochemical oxygen demand (mg/L)	Suspended solids (mg/L)
Beer spillage	90,000	4,000
Washings from process units	200–7,000	100–2,000
Cleaning solutions	1,000	100
Screen and press liquor[b]	15,000	20,000
Wort trubs[c]	50,000	28,000
Spent yeast	150,000	800
Precipitates from clarification	60,000	100

[a]Data obtained from Beszedits [17].
[b]From the pressing of wet spent grains prior to drying.
[c]Consists of haze, which forms either during wort boiling or cooling, and which is removed before fermentation.

the municipal sewerage system. The surcharge offsets the higher treatment costs imposed on the sewage treatment plant by these very high BOD wastes. This arrangement can allow discharge without treatment by the brewery. Or each of these waste streams my be treated individually to reduce the waste load to the sewerage system.

The spent grains and yeasts, for instance, can be recovered and sold as cattle feed supplements [17]. As long as the feed is used promptly these wastes can be sold wet. This has two advantages. It avoids the formation of the high BOD "screen and press liquor" stream resulting from filtration and the capital and energy costs of drying. If these waste materials cannot be consumed promptly they must be pressed and dried to improve keeping qualities and reduce shipping costs. If no feasible feed markets exist, these wastes may be dried and incinerated for disposal.

The beer spillage, washings, etc., of brewery wastewater streams have a lower BOD, but still high enough to have a sewerage impact. Conventional activated sludge methods are used by some breweries to reduce simultaneously the BOD and suspended solids content of these streams to acceptable values [17]. With added nutrient, BOD and suspended solids reductions of 95%–97% have been obtained, using this method [18]. Rotating biological contactors (RBCs or "biodisk units"), plastic media-filled bio-towers [19], and the Unox process developed by Union Carbide [17] have all been used by the brewing industry to accomplish these objectives. Use of oxygen with the Unox system, for example, gave a 95% reduction of a mixture of brewery wastes and domestic sewage of about 1000 mg/L BOD and 400 mg/L suspended solids.

16.3. WINEMAKING

The art of winemaking, like brewing, was developed well before the beginnings of recorded history. It is easy to speculate, again, that these skills were acquired by experimentation following the accidental fermentation of stored fruit or fruit juices by natural yeasts which were originally present on the surface of the fruit. Several theories exist. The theory that grape cultivation spread parallel with culture from East to West is most widely accepted. It is thought that viticulture originated in the lands south of the Caspian Sea since signs of grape cultivation and winemaking have been discovered in Mesopotamia dating back 4500 years. Certainly winemaking and aging skills were well known to the Greeks, some 2500 years ago, and very gradually spread from there, and probably other centers, in the following centuries [20].

From a chemical standpoint winemaking is similar to brewing since the essential alcohol-forming step still involves an anaerobic fermentation using yeast. But there are also some significant differences. The key distinction is that winemaking starts with a fruit, usually grape, which already contains sugar, the fermentation substrate. Thus many of the preliminary steps required in brewing to produce a sugar solution from starch-based raw materials are unnecessary for winemaking. This simplifies the early steps of the winemaking process.

A number of varieties of grapes form the dominant winemaking raw material but many fruits other than the grape, such as apples, loganberries, raspberries, strawberries, etc., are also used. Apart from the vested incentive to produce an appealing beverage product, there is also an economic value-added incentive to winemaking from grapes on a large scale. To illustrate this, in the mid-1970s grapes sold wholesale at some $190/tonne and retailed fresh at something like $660/tonne. The roughly 600 to 800 L of wine obtainable from the tonne of grapes (ca. 160 Imp. gal/ton grapes) was worth some $1600 at that time, a coarse measure of the profit potential of winemaking. Consequently, half or more of the grapes grown go into winemaking, and in some countries over 75%.

The types of grapes grown depend on the suitability of climate, microclimate, and soil types of an area, as well as the type of wine to be produced. Also the sugar content of any grape will also vary significantly with growing conditions, which will affect the type and quality of finished wine produced. These factors affect the development of the wine industry of any particular country (Table 16.7). It is possible to ship fruit from a grape-growing to a wine-producing area, such as may be necessary to supplement the sugar content of a local juice. However, because of the perishability and relatively low value of raw grapes this is only occasionally practiced. Thus, winemaking is

▆▆▆ **TABLE 16.7   Wine Production by a Selection of the Major Wine Producing Countries**[a,b]

	1960	1970	1980	1980	1990	1992
	Total (million L)	Total (million L)	Total (million L)	Per capita (L)	Total (million L)	Total (million L)
Argentina	1,675	1,836	2,300	84.2	1,404	1,150
Australia	154	287	414	28.3	445	459
Bulgaria	405[c]	409	445	50.2	228	250
Chile	485	401	570	51.4	398	320
France	6,311	7,540	7,155	133.2	6,553	6,522
Greece	290	453	440	46.3	353	450
Hungary	296	438	565	52.8	547	500
Italy	5,534	6,887	7,900	138.5	5,487	6,380
Portugal	1,146	1,150	943	95.0	1,137	724
Romania	88[c]	449	895	40.2	471	750
S. Africa	287	424	630	21.5	952	930
Spain	2,126	2,501	4,243	113.4	3,969	3,472
U.S.A.	1,033	969	1,729	7.6	1,600	1,545
U.S.S.R.	777	2,685	2,940	11.0	1,570	1,800
West Germany	684	1,012	399	6.5	949	1,340
Yugoslavia	335	548	678	30.3	517	523
Other	(361)[c]	841	872	0.3		
World total	24,300	30,199	33,921	7.8	28,226	28,825

[a]Selected from *U.N. Statistical Yearbooks* [9].

[b]To convert liters to Imperial gallons, divide by 4.55; to U.S. gallons divide by 3.79. A common international unit of wine production volume is the hectoliter, equivalent to 100 L.

[c]Averages for 1961–1965 period; 1960 not available. These estimates have not been included in the world total figure for 1960, hence the negative entry under "Other."

more or less restricted to areas which grow their own grapes. This reduces the number of countries engaged in commercial winemaking to about one-third of the number which produce beer [9].

Another natural factor which affects the volume of wine produced is the volume of available fruit. This factor alone can cause the volume of wine produced by any particular country to vary from year to year by a factor of two. However, world short-term production variations averaged over all wine producing countried are less than this (Table 16.7). Beer production is not so subject to year-to-year variations of available raw material as wine production, because of the variety of cereal crops which may be used. Therefore beer production shows far smaller annual variations in production volume from this source. Both beer and wine markets are dictated to some extent by changing beverage tastes, social customs, and even the weather!

16.3.1. Classification of Wines

A wider range of product variety and alcohol content is possible with wine than with beer. The "natural" wines, also called "table" or "dinner" wines, are those in which the alcohol content is derived entirely by natural fermentation. This limits the alcohol content to the range of about 7 to 14% alcohol, by volume. The upper end of this concentration range is only attainable by using special varieties of wine yeasts; however, all have a higher alcohol content than even the malt liquor types of beer.

Most natural wines are sold "still," that is, with no residual carbonation or effervescence. However, some are carbonated or "sparkling" varieties, which effervesce on opening from the release of dissolved carbon dioxide. Originally carbonation was accomplished by bottling the wine prior to complete fermentation so that the carbon dioxide produced during a secondary fermentation carbonated the product. This procedure is still used occasionally but because of its high cost is reserved for only a very few of the finest quality wines. Good sparkling wines are also made by fermentation in large pressure tanks. The larger scale of operation, lower labor requirement, and decreased risk of breakage using this method give a lower cost product. More often today the wine is carbonated at the time of bottling using purchased or fermentation carbon dioxide. Sparkling wines are classified according to their alcohol and carbon dioxide content. Thus, "crackling" wines are those carbonated to less than 2 atm pressure and contain 7 to 13.5% alcohol, by volume. "Sparkling" wines all contain about 7% alcohol, and may be carbonated to a varying extent. "Champagnes" may have an alcohol content in the 7 to 13.5% range and are carbonated to 2–6 atm pressure, depending on type.

An alcohol content of about 14% is the highest which can ordinarily be achieved directly by fermentation because at this concentration further sugar conversion by yeast is inhibited. Concentrations of up to about 18% by volume are possible with special strains of wine yeasts. A wine of higher alcohol content is possible by the addition of brandy (distilled from wine) or fermentation alcohol to a natural wine. Wines where the natural alcohol content is supplemented in this way to 14–20% by volume are referred to as "fortified," or dessert wines, and include the varieties commonly known as sherries and ports. Fortification of these wines improved the keeping qualities for shipment and export, by inhibiting further yeast action. Fortified wines are produced on about the same scale as natural wines.

Both natural and fortified wines may be further classified according to their residual sugar content. Dry wines have 1% or lower residual sugar content, i.e., as little unfermented sugar present in the finished wine as the vintner is able to achieve. This residual sugar usually consists of only unfermentable pentoses [21]. Red natural wines are obtained by fermenting the juice of special varieties of grapes in the presence of the skins to extract some of the color, or sometimes by heating the must (juice for fermentation) in the presence of the skins before fermentation. These wines tend to be drier, on the whole, than the white varieties of wines which are fermented in the absence of skins. Sweet wines, which are more often whites, are those which contain from 2.5 to 10% residual sugars. These are produced by starting with an initial sugar content which is higher than can be completely fermented by the yeast. Occasionally, however, the sweeter character is obtained by a stopped fermentation, by removing the yeast from the fermentation at the appropriate stage.

16.3.2. Principal Steps of Winemaking

Grapes are harvested in the fall, the exact timing depending on the grape variety, current weather conditions, and grape sugar content. On arrival at the winery the green stems are separated from the grapes, and the fruit is crushed. The fruit or juice is kept out of contact with air as much as possible to prevent oxidation, which can affect the flavor. If at all possible the time from picking the fruit to crushing is kept to less than 24 hours. When shipping grapes, or for temporary storage, a nitrogen or carbon dioxide atmosphere can be maintained above the chilled fruit, or the fruit may be treated with sulfur dioxide. For production of white wines the juice ("must") is only with the skins briefly at the juicing stage, after which it is separated. But red grapes, when destined for red wines, are crushed and kept with the juice for fermentation. Much of the color is extracted from the skins ("pomace") by the alcohol which forms in the fermenting mixture. This color extraction is promoted by the agitation induced by carbon dioxide evolution during the fermentation, assisted by occasional pressing of the thick mat of skins at the top of the fermenter under the surface of the must or by pumping the fermenting must over the skins.

Large numbers of natural yeasts (and other micro-organisms) are found adhering to the fruit in the field. Originally these were the yeasts which were used to carry out the fermentation of juice to wine with somewhat unpredictable results. Now, however, this initial juice, or must, is sterilized and inactivated by addition of 50 to 200 mg/L of "available sulfur dioxide" (SO_2 + H_2SO_3) before fermentation. Commercially, either liquefied sulfur dioxide itself or a 10% solution of sodium or potassium metabisulfite ($K_2S_2O_5$) may be used for this purpose. The 10% metabisulfite solution, equivalent to about 5% available sulfur dioxide (Eq. 16.10, 16.11), is a more convenient form for small-scale use.

$$Na_2S_2O_5 + H_2O \rightarrow 2 NaHSO_3 \qquad 16.10$$

$$NaHSO_3 + H_2O \rightarrow NaOH + H_2SO_3 \qquad 16.11$$

Sulfur dioxide sterilization inhibits the activity of the wild yeasts present, serves as a chemically reducing preservative, and lowers the pH of the must. The conditions obtained tend to favor the activity of the acid-tolerant yeasts added by the winemaker over any wild yeasts present. If boiling were used for sterilization in winemaking, as it is in brewing, much of the fruity elements of the product would be lost in the process, and these additional advantages would not be obtained. Before fermentation is begun, the must is analyzed for sugar and total acid content and the concentrations adjusted with added sweeter juices or sucrose, and tartaric or citric acids if necessary [12]. The product at this stage of the process will have an analysis roughly that given in Table 16.8 [22].

For fermentation cooled must is allowed to stand for a minimum of 2 hours, and sometimes up to 24 hours or so, for the antiseptic action of the sulfur dioxide to take effect, and for the sulfur dioxide content to be decreased somewhat by forming combined sulfur dioxide and by vaporization losses.

TABLE 16.8 Range of Concentrations of the Principal Constituents of a Must and a Dry Wine, in Percent by Weight[a,b]

Component	Must	Wine
Water	70–85	80–90
Total carbohydrates	15–25	0.1–0.3
Glucose	8–13	0.05–0.1
Fructose	7–12	0.05–0.10
Pentoses	0.08–0.20	0.08–0.20
Alcohols		
Ethanol	trace	8–15
Methanol[c]	0.0	0.01–0.02
C_3 and higher	0.0	0.008–0.012
Glycerol	0.0	0.30–1.40
Total organic acids	0.3–1.5	0.3–1.1
d-Tartaric	0.2–1.0	0.1–0.6
l-Malic	0.1–0.8	0.0–0.6
l-Citric	0.01–0.05	0.0–0.05
Succinic	0.0	0.05–0.15
Lactic	0.0	0.1–0.5
Acetic	0.00–0.02	0.03–0.05
pH	<3.3–3.6	
Tannins	0.01–0.10	0.01–0.30
Aldehydes	trace	0.001–0.050
Nitrogenous compounds	0.03–0.17	0.01–0.09
Inorganic compounds	0.3–0.5	0.15–0.40

[a]Selected from the comprehensive listings of Amerine [12] and Amerine *et al.* [22]

[b]Composition of the starting grape is about 80–90% pulp and juice, and 10–20% seeds, skin, and stems [12].

[c]Methanol is not a product of the fermentation but arises from the hydrolysis of naturally occurring pectins [22].

This period also allows heavy sediments to settle (fruit particles, seeds, sand, etc.) from which the juice is separated to give a better preservation of the natural fruitiness of the product.

After this standing period a starter of 1 to 2% of the must volume of the appropriate variety of yeast is added and fermentation started at close to 20°C (Eq. 16.8, 16.9). Fermentation temperatures for winemaking are less critical than for beermaking, so that many winemakers simply allow the process to take its course. However, better control of flavor and possible foaming problems is obtained by using chilled water to control white wine fermentations to within 2 or 3° of 15°C, a process which takes 1 to 2 weeks to complete [12]. High-quality wines may use lower temperatures to produce a wine with better varietal character (reflecting the species of grape fermented), higher residual sugar, and higher volatile acidity.

Red wines are fermented at about 20°C in the presence of the skins for 3 to 6 days, depending on the intensity of color (anthocyanins) and flavor (tannins) desired. The partially fermented must is then decanted and pressed from the skins, and a secondary slower fermentation carried out to the extent required. Acid and glycerin formation, small but important components of both white and red wines, tend to increase toward the end of the fermentation under the influence of the inhibiting ethanol already present from the primary fermention (Eq. 16.12).

$$2 \ C_6H_{12}O_6 + H_2O \xrightarrow{\text{yeast}} CH_3CH_2OH + CH_3COOH \qquad \qquad 16.12$$
$$+ \ 2 \ CH_2OHCHOHCH_2OH + 2 \ CO_2$$

Completion of fermentation for either type produces a "young" table wine which requires maturation before it becomes a palatable product.

As soon as the fermentation is stopped sulfur dioxide is added to bring the concentration to 25 ppm of free SO_2 throughout the subsequent clarification steps. This prevents oxidation, assists in biological stabilization, and binds aldehydes in the new wine. The binding of aldehydes (mostly acetaldehyde) improves the taste of the wine.

Clarification is needed after fermentation, primarily to remove the spent yeasts in the wine to prevent their autolysis in the product, and to remove any other suspended material (Fig. 16.3). Simple settling and racking (decantation), repeated several times, accomplishes much of this. Occasionally fining agents, such as isinglas (a fish protein), gelatin, a bentonite clay, or a highly refined diatomaceous earth, may be added to promote coagulation and precipitation of colloidal matter. Prolonged chilling to near freezing to promote tartrate crystallization or ion exchange methods are being employed by some wineries to remove excess tartrates, particularly before the bottling of young red wines [12]. This practice avoids the formation of haze from the crystallization of the residual tartrate in the bottle, which can occur particularly when the wine is chilled or is aged in the bottle. A final polishing step in clarification is filtration, used primarily to ensure that no yeast or other microorganism remains in the wine during the subsequent aging steps and to produce a brilliantly clear product [23]. At this stage the wine will have an analysis in the ranges quoted in Table 16.8.

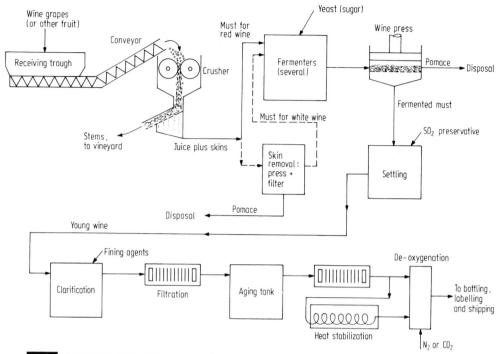

FIGURE 16.3 Flowsheet outlining the sequence of steps required to produce wine on a commercial scale.

Aging accomplishes a maturation of many of the flavor components of the young wine to give it a less harsh, more palatable flavor. A marrying of flavors and a development of bouquet (aroma) occurs, which we know happen through a variety of chemical steps. Probably the most important of these is esterification, the combining of acids and alcohols present to form esters. These contribute a more fruity, less harsh flavor and "nose" components than the initial precursors present in the young wine (Eq. 16.13).

$$\text{RCOOH} + \text{R}'\text{CH}_2\text{OH} \xrightarrow{\text{H}^+} \text{RCOOCH}_2\text{R}' + \text{H}_2\text{O} \qquad 16.13$$

Ethyl acetate is the most likely ester to form, because of the relatively higher concentrations of ethanol and acetic acid than other ester forming constituents present. It is also the most important for the progress of maturation. Malic to lactic acid conversion is also an important part of the aging process, particularly for red wines (Eq. 16.14).

$$\underset{\text{malic acid}}{\text{HOOC}-\text{CHOH}-\text{CH}_2-\text{COOH}} \xrightarrow{\text{Lacto-bacillaceae}} \qquad 16.14$$

$$\underset{\text{lactic acid}}{\text{HOOC}-\text{CHOH}-\text{CH}_3} + \text{CO}_2$$

Some North American white wines are bottled and marketed after only about 4 months aging following fermentation, and having undergone two or three rackings before bottling. Many are marketed in under a year [12]. Red wines are usually aged longer than this, a minimum of 24 months being normal and some as long as 4 to 6 years before bottling. Sherries and ports, the fortified wines, are generally aged 3 years or longer to allow full maturation and smoothness to develop before marketing. Sherries are subjected, during this aging process, to a partial oxidation with air, and a "baking" at about 50°C for 1 or 2 months to produce the characteristic sherry flavor. A sherry-like beverage may also be made from the by-products from domestic beekeeping, with or without supplementation from honey [24].

Prior to bottling, most commercial wines are stabilized and blended. Stabilization improves the keeping qualities of a wine. Filtration through a small pore filter removes yeast cells that might continue to ferment residual sugar when the wine is in the bottle, particularly with sweet wines. Filtration also prevents lysis (cell rupture) from contributing intracellular amino acids, etc., to the wine. Pasteurization, for 2 minutes at 86°C, may also be used, either in series with filtration or on its own, although it does not remove killed micro-organisms. Pasteurization also coagulates some proteins, thereby assisting in their removal. The wine is then gently sparged with carbon dioxide or nitrogen to dispel any dissolved oxygen which might be present and help preserve it from oxidation on storage. Residual sulfur dioxide, present at an average level of about 200 mg/L (as combined, and free) in Canadian wines, and at about twice this concentration in many European varieties, also provides further protection against wine oxidation or secondary fermentation on extended storage [25].

Finally the products of different fermentations based on similar or different grapes are blended to achieve consistent product flavor characteristics for a particular brand name of wine. Once the stabilized blend is bottled, particularly when closed with airtight caps, there is negligible further change on storage. Bottled wine, when kept in a cool, dark place, is a relatively stable product even on prolonged keeping. This stability is assisted by the practices of filtration and pasteurization ("hot bottling") described, and by the approximately 150–200 mg/L of total residual sulfur dioxide present. Government regulations in Canada and the U.S.A. permit up to 350 mg/L total sulfur dioxide in natural or dessert wines, still below the obtrusive level for the average palate. However, the sensitive palate of a trained wine taster can taste and roughly quantify concentrations of free (as opposed to "combined") sulfur dioxide above about 25 mg/L. Combined sulfur dioxide is present mostly as its bisulfite addition compound with acetaldehyde [22]. In the U.K. sodium or potassium sorbate, and ascorbic or erythorbic (isoascorbic) acid addition, is now also permitted to absorb oxygen. This helps to maintain a free SO_2 level for a long period of time. With these additives less sulfur dioxide is needed to maintain the required free SO_2 level for good keeping qualities, thus improving the taste.

Related to product quality control the concentration of asbestos fiber in all types of finished wines is now being minimized through better final product sampling and tests, and by the avoidance of asbestos-containing filtering materials [26]. The wide dispersion of asbestos fiber in the atmosphere means that harvested fruit of all kinds, not only grapes, inevitably contain a measurable adhering asbestos fiber count.

16.3.3. Utilization and Disposal of Winery Operating Wastes

Stems and pomace (skins and seeds) together amount to 10 to 20% of the weight of grapes crushed. These represent a significant disposal problem, particularly for large wineries producing upward of 36 million liters (ca. 8 million Imp. gallons) of wine per year. Some wineries return these directly to the soil of the vineyards, as an organic fertilizer [27]. At one time pomace was dehydrated for use as a component of cattle feed, the seeds being separated out first for oil recovery, but these are now less important [12]. This may be partly because of low nutritional value and partly because of storage difficulties. Pomace which is heap stored can start to ferment, and if fed to cattle in this state can be fatal. South Africa has made extensive use of winery by-products; alcohol, tartrates, cattle feed, methane, and fertilizer are being produced from pomace [28]. Many wineries recover the sugars from the pomace of white wines, and the residual sugars plus alcohol from the pomace of red wines by extraction with water. After fermentation of the sugars in the extracts, ethanol, useful for fortifying brandies, is recovered by distillation.

Alcohol may also be recovered from fresh lees, the spent yeast by distillation of the wine separated with the yeast. Some wineries pressure filter the lees for further wine recovery and produce a dry yeast cake, which may be dried and sold as cattle feed [23].

Monopotassium tartrate may be recovered from pomace, from lees, or from the virtually pure monopotassium tartrate obtained from tartrate removal processes. Monopotassium tartrate from this source provides most of the commercially marketed tartrates for use in such diverse products as baking powder and cream of tartar.

A poor batch of wine can sometimes be improved by chemical means, and then marketed as a blend. If it is not possible to correct the "off" wine, it may be distilled for recovery of alcohol, which again may be used for wine fortification. The aqueous residue, now significantly lower in BOD, is then discarded. An "off" wine may also be converted to wine vinegar by acetification (Eq. 16.15). This process is sufficiently profitable that one California winery has specialized in producing wine vinegar, rather than wine.

$$CH_3CH_2OH + O_2 \xrightarrow[\text{aceti}]{\text{Acetobacter}} CH_3COOH + H_2O \qquad 16.15$$

With appropriate local circumstances few of the by-products and waste materials of a winery operation need be wasted.

Dual aerated lagoons operated in series have been found to be an effective way of dealing with aqueous wastes. Better than 95% BOD removal has been obtained for influent BODs of 131 to 5000 mg/L when retention times averaged 30 days for the first lagoon and 100 days for the second.

16.4. BEVERAGE SPIRITS

Spirits are a class of alcoholic beverage with an alcohol content of 20% or more, above the range achievable solely by fermentation. Thus the spirits classification consists of beverages where distillation is used to raise the alcohol content above that reached by fermentation and where the distillate forms an essential and significant part of the final product.

To be able to produce beverage spirits requires a knowledge of both fermentation processes and of the art of distillation. Thus the preparation of spiritous liquors is historically much more recent than the arts of beer or wine production. Probably the earliest successful production of a distilled beverage from wine occurred in Greece or Egypt some 1800 years ago, most likely by alchemists [29]. The original discoveries did not become widespread because of the secretiveness of the early practitioners. Perhaps a thousand years elapsed before distilled beverages became at all widely available. Synonyms for whiskey, the Gaelic "uisge-beatha," the French "eau de vie," and the Russian "vodka," all came into use in the last 800 to 900 years. All can be liberally translated as "water of life," which gives some idea of the early recognition of its stimulant effects.

16.4.1. Specifying the Alcohol Content of Spirits

Much of the English-speaking world uses the proof system on a degrees Sikes scale, which originated in the U.K. in 1816, for specifying the alcohol content of beverage spirits. In this system 100% alcohol is equivalent to 175.2° proof. Proof alcohol, synonymous with 100° proof on this scale, is thus equivalent to 57.1% alcohol by volume. In Canada and the U.K. proof spirit (100° proof) is specified in statutes on a density basis as being exactly twelve-thirteenths the weight of an equal volume of distilled water at 51°F (ca. 10.5°C) [29]. Rumor has it that the early basis for deciding the alcohol content of spirits lay in the test that gunpowder moistened with an equal amount of proof (or higher alcohol content) spirits still ignited when touched with a flint or lighted taper. When moistened with underproof spirits it would not.

Under this proof system, spirits that are 20° overproof, which is the same as saying 120° Sikes or 120 proof, corresponds to a beverage which contains 120/175.2 × 100%, or 68.5% alcohol by volume (Table 16.9) [30]. Similarly, spirits that are "70° proof" are 70° Sikes or 30° underproof, and will contain 70/175.2 × 100%, or 40% (39.95%) alcohol by volume.

The most convenient way of determining ethanol concentration is by measuring the density and temperature of the solution and relating this to concentration by the use of appropriate tables. In turn the concentration by volume determined in this way can then be related to the British proof system, or to the Sikes scale for specification on the bottle.

The density system can be used to determine alcohol concentration for any country. The United States uses a simpler proof system based on 100% alcohol being equivalent to 200° proof. Fractions of this concentration, by volume, are specified by multiplying the concentration, in percent by volume, by two (Table 16.9). In France, alcohol concentrations in spirits are simply specified as percent alcohol by volume, sometimes referred to as the Gay-Lussac system. Percent alcohol by *weight* is used in Germany.

The excise taxes levied on distilled spirits is based on the volume of proof spirits produced, which corresponds to 57.1% ethanol (or ethyl alcohol) by

TABLE 16.9 **A Comparison of the Systems Used for Specifying the Alcohol Content of Beverages**

French, % alcohol by volume[a]	Approximate density at 15°C[b]	Canadian/ British proof[c]	American proof system	German, % alcohol by weight[a]
100	0.7939	175.2	200	100
97	0.8079	170	194	95.3
91	0.8306	160	182	87.0
86	0.8465	150	172	80.6
80	0.8639	140	160	73.5
74	0.8799	130	148	66.8
69	0.8925	120	138	61.4
63	0.9068	110	126	55.2
57.1	0.9199	100 (proof)[d]	114.2	49.3
51	0.9325	90	102	43.4
50	0.9344	87.5	100 (proof)[e]	42.5
46	0.9419	80	92	38.8
40	0.9519	70	80	33.4
34	0.9604	60	68	28.1
29	0.9666	50	58	23.8
23	0.9730	40	46	18.8
17	0.9791	30	34	13.8
10	0.9866	17.5[f]	20[f]	8.04[f]
5	0.9928	8.8[f]	10[f]	4.00[f]
0	1.0000	0	0	0

[a]System used in France, sometimes referred to as the Gay-Lussac system. In Germany, alcohol concentrations are specified on the basis of percent by *weight*.
[b]Related to the density of water, also at 15°C. Values will deviate slightly from this with sugar content, and at differing temperatures. Selected and calculated from the data of Lange [30].
[c]Values on this scale are sometimes referred to as "degrees Sikes." Entries are obtained by multiplying the percent alcohol by volume by 1.752; consequently these are not exact figures. Values are rounded downward for ease of comparison and to keep proof value cited conservative for the percent alcohol given. Values above "proof" are sometimes quoted as "overproof," using (the value −100). Values below proof are sometimes quoted as "underproof," using (100 minus the value) to refer to lower alcohol concentrations.
[d]Proof alcohol in the British system, or 57.1% alcohol, by volume.
[e]Proof alcohol in the American system, or 50.0% alcohol by volume.
[f]Correspondence between the different scales becomes inaccurate in this range.

volume in Canada and the U.K., and 50.0% alcohol by volume in the U.S.A. Since the concentration basis of proof alcohol for taxation in the former areas is higher and the Imperial gallon is equivalent to 1.20 U.S. gallons, the alcohol content of one British "proof gallon" is equivalent to the alcohol content of 1.37 U.S. "proof gallons."

16.4.2. Steps in Spirits Production

The initial steps for spirits production parallel those described for beer or wine in that one starts with either a cereal source for starch to be hydrolyzed to sugar, or an inexpensive source of sugar itself. Thus a variety of grains or molasses are common raw materials. Whiskey has been, and still is, the dominant product of the distilled beverage class, although there has been an increase in the proportions of vodka, cordials, and brandies in the U.S.A. over the last 20 years (Table 16.10) [31,32]. The discussion of distilled beverage production will be centered on whiskey because of its continued dominance.

The choice of the main cereal component used is based on the lowest cereal cost on a starch content basis, since the sugar ultimately available for fermentation depends on the amount of starch hydrolyzed. Small amounts of a particular cereal, for example rye for rye whiskey, corn for bourbon, may still be employed to provide the desired flavor element in the final beverage. Rye is used extensively by distillers in the areas where it is plentiful. Corn, wheat, millet (65–68% starch), rice, and potatoes also provide supplementary sources of starch for spirits production, depending on local cost and availability. Small amounts of hydrolyzed rye are sometimes used to favor yeast propagation (about 2% of the total starch input).

Whatever grain is employed, high quality and an accurate knowledge of the water content at the time of purchase are vital since every bushel of grain shipped to a distillery has to be accounted for in terms of the proof gallons produced. For any grain the standard "distillers bushel" is 56 lb. Since this is measured on a mass, and not on the usual volume basis, an unsuspected high water content can cost the distiller not only the lost raw material but also the excise tax on the proof gallons not obtained, for which it is still liable based on the quota of grain received. With a cereal start, the preliminary steps are very similar to the first steps of beer production, to obtain a sterile mash ready for inoculation. The mash is cooled to about 20°C. The yeast "starter," which

TABLE 16.10   Recent Trends in the U.S. Consumption of Classes of Distilled Beverages (Spirits)[a]

Class of spirits	Percent of U.S. consumption, by class and year				
	1949	1960	1966	1979	1990
Whiskeys[b]	87.8	74.4[b]	68.2[b]	53.5[b]	38.2
Gin	7.1	9.3	10.5	9.8	8.6
Vodka	0.0	7.8	10.4	18.0	22.7
Cordials	2.2	3.8	4.3	6.8	11.1
(Liqueurs) brandy	1.3	2.6	3.2	3.9	4.9
Rum	1.3	1.6	2.2	4.2	8.3
Other	0.3	0.5	1.2	3.8	6.2
	100	100	100	100	100
Total volumes:					
L × 10⁶	641.5	888.3	1169.2	1621.8	1386.0
Wine gallons[c] × 10⁶	169.5	234.7	308.9	428.5	366.2

[a]Selected from data of *Business Week* [31] and Kirk-Othmer [32].

[b]Includes blends, straights, bonds, Scotch, Canadian, and other types of whiskeys. In 1949, consumption of all types of whiskeys totaled 87.8% of all distilled beverage consumption in the United States.

[c]The wine gallon is a strictly volume measure; in the U.S.A., equivalent to a U.S. gallon (3.7854 L); in Canada and the U.K., equivalent to an Imperial gallon (4.5460L).

is propagated from a master culture in four to five stages to a volume of about 2 to 3% of the mash volume to be fermented, is then added. Fermentation temperatures are less critical for spirit production than for beer, since the important flavor elements are less affected by temperature. However, only a few minutes at 50°C is enough to kill the yeast, so sufficient cooling is applied to keep the closed fermenter temperature below about 38°C to avoid this. The higher temperatures complete the fermention in a period of only about 3 days instead of the 6 to 7 days required for beer. By the time 50% of the sugar in the mash has been converted there is no further yeast propagation. The remaining sugar to alcohol conversion is accomplished by the existing yeast population to a "beer" of 7.5 to 8% alcohol by volume. Beverage spirits are obtained from this intermediate fermentation product by distillation, blending, and aging procedures.

16.4.3. Distillation of "Beers"

The theoretical basis of the success of distillation as a method for enrichment of the alcohol content of aqueous solutions lies in the application of Raoult's law. This law quantitatively relates vapor composition to liquid composition, for mixtures of ideal liquids, A and B, according to Eq. 16.16.

$$P_A = X_A P_A^0 \qquad\qquad 16.16$$

where

P_A = partial pressure of the vapor of liquid A, in the vapor mixture

X_A = the mole fraction of A in the liquid phase

P_A^0 = the vapor pressure of pure liquid A, at the same temperature

While few real liquids behave quite like the ideal liquids required for strict adherence to Raoult's law, many mixtures of real liquids, including ethanol/water mixtures, approximately follow this relationship on heating (Fig. 16.4). Water has a vapor pressure of 327 mm Hg at 78.3°C. Ethanol, with a boiling point of 78.3°C, has a vapor pressure of 760 mm Hg at this temperature and consequently forms a higher mole fraction in the vapor space above a heated ethanol/water mixture than it does in the liquid phase. Condensation of the alcohol-enriched vapor mixture obtained in this way produces a solution of ethanol in water again, but now enriched in the concentration of ethanol over that present in the initial mixture.

In a laboratory batch distillation the process described above may be carried out very easily, but this only achieves a limited (by the liquid-vapor composition diagram) improvement in concentration of ethanol obtained with each repetition of the distillation (Fig. 16.5a) [33]. Also, as the distillation proceeds, the concentration of alcohol in the distilling vessel becomes depleted. Consequently there is also a gradual depletion in the alcohol concentration obtained in the vapor and the condensate. Despite these problems, many small distilleries still use batch distillation to raise the alcohol concentrations to the requirement of the product [34].

It is also possible to operate a distillation in a continuous mode. When an incoming beer stream is fed continuously to a distilling vessel, and a vapor

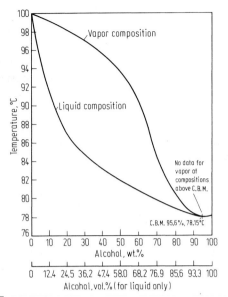

FIGURE 16.4   Liquid-vapor composition diagram for ethanol-water mixtures. (Reprinted from Shreve and Brink [3], courtesy of McGraw-Hill, Inc.)

stream and a distilled (alcohol depleted) liquid stream are continuously removed from the vessel, the mode of operation is referred to as continuous distillation. Continuous distillation simplifies quality control, because of the predictable and steady-state alcohol concentrations obtained from both product streams (Fig. 16.5b). The compositions of both the distillate and the residue streams, within the limits dictated by the liquid-vapor composition diagram, will depend primarily on the beer feed rate and the heat input to the system.

With a feed beer stream consisting of 7 to 8% alcohol, the increase in alcohol concentration which may be achieved with one stage of distillation under ideal conditions is close to 55% alcohol. This figure may be estimated by drawing a horizontal line from the 8% alcohol point of the liquid composition line to the vapor composition line. To allow more latitude in congener (flavor component) addition, blending, and the alcohol concentrations that are taken for aging, most distillers strive for higher alcohol concentrations than this. This may be achieved either by several repetitions of simple (batch) distillations or by several stages of continuous distillation (Fig. 16.5c).

Continuous distillation is the only way in which the large scale of a North American distillery, which can ferment over 1000 bushels of grain per day, can be conveniently handled. This is carried out in a distilling column in which a series of perforated plates or bubble cap plates separate the stages of distillation (Fig. 16.6). Using this arrangement, hot vapors from each stage heat the next higher stage as they partly condense, so that a preheated beers stream and only a single source of heat at the bottom of the column are sufficient to warm all the higher plates. As excess higher boiling fluid builds up on a plate, it overflows into a "downcomer" tube which carries it down to the next lower

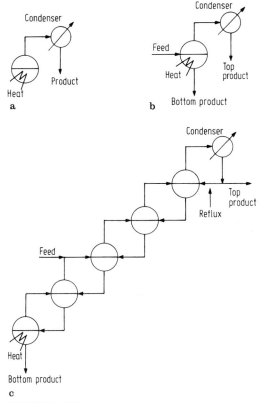

FIGURE 16.5 The relationship of (a) batch (simple) distillation to (b) single and (c) multistage continuous distillation. (Reprinted from Reuben and Burstall [33], courtesy of B.G. Reuben and M.L. Burstall and with permission of Addison Wesley Longman Ltd.)

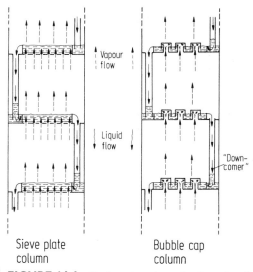

FIGURE 16.6 Horizontal sections of a sieve plate (left) and bubble cap (right) distilling columns.

plate. Any more volatile, lower boiling alcohol present in the fluid on a plate
has a tendency to be vaporized and carried up the distilling column in vapor
form as the fluid on the plate becomes heated by the upward moving vapors
passing through it. In these ways the liquid on each plate gradually comes to
equilibrium with the hot vapors continuously being forced through the liquid
on the plate.

Plate choice is a function of the variables the particular distilling column
is required to meet. Perforated plate columns are the most efficient for good
liquid-vapor contact but are rather sensitive to distillation rates. Too low a
distillation rate and the plates become dry from drainage through the the
perforations as well as via the downcomer. Too high a distillation rate and a
perforated plate column has a tendency to flood (become overloaded with
fluid) and hence lose distillation efficiency. A bubble cap column, in contrast,
can handle a wider range of distillation rates for a given column without the
drying out or flooding problems of a perforated plate column. But a bubble
cap column provides less efficient liquid-vapor contact than a perforated plate
column, and also costs more, for a given distilling capacity. Bubble cap plates
are usually specified when it is desired to obtain the principal advantage of
this system, flexibility in operating rates.

For a large-scale spirits producer fermenting over 1500 bushels/day, the
main beer distilling column can be shorter if the separation task is split among
several columns. The selective distillation, the aldehyde, and the fusel oil columns can all be significantly smaller than the main whiskey separating column
which has the job of alcohol enrichment for the whole of the beers stream
(Fig. 16.7) [35]. The first, and largest column, the beer distilling or whiskey
separating column, produces an overhead stream which contains 55 to 60%

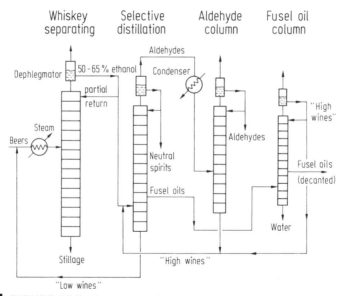

FIGURE 16.7 A suitable configuration of columns which can be used by a distillery for spirit
production. A more detailed flowsheet for continuous beverage spirits production is given by Kirk–
Othmer [35].

alcohol, together with the remaining volatiles, mainly aldehydes and fusel oils. This stream is the "high wines" stream of the distillery. At one time, with only some subsequent adsorption of impurities on charcoal, this stream comprised the whiskey product of early distilleries which was sent to aging [36]. The bottom stream from this column is "stillage," which is water from which all the alcohol has been stripped, but which will still contain some unfermented sugars and dextrins.

The smaller selective distillation column separates a neutral spirits stream (essentially 95% ethanol, i.e., the azeotropic composition) plus aldehydes (overhead) and fusel oils (bottom product) streams from the high wines stream of the beer column. The last two columns recover high wines streams from the aldehydes and the fusel oil fractions. The more concentrated streams of aldehydes and fusel oils obtained are about 8% and 7% of the volume of the high wine stream, respectively [34].

16.4.4. Distinct Distilled Beverage Products

The aldehydes and fusel oils recoverd from distillation are also referred to as congeners, or flavor components. Most of the aldehyde content is acetaldehyde, although other aldehydes are also present. The fusel oils (from the German "fuselol" meaning "bad spirits") are thought to arise via yeast fermentation of amino acids present in the original mash. They include amyl, and iso-amyl alcohols (about 50%), n-butanol (about 20%), and a residue of fatty acids and several other higher alcohols [37] (Eq. 16.17).

$$CH_3CH_2CH_2CH_2CH_2OH \quad CH_3CH(CH_3)CH_2CH_2OH \quad CH_3CH_2CH_2CH_2OH$$
$$\text{amyl alcohol} \qquad\qquad \text{iso-amyl alcohol} \qquad\qquad \text{n-butanol}$$

16.17

Once the congeners are separated from the high wines stream, varying proportions of each may be added back to the neutral spirits fraction as required for the type of product being made. In this way the distiller obtains good control of the whiskey flavor of the final product. Thus, many whiskey products start out as blends of various proportions of neutral spirits, aldehydes ("heads"), and fusel oils ("tails"), together with enough water to bring the alcohol content to the range required by statute. Bourbon-type whiskeys generally have the highest concentration of congeners, followed by Scotch whiskeys, though it can be seen that there is also a difference in the ratio of aldehydes and fusel oils (Table 16.11). Canadian blended whiskeys, in contrast, are generally lower in congeners than either the bourbon or Scotch counterparts, only some 0.5 to 1% of the heads and tails streams being added to neutral spirits before aging to produce these.

The newly mixed solution of neutral spirits, congeners, and sufficient water to bring the mixture to 60% alcohol is not a finished whiskey. It still must be matured to obtain a mellowing and smoothing of the initial blend to produce a more palatable product. The aging processes are carried out in charred oak barrels that are stored at a constant temperature as near 13–17°C as possible, and a relative humidity of about 65% for a period ranging from 4 to 18 years. If the conditions are cooler than this, the product doesn't mellow

TABLE 16.11 Congener Content of Some Widely Selling Types of Whiskeys[a,b]

Component	American blended whiskey	Bourbon whiskey	Canadian blended whiskey	Scotch blended whiskey
Fusel oil	83	203	58	143
Total acids (as acetic acid)	30	69	20	15
Esters (as ethyl acetate)	17	56	14	17
Aldehydes (as acetaldehyde)	2.7	6.8	2.9	4.5
Furfural	0.33	0.45	0.11	0.11
Total solids	112	180	97	127
Tannins	21	52	18	8
Total congeners, wt/vol %	0.116	0.292	0.085	0.160

[a]Selected examples from Kirk-Othmer [35].

[b]Grams per 100 L at 100° U.S. proof, density of about 0.914 g/mL, at 20° C. Thus, multiplying the numbers given by 10 gives a close approximation of respective concentrations in mg/L, and by 9.14 to give mg/kg.

well, and if warmer or drier it causes greater than acceptable annual evaporative losses, something the distiller tries to avoid. During the process, water and alcohol diffuse out through the wood, and air diffuses in. The air very gradually oxidizes the aldehydes and other congeners present. Flavoring and coloring is also picked up from the charred lining of the barrels, which at the same time adsorbs a certain amount of the fusel oil component from the mix. The complex changes in congener concentrations which occur during the aging process have been determined [37].

In Canada, the government requires a distiller to demonstrate production of at least 2.5 proof gallons (2.5 Imp. gal. of 57.1% alcohol by volume) for each bushel of grain mashed. If this is not achieved, the distiller is still taxed on this basis, hence the strong emphasis on pure, high-quality raw materials by the distiller. In practice a distiller can generally produce 3.5 to 4 proof gallons (Canadian) per bushel, which is still taxed at the same rate but allows some tax margin for error. The government also allows for an aging loss of 3% per year. If the loss rate exceeds this, the distiller is still taxed for product volume it should have had with a 3% per year loss rate. This is one incentive for care with the proper aging conditions.

After aging the product is bottled, sometimes after blending the contents of a number of barrels with differing flavor nuances, to yield a whiskey with consistent flavor characteristics. It is easy to see why an 8-, or 12-year old whiskey costs more per bottle than a 4-year-old whiskey, since there could be 10 to 20% less whiskey to bottle after these aging periods. There are also the inventory and inflation costs of keeping a product stored for any length of time. Once bottled there is very little, if any, change in the product because of the impermeability of the glass. Also stability is not a problem since whiskeys, like spirits generally, have a sufficiently high alcohol content to inhibit or kill any micro-organisms which may be present.

Vodka, presently second in sales volume to whiskeys in the United States, consists of straight neutral spirits (95% alcohol) diluted with pure water to the marketing concentration required. Ideally there should be no residual congeners in a first class vodka. This is ensured by some producers by passing the neutral spirits through activated charcoal, or by redistillation before water addition. Perhaps the growth in vodka sales can be attributed to the fact that it is a beverage without significant aroma, character, or taste. It readily lends itself to the preparation of many types of mixed drinks, punches, and homemade cordials.

Gin is produced entirely from neutral spirits too (i.e., without the fermentation congeners), but has an aromatic flavor element derived from a final distillation over juniper berries or from added coriander (a member of the carrot family).

Brandies have the special feature that they are prepared by distillation of a wine, generally a grape wine, rather than a "beer." The great variety in flavor and qualities of the various brandies originate from the different types of wines distilled, whether the distillation was batch or continuous, and the aging and blending details [33].

Rums are a distinctive beverage of this class in that they consist of the distillate from fermented molasses or other solution of sugarcane juice origin. So-called light or heavy rums are obtained depending on the type of sugar cane product that is fermented and the details of the distillation and maturation process used.

Cordials (liqueurs) may be made from any of the distilled spirits which are blended with various fruit juices, flowers, botanicals, and miscellaneous other flavoring materials [38]. Maceration, percolation, and distillation techniques are all used to place the blend of flavor components into the spirits base used for solution. Over the years many generic names have come into common use to enable recognition of the principal flavor elements, such as Creme de Cacao (chocolate), Creme de Menthe (peppermint), Triple Sec (citrus peel), Anisette (anis seed), etc. Also certain brand names have, by tradition, established a well-known flavor association with them, such as Benedictine (aromatic plant flavors) and Drambuie (whiskey flavor). Considering the wide range of flavors of the cordial class alone, the range of flavor variations possible from the distilled beverage class far exceeds that possible from either of the other two classes of alcoholic beverage. But a producer of distilled beverages, in turn, has to face a greater technological and investment input and far closer government regulation than the other two classes to produce a competitive consumer product for market.

16.4.5. Environmental Aspects of Distillery Operations

Similarities of raw materials and many of the procedures of distillery operations to those of brewing and winemaking mean that many of the emission control problems and solutions are similar. Much of the discussion of those sections is also appropriate here.

The higher fermentation temperatures used for whiskey production, over those used for beer or wine production, increase the vaporization rate of organics from the fermenting mash. While producing at the rate of 5.14 proof gallons (U.S.) per bushel of grain, one distillery had an average ethyl alcohol vapor emission rate corresponding to 182 g/m^3 of grain [39]. Vapors of ethyl acetate averaging 0.59 g/m^3, iso-amyl alcohol at 0.17 g/m^3, and isobutyl alcohol at 0.051 g/m^3 were also lost, all expressed per cubic meter of grain input. Activated charcoal may be used for vapor capture and analysis or for control and recovery of these organics, if necessary.

Distillery operations produce a stillage or "distillery slops" aqueous waste stream, as well as the usual brewery aqueous wastes. Beer still contains virtually all of the unfermented original ingredients, as well as the fermentation products. Distillery operations, however, since they are primarily interested in the alcohol and congener fractions produced by fermentation, are left with a residual wastewater stream from the distillation which contains 2 to 7% total solids [40]. The solids consist of mostly starches, dextrins, and unfermented sugars with beer-based spirits, or unfermented sugars with wine-based spirits. There may also be a significant nitrogen content present in the stillage when the distillery has fed the yeasts (lees) with the beer or wine stream to the still. Biochemical oxygen demands of 1500 to 5000 ppm have been cited for stillage in the absence of lees, and in the 7000 to 50,000 range for lees stillage [12]. Mass emission rates of 0.60 kg BOD/tonne grapes and 0.90 kg COD/tonne grapes, have been cited for a winery operating with stills.

If the distillery is operating near the sewerage system of a large center of population, liquid wastes can generally be discharged directly to this system. The composite waste stream will usually be amenable to the normal methods of sewage treatment. This disposal method may also require the payment of a surcharge to compensate for the additional treatment costs so imposed.

Stillage may be used as the organic feed to an anaerobic digestion unit in which bacteria convert the carbon content of the wastes into methane [41]. Combustion of methane rather than oil for a distillery's energy requirements, could save an estimated 116,000 Imp. gallons of oil annually.

If the distillery is operating in an area with available land, stillage may be discharged to an aerated pond for treatment or it may be used for irrigation [39]. Decomposition of the waste in soil is accelerated by disking or light harrowing at regular intervals.

An alternative is to recover dried yeast and dried spent grains from the stillage for sale as cattle feed. The more dilute stillage remaining is more economically treated for BOD removal. A trickling filter, or a biofiltration plant employing two aeration units in series, has demonstrated 60 to 90% BOD reductions [39].

16.5. INDUSTRIAL ETHYL ALCOHOL

Before 1945, most of the supply of ethyl alcohol for industrial solvent or feedstock uses was derived from fermentation sources (Table 16.12) [42–44]. Since this time, the reliability and low cost of petrochemical routes to the product caused a rapid replacement of fermentation sources by petrochemical routes in the United States. The early petrochemical sources were based on

TABLE 16.12   Trends in the U.S. Supply of Industrial Ethanol from Fermentation versus Synthetic Sources[a]

	Industrial ethanol (million L of 95%)			Percent by fermentation
	Fermentation[b]	Synthetic[c]	Total	
1930	364	3.4	367.4	99
1935	325	35	360	90
1940	364	122	486	75
1945	669	223	892	75
1950	232	394	626	37
1955	168	652	820	26
1960	95	939	1034	10
1965	229[d]	1126	1455	16
1970	241[d]	1164	1305	18
1975	316	850	1166	27
1980	246	574	810	30
1982	—	—	720[e]	—

[a]Compiled from Eveleigh [42,] Kirk-Othmer [43], and *Chemical and Engineering News* [44].
[b]Fermented from molasses, grain, sulfite liquors, cellulose, etc.
[c]From ethyl sulfate, and directly from ethylene, the latter predominating since 1970.
[d]Includes beverage alcohol, therefore, is higher than the straight industrial product volume.
[e]Preliminary estimate.

the formation and hydrolysis of ethyl sulfate, but in North America this has now been replaced by the direct gas phase hydration of ethylene (Eq. 16.18–16.20).

$$CH_2\!=\!CH_2 + H_2SO_4 \rightarrow CH_3CH_2OSO_3H \qquad\qquad 16.18$$

$$CH_3CH_2OSO_3H + H_2O \rightarrow CH_3CH_2OH + H_2SO_4 \qquad 16.19$$

$$CH_2\!=\!CH_2 + H_2O \xrightarrow{\;H_3PO_4\;} CH_3CH_2OH \qquad\qquad 16.20$$

Since 1975, however, increased costs and subsidies for fermentation alcohol have caused a reversal of this trend. Large new fermentation units based on a variety of substrates have recently been built, and distilleries formerly used for spirits production have been converted to industrial alcohol production [45].

Starches, fruit, and sugarcane sugars are used for industrial alcohol production using methods similar to those used for beverage products [46]. Industrial end uses have less stringent raw material quality requirements. For example a mobile, trailer-mounted fermentation distillation unit can usefully convert off-specification fruit and starch crops to fuel-grade ethanol on a contract basis. Sources of fermentable sugars that are inappropriate for beverage alcohol production, such as waste sulfite liquors, can yield as much as 80 L of ethanol per tonne of wood pulp produced [47]. Or cellulose may be deliberately hydrolyzed promoted by heat and hydrochloric (40%, at 20°C) or sulfuric (0.5%, at 130°C) acids in what are known as the Bergius and the Scholler processes, respectively (Eq. 16.21).

$$H^+$$
$$(C_6H_{10}O_5)n + n\ H_2O \longrightarrow n\ C_6H_{12}O_6 \qquad\qquad 16.21$$
$$\text{cellulose} \qquad\qquad\qquad \text{glucose}$$

Chemical hydrolysis of cellulose is not as economically attractive as some recently developed enzymatic routes [48]. Regardless of hydrolysis method cellulose opens up use of all kinds of farm wastes, straw, sawdust, waste paper, etc., to produce sugars for fermentation ethanol.

All fermentation processes strive to obtain a 6% or higher concentration of ethanol, the minimum which is presently economically attractive to recover. The energy costs for distillation become too high for lower alcohol concentrations. The azeotropic composition, 95% alcohol/5% water, is the maximum concentration of ethanol which can be obtained by distillation. However, this is a product already suitable for many industrial uses.

For the 10% alcohol-gasoline blends, a formula for the "gasohol" fuels for automotive use, as well as for some industrial applications, anhydrous ethanol is required. Anhydrous ethanol is miscible with gasoline. The presence of even a trace of water in the alcohol or the hydrocarbon component causes a water-rich phase to separate from the composite fuel.

Small amounts of 95% ethanol may be conveniently dried by addition of the correct amount of quicklime (calcium oxide, Eq. 16.22).

$$CaO + H_2O \rightarrow Ca(OH)_2 \qquad\qquad 16.22$$

No stable hydrate is formed so that a mole of calcium oxide is required for each mole of water to be removed. Since both lime components are slightly soluble in the ethanol the dried alcohol still requires distillation to obtain a high purity product.

Azeotropic distillation is the most economical method to dry large volumes of ethanol. A third component, usually benzene, is added to the 95% alcohol azeotropic composition. The boiling point of an azeotrope is always below that of the component with the lowest boiling point, and the vapor pressures of each of the components present at the boiling point are additive. Therefore heating the 95% ethanol benzene solution produces the ternary azeotrope of water, alcohol, and benzene, with a boiling point of 64.6°C, as the overhead product from the first distilling column (Table 16.13) [49]. Thus, if sufficient benzene is added to remove all of the water from the ethanol-water azeotrope in this way, the bottom product from the distilling column A, is pure, dry ethanol (Fig. 16.8) [50]. A slight excess of benzene is not critical to this result. Traces of benzene remain in this alcohol product, making it unsuitable as a beverage component.

Condensation of the top product from column A, which is the ternary azeotrope, then allows two phases to separate. The upper phase is rich in benzene (84.5%) with only a trace of water present (1.0%). The lower phase of the separator is rich in ethanol and water. The upper phase is returned directly to an upper plate of the first column amounts to directly recycle most of the benzene added to this column.

The lower phase from the separator is redistilled in a second column, column C, to recover the ternary azeotrope as the overhead product and water-rich ethanol as the bottom product. A third distilling column, column

Table 16.13 Boiling Points and Compositions of Azeotropes Appropriate to the Drying of Ethanol[a]

Components and % by vol. composition	Boiling point (°C)
Water/ethanol/benzene 7.4/18.5/74.1	64.6
Benzene/ethanol 67.6/32.4	67.8
Benzene/water 91.1/8.9	69.4
Ethanol/water 95.6/4.4	78.2
Pure ethanol	78.5
Pure benzene	80.1
Pure water	100.0

[a]Data selected from *Handbook of Chemistry and Physics* [49].

D, completes the recovery of ethanol as the ethanol-water azeotrope top product of this column, from the water-rich ethanol stream of the second column. The bottom product of the third column is the water originally present in the ethanol-water azeotrope feed to column A, now entirely freed of solvents.

Columns C and D are smaller than A since the volume to be distilled is less at these stages. The only raw material cost to dry ethanol by this method is for the benzene "make-up" (to replace process losses) which amounts to

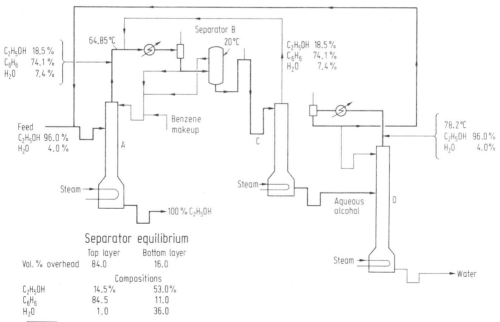

FIGURE 16.8 Drying of 95% ethanol by distillation with benzene. (Reprinted from Perry *et al.* [50], courtesy of McGraw-Hill, Inc.)

about 0.2% of the volume of ethanol produced. Some of this is lost in the dried ethanol, as evidenced by the strong benzene absorption seen in ultraviolet spectra of absolute alcohol. Thus, if one wishes to obtain an ultraviolet spectrum of a water stable compound free of this benzene absorption, it is better to use 95% ethanol as the solvent, to avoid this interference.

16.6. AEROBIC MICROBIOLOGICAL PROCESSES

Here the term "fermentation" is used in its more general sense to refer to a chemical conversion accomplished by micro-organisms, rather than in its strict sense of a microbiological conversion carried out in the absence of air. Most micro-organisms are aerobes or they are facultative, i.e., they are capable of functioning under either aerobic or anaerobic conditions. Therefore, to accomplish fermentations under aerobic conditions gives far greater scope to the types of chemical change possible than if one is limited to anaerobic transformations. But the wide range of aerobes and the variety of conditions to which they can adapt mean that the risk of contamination by stray organisms is much greater than with anaerobic fermentations. Also the very need of the aerobic system for oxygen or air provides a greater risk of exposure from entrained organisms. Nevertheless, the technology for aerobic fermentations has increased the number of options for microbiological products. Yeasts can function in an aerobic mode, as is used for bakers' yeast production, or for bread making. When increased cell mass or leavening are the primary objectives, aerobic function is valuable. However, the end products of aerobic fermentation of sugar by yeast are carbon dioxide and water, not a very useful chemical conversion.

16.6.1. Operating Details of Aerobic Processes

Neither oxygen nor air is very soluble in water so that without taking special measures to increase the rate of exchange of dissolved gases from a medium to and from air this rapidly becomes the limiting factor for aerobic conversions. In the laboratory, the rate of air exchange per unit volume of liquid is increased by use of a thin layer of the liquid fermentation medium in the bottom of an Erlenmeyer flask. Air exchange between the inside of the flask and the outside air is obtained using a tight, sterile cotton plug which also serves as a filter barrier to intrusion by outside organisms. Frequently the flask is also gently shaken to boost the air/liquid gas exchange rate. These methods function well for preliminary tests and process exploration, but are inappropriate for larger scale operation.

For larger scale operation, shallow trays made of corrosion-resistant metal or plastic and placed in portable trolleys may be used for efficient liquid-air contact. The trolleys are placed in a sterile room under closely controlled temperature and humidity and with frequent sterile air changes. Efficient air-liquid contact may also be obtained from a loosely packed, moist, solid medium permeated with air. For instance bran is used as the substrate for enzyme production by *Aspergillus oryzal*. Wine vinegar (acetic acid) is produced by trickling wine (basically ethanol in water) over a short tower of wood chips which are used both as a support and to provide air to *Acetobacter aceti* [12] (Eq. 16.15). In fact, by law, the acetic acid (3–5% in water) vinegar *must* be produced by fermentation, and not by petrochemical methods.

The most common procedure is to use deep tanks of the stirred liquid medium for large-scale aerobic fermentions. Efficient air exchange is obtained by continuous pumping of sterile air into crossed or coiled perforated pipe placed at the bottom of the tank at rates of up to one volume of air per unit of medium volume per minute [51]. This method promotes aerobic growth throughout the medium which accomplishes large volume production in a limited space. But it also has problems which relate to the maintenance of sterility and occasionally from foam formation. Overall, however, submerged aerobic fermentions are the most common method used by industry for aerobic conversions.

16.6.2. Aerobic Processes to Single Cell Protein

Without knowing the detailed metabolic pathways of many aerobic fermentative conversions several important products today are obtained via aerobic fermentation processes. The activated sludge, trickling filter, and aerated lagoon modes of industrial and domestic waste treatment are the largest scale, nonspecific versions of aerobic conversion.

Probably the largest scale, specific product aerobic fermentions are those now used to produce single cell protein (SCP) or yeast biomass from hydrocarbons [52]. Here, an aerated yeast suspension in water which contains trace nutrients is used to generate yeast biomass from gas oil, a mixture of straight chain hydrocarbons in the C_{12} to C_{20} range (Eq. 16.23).

$$\begin{array}{ccccc} \text{yeast} + & \text{water} & + \text{ gas oil} + & \text{air} & \rightarrow \text{yeast biomass} \\ & \text{(with} & (C_{12} \text{ to} & \text{(with} & \\ & \text{nutrients)} & C_{20}) & \text{ammonia)} & \end{array} \qquad 16.23$$

The yeast produced by continuous culture techniques is separated from the liquid medium and solvent washed by centrifugation or filtration techniques. After drying, a protein supplement is obtained which contains 65 to 68% protein and is suitable for addition to animal feeds. This protein content compares very favorably with that of dry fish meal, which contains about 65%, and dry skim milk powder with about 32%. SCP processes have operated on the thousands of tonnes per year scale in the U.K., France, and Italy, but regulatory problems with facilities operating on unpurified gas oil feedstocks have caused some shutdowns [53]. Nevertheless, because of the cell mass doubling time of 2.5 to 3 hours and the very efficient carbon conversion to protein of this technology, these developments deserve to be considered further. If problems involved with human consumption of SCP can be resolved this certainly represents a far more direct and less energy intensive route to protein than via meat or fish, and possibly even less energy intensive than plant proteins produced using high-technology (chemical fertilizer, tractors, herbicides, pesticides) agricultural methods.

It is also possible to convert methane, the lowest cost paraffin on a carbon content basis, to a mixed cell biomass in a similar type of fermenter [54]. But

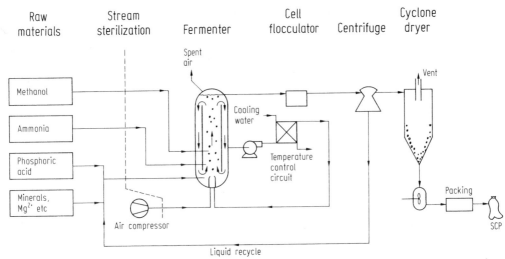

FIGURE 16.9 Details of the ICI single cell protein (SCP) process based on methanol. (Adapted from Prave and Faust [56], courtesy of The Royal Society of Chemistry.)

because of difficulties with obtaining sufficiently high concentrations of methane in the water phase, and the associated pressure required to keep it there, this system is not generally favored.

Methanol, relatively easily derived from methane and also readily purified, avoids some of the contamination problems of the higher, n-paraffin carbon sources. At the same time methanol is easily put into aqueous solution at any desired concentration. This is the basis of ICI's process to manufacture protein from methanol using *Methylophilus methylotrophus,* on a scale of 30,000 to 50,000 tonnes per year [55] (Fig. 16.9). Recombinant DNA technology was employed to raise the efficiency of methanol conversion by this organism. The dry product, trade named Pruteen, contains 72% protein and is suitable for animal feed supplementation.

16.6.3. Aerobic Processing of Hydrocarbon Substrates

Aerobic micro-organisms are capable of selective oxidation of specific hydrocarbon substrates which would be less efficient or impossible by direct chemical means. Strains of *Pseudomonas desmolytica,* for example, can carry out the specific oxidation of p-cymene to cumic acid [57], a process difficult to achieve by ordinary chemical means (Eq. 16.24).

$$\underset{\text{p-cymene}}{\overset{\displaystyle CH(CH_3)_2}{\bigcirc_{CH_3}}} \longrightarrow \underset{\text{cumic acid}}{\overset{\displaystyle CH(CH_3)_2}{\bigcirc_{COOH}}} \qquad 16.24$$

Some steroid transformations which are difficult to achieve by ordinary chemical means may also be accomplished readily by specific microbiological methods. Salicylic acid may be obtained in a 94% yield, on a weight for weight basis, by the action of *Pseudomonas aeruginosa* on naphthalene [58] (Eq. 16.25).

naphthalene salicylic acid "ASA" 16.25

Acetylation of the product with acetic anhydride or ketene nearly quantitatively converts this to acetylsalicylic acid (ASA, or aspirin), one of the most widely sold, nonprescription analgesics.

16.6.4. Microbiological Processes to Amino Acids

As the pressures on the world's available food supply become greater, the supply of first class (animal) protein will become more limited. This will place an increasing dependence on plant (vegetable) sources of protein for this necessary component of our diets. Most of the first class, animal-derived protein (meat, fish, eggs, milk, cheese, etc.) has a food demand to produce it of 2 to 20 times the weight of the protein obtained. Thus the amount of arable land required to support the production of animal protein is much larger than would be required to supply daily requirements directly from plant sources.

The problem with using plant sources of protein is that they do not provide the proper balance of the amino acids required for normal human nutrition [59]. Some of the deficient amino acids may be synthesized by the body. But some, the "essential amino acids," must be acquired through the diet for proper human growth and nutrition. Commercial production of amino acids has, thus far, been mainly for animal feed supplementation. However, cereals with synthetically produced essential amino acids could greatly improve the suitability of plant crops to provide the human protein requirement. Thus, a proportion of the crops now grown for animal feed could, with appropriate amino acid supplementation, be diverted to human consumption. In this way more people could be fed from the same area of arable land than would be possible using first class (i.e., animal) protein.

At the moment L-glutamic acid, a nonessential amino acid, is produced on the scale of about 340,000 tonnes per year, about a quarter of this in Japan. This is the largest volume of any commercially produced amino acid. Common starting materials for fermentation production of glutamic acid are glucose, molasses, corn or wheat glutens, or Steffen's waste liquor (from beet sugar mills) plus trace nutrients, although it is also possible to use purified fractions of n-paraffins (C_{12} to C_{20}) or acetic acid as substrates [46] (Eq. 16.26).

$$\text{glucose or molasses} \rightarrow \underset{\text{glutamic acid}}{\text{HOOCCH}_2\text{CH}_2\text{CH(NH}_2)\text{COOH}} \rightarrow \qquad 16.26$$

$$+$$

$$\underset{\text{zwitterionic form}}{\text{HOOCCH}_2\text{CH}_2\text{CH(NH}_3)\text{COO}^-}$$

Glutamic acid, as the end product of this fermentation, results from a tinkering of the normal cell metabolism of *Micrococcus glutamicus* in such a way that glutamic acid, as it is formed, "leaks" through the cell membrane and accumulates in the aqueous medium. Treatment of the glutamic acid product with sodium carbonate then yields monosodium glutamate (MSG). MSG is not an essential amino acid salt but is produced from an incentive from the large market for this product as a flavor enhancer. However, the experience gained from the fermentative production of glutamic acid on a commercial scale has served to pave the way for the commercial production of essential amino acids.

When monosodium glutamate is produced by a petrochemical route (Eq. 16.27), it yields the racemic mixture of D- and L-glutamic acid. This product is only about half as effective as the fermentative product on a weight basis.

$$\text{CH}_2\text{=CH—CN} \xrightarrow[\text{reaction}]{\text{oxo}} \text{OHCCH}_2\text{CH}_2\text{CN} \xrightarrow[\text{2. NH}_3]{\text{1. HCN}} \qquad 16.27$$

$$\text{NCCH(NH}_2)\text{CH}_2\text{CH}_2\text{CN} \xrightarrow[\text{H}_2\text{O}]{\text{hydrolysis}} \text{HO}_2\text{CCH(NH}_2)\text{CH}_2\text{CH}_2\text{CO}_2\text{H}$$

For this, and other economic reasons the petrochemical route is not as attractive as the fermentative route, so is not practiced commercially.

Of the two essential amino acids which are generally deficient in cereals, L-lysine and L-methionine, only L-lysine is currently produced by fermentation to any significant extent [42]. Again a carbohydrate source is a suitable substrate for cultures of *Corynebacterium glutamicum* to produce L-lysine (Eq. 16.28).

$$\text{carbohydrate source} \rightarrow \underset{\text{lysine}}{\text{H}_2\text{NCH}_2\text{CH}_2\text{CH}_2\text{CH}_2\text{CH(NH}_2)\text{COOH}} \rightarrow$$

$$+$$

$$\underset{\text{zwitterionic form}}{\text{H}_2\text{NCH}_2\text{CH}_2\text{CH}_2\text{CH}_2\text{CH(NH}_3)\text{COO}^-} \qquad 16.28$$

As with glutamic acid production, some modification of the normal metabolic pathway of *C. glutamicum,* imposed by the fermentation conditions, is necessary in order to accumulate lysine in the medium. At present, worldwide production of L-lysine by fermentation routes totals some 70,000 tonnes annually, about 40% of this by Japan. About 9000 tonnes of racemic lysine (DL mixture) per year is also produced synthetically from petrochemical sources. One interesting synthetic substrate for DL-lysine is caprolactam (or 6-aminocaproic acid), a starting material which is readily available on a tonnage scale.

As yet methionine, the other amino acid in which cereals are deficient, is only produced from pentaerythritol or acrolein by synthetic means, not by fermentation [60], which yields the DL product (e.g., Eq. 16.29 where

$$CH_3SH + CH_2{=}CHCHO \rightarrow CH_3SCH_2CH_2CHO \xrightarrow[\text{2. } (NH_4)_2CO_3]{\text{1. NaCN}}$$

$$\underset{\underset{CONH_2}{|}}{RCHNHCOO^-\overset{+}{N}H_4} \rightarrow \underset{\underset{NH{-}CO}{|}}{\overset{CH_3SCH_2CH_2CH{-}CO}{|}}{>}NH \rightarrow \qquad 16.29$$

$$\underset{\underset{NH_2}{|}}{CH_3SCH_2CH_2CHCOOH}$$

DL-methionine

$R = CH_3SCH_2CH_2-$). However, it is probably only a matter of time before the production of this and other essential amino acids yields to production via biochemical methods. This will come either from a detailed understanding of the metabolic processes of more types of suitable micro-organisms or by genetic engineering techniques to generate specific new types of micro-organisms with the required synthesis characteristics [61].

Plant genetic experiments presently being undertaken with corn (maize) and with rice may also help to improve the nutritional properties of these cereals. These may help to close the amino acid nutritional gap which exists between plant sources of proteins and first class proteins [59]. However, this will still not detract from the need for synthetic amino acids for their many other uses: e.g., for chelating agents, as dietary supplements, as building blocks for specialty proteins, and as components of intravenous feeding solutions. All of these uses will continue to provide incentives for large-scale biochemical and chemical synthesis of amino acids. Prospects and applications of amino acids have recently been surveyed [62].

16.6.5. Other Aerobic Fermentation Products

Since the original development and commercialization of penicillin production in about 1943, the antibiotics as a class and the penicillins in particular are still dominant in sales value in the aerobic fermentation products area [63]. The submerged culture methods used to prepare penicillin itself, which sold for \$330/g in 1943, were sufficiently improved and refined that by 1970 the price had dropped to about \$2/g (Eq. 16.30).

$$\begin{array}{c} H \qquad S \\ RCONH-C-CH \quad C(CH_3)_2 \\ \quad | \quad | \quad | \\ \underset{O}{\overset{\diagup}{C}}-N-CH(COOH) \end{array} \qquad 16.30$$

R = PhCH$_2$—	R = PhOCH$_2$—	R = PhCH(NH$_2$)—
"Penicillin"	Phenoxymethyl	Ampicillin
benzylpenicillin,	penicillin, or	(semi-synthetic)
or penicillin G	penicillin V	

Details of production methods and process flowsheets have been given by Shreve and Brink [3]. Despite the decreased cost of penicillin and of the related antibiotics such as streptomycin, neomycin, the tetracyclines, as their production methods were streamlined, the medical benefits of these drugs are so well established that the value of U.S. production of drugs in this class now exceeds a billion dollars annually [63].

Many vitamins too are produced by fermentation methods, although some are also produced by chemical synthesis procedures. Nevertheless those that are produced fermentatively represent another important class of submerged aerobic culture products. The total value of prescription vitamins alone that were marketed in the U.S.A. in 1980 exceeded $130 million [63]. In 1959, during the early development of the commercial fermentative product vitamin B-$_{12}$, it sold for $139/g. Again through processing improvements, the price of this product has been brought down dramatically so that it now sells for $8/g. Considering only the two B vitamins, riboflavin (vitamin B-$_2$), presently selling at $56/kg, and vitamin B-$_{12}$, these alone now account for U.S. sales of more than $20 million annually.

Citric acid is the largest scale pure organic acid produced by a microbiological process. World production in 1990 was about 550,000 tonnes. It is produced by *Aspergillus niger* from a medium containing sugars, obtained generally from low-cost cane or beet molasses, or occasionally a pure sugar solution. The stoichiometry of this process was originally established by Currie [64] (Eq. 16.31, 16.32).

$$C_6H_{12}O_6 + 3/2\ O_2 \rightarrow \qquad\qquad\qquad\qquad\qquad 16.31$$
glucose

$$HOOCCH_2COH(COOH)CH_2COOH + 2\ H_2O$$
citric acid

$$C_{12}H_{22}O_{11} + H_2O + 3\ O_2 \rightarrow \qquad\qquad\qquad\qquad 16.32$$
sucrose

$$2\ HOOCCH_2COH(COOH)CH_2COOH + 4\ H_2O$$
citric acid

To obtain citric acid as the metabolic product again requires interference with the normal Krebs tricarboxylic acid cycle in such a way that citric acid metabolism is blocked. Usually this is achieved by careful regulation of concentrations of trace metals available as coenzymes to the various enzyme pathways used by *A. niger,* so that some of these are rendered ineffective (are blocked).

About 40% of the citric acid produced is used as a synthetic flavoring agent, particularly for the tartness of fruit flavors in soft drinks. It is also a valuable chelating agent, helping to flavor and preserve foods, as well as contributing to the cleaning action of detergents. It may shortly be in large-scale use as a component of an aqueous sulfur dioxide scrubbing solution. No chemical synthesis of citric acid yet devised has been able to compete effectively with this microbiological method.

Several other organic acids are also made commercially on a large scale by microbiological techniques. Some of the better known examples of these are acetic acid (vinegar) via the metabolism of alcohol in water by *Acetobacter aceti,* and lactic acid (Eq. 16.33), produced by *Streptococcus lactis* fermentation of hexose sugars.

$$CH_3CH(OH)COOH \qquad HOCH_2(CHOH)_4COOH$$

lactic acid   gluconic acid

$$H_2C = C(COOH) - CH_2COOH$$

itaconic acid

16.33

Common sugar sources for lactic acid are molasses, hydrolyzed corn, and whey, a milk by-product of cheese production. Gluconic and itaconic acids are also microbiological oxidation products, obtained by methods similar to those used for citric acid [2]. Gluconic acid is formed, via simple oxidative fermentation, by *Aspergillus niger* or *Acetobacter suboxydans,* of the glucose aldehyde group.

16.6.6. Soluble and Immobilized Enzymes

A branch of the fermentation industry has pursued the isolation of enzymes, which are the catalytically active proteins of a cell. In the presence of the appropriate coenzyme, these proteins are the active principals of the cell which carry out the chemical changes from which the cell gains energy. Rennet, a milk curdling enzyme used in cheesemaking, and the amylases, used in food processing and for laundry stain removal, are among the earliest examples of isolated enzymes which have been used commercially [65]. The essence of the value of enzyme use is that the enormous rate enhancements obtained by microbiological means can be achieved in the absence of the organism, if the enzyme responsible is stable and can be isolated. A third of current enzyme production is also directed toward therapeutic uses. A direct result of these wide applications is an increase in the value of the world enzyme market from $300 million in 1979, $480 million in 1985, to over a billion dollars in 1990.

The advantage of using isolated enzymes rather than whole cells to carry out a chemical change is that the process does not have to cater to the special requirements of living cells. However enzymes, too, are sensitive entities so that care has to be exercised in the conditions under which they are used, in order to maintain catalytic activity. An additional problem is that use of the isolated enzyme is frequently a more expensive proposition, on a single-use basis, than use of propagated cells. There are other disadvantages to their use too, such as the need to remove enzyme from the product once the change desired is completed. Alternatively, long reaction times may be necessary if it is decided to use a low concentration of an expensive enzyme to decrease enzyme cost [66].

Some of these difficulties of enzyme utilization may be decreased or eliminated, and the full potential realized by fixing or immobilizing the enzyme in some way. The solution of substrate for conversion is then passed through a bed or tube of the immobilized enzyme, and the product(s) collected as the solution emerges after contact with, and conversion by, the enzyme.

A number of methods are being tested for enzyme immobilization, the actual method used in any particular case depending on the operating details of the enzyme system employed and the nature of the solvent to be used as the substrate carrier, which is usually water. Enzyme, or inactivated cells, may be encapsulated in a film, or encased in a gel which is permeable to both substrate and product, but not to enzyme [66]. Porous glasses, or insoluble polymers such as a derivatized cellulose may be used as a support onto which enzyme is adsorbed. Pendant functional groups of a polymer, such as those of the ion-exchange resins, can be used either to ionically bind the enzyme to the resin active sites or to covalently bond the enzyme to the resin [67]. Resin may also be used to form the backbone of a polymer chain using a linking bifunctional monomer such as glutaraldehyde to react with enzyme sites that do not affect its catalytic activity [68].

Hollow fiber technology may also be usefully wedded to immobilized enzyme technology by making the wall of a hollow fiber out of a gel in which the enzyme is already immobilized. The finished fiber may then be used simply as a large exposed surface area for enzyme activity, packed in small chopped pieces inside a larger tube. Or it may be used as a membrane into which substrate solution permeates from the outside, under a pressure differential. Passage through the tube wall allows contact of the substrate with the gelled enzyme, which converts substrate to product. Then, under the influence of the differential pressure across the wall of the tube, the product would flow through to the inside of the tube, eventually to be collected at a multi-tube header.

Immobilized enzymes have already been successfully tested for sugar conversion to ethanol [69,70]. Conversion of the lactose of whey to glucose and galactose is another confirmed application of immobilized enzymes. Preliminary tests have prompted predictions of 300-day operating cycles, when processing 6500 L/day (1800 U.S. gal/day) of whey containing 5% lactose [71]. In doing so, it converts the lactose in a cheese maker's waste stream into two monosaccharides of higher sweetening power and therefore greater food processing value than the lactose from which they are derived.

Enzymes, in both soluble and immobilized forms, are also increasingly being employed in the food processing industry to convert low sweetening power sugars into more potent sweeteners. As examples, glucose is isomerized to fructose, which is 2.5 times as sweet, by an immobilized glucose isomerase enzyme in a fixed or a fluidized bed [72]. Also a high-fructose corn syrup, to be produced from corn using two stages of immobilized enzymes, is likely to soon become a commercial reality. Starch hydrolysis to glucose would be accomplished by an immobilized glucoamylase in the first stage, in 6 minutes, followed by glucose isomerization to fructose, as outlined above.

This power of enzyme systems for specific interconversions of sugars has also been most usefully exploited as an additional fresh milk processing step, at present tested only in Milan. This process converts the normal lactose milk sugar, to which some people react unfavorably, into glucose and galactose which are much better tolerated [71]. Milk treated in this way can now be used as a beverage by those who could only tolerate milk substitutes previously.

Amino acid production too may extensively benefit from this immobilized enzyme technology. Optically active D($-$)-α-phenylglycine is one example which has been confirmed, but more are likely to follow.

The promise of fixed enzymes to process many food and waste products is only just beginning to be realized [73]. The prediction is that combining this technology with the developing field of synthetic enzyme preparations will lead to a wide range of feasible, clean, and specific chemical conversions never before thought to be possible [74].

REVIEW QUESTIONS

1. In the process of photosynthesis green plants use carbon dioxide and water to produce sugars according to the reaction:

$$H_2O + CO_2 \rightarrow C_{12}H_{22}O_{11} + O_2 \text{ (unbalanced)}$$

(a) What volume of CO_2 at 20°C and 760 mm Hg pressure is used by a plant to make 1 kg of sucrose, $C_{12}H_{22}O_{11}$?
(b) What volume of air is deprived of its normal amount of CO_2 in this process? (Air contains 0.033% CO_2 by volume.) Assume a barometric pressure of 760 mm Hg and a temperature of 20°C, and that photosynthetic CO_2 depletion is complete.

2. Normal brewing involves yeast fermentation of a wort initially containing 12% sugars (as $C_6H_{12}O_6$, 120 g/L solution) down to a beer containing 2.4% sugars plus ethanol, with evolution of carbon dioxide. Assume quantitative sugar conversion by zymase, negligible volume change on conversion, and unit density of starting and product solutions.

(a) What percent concentration of alcohol, weight per liter, should this produce?
(b) What mass of carbon dioxide would be coproduced per liter of wort fermented?
(c) If 0.029 L of carbon dioxide at 5°C and 1 atm pressure are required to carbonate one liter of beer, what proportion of the carbon dioxide produced is required for carbonation?

3. Winemaking essentially involves fermentation of a must containing sugars and flavor elements to a solution of ethanol plus flavor elements still containing some residual sugar.

(a) If a must containing 185 g/L sugar is fermented to a green wine with a sugar content of 45 g/L, what percent concentration of alcohol by weight should this produce? Assume quantitative conversion, of $C_6H_{12}O_6$, glucose, negligible volume change on conversion, and a density of 1.075 g/mL both before and after fermentation.
(b) What would be the technical classification of this wine, if the carbon dioxide produced was allowed to escape?
(c) What mass of carbon dioxide would be produced at the same time per liter of must fermented?
(d) What fraction of the cleaned-up carbon dioxide would be required for reinjection into the wine on bottling to produce a champagne? Bottling conditions are 1.0 L CO_2 added per L of wine, at 5 atm absolute pressure, and 0°C.

4. White vinegar is a solution of acetic acid in water. Knowing only that acetic acid contains C, H, and O atoms in the ratio of 1:2:1, respectively, and that a sample of vinegar gives an analysis of 2% carbon by weight, what percentage (by weight) of the sample is actually acetic acid?

5. An American 20° overproof navy rum is to be sold in Canada.
(a) What is the percent alcohol (ethanol) by volume in this product?
(b) What would be the proper designation of this product using the Canadian proof system (which is equivalent to degrees Sikes)?
(c) How should a brandy containing 45.7% alcohol by volume be specified on the Canadian proof system, i.e., degrees Sikes.

6. Corn contains 82% starch (dry weight basis) and an unfermentable residue of corn oil, gluten, fiber, etc.
(a) What dry weight mass of corn would be required to produce 1000 L of 100% ethanol (density 0.789 g/mL) assuming quantitative conversions of starch to fermentable sugars, and sugar fermentation to ethanol?
(b) Ethanol (100%) currently sells at $1.50 Imp. gallon. At what price would corn have to be available, allowing 50% of the price of the ethanol for operating overheads (capital costs, energy, labor, etc.) to enable this to be an economic source of the starch? 1 Imp. gal. = 4.546 L.
(c) At a straight gasoline price of 42¢/L, what would be the price of gasohol at 10% ethanol by volume?
(d) Name any two other potential sources of fermentable sugars for fuel alcohol production which are likely to be significantly less expensive than corn.

7. Vinegar is produced by the oxidative microbiological conversion of a 6.0% by weight solution of ethanol in water to acetic acid in water as the sole carbon-containing product.
(a) What concentration of acetic acid by weight would be present in the product, assuming quantitative conversion of ethanol and unit density for the starting and product solutions?
(b) What volume of air (assume ½ O_2:2 N_2) at 1 atm and 0°C would be required to carry out the conversion of 2.0×10^{-3} L of the starting solution?

8. At one time a test for the detection of adulteration of tobacco by sugar involved extraction of a weighed dry tobacco sample with hot water, and fermentation of the extract by yeast [P. Hammond, *Chem. in Britain* **28**(9): 796–798 (1992)]. Unadulterated tobacco contains 1–2% by weight fermentable sugars. Fermentation of the water extract from a 100-g sample of tobacco gives 6.6 g of pure ethanol and 0.13 mole of carbon dioxide. Assume quantitative fermentation using $C_6H_{12}O_6$.
(a) What is the percent sugar by weight in the tobacco sample from the ethanol recovered?
(b) What is the percent sugar by weight in the tobacco sample calculated from the moles of carbon dioxide collected?

(c) Is the tobacco sample adulterated or not, with consideration of the possibility of experimental error in each of the two results?

FURTHER READING

American Chemical Society, "Chemistry of Winemaking," ACS Adv. Chem. Ser., No. 137. American Chemical Society, Washington, DC, 1974.

American Chemical Society, "Wine Production Technology in the United States," ACS Symp. Ser., No. 145. American Chemical Society, Washington, DC, 1981.

M.E. Bushell, "Computers in Fermentation Technology." Elsevier, Amsterdam, 1988.

A.C. Olson and C.L. Cooney, eds., "Immobilized Enzymes in Food and Microbial Processes." Plenum, New York, 1974.

C.S. Pederson, "Microbiology of Food Fermentations," 2nd ed. AVI Publ. Co., Westport, CT, ca. 1979.

O.P. Ward, "Fermentation Biotechnology: Principles, Processes, and Products." Prentice-Hall, Englewood Cliffs, NJ, 1989.

REFERENCES

1. P. Dunnill, *Chem. Ind. (London)*, p. 204, April 4 (1981).
2. J.A. Kent, ed., "Riegel's Handbook of Industrial Chemistry," 7th ed., p. 156. Van Nostrand-Reinhold, New York, 1974.
3. R.N. Shreve and J.A. Brink, Jr., "Chemical Process Industries," 4th ed., p. 525. McGraw-Hill, New York, 1977.
4. H.J. Phaff, *Sci. Amer.* **245**(3), 76, Sept. (1981).
5. W.N. McCutchan and R.J. Hickey, *in* "Industrial Fermentations" (L.A. Underkofler and R.J. Hickey, eds.), Vol. 1, p. 347. Chem. Publ. Co., New York, 1954.
6. A. Mulchandani and B. Volesky, Production of Acetone-butanol-ethanol by *Clostridium acetobutilicum* using a Spin Filter Perfusion Bioreactor. *J. Biotechnology* **34**, 51–60, April 30 (1994).
7. F.W. Salem, "Beer, Its History and Economic Value as a National Beverage." Arno Press, New York, 1972, reprint of 1880 edition.
8. T. Lom, *Can. Chem. Process.* **58**(12), 14, Dec. (1974).
9. "United Nations Statistical Yearbook 1993," 40th ed., United Nations, New York, 1995, and earlier years.
10. "Kirk-Othmer Encyclopedia of Chemical Technology," 3rd ed., Vol. 3, p. 692. Wiley, New York, 1978.
11. R. Stevens, *Chem. Rev.* **67**, 19 (1967).
12. M.A. Amerine, *Sci. Amer.* **211**(2), 46, Aug. (1964).
13. G. Donaldson and G. Lampert, eds., "The Great Canadian Beer Book." McLelland and Stewart, Toronto, 1975.
14. R.I. Tenney, The Brewing Industry. *In* "Industrial Fermentations" (L.A. Underkofler and R.J. Hickey, eds.), Vol. 1, p. 172. Chem. Publ. Co., New York, 1954.
15. A. Vogler and H. Kunkely, *J. Chem. Educ.* **59**(1), 25 (1982).
16. Avoiding an Environmental Hangover, *Can. Chem. Process.* **64**(2), 29, March (1980).
17. S. Beszedits, *Water Pollut. Control* **115**(10), 10, Oct. (1977).
18. H.M. Malin, Jr., *Environ. Sci. Technol.* **6**, 504 (1972).
19. J.F. Walker, *Water Pollut. Manage.*, p. 174 (1972), cited by *Water Polln Abstr. (U.K.)* **45**, 2953 (1972).
20. H.W. Allen, "A History of Wine." Horizon Press, New York, 1961.
21. P. Duncan and B. Acton, "Progressive Winemaking." Amateur Winemaker, Andover, Hants, 1967.

22. M.A. Amerine, H.W. Berg, and W.V. Cruess, "The Technology of Wine Making," 3rd ed. The Avi Publishing Co., Westport, CT, 1972.

23. L. Trauberman, *Food Eng.* **39**, 83, Nov. (1967).

24. B. Hocking, Mead—A Silk Purse from a Sow's Ear. *Can. Bee J.* **63**(12), 4–6,18, Dec. (1955).

25. L. Morisset-Blais, *Can. Chem. Process.* **57**(12), 31, Dec. (1973).

26. H.M. Cunningham and R. Pontefract, *Nature (London)* **232**, 332 (1971).

27. W.H. Detwiler, *Chem. Eng. (N.Y.)* **75**, 110, Oct. 21 (1968).

28. A. Agostini, Wynboer **32**(395), 13 (1964); cited by Amerine *et al.* [22].

29. "Kirk-Othmer Encyclopedia of Chemical Technology," 2nd ed., Vol. 1, p. 501. Wiley-Interscience, New York, 1963.

30. N.A. Lange, ed., "Lange's Handbook of Chemistry," 10th ed., p. 1183. McGraw-Hill, New York, 1969.

31. Business Week, issues of Feb. 17, 1962 and March 21, 1977. Cited by Kirk-Othmer [10], p. 830.

32. Kirk-Othmer Encyclopedia of Chemical Technology," 4th ed., Vol. 4, p. 160. Wiley, New York, 1992.

33. B.G. Reuben and M.L. Burstall, "The Chemical Economy," p. 428. Longman Group Ltd., London, 1973.

34. J.S. Swan and S.M. Burtles, *Chem. Soc. Rev.* **7**(2), 201 (1978).

35. "Kirk-Othmer Encyclopedia of Chemical Technology," 3rd ed., Vol. 3, p. 830. Wiley, New York, 1978.

36. W.F. Rannie, "Canadian Whisky, The Product and the Industry." W.F. Rannie Publishing Co., Lincoln, Ontario, 1976.

37. "McGraw-Hill Encyclopedia of Science and Technology," 5th ed., Vol. 4, p. 346. McGraw-Hill, New York, 1982.

38. P.V. Price, "Penguin Book of Spirits and Liqueurs," Handbook Series. Penguin Books, New York, 1981.

39. R.V. Carter and B. Linsky, *Atmos. Envir.* **8**(1), 57 (1974).

40. N.L. Nemerow, "Industrial Water Pollution, Origins, Characteristics, Treatment." Addison-Wesley, Reading, MA, 1978.

41. Methane Pilot for Distillery, *Can. Chem. Process.* **66**(1), 8, Feb. 19 (1982).

42. D.E. Eveleigh, *Sci. Amer.* **245**(3), 154, Sept. (1981).

43. "Kirk-Othmer Encyclopedia of Chemical Technology," 3rd ed., Vol. 9, p. 338. Wiley, New York, 1980.

44. No Recovery in Sight for Major Solvents, *Chem. Eng. News* **59**(45), 12, Nov. 9 (1981).

45. National Distillers to Convert Plant for Gasohol, *Chem. Eng. News* **58**(19), 8, May 12 (1980).

46. W.L. Faith, D.B. Keyes, and R.L. Clark, "Industrial Chemicals," 3rd ed. Wiley, New York, 1965.

47. C.R. Noller, "Textbook of Organic Chemistry," 3rd ed. Saunders, Philadelphia, 1966.

48. W. Worthy, *Chem. Eng. News* **59**(49), 35, Dec.7 (1981).

49. R.C. Weast, ed., "Handbook of Chemistry and Physics," 56th ed. CRC Press, Cleveland, OH, 1975.

50. R.H. Perry, C.H. Chilton, and S.D. Kirkpatrick, eds., "Chemical Engineers' Handbook," p. 13–50. McGraw-Hill, New York, 1969.

51. B. Kristiansen, *Chem. Ind. (London),* p. 787, Oct. 21 (1978).

52. G. Hamer and I.Y. Hamdan, *Chem. Soc. Rev.* **8**(1), 143 (1979).

53. BP Contests Petroprotein Plant Health Issue, *Chem. Eng. News* **56**(12), 12, March 20 (1978).

54. J.C. Mueller, *Can. J. Microbiol.* **15**(9), 1047 (1969).

55. ICI's Bugs Come on Stream, *Chem. Br.* **17**(2), 48, Feb. (1981).

56. P. Prave and U. Faust, *Chem. Br.* **14**(11), 552 (1978).

57. K. Yamada, S. Horiguchi, and J. Takehashi, *Agric. Biol. Chem.* **29**, 943 (1965).

58. T. Ishikura, H. Nishida, K. Tanno, M. Mujachi, and A. Ozaki, *Agric. Biol. Chem.* **32**, 12 (1968).

59. M. Hamdy, *CHEMTECH* **4**(10), 616 (1974).

60. R.F. Goldstein and A.L. Waddams, "The Petroleum Chemicals Industry," 3rd ed., p. 345, 407. Spon, London, 1967.

61. D.A. Hopwood, *Sci. Am.* **245**(3), 90, Sept. (1981).

62. "Kirk-Othmer Encyclopedia of Chemical Technology," 4th ed., Vol. 2, pp. 504–579. Wiley, New York, 1992.

63. Y. Aharanowitz and G. Cohen, *Sci. Am.* **245**(3), 140, Sept. (1981).

64. J.N. Currie, *J. Biol. Chem.* **31**, 15 (1917); *Chem. Abstr.* **11**, 2814 (1917).

65. J.L. Meers, *Chem. Br.* **12**(4), 115, April (1976).

66. C. Bucke and A. Wiseman, *Chem. Ind. (London)*, p. 234, April 4 (1981).

67. C.J. Suckling, *Chem. Soc. Rev.* **6**(2), 215 (1977).

68. W.R. Vieth and K. Venkatasubramanian, *CHEMTECH* **4**(1), 47 (1974).

69. Y. Shabtai, S. Chaimovitz, and A. Freeman, Continuous Ethanol Production by Immobilized Yeast Reactor coupled with Membrane Pervaporation Unit. *Biotechnol. Bioeng.* **38**, 869–876 (1991).

70. Ethanol Via Fixed Bed, *Can. Chem. Process.* **65**(1), 18, Feb. (1981).

71. R. Greene, *Chem. Eng. (N.Y.)* **85B**, 78, April 10 (1978).

72. Enzymes Now Winning New Applications, *Can. Chem. Process.* **61**(12), 20, Dec. (1977).

73. D. Perlman, *CHEMTECH* **4**(4), 210 (1974).

74. C.J. Suckling and K.E. Suckling, *Chem. Soc. Rev.* **3**(4), 387 (1974).

Bucket excavator used for the large-scale shifting of surface materials required for the tar sands processing, Syncrude, Fort McMurray, Alberta.

17
PETROLEUM PRODUCTION AND TRANSPORT

Energy is beauty—a Ferrari with an empty tank doesn't run.

—Elsa Peretti (1940–)

17.1. PRODUCTION OF CONVENTIONAL PETROLEUM

Early use of petroleum or mineral oil, as opposed to animal or plant oils, was by direct harvesting of the crude product from surface seeps and springs. For example, tar obtained from the Pitch Lake (La Brea) area, Trinidad, has been used for the caulking of ships since the Middle Ages and is still marketed to the extent of about 142,000 tonnes per year [1]. Tar from the Alberta tar sands was used in the 1700s by Cree Indians of the Athabasca river area to seal their canoes, as recorded by Peter Pond. Also a thick bituminous gum was collected from the soil surface in the vicinity of the St. Clair River, Southern Ontario [2], and from Guanoco Lake, Venezuela [3], and these too were marketed for a range of purposes.

It was a relatively small but important step to advance from opportunistic oil recovery and utilization from surface seeps of this kind to the placement of wells in or near these surface deposits in an organized attempt to try to increase production. Probably the first wells drilled in the Western world with the specific intent of oil recovery were placed during the period 1785 to 1849 in the Pechelbronn area of France [4]. None of these wells, which ranged in depth from 31 to 72 m, was a prolific producer but they did serve to establish the practice of drilling for oil in Europe. In North America, "Colonel" Edwin H. Drake drilled a well to about 19 m (60 feet) in depth at Titusville, Pennsylvania, in 1859, where he struck oil. This well initially produced oil at the rate of about 20 barrels per day. Some historical accounts of the development of the North American petroleum industry credit this event as being the first instance of crude oil production from a well [5]. But more comprehensive accounts also describe the accomplishments of James M. Williams in Canada who, at about the same time, had already marketed some 680 m^3 (150,000 gallons) of oil produced from his wells near Black Creek in the vicinity of Petrolia and Oil Springs in Southern Ontario [6]. Some of this production was also partially refined, on site. The first of Williams's wells, several of which were placed by digging rather than by drilling, produced oil at a depth of about 18 m (59 ft) in late 1857 to early 1858. By 1860 each of Williams's five successful wells were producing on the average about 3.2 m^3 (20 barrels) of oil a day netting an annual production rate for Ontario of some 1600 m^3 (10,000 barrels) in that year. American production rapidly made up for the marginally later start in the production of crude oil with a gross of something more than 40,000 m^3 (250,000 barrels) in the same year.

17.1.1. Modern Exploration and Drilling for Oil and Gas

Prospecting for petroleum (literally "rock oil") draws significantly on the present state of knowledge regarding its genesis. One requirement for oil formation is the presence, millions of years ago, of shallow seas having a rich microscopic and/or larger animal life and probably also plant life. For the effective formation of oil deposits this condition must have been accompanied by a large sediment source such as a turbid river. Thus, as organisms died or were trapped in the accumulating sediments anaerobic decomposition occurred with the assistance of bacteria. As further sediments accumulated the transformation process was further assisted by the contributions of heat (geothermal and decompositional) and possibly also pressure. Pressure of accumulating sediment eventually caused expression of the fluid and gaseous decomposition products which then migrated through a porous deposit to formations of lower pressure, generally upward or sideways (horizontally), until the mobile products reached impermeable strata. If this type of hydrocarbon trapping occurred a petroleum reservoir or deposit was formed (Fig.17.1). Or there may have been no nonporous barrier between the formation zone and the surface, or a fault line may have subsequently permitted migration in which case a large part of the hydrocarbon content either moved to a fresh entrapment zone or was lost to the atmosphere and surroundings by evaporation and weathering. Losses of the latter type probably combined with some further bacterial action are what have given rise to surface deposits of viscous, high molecular weight hydrocarbons such as those that occur at Pitch Lake, Trinidad, and the other surface seeps and springs mentioned earlier.

While the harvesting and tapping of more or less self-evident oil deposits was sufficient to provide for the early demand for petroleum products, the incentive very soon became strong to devise methods to explore areas without any such obvious signs, to improve the prospects of success on drilling. For a new area this initially involves large-scale aerial strip photography, with overlap of adjacent frames and strips to allow stereoscopic viewing to determine the topography and geologic features of the area of interest. From study of these photographs, maps are made which include accurate placement of the

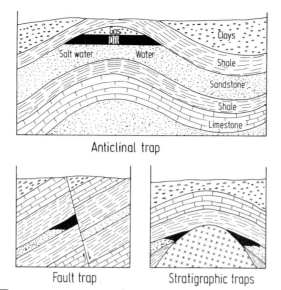

Anticlinal trap

Fault trap   Stratigraphic traps

FIGURE 17.1   Three types of structural traps which can provide a situation favorable to the formation of oil accumulations. Other types of structural traps such as salt domes are also recognized.

main physical features of importance and also the principal aerially observable geological features of interest. Aerial surveys may be continued via gravity meter measurements, in which the small differences in the force of gravity from variations in rock distribution may be measured, and/or by magneto-meter surveys, from which local variations in the earth's magnetic field are used to estimate the thickness of underlying sedimentary rock formations. More recently, geochemical methods which rely on determining hydrocarbon gas content above surface soils are also being employed [7]. The presence of a hydrocarbon "halo" at the surface, from the diffusion of gases from an underground deposit, is taken as promising evidence of an accumulation.

More detailed ground surveys are conducted in the areas revealed to be of prime interest from the large-scale methods. Surface study draws on a variety of procedures of which the first essential is geologic mapping, a step which may include fossil collecting and analysis. This is commonly followed by seismic surveys to try to establish evidence of an anticlinal structure or other rock structure in which a petroleum reservoir may occur. Evidence of the presence of a thick sequence of marine sedimentary rock, including some rich in kerogen, would be a good sign of hydrocarbon formation.

Seismic study employs a small explosive charge or several charges to set off shock waves which are transmitted through the rock formations. Placement of 12 or more sensitive geophones (listening devices) in a pattern around the site of the explosive charge allows the travel time and frequency of the shock waves to be traced on a high-speed chart. Study of these charts, in turn, provides geologists with sufficient shockwave intensities and times, including reflections from deep-seated formations and other data, to permit assembly of an estimate of the shape of the underlying formations.

Test drilling will follow these preliminary procedures, if the site area still looks promising. Core samples (clean cylinders of rock) are taken using a special circular cutting drilling bit. The cores are carefully stored at the surface in the sequence in which they were obtained. Study of these permits determination of formations or strata at this drilling site which are parallel with those previously collected at other drilling sites. It also enables any fossils present to be identified and from a prior knowledge of these, the approximate age of the formations may be determined. Stratum dating is an important feature of core analysis because of the good correspondence observed between formation age and the geological periods during which many oil deposits were laid down [8]. It is also possible from core samples to establish whether the underlying strata at the test site are appropriate for petroleum entrapment. The presence of vugs, or holes in a limestone formation, for instance, signals the possibility of that formation serving as an oil reservoir. Evidence of a porous and permeable zone of the appropriate geological age overlain by an impermeable shale would also be promising.

The high cost of coring, in both time and money, means that this method of obtaining information about underlying strata is usually held to the minimum. Much subsurface geologic information is derived from a detailed study of the rock cuttings (chips) brought up with the flow of drilling mud from the operation of the regular rotary bit. More coring is done in areas which are being newly investigated, usually when the bit has just entered an oil-stained and/or porous zone as evidenced by the cuttings. Cuttings can also be used for hole-to-hole comparisons, again decreasing the extent to which coring has to be used for this purpose. Coring, however, remains the only method by which precise data on porosity and permeability of a formation can be obtained, from physical tests conducted at the surface on the comparatively larger volumes of rock provided by core samples.

Drilling success rates, that is, wells drilled which show any evidence of the presence of oil or gas, are markedly improved with the sophistication of the initial exploration techniques used. Ratios of 1 success in 30 or more holes is about all that can be expected without the application of the methods outlined whereas success ratios of approximately 1 in 5 to 1 in 10 is achievable when using evidence gained from these procedures [9]. Since to drill and complete (bring into production) a typical oil well in an accessible region can cost upwards of half a million dollars, and $10 million or more in frontier regions, it is important to assemble as much favorable evidence as feasible before drilling. For a well to be worth bringing into production it must be capable of a significant initial rate of oil or gas production. Otherwise the cost of the associated surface collection equipment becomes too high in relation to the anticipated value of the product, so the well is capped. The success rate of producing wells, compared to dry holes or nonproducers, runs to about 1 in 9 or 10 wells completed in North America, hence the high exploration investment required by the oil industry.

Early drilled wells were made by a cable-tool system in which a heavy tool, about 10 cm in diameter by 1.5 to 2 m long and with a sharpened end, was alternately lifted and then let drop in the hole by a cable and winch system. At intervals the loosened material, suspended in a few centimeters of water at the base of the well, was lifted out with a bailer end to the cable system. The bailer consisted of a length of pipe fitted with a weighted valve at the lower end which opened when the pipe was lowered to the end of the hole, and which reclosed on lifting. Then the percussion drilling could be resumed. This method, which had been used by the Chinese for water and brine wells for more than a thousand years, was effective, simple, inexpensive, and could reach depths as great as 1100 m [4]. It was virtually the only method used to drill oil wells in the nineteenth century, and the predominant method for the first two decades of the twentieth century. But cable-tool systems were slow relative to other developing methods, particularly in softer formations. They had a depth limitation too and, using this system, it was not possible to provide an effective safeguard for pressure containment in the event that the drill penetrated formations under high gas pressures.

Rotary drilling, which gradually became the dominant well drilling method in the first quarter of this century, enabled more rapid well completion and permitted working to depths as great as 8000 m but required a greater equipment and labor investment to achieve this. First, a larger derrick or support structure about 40 m high is required above ground to enable lifting, lowering, and guidance of the drill string and to act as a temporary storage rack for lengths of drill pipe during bit changes, etc. (Fig. 17.2). The actual drilling is accomplished by special types of bit, about 16 cm in diameter, threaded to the end of 9-m lengths of about 8-cm-diameter special high-strength steel pipe sufficient to reach the bottom of the hole from the surface. A lifting system from the top of the derrick allows control of the weight or force on the bit and hence the torque required to rotate it as the bit cuts into the rock formations being penetrated. Since three 9-m lengths (the working unit kept stacked inside the derrick) of even 12-cm-diameter drill pipe weighs some 730 kg, sufficient pipe to reach a 2000-m depth would weigh some 54 tonnes. Thus this control task using massive equipment is a delicate one, assigned to the head driller of each operating crew to ensure optimum drilling rates without breakage of the drill string from the application of excessive vertical pressure on the bit. A rotating table at the base of the derrick, driven by 3000 to 4000 hp of tandem diesel engines, is used to grip and rotate the pipe string from the top of the string at 100 to 250 rpm [9].

A specially formulated drilling mud pumped down the inside of the pipestring and returning between the pipe and the side of the hole, provides lubrication and cooling for the pipe string and drill bit. It also carries rock chips up to the surface for removal from the mud by screening, assists in sealing any somewhat porous formations which are being penetrated, and, most importantly, provides a safety feature in the event that a formation under high gas pressure is reached. The density of the drilling mud used will be deliberately increased by the addition of powdered barytes ($BaSO_4$; density 4.15 g/cm^3), iron oxide, or the like when it is anticipated that a high-pressure formation is likely to be encountered. This can be predicted, for instance, when further developing an oil reservoir with known high formation pressures which has already been tapped by exploratory wells.

Blowouts, or the loss of drilling fluid and sometimes the whole of the drill pipe and derrick as well when unanticipated high pressures are encountered,

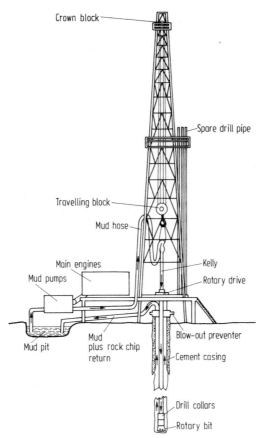

FIGURE 17.2 Schematic diagram of a rotary drilling rig showing hoisting derrick and routing of drilling mud.

only rarely occur today. Warning signs such as a rapid unexplained increase in drilling mud volume, sometimes accompanied by a frothy appearance, are taken seriously and the drill mud exit stream flow rate may be restricted to increase the drilling mud pressures in the hole. Also modern drill rigs are equipped with blowout preventers, massive valves placed immediately under the derrick floor and cemented into the top of the hole, as well as additional valves which may be placed at one or two intermediate depths for deeper wells, and can be closed at a moment's notice on the threat of a blowout. In these ways both the hazards and the significant hydrocarbon and formation pressure losses which can result from a blowout are generally avoided.

One further mode of drilling, in addition to the cable-tool and rotary methods already described, is the turbo drill. This method, developed significantly by the Russians, uses much of the same equipment such as derrick, draw works (lifting system), and mud pumps of the rotary drilling method. But instead of imparting the turning force to the bit by rotating the string of pipe from the top of the string, in the manner of a conventional drill, this system uses a turbo-powered unit located in the drill string at the bottom of the hole and just above the drill bit itself to rotate the bit. The pipe string remains stationary, except for vertical movement during bit changes, etc. Power is carried to the turbo unit and thence to the bit via the high-pressure mud stream itself, flowing down the inside of the string of pipe. Since the drill string does not need to provide the rotating torque to the drill bit with this method, the pipe stresses are much less. For this reason turbo-drilling is of primary value for the making of very deep holes of 5000 m or more.

17.1.2. Petroleum Production

Conventional petroleum accumulations can generally be classified into one or the other of four types of oil reservoirs (Fig. 17.3) [10]. Production from a water drive reservoir involves the movement of petroleum upward into the producing well by displacement of petroleum from the lower portions of the reservoir into the producing zone by the hydrostatic pressure of underlying water, also present in the formation. Conversely a gas cap drive reservoir relies on the pressure of a separate natural gas phase within the formation to move oil downward toward a producing well. In some reservoirs containing both oil and gas, however, the two components are more or less homogeneously distributed throughout the formation, a production situation described as a dissolved gas drive. When this occurs, both components move into producing wells. The fourth type of reservoir, where little or no production assistance is derived from the natural pressure of the formation contents, is described as a gravity drainage reservoir. In some petroleum producing situations combinations of these reservoir types may be encountered.

The fraction of oil-in-place which is recoverable from conventional petroleum reservoirs varies greatly with the reservoir type, oil viscosity, formation pressure, production rate, and finesse employed. Water drive reservoirs,

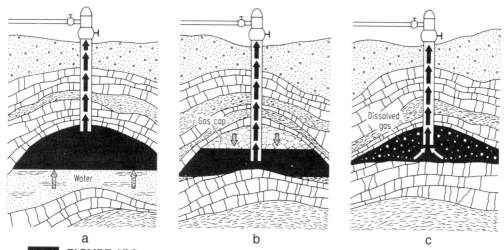

| a | b | c |

FIGURE 17.3 Diagrammatic representations of three of the four common types of oil reservoir: (a) Water drive. (b) Gas cap drive. (c) Dissolved gas drive. (From Purdy [10], courtesy G.A. Purdy.)

because of the positive displacement aspect, generally give the highest ultimate petroleum recoveries, or up to about 70% of the oil-in-place [10]. Estimates of ultimate recoveries possible from gas cap drive and dissolved gas drive types of reservoirs are generally much lower, in the 25–50% range for the former and 10–30% for the latter. Recovery from gravity drainage reservoirs will also be low, at the lower end of the recovery ranges of the two gas drive types of reservoirs.

When a new oil formation of one of the first three types is first tapped, the formation pressures of up to 20 mPa (3000 lb per square inch) or more which may be encountered are sufficient to force the oil being produced to the surface under formation pressure. Thus, artesian well fashion, the crude oil (plus associated gas) simply requires piping to a temporary holding area or to a well-head separation manifold. In the early days of the oil industry, when competing interests were all vying for product from the same oil reservoir under little or no government regulation, it became a race to access as many producing wells into the deposit as soon as possible to obtain the maximum share of the product. At the time there was also little application for the coproduced natural gas so that any which did separate from the oil product during production was simply flared at the well site. While these practices may well have given one company a larger share of the product recovered from a particular oil field than another company, they also served to markedly decrease the ultimate total recovery of oil possible from the field, to the detriment of all producers. This yield decrease occurred as a net result of the premature loss of formation pressure, an increase in the average viscosity of residual formation oils still in place, and from discontinuities in fluid-filled voids in the rock. All of these production problems were a direct result of the rapid production rates, sometimes complicated by reservoir structural features.

For some newly developed oil fields and for most oil fields when they reach a mature stage of production, formation pressures are inadequate to raise the oil to the surface although they may be sufficient to raise the oil some distance toward the surface of a producing well. In these instances pumping with specially designed versions of deep well lift pumps is used to bring the oil to the surface. When gas is present with the oil, gas separators are placed in the well below the pumping zone to avoid the interference of vapor lock which would otherwise severely affect pumping efficiency.

17.1.3. Economizing Techniques

Today most oil producing areas have some framework of drilling and production regulations in place designed to put the development of oil fields and the production of petroleum on a more orderly and less wasteful basis. For instance, in British Columbia not more than one well may be drilled per 4 hectares (10 acres). Regulations also limit the production rate per well based on a formula which takes into account the unrestricted production rate for a well on initial testing, estimated reserves present in the particular oil field, product properties, and the like. Production rate of a particular well is controlled by installation of a choke, a plate with a hole of an appropriate size for the required production rate, or by appropriately adjusted valves installed at the well head. One estimate of an efficient production rate for a water drive reservoir is that the total annual production of all wells producing from the reservoir be not more than 3 to 5% of estimated ultimate yield (not oil-in-place) of the reservoir [11].

Despite conservation measures such as these taken during the early life of mature oil fields, and sometimes even for new oil fields which have little or no formation pressure when initially brought into production, pumping has to be used to maintain or initiate petroleum production. This is a way in which the producing life of an oil field, even though slowed, may be extended for some years.

Water flooding, by pumping water or water plus additives into the formation via one well, while producing displaced oil plus water from surrounding wells, is a further method of improving the ultimate recovery of oil [12]. These methods, which are variously called secondary or tertiary oil recovery techniques, represent two types of "enhanced oil recovery" (EOR) methods [13].

Some recovery of reservoir pressure and a decrease in viscosity of the residual petroleum in the reservoir are both obtained by returning the natural gas to the formation. High formation pressures contribute to the solubility of methane, ethane, propane, etc., in the residual petroleum which brings about the decreased viscosity. Carbon dioxide injection is also used which accomplishes similar objectives as natural gas return [14]. Nitrogen has also been used for this purpose [15].

An effective variant of the gas injection EOR procedure is the proposal to inject molasses or some other low-cost fermentable material plus a species of bacterium such as *Clostridium acetobutylicum* into the formation. Fermentation products such as acetone, butanol, and lower molecular weight acids are produced. These and the carbon dioxide which comprises the balance can augment production by 200 to 250%.

Direct application of heat via *in situ* combustion or via superheated steam generation at the surface and injection are other effective methods to boost production, either in mature oil fields or in "heavy oil" fields where the petroleum is naturally quite viscous. While both formation heating methods achieve production rate improvements by viscosity reduction, the apparent simplicity of the *in situ* combustion concept is offset by the difficulty of separation of recovered oil from an aqueous solution containing nitrogen oxides, sulfur oxides, and other combustion products. The acids present in the aqueous phase contribute to the stability of the emulsions obtained from the producing wells of an *in situ* combustion project, and are highly corrosive to steel pipes and tanks.

Sometimes a measure is needed to increase the porosity of the formation in the producing zone, either to initiate production in a new oil field or to stimulate production in a mature one. This may be achieved hydraulically ("fracturing") by pumping a mixture of water and coarse sand into the well at a high flow rate and pressure. Under the impetus of pumps driven by 2000 to 3000 horsepower, the injected high-pressure stream of water and sand opens up cracks in the formation extending radially from the well being stimulated. When pumping is stopped the coarse sand present jams in the cracks and holds them open. Otherwise they would tend to close again from the weight of the overlying rock when the pressure from the injected water is removed. These propped open cracks now serve as channels through which oil may flow to the producing well.

If the producing zone of a well is in a limestone formation, acidizing techniques may be used to stimulate production. Corrosion-inhibited hydrochloric acid is pumped down the well, which dissolves the limestone with which it comes into contact and in this way generates new channels for oil flow (Eq.17.1).

$$CaCO_3 + 2\,HCl \rightarrow CaCl_2 + H_2O + CO_2 \qquad\qquad 17.1$$
$$\text{insol.} \qquad\qquad\quad \text{soluble}$$

Explosives may also be used in various ways to augment oil flow. For old or new wells, any deposits of wax or bitumen accumulated in the casing at one time were removed and at the same time the producing zone enlarged by setting off a delicately placed charge of nitroglycerin in a "torpedo," a special device designed for this purpose. More often today casing cleaning functions are conducted with better control by using a combination of solvent and scraping tools. A more specialized well servicing unit, the "gun perforator," may be used to boost production of a sluggish producing zone. The unit is lowered to the required level where high powered, steel-piercing shells previously loaded into the device are fired. The paths cut into the rock by the high-velocity shells provide new channels for enhanced oil flow. A gun perforator may also be lowered to a higher or lower section of the steel casing of the well, which was not originally brought into production but which was noted as a potential producing zone at the time the well was originally drilled and logged (formation details recorded). Firing the shells at this new perforator position both penetrates the steel casing at this point and provides short flow channels in the rock at this new producing zone to improve net production from the well.

Of all the economizing techniques described water flooding is generally considered to be the most effective, particularly in sandstones. Frequently, however, a combination of techniques will provide the optimum improvement in production rates.

17.1.4. Supply Prospects of Conventional Petroleum

World crude oil production and consumption has been growing fairly steadily during the last 50 years [16]. From production values of 177 million metric tonnes and 262 million metric tonnes in 1930 and 1940, respectively, this has grown to a world production figure of 3.16×10^9 metric tonnes for 1993 (Table 17.1) [17,18]). Half the cumulative total oil produced took from 1857 to 1960, a 103-year period, and the other half took only 10 years, from 1960 to 1970 [16]. Although generally growth has continued throughout this period the rate of growth has varied widely. The doubling time, or the time in years to obtain double the initial annual production rate stood at 21 years for the 1930 to 1940 period. But the doubling time rapidly shrank from this value to 11.8 years for 1940 to 1950, 8.5 years for 1950 to 1960, and 8.7 years for 1960 to 1970, a trend which was certainly reason for concern about the

TABLE 17.1 Trends in Annual Petroleum Production by the Present Major World Producers of Crude Oil[a]

	Density (g/cm³)	Production (in millions of metric tonnes)[b]					
		1950	1960	1970	1978	1990[c]	1993[c]
Algeria	0.80	<0.1	8.8	49.0	54.0	52.9	52.1
Canada	0.85	3.7	26.0	62.0[d]	64.3[d]	81.9	88.4
China	0.86	—	3.7[e]	23.9	104.1	138.3	145.2
Indonesia	0.85	6.4	20.6	42.6	80.5	74.9	77.0
Iran	0.86	32.3	52.2	191.3	262.8	155.9	182.9
Iraq	0.85	6.5	47.5	76.5	125.6	99.2	23.6
Kuwait	0.86	17.3	81.9	150.6	126.0	54.2	87.5
Libya	0.83	—	—	159.8	95.4	65.8	66.5
Mexico	0.89	10.3	14.4	21.5	60.8	147.9	155.3
Nigeria	0.85	—	0.9	54.2	94.9	88.3	95.8
Norway		—	—	—	—	82.1	114.5
Saudi Arabia	0.86	26.9	62.1	188.4	476.3	330.6	424.7
United Arab Emirates	0.85		—	37.7	89.6	110.9	113.9
U.K.	0.86	<0.1	ca. 0.1	ca. 0.1	52.9	91.6	100.1
U.S.A.	0.85	270.1	348.0	475.3	429.2	411.8	389.0
U.S.S.R.	0.86	—	147.9	353.0	572.5	570.8	411.5
Venezuela	0.90	78.2	152.4	194.3	113.5	114.4	130.4
Other		33.3	90.3	196.6	284.2	475.3	499.5
World total		485.0	1056.8	2276.8	3086.6	3146.8	3157.9

[a] Data for those countries producing more than 50 million metric tonnes in 1978, from *U.N. Statistical Yearbooks* [17] and OECD [18]. Does not include natural gas or natural gas liquids.

[b] A close estimate of the production volume in petroleum industry "barrels" may be obtained by dividing the number of metric tonnes by the appropriate density figure and then multiplying by 6.2898. The standard petroleum industry barrel is equivalent to a volume of 0.159 m³, or 35 (34.97) Imperial gallons, or 42 U.S. gallons.

[c] Includes natural gas liquids.

[d] Includes synthetic crude oil produced from tar sands: 1.63 million tonnes in 1970 and 2.70 million metric tonnes in 1978. Very little was produced from tar sands in 1960, and none in 1950.

[e] Data for 1959.

projected life of the petroleum resource. In fact, Hubbert, in 1973, estimated that world petroleum production would peak at about the year 2000, with new finds failing to keep up with increases in demand. However, since this estimate, the doubling time for the 1970 to 1978 production interval has actually showed a dramatic reversal of the earlier growth trend by increasing to a period of 28.1 years. Undoubtedly significant factors were the imposition and continuance of the oil embargo and price increases by the Organization of Petroleum Exporting Countries (OPEC) in 1974.

Since 1978 production has generally leveled off and the production of Indonesia, Iran, Iraq, Saudi Arabia, and the USSR has actually declined. This decline was not the consequence of decreased reserves (Table 17.2) [19]. In fact their known reserves increased and this, combined with their reduced production rates, has caused the reserves to production ratios of all of these countries to go up. The anomalously high values seen for Iraq and Kuwait are largely the consequence of the Gulf War and its aftermath. Worldwide resource life projections of this kind show a similar trend moving from a value of 29.7 "years" in 1970, 25.2 "years" in 1978, and encouragingly to 42.5 "years" in 1992. It is the impact of the large-scale consumers like the United States, Canada, and more recently the United Kingdom, whose consumptions are large fractions of their reserves, which has tended to depress the world petroleum life projections. The conventional petroleum production rates for Canada and the U.S.A. both peaked some time ago, Canada in 1973 and the United States in 1970 (Fig. 17.4).

Some countries with a large petroleum consumption, most notably China, Norway, and the United Kingdom, have been able to continue to increase their petroleum production from either diligent exploration, late resource development, or a combination of these factors (Tables 17.1, 17.2). The U.S.S.R., until recently showing increased production, was a net exporter of oil, mainly to other socialist bloc nations. But recent lower production levels perhaps because of political changes have apparently changed this. Norway and the U.K., from their first discoveries of natural gas and oil in the North Sea in the late 1960s and early 1970s, were both able to dramatically increase their domestic petroleum production by a factor of more than 600 from 1970 to 1980. Since then their production has roughly doubled again.

The large-scale petroleum consumers are exploring ways to supplement the convenient liquid fuel and chemical feed stock aspects of a conventional petroleum resource. Coal can be used in this way. From current consumption projections the known reserves of hard coal have an estimated life of more than 200 years, considerably more than that of petroleum. But extraction of coal from deposits is generally more difficult than extraction of conventional petroleum, equipment to consume it directly needs to be more complex, and efficient emission control is more complicated than for consumption of petroleum.

Coal may be partially converted to a liquid fuel by an aerobic distillation (Table 14.13), or may be made to yield a higher return of liquid fuel by gasification followed by catalytic hydrogenation of the primary liquid gasification products using Fischer-Tropsch technology [20]. This technology was developed and operated on a large scale in plants in Germany when the

TABLE 17.2 Trends in Petroleum Reserves and the Anticipated Reserve Life at Current Production Rates for the Present Major Oil Producing Countries[a]

	1970		1978		1992	
	Reserves (million tonnes)	Ratio of reserves/ production	Reserves (million tonnes)	Ratio of reserves/ production	Reserves (million tonnes)	Ratio of reserves/ production
Algeria	1,056	22.3	1,309	24.2	—	—
Canada[b]	1,157	19.1	791	12.8	893	10.1
China	—	—	2,738	26.3	3,340	23
Indonesia	1,367	32.5	1,070	13.3	—	—
Iran	8,204	42.8	6,148	23.4	13,535	74
Iraq	3,873	50.7	4,702	37.4	13,405	568
Kuwait	10,443	76.0	10,184	80.8	25,040	286
Libya	3,959	24.5	3,719	39.0	—	—
Mexico	779	35.5	3,884	63.8	—	—
Norway	—	—	—	—	1,271	11.1
Nigeria	757	14.0	1,678	17.7	—	—
Saudi Arabia	12,041	68.1	15,911	33.4	34,400	81
United Arab Emirates	2,161	56.9	4,318	48.2	10,707	94
U.K.	132	1590[c]	1,393	26.3	581	5.8
U.S.A.	5,270	11.1	3,801	8.9	3,851	9.9
U.S.S.R.	7,930	22.5	7,990	14.0	—	—
Venezuela	2,009	10.4	2,492	22.0	8,998	69
Other	6,562		5,559			
World	67,700	29.7	77,687	25.2	134,211	42.5

[a]Data from *U.N. Statistical Yearbooks* [17] and Kirk-Othmer [19]. Reserve data may be converted to petroleum industry barrels by dividing the number of metric tonnes by the appropriate density figure (obtained from Table 17.1) and multiplying by 6.2898. The petroleum industry barrel is equivalent to 0.159 m^3, ca. 35 (34.972) Imperial gallons or 42 U.S. gallons (exactly).

[b]Does not include synthetic crude oil produced from the tar sands nor the reserves represented by the tar sands.

[c]Anomalously high values for the ratios of reserves to production are obtained here during the early stages of development of a new large petroleum producing area, North Sea oil (see text).

a

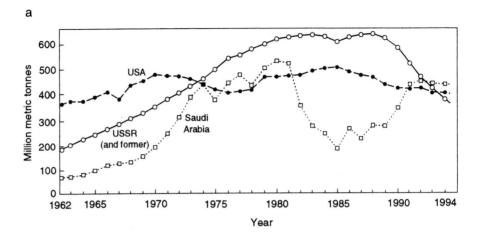

b

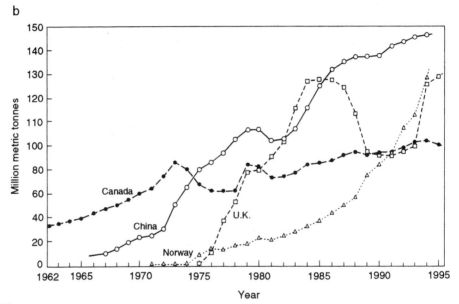

FIGURE 17.4 Trends in annual conventional petroleum production volumes for examples of (a) large producers where this appears to have peaked and (b) smaller scale, more recent producing countries that have shown dramatic growth.

country was cut off from sources of conventional petroleum. The plants were dismantled at the end of the war. The plants operated today by the South African Coal, Oil, and Gas Company (SASOL) are also not profitable, but they continue to be run for strategic and political reasons. Second-generation, more direct coal liquefaction plants may bring down the cost of liquid fuel from this source.

Using technology that bears some resemblance to production of liquid fuels from coal, hydrogenated urban refuse has also been studied as a potential source for liquid fuel [21]. Refuse hydrogenation appears to be too costly and to provide too small a return of liquid fuel to justify the large capital cost, except perhaps for very large urban centers. Use of refuse pyrolysis to produce liquid fuels from municipal waste has been found to be economically feasible by some cities [22]. The mixture of oils which is produced may be used directly as a fuel for space heating, if not for vehicles. The gross volume of municipal waste available is large, but not large enough to provide more than a small fraction of our liquid fuel requirements. Thus municipal waste can never make a significant contribution to the world liquid fuel supply.

Liquid fuel, more specifically ethanol, may also be produced by fermentation or other processing of carbohydrates. Technology for this already exists in the form of alcoholic beverage and spirits production, and has been widened to utilize lower cost cellulosic sources of carbohydrate and enlarged in scale to increase the volumes of alcohol produced (see Chap. 16). Brazil, for instance, has carried the development of this technology to the point where automobiles are produced that are capable of operating entirely on ethanol produced from sugarcane. In general, however, this renewable fuel technology is being employed as a supplement to conventional automotive fuels, as a "gasohol" blend with them, and as an aid to conserving conventional petroleum supplies rather than as a means of entirely replacing them.

Probably the more important and more significant developments to extend the available supply of conventional petroleum-based liquid fuels lie in the processes being introduced to recover oil from the tar sands, and in pilot plant projects involving oil shale pyrolysis experiments to liquid fuels.

17.2. LIQUID FUEL FROM NONCONVENTIONAL SOURCES

17.2.1. Petroleum Recovery from Tar Sands

Major deposits of tar sands, also called bituminous sands or oil sands, occur in several areas of the world, and represent considerable hydrocarbon reserves (Table 17.3) [23,24]. The largest of these in both area and reserves in place occurs in Canada in the Fort McMurray area of Alberta. While the reserves in place for this deposit are estimated to be of the order of 120×10^9 metric tonnes it is thought that only about one-quarter of this, or about 30×10^9 metric tonnes, is likely to be eventually recoverable [23]. Nevertheless the potential recoverable synthetic crude oil from the tar sands of this one locale represents an enormous hydrocarbon potential since even the recoverable oil estimate is equivalent to about 95% of the estimated reserves in the Middle East.

The second largest tar sand deposits occur in the Oficina-Tremblador area of Venezuela, and are also sufficiently large to have a significant development potential for the production of synthetic crude oil. Smaller deposits also occur in Malagasy, U.S.A., Albania, and Trinidad, all of which are less significant from a world reserve standpoint but still are important for the locales in which they occur. Other oil sand deposits are known, such as occur in the Lower Triassic Bjorne formation of Melville Island in the Canadian Archipelago, but these are generally less well defined.

TABLE 17.3 Location and Significance, in Millions of Metric Tonnes, of the Known Major Tar Sands Deposits of the World[a]

Location and name of deposit	Areal extent [km² (10³ha)][b]	Overburden (m)	Reserves in place[c]
Canada, Alberta			
Athabasca	2323	0–580	99,510
Cold Lake	486	210–790	8,188
Peace River	456	90–430	5,310
Wabasca	714	76–762	5,500
Totals	3979		118,508
Venezuela			
Oficina-Tremblador	2323	0–900	31,790
Malagasy			
Bemolanga	39	0–30	278
U.S.A.			
Utah deposits	19.8	0–610	276
California deposits	2.7	0–180	34
New Mexico, Santa Rosa	1.9	0–12	9
Kentucky, Asphalt	2.8	2–15	8
Totals	27.2		327
Albania			
Selenizza	2.1	shallow	59
Trinidad			
LaBrea	0.05	0	10

[a]From the considerable range of data of Berkovitz and Speight [23] and Bridges [24].
[b]To convert to acres multiply by 2.47.
[c]To convert into units of petroleum industry barrels, since the density of these heavy oils is close to 1.0, mulitply the number of tonnes given by 6.290

The tar sands situated in Alberta consist of deposits of oil-bearing sand-stones. Surface exposures occur in parts of the Athabasca deposit, but much of the deposit lies 100 m or more beneath the surface. Oil present in these Lower Cretaceous sandstone deposits is very viscous and thick and partially oxidized so that it cannot be recovered by simple pumping alone. Hence, the terms "tar" or "bitumen" are used to describe this heavy oil fraction, which averages about 12% ranging up to as much as 18% of the deposit by weight. Deposits with less than 2–3% bitumen have been excluded from the reserve data given in Table 17.3.

The surface exposures and the near surface deposits (<45 m of cover) of the Athabasca region may be profitably surface mined for bitumen recovery. These shallow deposits, which are accessible by surface strip mining techniques, amount to a total of about 10% of the Athabasca deposit. The remainder of the Alberta tar sands, which lie under 75 m or more of overburden, is uneconomical to surface mine and at the same time is too poorly consolidated for underground mining. These deeper deposits are yielding bitumen to the surface via various *in situ* techniques.

To obtain bitumen free of sand from the strip mined bituminous sands requires treatment with hot water and steam, together with a small amount of sodium hydroxide (as a surfactant), in a process originally devised by K.A.

Clark [25,26]. In primary separation cells, the hot frothy bitumen rises to the surface and is skimmed off for further processing (Fig. 17.5). The separated sand sinks to the bottom of the cell and is moved hydraulically to the tailings disposal area. By directing the discharge piping carrying the sand tailings onto the walls of the tailings pond, the walls of the pond are continuously raised and made thicker by the precipitated sand. The tailings pond is used by a hot water process tar sands extraction plant to retain wastewaters which contain about 20% of accumulated mineral fines plus 1–2% bitumen in a very stable suspension [27,28].

There is also a middle, less well-differentiated fraction from the primary separation cell which contains a mixture of water, clay fines, and bitumen. This "middlings" fraction is processed through scavenger cells in which more vigorous aeration and flotation techniques recover some further bitumen froth, which is eventually combined with the bitumen layer from the primary separation cell for further processing. It is the wastewater stream from these cells that contributes much of the mineral fines fraction to the tailings pond.

The bitumen layer from these two separation processes is a similar density to water and also quite viscous. It contains some occluded water plus a small amount of mineral fines. By diluting this fraction with naphtha, in which the bitumen is soluble, the viscosity is decreased sufficiently to allow phase separation to take place. Phase separation, accelerated by centrifuging, produces separate streams of the wastewater (plus some mineral fines) and the naphtha

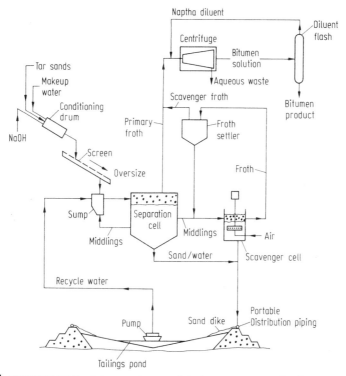

FIGURE 17.5 Schematic diagram of the hot water process for the extraction of bitumen from tar sands. (Reprinted from Hocking [30], with permission.)

solution of bitumen. Flash distillation of the naphtha solution then yields the crude bitumen product and permits recovery of the naphtha for recycle.

Crude bitumen is a black, very viscous (about 100 centipoise at 38°C) rather intractable material as initially obtained. This crude bitumen is not very useful directly, since the temperature at which it softens is too low to make it useful as a component of road building materials. And with its high viscosity and relatively high pour point (minimum temperature at which it will still just pour) of about 10°C it would have to be heated to move it via pipeline to another location for further processing, which would be expensive. So, right at each extraction plant, hot crude bitumen obtained directly from the naphtha recovery unit is fed to a coker. Here, by applying heat at temperatures up to about 500°C, gases and the volatile liquids present are distilled off and the heavier asphaltene fraction (Eq. 17.2), consisting of higher molecular weight polyaromatics, is cracked to more volatile hydrocarbons and coke (Table 17.4 [29]; Fig. 17.6).

Examples of hypothetical asphaltene skeletons [31]:

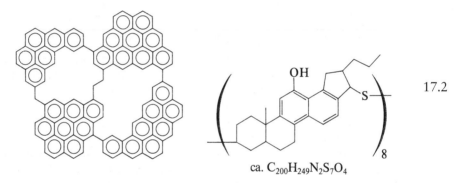

ca. $C_{200}H_{249}N_2S_7O_4$ 17.2

Eq. 17.3 gives an example of the kinds of chemical changes occurring during this processing step.

$$C_{42}H_{56} \rightarrow 2\ C_8H_{18} + C_4H_{10} + C_2H_6 + C_2H_4 + 18\ C$$

A typical   octane   butane   ethane   ethylene   coke   17.3
asphaltene

TABLE 17.4 Approximate Breakdown of Products on the Delayed Coking of Bitumen Recovered from the Alberta Tar Sands by Hot Water Extraction[a]

Component	% by weight	Boiling range (°C)	Density (g/cm³)
Gases[b]	8	—	—
Naphtha	12.7	95–190	0.775
Kerosene	15	190–260	0.832
Gas oil + heavy fuel oil	42.1	260–460	0.884
Coke	22.2	>460	—

[a]Compiled from Berkovitz and Speight [23] and Bachman and Stormont [29].
[b]Breakdown given in Table 17.5.

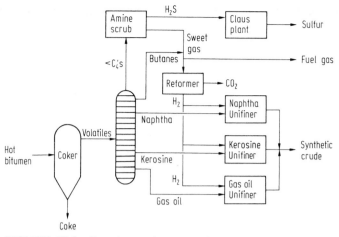

FIGURE 17.6 Flow diagram for the purification and stabilization of the products of bitumen pyrolysis. (Reprinted from Hocking [30], with permission.)

The gases obtained from volatilization and asphaltene pyrolysis comprise about 8% by weight of bitumen coked, more than half of which consists of hydrogen and methane (Table 17.5). Thus, the crude gaseous product is a valuable fuel and can also be used for the production of hydrogen. Volatile liquid products amount to some 68% of the bitumen feed, leaving a residue of about 20–22% coke. Overall mass balances and the properties of the products have been published [30].

The coke residue is the result of the hydrogen-deficient stoichiometry of the process. With a proximate analysis (determination of the compounds, moisture, ash, etc., present) of carbon 80%, volatiles 10%, sulfur 6%, and

TABLE 17.5 Composition of Gas Stream from Delayed Coker, Before and After Sweetening[a]

Component	Mole fraction (%)	
	Crude	Sweetened
Hydrogen	24.20	31.7
Carbon monoxide (+ N_2)	0.9	1.7
Carbonyl sulfide	0.1	0.0
Hydrogen sulfide	15.2	0.0
Carbon dioxide	1.2	0.0
Methane	32.9	37.9
Ethane	11.6	13.1
Ethylene	2.3	2.6
Propane	7.6	8.6
Propylene	3.4	3.8
Butane	0.6	0.6
	100.0	100.0

[a]From Bachman and Stormont [29].

ash 4%, the coke has a fuel value near that of high rank coals. It is burned in the site power plants to provide steam and electrical power for oil sands processing. However, the high sulfur content detracts from its wider utility as a fuel. Any coke in excess of the current fuel requirement has been finely powdered and incorporated into the dyke walls to help trap any hydrocarbons present in water seepage through the wall.

The crude liquid product contains alkenes and other unstable components resulting from the pyrolysis reactions and is also high in bound sulfur. These interact with air to cause accumulation of gummy material on storage. Also the presence of sulfur compounds is undesirable in fuels. High-pressure hydrogenation (Unifining) is used on site to produce more stable alkanes from alkenes, and volatile hydrogen sulfide from the sulfur bound to organic compounds (Eq. 17.4, 17.5).

$$H(R)C=CH_2 + H_2 \xrightarrow{\text{catalyst}} H_2(R)C-CH_3 \qquad 17.4$$

$$R_2S + 2\ H_2 \xrightarrow{\text{high pressure}} 2\ RH + H_2S \qquad 17.5$$

The hydrogen sulfide is now easily separated from the liquid hydrocarbon stream by distillation, and is then converted to elemental sulfur, another product of tar sands operations, via the Claus process. The stabilized liquid hydrocarbon stream is the synthetic crude oil product of tar sands extraction plants.

In this way the useful petroleum fractions are recovered from the surface or near surface exposures of tar sand by the two currently operating hot water process extraction plants in Alberta. Together Suncor (since 1968) and Syncrude (since 1978) now produce about 10^7 metric tonnes of synthetic crude per year, and between them supply about 12% of Canada's current crude oil requirements. Other processes for bitumen recovery from minable sands, such as preliminary partial sand removal with the help of cold water, followed by direct coking of the whole of the bitumen/solid residue, and solvent extraction methods have both been tested but are apparently not attractive for commercial development [32].

Various measures are being developed for recovery of the 90% of bitumen which is in deposits too deep for surface mining. Among these experimental *in situ* methods being tested are formation heating using steam, *in situ* bitumen combustion [33], or electrical resistance heating [34] to increase bitumen temperatures sufficiently to make it flow and allow recovery by pumping. Hot water flooding with a polymer or sodium hydroxide additive content to enhance the interfacial interactions, in the presence of heat, has also demonstrated some positive results [35]. Mining integrated with one or more of the above methods is also under consideration as an *in situ* recovery method. As one or more of these *in situ* combinations is refined to efficient full-scale production the petroleum recovery potential of the deeper deposits of the Alberta tar sands should be realized.

17.2.2. Petroleum from the Oil Shales

The world oil shale deposits, which are distributed among a number of different countries, represent a considerable potential source of oil (Table 17.6

■ **TABLE 17.6 Estimates of the Distribution of Potential Oil-in-Place by Country, from the Major Oil Shale Deposits of the World[a]**

	Potential oil-in-place (10^9 m^3)
Brazil	127
Canada	8.0
China	4.5
Sicily	5.6
Congo	15.9
U.S.A.	350
U.S.S.R.	16.7
Other	3.4
Total	531.1

[a]Corresponds to 1969 estimates from Kirk-Othmer [36]. Data published in 1981 give a world total potential of 332.8 $\times$ 10^{12} m^3, considerably more than quoted here, distributed by grade as follows: 83%, 21–42 L/tonne shale; 16%, 42–104 L/tonne; and 1%, 104–417 L/tonne [37].

[36,37]). But realizing this potential with an oil shale is a significantly different proposition than with the tar sands. First, the organic fraction present in the rock is not an oil or bitumen, but kerogen. Kerogen is a complex, high molecular weight (>3000), three-dimensional polymeric solid which is intimately distributed through the rock. It is insoluble in water or in common organic solvents. As an indication of this, less than 1% of the organic content is recovered on extended Soxhlet extraction of oil shale with boiling toluene, and this extract is mostly traces of bitumen commonly associated with the kerogen. Thus, in contrast to the tar sands, the organic component of oil shales in place is not in fact an oil. However, on vigorous heating of the shale in the absence of air to temperatures of the order of 500°C, a process called retorting, it may be made to yield from 21 L/tonne shale (5 U.S. gal/short ton) to occasionally 417 L/tonne (100 U.S. gal/ton) of a dark viscous oil (Tables 17.7 [39,40], 17.8).

A second misnomer of the oil shale resource is the matrix in which the kerogen resides. This is not a true shale but a marlstone composed of dolomite ($CaCO_3 \cdot MgCO_3$; ca. 32%), calcite ($CaCO_3$; 16%), quartz (SiO_2; 15%), illite (a silica clay; 19%), and an albite ($Na_2O \cdot Al_2O_3 \cdot 6SiO_2$; ca. 10%), and smaller amounts of several other minerals [40]. The carbonate constituents are susceptible to thermal decomposition, e.g., dolomite at 600–750°C, so

■ **TABLE 17.7 Percent Elemental Composition by Weight of Kerogen and the Products from the Retorting of Oil Shale**[a]

	Raw shale	Kerogen	Retorted shale	Crude oil
Organic carbon	16.5	80.5	4.94	84.68
Hydrogen	2.15	10.3	0.27	11.27
Nitrogen	0.46	2.39	0.28	1.82
Sulfur	0.75	1.04	0.62	0.83
Oxygen	—	5.75	—	—

[a]Data from Atwood [39] and Bozak and Garcia [40].

■ **TABLE 17.8 Properties of Typical Crude Shale Oil, and Composition of Gases from a Surface Retorting of Oil Shale**[a]

Shale oil composite properties		Typical composition of gases (mol %)	
Gravity, °API[b]	22	Hydrogen	22.4
Pour point, °C	−1.0	Carbon dioxide	21.4
Carbon, %	84.7	Carbon monoxide	3.6
Hydrogen, %	11.3	Methane	15.2
C:H ratio	7.5	Ethane	10.3
Nitrogen[c] %	1.8	Ethylene	5.4
Sulfur, %	0.8	Propane	4.0
Boiling ranges, % by wt.:		Propylene	3.7
Initial to 204°C	18	Butane	1.6
205–316°C	24	Butenes	2.7
317–482°C	34	C₅, and higher	5.4
483°C and above	24	Hydrogen sulfide	4.3
	100		100.0

[a]Compiled from Atwood [39].
[b]°API gravity, short for degrees. American Petroleum Institute gravity is a density scale used to relate this property of different crude oils and distillate fractions.

$$°API \text{ gravity} = \left(\frac{141.5}{\text{specific gravity } 60°/60°F} \right) - 131.5$$

Thus, water, with specific gravity of 1.00 has an °API of 10, crude oils run the range from about 5 to about 65, lubricating oils run about 26 to 35, and gasolines run about 60. The less dense the crude or distillate fraction, the higher the °API.
[c]Nitrogen-containing constituents present are mainly pyridines, quinolines, pyrroles, and carbazoles [35].

their presence influences the composition of the pyrolysis gases, particularly at higher retorting temperatures (Eq. 17.6; Table 17.9) [37].

$$CaCO_3 \cdot MgCO_3 \xrightarrow{>600°C} CaO + MgO + 2\ CO_2 \qquad 17.6$$

The Fischer assay, which is a standardized laboratory test in which an oil shale sample is retorted at 500°C to determine its oil yield, provides a measure

TABLE 17.9 Effect of Retorting Temperature of Colorado Oil Shale on Product Distribution[a]

Retorting conditions	Distillate hydrocarbon yield (volume percent)[b]		
	Saturates	Olefins	Aromatics
537°C	18	57	25
649°C	7.5	39.5	53
760°C	0	2.5	97.5
871°C	0	0	100
Simulated *in situ*	41	37	22
Actual *in situ*	59	16	65

[a]Data selected from Kirk-Othmer [37].
[b]Volume percents quoted for product test distillation to a 300°C boiling point.

of the grade of oil shale being processed. Commercial processes such as The Oil Shale Corporation's TOSCO II process give oil recoveries (synthetic crude or syncrude) of up to 100% of the Fischer assay. Oil and gas formation from the kerogen occur by what is now believed to be two successive first-order processes to give a product distribution of about 70% oil, 10% gas and light oils, and 20% coke which remains on the solid residue [41] (Eq. 17.7, 17.8).

$$\text{kerogen} \xrightarrow{k_1} \text{bitumen} + \text{gas} + \text{coke} \qquad\qquad 17.7$$

$$\text{bitumen} \xrightarrow{k_2} \text{oil} + \text{gas} + \text{coke} \qquad\qquad 17.8$$

Thus, the observed products arise both directly from kerogen pyrolysis, and indirectly from kerogen via a bitumen intermediate formed from the kerogen. The product distribution is similar to that obtained on the coking of tar sands bitumen. Heat for the pyrolysis is provided from combustion of a part of the kerogen in the shale feed, and sometimes by burning a portion of the gases and/or the residual coke in the pyrolyzed shale (e.g., Fig. 17.7) [38]. Higher overall thermal efficiencies are obtained for those processes which utilize the residual coke as well as pyrolysis gases for shale heating.

The key difference between the methods used for oil recovery from oil shales and that used for tar sands is in the methods used for separation of the organic constituent from the naturally occurring material. Oil shale processing requires the whole of the mined material to be heated up to the pyrolysis temperatures of 500°C or more, whereas the hot water process for tar sands extraction requires the mined tar sand (plus process water) to be heated to only around 70–80°C. Only the extracted bitumen from the tar sand, some 10–12% of the mined mass, has to be heated up to ca. 500°C during the coking step to obtain synthetic crude. Because all the oil shale must be heated to pyrolysis temperatures to effect oil recovery, efficient heat transfer and reclamation from hot spent shale is crucial to the commercial success of oil recovery from oil shales. This is achieved in the TOSCO II process by recirculating steel or ceramic balls (Fig. 17.7). A number of variations of this

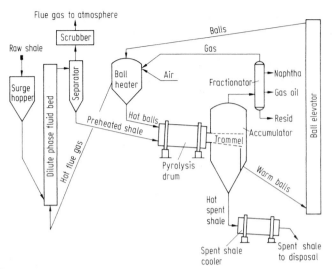

FIGURE 17.7   Diagram of the components of The Oil Shale Corporation's TOSCO II process for the retorting of mined oil shale for the production of synthetic crude. (From Hall and Yardumian [38], courtesy L. Yardumian.)

process are being tested to determine the most efficient heat transfer methods and to confirm product quality. One recent procedure borrowed from coal gasification technology uses the hot spent shale itself to transfer heat to the incoming pulverized raw shale. This Lurgi-developed process obtains oil recoveries equivalent to 100% of the Fischer assay and at thermal efficiencies of 92.7% [42].

Whatever pyrolysis process is used the crude black, viscous pyrolysis oil requires an upgrading step to make it suitable as a refinery feedstock. This is accomplished by high-pressure hydrogenation in a manner very similar to the upgrading step used for the coking products in tar sands processing.

As required for deeper tar sands deposits, *in situ* methods are required to economically exploit the deeper oil shale deposits. The nature of the organic fraction of oil shales requires methods to conduct underground kerogen pyrolysis which permit simultaneous or subsequent recovery of the oil. The most promising methods to date use explosives to fracture the oil shale formation to initially loosen the material sufficiently to permit underground combustion and flow of hot gases to take place [35]. Air is then injected in the presence of an ignition source located at the top of the loosened rubble. This heats the formation sufficiently to cause kerogen pyrolysis and oil generation. As the oil forms, it trickles to the bottom of the fractured rubble from where it is recovered by conventional pumping.

At the moment high projected prices for the production of oil from oil shale have stalled the construction of commercial-scale oil shale processing plants, although several pilot-scale test plants have been built. One recent venture, that of The Oil Shale Corporation (TOSCO) constructed near Grand Valley, Colorado, was designed to process 100 short tons of shale per day (ca. 36,000 U.S. gal/day) and was operated close to this scale of production at intervals from 1965 to 1967. Increases in the world price of conventional petroleum have reawakened an interest in the development of U.S. oil shale deposits. Commercial-scale plants capable of producing 1600 m^3/day (10,000 barrels/day) to be constructed by 1982, to be enlarged to 8000 m^3/day (50,000 barrels/day) by 1990, were announced [43].

17.3. ENVIRONMENTAL ASPECTS OF PETROLEUM PRODUCTION

17.3.1. From the Exploratory Geology Phase

The significance of any environmental impact of petroleum production is to a great extent affected by the particular stage in the producing process. Land exploration by aerial surveying, or by the use of aerial or surface based geophysical techniques such as seismic, gravimetric, magnetic, or electrical methods will have relatively little direct effect on the land surface. Disturbances caused by the placement of seismic shots and travel requirements of a field crew will need to be considered. Surface travel may pose a significant local impact, however, in tundra, permafrost, or other ecologically delicate regions. Summer passage of exploration teams in the Arctic can leave thermokarst line scars which remain and in some cases worsen for decades later because of the low temperatures, limited plant species, and short growing season. Impacts on these polar and subpolar "frontier regions" can be minimized, however, through use of proper vehicles such as hovercraft, helicopters, and vehicles such as the Rolligon which use wide, ultra-low-pressure pneumatic tires [44]. Timing this activity to coincide with a frozen surface layer can also help to minimize impact on the tundra surface, though it may introduce field work difficulties.

Offshore exploration has risks of sediment or underwater installation disturbance, but the overlying water column serves to minimize any visual impacts of these activities.

17.3.2. Impacts of Drilling Activities

After the surface and seismic exploration phases have been conducted to gain information to assess the prospects of a drilling site, wells are drilled to test this information. These wells are placed with little knowledge of the conditions to be expected in the penetrated formations. The greatest environmental risk faced by these operations comes from the threat of a "blowout" or an uncontrolled release of drilling fluid mixed with oil, gas, and brine under high pressure, which can occur when the drill rig penetrates a high-pressure reservoir. The on-site hazards of such an event comprise not only the mud and debris from the well itself but can occasionally also include the forcible ejection of the whole of the drill string (one or more kilometers of heavy pipe). There is also the risk of a well-head explosion or fire from the release of flammable gases at the surface. A blowout may also have an impact from the release of hydrogen sulfide gas which can occur in natural gas. Fortunately the rotten egg smell of hydrogen sulfide serves as a warning signal. These collective risks in the drilling of exploratory wells has led to them being termed "wildcats" by drillers, as recognition of their unpredictable nature.

To be able to prevent a blowout has favorable environmental consequences and can reduce the risk of injury to personnel or damage to equipment. The immediate large-scale losses of the valuable resource, the longer term potential savings in ultimate hydrocarbon recovery, and reduced fire risk give blowout control a high priority in the drilling phase.

Increased density of drilling mud and restricting its outflow from the mud circuit are primary blowout control measures. These are accompanied by at least a pair of blowout preventers. These consist of stout large-bore valves, actuated by a pair of hydraulic rams. The upper valve, which is used for control when a drill string is in place, has semicircular cutouts in the two faces of the valve corresponding to half the cross-sectional area of the pipe used in the drill string. When activated the two halves close tightly around the drill string, controlling both movement of the pipe itself and formation pressures. There is also a lower blowout preventer, without the semicircular cutouts, which may be used to close off the well when there is no drill string present. In deeper wells, when the size of the well casing (pipe lining) is decreased to a smaller size for the lower reaches of the well, an additional one or more blowout preventers will be installed at the junction. This measure increases the capability to control very high pressures which are more likely to be encountered at depth. Blowouts are now rare, but they still do occur occasionally, despite the precautions. Shielding of potential ignition sources reduces the risk of fire from the dissipation of gas captured in the drilling mud, should there be a blowout.

An unsuccessful wildcat is called a "dry hole," and a successful wildcat is called a "discovery well." When drilling equipment is removed and the well is brought into production there is very little impact on the surrounding area during the producion phase. "Appraisal wells" may be drilled to better determine the size and extent of the initial discovery. These will have a much lower risk of the occurrence of a blowout because the depths at which formations under high pressure may be encountered will already have been established. "Development wells" may or may not be drilled following completion of the appraisal wells depending on the results of the appraisal.

Exploratory drilling in tundra or permafrost areas requires greater care because of the lower natural recovery capability [45]. On completion of wildcat wells in these areas greater care is necessary for disposal of waste drilling mud, etc., to avoid terrain collapse from the introduction of a thaw-susceptible area into the permafrost [46].

Offshore drilling introduces several complicating factors simultaneously. Drilling from a ship is accomplished by radial placement of four (or more) anchors or sometimes by computer-controlled propellors. In rough weather drilling may have to be curtailed. A sea-based drilling platform uses ballasted stabilizing legs extending well into ocean depths not subject to wave action, or to the sea floor. Drilling from these, or from artificial islands constructed from material dredged from the surrounding area, is positionally stable but severe weather can introduce risks to all of these situations. Add the problem of floating massive sea ice to northern offshore drilling and one can see why oil exploration in frontier areas is so expensive.

Blowout control methods for offshore drilling are still basically the same as those used on land. However, in the event of the failure of blowout prevention equipment, there is an oil pollution risk to a much wider area than would be the case on land.

Oil production, particularly in offshore areas, is facing increasing difficulty in meeting the tighter regulations for oil discharges at sea. Water that is received at the surface with the oil, "produced water" is now required to be processed through three or more stages of treatment before discharge to control these losses [47].

17.3.3. Emission Problems of Synthetic Crude Production

One of the problem areas of the hot water process for tar sands extraction arises from the clay mineral fines which comprise from less than 1% to over 15% of the mined material. These mineral fines interfere with efficient bitumen separation in the primary separation cell and require the backup scavenger cell to maintain bitumen recovery efficiencies. Selective mining could be used to avoid the problem by leaving high fines tar sands in the deposit. But this procedure would raise mining complexity and cost and would waste bitumen present in the fines, so it is not practiced. As a result of this, the spent water discharged from the scavenger cells still contains much of the mineral fines plus traces of bitumen. This is accumulated in large holding ponds, totaling 30 km^2 (about 11 square miles) in extent for the larger of the two extraction plants which are presently operating.

Prolonged settling in the holding ponds allows the supernatant water stream to be recycled into the extraction process. A sludge consisting of about 78% water, 20% mineral fines, and 1–2% bitumen gradually accumulates in the holding pond. Ordinarily, the residual sand after the bitumen has been extracted could be accommodated in the mined-out areas as mining proceeds. But if the extracted sand is combined with the large volume of sludges the gross volume of waste is greater than the volume of available mined-out areas. An economic method of sludge dewatering could allow the spoils to be accommodated [48]. Spherical agglomeration methods which are presently being tested for direct solvent extraction of bitumen from tar sands may be usefully applied to this problem area [49].

Sulfur gases arising during synthetic crude production from the bitumen and from the high-pressure hydrogenation process for synthetic crude stabilization are captured in amine scrubbers, and are subsequently converted to sulfur via the Claus process (Chap. 9). About 1500 tonnes of sulfur is produced daily from these sources by the two presently operating hot water extraction plants. Occasionally these control measures have been inadequate to maintain low ambient air sulfur dioxide concentrations, particularly during an inversion episode. The potential of vanadium and nickel recovery from fly ash, which is possible on the scale of 1600 and 3900 tonnes per year, respectively, has also been considered.

In situ bitumen recovery from tar sands promises to have a lower environmental impact than surface mining and extraction. Water recovered with bitumen from steam drive *in situ* tar sands processing requires treatment before reuse or discharge.

17.3.4. Impact Control of Oil Shale Processing

Production of synthetic crude oil from oil shale has three main environmental problem areas. These center on a scarce water resource in the operating area (at least in the United States), disposal of spent shales and reclamation of the disposal areas, and sulfur gas containment. Even though this process does not require water for extraction, the water needs are significant, for pretreating shale, for condensing the crude and upgraded shale oils, and for the operation of any scrubber-based emission control devices. Thus, while the present water supply is adequate for pilot-scale recovery plants, provision of an adequate water supply for large-scale shale oil producing units is a major concern [50]. The "burned" shale contains little in the way of plant nutrients and has a high salt content so is not easily reclaimed. This is complicated by the dry conditions and the large amounts involved, though because the waste is dry its volume is a closer match to the volume of the mined-out areas. However, an oil shale industry that produces a million barrels of oil per day would produce over 1.25 million tonnes of burned shale each day, so the areas potentially involved for production-scale units are large [51]. However, use of oil shale solid waste for soil stabilization [52] or reclamation activity is not likely to be complicated by health problems from either the raw oil shale rock or from the spent shale [53].

Sulfur gas containment from either oil shale pyrolysis or oil upgrading can be accomplished using the largely existing technology already outlined for tar sands processing. Environmental aspects of oil shale processing have been reviewed [37].

Recovery of shale oil from *in situ* shale pyrolysis processes is likely to have a lower environmental impact than the methods based on surface mining. Small volumes of formation water which will be mixed with the shale oil recovered by pumping from the lower reaches of the fractured deposit will require treatment before being recycled.

17.3.5. Loss Prevention during Petroleum Shipment

The major petroleum producing areas of the world do not coincide very closely with the major consuming areas, so that much ocean shipping activity today is involved with tanker movement of oil. When the producing area is accessible to the consuming area by a land route, or for the transport of oil to or from an oil port to a refinery, a pipeline is normally used. And for smaller quantities, segregated products, or shorter distances, small tanker ships, tank trucks, or rail tank cars are generally used.

Since the volumes of oil shipped long distances are very large, and transportation costs per tonne of oil shipped are lowest when using large ocean tankers or pipelines, more than half the present world's ocean cargo weight is now oil, and has been since the mid 1960s. More than 75% of all this oil traffic is now crude oil, which means that increased shipping economy is realized by the use of very large tankers [54]. This has led to a rapid escalation in the size of the largest tankers used, from the 5000 to 15,000 deadweight tonnes (dwt) size range in the 1939–1945 period to more than 85 vessels of over 320,000 deadweight tonnes operating in the mid 1980s. More than 80% of the tonnage was vessels of over 72,000 dwt and several were 430,000 dwt.

This escalation in tanker size caused a decrease in vessel maneuverability. The standard World War II tanker of 15,000 dwt had a loaded draft of about 9 m and a length of 170 m, whereas the loaded 430,000 dwt very large crude carrier (VLCC) requires a draft of about 27.5 m and has a length of more than 400 m, both adversely affecting maneuverability. In the event of a serious accident involving a VLCC the polluting potential is also enormous, but the probable number of accidents for the same volume of oil transported goes down. It would take some 30 or more World War II vintage tankers to move the same volume of oil as one VLCC.

These factors combine to result in only about 9% of the world oil loss to the oceans being via accidental tanker spills (Table 17.10) [55,56]. This represents 100,000 to 270,000 tonnes per year, about one major VLCC accident every 2 years or 10 accidents involving smaller vessels every year. Much larger quantities than this, from 530,000 to 670,000 tonnes per year are estimated to be discharged to the sea as a result of normal tanker operations. Direct discharge to the sea of water used for tank cleaning operations or used as seawater ballast for vessel stability on the return trip were the oil discharge practices which contributed to this total. Today, most tanker facilities are equipped with shore lagoon storage facilities to accept large volumes of oily ballast and tank cleaning wastewaters which are cleaned using oil/water separation units before the saltwater is discharged.

Waste oils and used oils are being re-refined either to produce petroleum products or for energy recovery [57,58]. These measures are decreasing the oil lost to the oceans by poor used oil disposal practices. The stimulus for these actions has come from the increase in value of petroleum and from an increased recognition of the harm done to oceans by continual discharge at this rate.

Oil loss may occur from pipeline rupture caused by corrosion, subsidence or land slip, or from seismic disturbances [59]. Regularly scheduled inspections of pipeline rights of way, and efficient containment and cleanup operations

TABLE 17.10 Estimates of Direct Petroleum Losses to the World's Oceans[a]

Source	Thousands of metric tonnes	Percent of total
Waste oil disposal of industrial and motor oils	745–3300	25–67
Tankers, normal operations, tank cleaning, ballasting	530– 725	11–25
Other ships, normal operations, cleaning of bilges	500– 635	10–22
Refineries, petrochemical plants	300– 405	6–14
Tanker and other ship accidents, other accidental spills	200– 300	4–9
Offshore petroleum production, normal operations	100– 144	2–5
Range of total	2954–4930[b]	

[a]Compiled from Pryde [55] and Matthews et al. [56].
[b]Amounts to 0.09–0.15% of current world production.

generally keep these losses small. A useful precaution is to dike any pipeline rights of way which run alongside a lake, stream, or maritime coast, at the time of construction. This can be aesthetically blended into natural contours, and seeded for erosion protection.

When an oil spill does occur, a first priority is containment to limit the spread as much as possible, using sandbags or earth dikes on land or floating booms with underwater skirts on water (Fig. 17.8) [60]. Then as much of the oil as possible is picked up by pumping from land-based spills, or by using one of the commercially available oil/water separation devices (Fig. 17.9) [61]. Final cleanup is by oil sorbents, detergents, and high-pressure water jets. Residues in soils may be biodegraded with the assistance of the correct balance of micronutrients or maybe removed to a pyrolysis device and the oil content burned.

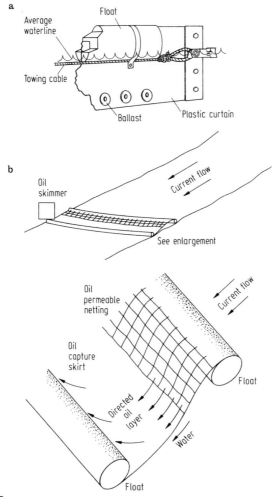

FIGURE 17.8 (a) Important components of an oil boom constructed to contain an oil spill under open water conditions. (b) A boom specifically designed to capture spills in moving water. (Reprinted from Smith [60], courtesy of *Pollution Engineering*.)

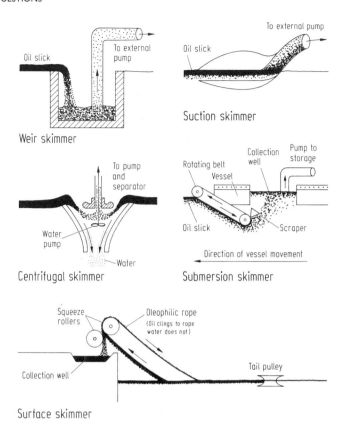

FIGURE 17.9 Illustrations of the operating principles of different types of recovery equipment for oil spills on water. (From Shell Canada [61], with permission.)

More novel proposals for frontier oil shipment have included dirigibles having a 450-tonne payload. Methane could be used as a lifting gas which would allow the craft to serve a dual role, movement of oil and natural gas. Also Boeing officials conducted a feasibility study in 1972 in which 12-engined jet aircraft having a 154-m wingspan and capable of one million kilogram payloads were judged to be a competitive option. Perhaps accident risk perceptions caused these alternatives to be abandoned.

REVIEW QUESTIONS

1. (a) Compare the structures and properties of the material in place for conventional petroleum, "heavy oil," tar sands, and oil shale deposits.
(b) How do the recovery methods and the recovery rates (as a proportion of the resource in place) compare for the hydrocarbon sources of part (a)?
(c) What postproduction measures are necessary for the reclamation of exhausted production areas for the hydrocarbon sources of part (a)?

2. (a) What production practices are used to help maximize the life and the ultimate hydrocarbon recovery from a newly discovered conventional oil or gas field?

(b) Outline the details of any four measures (secondary and tertiary recovery methods) that may be used to maintain production of an aging oil field. Explain how each of these functions.

3. (a) What is the major constituent of natural gas, and what are the minor constituent(s) that may be present?

(b) Why do these minor constituents have to be removed before pipeline delivery, or use, and how is the well-head composition cleaned up for these purposes? (See Section 9.5.)

4. Qualitatively rank the energy requirement for the production of conventional petroleum in comparison to the requirement for oil production from the tar sands and from the oil shales. Explain the reasons for this ranking.

5. (a) Define the meaning of °API and explain how it is evaluated.

(b) What general composition information does this quantity convey with respect to petroleum fractions?

(c) What range of °API values could be expected for crude oils, for gasolines, and for lubricating oils?

6. (a) Fully define the petroleum industry volume units: the barrel (bbl) and, in relation to natural gas, the "standard cubic foot."

(b) Make the following unit conversions: 1000 bbl/day of 0.85 kg/dm^3 density, to metric tonnes/day; 30 metric tonnes/day of crude oil of density 0.90 kg/dm^3, to bbl/day; 22,000 ft^3/day of natural gas at 1200 lb/in.^{2} pressure and 20°C to std ft^3/day (state conditions) <u>and</u> to m^3/day at 0°C and 1 atm pressure. (Assume ideal gas relations for gas conversions.)

FURTHER READING

J.W. Bunger and N.C. Li, eds., "Chemistry of Asphaltenes," Adv. Chem. Ser., No. 195. American Chemical Society, Washington, DC, 1981.

J.J. Fitzgerald, "Black Gold With Grit." Gray's Publishing Ltd., Sidney, B.C., 1978.

J.H. Gary, ed., "15th Oil Shale Symposium Proceedings." Colorado School of Mines Press, Golden, 1982.

H.J. Lamp'l, Spill Control Systems and Standards. ASTM Stand. News 5, 11, Nov. (1977).

J. Milgram, Clean-up of Oil Spills from Unprotected Waters. Oceanus 20, 86, Fall (1977).

Petroleum Publishing Co., "International Petroleum Encyclopedia." Petroleum Publishing Co., Tulsa, OK, 1978.

F. Zurcher and M. Thuer, Rapid Weathering Processes of Fuel Oil in Natural Waters. Environ. Sci.Technol. 12, 838, July (1978).

REFERENCES

1. J.F. van Oss, "Chemical Technology: An Encyclopedic Treatment," Vol. 4, p. 34. Barnes and Noble, New York, 1972.

2. R.M. McPherson and R.W. Ford, "A History of the Chemical Industry in Lambton County." Chemical Institute of Canada, Sarnia, 1964.

3. "Encyclopedia Brittannica," 15th ed., Vol. 18, p. 712. Macropaedia, Encyclopaedia Brittannica Inc., Chicago, 1974.

4. J.E. Brantly, "History of Oil Well Drilling." Gulf, Houston, TX, 1971.

5. P.H. Giddens, "The Birth of the Oil Industry." Arno Press, New York, 1972 (reprint edition).

6. "The Canadian Petroleum Industry," p. 31. Chemical Division, Shell Oil Co. of Can., Ltd., Ryerson Press, Toronto, 1956.

7. Fresh Look at Geochemistry, *Chem. Eng. News* **59**(15), 60, April 13 (1981).

8. E.J. Lynch, "Formation Evaluation." Harper & Row, New York, 1962.

9. "The Petroleum Handbook," 4th ed. Shell International, London, 1959.

10. G.A. Purdy, "Petroleum, Prehistoric to Petrochemicals." McGraw-Hill, New York, 1958.

11. L.M. Fanning, ed., "Our Oil Resources," 2nd ed., McGraw-Hill, 1950.

12. W. Worthy, *Chem. Eng. News* **57**(16), 40, April 16 (1979).

13. M.M. Schumacher, ed., "Enhanced Recovery of Residual and Heavy Oils," 2nd ed. Noyes Data Corp., Park Ridge, NJ, 1980.

14. Alberta Stimulated Enhanced Oil Recovery, *Can. Chem. Process.* **66**, 4, Nov. (1982).

15. Nitrogen Scheduled For Oil Recovery, *Chem. Eng. News* **58**(32), 7, Aug. 11 (1980).

16. M.K. Hubbert, *Can. Min. Metall. Bull.* p. 37, July (1973).

17. "United Nations Statistical Yearbook 1979/80." United Nations, New York, 1981, and earlier years.

18. Organization for Economic Cooperation and Development, "IEA Statistics, Oil and Gas Information, 1994," p. 77. OECD, Paris, 1995.

19. "Kirk-Othmer Encyclopedia of Chemical Technology," 4th ed., Vol. 18, p. 471. Wiley, New York, 1995.

20. Liquid Fuels from Coal, *Hydrocarbon Process.* **54**(5), 119, May (1975).

21. Work Intensifies on Biomass, *Chem. Eng. News* **57**(40), 34, Oct. 1 (1979).

22. M.B. Hocking, *J. Environ. Syst.* **5**(3), 163 (1975).

23. N. Berkovitz and J.G. Spreight, *Fuel* **54**, 138 (1975).

24. L. Bridges, *Sci. Affairs* **9**(3), 4 (1976).

25. K.A. Clark, *Can. Min. Metall. Bull.* **212**, 1385 (1930).

26. K.A. Clark, *Trans. Can. Inst. Min. Metall. Min. Soc. N.S.* **47**, 257(1944).

27. M.B. Hocking, *Fuel* **56**, 334 (1977).

28. M.B. Hocking and G. W. Lee, *Fuel* **56**, 325 (1977).

29. W.A. Bachman and D.H. Stormont, *Oil Gas J.* **65**, 69 (1967).

30. M.B. Hocking, *J. Chem. Educ.* **54**(12), 725 (1977).

31. T. Ignasiak, A.V. Kemp-Jones, and O.P. Strausz, *J. Org. Chem.* **42**, 312 (1977).

32. M.A. Carrigy, ed., "The K.A. Clark Volume." Research Council of Alberta, Edmonton, 1963.

33. R.E. McRory, "Oil Sands and Heavy Oils of Alberta." Alberta Energy and Natural Resources, Edmonton, 1982.

34. D.E. Towsonin, The Future of Heavy Crude and Tar Sand. *Proc. 1st Unitar Conf.*, p. 410 (1979).

35. T.F. Yen, ed., "Shale Oil, Tar Sands and Related Fuel Sources," Adv. Chem. Ser., No. 151. American Chemical Society, Washington, DC, 1976.

36. "Kirk-Othmer Encyclopedia of Chemical Technology," 2nd ed., Vol. 18, p. 1. Wiley, New York, 1969.

37. "Kirk-Othmer Encyclopedia of Chemical Technology," 3rd ed., Vol. 16, p. 333. Wiley, New York, 1981.

38. R.N. Hall and L.H. Yardumian, The Economics of Commercial Shale Production by the Tosco II Process. *61st Annu. Meet., Am. Inst. Chem. Eng.*, Los Angeles, 1968 (unpublished).

39. M.T. Atwood, *CHEMTECH* **3**, 617 (1973).

40. R.E. Bozak and M. Garcia, Jr., *J. Chem. Educ.* **53**, 154 (1976).

41. H.C. Stauffer, ed, "Oil Shale, Tar Sands, and Related Materials," ACS Symp. Ser., No. 163. American Chemical Society, Washington, DC, 1981.

42. Shale Oil Processes Ready, *Chem. Eng. News* **60**(15), 60, April 12 (1982).

43. Production, Engine Tests Start for Shale Oil, *Chem. Eng. News* **58**(25), 39, June 23 (1980).

44. W.E. Rickard, Jr. and J. Brown, *Environ. Conserv.* **1**(1), 55 (1974).

45. J.P. Schumacher, E. Malachosky, D. M. Lantero, and P.D. Hampton, Minimization and Recycling of Drilling Waste on the Alaskan North Slope. *J. Pet. Technol.* **43**, 722–729 (1991).

46. H.M. French, *Arctic* **3**(4), 794 (1980).

47. S. Davies, Recycling, Water Output Limitation Challenge Set Treatment Methods, *Offshore Int.* **54**, 73–74, Sept. (1994).

48. N.N. Bakshi, R.G. Gillies, and P. Khare, *Environ. Sci. Technol.* **9**(4), 363 (1975).

49. W. Campbell, *Sci. Dimens. (Ottawa),* **8**(1), 10 (1976).

50. D.L. Klass, *CHEMTECH* **5**, 1, Aug. (1975).

51. R.N. Heistand, Retorted Oil Shale Disposal Research. *In* "Oil Shale, Tar Sands, and Related Materials" (H.C. Stauffer, ed.), ACS Symp. Ser., No. 163. American Chemical Society, Washington, DC, 1981.

52. J.P. Turner, Soil Stabilization Using Oil-Shale Solid Waste. *J. Geotech. Eng.* **120**, 646–660 (1994).

53. Shale Oil Materials Pose Few Health Problems, *Chem. Eng. News* **56**(41), 13, Oct. 9 (1978).

54. "BP Statistical Review of World Energy 1981." British Petroleum Company, London, 1982.

55. L.T. Pryde, "Environmental Chemistry." Cummings, The Philippines, 1973.

56. W.H. Matthews, F.E. Smith, and E.D. Goldberg, eds., "Man's Impact on Terrestrial and Oceanic Ecosystems." Massachusetts Institute of Technology, Cambridge, MA, 1971, cited by H.S. Stoker and S.L. Seager, "Environmental Chemistry: Air and Water Pollution," 2nd ed. Scott Foresman, Glenview, IL, 1976.

57. D.J. Skinner, "Preliminary Review of Used Lubricating Oils in Canada," Rep. No. EPS 3-WP-74-4. Environmental Canada, Ottawa, 1974.

58. N.J. Weinstein, *Hydrocarbon Process.* **53**(12), 74, Dec. (1974).

59. R.O. Van Everdingen, "Potential Interactions Between Pipelines and Terrain in a Northern Environment," Pap. No. 8. National Hydrology Research Institute, Ottawa, 1979.

60. M.F. Smith, *Pollut. Eng.* **2**(5), 24, Nov/Dec (1971).

61. "Perspective on Oil Spills." Shell Canada, Montreal, 1981.

18

PETROLEUM REFINING

*No fewer than ten substances are obtained from pe-
troleum by the refining process . . . 2nd, gasolene,
used in artificial gas machines.*
                         Century Magazine, *July 1883*

18.1. COMPOSITION OF CONVENTIONAL PETROLEUM

It is commonly thought that crude oil from conventional oil wells is quite
similar in composition, regardless of the source. This is not so. Both the physical characteristics and the composition of crude oils vary widely with the
stage of production (new, or mature) and the location of the oil field.

Conventional crude oil ranges from green to brown or black in color
depending on the petroleum type and the mineral matter present, and is composed of a heterogeneous mixture of liquids, solids, and gases. Some components of the crude oil are dissolved in others and some are not. Water may
occur as a readily separated phase with the petroleum produced or it may
occur as an emulsion containing as much as 80–90% water. The "pour point"
is used as one indicator of the low-temperature viscosity or flow characteristics
of an oil. It is defined as being 3 C° (or 5 F°) above the setting temperature
(maximum temperature at which no observable flow occurs) of the oil. Pour
points of some viscous conventional crudes can lie above 5°C, while the pour
points of the less viscous, or lighter, crudes can be less than −15°C (Table
18.1) [1–3].

The industry standard for bulk measurement for both crude oil and liquid
products was the barrel (abbreviated bbl) but now the metric tonne and the
cubic meter are more commonly used, especially in international trade. The
metric tonne may be converted to the barrel volume unit by dividing the mass
unit by the density (specific gravity) of the particular oil being measured and

■■■ **TABLE 18.1 Specifications of Examples of Typical Conventional and Synthetic Crude Oils**[a]

	Conventional crude oils			Synthetic crude oils	
	Paraffin base oil (wax bearing)	Intermediate naphthene base oil (wax bearing)	Naphthene base oil (wax free)	From tar sands[b]	From oil shales[c]
Density (specific gravity), g/cm^{3d}	0.781	0.964	0.910	0.84	0.93
Pour point, °C	<−15	5	<−15	−35	5
Saybolt universal viscosity at 38°C, seconds[e]	34	4000	55	34	78
Color	green	brown-black	green	pale yellow	black
Sulfur content, %	0.10	3.84	0.14	0.03	0.72
Distillation:					
1st drop, °C	34	138	157	—	—
Fractions, as %					
Gasoline + naphtha	45.2	2.9	1.1	30	18
Kerosine	17.7	4.5	0.0	20	24
Gas oil	8.3	10.6	55.5	50	34
Nonviscous lube	9.8	8.6	14.2	—	—
Medium lube	3.8	6.7	4.7	—	—
Viscous lube	0.0	1.0	11.6	—	—
Residue, %	14.7	58.4	12.7	—	24
Distillation loss, %	0.9	1.9	0.2	—	—
Carbon in residue, %	1.1	18.2	4.5	—	—
Carbon in crude, %	0.2	10.6	0.6	(22)[b]	—

[a]Selected from Lane and Garton [1]. Distillate and residue breakdowns are quoted in percent by weight.
[b]Data calculated from that for composite synthetic crude obtained after coking and Unifining of extracted bitumen, from Bachman and Stormont [2]. Before Unifining (hydrogenation) mean density of the composite stream would be somewhat higher, and sulfur content would be about 3%. Proportions of distillate components are approximate, carbon content quoted is the coke residue on pyrolysis of bitumen.
[c]Data calculated from that given for crude Fischer assay oil (the crude pyrolysate) from Atwood [3]. Proportions given on distillation are approximate. The percent residue quoted corresponds to the fraction of the pyrolysate having a boiling point higher than 482°C.
[d]Density is a necessary property for the conversion of volume units such as m^3, bbl (barrels), or L to mass units such as tonnes. It is also a useful indicator of the composition of the crude oil (see text). °API is a petroleum industry density unit obtained from the specific gravity at 16°C: °API = (141.5/specific gravity) − 131.5.
[e]Scale of viscosity, measured by the number of seconds required for a sample to pass through a standard orifice in a Saybolt viscosimeter, usually specified at 100°F.

multiplying by 6.2898. For example 1 tonne of Canadian crude, of density 0.85 g/cm^3, equates to 7.4 bbl of oil (Eq. 18.1).

$$1 \text{ tonne} \div 0.85 \text{ tonne/m}^3 \times 6.2898 \text{ bbl/m}^3 = 7.4 \text{ bbl} \qquad 18.1$$

In all bulk measurements of petroleum and especially of crude oil it is necessary to specify the gross concentration of solids and nonpetroleum liquids present. This is normally stated as percent bottom sediment and water (% BS&W), and is determined by centrifuging of a representative sample.

Light petroleum gas is almost always present in solution in conventional crude oil and some wells are brought into production to produce solely natural gas. But even when the objective is to produce oil this may contain as much as 50 m^3/m^3 (ca. 300 ft^3/bbl) of dissolved gas. High formation pressures help to dissolve the gas in the oil. Gas is usually separated at well gathering stations by the controlled release of gas from solution. Depending on the producing temperature and pressure of the oil, gas may be released from the oil in several stages and flared. Or it or may be compressed for pipeline transport or for injection back into the formation when oil, rather than gas, is the product desired.

Formation water is produced along with the crude oil at rates which average about 10% by volume. Much higher volumes are obtained from oil fields which are using water flooding for enhanced recovery. The water (and any dissolved salts) is separated from the oil by gravity settling in field tank batteries after degassing, which leaves the crude oil ready for delivery. Shipped oil usually contains less than 0.5% by volume sediment and water. Sediment and salts, mainly sodium chloride, can occur either dissolved in the aqueous phase of the oil or as a fine particulate suspension in the oil phase. Other entrained insoluble mineral matter such as sands and silts may also be present to the extent of 15–30 g/tonne oil (5–10 lb/1000 bbl) although the salt content alone can run as high as 300–1500 g/tonne (100 to 500 lb/bbl). This mineral matter, if not removed before distillation, can concentrate and accumulate on heat transfer surfaces during distillation and in the process decrease thermal efficiency of the distillation. Thus, nondissolved material has to be considered when designing refinery process sequences.

The petroleum itself consists of a mixture of hydrocarbons and heteroatom-substituted hydrocarbons. Even the natural gas, which is mostly methane, usually also contains small amounts of four or five other hydrocarbon components. The paraffin hydrocarbon component of crude oil has a generic formula C_nH_{2n+2}, and may be gases, liquids, or solids depending on their molecular weight (or the value of n). The simplest example of this group is methane, CH_4, the chief constituent of natural gas. It has a boiling point of −164°C and a melting point of −183°C. Propane and butane, C_3H_8 and C_4H_{10}, are also both gases under ordinary conditions, but with boiling points of −42 and −1°C, respectively, are both relatively easily liquefied. Butane and the higher (larger carbon number) members of this series occur not only in the straight chain form, referred to as the normal or "n" form, but also in various branched chain structures of the same molecular formula but which have different physical and chemical properties (e.g., Eq. 18.2).

$$\begin{array}{ccc} & \overset{\displaystyle CH_3}{\underset{\displaystyle |}{}} & \\ CH_3CH_2CH_2CH_3 & CH_3-CH-CH_3 & \quad 18.2 \\ \textit{n}\text{-butane (or butane)} & \text{isobutane} & \\ \text{b.p. } -0.5°C & \text{b.p. } -12°C & \end{array}$$

Pentane isomers, C_5H_{12}, have boiling points in the range of normal ambient conditions, and represent the approximate borderline between gases and liquids in the paraffin series (Eq. 18.3).

$$CH_3CH_2CH_2CH_2CH_3 \qquad CH_3\overset{\overset{\displaystyle CH_3}{|}}{C}HCH_2CH_3 \qquad CH_3-\overset{\overset{\displaystyle CH_3}{|}}{\underset{\underset{\displaystyle CH_3}{|}}{C}}-CH_3 \qquad 18.3$$

n-pentane	isopentane	neopentane
b.p. 36°C	b.p. 28°C	b.p. 9.5°C

As the carbon number gets larger in the paraffin series the number of possible structural isomers also gets larger so that hexane (C_6H_{14}), for example, has five structural isomers, and heptane (C_7H_{16}), six. All 18 isomers of octane have been isolated or synthesized, as have the 35 isomers of nonane. Beyond this, however, little is established about the natural occurrence of the 75 possible structural isomers of decane ($C_{10}H_{22}$) or the over 4000 isomers possible with pentadecane ($C_{15}H_{32}$). Nevertheless, many of the possible paraffin isomers have been found and isolated from exhaustive separations of natural petroleum [4].

Normal octane, C_8H_{18}, with a melting point of $-57°C$ and boiling point of 126°C, lies near the upper end of the liquid paraffinic constituents of gasoline. n-Eicosane $C_{20}H_{42}$ ($CH_3(CH_2)_{18}CH_3$), with a melting point of 36°C and a boiling point of >340°C, is the first of the higher (longer carbon chain or larger molecular weight) paraffins which is isolated in the solid state (a wax) under ordinary conditions. These larger saturated hydrocarbons occur dissolved in the lower molecular weight liquid hydrocarbons which comprise the bulk of light natural petroleums.

Saturated hydrocarbons also occur in petroleum in cyclic form, and will have molecular formulas that fit C_nH_{2n} (if monocyclic). These cycloparaffins are referred to as naphthenes in the petroleum industry and occur primarily as five, six, and seven-membered rings, with and without alkyl substituents. They also occasionally occur as various combinations of two of these ring systems linked or fused together (e.g., Eq. 18.4).

			18.4
cyclohexane	ethylcyclopentane	bicyclo[4.3.0]nonane	
b.p. 81°C	b.p. 103.5°C	(hexahydroindane) b.p. 161°C	

Aromatic hydrocarbons occur to a varying extent in petroleum and have a higher ratio of carbon to hydrogen than any of the commonly occurring paraffins or naphthenes. They correspond to a molecular formula of C_nH_{2n-6} if they are mononuclear (single ring only). Benzene, toluene, and cumene are mononuclear examples of this series, and naphthalene is a dinuclear example (Eq. 18.5).

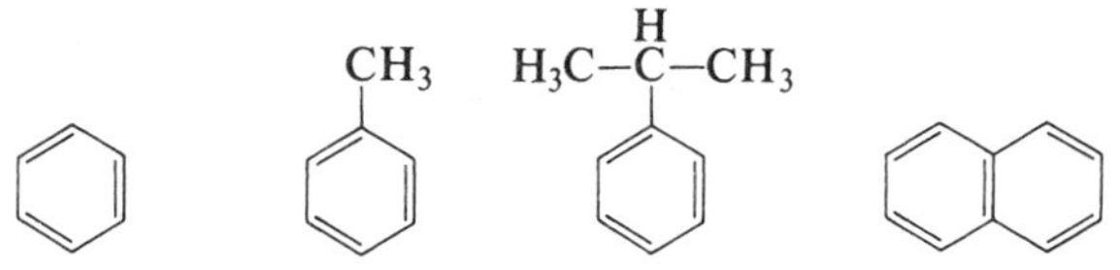

benzene
b.p. 80 °C

toluene
b.p. 111 °C

cumene
b.p. 152 °C

naphthalene
b.p. 218 °C
m.p. 81 °C

18.5

These and many other aromatics have been isolated from petroleum fractions. The aromatic content of crude oils can vary widely but an aromatic content of a third or more of the total, as has been noted for some Borneo crudes, is not unusual [5]. The density (or °API) of a crude oil is an indicator of the aromatic content since the high C:H ratio of aromatic components tends to make these the densest constituents present. This is particularly true with polynuclear aromatic constituents since each additional ring further reduces the hydrogen count by two, increasing the already high C:H ratio and therefore also the density.

While the bulk of the hydrocarbon content of crude oils is represented by the paraffins, naphthenes, and aromatics, small percentages of several other types of compounds are also present. Olefinic hydrocarbons, unsaturated chain compounds having a carbon-carbon double bond and a type formula C_nH_{2n}, also occur in natural petroleum but only to a very small extent since they are quite reactive compounds. They are mentioned here, however, since they are formed to a significant extent by some refinery processes, particularly those involving cracking.

A wide variety of compounds which contain a heteroatom as well as carbon and hydrogen also occur in petroleum, but generally only to a limited extent. Hydrogen sulfide and a variety of thiols, sulfides, and thiophenes are some examples of the sulfur compounds present [6] (e.g., Eq. 18.6).

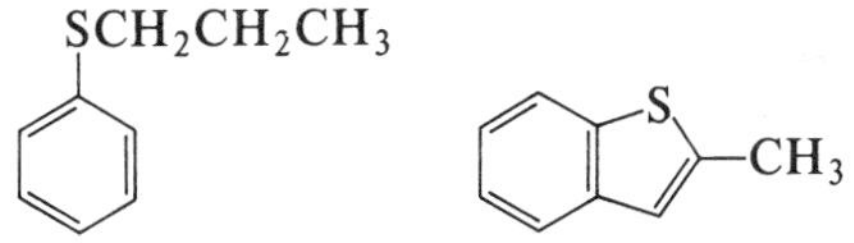

$CH_3(CH_2)_5SH$

1-hexanethiol

1-thiapropyl-
benzene

2-methylbenzo-
(b) thiophene

18.6

The sulfur content of petroleum containing these compounds is usually quite low but it can be as high as 6% of the total. The more stable oxygen derivatives of hydrocarbons such as the paraffinic acids, ketones, and phenols (e.g., Eq. 18.7), also occur in crude oils.

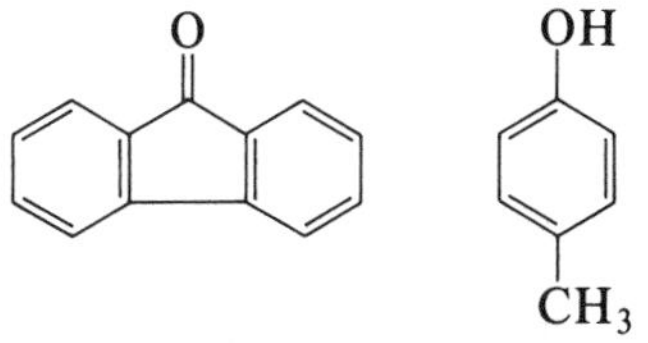

$CH_3(CH_2)_2CHCH_2CO_2H$

3-methylhexanoic acid

fluorenone

p-cresol

18.7

And nitrogen compounds, either on their own or complexed to a transition metal such as vanadium, also occur in petroleum to a small extent. Pyridine, quinoline, and many other heteroaromatics (e.g., Eq. 18.8) have also been found.

pyridine   quinoline   carbazole   benzonitrile   18.8

The first two of these compounds occur to a sufficient extent to be produced from petroleum. Benzonitrile has also been detected [6]. Much of the trace metal content of petroleums, in particular vanadium and nickel, is present in association with petroporphyrins, which are polycyclic pyrroles closely related in structure to the hemes and chlorophylls. These materials are examples of the more complex nitrogen heterocycles to be found in petroleum. The presence of these particular heterocycles, with their complexed metal atoms, contributes much to our present knowledge of the original biogenesis of the petroleum hydrocarbons [6].

In the composite of molecular types found in a particular petroleum reservoir a light, fluid (low-viscosity) crude is obtained if the proportion of low molecular weight hydrocarbons (low carbon number, small molecules) to high molecular weight hydrocarbons is large. If, however, high molecular weight paraffins or polynuclear aromatics (asphaltenes) predominate, then this is a viscous, high pour-point crude such as occurs in oil fields which produce the so-called "heavy crudes." The composition of crude oil produced from a particular oil field will vary somewhat too, with the stage of production. The proportion of lighter (lower molecular weight) hydrocarbons will generally be higher in the early stages of production of a new oil field. The ratio of the percentage of paraffins to the proportion of naphthenes and aromatics present will be relatively consistent from the different producing wells of a particular oil field but can vary widely from one oil field to another. With all these variables in mind, refineries seldom rely on crude oil from a single oil field for the whole of their production but will employ a selection of crudes depending on price, availability, operational processing equipment, and the proportions of the various products that are desired.

18.2. DESALTING AND DISTILLATION

18.2.1. Crude Oil Desalting

Crude oils delivered to the refinery frequently contain significant quantities of water, silts, sand, extraneous salts, etc. "Desalting" involves removal of most of these impurities before further processing. If not removed they can increase corrosion rates, and also the scaling (buildup of deposits) or blockages of refinery equipment such as heat exchangers. Two desalting methods are in common use, each of which uses several processing units operating in series.

Chemical desalting is accomplished by addition of water equivalent to about 10% of the volume of the oil to be treated, plus sulfuric acid or sodium hydroxide as necessary for crude pH adjustment. Reagents are added just before the oil enters the desalting system charging pump to obtain good mixing (Fig. 18.1) [7]. The acid or base addition may be sufficient by itself to cause rapid demulsification after mixing or a small concentration of a proprietary demulsifier such as Tretolite (a polyethylenimine) may have to be added with the water to assist the process. The mixture is then heated to 65 to 180°C (depending on oil viscosity) and mixing is assured by passage through a mixing valve [7]. After treatment, the mixture is held for a sufficient residence time to allow separation of the oil and aqueous phases. Cleaned oil is drawn from the top of the vessel and the aqueous phase containing the salts, sand, silt, and other extraneous material is drawn from the bottom.

Electrostatic desalting is also common [8]. As in chemical desalting, water is added to the oil stream (to dissolve any suspended salts) and the stream passed through a mixing valve into the desalter. Here oil/water separation is mainly accomplished with a high potential electrostatic field instead of with demulsifying chemicals. The electrostatic field induces rapid coalescence and settling of water droplets together with any other aqueous-associated impurities and "creaming" of the oil phase. The water phase is drawn continuously from the bottom of the desalter vessel and passed to the refinery effluent treatment plant. The oil phase is skimmed from the top for further processing. Desalting by one or the other or both of these two procedures is now conducted as a routine preliminary step for most crudes processed in the United States and Canada.

18.2.2. Petroleum Distillation

Distillation of crude oil accomplishes a rough grouping of the components on the basis of boiling point differences. This was originally carried out on a batch basis, where the crude oil to be distilled was entirely in place in the distilling vessel at the start of heating. With a batch still the components of the crude are obtained in vapor form in sequence as heat is applied, the lighter

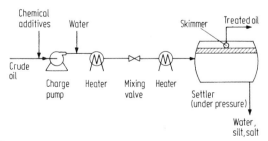

FIGURE 18.1 Flowsheet for the main steps involved in chemical desalting. (From *Hydrocarbon Processing* [7], with permission.)

(lower boiling point) components first. Midrange and heavier, less volatile constituents are distilled later in the sequence, eventually leaving behind a viscous, high boiling point asphaltic residue in the distilling vessel. Batch distillations are inefficient because of the need for separate steps for filling, heating, stopping, pumping out the residue, and refilling of the distilling vessel again. For this reason, they tend to be economic for distillations on a scale not exceeding about 300 bbl/day (ca. 12,000 U.S. gallons, 10,000 Imp. gallons, or 45 m^3 per day). This scale of production was only appropriate for local, very small distillate requirements from a simple refinery which also happened to be close to a petroleum producing area, not a common situation today.

Even a small modern refinery distills 2000 to 10,000 bbl/day and the largest American refineries process 175,000 bbl (27,800 m^3) or more crude oil per day [9]. These are all scales of operation which demand continuous, rather than batch, distillation to be accomplished efficiently and economically.

For continuous distillation the crude oil is first heated to 400–550°C while it is continuously flowing through a pipe still, using natural gas, "light ends" (miscellaneous low boiling point hydrocarbons), or fuel oil for fuel. The heated crude is then passed into a fractionating tower near the bottom, the hotter end of the tower. In the fractionating tower, a unit 2–3 m in diameter and 30–40 m high for a large modern refinery, the lower boiling components move up as vapors (Fig. 18.2). As the vapor moves upward past each plate of the column it is forced, via the bubble caps ("bell caps") of that level, to pass through and come to thermal equilibrium with the liquid on that plate or tray. Hydrocarbons having a boiling point lower than the temperature of the liquid on the plate will continue to move up the column, in vapor form. Components of the heated crude having a boiling point higher or the same as the temperature of the liquid on the plate will tend to condense in the liquid on the plate. As liquids of similar boiling points accumulate on the plate they either overflow into the projecting end of a "downcomer" pipe to the next lower plate, or they are drawn off the plate and the column as one of the product streams of the crude distillation. The lower end of each downcomer is of sufficient length that it dips some distance into the liquid on the lower tray to prevent vapor movement up the downcomer.

In this way the lowest boiling components, the petroleum gases (mostly C$_3$ and C$_4$ hydrocarbons, and commonly referred to as liquefied petroleum gas, or LP gas) and frequently some water, are collected as vapors from the top of the fractionating column. The vapors pass through a dephlegmator (a partial condenser) which condenses naphtha, gasoline, and water vapor components out of this vapor stream, and allows passage of butane and lighter fractions through as vapors (Table 18.2) [10]. Water is phase-separated from the condensed liquids for removal in a unit outside the fractionating column, and most of the condensed naphtha/gasoline fraction is withdrawn from this as a product stream.

A part of the phase-separated hydrocarbon stream is returned to the top plate of the fractionating column as a reflux for temperature control. This return of condensed light distillate to a top plate of the column is what controls the temperature of the cool end of the fractionating column to achieve

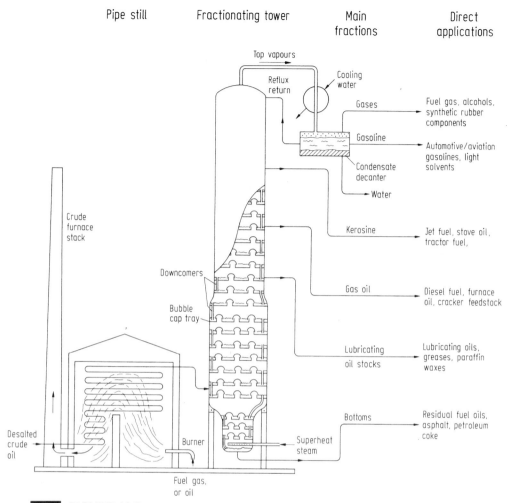

FIGURE 18.2 Partial cut-away view of continuous crude oil distillation via a pipe still and fractionating column. Frequently the heat content of the distillate fractions is employed to preheat the incoming crude oil (not shown here).

the thermal gradient of 12–15 C°/m required. The bottom and hot end of the column is kept at about 500°C via heat brought in through the entry of preheated crude oil, already mostly vaporized, and also occasionally by means of a reboiler operating at the bottom of the column indirectly heated by heat exchange fluid. To decrease the thermal gradient per meter of height and sharpen (narrow) the boiling range of the fraction separated from crude oil, the fractionating column will occasionally be split into two columns, each taking over one-half of the fractionation duty of the single column they replace. One of these will normally operate over the lower temperature range of about 40°C to 350–370°C. The bottom, higher boiling point stream from this column, which includes all components boiling above about 370°C, is then fed hot to a second column, which will normally operate under reduced

■■■■■ TABLE 18.2 Atmospheric Pressure Boiling Point Ranges, Trends in Properties, and Approximate Composition of Representative Fractions from Crude Oil Distillation[a]

Fraction	Boiling range (°C)	Density (g/cm³)	Viscosity[b] (cSt at 38°C)	Range of n-paraffins[c]	No. of compounds[d]	Sulfur (wt. %)
Gas	<40	0.4–0.6[e]	—	C_1–C_5	7	0.001
Naphtha/gasoline	40–180	0.70	low	C_6–C_{10}	>100	0.011
Kerosine	180–230	0.79	1.1	C_{11}, C_{12}	>>40	0.20
Light gas oil	230–320	0.85	3.7	C_{13}–C_{17}	>>12	1.40
Heavy gas oil	320–400	0.90	50	C_{18}–C_{25}	>>10	2.0
Lubricant	400–520	0.91	60	C_{26}–C_{38}	>>7	3
Residue	>520	1.02	solid	>C_{38}	—	4–6

[a]Compiled from Rossini [4], Kirk-Othmer [9], and Guthrie [10]. The values of the data given vary somewhat with source so the values given are approximate. The uncertainty is roughly indicated by the number of significant figures quoted. The trends seen here are realistic.

[b]Viscosity in centistokes is equivalent to units of mm²/sec. For the lubricant fraction the value would be about 7 cSt at 99°C and for the residue about 13,400 at 99°C.

[c]Carbon numbers quoted are for normal (i.e., straight chain) paraffins. For branched paraffins the carbon numbers for the components included in this boiling range will tend to be somewhat higher, and for cyclic paraffins (naphthenes) and aromatics somewhat lower.

[d]Numbers given are the number of individual compounds actually isolated from a sample in the boiling range given. This is preceded by a single carat if the compounds isolated comprised most of the sample of that boiling range, and a double carat if this comprised only a small fraction of the sample.

[e]A range since this is a composite of the densities at 15°C of methane, 0.3; ethane, 0.37; propane, 0.51; isobutane, 0.56; butane, 0.58; and pentane/isopentane, 0.63, and is therefore significantly influenced by the properties of these hydrocarbons present.

pressure, to provide sharper fractions of the higher boiling constituents of the crude.

Whether the crude distillation is conducted in a single or dual column arrangement the principle of operation is the same. Hydrocarbon vapors plus a naphtha/gasoline fraction boiling in the 40–180°C range is the top take-off product of the crude still. A kerosine/jet fuel fraction boiling at approximately 180–230°C is taken off a plate further down the column, followed by fractions of light gas oil, b.p. 230–300°C, and heavy gas oil, b.p. 300–400°C, from appropriate lower plates of the column (Fig. 18.2). The highest boiling distillate (i.e., volatile) fraction from the crude still is the lubricating oil stream, boiling in the range 400–520°C, which also contains much of the grease and wax yield of the crude oil. The residue or the bottom stream of the crude fractionating tower includes all the crude components not vaporized below about 520°C, and consists of mostly asphalt and suspended petroleum coke. This is the highest distillation temperature normally used for crude distillation because at temperatures in this range the decomposition or breakdown of the residual large hydrocarbon constituents begins to become significant.

Since the boiling point ranges of the petroleum cuts (fractions) obtained from the crude distillation column(s) are quite wide, many, if not all, of these crude streams will be subjected to redistillation and/or stripping to sharpen (narrow down) the boiling point ranges. For instance the gas stream, composed entirely of components boiling at below 40°C, is normally redistilled under pressure. Pressure distillation increases the boiling and condensation temperatures of the components present. This enables condensation of many of the separated compounds from the original mixture and collection of these as liquids. Refrigeration would be required to achieve condensation if distillation of these components was conducted at atmospheric pressure. Distillation under pressure also increases the boiling point differences of the components being distilled which makes sharper separations possible in fractionating columns of equivalent separation capability. A separate column is generally used for the final separation of each component.

The naphtha/gasoline, kerosine, and higher boiling streams will usually be sharpened to narrower boiling ranges than obtained directly from the crude still by heating with steam in a reboiler to strip (remove by vaporization) excess volatiles. Or these streams may be redistilled in small columns to separate each of them into two or three separate sharper fractions.

The trends in density, viscosity, sulfur content, etc., that are obtained as one proceeds from the low boiling point to higher boiling point fractions are quite informative as to content (Table 18.2). For instance, as the carbon to hydrogen ratio increases there is a corresponding increase in density. Viscosity also increases steadily with an increase in molecular size. The sulfur also tends to occur in the larger molecules from the increased sulfur content observed with increasing boiling point. The content of transition metals such as vanadium, nickel, and iron also tends to predominate with the high boiling and residual fractions of crude oil distillation [9]. Up to 250 ppm of vanadium is not uncommon in the residue [11].

Simple distillation, secondary distillation, and stripping processes separate crude petroleum into useful hydrocarbon fractions of similar physical properties. Each fraction can contain from 5 to over 100 different compounds (Tables 18.2, 18.3). But the proportions of the main fractions separated vary widely depending on the origin of the crude oil (Table 18.1). Some crude oils contain as little as 1 or 2% of a gasoline fraction. The minimum demand for gasoline in Japan and Western Europe is from 16 to 20% of the petroleum refined, and in the United States, the gasoline demand is about double these

TABLE 18.3 Some Primary and Secondary Uses of the Gas Stream from Petroleum Refining

Component	Direct uses	Primary products	Secondary products
Methane	natural gas, heating	hydrogen	ammonia, methanol
Ethane	natural gas, heating	ethylene	polyethylene, ethanol, styrene, ethylene glycol
Propane	bottled gas (liquefied petroleum gas, LPG)	propylene	polypropylene, propylene glycol, cumene
Butane	bottled gas, natural gasoline (winter blends)	butadiene, polymer gasoline	synthetic rubber, adiponitrile (nylon polymers)
Pentane	natural gasoline	1-, 2-pentenes, cyclopentane	*sec*-amyl alcohols (oxo process)

████ **TABLE 18.4** **Principal Uses and Demand Distribution for the Primary Liquid Refinery Products**[a]

Refining stream	Components	Direct uses	Demand distribution (wt. %)		
			Japan	U.S.A.	W. Europe
Gases	C_1–C_5	natural gas, LPG,[b] etc.	1–2	1–2	1–2
Naphtha/gasoline	C_6–C_{10}	aviation gasolines, motor gasolines, light weed control oil, dry cleaning, and metal degreasing solvents	20	42–45	16–21
Kerosene/jet fuels	C_{11}, C_{12}	diesel fuels, jet fuels, illuminating and stove oils, light fuel oils	8–9	9–11	4–8
Light gas oil	C_{13}–C_{17}	gas turbine fuels, diesel fuels, cracking feedstock, furnace oil	7–8	20–22	33–35
Heavy gas oil	C_{18}–C_{25}	cracking feedstock, fuel oils	48–50	8–10	35–37
Lubricant fraction	C_{26}–C_{38}	gear and machinery lubricating oils, cutting and heat-treating oils, lubricating greases, medicinal jelly, paraffin waxes	2–3	1–2	1–2
Residue	>C_{38}	roofing, waterproofing, and paving asphalts, residual fuel oils	—	5–6	2
Refinery consumption and losses			—	20–22	11–13

[a]Compiled from data of Kirk-Othmer [9,12], Guthrie [10], and Garner [13].
[b]Liquefied petroleum gas, i.e., propane, butane.

figures (Table 18.4) [12, 13]. In northern countries there is also a seasonal swing in gasoline demand from 27% of refinery output in winter to about 35% in summer. These poor matches between the proportions of constituents actually present in crude oil and the proportions demanded of refineries means that molecular modification of crude distillation fractions is necessary to achieve a better correspondence of product to demand.

18.3. MOLECULAR MODIFICATION FOR GASOLINE PRODUCTION

The naphtha/gasoline fraction from crude distillation is not a large enough fraction of the crude petroleum being refined to supply the demand. Thus, gasoline, as sold at the pump, contains not only desulfurized straight-run gasoline, which is the component obtained directly from the distillation of crude, but also cracked gas oil, reformate, polymer gasoline, and occasionally natural gasoline as well. Natural gasoline, also referred to as casinghead gasoline, is the liquid fraction of C_5 and higher hydrocarbons which is condensed at the well head from natural gas in field installation. Natural gasoline is added to the crude oil before primary distillation. The gasoline component of cracked gas oil comprises the C_6 to C_{10} molecular fraction recovered from the partial thermal breakdown of the larger molecules present in gas oil. Polymer gasoline is produced by the fusing of smaller hydrocarbon molecules, primarily C_3 and C_4 hydrocarbons, to produce larger molecules of an appropriate size range for blending into gasoline. Reformate is straight-run gasoline which is desulfurized and upgraded to a higher octane rating by thermalor catalytic reforming. So by a combination of decreasing the molecular size range of crude oil components containing more than 10 carbon atoms, and increasing the carbon number of C_3 and C_4 fractions of crude oil, the gasoline yield may be increased beyond that originally present to better match the supply to suit the demand. The direct separation of the first two of the four primary gasoline components from the crude oil has already been discussed. Details of the methods of production of the other two components follow.

18.3.1. Thermal Cracking of Gas Oils

Cracking is a process by which high temperatures and moderate pressures are used to break down, or decompose, the larger molecules of gas oil into smaller ones suitable for incorporation into gasoline. The various cracking processes in regular use for production of gasoline constituents represent the most important of the petroleum modifying processes used by refineries today, with a capacity in North America that exceeds 50% of the current crude distillation capacity. If the thermal decomposition is relatively mild, and is conducted on a residual feedstock to decrease its viscosity or to produce a heavy gas oil from it, the process is called viscosity-breaking, or "vis-breaking" for short. If the decomposition is sufficiently severe that coke is the residual product, the process is termed coking. If, however, the feedstock being treated is already in the naphtha/gasoline boiling range and is only subjected to mild decomposition by heat to decrease the boiling point and increase the octane number slightly, then the process is referred to as thermal reforming [12]. Vis-breaking, coking, and thermal reforming are all related to, but are far less important than, the processes which are used to crack gas oil directly to gasoline fractions.

There are two broad categories of methods for cracking gas oil to gasoline. These are the straight thermal cracking and the catalytic cracking processes. Thermal cracking, or the application of heat and pressure alone to modify gas oil refinery streams, was the original method by which the gasoline yield from a crude oil was increased. Today straight thermal cracking is only used in "simple" refineries where the available petroleum modification equipment is modest and where the gasoline requirement is small. Gas oil vapor is heated to 500–600°C under a pressure of about 3.5×10^6 Pa (N/m^2; or about 34 atm) for a thermal contact time of the order of 1 minute. Temperatures this high are sufficiently energetic to cause homolysis (rupture to two radicals) of carbon carbon bonds to form two highly reactive, but uncharged, radical fragments (Eq. 18.9).

$$\begin{array}{ccccc} & \text{H} & \text{H} & & \text{H} & \text{H} \\ & | & | & & | & | \\ \text{RC} & : & \text{CCH}_2\text{CH}_2\text{R}' & \rightarrow \text{RC} \cdot & \cdot\text{CCH}_2\text{CH}_2\text{R}' \\ & | & | & & | & | \\ & \text{H} & \text{H} & & \text{H} & \text{H} \end{array} \qquad 18.9$$

Each radical can then either attack another hydrocarbon molecule, by collision and abstraction of a hydrogen atom from the new molecule to form a new stable species from itself and another radical species from the other molecule (Eq. 18.10).

$$R\dot{C}H_2 + RCH_2CH_2CH_2CH_2CH_2CH_2CH_3 \rightarrow$$
$$RCH_3 + RCH_2CH_2CH_2CH_2\dot{C}HCH_2CH_3 \qquad 18.10$$

This process is thermodynamically favorable since a secondary (or internal) radical is more stable than a primary (terminal) one. Or in a further option the fragment can undergo β-fission to lose ethylene and form a new smaller primary radical (Eq. 18.11).

$$R'CH_2CH_2\dot{C}H_2 \rightarrow R'\dot{C}H_2 + CH_2{=}CH_2 \qquad 18.11$$

It is β-fission that produces much of the ethylene obtained from thermal cracking.

The secondary radical can also abstract a hydrogen atom from another hydrocarbon molecule, but this process will not be as favorable as the process which produced it since neutralization and formation of radicals will be energetically similar. Also there would be no reaction progress. But this radical center can undergo β-fission too, forming a smaller hydrocarbon radical and olefin products (Eq. 18.12). This process can continue (Eq. 18.13).

$$RCH_2CH_2CH_2CH_2\dot{C}HCH_2CH_3 \rightarrow$$
$$RCH_2CH_2\dot{C}H_2 + CH_2{=}CHCH_2CH_3 \qquad 18.12$$

$$RCH_2CH_2\dot{C}H_2 \rightarrow R\dot{C}H_2 + CH_2{=}CH_2 \qquad 18.13$$

Some of the larger olefins formed by these processes are useful gasoline constituents. The ethylene also formed is used for other purposes.

There is another process open to a terminal radical on a long hydrocarbon chain which can give rise to useful gasoline constituents; it is called "backbiting." The radical on the terminal carbon can bend back on itself to abstract a hydrogen atom from the carbon atom six from the end of the chain, forming a more stable secondary radical from a primary one (Eq. 18.14). Progression of the resulting secondary radical center then gives heptene plus a primary radical fragment (Eq. 18.15).

$$\begin{array}{ccc} & \overset{\text{H}_2}{\underset{}{C}} & \\ \text{H}_2\dot{C} & & \text{CH}_2 \\ & & | \\ \text{RH}_2\text{C}(\text{H}_2)\text{C} & & \text{CH}_2 \\ & \underset{\text{H}_2}{C} & \end{array} \longrightarrow \begin{array}{ccc} & \overset{\text{H}_2}{\underset{}{C}} & \\ \text{H}_3\text{C} & & \text{CH}_2 \\ & & | \\ \text{RH}_2\text{C}(\text{H})\dot{C} & & \text{CH}_2 \\ & \underset{\text{H}_2}{C} & \end{array} \qquad 18.14$$

$$RCH_2CH_2\dot{C}HCH_2CH_2CH_2CH_2CH_3 \rightarrow$$

$$RCH_2\dot{C}H_2 + CH_2{=}CHCH_2CH_2CH_2CH_3$$
18.15

Thus, by the occurrence of many variations of processes such as these, thermal cracking produces a mixture of gaseous and liquid hydrocarbons of paraffinic (saturated) and olefinic (unsaturated) types, plus coke (Eq. 18.16).

$$\text{gas oil vapor} \xrightarrow[\substack{34 \text{ atmos,} \\ \text{ca. 60 sec}}]{500{-}600°C} \begin{cases} \text{--gases} \\ \text{--naphtha/gasoline} \\ \text{--cracked gas oil (heavier fractions)} \\ \text{--coke} \end{cases}$$
18.16

The gasoline component is separated from the product mixture by distillation (Fig. 18.3). Of course, the objective of the cracking process is to produce as much hydrocarbon in the C_5 to C_9 range appropriate for gasoline blends as possible (e.g., Eq. 18.17).

$$\underset{\text{pentadecane}}{C_{17}H_{36}} \rightarrow \underset{\text{octane}}{C_8H_{18}} + \underset{\text{nonene (olefinic)}}{C_9H_{18}}$$
18.17

Larger paraffinic hydrocarbons cannot be cracked into two smaller paraffins because of the hydrogen-deficient nature of cracking stoichiometry. This is why the best that can be done is to obtain a paraffin plus an olefin. This process is a sensitive one. At too high a temperature or too long a contact time, the coke-forming reactions become more significant at the expense of hydrocarbons of the desired size range (e.g., Eq. 18.18).

$$\underset{\text{pentadecane}}{9\ C_{15}H_{32}} \rightarrow \underset{\text{octane}}{16\ C_8H_{18}} + 7\ C$$
18.18

Thus care is required to select suitable cracking conditions to minimize the extent of coke formation.

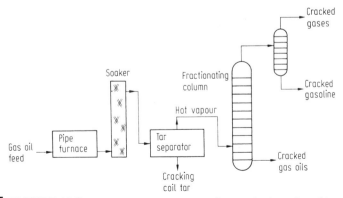

FIGURE 18.3 Tube and tank cracking unit for straight thermal cracking.

18.3.2. Catalytic Cracking

In about 1940 it was discovered that when a suitable catalyst was used for cracking to gasoline the process was accelerated by several hundred to a thousand times that of straight thermal cracking [12]. This discovery firmly established the dominance of "cat-cracking" for the production of gasoline from gas oils (Table 18.5) [14, 15].

Catalysts used in this process are of two types. Special acid-washed clays of particle diameters in the 2- to 400-μm range are used in fluidized bed versions of cracking [16]. In these units the catalyst is kept suspended or "fluffed up" on an upward moving stream of hot gas oil vapors, which ensures continuous exposure of all catalyst faces to the raw material and provides continuous turnover. Synthetic catalysts are prepared from a mixture of 85–90% silica plus 10–15% alumina, or from synthetic crystalline zeolites (molecular sieves) [16]. The synthetic catalysts are either used in a small particle size suitable for use in a fluidized bed, or can be formed into 3- to 4-mm-diameter pellets appropriate for crackers which use a moving bed for catalyst cycling.

A common feature of these catalysts is their acidic nature, i.e., they all act as solid phase acids in the hot gas oil vapor stream. Synthetic silica/alumina catalyst composites, for example, have an acidity of 0.25 mEq/g distributed over an active surface area of some 500 m^2/g. This acidity is the key feature that distinguishes catalytic cracking from straight thermal cracking.

Among the various possibilities open to acid catalyst action is the possibility that it may donate a proton to an olefin, generating a carbocation. The carbocation or carbonium ion may simply lose a proton again but this time from an internal position along the chain, which yields an olefin, a stable product (Eq. 18.19).

$$RCH_2CH_2CH{=}CH_2 \xrightarrow{H^+} RCH_2CH_2\overset{+}{C}HCH_3 \rightarrow RCH_2CH{=}CHCH_2 + H^+$$

18.19

Or a carbonium ion might migrate to a position further along the chain to move the cationic center to a position which then favors alkyl group migration toward a more stable carbonium ion, rather than olefin formation (Eq. 18.20).

TABLE 18.5 Trends in the Utilization of Cracking Processes for Petroleum Modification in the U.S.A.[a]

Cracking process	Cracking capacity as % of crude capacity						
	1930	1940	1950	1960	1970	1979	1993
Thermal	42	51	37	16	12	n/a	1.5
Catalytic	0	3	23	36	37	33	35
Hydrocracking	0	0	0	0	5	5	9

[a]Data from Kirk-Othmer [9], Considine [14], and OECD [15].

$$RCH_2CH_2\overset{+}{C}HCH_3 \rightarrow RCH_2\overset{+}{C}HCH_2CH_3 \rightarrow RCH_2\underset{\underset{+}{|}}{\overset{\overset{CH_3}{|}}{C}}H-CH_2 \qquad 18.20$$

Migration of a hydride ion from the branching carbon of the primary (terminal) carbonium ion so formed can then produce a relatively stable tertiary carbonium ion. An unchanged olefinic product may be formed from this by loss of a further proton (Eq. 18.21).

$$RCH_2\underset{+}{\overset{\overset{CH_3}{|}}{-}}CH-CH_2 \rightarrow R-CH_2\underset{+}{\overset{\overset{CH_3}{|}}{-}}C-CH_3 \rightarrow RCH_2\overset{\overset{CH_3}{|}}{-}C{=}CH_2 + H^+$$

$$18.21$$

Processes of these kinds lead to the internal olefinic (alkene) and the branched paraffinic (alkene) and other branched olefinic products which are characteristic of catalytic cracking processes.

When a carbonium ion is formed several carbons from the end of a large hydrocarbon molecule, either by withdrawal of a hydride ion by another carbonium ion species in the gas phase or by an electron-deficient area on the surface of the catalyst, it can form a neutral product by one of the processes just outlined. If so, the properties will change but there will be no significant decrease in average molecular weight of the product. Or it can undergo β-fission via either of two possible routes each yielding two smaller fragments, one with a terminal carbonium ion and the other with a terminal olefin (Eq. 6.22)

$$\overset{1}{RCH_2CH_2CH_2}\underset{}{\overset{+}{-\!\!-\!\!-}}\overset{2}{CH_2CHCH_2}-\!\!-\!\!-CH_2CH_2CH_2R'$$

Fission at: 18.22

$$1. \rightarrow RCH_2CH_2\overset{+}{C}H_2 + CH_2{=}CHCH_2CH_2CH_2CH_2R'$$

$$2. \rightarrow RCH_2CH_2CH_2CH_2CH{=}CH_2 + \overset{+}{C}H_2CH_2CH_2R'$$

These are the types of processes which give the molecular size reduction activity of catalytic cracking processes.

These acid-catalyzed changes are brought about by passing hot gas oil vapors either through a mechanically circulated bed of the larger pellets or upward through a fluidized bed of more finely divided catalyst particles. The upflow of vapors provides the fluidizing action on the catalyst, hence, the units are referred to as "fluidized cat crackers" or FCC units. To avoid loss of catalyst as the vapors of cracked gas oil leave the catalyst bed, the vapors from fluidized bed crackers are passed through two cyclone separators in series. These retain any entrained solid particles in the vapor stream from the catalyst bed and return these to the bed (Fig. 18.4) [17]. The various products formed are then separated from the composite vapor stream on a distilling column.

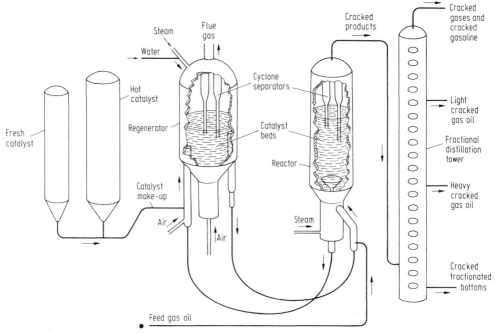

FIGURE 18.4   Main operating components of a fluid catalytic cracking unit. (From Purdy [17], courtesy of G.A. Purdy.)

Overall, catalytic cracking processes achieve a far higher cracking rate than is possible from straight thermal cracking. Furthermore this is achieved at somewhat lower temperatures and much lower pressures than is possible with thermal cracking (Table 18.6) [18]. A further benefit of catalytic cracking is that a much higher proportion of gasoline-suitable products is ob-

TABLE 18.6   Comparison of the Conditions Required for Straight Thermal and Catalytic Cracking of Gas Oil to Gasoline[a]

	Cracking conditions	
	Straight thermal	**Fluidized bed**
Temperature, °C	450–565	465–540
Pressure,[b] N/m^2	$1.8–6.2 \times 10^6$	$6.9–14.5 \times 10^4$
(atm)	(14–60)	(0.7–1.4)
Contact time, seconds[c]	40–300	100–300
Carbon removal method, specifications	periodic shutdown	on spent catalyst, 0.5–2.6% C on regenerated catalyst, 0.4–1.6% C

[a]Selected from the data of Kent [16,18], and converted to appropriate units.
[b]Above atmospheric pressure, i.e., gauge pressure, in pascals (N/m^2).
[c]Estimated from a space velocity range of 0.5 to 3.0. Space velocity here defined as (weight of oil feed per hour)/(weight of catalyst in reactor), since catalyst volume is not fixed in the fluid bed mode of operation.

tained than from straight thermal cracking, almost 50% of the total. In fact, modern fluid catalytic cracking units can give as much as 75 to 80% gasoline fractions.

Continuous decoking of the catalyst is necessary at the high cracking rates of these systems, in order to maintain catalytic activity. This is accomplished by continuously moving a portion of the catalyst from the catalyst bed to a regenerator, using pneumatic conveying or gravity (Fig. 18.4). Here, the still hot catalyst particles are suspended in a current of air, which forms a second fluidized bed in the regenerator and effectively burns off much of the carbon (Table 18.5). Burning off the carbon also reheats the catalyst particles before they are returned to the endothermic processes going on in the cracker. Catalyst return is accomplished using hot gas oil vapor as the driver. This method of catalyst regeneration allows an easy means to maintain high recycle rates and also provides an efficient solid heat transfer medium between the regeneration unit and the cracker. This catalyst regeneration system is also a substantial petroleum engineering feat since a fluid bed cat cracker processing 100,000 barrels of gas oil per day moves catalyst between the cracker and the regenerator at the rate of a boxcar load per minute, all without any moving parts [16]. The waste combustion gases from the regenerator exit through two cyclones in series to avoid catalyst loss. These are followed by heat exchange units for energy recovery as steam, finally passing through an electrostatic precipitator for fine particulate emission control. It has been conventional wisdom for refiners to keep the catalyst level in the FCC unit as low as possible to minimize catalyst inventory. However, it has been found that increasing this results in increased catalyst activity, lower catalyst consumption rates, and lower stack losses [19].

18.3.3. Polymer and Alkylate Gasoline

The volume of polymer gasoline produced is only 2–5% of the volume contributed by catalytic cracking. Nevertheless, the preparation of polymer gasoline by joining together two (or more) light hydrocarbons forms an important part of integrated refinery operations. Polymer gasoline production uses relatively large amounts of the C_3 and C_4 hydrocarbons formed as by-products of cracking processes. Thus, it further incrementally increases the ultimate yield of gasoline possible from gas oil, and in the process generates a high octane component for gasoline blends.

Polymerization is carried out on a $C_3 + C_4$ hydrocarbon stream containing a high proportion of olefins under heterogeneous conditions. This process is usually catalyzed by phosphoric acid on a solid support. At pressures of $2.8–8.3 \times 10^6$ N/m^2 (27–82 atm), which are high compared to the conditions used for cracking, and more moderate temperatures in the 176–224°C range, carbon carbon bond formation occurs, aided by the acid catalysis and close molecular proximity. With isobutylene (2-methylpropene), for example, initial protonation gives t-butyl carbonium ion, which then adds to a further molecule of isobutylene to give a new branched eight-carbon skeleton containing a different tertiary carbonium ion (Eq. 18.23).

$$
2 \; CH_3\underset{\underset{CH_3}{|}}{C}\!=\!CH_2 \rightarrow CH_3\underset{\underset{CH_3}{|}}{\overset{\overset{CH_3}{|}}{C^+}} + CH_2\!=\!\underset{\overset{CH_3}{|}}{C}CH_3 \rightarrow \qquad 18.23
$$

$$
\underset{\underset{CH_3}{|}}{CH_3}\underset{+}{\overset{\overset{CH_3}{|}}{C}}CH_2\underset{}{\overset{\overset{CH_3}{|}}{C}}CH_3 \quad\xrightarrow{-H^+}\quad CH_3\underset{\underset{CH_3}{|}}{\overset{\overset{CH_3}{|}}{C}}CH_2\overset{\overset{CH_3}{|}}{C}\!=\!CH_2
$$

Loss of a proton back to the catalyst surface again releases a molecule of 2,4,4-trimethylpentene-1, a branched C_8 olefin, from the catalyst surface. Processes analogous to this can occur not only between two branched four-carbon olefins but also between straight chain olefins, propylene and C_4 olefins, and two propylene molecules. The carbonium ion formed by coupling of any two of these olefin units may also, occasionally, add a third olefin. The composite result of all of these processes is that a stream of C_6, C_7, and C_8 olefins, plus much still unreacted material, is obtained from the predominantly C_3 plus C_4 feed stream (Eq. 18.24).

Polymerization:

$$
C_3,\ C_4 \text{ olefins} \xrightarrow[\text{catalysis}]{H_3PO_4} \underset{\text{olefins}}{C_6,\ C_7,\ C_8} \qquad 18.24
$$

Components appropriate for blending into gasoline are separated from this product stream by fractionation.

While polymerization was the first refining process to produce larger molecules appropriate for gasoline, alkylation, a chemically similar process, is more prominent in importance today [12]. Alkylation uses either concentrated sulfuric acid or hydrofluoric acid as liquid phase catalysts to form dimers or trimers from C_3 and C_4 olefins reacted with paraffinic hydrocarbons [20]. The yield of the alkylation process, based on the olefin feed, is about twice that of the polymerization process. This factor, and a product that has as good or better octane rating as polymer gasoline, has made alkylation a dominant synthetic route to superior gasoline components.

Alkylation conditions are quite mild with both common catalysts. Processes employing concentrated sulfuric acid operate at 2–12°C (requiring refrigeration) and 4.1–4.8×10^6 N/m^2 (4.1–4.8 atm) and use a gas contact time of 5 minutes although commercial processes normally allow 15–30 minutes. Catalysis with anhydrous hydrogen fluoride requires slightly higher temperatures (25–45°C) and pressures than used with sulfuric acid, but functions effectively with contact times of 20–40 seconds. Both catalysts operate by initial protonation of the double bond of an olefin. With isobutylene, for example, t-butylcarbonium ion is formed. This t-butylcarbonium ion can add to a further molecule of isobutylene in the same manner as occurs during polymerization (Eq. 18.25).

$$(CH_3)_3C^+ + CH_2{=}C(CH_3)_2 \rightarrow (CH_3)_3C{-}CH_2{-}\overset{+}{C}(CH_3)_2 \qquad 18.25$$

But rather than lose a proton, as would normally occur under polymerization conditions, the dimeric ion obtained abstracts a hydride ion (H^-) from a neighboring isobutane to give the saturated dimer 2,2,4-trimethylpentane, and regeneration of a tertiary butyl carbonium ion (Eq. 18.26).

$$(CH_3)_3C{-}CH_2{-}\overset{+}{C}(CH_3)_2 + HC(CH_3)_3 \rightarrow \qquad 18.26$$
$$(CH_3)_3C{-}CH_2CH(CH_3)_2 + {}^+C(CH_3)_3$$

The carbonium ion regenerated in this way can of course continue the process, a key feature being that under alkylation conditions this continuing active species is formed from saturated alkane, not an olefin as required by polymerization. Different alkenes such as propylene, 1-butene, or the 2-butenes may also form carbonium ions in a similar manner to the process of Eq. 18.25, but neither *n*-butane nor *n*-pentane can replace an isoalkane for the hydride transfer. An *n*-alkane is not capable of forming a sufficiently stabilized carbonium ion so this process is thermodynamically unfavorable. Nevertheless, this is one advantage that the alkylation process offers over polymerization as a route to gasoline; it is able to use both light hydrocarbon alkanes (as long as they are branched) and alkenes. Alkylation and polymerization both produce branched products, but the alkylation products are saturated (Table 18.5) whereas the polymerization products are alkenes.

18.3.4. Upgrading of Gasoline Components

Polymer and alkylate gasoline components have characteristically high octane numbers because of the high degree of branching of these products (Table 18.5). Cat-cracked gasoline also has a high octane, 92 or better research octane number (RON), mainly contributed by its high olefin content. In contrast, straight-run gasolines generally give low octane numbers because they are mainly composed of normal alkanes. The octane number of a gasoline blend may be improved, within limits, by the addition of lead-containing or other types of antiknock compounds. But with increasing incentives to reduce lead dispersion into the atmosphere there is a growing trend to upgrade the octane number of a gasoline by raising the octane number of the various hydrocarbon feedstocks being blended into it or by using nonleaded antiknock additives, rather than using lead-containing additives.

Catalytic thermal reforming, under a variety of trade names such as Platforming, Ultraforming, Renoforming, etc., and using similar but somewhat milder conditions than catalytic cracking, accomplishes many octane improving changes of normal (straight chain) alkane streams [21]. For example, naphthenes (cycloalkanes) are dehydrogenated to benzenes, and alkyl naphthenes undergo dehydroisomerization (e.g., Eq. 18.27, 18.28).

$$\qquad + 3\,H_2 \qquad\qquad 18.27$$

$$18.28$$

In the same processing step alkanes are also isomerized to branched alkanes (e.g., Eq. 18.29).

$$CH_3CH_2CH_2CH_2CH_2CH_3 \rightarrow CH_3-\underset{\underset{\displaystyle CH_3}{|}}{CH}-CH(CH_3)_2 \qquad 18.29$$

Since aromatics, such as benzene and toluene, and branched alkanes all possess higher octane numbers than their precursors, the octane number of the fuel is markedly improved by these structural modifications.

It should be noted that since sulfur is a poison for platinum-based catalysts, the feed for the catalytic reformer has to be essentially sulfur free. Sulfur is removed by passing the feedstock through a cobalt/molybdenum catalyst bed in the presence of hydrogen, normally generated from the catalytic reformer. Carbon-bound sulfur is converted to hydrogen sulfide (Eq. 18.30).

$$+\ 2H_2 \longrightarrow CH_3CH_2CH_2CH_2CH_3 + H_2S \qquad 18.30$$

Hydrogen sulfide is easily separated from the other constituents after hydrogenation by stripping or fractionation. Hydrogen sulfide and mercaptans (thiols) may also be removed from refinery streams by washing with aqueous sodium hydroxide (lye treating, Eq. 18.31, 18.32).

$$2\ NaOH + H_2S \rightarrow Na_2S\ (water\ soluble) + 2\ H_2O \qquad 18.31$$

$$NaOH + RSH \rightarrow NaSR\ (water\ soluble) + H_2O \qquad 18.32$$

Sodium hydroxide may be regenerated, and mercaptan recovered from the alkaline scrubber effluent by steam stripping (Eq. 18.33).

$$NaSR + \underset{(as\ steam)}{H_2O} \rightarrow NaOH + \underset{``mercaptan\ oil"}{RSH\uparrow} \qquad 18.33$$

Any mercaptans boiling below 80°C are also readily dissolved in alkaline solutions (Eq. 18.32). A common process for the removal of dissolved mercaptans of this kind, especially from catalytically cracked gasoline and liquefied petroleum gas, is the Universal Oil Products (UOP) Merox process which also uses caustic soda for extraction. In this process, however, the mercaptans are then oxidized to disulfides using air assisted by a metal complex catalyst dissolved in the caustic soda (e.g., Eq. 18.34).

$$4\ NaSC_2H_5 + O_2 + 2\ H_2O \rightarrow 2\ C_2H_5SSC_2H_5 + 4\ NaOH \qquad 18.34$$
$$\text{diethyldisulfide}$$

In this way the sodium hydroxide is regenerated for further use, and the disulfides, which are not soluble in sodium hydroxide, form an oily layer that can be removed. This is one way in which the sulfur content of gasolines is kept below the 0.1% limit required for marketing.

18.3.5. Gasoline Blending

For optimum use of the various refinery fractions in gasoline, these must be blended to achieve the correct volatility and rate of combustion for the engines in which they are to be employed. Winter conditions require more volatile gasolines, which are obtained by increasing the proportions of butane and pentane in the blend. To decrease standing losses (vaporization from the carburation systems, fuel tank, etc.) in summer the proportions of these more volatile constituents is decreased.

Control of the rate of combustion of a gasoline is necessary to obtain smooth engine operation over a range of loads and operating speeds. Rough engine operation, or knocking, is caused by too rapid combustion of the gasoline vapor/air mixture in the cylinder. This condition is generally aggravated when the engine is operating under a heavy load. Rough operation may also be the consequence of preignition (too sensitive ignition) which might in turn be caused by hot carbon deposits on the cylinder head, by an overly hot exhaust valve, or by compression ignition of the fuel vapor/air mixture in the cylinder.

The components of commercial gasolines are always complex mixtures of hydrocarbons. Hence, these are best tested, individually or in already-mixed blends, in an operating engine to determine their resistance to knock. An octane rating is then assigned to the particular fuel being used by comparison of its knock, or preignition resistance as compared to the knock resistance of a standard fuel of known octane rating. Octane numbers have been defined for some particular hydrocarbons. Pure 2,2,4-trimethylpentane, for example, which burns smoothly in a high compression engine, is assigned an octane number of 100. *n*-Heptane, which has a strong tendency to knock, is assigned a value of zero. Octane numbers of other hydrocarbons with knock resistance values intermediate between these two are determined experimentally and also can be obtained from suitable blends of the pure constituents [22]. As a general rule, the octane number of a straight chain hydrocarbon is inversely proportional to molecular weight and directly proportional to the degree of branching or unsaturation (Table 18.7). Rather than using the expensive pure hydrocarbons for test purposes, commercial blends of hydrocarbons of known standardized octane ratings are used for comparison with gasoline blends of current production to determine their market suitability, and adjustments are made if necessary.

If the octane number of the blend is not high enough, it may be raised by increasing the proportion of high octane alkylate, catalytically reformed product, or aromatics in the blend [23]. Or an antiknock compound may be added to accomplish inexpensively the same thing. Use of tetraethyl, or tetramethyl lead with other additives, has been discontinued in Canada and the United States for a decade and these products are gradually being phased out in other areas. However, they are still used to a significant extent to increase the octane number of "leaded" gasolines. Treatment with 1.8 mL of tetraethyl

TABLE 18.7 Motor Octane Numbers Related to Molecular Weight and Degree of Branching of Some Representative Gasoline Constituents[a]

Component	Molecular weight	Octane number
n-Butane	58.1	89
1-Butene	56.1	92
trans-2-Butene	56.1	95
Isobutane	58.1	97
n-Pentane	72.2	63
IsoPentane	72.2	90
n-Hexane	86.2	26
2-Methylhexane	100.2	47
3-Methylhexane	100.2	55
2,2-Dimethylpentane	100.2	96
2,2,3-Trimethylbutane	100.2	101

[a]Data selected from Hutson and Logan [22]. Motor octane numbers (MON) are determined under more rigorous test conditions than research octane numbers (RON) and hence are generally somewhat lower than the latter.

lead per Imperial gallon can increase the octane number of a paraffinic gasoline from the 75 range up to about 85. But there is a decreasing rate of return on increasing the additive concentration. Doubling the amount added to 3.6 mL/Imperial gallon, the maximum permitted by law because of the toxicity of these organic lead compounds, only increases the octane number from 85 to about 88. Gasolines with a high aromatic content have a better octane number response with tetramethyl lead than with tetraethyl lead, which is the additive normally used with high paraffinic and high naphthenic gasolines. Both are thought to function by breaking up (interrupting) the preflame chain reactions leading to combustion, which decreases the tendency of the fuel/air mixture to preignite [24]. The routine levels of tetraethyl lead addition in the United States used to be in the 0.48–0.83 mL/L range (1.8–3.15 mL/ U.S. gal; 2.2–3.8 mL/Imperial gallon). By 1980 this was reduced to not more than 0.5 g of contained lead per U.S. gallon, even for leaded gasolines [25], and was phased out by 1985. Today methyl *t*-butyl ether is the dominant additive used to raise the octane rating of gasolines in Canada and the United States. Much higher addition rates are necessary for an equivalent octane response, but it is relatively inexpensive (Chap. 19).

In addition to the tetraethyl or tetramethyl lead, both types of antiknock fluids also contain 1,2-dichloroethane and 1,2-dibromoethane (ca. 35% of the fluid additive by weight) to react with the lead released on combustion to form lead bromide and lead chloride. These lead halides are volatile at the cylinder combustion temperatures of 800–900°C, and leave the combustion chamber with the exhaust. Without these components, buildup of lead deposits in the combustion chamber could eventually interfere with engine operation. An identifying dye is also added to these antiknock fluids so that leaded gasolines may be immediately recognized.

The toxicity of leaded gasolines and their combustion products has stimulated the testing of a number of other compounds as antiknock additives. The organometallics methylcyclopentadienyl manganese tricarbonyl (MMT), ferrocene, and other related compounds have antiknock activity, but are not as compatible in fuels as the lead alkyls [26]. Many oxygen-containing compounds such as *t*-butanol, *t*-butyl acetate, and in particular methyl *t*-butyl ether (MTBE) also have antiknock capabilities [27]. A higher proportion of these compounds is needed to obtain equivalent antiknock activity. Ethanol also has antiknock activity, and if produced by fermentation provides a second advantage in that it is a renewable fuel component [28].

The final step in gasoline production is the addition of antioxidants. Mono-, di-, and triolefins such as are contributed by catalytically cracked gasoline components, are highly susceptible to gum formation on exposure to air. Contact with air tends to peroxidize the labile allylic hydrogens of the olefins present to produce free radical centers. These radical centers then initiate polymerization to give the observed gums in the gasoline. However, addition of one, or a combination of p-phenylenediamine (PDA), N,N′-diisobutyl-*p*-phenylenediamine (DBPDA, Eq. 18.35)

$$H_2N-\langle\bigcirc\rangle-NH_2 \quad (CH_3)_2CHCH_2NH-\langle\bigcirc\rangle-N(H)CH_2CH(CH_3)_2$$

$$\text{PDA} \qquad\qquad\qquad\qquad\qquad\qquad \text{DBPDA} \qquad\qquad 18.35$$

or 2,6-di-*t*-butylphenol as antioxidants effectively trap radical centers to prevent polymerization [24]. The antioxidant product is a stable free radical, too sluggish to initiate polymerization.

Other minor additives such as corrosion inhibitors, anti-icers, carburetor detergents, and intake valve deposit control additives may also be placed in various gasoline blends. These help to maintain smooth engine operation and long service intervals. Thus gasoline, an apparently simple product, is actually a highly complex carefully engineered product of the petroleum refining operation.

18.4. MANUFACTURE OF LUBRICATING OILS

There are four main steps to produce a lubricating oil from the appropriate high boiling fractions separated on the crude distillation column. These are vacuum fractionation, to decrease the boiling point range of the fractions used, solvent dewaxing, to lower the pour point (gelling point) of lubricating oils, followed by decolorization and formulation steps.

18.4.1. Vacuum Fractionation

Redistillation of the appropriate lube oil fractions obtained from crude distillation is required to remove "lighter ends" (lower molecular weight components such as gas oils), which are poor lubricants, and most of the high molecular weight asphaltic and wax constituents from the lube oil base stock. The boiling point of these oils under ambient conditions is high, close to the temperatures at which cracking and pyrolysis occur. For this reason distillation is conducted under reduced pressures in the range 6 to 9 kPa (0.06–0.09 atm) [12]. This decreases the boiling points by 50°C or so, sufficient to avoid thermal degradation of the oil on distillation.

Two to three steam ejectors (similar to laboratory water aspirators) operating in series with interstage condensers provide the pressure reduction necessary and are connected to the top end of the vacuum distilling column (Fig. 18.5). Each take-off point on the main vacuum column is also passed through a stripper unit, a smaller version of the main column, to ensure thorough removal of volatiles before the lubricating oil proceeds to finishing stages. Fractions more volatile than lubricating oils a rereturned to the column.

18.4.2. Solvent Dewaxing

Waxes are mainly paraffinic hydrocarbons in the C_{20} to C_{30} molecular weight range. Their presence in lubricating oils tends to raise the pour point roughly proportional to the wax content. A lubricating oil with a high pour point can make a cold engine impossible to start under winter conditions. Thus, wax removal is a necessary part of lubricating oil manufacture. It is also one of

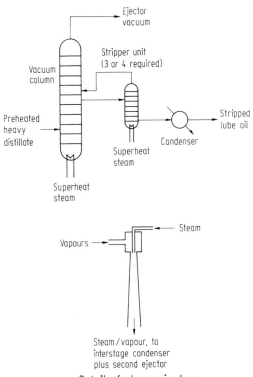

FIGURE 18.5 Simplified flowsheet of vacuum distillation for lubricating oil refining.

the most complicated and expensive of lube oil stock preparation steps [29]. Solvents in use for wax removal include propane, which also serves as an autorefrigerant for the process by solvent evaporation, methyl ethyl ketone (MEK)-toluene mixtures, and methyl isobutyl ketone (MIBK). The last two systems are in widest use at present. MEK-benzene was once a common solvent combination for dewaxing but has now been abandoned from benzene toxicity considerations.

For dewaxing with an MEK-toluene mixed solvent, the solvent is mixed with 1 to 4 times its weight of vacuum-fractionated lube oil stock. The intimate mixture is then chilled to somewhere in the -10 to $-25°C$ range, depending on the degree of dewaxing required (Fig. 18.6). Ammonia or propane evaporation is used as the refrigerant. Scrapers operate continuously on the oil side of the exchanger to prevent the accumulation of adhering wax on the chiller exchange surfaces. The emerging stream from the chillers is a slurry consisting of crystallized wax suspended in a mixture of dewaxed oil and solvent. While the mixture is still cold, the wax is filtered from it and then washed with a small amount of fresh chilled solvent to recover traces of oil still adhering to the wax. The dewaxed oil is recovered from the oil/solvent filtrate in a solvent stripper, which removes the volatile solvent(s).

Waxes, in the 200 to 400 molecular weight range covering several melting point ranges, are recovered in finished form after further processing. The 50,000 tonne/year wax market in Canada of the 1970s was distributed as follows: packaging, 32%; tires and molded rubber, 10%; wire and cable insulation materials, 10%; candles, 8%; carbon paper, 8%; other uses, 32% [30].

18.4.3. Lubricating Oil Decolorization

Vacuum fractionation and dewaxing helps to refine lubricating oil stock, but color, caused by olefin and asphaltic residues, still must be removed for long lubricant life. Originally these impurities were removed by the formation of the acid-soluble alkyl hydrogen sulfates and sulfonic acids (Eq. 18.36–18.38),

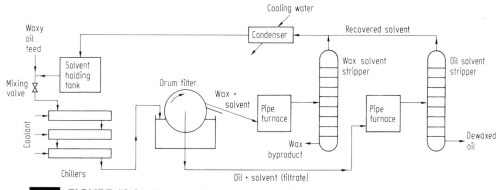

FIGURE 18.6 Flowsheet of the steps required for solvent dewaxing of a lubricating oil stock.

$$RCH{=}CH_2 + HOSO_3H \rightarrow RCH(SO_3H)CH_3 \qquad 18.36$$

$$2\ H_2SO_4 \rightarrow SO_3 + H_3O^+ + HSO_4^- \qquad 18.37$$

$$C_6H_6 + SO_3 \rightarrow C_6H_5SO_3H \qquad 18.38$$

obtained by treating the lubricating oil stocks with concentrated sulfuric acid. Waste disposal and other problems have led to its gradual abandonment.

Today, clay adsorption or hydrogen treating (catalytic hydrogenation) are used to accomplish decolorization [31]. Passage of oil through a clay coating on a continuous filter is quite effective and is suitable for small refineries. Larger refineries with a wider product range that can justify hydrogenation facilities use catalytic hydrogenation. Olefins and aromatics are reduced to their saturated equivalents, paraffins and naphthenes. This method of decolorization decreases the number and extent of conjugated double bonded species and stabilizes the lubricating oil stocks toward oxidation.

The product at this stage is suitable for simple lubricating functions. However, engine lubricants are called on to meet many other demands which require blending and several additives.

18.4.4. Formulation of Lubricating Oils

Paraffinic and naphthenic (cycloparaffinic) stocks may be used for the formulation of lubricating oils, each with favorable characteristics for particular uses. Paraffinic stocks are generally preferred for their superior lubricating power and oxidation resistance. Naphthenic stocks, on the other hand, have naturally lower pour points, that is they maintain flow characteristics at lower temperatures than paraffinics (Table 18.8) and are better solvents, features which are more important for applications such as heat transfer, metal working, and fire-resistant hydraulic fluids [29]. Any residual aromatics in the lubricating base stock will have been removed before formulation by solvent extraction, using N-methylpyrrolidone, furfural, or less frequently today, phenol (Eq. 18.39).

N-methylpyrrolidone   furfural   phenol   18.39

When an oil is used for lubrication of an internal combustion engine it has to reduce the friction of moving parts and has to accomplish this efficiently under a wide range of operating temperatures and loads. At the same time it performs heat transfer functions and has to deal effectively with carbon and dust particles, gum deposits, water and corrosive gases, and accomplish all of this without decomposing in the process. Oil-soluble detergents, added to the oil to the extent of 0.02 to 0.2% by weight, capably handle many of these problems. Various sulfonates or phenolates are used as their calcium or bar-

TABLE 18.8   A Comparison of the Properties of Lubricating Oil Stocks[a]

	Paraffinic	Naphthenic
Density, g/mL	0.887	0.905
Pour point, °C	7	−29
Flash point, °C	246	224
Viscosity, cS		
38°C	107	120
100°C	11.3	10.0
Viscosity index	100	57

[a]Calculated from the data of Gillespie *et al.* [29].

ium salts, CaX_2 where X is a sulfonate or phenolate ligand, to confer oil solubility to the dispersant (e.g., Eq. 18.40).

$$CH_3(CH_2)_{11}SO_3^-, \quad \text{wax}-\!\!\bigcirc\!\!-SO_3^-, \quad CH_3(CH_2)_7-\!\!\bigcirc\!\!-O^- \qquad 18.40$$

Basic alkyl sulfonates, $Ca(OH)SO_3R$), which contribute both acidity neutralizing and dispersing activity to the oil, are also used. These detergents, or dispersants effectively keep particulate impurities such as carbon, dust, or metal fines (e.g., lead or other metal oxides, and salts) and water or acid droplets suspended in the oil, preventing deposition on or attack of critical moving parts. At the appropriate oil change interval the bulk of this suspended material is removed from the engine with the drained oil.

Mineral oils and in fact most liquids show a decrease in viscosity with an increase in temperature, i.e., an inverse relationship exists between the viscosity of an oil and its temperature. At higher operating temperatures the viscosity of the oil decreases (the oil becomes "thinner"). If this decrease of viscosity with increased temperature is too large, lubrication is impaired under hot conditions which can cause excessive engine wear and at the same time increase oil consumption. Thus, it is important that this change in viscosity of an oil with temperature be kept within reasonable limits. The ratio of high to low temperature viscosity is an important property of a lubricating oil and is referred to as its viscosity index. The viscosity index is an empirical property determined by measuring the viscosities of an oil at 40°C and at 100°C and comparing these values with ASTM (American Society for Testing Materials) tables. Under this system Pennsylvania oils, which have a relatively small change in viscosity with temperature, are assigned a viscosity index of 100. U.S. Gulf Coast oils, which have a relatively larger change in viscosity with temperature, are assigned a viscosity index of zero.

A few percent of a synthetic polymeric material, such as polymethyl methacrylate, polyisobutylene, or an ethylene-propylene copolymer added to an oil improves its viscosity index [10]. In this way the viscosity of a low pour point oil, which remains fluid at low temperatures, although it is still reduced when hot, is still kept acceptably high because of the presence of the polymer additive (Fig. 18.7) [32]. The additive is thought to function by the individual

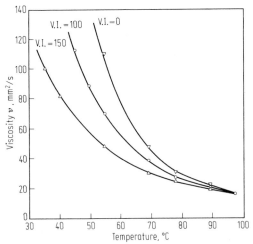

FIGURE 18.7   Relationship of viscosity indices to viscosity change with temperature at different polymer loadings. (Reprinted from *Tribology International*, Vol. 11, H. G. Muller, p. 189, Copyright © 1978, with kind permission from Elsevier Science Ltd., The Boulevard, Langford Lane, Kidlington 0X5 1GB, UK.)

polymer molecules assuming compact spherical shapes at low temperatures, because of low solvent power of the base oil under these conditions. At higher temperatures, when the oil is a better solvent for the polymer, the polymer molecules are well solvated, become extended, and undergo intermolecular intanglements. The presence of these extended polymer molecules has the effect of partially offsetting the tendency of the base oil to "thin out" [32] (Fig. 18.8). This method of viscosity stabilization enables many base oils, which would be poorly suitable on their own, to be used for engine lubrication. These additives also contribute to the formulation of the multi-grade oils which may be used safely summer and winter, without requiring seasonal oil changes. In part, multi-season use is achieved because the same or similar additives are also efficient pour point depressants. These decrease the temperatures at which any residual wax may crystallize out which reduces the risk of oil gelation [33].

Rapidly moving engine parts tend to whip up a froth in the lubricant as they operate, which is promoted by detergent additives in the oil. An oil foam can cause air to enter the circulating pump and cause temporary failure of the lubricant circulation system in an engine. Addition of 1 to 10 ppm of silicone antifoam agents can prevent this. Silicones are also effective lubricants so that this addition does not harm function.

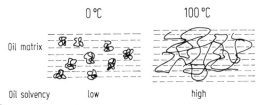

FIGURE 18.8   Mechanism of action of polymeric viscosity index improvers.

Many other additive materials are formulated into lubricating oils to improve their function. Some of these "oiliness" additives are polar compounds which tend to form a stable film on metal surfaces by plating out on them, and extreme-pressure additives which are commonly lead soaps. Greases are used for lubricating functions where less viscous materials would be forced out of or lost from the friction zone. They consist of a thickening agent combined with a lubricating oil. Common thickening agents for greases are the soaps (metal salts of a fatty acid) of aluminum, barium, calcium, lithium, or sodium.

Lubricating oils represent only about 2 to 3% of the volume of crude oil processed. They are a high value component, however, and since their use is only partially consumptive recycling of this petroleum fraction is possible. Increased recycling could also decrease the environmental mutagenic burden that can occur from some disposal methods. Almost half of the 850 million liters of lubricating oil sold in Canada each year is potentially reclaimable, and yet less than 40 million liters are currently recycled [34]. If reclaimed lubricating oil is downrated to a fuel oil after treatment, only water and sludges have to be removed in a relatively simple process [35]. Full re-refining requires six or more stages of upgrading. This method of processing is favored, primarily because of the higher value of the lubricating oil stock as a product rather than as a fuel. In 1974, capital costs for a 19 million liter (4.2 million imperial gallon) per year re-refining to lube oil plant was estimated to be $0.9 to 1.4 million, and operating costs 3.9 to 5.8 cents per liter [36]. Reclaimed oil has also been reprocessed by delayed coking to yield gasoline, fuel oil, and petroleum coke [37]. Areas with a high population density could make the economics of reprocessing or re-refining more favorable.

18.4.5. Synthetic Engine Oils

Many special lubricant properties such as high viscosity indices, very low pour points, high thermal stability, and greater oxidation resistance can all be achieved by the use of certain synthetic lubricating fluids [38]. Original developments of synthetic lubricants used alkylated aromatics as the bulk constituent, but these had worse low temperature viscosity characteristics than refined petroleum stocks and were still susceptible to oxidation. Olefin oligomers (low molecular weight polymers) which were developed subsequently were much better in these respects [39]. For automotive use, blends of olefin oligomer and high molecular weight esters such as di-2-ethylhexyl sebacate or trimethylolpropane triheptanoate are presently favored from cost and performance considerations. These synthetic lubricant blends provide reduced wear, better engine cleanliness, longer oil change intervals, and slightly improved fuel economy, and all at lower oil consumption rates than are experienced with conventional petroleum-based lubricants. The current much greater cost of this class of lubricant is likely to continue to limit their use to a fraction of the total automotive market, where their advantages outweigh their higher cost [40]. Much additional information relating to the properties of synthetic materials which may be used in this application has been published [37,41].

18.5. FUEL OILS, ASPHALTS, AND PITCHES

The fuel oils include a wide variety of petroleum end uses, from jet fuels at the more volatile low molecular weight end of the scale, through kerosene, stove oil, several grades of diesel fuel, to industrial heating fuel and ultimately to asphalt for road cements at the high molecular weight end. A convenient classification to consider this wide range of products is related to their distillate or residual origin, at the crude distillation step.

18.5.1. Distillate Fuel Oils

There is considerable overlap in the boiling ranges of the diesel, jet, kerosene, and light fuel oils represented by the distillate fuel oil class. But all, if desulfurized, have the capability of burning relatively cleanly with little more than a trace of ash which consists of transition metal oxides (mainly vanadium).

Jet fuels are a blend of the higher boiling constituents of the gasoline/naphtha fraction together with a proportion of the C_{11}, C_{12}, and higher components of the kerosene fraction. These are the lowest boiling range of the fuel oil product categories.

Kerosene is the lightest straight fuel oil in the distillate category, and has uses which range from lamp oils, to light stove oils, and diesel fuels for use in Arctic service. Pour points of below −50°C are the attraction for low temperature diesel fuel applications. A small addition of lubricating oil stocks is made to diesel fuel in this service.

Conventional diesel fuel is made from crude oil fractions boiling in the range 190–385°C. It is sometimes also referred to as No. 1 fuel oil. Narrower fractions of this boiling range are used depending on the class of diesel fuel being produced. These differ in volatilities and pour points. Diesel fuels have 40–50°C higher flash points than gasoline or jet fuels which makes them safer [12].

After physical properties, the most important specification of a diesel fuel is its cetane number. The cetane number is a measure of the ease of autoignition of a diesel fuel and is important because this is the means of ignition used in diesel engines. Cetane (n-hexadecane, $C_{16}H_{34}$) itself performs very well in a diesel engine, meaning that there is negligible delay between the time of injection of the fuel into the cylinder and fuel combustion. Pure cetane is assigned a cetane number of 100. In contrast to this α-methylnaphthalene ($CH_3C_{10}H_7$), which performs very poorly (does not readily ignite) in a diesel engine, is assigned a cetane number of 0. Cetane numbers intermediate between these two extremes correspond to the percentage of cetane in a blend of these components. A commercial diesel fuel is composed of a large number of hydrocarbons with physical properties in the right range. It is blended to produce a match of its ignition properties to those of the desired percentage of cetane in a research fuel.

Diesel fuels of low cetane number, the 25–40 range, are appropriate for operation of slow-speed engines such as are used for the propulsion of ships and some stationary engines. Medium- and high-speed engines require cetane numbers of about 40, and 50 to 65, respectively, for smooth operation. Another factor which may be important at low temperatures is that the higher the cetane number the lower the operating temperature at which a diesel engine may be successfully started. In general, the higher cetane number fuels also tend to give smoother engine operation, once it reaches operating temperature, and reduce the formation of engine deposits and smoke.

The higher boiling No. 2 and No. 3 distillate fuel oils are largely used for space heating (homes and buildings), and can also form part of the fuel composition used in slow-speed diesels. Also the No. 2 and No. 3 distillate fuel oil fractions, depending on availability, are used as the solvent in the formulation of residual fuel oils.

18.5.2. Residual Fuel Oils

As the name implies, these lower cost fuels are formulated from the residues of the refining and distillation processes. Pitchy residues of the crude distillation are dissolved in ("cut with") varying amounts of the residual heavy oil fractions remaining after catalytic cracking, so-called cat-cracked gas oil, or cycle oil (recycle stock). Thus, this group of fuels contains a relatively high concentration of all the less volatile and involatile impurities present in the original crude being distilled. These impurities include trace metals, heavy sulfur-containing compounds, asphalts, coke, and any residual incombustibles such as salt, silt, etc. They still possess considerable fuel value, however; in fact this is 5 to 10% higher than the same volume of distillate fuel oils [42]. However, they require combustion equipment that can cope with the higher viscosities and higher ignition temperatures that are characteristic of these fuels.

Bunker A, or No. 4, fuel oil, so-called because of its use in the fueling (bunkering) of ships, is the lightest grade of residual fuel oil. It has the highest proportion of cracked gas oil to pitch, and is designed to stay fluid at ambient temperatures above freezing (pour point of ca. −7°C) without difficulty. Its lower viscosity and hence greater versatility command a price somewhat higher than the other residual fuels. It is an attractive fuel choice for smaller combustion units operating under colder than average conditions, such as ships in Arctic service and the like. The use of especially low pour point (−40°C) No. 2 light diesel oil is sometimes necessary in Arctic conditions for small ships, such as those of northern fishing fleets.

Bunker B, or No. 5, fuel oil has a lower proportion of cat-cracked gas oil to pitch than No. 4 fuel oil, but still remains relatively fluid down to storage temperatures of about 10°C. Combustion equipment must have some provision to heat the fuel prior to atomization, in order to obtain efficient combustion. Outside storage at temperatures much below 10°C would require insulated tanks fitted for heating, to enable the fuel to be pumped under cold conditions.

Bunker C, or No. 6, fuel oil has the lowest proportion of cracked gas oil to pitch of all the residual fuels. Consequently storage tanks for Bunker C require heating even under ordinary ambient conditions to enable the fuel to be pumped and to assist in atomization for burning. Heated storage is not difficult to arrange in return for the lowest in cost and the highest in fuel value among the fuel oils. The fuel value of No. 1 fuel oil is in the range 38,000 to 39,000 kJ/L (135,800–138,800 Btu/U.S. gal) whereas No. 6 fuel oil produces some 42,200 to 42,600 kJ/L (150,700–152,000 Btu/U.S. gal) on combustion.

An occasional problem with large thermal combustion sources coupled with little air movement is that excessive concentrations of sulfur dioxide may accumulate in the ambient air (Section 18.6.1). This type of problem is aggravated by the combustion of high sulfur content residual fuels. However, the sulfur content of residual fuels may be decreased by using hydrodesulfurized cutting solvents, and by selecting the residual fractions from low sulfur crude oils to be made into residual fuels. The residues from high sulfur content crudes, if properties are appropriate, could then go into the production of pitches and asphalts where the sulfur content is not subjected to combustion and discharge. A less desirable option for the residues from high sulfur content crudes is to formulate these into residual fuels to be marketed outside regulated areas, that is, in areas where thermal combustion sources are smaller and less numerous. If required the formulated residual fuel itself may be hydrodesulfurized although the processing costs to do this are much higher than for the desulfurization of distillate fractions [43]. When this level of sulfur reduction is necessary in residual fuels a product price increase is generally necessary to offset the higher production costs.

18.5.3. Asphalts and Pitches

Usually the residue from crude distillation in North America will be processed to an asphalt product as much as possible, since as a component of residual fuel oils it generally fetches a lower price than as asphalt. The residual fuel oil market has quite a competitive international supply network, hence ships and tankers can choose to load at centers which offer the lowest cost bunkering and bulk fuels. Also the cost of the heavier grades of bunker fuel used to supply power stations has to be comparable to that of coal, in order to remain a competitive energy source.

Asphalts are marketed in liquid, and low, medium, and high melting point solid grades, depending on the intended end use. The melting point and degree of hardness of an asphalt is affected by the completeness with which the volatile fractions have been removed. If the melting point or hardness of a crude separated asphalt product is too low, it is oxidized by blowing air through the heated asphalt until the melting point is raised to the extent desired.

Low melting point asphalts are generally used in such applications as the waterproofing of flat, built-up roofs and the like, where the self-sealing qualities are an attraction. For sloping built-up roofs, higher melting asphalts are required. For asphalt shingles, roll roofing, and similar sheet roofing products, the melting point has to be high enough to give an essentially nonsticky surface after fabrication, under ordinary ambient conditions. Road asphalts normally have to pass a melting point specification, a penetration test ("pen test"), and other requirements. The penetration test is conducted on an asphalt sample at 25°C. It is the extent to which the sample is penetrated by a steel ball or pin of standard dimensions while a force of 100 g is applied for 5 seconds. This test is an important measure of the suitability of the asphalt for various paving applications. For application as road surfacing a narrow size distribution of crushed gravel, the aggregate, is blended into hot asphalt, which serves as the binder.

By the variety of processing procedures just outlined very little of the crude oil processed is wasted. From the high volatility dissolved gases present, all the way to the nearly involatile asphaltenes and the suspended carbon, etc., present in the original oil, virtually all are converted to one or the other of a host of useful products. Any combustible components that are not directly used toward the formulation of a product are used to generate energy for the refinery itself.

18.6. REFINERY EMISSION CONTROL

The overall interest in producing useful products from the feedstocks entering a refinery, and recognition of the potential hazards to operating personnel in the event of process losses, have meant that good control of process streams has always been maintained. Some aspects of refinery operations require closer attention than others to ensure safe and environmentally acceptable production. For the convenience of discussion the areas of concern have been grouped as they relate to atmospheric emission control, aqueous emission control, and waste disposal practices.

18.6.1. Atmospheric Emission Control

Hydrocarbon vapor losses from refineries are about 5 to 10% of the total hydrocarbons discharged to the atmosphere, from pollutant inventories [44]. However, refinery losses are highly localized so that control is important to minimize local ambient air concentrations of hydrocarbons. Any volatile sulfides, because of their intense odors, need attention to their control out of proportion to the potential losses.

Losses from pump seals, flanges connecting lengths of pipe, and operating vessels are best controlled by properly scheduled maintenance procedures [45]. Losses of methane or ethane from crude or process stream preheat furnaces using these hydrocarbons as fuels used to occur occasionally, when there was inadequate combustion air. Control is now maintained using a slight excess of combustion air. An excess is ensured with a continuous oxygen analyzer which is used to monitor the flue gases of these units. Introduction of flue gas combustion control has also contributed a fringe benefit in that more energy is recovered using a slight excess of air rather than using large excesses or deficiencies of air.

Potential hydrocarbon losses from the overpressuring of operating vessels is controlled first via staged computer alerts and/or manual alarms to provide for correction of the condition. If the overpressure exceeds a second set point, pressure relief valves vent the contents of the vessel to a flare release system. The flare system provides a means of controlled burning of hydrocarbon vapors at a nonhazardous point to avoid fire or explosion risks. Smoke problems from flares are avoided by more efficient designs which use the Coanda effect, or steam injection at appropriate points in the flare streams, to promote good fuel-air mixing [46]. Use of a steam to hydrocarbon ratio of about 0.3:1 for flaring has also been shown to reduce soot formation by the water gas reaction of carbon particles with steam [47] (Eq. 18.41).

$$C + H_2O \rightarrow CO + H_2 \text{ (at high temperatures)} \qquad 18.41$$

The carbon monoxide and hydrogen formed by this process both burn cleanly. Energy recovery or recycle may be possible more often for flared hydrocarbons.

Storage tanks which hold the more volatile liquid products can lose a significant amount of product as vapor, particularly with short cycle times. Storage tanks are always vented to the air to prevent overpressuring from vapor inside on a hot day, which could cause tank rupture, or collapse from vapor condensation and atmospheric pressure acting on the outside of the tank during a cold night. A single storage tank can lose as much as 190,000 L of gasoline per year from this source [48]. These losses may be controlled by fitting floating roofs to tanks used to store products of high vapor pressures, which eliminates the vapor space responsible for the losses from a fixed roof tank [49]. Or hollow polyethylene spheres, 2–3 cm in diameter and two to three layers deep, may be floated on the surface of the liquid inside the tank to decrease losses from the vapor space by 90+%. A slight positive pressure can also help to reduce vapor loss. These measures not only decrease emissions but are also generally cost effective, because of the conserved product. Floating roofs also decrease the explosion risk in fixed roof tanks that exists from the air-vapor mixture above volatile hydrocarbon products such as gasoline.

Crude oils all contain some sulfur, mostly in the form of thiols, sulfides, and hydrogen sulfide. The sulfur content contributed by these compounds ranges from about 6% for light Arabian crudes, to 1–2% for California crudes, down to as little as 0.4% for some Canadian Peace River crudes [50]. Volatile sulfides of stored crude oil are also controlled using floating roof storage tanks.

The hydrogen sulfide content of gaseous hydrocarbon streams is removed during refining by either scrubbing with organic bases (details in Chap. 7) or by physical absorption, for example by the Sulfinol process. Liquid process streams in the middle boiling point range require catalytic hydrogenation first (Hydrofining or Hydrotreating) to convert less volatile sulfides to hydrogen sulfide and paraffinic hydrocarbons (Eq. 18.42).

$$R_2S + 2 H_2 \rightarrow 2 RH + H_2S \qquad 18.42$$

The more volatile hydrogen sulfide may then be readily separated from the hydrocarbon stream by stripping (sparging) with air or steam. Separation of the sulfur-containing compounds present in higher boiling point streams and involatile residues requires similar chemistry to this to convert high boiling sulfides to the more volatile hydrogen sulfide and thiols. Following the catalytic conversion they may be separated by fractionation or stripping. Sulfur removal from the residue stream is the more difficult since the wide variety of metallic and other impurities present slow down or rapidly deactivate even the best catalysts. This problem makes frequent catalyst replacement necessary and makes desulfurization of residues more expensive. This is why residues are less often desulfurized. The sulfur content of a residual fuel oil may be minimized without having to hydrodesulfurize the distillation residues by thorough desulfurization of the catalytically cracked gas oil diluent before it is blended with the residue.

Hydrogen sulfide accumulated from desulfurization processes and any on-site sulfur dioxide are converted to elemental sulfur using two or three stages of Claus reactors [51] (Chap. 9). However, Claus plant operation is seldom a profitable operation for a refinery since the sulfur yield frequently amounts to only a few tonnes per day. But the improved ambient air quality achieved in the vicinity of the refinery makes this measure worthwhile.

Carbon monoxide is a potential emission problem. It can arise from the catalyst regenerators of some recent models of catalytic cracking units which run at temperatures too low to obtain complete oxidation of carbon to carbon dioxide. Fortunately, carbon monoxide is a valuable industrial fuel and may be burned in a "CO boiler" to recover energy as steam and discharge carbon dioxide. The most recent designs of catalyst regenerators operate at somewhat higher temperatures and achieve complete conversion of carbon to carbon dioxide, without the need for a CO boiler.

Control of catalyst particle losses from both the cracker and regenerator of fluid catalytic cracking units is achieved by two cyclones operating in series right inside each unit. This is usually followed by an electrostatic precipitator for fine particle control, working on the exhaust side of the catalyst regenerator [52]. The metal content of spent catalysts may be recovered for reuse [53].

Nitric oxide can form to the extent of 270 ppm or more at the hot metal surfaces of the refinery combustion units [54] (Eq. 18.43).

$$N_2 + O_2 \xrightarrow[\text{metal}]{\text{hot}} 2\,NO \qquad\qquad 18.43$$

Its formation can be kept to a minimum by keeping the excess air supplied to combustion units to a minimum value for safe complete combustion [45]. Burner designs which produce a more diffuse flame front (large flame volume) achieve lower peak combustion temperatures which also helps to decrease the formation of nitric oxide. Injection of ammonia into the flue gas, while it is still hot, can decrease NO_x concentrations down to 80–120 ppm, one-third to one-half that of uncontrolled concentrations [54].

18.6.2. Aqueous Emission Control

Standards for aqueous effluent from refineries have been set, both in Canada and the United States (Table 18.9) [55–57]. For parameters such as total suspended solids, oil and grease, and ammonia nitrogen, American standards appear to be more strict than Canadian federal standards. Also provincial standards in Canada are generally more restrictive than the federal standards, perhaps because of the more numerous very large refineries operating in the United States.

TABLE 18.9   Aqueous Effluent Standards for Petroleum Refining[a]

Effluent component or property	Average of daily values for one month, max. permitted (g/m³ crude)			Maximum values, any one day (g/m³ crude)	
	U.S.A.	**Canada**[b]	**B.C.**[b,c]	**U.S.A.**	**Canada**[d]
Total suspended solids	2.0	20.6	20[d]	2.4	42.8
BOD	2.0	—	6.6	2.5	—
COD	8.0	—	—	10.0	—
Oil and grease	0.4	8.6	1.7	0.5	21.4
Phenols	0.0060	0.86	0.066	0.012	2.1
Sulfide	0.035	0.29	0.031	0.05	1.4
Ammonia nitrogen	0.51	10.3	1.65	0.68	20.6
Total chromium	0.105	—	0.57	0.124	—
pH	—	6–9.5	6.5–8.5	—	6–9.5

[a]Derived from Water Resources Service [55], Greenwood *et al.* [56], and Water Pollution Control Directorate [57].

[b]Recalculated from units of pounds per thousand barrels to grams per cubic meter of crude by multiplying by 2.855.

[c]Level A standards, test intervals vary from daily to monthly.

[d]Units of mg/L, not g/m³.

Aqueous desalter effluent contains sediments, oil, dissolved salts, and sulfides. It is treated initially by using an American Petroleum Institute (API) separator for residual oil removal and recovery (Fig. 18.9). Any oil droplets larger than about 150 μm (0.15 mm) in diameter are collected as an oil phase, which is routed to the refinery "slop oil" stream to join any other off-specification liquid oil streams for reprocessing. Inclined plastic plates inserted into an oil/water separator of this type greatly increase the speed and com-

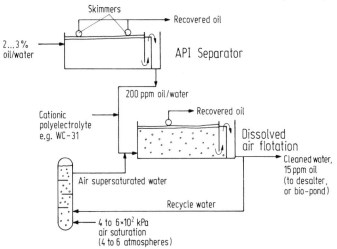

FIGURE 18.9 Operating details of an American Petroleum Institute separator, followed by a dissolved air flotation unit for oil/water separation.

pleteness of creaming, and reduces the average droplet size collected to about 60 μm (0.06 mm) [58]. After oil separation the water phase of the desalter is passed through a sour water stripper (Fig. 18.10), where hydrogen sulfide is removed by blowing with low-pressure steam, 140 kPa (ca. 130°C, 20 lb/in.²) [59]. The separated hydrogen sulfide is processed to elemental sulfur in a Claus unit. The sweetened water obtained as the bottom stream from the stripper and contains <10 ppm hydrogen sulfide and <50 ppm ammonia [60]. It may be recycled to feed the incoming desalter water, or for other purposes to help control losses of volatile compounds [61]. In case of excess water buildup in the desalter circuit, it is bled from this point in the recycle to aerated bio-ponds, etc., for further treatment before discharge.

Sour water (hydrogen sulfide in water) is also obtained as aqueous condensate from the fractionation of the volatile products of refinery operations such as the fluid catalytic cracker, hydrotreater, hydrocracker, or coker. In fact, in a refinery which is endeavoring to minimize water use, sour waters may make up as much as 25% of the total aqueous effluent discharged. All of this will be treated for hydrogen sulfide (and ammonia) removal in manner similar to the procedure described for desalter effluent [59].

Site drainage waters from surface runoff invariably include some oil picked up from small spills and process leaks. These will also be routed through an API separator first, sometimes followed by a dissolved air separator (Fig. 18.8). Oil separated from this drainage water stream will join the slop oil circuit for recycling. During times of desalter operation when water is short, treated site drainage waters may be used for this purpose. Any water in excess of desalter requirements will proceed to an artificially aerated bio-pond for further BOD reduction before discharge [62].

Phenols are removed from refinery hydrocarbon streams by treating with a solution of sodium hydroxide in water. This produces a waste stream which is difficult to treat in a bio-pond [63]. If feasible, this stream will be sold (or given) to a neighboring petrochemical plant if it can make use of the phenol content. Phenol is recovered from this stream by economical acidification using flue gases, followed by solvent extraction (Eq. 18.44).

$$NaOPh + H_2CO_3 \rightarrow NaHCO_3 + PhOH \qquad 18.44$$

Caustic in water is also frequently used for removal of sulfides from certain liquid hydrocarbon refinery streams (Eq. 18.45, 18.46).

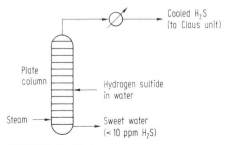

FIGURE 18.10 Details of hydrogen sulfide removal from water in a sour water stripper.

$$2 \text{ NaOH} + \text{H}_2\text{S} \rightarrow \text{Na}_2\text{S} + 2 \text{ H}_2\text{O} \qquad\qquad 18.45$$

$$\text{NaOH} + \text{CH}_3\text{SH} \rightarrow \text{NaSCH}_3 + \text{H}_2\text{O} \qquad\qquad 18.46$$

Sodium sulfide, at least, is of value to pulp mills which use the kraft process for pulping and, if near by, can provide an outlet for this waste product. However, if this disposal method is not feasible a "caustic oxidizer" may be used to convert the sodium sulfide to the more environmentally acceptable sodium sulfate, for ultimate disposal by landfill (Eq. 18.47).

$$\text{Na}_2\text{S (aqueous)} + 2 \text{ O}_2 \xrightarrow[\text{heat}]{} \text{Na}_2\text{SO}_4 \qquad\qquad 18.47$$

Or, alternatively, aqueous sodium sulfide may be acidified to produce a sour water stream (Eq. 18.48).

$$\text{Na}_2\text{S} + \text{H}_2\text{SO}_4 \rightarrow \text{Na}_2\text{SO}_4 + \text{H}_2\text{S} \qquad\qquad 18.48$$

This can then be routed to the sour water stripper for hydrogen sulfide recovery and conversion to sulfur. In this case the more economical method of acidification, flue gas contacting, cannot be used since this would liberate highly volatile hydrogen sulfide into the flue gases contacted.

When operating in areas with a restricted water supply, or on municipal water services, significant economies in water usage may be achieved by recycling once-used and warmed cooling water through direct or indirect cooling towers for reuse. Process cooling directly with air, by fan forcing against finned tubes, is also a measure which can reduce overall refinery water requirements, particularly when operating in cold climates.

The potential impact of inadvertent losses of some constituents of refinery liquid wastes has been reviewed by Cote [64], and a summary of aqueous treatment processes has been presented by Baker [65].

18.6.3. Refinery Waste Disposal Practices

The components of most refinery liquid waste streams are recovered and reused, whenever feasible. But some of these, such as aqueous caustic phenolic or caustic sulfidic wastes, do not lend themselves readily to reuse in many refinery situations. Deep well disposal, incineration, or precipitation in some manner and landfilling of the separated solids are the measures used in these instances. Preliminary concentration of brine streams by reverse osmosis can help decrease final disposal costs by decreasing the waste volume [66].

Sludges, such as those accumulated in the monoethanolamine or diethanolamine streams used for hydrogen sulfide removal by scrubbing, comprise another intractable waste material. Sludges may be reprocessed to recover valuable materials [67], or failing this they may be safely discarded by incineration. Others use landfilling (dry) or landfarming as disposal methods. Biotreatment or deep well disposal are also used by some Canadian refineries, but to a more limited extent.

Spent acids are used where feasible to neutralize alkaline waste streams, or are neutralized with purchased lime, or caustic, and are then routed to the normal effluent treatment system for further cleanup before discharge [51].

Waste solids from refinery operations include such materials as spent clay from decolorization of lubricating oils or waxes, sand used as a catalyst support, filter aid, or filter base, and exhausted Claus catalysts (primarily ferric oxide on alumina). These are disposed of by landfilling by the majority of Canadian refineries. However, some use landfarming for disposal of these materials. One refinery recovers spent Claus catalysts for regeneration into new catalyst.

Landfarming of oily sludges for degradation by soil micro-organisms has been found to be an efficient method of disposal, providing proper conditions are maintained to achieve optimum oil decomposition rates within the soil. Oil degradation rates in outside experimental plots of soil are high (Table 18.10) [68]. Decomposition rates range from about 150 kg oil carbon/hectare/ day in the temperate climate of the area of Edmonton, Alberta [68,69], up to as high as 569 kg oil carbon/hectare/day for landspreading operations in Texas [70]. Experiments that examined the rates of degradation of crude oil spills on land showed that the rate was significantly enhanced by the application of fertilizer and helped somewhat by the addition of a suitable mixed culture of bacteria (Table 18.11) [71,72].

The methods outlined above may also be used for other biodegradable waste products of refinery operation such as amine sludges and spent oily clay and sand. With only light sludge loadings the land can be returned to agricultural use after one season. However, landfarming for sludge disposal has been regulated against in some areas, e.g., the State of Minnesota. The Water Quality Programs Committee of the International Joint Commission (U.S.A. and Canada) recommends that groundwater pressure and quality be monitored at different depths, and that routine soil samples be taken for oil content and trace metal analysis from any operating landfarm disposal system [73]. And when the land employed for landfarming is retired from this use it is also recommended that these soils not be used for the growing of crops or for grazing because of the risk of contamination of foodstuff from metals or other materials that could have accumulated in the soil [74]. Concern with land disposal of petroleum processing wastes has led to the use of these materials as fuel for cement kiln operation [75].

TABLE 18.10 Decomposition Rates of Crude Oil in the Field, Edmonton, Alberta[a]

Elapsed time (days)	Application rate/oil content (kg/m²)[b]		
	Light	Medium	Heavy
0	5.5	13.7	24.9
43	3.7	10.7	17.2
340	2.1	8.5	11.5
460	2.0	7.2	11.0
735	1.5	4.2	7.8

[a]Calculated from the data of Toogood [69].
[b]Cultivated to a 30-cm depth.

TABLE 18.11   Percent Composition of Crude Oil after 308 Days in Contact with Soil Plus Various Additives, Swan Hills, Alberta[a]

Crude oil fraction	Percent composition of oil[b]			
	In barrel	Soil alone	Soil plus fertilizer[c]	Soil + bacteria[d] + fertilizer
Asphaltenes, soluble	2.77	1.35	2.01	2.58
Asphaltenes, insoluble	4.19	12.7	20.7	21.3
Saturates	59.7	54.7	39.3	37.9
Aromatics	23.6	17.9	20.9	21.4
NSOs,[e] soluble	5.64	6.85	9.06	9.98
NSOs,[e] insoluble	5.03	6.42	8.04	7.27

[a]Data selected from that of Cook and Westlake [71,72]. Little change in composition was noted in these experiements after 12 days, but significant degradation was already evident after 66 days and a further progression of degradation was observed after 433 days, the length of the experiment.
[b]Local crude oil was distributed at the rate of 60 liters per 9 m² plot; results averaged over four replicate plots.
[c]Fertilizer (27-27-0) was applied at the rate of 60 grams of nitrogen per square meter of plot.
[d]A mixed culture of suitable bacteria was applied at the rate of 10^6 per square centimeter of plot.
[e]NSOs = nitrogen-, sulfur-, and oxygen-containing compounds.

REVIEW QUESTIONS

1. Why is 520°C approximately the upper limit for atmospheric pressure distillation of crude oil? How does this factor relate to the need to use reduced pressure for the distillative refinement of lubricating oils?

2. (a) Describe the similarities and the differences between straight thermal cracking and catalytic cracking as oil modification procedures.
(b) How do the details of the process parameters affect the types and proportions of the observed products from the two processes?

3. (a) Explain the chemical basis for the inevitable formation of olefins and/or coke (as well as saturated paraffins) from cracking reactions.
(b) Write two plausible equations for cracking reactions to useful products from n-dodecane ($C_{12}H_{26}$), and from n-hexadecane ($C_{16}H_{34}$).
(c) What mass yield of each of 1-heptene and the other expected product would be obtained from one pass of 1000 kg of n-dodecane through a catalytic cracker for a conversion of 52% and a selectivity of 86%?
(d) Name and explain at least two other products that could have resulted from the 14% of dodecane that does not form the products given in part (c).

4. (a) Explain and give equations to illustrate your explanation of the mode of formation of a high proportion of branched, rather than straight chain, products formed during the formation of polymer alkylate for use in gasoline.

(b) What property is contributed by this component of gasoline that is valued by refiners?

(c) Explain the relationship between the typical structures of polymer alkylate and the valuable property that this fraction contributes to gasoline. Equations may assist in interpretation and presentation.

5. (a) Assuming that a statistical distribution of products is obtained from a polymerization process to a gasoline fraction, what proportion of C_6, C_7, and C_8 products would result from the polymerization of 1000 kg of isobutylene and 1000 kg of propane?

(b) For a 60% conversion, 80% yield on polymerization of 1000 kg of isobutylene, what total mass of octanes and octenes would be expected, theoretically?

FURTHER READING

J.L. Allen, Evaluating a Waste-oil Reclamation Program. *Plant Eng.* **30,** 255 (1976).

P.J. Bailes, Application of Solvent Extraction to Organic and Petrochemical Industries. *Chem. Ind. (London)*, p. 724, Sept. 3 (1977).

D. Burris, Field Desalting, a Growing Producer Problem Worldwide. *Pet. Eng.* **46,** 36 (1974).

CanTest Ltd./IEC/PAC, "Survey of Trace Substances in Canadian Petroleum Industry Effluents," by CanTest Ltd. and IEC International Environmental Consultants Ltd., Petroleum Association for Conservation of the Canadian Environment, Ottawa, 1981.

J.G. Speight, "The Chemistry and Technology of Petroleum." Dekker, New York, 1980.

REFERENCES

1. E.C. Lane and E.L. Garton, Rep. Invest.—*U.S., Bur. Mines* **RI-3279,** (1935), cited by J.A. Kent, ed., "Riegel's Handbook of Industrial Chemistry," p. 406. Van Nostrand-Reinhold, New York, 1974.

2. W.A. Bachman and D.H. Stormont, *Oil Gas J.* **65,** 69 (1967).

3. M.T. Atwood, *CHEMTECH* **3,** 617 (1973).

4. F.D. Rossini, *J. Chem. Educ.* **37**(11), 554 (1960).

5. L.F. Fieser and M. Fieser, "Advanced Organic Chemistry."Reinhold, New York, 1961.

6. G. Eglinton and M.T.J. Murphy, eds., "Organic Geochemistry," pp. 640, 676. Longman-Springer Verlag, Berlin and London, 1969.

7. Chemical Desalting, *Hydrocarbon Process.* **49**(9), 235, Sept. (1970).

8. Electrostatic Desalting, *Hydrocarbon Process.* **49**(9), 237, Sept. (1970).

9. "Kirk-Othmer Encyclopedia of Chemical Technology," 3rd ed., Vol. 17, p. 183. Wiley-Interscience, New York, 1982.

10. V.B. Guthrie, ed., "Petroleum Products Handbook." McGraw-Hill, New York, 1960.

11. E.M. van Z. Bakker and J.F. Jaworski, "Effects of Vanadium in the Canadian Environment." National Research Council of Canada, Ottawa, 1980.

12. "Kirk-Othmer Encyclopedia of Chemical Technology," 2nd ed., Vol. 15, p. 1. Wiley-Interscience, New York, 1968.

13. P.J. Garner, *Chem. Ind. (London)*, p. 131, Feb. 16 (1974).
14. D.M. Considine, ed., "Chemical and Process Technology Encyclopedia." McGraw-Hill, New York, 1974.
15. Organization for Economic Cooperation and Development, "IEA Statistics, Oil and Gas Information 1994," p. 81. OECD, Paris, 1995.
16. J.A. Kent, ed., "Riegel's Handbook of Industrial Chemistry," 7th ed. Van Nostrand-Reinhold, New York, 1974.
17. G.A. Purdy, "Petroleum Prehistoric to Petrochemicals." McGraw-Hill, Toronto, 1958.
18. J.A. Kent, ed., "Riegel's Industrial Chemistry," 6th ed. Reinhold, New York, 1962.
19. R.F. Wong, Increasing FCC Regenerator Catalyst Level, *Hydrocarbon Process.* **72**(11), pp. 59–61, Nov. (1993).
20. L.F. Albright, *Chem. Eng. (N.Y.)* **73**, 143, Aug. 15 (1966), cited by Kirk-Othmer [12].
21. S.R. Tennison, *Chem. Br.* **17**(11), 536 (1981).
22. T. Hutson, Jr. and R.S. Logan, *Hydrocarbon Process.* **54**(9), 107, Sept. (1975).
23. G.C. Ray, J.W. Myers, and D.L. Ripley, *Hydrocarbon Process.* **53**(1), 141, Jan. (1974).
24. P. Polss, *Hydrocarbon Process.* **52**(2), 61, Feb. (1973).
25. EPA Mulls Over Lead-in-Gasoline Rules, *Chem. Eng. News* **60**(18), 28, May 3 (1982).
26. Manganese Rather than Lead, *Can. Chem. Proc.* **56**(6), 48, June (1972).
27. MTBE Can be Made from Natural Gas, *Chem. Eng. News* **58**(28), 26, July 14 (1980).
28. DOE Endorses Use of Alcohol Fuels, *Chem. Eng. News* **57**(29), July 16 (1979).
29. B. Gillespie, L.W. Manley, and C.J. Di Perna, *CHEMTECH* **8**, 750 (1978).
30. L. Morriset-Blais, *Can. Chem. Process.* **57**(8), 48, Aug. (1973).
31. "Kirk-Othmer Encyclopedia of Chemical Technology," 3rd ed., Vol. 14, p. 477. Wiley-Interscience, New York, 1991.
32. H.G. Muller, *Tribol. Int.* **11**(3), 189, June (1978).
33. R.R. McCoy and D.S. Taber, A Cold Look at Lubricants. *SAE Trans.* **80**, Paper 719716 (1971).
34. D.J. Skinner and W.A. Neff, "Preliminary Review of Used Lubricating Oils in Canada," Rep. EPS 3-WP-74-4. Environment Canada, Ottawa, 1974.
35. E. Baumgardner, *Hydrocarbon Process.* **53**(5), 129, May (1974).
36. N.J. Weinstein, *Hydrocarbon Process.* **53**(12), 74, Dec. (1974).
37. J. Sternberg, Used Lubricants to Gasoline. *Biocycle* **33**, 80, May (1992).
38. E.E. Klaus and E.J. Tewkesbury, *Hydrocarbon Process.* **53**(12), 67, Dec. (1974).
39. D.B. Barton, J.A. Murphy, and K.W. Gardner, Synthesized Lubricants Provide Exceptional Extended Drain Passenger Car Performance. *SAE Tech. Pap. Ser.* 780951 (1978).
40. Synthetic Lubricants Poised for Big Growth, *Chem. Eng. News* **58**(13), 12, March 31 (1980).
41. M.W. Ranney, ed., "Synthetic Oils and Additives for Lubricants—Advances Since 1977," Chem. Technol. Rev. No. 145. Noyes Data Corp., Park Ridge, NJ, 1980.
42. R.H. Perry, ed., "Chemical Engineers Handbook," 4th ed., p. 9–6. McGraw-Hill, New York, 1963.
43. F. Tamburrano, Disposal of Heavy Oil Residues. *Hydrocarbon Process.* **73**, 79–84, Sept. (1994).
44. State of the Environment Reporting, "The State of Canada's Environment." Government of Canada, Ottawa, 1991.
45. H.F. Elkin and R.A. Constable, *Hydrocarbon Process.* **51**(10), 113, Oct. (1972).
46. W. Lauderback, *Hydrocarbon Process.* **51**, 127, Jan. (1972).
47. S.H. Tan, "Flare System Design Simplified, in Waste Treatment and Flare Stack Design Handbook," p. 81. Gulf Publ. Co., Houston, TX, 1968.
48. P.L. Magill, F.R. Holden, and C. Ackley, eds., "Air Pollution Control Handbook." McGraw-Hill, New York, 1956.
49. Control of Atmospheric Emissions from Petroleum Storage Tanks, *J. Air Pollut. Control Assoc.* **21**(5), 263 (1971).
50. C.M. McKinney, *Hydrocarbon Process.* **51**(10), 117, Oct. (1972).
51. W.J. Racine, *Hydrocarbon Process.* **51**(3), 115, March (1972).
52. C.S. Russell, "Residuals Management In Industry, A Case Study of Petroleum Refining." Johns Hopkins University Press, Baltimore, 1973.

53. G. Berrebi, P. Dufresne, and Y. Jacquier, Recycling of Spent Hydroprocessing Catalysts: EURECAT Technology. *Environ. Prog.* **12**(2), 97–100, May (1993).

54. Pollution Control Process Demonstrated, *Chem. Eng. News* **56**(8), 23, Feb. 20 (1978).

55. "Pollution Control Objectives for the Chemical and Petroleum Industries of British Columbia." Water Resources Service, Queen's Printer, Victoria, 1977.

56. D.R. Greenwood, G.L. Kingsbury, and J.C. Cleland, "A Handbook of Key Federal Regulations and Criteria for Multimedia Environmental Control," Rep. 600/7-79-175. Environmental Protection Agency, Washington, DC, 1979; cited by Kirk-Othmer [9].

57. "Petroleum Refinery Effluent Regulations and Guidelines," Rep. EPS 1-WP-74-1. Water Pollution Control Directorate, Ottawa, 1974. From the Canada Gazette Part 2 of April-June 1982, which refers to the Consolidated Regulations of Canada, Vol. VII, c. 828, p. 5225, Ottawa, 1978. These figures represent Canadian standards to date.

58. J. Wardley-Smith, "Prevention of Oil Pollution." Graham and Trotman Ltd., London, 1979.

59. R.J. Klett, *Hydrocarbon Process.* **51**(10), 97, Oct. (1972).

60. Fina Recovers H2S from Wastewaters, *Can. Chem. Process.* **60**(10), 28, Oct. (1976).

61. A. Hasbach, Closed Loop System Recycles VOC's from Refinery Waste Water. *Pollut. Eng.* **24**, 69(1992).

62. J.F. Ferrel and D.L. Ford, *Hydrocarbon Process.* **51**(10), 101, Oct. (1972).

63. J.C. Hovious, G.T. Waggy, and R.A. Conway, "Identification and Control of Petrochemical Pollutants Inhibitory to Anaerobic Processes," EPA-R2-73-194. Environmental Protection Agency, Washington, DC, 1973.

64. R.P. Cote, "The Effects of Petroleum Refinery Liquid Wastes on Aquatic Life, with Special Emphasis on the Canadian Environment." National Research Council, Ottawa, 1976.

65. D.A. Baker, *J. Water Pollut. Control Fed.* **46**(6), 1298 (1974).

66. R.W. Newkirk and P.J. Schroeder, *Hydrocarbon Process.* **51**(10), 103, Oct. (1972).

67. W.J. Hahn, High Temperature Reprocessing of Petroleum Oily Sludges. *SPE Prod. Facilities* **9**, 179–182, Aug. (1994).

68. J.A. Toogood, ed., "The Reclamation of Agricultural Soils After Oil Spills," AIP Publication No. M-77-11. Dept. of Soil Science, University of Alberta, Edmonton, 1977; cited by Beak Consultants Ltd., "Landspreading of Sludges at Canadian Petroleum Facilities," Rep. No. 81-5A. Petroleum Association for Conservation of the Canadian Environment, Ottawa, 1981.

69. W.K. Mann, H.B. Shortly, R.M. Skallerup, eds., "Industrial Oily Waste Control." American Petroleum Institute/American Society of Lubrication Engineers, New York and Baltimore, ca. 1970 (no date specified).

70. C.B. Kincannon, "Oily Waste Disposal by Soil Cultivation Process," EPA-R2-72-110. Environmental Protection Agency, Washington, DC, 1972; cited by Beak Consultants Ltd., "Landspreading of Sludges at Canadian Petroleum Facilities," Rep. No. 81-5A. Petroleum Association for Conservation of the Canadian Environment, Ottawa, 1981.

71. F.D. Cook and D.W.S. Westlake, "Biodegradability of Northern Crude Oils." Task Force on Northern Oil Development Rep. No. 73-20, Information Canada Cat. No. R72-8373, Ottawa, 1973, Ottawa, 1973.

72. F.D. Cook and D.W.S. Westlake, "Microbiological Degradation of Northern Crude Oil." Task Force on Northern Oil Development Rep. No. 74-1, Information Canada Cat. No. R72-12774, Ottawa, 1974.

73. "A Review of the Pollution Abatement Programs Relating to the Petroleum Refinery Industry in the Great Lakes Basin." Great Lakes Water Quality Board, Windsor, Ontario, 1982.

74. "Manual for Landspreading of Petroleum Industry Sludges." Rep. No. 81-5A, Beak Consultants Ltd., Petroleum Association for Conservation of the Canadian Environment, Ottawa, 1981.

75. D. Gossman, The Reuse of Petroleum and Petrochemical Waste in Cement Kilns. *Environ. Prog.* **11**(1), 1–6, Feb. (1992).

Skyline of the styrene production facilities of the former Polymer Corporation, Sarnia. Shorter stout tower near the center is the styrene distillation column.

19
■ PETROCHEMICALS

Take Carbon for example then
What shapely towers it constructs
To house the hopes of men!
            —A.M. Sullivan (1896–1980)

19.1. BACKGROUND

Organic chemicals such as dyestuffs and derivatives of benzene and hetero-cyclic compounds were first produced from coal-based raw materials. Since the mid-nineteenth century these products were obtained by the further processing of discrete volatile fractions recovered during the coking of coal. A large fraction of the coke product was destined for the reduction of iron ore to produce pig iron (Chap. 14). Although volatile fractions from the coking of coal, such as benzene, toluene, and the xylenes, are still important items of commerce, petroleum-based sources for these materials have become much more important than coal in recent years [1]. The relatively greater attractiveness of petroleum-based feedstocks grew as both the reliability and scale of petroleum production increased. These aspects, plus the gradually increased sophistication of refining methods of the oil industry of the early twentieth century contributed to the increased appreciation of petroleum as a resource for chemicals production.

As a consequence of the recent occurrence of these developments the petrochemical industry is relatively much younger than the industry and the processes of production of the major inorganic chemicals. But one should be careful in the assessment of products that are termed "petrochemical." Basically a petrochemical is derived directly or indirectly from a petroleum or natural gas fraction. It maybe organic, such as ethylene, benzene, or formaldehyde, or it may be inorganic such as ammonia, nitric acid, and ammonium nitrate

(Chap. 11). So "petrochemical" is not synonymous with organic chemical, although most petrochemicals are also organic chemicals.

What provides the incentive for a producer to construct facilities for the production of petrochemicals? It is the accessibility to the producer to further profit from the additional value-added stages that are provided by the conversion of a refined petroleum or gas fraction into a petrochemical, compared

▪ TABLE 19.1   The 30 Largest Volume Chemicals and Their Scale of Production in the U.S.A. and Canada in 1995[a]

U.S.A. rank 1995	Petro-chemical	Product	Production (million tonnes)	
			U.S.A.	**Canada[b]**
1		Sulfuric acid	43.25	4.22
2		Nitrogen	30.86	
3		Oxygen	24.26	
4	√	Ethylene	21.30	3.13
5		Lime	18.70	
6	√	Ammonia	16.15	4.66
7		Phosphoric acid	11.88	
8		Sodium hydroxide	11.88	1.18
9	√	Propylene	11.65	0.72
10		Chlorine	11.38	1.13
11		Sodium carbonate	10.11	
12	√	Methyl t-butyl ether	7.99	
13	√	Ethylene dichloride	7.83	
14	√	Nitric acid	7.82	0.99
15	√	Ammonium nitrate	7.25	1.03
16	√	Benzene	7.24	0.78
17	√	Urea	7.07	3.33
18	√	Vinyl chloride	6.79	
19	√	Ethylbenzene	6.20	
20	√	Styrene	5.17	
21	√	Methanol	5.12	
22	√	Carbon dioxide	4.94	
23	√	Xylene	4.25	
24	√	Formaldehyde	3.68	
25	√	Terephthalic acid	3.61	
26	√	Ethylene oxide	3.46	
27		Hydrochloric acid	3.32	0.14
28	√	Toluene	3.05	0.28
29	√	p-Xylene	2.88	0.43
30	√	Cumene	2.55	
		Total inorganic chemicals (9)	165.64	[b]
		Total petrochemicals (21)	140.00	[b]

[a]Calculated from data of *Chemical and Engineering News* [2].
[b]Some data unavailable because of Statistics Canada confidentiality regulations.

to the sales value of the raw petroleum fraction. Depending on the complexity and number of stages involved, a petrochemical product may command a price that is 5 or 6 times the price of the oil used to produce it. If the processor incorporates several more value-added stages to convert a petrochemical into a consumer product, this may fetch a price as high as 20 times that of the oil from which it is derived. In short, the incentive is that oil (or gas) is worth far more as chemicals than as fuel.

By considering the scale of production of the top 30 industrial chemicals without regard to source, it is possible to obtain an impression of the relative importance of the petrochemical sector to the chemical industry as a whole. Fully 21 of these top 30 chemicals are classified as petrochemicals with a gross production of the group in the United States in 1995 of 140 million metric tonnes (Table 19.1) [2]. Only 9 are nonpetrochemical inorganic compounds. However, the gross volume of production of these 9 compounds is still larger at 166 million metric tonnes annually. The product distribution for Canada and other developed nations is similar. Sulfuric acid, an important inorganic commodity chemical, has been and probably will continue to be the chemical produced on the largest scale in developed nations. But in Canada, where the raw sulfur for sulfuric acid production is derived largely from the sweetening of sour natural gas (Chap. 9), it could be argued that sulfuric acid should be classified as a petrochemical under these circumstances.

Not only are petrochemicals produced on a vast scale but the diversity of materials represented by this class is also great, more than 3000 of these being in current commercial production. Even with their significant commercial importance and scale, however, only 5–6% of current oil and gas production is consumed to produce organic petrochemicals and another 4–5% for inorganic petrochemicals. It seems to be an extravagant use of a key flexible energy resource to apply only 10% of it for high value-added products, many of them in recoverable or durable applications, and simply burn the remaining 90% for energy production.

19.2. CARBON BLACK

The first chemically simple product from hydrocarbon sources was carbon black, which has been produced from natural gas (predominantly methane) since 1872. Normal combustion of natural gas with adequate air produces carbon dioxide and water together with a flame temperature of 1000–1200°C. However, if the combustion is carried out in a deficiency of air with strong cooling of the flame a large proportion of the "combustion" proceeds to yield finely divided carbon (carbon black) and water vapor (Eq. 19.1).

$$CH_4 + O_2 \xrightarrow{\text{ca. } 500°C} C + 2\,H_2O \qquad\qquad 19.1$$

The cooling of the flame was accomplished by cold iron bars or channels placed into the flame and onto which the carbon deposited. The product, called channel black, was periodically scraped from the iron channels for further processing and sale. Yield of carbon from this process was about 1 tonne per 50,000 m^3 (at 1 atm, 15.6°C) of natural gas, or about 5% [3]. The low yields from this process and the increasing concerns about the traces of particulate lost from channel black plants have contributed to their obsolescence.

Better yields of carbon black are obtained by better separation of the combustion function from the carbon-forming function, as is accomplished in the newer furnace black processes. This approach enables "gas oil," relatively high boiling point liquid petroleum fractions, or natural gas to be used to produce carbon black. The cooling function from an initial 1400°C to about 200°C is accomplished by direct water sprays. The product is removed from the gas stream via a combination of cyclone collectors and glass fiber, or Teflon bag filters. One tonne of furnace black is obtained from 5300–7000 m^3 (1 atm 15.6°C) of natural gas, or 1400–2800 L of gas oil corresponding to 50–70% yields. Oil-based furnace black now supplies 90% or more of the current carbon black market, although the special features of the product from the other smaller scale processes still lead to product contributions from these [4].

"Thermal black" is produced by another variant in which natural gas is cracked to carbon and hydrogen by the direct application of heat via firebrick checker work in a cyclical process (Eq. 19.2) [5].

$$CH_4 \xrightarrow{>870°C} C + 2\ H_2 \qquad\qquad 19.2$$

Elemental analysis of carbon blacks gives a carbon content of 90+%, and in some grades as high as 97%. Although the chemical composition is simple, the structures of carbon black products are complex and vary widely depending on the feedstock used, the process and conditions employed, and the effects of minor amounts of feedstock additives that may be used [6]. Since the discovery of the rubber reinforcement properties of carbon blacks in about 1912, this industry has grown to a scale of the order of 400,000 and 1.5 million tonnes per year in Canada and the United States, respectively, and its volume currently ranks in the mid-1930s in comparison to other large-scale commodity chemicals. Its chief enduse (ca. 90%) is in rubber reinforcement, for which the oil-based furnace blacks are nearly ideal. Automobile tires alone account for about 70% of the total carbon black market.

Other forms of carbon are also of significant industrial interest, among them the various forms of activated carbon that are used in many air and water emission control applications. Carbon fibers used for reinforcement of composite materials in aerospace and sporting goods applications can sell from $22 to more than $220 per kilogram, an attractive value added [7–9]. Industrial diamonds used for the cutting and grinding of very hard materials, for optical devices such as lasers, for heat dissipation, and for diamond coatings, which confer enhanced wear resistance, also represent small-scale, high-technology carbon products [10]. All except the first of these are relatively low-volume, but high-value specialty products.

19.3. ISOPROPANOL (ISOPROPYL ALCOHOL)

The first petrochemical to be produced commercially which required more than one step was basically a scaled-up laboratory procedure adapted by Standard Oil in 1920. By 1925 the scale of production of isopropanol by this process had reached about 68 tonnes per year. By 1995 volumes of the order of a million tonnes per year were produced in the United States alone and still mainly by the same process.

The initial step is the absorption of propylene-rich gases into 85% sulfuric acid at temperatures of 20–25% to form isopropyl hydrogen sulfate via addition of the acid across the double bond (Eq. 19.3). Control of acid temperature and concentration is important to provide selectivity and avoid absorption of olefins, which require higher concentrations or temperatures to react.

$$CH_3 - CH = CH_2 + H_2SO_4 \cdot H_2O \rightarrow CH_3C(H)(SO_3H)CH_3 \quad 19.3$$

Hydrolysis of the intermediate monoester with water in a lead-lined vessel then gives isopropanol and diluted sulfuric acid (Eq. 19.4).

$$CH_3C(H)(OSO_3H)CH_3 + H_2O \text{ (excess)} \rightarrow CH_3CHOHCH_3 \quad 19.4$$
$$+ H_2SO_4(\text{dil.})$$

The heat of reaction plus additional heat from added steam drives off the crude isopropanol from the spent sulfuric acid, now about 20% acid. The acid is reconcentrated to 85% by boiling off the excess water, after which it is ready for reuse. Since it plays an active part in the process, the sulfuric acid is a recycled reactant rather than a true catalyst in this instance.

The condensed, crude isopropanol-water solution is concentrated and purified by distillation to the 91% isopropanol-water azeotrope (boiling point 80.4°C, Fig. 19.1). Any higher alcohols which may have formed in the process

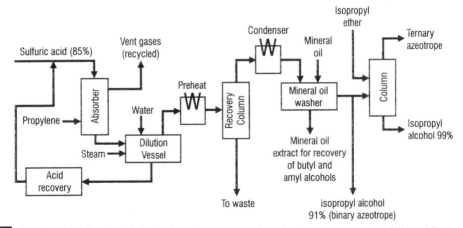

FIGURE 19.1 Simplified flowsheet for isopropanol production from propylene. (Adapted from *Faith Keyes and Clark's Industrial Chemicals*, F. A. Lowenhem and M. K. Moran. Copyright © 1975, reprinted by permission of John Wiley & Sons, Inc.)

■ TABLE 19.2   Important Parameters for the Drying of 91% Isopropanol Via the Ternary Azeotrope

| Component | Boiling point (°C) | | Percent composition | | |
| | Pure | Azeotrope | Azeotrope | Condensate layer | |
				Upper	Lower
Isopropyl ether	67.5		91.0	94.7	1.0
Isopropanol	82.3	61.8	4.0	4.0	5.0
Water	100.0		5.0	1.3	94.0

from traces of higher olefins present in the propylene feed are absorbed from the azeotrope into mineral oil, in which the isopropanol is insoluble. Pure isopropanol is obtained by ternary distillation of the cleaned water azeotrope with the appropriate proportion of added di-isopropyl ether. The ternary azeotrope (Table 19.2) is the top product from the column, and pure isopropanol is removed from the bottom (see Section 16.4 for related information).

About two-thirds of the isopropanol produced is used as an intermediate to form other chemicals, mostly for acetone production. The remainder is used as a solvent in paints and coatings or for other chemical processes.

From the chemically simple process of methane or gas oil to carbon black to the first more complex petrochemical process to isopropanol the scope of this industry has expanded to the production of a truly vast range of large-scale chemicals. Here we will cover outlines of representative samples of the more important of these chemicals. For more detailed information consult the items listed in the Further Reading section at the end of this chapter.

19.4. ALKENE AND AROMATIC PRODUCTS

The majority of the chemical conversion processes of the modern petrochemical industry are reliant on just three refinery streams, methane (or natural gas), olefins, and aromatics comprising just seven hydrocarbons: methane, ethylene, propylene, the butylenes, benzene, toluene, and the xylenes (dimethyl benzenes). Details of the processes used to isolate naturally occurring methane from raw oil or gas streams have been discussed (Chap. 18). A few of the other hydrocarbons listed also occur to a varying extent in crude oil streams but in proportions that are far from adequate to meet the demand. Thus molecular changes are necessary to the hydrocarbon streams that are available, to convert these to the materials required. Section 18.3 outlines some of the procedures used with the larger molecules that form components of gasoline. Some of the procedures used here, such as cyclo-alkylation and aromatization, are the same as those already described. Others differ in important respects but still employ the same principles to obtain the desired product, namely, sufficiently high temperatures to cause carbon-carbon and carbon-hydrogen bond homolytic cleavage and to supply the endotherm of the reactions, plus conditions (contact time, catalyst, concentration, etc.) adequate to achieve the optimum product result from the energy expended. Since many refinery operations lead to several of these products simultaneously it is common for modern operations to employ computer programs to aid in the optimization of their processes to produce the particular ratio of products that they currently require. The tendency to get several simultaneous products from a single process will be evident from the following discussion of related product groups from the seven primary hydrocarbon products of the petrochemical industry.

19.4.1. Ethylene (Ethene)

Ethylene, the largest scale petrochemical, is probably also the oldest to have had confirmed preparation in the laboratory. The initial method used, catalytic dehydration of ethanol, was reported as early as 1797 (Eq. 19.5).

$$CH_3CH_2OH \xrightarrow[\text{catalyst}]{300-400°C} CH_2{=}CH_2 + H_2O \qquad 19.5$$

Although this method is capable of well over 90% yields of ethylene, and may be used when the local availability and price of ethanol (e.g., occasionally fermentation sources, Brazil) are attractive, other methods are generally more important.

Refinery ethylene is generally made by the catalytic cracking of ethane, propane, or a mixed hydrocarbon stream such as recovered natural gas liquids, naphthas, or gas oil. Cracking conditions are quite severe: 750–900°C and 0.1–0.6 second residence time for a low partial pressure hydrocarbon-stream. A number of metal oxide catalysts have recently been evaluated for this purpose [11]. The usual diluent is steam, used at a weight ratio of steam to hydrocarbon of 0.2:1 for ethane feed, to progressively higher ratios with the higher molecular weight hydrocarbons of up to 2.0:1 for gas oil.

With an ethane feed the dominant cracking reaction is to ethylene and hydrogen, together with traces of methane and higher hydrocarbons. The initiation step involves the homolysis of the carbon–carbon bond (Eq. 19.6).

$$CH_3CH_3 \rightarrow 2H_3C^{\cdot} \qquad 19.6$$

Carbon hydrogen bonds, which are stronger, are much less affected. The methyl radicals formed can then abstract hydrogen atoms from ethane in the first of series of rapid propagation reactions (e.g., Eq. 19.7–19.9).

$$H_3C^{\cdot} + CH_3CH_3 \rightarrow CH_4 + {}^{\cdot}CH_2CH_3 \qquad 19.7$$

$${}^{\cdot}CH_2CH_3 \rightarrow CH_2{=}CH_2 + H^{\cdot} \qquad 19.8$$

$$H^{\cdot} + CH_3CH_3 \rightarrow H_2 + {}^{\cdot}CH_2CH_3 \qquad 19.9$$

Termination reactions are one of the ways in which the higher hydrocarbons are formed in the cracking of ethane (e.g., Eq. 19.10, 19.11).

$$2\ {}^{\cdot}CH_2CH_3 \rightarrow CH_3CH_2CH_2CH_3 \qquad 19.10$$

$$H^{\cdot} + {}^{\cdot}CH_2CH_3 \rightarrow CH_3CH_3 \qquad\qquad 19.11$$

Propane cracks both to ethylene and methane, and to propylene and hydrogen (Eq. 19.12, 19.13) with product proportions which change with changes in the cracking temperature (Fig. 19.2) [12].

$$CH_3-CH_2-CH_3 \rightarrow CH_2{=}CH_2 + CH_4 \qquad\qquad 19.12$$

$$CH_3-CH_2-CH_3 \rightarrow CH_2{=}CH-CH_3 + H_2 \qquad\qquad 19.13$$

The initial step in the process with propane is still predominantly carbon carbon bond cleavage, with methyl and ethyl radicals and to a lesser extent hydrogen atoms serving in the propagation steps which lead to the products.

Naphtha, or gas oil, may also be efficiently cracked to ethylene and propylene, although the proportion of ethylene in the product stream is generally lower for these feedstocks than when using the lighter feedstocks ethane or propane. However, this flexibility of possible feedstocks to ethylene enables a refinery to process those streams in current excess in order to produce the olefins that it may currently need.

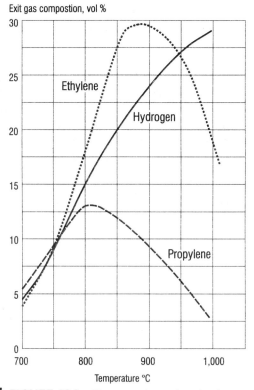

FIGURE 19.2 Changes in the product distribution with temperature on the thermal cracking of propane. (Drawn from selected data of Frolich and Wiezevich [12].)

Recovery of ethylene (b.p. $-103.7°C$) from the mixed hydrocarbon product vapors is accomplished by compression, condensation, and then fractionation at progressively lower temperatures. Ethane (b.p. $-88.6°C$) is the most difficult constituent to separate from ethylene, but even this is technically feasible.

The single largest use of ethylene is for the production of all types of polyethylene, which consumes about half of the total. Other major uses are for the production of ethylene oxide, ethylene dichloride (1,2-dichloroethane), and ethylbenzene enroute to styrene. The remaining 10–15% of production is distributed among a multiplicity of smaller scale products, including vinyl acetate, ethanol, and acetaldehyde.

19.4.2. Propylene (Propene)

Propylene, or propene by IUPAC nomenclature, is probably the oldest petrochemical feedstock, employed as it was in the early processes to isopropanol. It is produced by the cracking of propane or higher hydrocarbons in the presence of steam under conditions very like those outlined for ethylene. Propane is the feedstock of choice for a product stream to contain the maximum concentration of propylene. The yield of propylene rises as the cracking severity of the propane is increased, but reaches a maximum before the maximum conversion of ethylene is achieved (Fig. 19.2). Thus, even with use of propane as the feedstock for propylene production, it is common to have more moles of ethylene than propylene product in the exit stream for optimum cracking economics.

Propylene is also recovered as a by-product of other refinery operations, principally from the fluid catalytic cracking (FCC) of gas oils and to a lesser extent from the volatile products of coking, when practiced. All refinery streams containing recoverable fractions of propylene will be combined into a mixed C_3 stream for propylene separation. Distillation of this combined stream then gives propylene (b.p. $-47.7°C$) as the overhead product and propane (b.p. $-42.1°C$) plus traces of other higher boiling point products as the bottom fraction.

About 45% of the propylene produced is used for the production of isotactic polypropylene. About 10% fractions of the total are consumed for acrylonitrile production, in the preparation of the "oxo-alcohols," for propylene oxide, and for cumene (isopropylbenzene) production.

19.4.3. The Butylenes (Butenes)

There are four C_4H_8 mono-olefins because of the various possible geometric isomers. These are 1-butene, *cis-* and *trans-*2-butene, and isobutylene (2-methylpropene). All occur in many of the catalytically processed refinery streams, and predominantly in the C_4 fraction separated from the products of thermal or fluid catalytic cracking. Some may also be produced in the refinery processes designed to raise the octane rating of gasolines. Including the butanes, which are also present in this stream, we get a total of six compounds

■■■ **TABLE 19.3 Boiling Points of
the Butanes and Butylenes
(the "BB Fraction") of the
C$_4$ Refinery Stream^a**

Hydrocarbon	b.p. (°C)
2-Methylpropane (isobutane)	−11.7
2-Methylpropene (isobutylene)	−6.9
1-Butene	−6.3
n-Butane	−0.5
trans-2-Butene	+0.88
cis-2-Butene	+3.7

^{a}1,3-Butadiene, which may occasionally be present in this fraction, boils at −4.4°C.

present in this C$_4$'s fraction, all boiling in the range −11.7 to +3.7°C. This presents a formidable separation task (Table 19.3). Fractional distillation is capable of separating the isobutane, isobutylene, and 1-butene fraction from the n-butane and the 2-butenes. Extractive distillation, using acetonitrile or aqueous acetone, is used to separate the 1-alkenes from isobutane. Extractive distillation with furfural separates the components of the bottom fraction of the distillation.

The production of methyl *t*-butyl ether (MTBE) is the single largest consumer of the isobutylene fraction of the butylenes. The next largest market for the butylenes (both 1-butene and the 2-butenes) is for the production of 1,3-butadiene, destined for the production of various types of synthetic rubber.

19.4.4. Benzene, Toluene, and the Xylenes (BTX)

These monoaromatic products, usually produced in several different refinery operations, all rank in the top 30 chemicals produced in the United States by volume. They have reached this significance because of their large demand as feedstocks for many of the important plastics. Preparation of these materials is discussed together because several are produced at the same time from many of the processes from which they arise. The chief sources of these aromatics may be conveniently grouped into three classes of related processes.

The first class, production of aromatics as a complementary objective to the refinery processing of gasoline fractions to raise the aromatic content, evidently links these refining functions. Catalytic reforming processes are used to convert paraffins to naphthenes (cycloparaffins) to be followed by dehydrogenation of naphthenes to aromatics (Chap. 18). Since aromatization of naphthenes is an easier process to accomplish than cyclo-alkylation, the emphasis in refinery operations is on maximization of the second step in this sequence, in the presence of an adequate supply of naphthenes. The demand for the aromatics component of gasoline will compete with the feedstock aromatic need from this source.

Some of the processes used for ethylene production, particularly those using naphtha, or gas oil as feedstocks, may also produce large amounts of BTX. The conditions used (temperature, pressure, feedstock ratios, etc.) to operate these processes provide some flexibility to alter the product distribution slightly in response to the distribution of the demand.

The third class of processes to benzene, toluene, and xylenes, or *p*-xylene is discretionary and depends on the proportion of these products that are available from the first two classes of processes incomparison to the current demand. Benzene consumption, which is more than twice that of toluene, may be conveniently produced in almost 90% yields by the catalytic hydrodealkylation of toluene (Eq. 19.14). Or if both benzene and xylenes are in short supply, toluene may be catalytically transalkylated to mainly these products.

$$C_6H_5CH_3 + H_2 \xrightarrow[H_2, \text{ catalyst}]{540-800°C} C_6H_6 + CH_4 \qquad 19.14$$

Ethylbenzene, too, may be isomerized to xylenes. Xylenes are normally recovered from a mixed aromatics stream by extraction with sulfolane or a glycol. If the market for one xylene isomer is greater than that available by separation of the mixture the low demand isomers may be isomerized to raise the proportion of the desired isomer(s).

About half of the benzene produced as a chemical feedstock is for styrene production, followed by large fractions for phenol and cyclohexane-based products. As much as half of the toluene produced is converted to benzene, depending on the price and demand differential. The largest use as toluene itself is as a component of gasoline. Much smaller amounts are used as a solvent, or in the manufacture of dinitrotoluene and trinitrotoluene for military applications.

Xylenes are also used in gasoline formulations and function like toluene in this application as octane improvers. *para*-Xylene and o-xylene are the dominant isomers of value as chemical feedstocks, for the production of terephthalic acid (and dimethyl terephthalate) and phthalic anhydride, respectively. Polyester and the synthetic resin markets, in turn, are major consumers of these products. *meta*-Xylene is oxidized on a much smaller scale to produce isophthalic acid, of value in the polyurethane and Nomex aramid (poly (*m*-phenylene isophthalamide)) technologies.

19.5. PRODUCTS FROM METHANE

Production of ammonia and the downstream products nitric acid, urea, and ammonium nitrate, which currently rank 6th, 14th, 15th, and 17th in American volume of production, are discussed in Chapter 11. Two other petrochemicals derived from methane, methanol and formaldehyde, currently rank 21st and 24th in volume in the United States at about 4 million tonnes per year each. Details of their production will be outlined here.

19.5.1. Methanol (Methyl Alcohol)

Methanol is made from a mixture of carbon monoxide and hydrogen. The feed gases required are produced by the reforming of natural gas (Section 9.1). At about 800°C in the presence of a promoted nickel catalyst a gas mixture of close to the correct proportions of carbon monoxide and hydrogen is obtained (Eq. 19.15).

$$3\ CH_4 + CO_2 +\ \ 2\ H_2O\ \ \rightarrow 4\ CO + 8\ H_2 \qquad\qquad 19.15$$
$$\text{(as steam)}$$

Similar gas mixtures may also be made by the old water-gas reaction using coal, which gives approximately 1:1, carbon monoxide to hydrogen. The hydrogen may be supplemented to the correct proportions from refinery sources or from hydrogen obtained by the electrolysis of water or brine.

The methanol synthesis reaction requires compression of the gas mixture to 200–330 atm, temperatures in the region of 300°C, and a catalyst such as a chromium oxide-promoted zinc oxide, to result in an approximately 60% conversion to methanol (Eq. 19.16).

$$CO + 2\ H_2 \rightarrow CH_3OH \qquad \Delta H = -103\ kJ\ (-24.6\ kcal) \qquad 19.16$$

Methanol is condensed from the exit gases, still at converter pressures, and the unreacted gases are recycled after a partial purge and recompression. As with ammonia production, the purge is required to avoid accumulation of inert gases in the recycle stream. The product under these conditions is about 99% pure methanol but contains 1 or 2% of dimethyl ether and traces of higher alcohols.

The process just described may also be operated at somewhat lower pressures and temperatures than outlined, by using some of the more recent copper-based catalysts. Energy saved by doing this achieves some production economies. It is also possible, with an extra mole of hydrogen, to produce methanol from carbon dioxide (Eq. 19.17).

$$CO_2 + 3\ H_2 \rightarrow CH_3OH + H_2O \qquad\qquad 19.17$$

In this instance distillation in a series of columns is necessary for methanol recovery [5].

Major end uses for methanol are for the production of formaldehyde, about 30%, which is used for the preparation of phenol-formaldehyde resins. About 20% is used for the production of methyl *t*-butyl ether, which is used as an additive alone, and in blends with methanol as a fuel component. Further uses are for esterification of terephthalic, and acrylic acids, and for acetic acid preparation, about 10% each.

19.5.2. Acetic Acid

The Monsanto process to acetic acid uses a rhodium-iodine-containing catalyst to convert a mixture of methanol and carbon monoxide under mild conditions and ambient pressure to very high (99+% based on methanol) yields of the product (Eq. 19.16). Several stages of distillation are required to recover the acetic acid (b.p. 118°C) in purities of 99.8% or better. A substantial fraction of acetic acid is also made commercially via the liquid or gas phase oxidation of acetaldehyde or the liquid phase oxidation of butane, in each case using air plus a catalyst.

The largest single use for acetic acid, about half the total, is for the manufacture of vinyl acetate for use in the production of poly(vinyl acetate) (Section 19.10.4). Cellulose acetate destined for fiber spinning consumes a further quarter.

19.5.3. Formaldehyde

Catalytic oxidation of methanol in the presence of a catalyst based on transition metal oxides produces good yields of formaldehyde. Both oxidation and dehydration reactions participate in this process (Eq.19.18, 19.19).

$$CH_3OH + \tfrac{1}{2} O_2 \text{ (from air)} \rightarrow CH_2O + H_2O \qquad 19.18$$

$$CH_3OH \rightarrow CH_2O + H_2 \qquad 19.19$$

Evidence for the contribution of the second reaction is obtained by the detection of hydrogen in the reactor exit gases. Formaldehyde (b.p. −21°C) is recovered from these by fractionation.

Most of the formaldehyde produced is consumed in the production of urea-formaldehyde resins and phenol-formaldehyde resins. These cross-linked polymer products are in turn used in adhesive and laminate applications.

19.6 PRODUCTS FROM ETHYLENE

Ethylene oxide is the next most significant product to consume ethylene, after the various types of polyethylene which together consume close to half of the total. It used to be produced via the chlorohydrin process, as propylene oxide still is. However, now it is made by the direct oxidation of ethylene with air in the presence of a silver catalyst (Eq. 19.20).

$$CH_2 {=\!=} CH_2 + \tfrac{1}{2} O_2 \rightarrow \underset{\underset{O}{\diagdown\!\diagup}}{CH_2\text{-}CH_2} \text{ (+ some } CO_2, H_2O) \qquad 19.20$$

Temperatures in the 270–290°C range, a pressure of about ambient, and a 1-second contact time give conversions to ethylene oxide of about 60%. The ethylene oxide is absorbed in water from the reactor exit gases, and the aqueous solution is then fractionated for ethylene oxide (b.p. 13.5°C) recovery. The nonabsorbed ethylene and the absorption water recovered from the bottom of the ethylene oxide fractionator are both recycled.

Close to half of the ethylene oxide produced is directly converted to ethylene glycol (1,2-ethanediol) by acid-catalyzed or pressure hydration. Roughly half the ethylene glycol is used in automotive antifreeze and half in polyester (poly(ethylene terephthalate)) fiber production (Chap. 21). Smaller amounts are consumed for alkyd resins production and for the formulation of latex paints.

19.6.1. Ethylene Dichloride (1,2-Dichloroethane) and Vinyl Chloride

Direct reaction of ethylene and chlorine in the presence of a catalyst, and in either vapor phase or liquid phase reactors under mild conditions produces a 96+% yield of ethylene dichloride (Eq. 19.21).

$$CH_2{=}CH_2 + Cl_2 \rightarrow ClCH_2CH_2Cl \qquad\qquad 19.21$$

Some producers with a good supply of hydrogen chloride available as a by-product from the production of chlorinated solvents may prefer to use the oxychlorination route to ethylene dichloride (e.g., Eq. 19.22, 19.23).

$$CH_4 + Cl_2 \rightarrow CCl_4 + 4\ HCl \qquad\qquad 19.22$$

Oxychlorination:

$$CH_2{=}CH_2 + 2\ HCl + \tfrac{1}{2}\ O_2\ \text{(from air)} \xrightarrow[\text{ClCH}_2\text{CH}_2\text{Cl} + \text{H}_2\text{O}]{\text{CuCl}_2\ \text{cat.}} \qquad 19.23$$

Small amounts of ethylene dichloride (b.p. 83.5°C) are also recovered as a by-product from the direct chlorination of ethane to ethyl chloride (chloro-ethane).

Whatever the source, 80% or more of the ethylene dichloride produced is directly converted into vinyl chloride. Target conversions are 50% under the usual operating conditions of about 3–4 atm pressure and 500°C to give better than 94% yields of vinyl chloride (Eq. 19.24).

$$ClCH_2CH_2Cl \rightarrow CH_2{=}CHCl + HCl \qquad\qquad 19.24$$

Gas exposure times to furnace temperatures are kept short by a direct spray of ethylene dichloride at ambient temperatures onto the exit gases and by an indirect heat exchanger. Hydrogen chloride may be recovered by scrubbing with water, or by some form of fractionation if it is to be fed in anhydrous form to an oxychlorination unit. Many vinyl chloride producers operate an oxychlorination unit solely to convert the by-product hydrogen chloride from the ethylene dichloride cracker to produce additional ethylene dichloride feed-stock. Vinyl chloride (b.p. −13.4°C) is almost entirely consumed for the production of poly(vinyl chloride), PVC.

19.6.2. Ethylbenzene and Styrene (Vinylbenzene)

Ethylbenzene, a product with a smaller demand for the ethylene stream than the products already described, currently stands 19th in volume of production. It is made by both liquid phase processes under moderate conditions employing aluminum chloride catalysis and by vapor phase processes at 150–250°C and 30–50 atm in the presence of a supported boron trifluoride catalyst (Eq. 19.25).

$$C_6H_6 + CH_2{=}CH_2 \xrightarrow{\text{catalyst}} C_6H_5CH_2CH_3 \qquad\qquad 19.25$$

Ethylbenzene (b.p. 136°C) is recovered from the excess unreacted benzene and the more heavily alkylated material by fractionation. The polyalkylated residues may be partly recycled to take advantage of transalkylation reactions to produce more ethylbenzene, or may be burned as fuel for the process.

Virtually all of the ethylbenzene produced is destined for styrene production, ultimately leading to polystyrene and other polymers. A preheated mixture of ethylbenzene vapor and superheated steam passed over one of several possible dehydrogenation catalysts preheated to 620–640°C cracks 30–40% of the ethylbenzene in one pass (Eq. 19.26).

$$C_6H_5CH_2CH_3 \xrightarrow{\text{catalyst}} C_6H_5CH{=}CH_2 + H_2 \qquad 19.26$$

The exit stream is condensed and, after addition of a polymerization inhibitor (usually sulfur), is successively distilled under reduced pressure to a styrene purity of better than 99.7%. A few ppm of 4-t-butylcatechol inhibitor is added to the purified product after which it is ready to be shipped. Industrial yield of styrene from ethylbenzene is about 90%.

19.6.3. Vinyl Acetate

Currently ranked 41 in American volume of production, vinyl acetate is made by the vapor phase reaction of ethylene with acetic acid at 170–200°C. Pressures of 5–8 atm and a palladium catalyst supported on carbon are required to give conversions of 10–15% on ethylene and 15–30% on acetic acid. Equation 19.27 is an overall representation of the stoichiometry of the complex process.

$$CH_2{=}CH_2 + CH_3COOH + \tfrac{1}{2} O_2 \rightarrow \qquad\qquad 19.27$$
$$CH_3COOCH{=}CH_2 + H_2O$$

The reactor exit gases are cleaned up by various washes and the vinyl acetate (b.p. 72.2°C) separated from unreacted starting materials, water, and a small amount of polymeric material by distillation. Industrial yields based on ethylene are 92–95%. The product purity is 99%, to which traces of inhibitor such as hydroquinone will be added to stabilize it during storage and shipment.

All of the product vinyl acetate is consumed in the production of homopolymer (one monomer) and copolymer products, among them poly(vinyl acetate) latices and resins, poly(vinyl alcohol) by postpolymerization hydrolytic removal of the acetate, and copolymers with vinyl chloride and ethylene.

19.6.4. Ethanol (Ethyl Alcohol)

Production of ethanol by fermentation processes probably represents among the earliest processes operated by humankind. Certainly until the late 1940s it was the dominant process used to produce industrial alcohol (Fig. 19.3). The world market for industrial ethanol in 1991 was about 19×10^9 L. Most of this was for fuel use, Brazil alone consuming about 12×10^9 L, more than half of the total, and the United States a further 3.5×10^9 L. Most of the Brazilian market was supplied by the domestic fermentation of sugar solutions plus distillation, although the proportion produced in this way varies inversely with world sugar prices (Chap. 16). Any Brazilian supply shortfall which may result is supplied from imports of industrial alcohol. In the United States fermentaion supplied almost all of the industrial ethanol in 1930. Fermentation sources provided only about a tenth of the total in 1960 but with the stimulus of subsidies had grown to about one-third of a larger total by 1990. The balance of the American industrial ethanol market was supplied from petrochemical sources. Today fermentation of sugar cane juice provides most of the world's ethanol, thanks to the massive contribution to this source from Brazil.

The earliest petrochemical process to convert ethylene to ethanol was first practiced in about 1930, and was indirect. In this process ethylene, under pressure, is absorbed by countercurrent passage against sulfuric acid (90–98%) at about 80°C in an absorber to form a mixture of the monoesters, and diesters (Eq. 19.28, 19.29).

$$CH_2{=}CH_2 + H_2SO_4 \rightarrow CH_3CH_2OSO_3H \qquad\qquad 19.28$$

$$2\ CH_2{=}CH_2 + H_2SO_4 \rightarrow (CH_3CH_2O)_2SO_2 \qquad\qquad 19.29$$

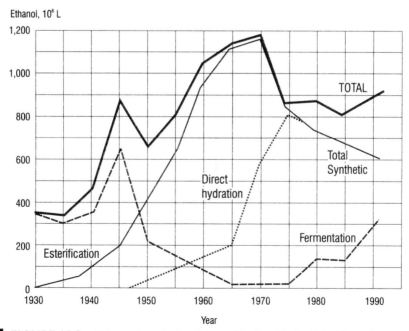

FIGURE 19.3   Volume of production of industrial ethanol in the United States, by process.

Both the acid concentration and the temperature have to be higher for ester formation from ethylene than for ester formation from propylene because of the less reactive carbon carbon double bond. The monoester is preferred for its easier hydrolysis and its lower tendency to form diethyl ether. After reaction, the mixed esters are hydrolyzed with just sufficient excess water to complete the hydrolysis, which takes about 2 hr at 60–70°C (Eq. 19.30).

$$CH_3CH_2OSO_3H + (CH_3CH_2O)_2SO_2 + 3\ H_2O \rightarrow \qquad\qquad 19.30$$
$$3\ CH_3CH_2OH + H_2SO_4\ (dil.)$$

The ethanol product (b.p. 78.3°C) is recovered by distillation and the diluted acid is recovered as the bottom stream of the column. Yields on ethylene are 90–95%. Diluted acid is reconcentrated in several stages, the last stages under reduced pressure, for recycling to the absorption stage. The volume of ethyl alcohol produced in the United States by this process peaked in about 1960. Since the late 1940s it has gradually been supplemented and then largely replaced by the process which uses direct hydration of ethylene (Fig. 19.3).

The direct hydration process began to be important for American industrial alcohol production in the late 1940s. It has grown from this beginning to produce roughly 70% of the total industrial alcohol in 1975 and certainly has provided most of the supply since then. Like many catalytic petrochemical process high pressures (ca. 70 atmos) and temperatures of around 300°C are again required to converta 0.6:1 mole ratio of steam:ethylene to ethanol. In the presence of a supported phosphoric acid catalyst using a space velocity of 1800 hr^{-1} gives conversions of 4.0–4.2% and yields on ethylene of 95+% (Eq.19.31).

$$CH_2{=}CH_2 + H_2O \rightleftharpoons CH_3CH_2OH$$
$$\Delta H = -40.2\ kJ\ (-9.6\ kcal) \qquad\qquad 19.31$$

The exit stream is cooled and the condensate separated in a high-pressure separator. After scavenging any remaining product from this stream, the unreacted ethylene is recycled.

Ethanol is recovered from the condensate of the direct hydration process by distillation. Traces of acetaldehyde, which probably arise from the hydration of a small amount of acetylene present in the feed ethylene, are catalytically reduced to ethanol with hydrogen (Eq. 19.32, 19.33).

$$HC{\equiv}CH + H_2O \rightarrow CH_3CHO \qquad\qquad 19.32$$

$$CH_3CHO + H_2 \xrightarrow{\ Ni\ cat.\ } CH_3CH_2OH \qquad\qquad 19.33$$

Traces of diethyl ether impurity are hydrated back to ethanol using conditions similar to those used for the ethylene hydration reaction.

Ethanol is used in diverse fuel, solvent, and chemical applications in industry. Chemical applications have dropped sharply in the last 20 years as more direct routes to products such as acetaldehyde, acetic acid, ethyl acetate, and ethyl chloride, which were formerly made from ethanol, came on stream. The automotive fuel component is anticipated to have the highest potential for growth.

19.7. PRODUCTS FROM PROPYLENE

Almost half of the propylene produced is used to prepare polypropylene (Chap. 23). Most of the remainder is consumed for the production of the following products.

19.7.1. Acrylonitrile (Vinyl Cyanide)

This versatile monomer is produced by the catalytic ammonoxidation of propylene. A mixture of propylene, ammonia, and air in the volume ratio 1:1:10 provides roughly two volumes (moles) of oxygen per volume (mole) of the other reactants, or a 30% excess to that required by the process. The gas mixture at near ambient pressures is given a 10–20 second contact time over a molybdenum-based catalyst on a silica support kept at about 400°C. Better than 70% conversion to acrylonitrile is obtained, together with traces of by-product acetonitrile and hydrogen cyanide (Eq. 19.34).

$$CH_3-CH{=}CH_2 + NH_3 + 3/2\ O_2 \rightarrow CH_2{=}CHCN + 3\ H_2O \qquad 19.34$$

Reactor exit gases are cooled and washed with water, from which the acrylonitrile is recovered and purified by distillation.

A recently developed one-step ammonoxidation process to acrylonitrile developed by BP Chemicals uses propane instead of propylene feed [13]. Key to the feasibility of this lower cost option, now at the commercial demonstration stage, was the development of the new catalyst system. Almost all the acrylonitrile production goes into synthetic polymers and copolymers mostly for applications as fibers, some for plastics applications and with a small percentage to elastomer markets (the nitrile rubbers).

19.7.2. Isopropyl Alcohol (Isopropanol)

The indirect, sulfate ester-based process, used to introduce this chapter, was one of the first petrochemical processes (Section 19.3). This method still has a place since its capability of selective absorption of propylene enables it to be used with refinery gas streams containing low concentrations of propylene, unlike alternative processes. The direct hydration of propylene to isopropanol is also possible, but this requires a refinery stream containing a much higher concentration of propylene to be competitive. For this reason both petrochemical processes are still viable in their respective feedstock niches.

With the direct hydration process to ethanol, the Le Chatelier principle would suggest that high pressures and low temperatures would tend to favor product formation. This is also true with propylene. Pressures of 100–250 atm and temperatures of 150–270°C are used because, unfortunately, all known catalysts (supported sulfonic acids, or phosphoric acid) for this process require high temperatures to be effective (Eq. 19.35).

$$CH_3CH{=}CH_2 + H_2O \rightleftharpoons (CH_3)_2CHOH$$
$$\Delta H = -210\ kJ\ (-50\ kcal) \qquad 19.35$$

Isopropanol formation is thought to proceed via an intermediate isopropyl carbocation generated by interaction with water and the acid catalyst, which is then hydrated with water. Recovery is by distillation, to the 91% azeotrope, or by ternary distillation of the azeotrope to 100% isopropanol.

19.7.3. The Oxo Alcohols: *n*-Butanol and Isobutanol (2-Methylpropanol-1)

The oxo reaction is used to produce aldehydes or alcohols containing one additional carbon atom, from the corresponding α-olefin. Propylene, in the presence of a 1:1 mixture of carbon monoxide and hydrogen (synthesis gas) all raised to 100–250 atm and temperatures of about 170°C, gives 70–80% yields of the initial products butyraldehyde and isobutyraldehyde (Eq. 19.36).

$$CH_3CH{=}CH_2 + CO + H_2 \rightarrow \qquad\qquad\qquad 19.36$$
$$CH_3CH_2CH_2CHO + (CH_3)_2CHCHO$$
$$\text{60–40\%} \qquad\qquad \text{40–60\%}$$

An organocobalt derivative such as cobalt naphthenate is used as a catalyst. In a second step, the mixed aldehydes are reduced with hydrogen in the presence of a nickel-based catalyst to give *n*-butanol (b.p.118°C) and isobutanol (b.p. 108.1°C) (Eq. 19.37).

$$CH_3CH_2CH_2CHO + H_2 \rightarrow CH_3CH_2CH_2CH_2OH \; (+ \; iso) \qquad 19.37$$

The products are recovered and purified by distillation. It should be mentioned that this process is not limited to the oxygenation of propylene, but may be applied to α-olefins generally.

19.7.4. Propylene Oxide

Two process routes to propylene oxide are commercially practiced; hydroperoxide formation and then use of this to oxidize propylene, and formation of propylene chlorohydrin followed by treatment with a base to form propylene oxide [14, 15]. It has not been possible to produce adequate yields of propylene oxide via the direct oxidation of propylene with air in the manner in which ethylene oxide is now produced, although attempts to come close to this continue [16].

Epoxidation of propylene by a hydroperoxide has been found to be commercially viable for those hydroperoxides which, when spent, can easily give another commercial product. One option using this concept is to produce t-butyl hydroperoxide (plus some *t*-butanol) by the air oxidation of isobutane at 90°C and ca. 30 atm in the presence of molybdenum naphthenate (Arco and Texaco). The separated hydroperoxide is then used to oxidize propylene to propylene oxide with yields on hydroperoxide of over 90% (Eq. 19.38, 19.39).

$$(CH_3)_2CHCH_3 + O_2 \rightarrow (CH_3)_3COOH \qquad\qquad 19.38$$

$$(CH_3)_3COOH + CH_3-CH=CH_2 \rightarrow \qquad 19.39$$

$$CH_3\text{-}\underset{\underset{O}{\diagdown\diagup}}{CH}\text{-}CH_2 + (CH_3)_3COH$$

Coproduct t-butanol, produced at the rate of about 2.2 kg per kilogram of propylene, is converted to methyl t-butyl ether for sale as an automotive fuel additive.

Ethylbenzene hydroperoxide from ethylbenzene provides the other viable option to oxidize propylene (Arco, Eq. 19.40, 19.41).

$$PhCH_2CH_3 + O_2 \rightarrow PhC(OOH)CH_3 \qquad 19.40$$

$$PhC(OOH)CH_3 + CH_3CH=CH_2 \rightarrow \qquad 19.41$$

$$CH_3\text{-}\underset{\underset{O}{\diagdown\diagup}}{CH}\text{-}CH_2 + PhCH(OH)CH_3$$

In this case the spent oxidant is α-methylbenzyl alcohol, produced at almost twice the rate per unit of propylene oxide as t-butanol. This large amount of coproduct is anticipated, however. Easy dehydration of α-methylbenzyl alcohol converts the coproduct to styrene, for which a large market exists (Eq. 19.42).

$$PhCHOHCH_3 \rightarrow PhCH=CH_2 + H_2O \qquad 19.42$$

The chlorohydrin process, practiced by Dow Chemical in the United States, weds the chlorine component of chloralkali technology to propylene oxide production. Chlorine added to water produces hypochlorous acid and hydrochloric acid. (Eq. 19.43).

$$Cl_2 + H_2O \rightleftharpoons HOCl + HCl \qquad 19.43$$

Contacting this solution countercurrently with propylene converts some of the propylene to propylene chlorohydrin which dissolves in the aqueous phase leaving the bottom of the contacting tower. Both possible isomers are formed (Eq. 19.44).

$$CH_3-CH=CH_2 + HOCl \rightarrow \qquad 19.44$$
$$\underset{90\%}{CH_3-CH(OH)CH_2Cl} + \underset{10\%}{CH_3-CH(Cl)CH_2OH}$$

The rate of reaction of chlorine with propylene is much slower, though 7–8% by weight propylene chloride (1,2-dichloropropane) is also formed. In a second stage the propylene chlorohydrin is contacted with a lime slurry in water which causes cyclization of the chlorohydrin to the epoxide (Eq. 19.45).

$$2\ CH_3-CH(OH)CH_2Cl + Ca(OH)_2 \rightarrow \qquad 19.45$$
(or other isomer)

$$2\ CH_3\text{-}\underset{\underset{O}{\diagdown\diagup}}{CH}\text{-}CH_2 + CaCl_2 + 2\ H_2O$$

To minimize hydrolysis of the propylene oxide (b.p. 34.2°C), as it is formed, it is flashed (rapidly removed) from the reactive lime slurry. Yields of propylene oxide are 75% or better based on propylene. The advantage of the chlorohydrin route to propylene oxide over the two hydroperoxidation processes is that it yields essentially a single product to market. The disadvantage is the large quantities of coproduced aqueous calcium chloride that has to be dealt with safely. The small amount of by-product 1,2-dichloropropane may be pyrolyzed to allyl chloride, or may be formulated with 1,3-dichloropropene, as an effective soil fumigant.

Close to two-thirds of the propylene oxide produced goes to the production of polypropylene glycol and polyester glycols destined for use in polyurethanes. The polyol diols are made by the base-catalyzed reaction of propylene glycol with propylene oxide (Eq. 19.46). Replacing propylene glycol by glycerol in this process produces polyol triols.

$$CH_3 - CH(OH)CH_2OH + n \ CH_3\text{-}\underset{\underset{O}{\diagdown\diagup}}{CH\text{-}CH_2} \xrightarrow{HO^-}$$

$$HOCH_2CH(CH_3)O[CH_2CH(CH_3)O]_{n-1}CH_2CH(CH_3)OH \qquad\qquad 19.46$$

Most of the rest of the propylene oxide is hydrolyzed in water using a trace of mineral acid to give propylene glycol (1,2-propanediol, b.p. 187°C) of utility in the manufacture of polyester resins.

19.7.5. Cumene (Isopropylbenzene) and Phenol

About 10% of propylene is consumed for the alkylation of benzene to produce cumene. Use of a supported phosphoric acid catalyst and a high-purity stream of propylene in the presence of a large excess of benzene (to minimize dialkylation and oligomer formation) gives cumene with a 90+% yield based on ethylene, 96+% yield on benzene (Eq. 19.47).

$$C_6H_6 + CH_3 - CH{=}CH_2 \rightarrow C_6H_5CH(CH_3)_2 \qquad\qquad 19.47$$

Reaction conditions are about 200°C at 20–30 atm pressure. Product cumene (b.p. 152°C) is recovered and purified by distillation.

Almost all of the cumene produced is consumed by the cumene peroxidation process to phenol and acetone. Aspects of this process resemble the peroxidation processes to propylene oxide already described, except that in this case the final products are formed by an intramolecular rearrangement of the same molecule that is peroxidized rather than by reaction of the peroxidized molecule with a second component, propylene. In the first stage of the cumene to phenol process a suspension of purified cumene in a dilute solution of sodium carbonate in water is heated to about 110°C under slight pressure. Air is blown through this suspension until about 25% of the cumene has formed the hydroperoxide (Eq. 19.48).

$$PhCH(CH_3)_2 + O_2 \ \text{(from air)} \rightarrow PhC(CH_3)_2OOH \qquad\qquad 19.48$$

The dissolved sodium carbonate helps to maintain the system to near neutral pH to avoid prematurerearrangement of the hydroperoxide. The partially converted cumene from the first stage is carefully concentrated to about 80% cumene hydroperoxide before proceeding to the rearrangement step.

$$\text{PhC(CH}_3)_2\text{OOH} \underset{-\text{H}^+}{\overset{\text{H}^+}{\rightleftharpoons}} \text{PhC(CH}_3)_2\text{OOH}_2^+ \rightarrow \text{PhC(CH}_3)_2\text{O}^+ \xrightarrow{\text{1,2 shift}}$$

$$\text{PhOC}^+(\text{CH}_3)_2 \xrightarrow{\text{H}_2\text{O}} \text{PhOC(CH}_3)_2\text{OH}_2^+ \underset{\text{H}^+}{\overset{-\text{H}^+}{\rightleftharpoons}} \text{PhOC(CH}_3)_2\text{OH} \xrightarrow{\text{H}^+}$$

$$\text{PhOH} + (\text{CH}_3)_2\text{CO}$$

FIGURE 19.4   Mechanistic details of the acid catalyzed rearrangement of cumene hydroperoxide to phenol and acetone.

Dilute sulfuric acid at 70–80°C is used to rearrange the cumene hydroperoxide to phenol and acetone, for which the stoichiometry is represented by Eq. 19.49.

$$\text{PhC(CH}_3)_2\text{OOH} \rightarrow \text{PhOH} + (\text{CH}_3)_2\text{CO} \qquad\qquad 19.49$$

The details of the complicated mechanism of this rearrangement were only discovered some time after the startup of commercial facilities (Fig. 19.4). Distillation is used to recover the 90+% yield of the rearrangement products. Unreacted cumene is recycled after careful purification to remove any traces of phenol, which would be a powerful inhibitor of hydroperoxide formation, and α-methylstyrene (a by-product), which would reduce the yield of the peroxidation stage. Any separated α-methylstyrene may be sold as such, or may be catalytically reduced with hydrogen back to cumene again.

Early petrochemical routes to phenol involved sulfonation followed by hydrolysis, or catalytic chlorination in the presence of iron followed by drastic hydrolysis (Eq. 19.50, 19.51).

$$\text{C}_6\text{H}_6 + \text{H}_2\text{SO}_4 \rightarrow \text{C}_6\text{H}_5\text{SO}_3\text{H}, + \text{H}_2\text{O} \rightarrow \text{C}_6\text{H}_5\text{OH} \qquad 19.50$$

$$\text{C}_6\text{H}_6 + \text{Cl}_2 \rightarrow \text{HCl} + \text{C}_6\text{H}_5\text{Cl} \rightarrow \text{C}_6\text{H}_5\text{OH} \qquad 19.51$$

Improved yields under less drastic conditions and the decreased waste stream problems of the cumene process have led to phasing out of these older processes.

A large fraction of the phenol product, currently ranked 34th, is directed toward the production of several types of phenol-formaldehyde resins of utility as adhesives and as components of laminates. Some is destined to the preparation of *bis*-phenol A enroute to epoxyresin production.

19.8. PRODUCTS FROM ISOBUTYLENE (2-METHYLPROPENE)

Isobutylene itself ranks 38th in volume in the American chemical industry. Its two major uses are for the production of methyl *t*-butyl ether and 1,3-butadiene, which currently stand 12th and 36th in volume of production, respectively.

19.8.1. Methyl *t*-butyl Ether (MTBE)

Commercial production of methyl *t*-butyl ether began in 1979, shortly after the discovery of its octane-improving capability for motor fuels. Although a higher proportion of this additive was required for equivalent octane enhancement, it was less costly and eliminated the lead particulate discharges associated with the tetraethyl lead-based formulations previously used for this purpose (Chap. 18). By 1984 it ranked 49th in American volume of production and jumped to 12th by 1995, which represents an average compound growth rate of almost 50% per year during that period. It has been predicted that by the year 2000 MTBE production will rank only behind ethylene and second in volume among American organic chemicals.

Methyl *t*-butyl ether production is the single largest consumer of butylenes. Liquid phase reaction of methanol with isobutylene in the presence of an acidic ion-exchange resin catalyst at temperatures of below 100°C and moderate pressures give excellent yields of this novel, oxygenated gasoline additive (Eq.19.52).

$$CH_3OH + CH_2{=}C(CH_3)_2 \rightarrow CH_3OC(CH_3)_3 \qquad 19.52$$

Purification of MTBE (b.p. 55°C) for general solvent use is by distillation. Its use in general solvent applications is still small. However, since MTBE has no secondary or tertiary hydrogens it is very resistant to oxidation and peroxide formation. This makes it an attractive replacement for many of the applications of the more traditional diethyl and diisopropyl ethers.

19.8.2. 1,3-Butadiene

The other major use for the *n*-butene fraction (both 1-butene and the 2-butenes) of the butylenes is as a feedstock for 1,3-butadiene manufacture, ultimately for the production of synthetic rubber. Careful thermal catalytic dehydrogenation in the presence of steam gives a 75–86% yield of 1,3-butadiene from a 25–30% butene conversion per pass (Eq. 19.53).

$$n\ C_4H_8 \xrightarrow{\text{ca. 650°C}} CH_2{=}CH{-}CH{=}CH_2 + H_2 \qquad 19.53$$

In 1997 1.7 million metric tonnes of *n*-butenes were consumed in the United States for this purpose, but this was still only about one-third of that consumed to produce MTBE.

Butane may also be used as the raw material for butadiene production in a different, though related, process. The need to remove four hydrogens requires process operation at lower conversions of 10–11% per pass to maintain reasonable yields, and results in ultimate yields of 1,3-butadiene from this process of about 60%.

19.9. PRODUCTS FROM BENZENE, TOLUENE, AND THE XYLENES (BTX)

The number of derivatives commercially produced from this group of aromatic substrates is extensive, many of these targeted as monomers for polymerization to the large-volume fibers and plastics. Some of these, the ethylbenzene to styrene sequence and cumene to phenol, have already been discussed in connection with ethylene and propylene derivatives. Details of some further representative examples of major products will be outlined here.

19.9.1. Benzene to Cyclohexane

Benzene is the aromatic petrochemical used as such to ultimately produce styrene and phenol, both very large-scale products in their own right. Other major products from benzene use it in hydrogenated form, as cyclohexane. Adipic acid (48th in volume), for example, is almost entirely derived from cyclohexane (47th in volume) as is much of the caprolactam and hexamethy-lenediamine (1,6-diaminohexane) production. The production sequences to all of these products will be outlined.

Cyclohexane is made by the catalytic reduction of benzene with hydro-gen, usually in a train of several reactors with partial conversion conducted in each. This process uses a nickel, or platinum catalyst and requires tem-peratures of about 200°C, 25–40 atm pressure, and the presence of a cyclo-hexane recycle diluent to help absorb the exotherm of the hydrogenation (Eq. 19.54).

$$C_6H_{6(g)} + 3\ H_{2(g)} \rightarrow C_6H_{12(g)} \qquad \Delta H = -151\ \text{kJ}\ (-36\ \text{kcal}) \qquad 19.54$$

Cyclohexane recovery (99+% yields) is by distillation.

19.9.2. Cyclohexane to Adipic Acid (1,6-Hexanedioic Acid)

Oxidation of cyclohexane, usually in two stages, gives adipic acid. In the first stage cyclohexane isoxidized with air in the presence of a cobalt naphthenate catalyst under moderate conditions to give a mixture of cyclohexanol and cyclohexanone (Eq. 19.55, stoichiometry only).

$$2\ C_6H_{12} + O_2 \rightarrow \quad \bigcirc\!\!=\!\!O \quad + \quad \bigcirc\!\!-\!\!OH + H_2O \qquad 19.55$$

On completion, water is added to the mixture after which it is fractionated. Cyclohexane (b.p. 81°C) containing some benzene is collected from the top of the column, and after hydrogenation of the benzene, is recycled. The cyclohexanol-cyclohexanone mixture, which consists of approximately equal volumes of cyclohexanol (b.p. 161°C), cyclohexanone (b.p. 156°C), plus a mixture of several esters and ethers, is collected from the bottom with 80+% yields on cyclohexane. An alternative route to cyclohexanol used by some plants is to catalytically hydrogenate phenol.

The cyclohexanol-cyclohexanone mixture isolated from air oxidation, without separation, is oxidized with 5 volumes of 50% nitric acid plus catalyst at 50–90°C, and pressures only slightly above ambient for 10 to 30 min, depending on the temperature used (Eq. 19.56).

$$C_6H_{11}OH + \quad \bigcirc\!\!=\!\!O + 4\ HNO_3 \rightarrow \qquad 19.56$$

$$2\ HO_2C(CH_2)_2CO_2H + 2\ H_2O + 2\ N_2O \uparrow$$
(approximate stoichiometry)

Adipic acid (m.p. 152°C) is crystallized from the reaction mixture by cooling this to ca. 5°C.

Over 90% of the adipic acid produced is consumed for the production of nylon fiber and for nylon engineering resin applications. A large fraction of this consumption is direct, as adipic acid in the production of nylon 6,6, but a substantial fraction of the adipic acid is further processed to give hexamethylene diamine, the other monomer required. A further small fraction of the adipic acid is converted into di-octyl (di-2-ethylhexyl) or di-hexyl esters for use as plasticizers in flexible grades of PVC, etc., or as a high boiling point component of synthetic motor oils.

19.9.3. Hexamethylenediamine (Hexane-1,6-diamine)

Vapor phase reaction of adipic acid with ammonia at about 400°C and 270–410 atm in the presence of boron phosphate catalyst gives good yields of adiponitrile (Eq. 19.57).

$$HO_2C(CH_2)_4CO_2H + 2\ NH_3 \rightarrow NC(CH_2)_4CN + 4\ H_2O \qquad 19.57$$

Alternatively, adiponitrile may also be made from 1,3-butadiene via chlorination, followed by treatment with sodium cyanide. Catalytic reduction of adiponitrile with hydrogen then produces hexamethylenediamine (b.p. 205°C, Eq. 19.58).

$$NC(CH_2)_4CN + 4\ H_2 \rightarrow H_2N(CH_2)_6NH_2 \qquad 19.58$$

Product recovery is by distillation.

Production of this petrochemical is dedicated to a single end use, the production of nylon 6,6.

19.9.4. ε-Caprolactam (2-Oxohexamethyleneimine)

This cyclic amide is of major importance as the single monomer used to produce nylon 6, which rivals nylon 6,6 in utility and in volume of production. Cyclohexanone is required as the starting material for ε-caprolactam production. The mixture of cyclohexanol and cyclohexanone obtained initially from the air oxidation of cyclohexane for the second stage en route to adipic acid may be used. Subjecting the mixture to dehydrogenation converts the cyclohexanol component to cyclohexanone producing suitable starting material for this process (Eq. 19.59).

$$19.59$$

Alternatively, pure cyclohexanone may be made by the catalytic hydrogenation of phenol in the presence of palladium on charcoal.

Cyclohexanone oxime, an intermediate, is made by reacting aqueous hydroxylamine sulfate with cyclohexanone (Eq. 19.60).

the oxime   caprolactam   19.60

A Beckmann rearrangement of the oxime in hot concentrated sulfuric acid then gives the desired seven-membered cyclic amide, ϵ-caprolactam. The crude product forms a separate oily phase which is separated from the reaction mixture and purified by distillation under reduced pressure (b.p. 136–138°C at 10 mm; m.p. 72°C). There are also several alternative routes to produce caprolactam [17]. Caprolactam production like that of hexamethylenediamine, is dedicated to the production of a single product, in this case the various forms of nylon 6.

19.9.5. Toluene to Toluenediisocyanate (TDI)

Toluene output at one time used to be almost entirely directed to trinitrotoluene (TNT) production for industrial and military uses. Cheaper industrial explosives such as various ammonium nitrate formulations have now become available, making TNT a less important product. Uses of TNT are now almost entirely confined to military applications. However, the first stage in the preparation of toluenediisocyanate still involves the mononitration and dinitration of toluene using a mixture of nitric and sulfuric acids, as in the first stages of TNT production. Depending on the nitration strategies employed, one of two isomer mixtures of 2,4-dinitrotoluene and 2,6-dinitrotoluene, an 80:20 ratio and a 65:35 ratio, or the pure 2,4-dinitrotoluene may be obtained (Eq. 19.61) [18].

$$C_6H_5CH_3 + HNO_3/H_2SO_4 \rightarrow \qquad\qquad 19.61$$
$$\text{2,4-, and 2,6-}PhCH_3(NO_2)_2 + \text{dil. acids}$$

All three possible toluene nitration outcomes are separately carried through the reduction and phosgenation stages to give the corresponding diisocyanate mixtures, or pure 2,4-toluene diisocyanate. Reduction to the diamines is conducted catalytically (Raney nickel) with hydrogen (ca. 50 atm) in the liquid phase at about 90°C, to avoid the considerable explosion risk of gas phase operation (Eq. 19.62).

$$\text{2,4-, or 2,6-}PhCH_3(NO_2)_2 + 6\ H_2 \rightarrow \qquad\qquad 19.62$$
$$\text{2,4-, or 2,6-}PhCH_3(NH_2)_2 + 4\ H_2O$$

Before the catalytic route had been developed, iron in hydrochloric acid was used to reduce the dinitrotoluenes to the diamines.

The second stage, phosgenation of the mixed diamines, or the 2,4-diaminotoluene is conducted in an inert (aromatic, or chlorinated) solvent at 20 to 50°C. This uses an excess of phosgene in one or preferably two (for better yields) stages (Eq. 19.63).

$$R(NH_2)_2 + 2 \ ClCOCl \rightarrow R(NHCOCl)_2 \rightarrow \qquad\qquad 19.63$$
$$+ 2 \ HCl$$
$$R(N{=}C{=}O)_2 + 2 \ HCl$$

The intermediate carbamoyl chloride is dissociated to the diisocyanate with the aid of second stage heating of the reaction mixture to 100–110°C plus sparging with natural gas or nitrogen to remove the hydrogen chloride as it forms. Material for sale is produced by reduced pressure distillation; pure toluene-2,4-diisocyanate has a b.p. of 120°C at 10 mm Hg pressure.

The largest volume product, and the least expensive from this process, is the 80:20 2,4,- to 2,6-toluenediisocyanate blend which is the composition obtained without any intermediate separations of nitrotoluenes. The 65:35 2,4- to 2,6-TDI blend, and the pure 100% 2,4-toluenediisocyanate are progressively more expensive and are produced in lower volumes than the 80:20 blend since they each require one nitrotoluene isomer separation step. However, regardless of composition essentially all commercial TDI is consumed in the production of a wide variety of polyurethanes (Chap. 21). Diphenylmethane-4,4'-diisocyanate (MDI), with a somewhat lower volatility than TDI, is also used in some coating, elastomer, and fiber polyurethane applications.

19.9.6. p-Xylene to Dimethyl Terephthalate (Dimethyl Benzene-1,4-dicarboxylate, DMT)

Catalytic oxidation of p-xylene with air is the chief commercial method used to produce terephthalic acid. A solution of p-xylene in acetic acid, together with a manganese or cobalt derivative and heavy metal bromides which serve as cocatalysts, is fed to a continuous reactor, vigorously stirred, and heated to 200°C while under about 25 atm pressure. Air is continuously fed into the reactor at the same time as a small stream of partially reacted solution is removed (Eq. 19.64).

$$H_3C - \text{(benzene ring)} - CH_3 \ + \ 7/2 \ O_2 \ \longrightarrow \qquad 19.64$$

$$HO_2C - \text{(benzene ring)} - CO_2H \ + \ 2 \ H_2O$$

The pressure is released on the exit stream which simultaneously flashes off much of the excess p-xylene and the acetic acid, and cools the residual solution causing terephthalic acid to crystallize out. Residual acetic acid and p-xylene are removed from the crystals by centrifugation. The crude product acid is then slurried in hot water first for washing, and then for hydrogenation to decolorize any residual traces of colored impurities. It is then recrystallized and dried to give "fiber-grade" material (m.p. >300°C) in yields of about 90%. p-Xylene is also oxidized to terephthalic acid by systems which replace the bromide component of the catalyst system by acetaldehyde or methyl ethyl ketone. These systems function at lower temperatures and pressures, presumably via peroxide-derived radicals.

Terephthalic acid is usually directly converted to the dimethyl ester for polymer synthesis. This is prepared by superficially conventional technology using methanol and a mineral acid under moderate conditions (Eq. 19.65).

$$HO_2C-\langle\bigcirc\rangle-CO_2H \quad + \quad 2\ CH_3OH \quad \longrightarrow \qquad\qquad\qquad 19.65$$

$$H_3CO_2C-\langle\bigcirc\rangle-CO_2CH_3 \quad + \quad 2\ H_2O$$

Proportions and exact conditions required for optimum conversions and yields of diester are proprietary. Recovery of the dimethyl terephthalate product from unreacted starting materials, etc., is by distillation using a series of four or five columns. Polymer-grade material (m.p. 142°C; b.p. 288°C) is obtained by distillation under reduced pressure from the top of the last fractionating column.

Almost all of the dimethyl terephthalate (DMT) output is consumed for the production of poly(ethylene terephthalate), PET, by transesterification with ethylene glycol (ethane-1,2-diol), mostly for fiber and film applications.

19.9.7. m-Xylene to Isophthalic Acid

Isophthalic acid is produced from m-xylene by processes analogous to those used for terephthalic acid except that five or six times higher pressures are required. The product (m.p. 342°C) is chiefly used forthe preparation of the high thermal stability aramid polymer, Nomex (Dupont), used for electrical insulation in thermally aggressive environments.

19.9.8 o-Xylene to Phthalic Anhydride

Air is sufficient to oxidize the methyl groups of o-xylene, under the right conditions, like it is with p- or m-xylene just described. However, here the similarity ends since commercial o-xylene oxidation is a vapor phase process. o-Xylene vapor, mixed with a large excess of air to ensure operation outside the explosive range, is fed to a reactor containing a supported vanadium pentoxide catalyst and heated toabout 550°C. Using about a 0.1-second contact time under these conditions produces exit gases composed of phthalic anhydride, water, and carbon dioxide (Eq. 19.66).

$$\begin{array}{c}\text{o-xylene (CH}_3)_2\end{array} \quad + \ 3\ O_2 \quad \text{(from air)} \quad \longrightarrow \quad \text{phthalic anhydride} \quad + \quad 3\ H_2O \qquad 19.66$$

Cooling the exit gases causes condensation and crystallization of phthalic anhydride, which is collected and recrystallized to yield >99.5% pure product (about 75% yield). Phthalic anhydride is also made by catalytic oxidation of naphthalene with air.

About one-half of the phthalic anhydride production is consumed for the preparation of plasticizers, mostly for the various flexible grades of poly(vinyl chloride). The remainder is roughly split between alkyd resin preparation used for many types of surface coatings, and for polyester resin composites with fiberglass reinforcement, the so-called "fiberglass" resins used in boats and other sporting equipment as well as for corrosion-resistant vessels and ducts used in chemical processing, some automotive parts, and as a convenient means of field repair of many of these items.

19.10. ENVIRONMENTAL CONCERNS

Principal objectives of a modern petrochemical complex are good to excellent yields to minimize waste production and reuse or disposal costs, and good containment throughout to minimize losses of raw materials and products to air, water, or soil. Many newer facilities are now able to operate with only make-up water requirements and no aqueous waste streams [19]. Aqueous streams containing volatile organics may be sparged with air to remove these, and the organic vapor-air stream produced used for feed air to a boiler to eliminate air pollution from this source. Bacterial treatment of the residual wastewater stream under anaerobic and/or aerobic conditions can do much to consume and remove the residual impurities [21,21]. By-product organics that are not amenable to recycle or to be burned for energy recovery on site may be usefully consumed for this purpose in cement kilns [22].

Let us use as an example one recent environmental case history relevant to petrochemicals production. Nitrous oxide (N_2O) is implicated as a contributor to stratospheric ozone damage, and is also a potent greenhouse gas (Table 2.7). Recent work indicated that about 10% of the nitrous oxide contributions to the atm was from the world's adipic acid plants, slightly more than the fraction contributed by biomass burning [23,24]. It has been estimated that about 1 mol of N_2O is produced per mole of adipic acid, or about 0.3 kg of N_2O per kg of adipic acid.

Nitrous oxide control options at various stages of development included thermal catalytic reduction of N_2O in the presence of methane, conversion of N_2O to recoverable NO and use of this to prepare nitric acid, and catalytic dissociation to nitrogen and oxygen (Eq. 19.67–19.71) [25].

$$CH_4 + 4\,N_2O \rightarrow 4\,N_2 + CO_2 + 2\,H_2O \qquad 19.67$$

$$CH_4 + 4\,NO \rightarrow 2\,N_2 + CO_2 + 2\,H_2O \qquad 19.68$$

$$N_2O + M \rightarrow N_2 + O + M \qquad 19.69$$

$$N_2O + O \rightarrow 2\,NO \quad \text{(to use for nitric acid, Chap. 11)} \qquad 19.70$$

$$N_2O + O \rightarrow N_2 + O_2 \qquad 19.71$$

It could also be recovered as nitrous oxide for sale, but this has not been economically attractive. The major adipic acid producers, worldwide, have agreed to implement N_2O abatement measures using one or the other of these options by 1998. Feasibility studies of a possible longer range solution have been conducted of an alternate enzymatic route that converts glucose to muconic acid, which could then be catalytically reduced to adipic acid, but this has not as yet been tested on a commercial scale (Eq. 19.72) [26].

$$C_6H_{12}O_6 \xrightarrow{\text{enzym.}} \underset{\text{cis, cis-muconic acid}}{HO_2C-CH=CH-CH=CH-CO_2H} \xrightarrow{H_2,\ Pt} \qquad 19.72$$

$$\underset{\text{adipic acid}}{HO_2C(CH_2)_4CO_2H}$$

REVIEW QUESTIONS

1. (a) What would be the mass (kg) of carbon theoretically possible from the processing of $1.00 \times 10^4\ m^3$ (standard conditions) of natural gas (standard conditions are 15.6°C, 1 atm pressure)?
   (b) Compare the percent yields reported for channel black and thermal black in this chapter with the percent yields calculated using the mass ratio determined in part (a)?

2. (a) What mass of isopropanol (100% basis) would be expected from the passage of 1000 kg of propylene through the two-step process, ester formation and hydrolysis, assuming a 52% conversion and 71% selectivity (industrial yield)?
   (b) Assuming the same conversion and yield as in part (a), what mass of propylene would be required to produce 1000 kg of isopropanol (100% basis)?
   (c) What happens to the unreacted propylene?
   (d) Explain some of the possible causes of the relatively low selectivity.

3. (a) By using plausible reactions of ethyl radicals with ethylene and related reactions, sketch reasonable schemes leading to the various possible C_4 hydrocarbon products from butane which applies the basic mechanisms to ethylene outlined in Section 19.5.
   (b) Sketch plausible propagation and termination reactions for the products observed from the cracking of propane.

4. (a) What pentanol isomer(s) would be possible from the oxo reaction using isobutylene $(CH_2=C(CH_3)_2)$ as the feed olefin? Outline the equations for the steps involved.
   (b) Describe and explain plausible methods which could be used to isolate and purify the pentanol product(s).

5. (a) Calculate the research (academic) yield and the conversion and selectivity (industrial yield) on ethylene from the following data collected from an autoclave (pressure vessel) experiment on the direct hydration of ethylene.

Feed to reactor:		Recovered from reactor:	
Ethylene:	280.0 g	Ethylene:	268.3 g
Water:	300.0 g	Water:	292.6 g
Acid catalyst:	trace	Ethanol:	19.17 g

(b) Based on the research yield, would you authorize construction of a plant to use this process? Explain.

(c) Consider how knowledge of the selectivity of the process might affect your decision given in part 2 (b) and explain your answer.

(d) Outline positive and negative selection features of the direct hydration ethylene process in comparison to the esterification (indirect) route (Section 19.10.5).

6. (a) Calculate the theoretical masses of propylene, chlorine, and calcium hydroxide required to produce 1000 kg of propylene oxide via the chlorohydrin process, and the mass of calcium chloride (100% basis) that would be coproduced.

(b) What actual masses of starting materials would be required and mass of 12% by weight calcium chloride in water that would be produced for each 1000 kg of propylene oxide when the yield (selectivity) on each of the starting materials is 75%?

(c) What would be the theoretical, and actual (same selectivity), base requirement for 1000 kg of propylene oxide if the process was switched to use sodium hydroxide instead of calcium hydroxide as the base? What mass of sodium chloride (100% basis) would also result?

(d) Using the theoretical raw material requirements, at what price would sodium hydroxide have to be available to equal the base cost when slaked lime ($Ca(OH)_2$) is purchased at $66 per metric tonne (1000 kg).

(e) What environmental factor(s) might influence the base selection decision, apart from equivalent raw material cost?

7. Compare the contact times (in seconds) and the respective space velocities for each of the three following process conditions.

Ethylene oxidation: 1.0-second contact time at 280°C, 1.0 atm
Ethylene hydration: 1800 hr^{-1} space velocity at 300°C, 70 atm
Propylene ammonoxidation, 20-second contact time at 400°C, 1 atm.

FURTHER READING

C.G. Bertram, Minimizing Emissions from Vinyl Chloride Plants. *Environ. Sci. Technol.* **11**, 864–868 (1977).

A. Chauvel and G. Lefebvre, "Petrochemical Processes," Vols. 1 and 2, Gulf Publ. Co., Houston, TX, 1989.

Organization for Economic Cooperation and Development, "OECD Petrochemical Industry: Energy Aspectsof Structural Change." OECD, Paris, 1985.

P.H. Spitz, "Petrochemicals: The Rise of An Industry." Wiley, New York, 1988.

P. Wiseman, "An Introduction to Industrial Organic Chemistry, 2nd ed." Appl. Sci. Publ., London, 1979.

H.A. Wittcoff and B.G. Reuben, "Industrial Organic Chemicals in Perspective. Part One: Raw Materialsand Manufacture." Wiley, New York, 1980.

REFERENCES

1. G.T. Austin, "Shreve's Chemical Process Industries," 5th ed., p. 75. McGraw-Hill, New York, 1984.
2. Facts and Figures for the Chemical Industry, *Chem. and Engin. News* **74**(26), 38–80, June 24 (1996).
3. W.L. Faith, D.B. Keyes, and R.L. Clark, "Industrial Chemicals," 3rd ed., p. 205. Wiley, New York, 1966.
4. A.M. Thayer, Carbon Black Industry Rattled by Exit of Two Established Producers. *Chem. Eng. News* **73**(29), 33–40, July 17 (1995).
5. F.A. Lowenheim and M.K. Moran, "Faith Keyes and Clark's Industrial Chemicals," 4th ed., p. 210. Wiley, New York, 1975.
6. G.D. Ulrich, Flame Synthesis of Fine Particles. *Chem. Eng. News* **62**(32), 22–29, Aug. 6 (1984).
7. M.S. Reich, Demand for Carbon Fibers Rebounds, *Chem Eng. News* **72**(48), 23–24, Nov. 28 (1994).
8. J.E. Bailey and A.J. Clarke, Carbon Fibers, *Chem. Br.* **6**(11), 484–489, Nov. (1970).
9. P. Beardmore, J.J. Harwood, K.R. Kinsman, and R.E. Robertson, Fiber-Reinforced Composites: Engineered Structural Materials. *Science* **208**, 833–840 (1990).
10. P.K. Bachmann and R. Messier, Emerging Technology of Diamond Thin Films. *Chem. Eng. News* **67**(20), 24–39, May 15 (1989).
11. A.A. Lemonidou, I.A. Vasalos, E.J. Hirschberg, and R.J. Bertolacini, Catalyst Evaluation and Kinetic Study for Ethylene Production. *Ind. Eng. Chem. Res.* **28**, 524–530 (1989).
12. P.K. Frolich and P.J. Wiezevich, Cracking and Polymerization of Low Molecular Weight Hydrocarbons. *Ind. Eng. Chem.* **27**(9), 1055–1062 (1935).
13. J. Krieger, Propane Route to Acrylonitrile Hods Promise of Savings. *Chem Eng. News* **74**(39), 18, 19, Sept. 23 (1996).
14. S.J. Ainsworth, Propylene Oxide Producers Look for Ways to Counter Sluggish Market. *Chem. Eng. NEws* **70**(9), 9–12, Mar. 2 (1992).
15. Anonymous, Arco Expands PO and SM. *Chem. Br.* **32**(8), 9, Aug. (1996).
16. T. Hayashi, L.-B. Han, S. Tsubota, and M. Haruta, Formation of Propylene Oxide by the Gas-Phase Reaction of Propane and Propene Mixture with Oxygen. *Ind. Eng. Chem.* **34**, 2298–2304 (1995).
17. F.A. Lowenheim, and M.K. Moran, "Faith, Keyes and Clark's Industrial Chemicals." 4th ed., p. 201, Wiley-Interscience, New York, 1975.
18. K.J. Saunders, "Organic Polymer Chemistry," 2nd ed. Chapman & Hall, London, 1988.
19. D. Dembicki and K. Tsang, Off the River. *Hazard. Mater. Manage.* **8**(2), 4–9, April/May (1996).
20. J.E. Rucker, Wastewater Control Facilities in a Petrochemical Plant. *Chem. Eng. Prog.* **66**(11), 63–66, Nov. (1970).
21. D. Dempster, Bacterial Bed Consumes Phenol. *Can. Chem Process.* **65**(1), 36–37, Feb. (1981).
22. D. Gossman, The Reuse of Petroleum and Petrochemical Waste in Cement Kilns. *Environ. Prog.* **11**(1), 1–6, Feb. (1992).
23. M.H. Thiemens and W.C. Trogler, Nylon Production: An Unknown Source of Atmospheric Nitrous Oxide. *Science* **251**, 932–94 (1991).
24. W.R. Cofer, III, J.S. Levine, E.L. Winstead, and B.J. Stocks, New Estimates of Nitrous Oxide Emissions from Biomass Burning, *Nature (London)* **349**, 689–691 (1991).
25. R.A. Reimer, C.S. Slaten, M. Seapan, M.W. Lower, and P.E. Tomlinson, Abatement of N_2O Emissions Produced in the Adipic Acid Industry. *Environ. Prog.* **13**(2), 134–137, May (1994).
26. J.W. Frost and K.M. Draths, Sweetening Chemical Manufacture. *Chem. Br.* **31**(3), 206–210, March (1995).

20

CONDENSATION (STEP-GROWTH) POLYMER THEORY

Mr. McGuire: I want to say just one word to you
. . . just one word.
Benjamin Braddock: Yes, sir.
Mr. Mcguire: . . . Plastics.
                —The Graduate, *Embassy Pictures, 1967*

. . . each primordial Bean
Knew cellulose by heart: Nature alone of Collagen
And Apatite compounded Bone.
                —*John Updike, 1968*

20.1. BACKGROUND

Natural polymers with a wide variety of fascinating properties have been around for a very long time. Materials such as wood, cotton, hides, wool, muscle, collagen, and starch all comprise examples of natural polymers, most of which have been used by humans since before historical records were kept, for heat, shelter, tools, food, and other purposes. For all of these applications the natural materials were used in more or less the chemical form in which they originally occurred. It was not until the nineteenth century, when our understanding of the nature of very basic chemical transformations began to develop, that technologists began to experiment with the modification of natural polymers to yield products with more versatile target properties. By these means the applications possible with natural polymeric materials were gradually expanded beyond those possible with the native materials.

Natural rubber began public awareness as a scientific curiosity, when it was discovered by Priestley that it could be used to "rub out" pencil marks. This novel capability gave the material its name, but still gave rubber little utility because of its dimensional instability and poor durability. The discovery of vulcanization by Charles Goodyear in 1839 changed all this. Vulcanization, or the heating of natural rubber mixed with small amounts of sulfur, permitted the stabilization of formed rubber shapes and rapidly expanded the uses of this modified natural polymer.

Chemically modified cellulose, as cellulose nitrate or nitrocellulose, was made and tested for commercial applications in Britain in the 1855–1860 period without much success. The discovery by Hyatt, in 1863, that cellulose nitrate could be plasticized with camphor to give moldability to the combination, made this material much more useful. By 1870, celluloid, the commercial name for plasticized cellulose nitrate, was being produced into a variety of commercial products such as billiard balls, decorative boxes, and combs. Nitrocellulose was also soluble in organic solvents, unlike cellulose, and so could be applied to surfaces in solution to form a coating, as in airplane dopes and automobile lacquers. It could also be solution spun into fibers (synthetic silk) and formed into photographic film, or used as a laminating layer in early auto safety glass. It was also used as an explosive. The hazard introduced to many of these uses of nitrocellulose by its extremely flammable nature resulted in an interest in discovering other cellulose derivatives that could still be easily formed, like nitrocellulose, but without its extreme fire hazard.

Cellulose acetate, first prepared in 1865, was a suitable candidate to replace nitrocellulose. Cellulose triacetate, however, had lower solvent solubility than nitrocellulose and also was difficult to dye. Lesser levels of acetylation gave more tractable, fiber-formable products which could also be more easily dyed, and could be formed into film suitable for photographic use. These were the so-called acetate rayon fibers and plastics, which were also considerably safer to use than nitrocellulose.

Later still, methods were discovered by which cellulose could be dissolved unchanged, and then reprecipitated as fiber or film by subsequent chemical treatment. This process, which is still in use, yields viscose rayon fiber and cellophane film from the reconstituted cellulose, more widely useful products than were available from the early acetate rayon.

All of these early attempts to produce useful polymer products were based on modified, or reconstituted ("semi-synthetic") natural polymers, but many of these are still in use today. The first of the purely synthetic commercial polymers came with the small-scale introduction of Bakelite in 1907. This phenol-formaldehyde resin product was developed by Leon Baekeland. It rapidly became a commercial reality with the formation of The General Bakelite Company by Baekeland, and construction of a larger plant at Perth Amboy, New Jersey, in 1910. At about this time styrene was being combined with dienes in the early commercialization of processes to synthetic rubber. Polystyrene itself was not a commercial product in Germany until 1930 and in the United States in 1937. The only other purely synthetic polymers that made a commercial appearance during this early development period were polyvinyl chloride and polyvinyl acetate, both in the early 1920s.

From these relatively recent beginnings, while the polymer industry learned much from the modification of natural polymers, production figures were small. As the purely synthetic polymers came on stream the range of valuable applications for plastic materials rapidly increased. Today the annual mass of polymeric materials produced greatly exceeds the scale of aluminum production, and rivals pig iron production, both in the United States and the world. The generally low density of organic polymeric material means that on a volume basis production of these products already exceeds the combined *volume* of production of these two important fabrication metals. By substituting the low-density, high-strength plastics for metals, corrosion resistance and substantial weight savings are being realized in the construction of automobiles, buses, aircraft, etc. This, in turn, is resulting in a substantial saving in the energy required to transport people and goods. A further factor of relevance, particularly to those who study polymer chemistry, is that it is estimated that one out of every two chemists is employed in some aspect of the polymer industry.

20.2. BASIC POLYMER CONCEPTS

What comprises a polymer? A general definition which can include natural, modified natural (semi- synthetic), and purely synthetic polymers of all types is that a polymer is a large molecule built up of small structural units combined in any conceivable pattern. Staudinger, a major contributor to the early development of polymer theory, set as an arbitrary guideline that a "polymer" was a molecule with a molecular weight of more than 10,000, or that consisted of a total of more than 1000 carbon atoms. While there are also a number of important polymers with an inorganic backbone, such as silica, the silicones, and phosphonitrilic compounds, where the second criterion would not apply, they would still qualify under the first.

The repeating unit of a polymer, the "-mer" or monomer, is the basic building block required for polymer preparation. To be capable of polymerizing it must have two or more "bonding units," i.e., it must have a functionality, or a potential functionality, of two or more.

$$8 \times \cdot A \cdot \rightarrow A \cdot A \cdot A \cdot A \cdot A \cdot A \cdot A \cdot A$$

The "degree of polymerization" is the number of monomer units that the polymer contains, 8 in the example above. This is sometimes expressed as the DP, X_n, or X, for short. If the monomer used has a functionality of 2 we normally obtain polymers of long continuous chains, but we may get occasional short branches from impurities in the monomer or irregularities in the polymerization process (Eq. 20.1):

20.1

Polymerization of a monomer with a functionality of 3 gives more possible structures. Short or long branches will be numerous, as will be branching branches and links between adjacent chains to form a cross-linked, three-dimensional polymer (Eq. 20.2).

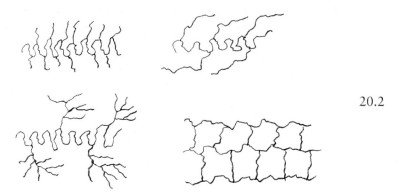

20.2

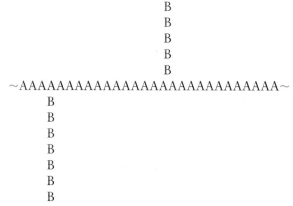

Functionalities of greater than 3 may be used for some purposes, which will further increase the complexity of the product. Thus, the functionality of the monomer has a significant influence on polymer structure and also on the resultant properties of articles made from that polymer.

Another factor which differentiates polymers arises from the number of monomers used in their preparation. The functionality discussion above implied the use of a single monomer in the polymer synthesis, which would give a "homopolymer" product, or simply "polymer" for short. A polymer composed of two types of monomers, A and B, would be called a "copolymer." If A and B are only bifunctional, the copolymer possibilities are as follows:

(a) Random copolymer, ~AAABABBAABABBB~, in which the monomer elements are present in a linear polymer in a statistically random order

(b) Alternating copolymer, ~ABABABABABAB~, in which the monomer elements occur in a regular alternating sequence

(c) Block copolymer, ~AAAABBBBBBAAAA~, in which predictable blocks of first one monomer and then the other occur regularly in the polymer chain

(d) Graft copolymer,

$$
\begin{array}{c}
B \\
B \\
B \\
B \\
B \\
\text{\textasciitilde}AAAAAAAAAAAAAAAAAAAAAAAAAAAAAA\text{\textasciitilde} \\
B \\
B \\
B \\
B \\
B \\
B \\
B
\end{array}
$$

in which blocks of one monomer are grafted, or bonded, to a backbone of polymer formed from the other monomer.

669

There are also the possibilities of using three monomers to produce a terpolymer, and four (or more), to produce a tetrapolymer and so on. Of course if the monomers used to prepare a copolymer have functionalities greater than 2, the structural complexity of the various possible copolymers obtained also rises.

20.3. POLYMERIZATION PROCESSES

Monomers can be joined by means of two principal methods to form polymers, and these methods are used as a broad basis for classification of synthetic polymers. The first of these, condensation, or step-growth polymerization, involves the use of functional group reactions such as esterification or amide formation to form polymers. When each of the molecules involved has only one functional group then the reaction between a carboxylic acid and an alcohol gives an ester (Eq.20.3). In this equilibrium reaction water removal will help drive the reaction to the right.

$$CH_3CO_2H + C_2H_5OH \overset{H^+}{\rightleftharpoons} CH_3CO_2C_2H_5 + H_2O \qquad 20.3$$

A functionality of 2 is required in each of the monomer units in order to obtain a linear condensation polymer from esterification reactions. The initial dimeric product still has residual, potentially reactive functionalities (Eq. 20.4).

$$HO_2C - C_6H_4 - CO_2H + HOCH_2CH_2OH \rightleftharpoons$$
$$\text{terephthalic acid} \qquad \text{1,2-ethanediol} \qquad 20.4$$
$$HO_2C - C_6H_4 - CO_2CH_2CH_2OH + H_2O$$
$$\text{ethylene terephthalate}$$

Further reaction and water removal from this system will ultimately give polyethylene terephthalate, well known commercially in fiber form as Dacron, or Terylene, and as a film by the Mylar trade name (Eq. 20.5).

$$n\ HO_2C - C_6H_4 - CO_2CH_2CH_2OH \rightleftharpoons \qquad 20.5$$
$$HO + CO - C_6H_4 - CO_2CH_2CH_2O +_n H + (n - 1)\ H_2O$$
$$\text{polyethylene terephthalate}$$

Features that generally serve to recognize condensation polymerizations are the loss of a small molecule such as water, methanol, or hydrogen chloride during the polymerization process, and a relatively slow progression toward high molecular weight material which is aided by the removal of the small molecule from the system.

In contrast with the usually slow progress of condensation polymerization the second major classification, addition, or vinyl-type polymerizations, usually proceeds very rapidly, so rapidly that they are sometimes referred to as chain reaction polymerizations. This method of producing synthetic polymers applies the potential dual functionality present in a carbon-carbon double bond. The process is initiated by the use of radical or charged initiator species to form new sigma bonds from the carbon-carbon double bonds of the monomer, to link the monomer units (Eq. 20.6).

$$n\ CH_2=CHX \rightarrow + CH_2-CHX +_n$$
$$e.g.,\ for\ X = H:\ \ n\ CH_2=CH_2 \rightarrow + CH_2-CH_2 +_n,$$
$$\text{polyethylene, or PE}$$

for X = Cl: $n\ CH_2=CH \rightarrow + CH_2-CH_2 +_n,$ 20.6
$$\ ||$$
$$ClCl$$

$$\text{polyvinyl chloride, or PVC}$$

for X = phenyl: $n\ CH_2=CH \rightarrow + CH_2-CH +_n,$ polystyrene, or PS
$$\ ||$$
$$PhPh$$

Addition polymerization is usually such a rapid process that only monomer and final polymer chains are present. Very little of the active material in the system is oligomeric, that is, consisting of only a small number of linked monomer units en route to polymer, at any one point in time. Also, in addition polymerization the whole monomer molecule adds to form polymer. No small molecule is lost in the process. Further details of the polymerization processes involved and the structural properties of the products obtained from addition polymerization are discussed in Chapters 22 and 23. This chapter will focus on the background theory of synthetic polymers, and condensation, or step-growth, polymerization.

20.4. POLYMER MOLECULAR WEIGHTS

With either type of synthetic polymerization we are dealing with a series of completely random events. As a result the molecular weights of the polymer products that are obtained will be within a range, rather than an exact, single molecular weight. Synthetic polymer samples are therefore described as being polydisperse with regard to molecular weights; any particular sample will consist of a composite of chains of many different molecular weights. Some natural polymers are made in a template fashion and can have a single, specific (and characteristic) molecular weight as, for example, some of the proteins. These polymers are described as being monodisperse, in contrast to the usual spread of molecular weights of synthetic polymers.

   In work with the molecular weights of the polydisperse synthetic polymers, averages are used a great deal. The number average molecular weight is one of the important averages in common use. Number average molecular weight M_n, $\overline{M_n}$, or $<M_n>$, is defined as:

$$M_n = \frac{w}{\sum\limits_{i=1}^{\infty} N_i} = \frac{\sum\limits_{i=1}^{\infty} N_i M_i}{\sum\limits_{i=1}^{\infty} N_i}$$ 20.7

where

w = mass of polymer sample measured
N_i = the number of molecules (moles) of molecular weight i in the sample
M_i = the molecular weight.

Since the numerator of Eq. 20.7 represents the mass of the polymer sample measured, and the denominator the total number of polymer molecules present, the result of this calculation is simply the average molecular weight per polymer molecule in the sample.

The number average molecular weight of a polymer sample is experimentally determined by any method which determines colligative properties of polymer solutions. Any measurements such as osmotic pressure, boiling point elevation, or freezing point depression, where the result is affected by the *numbers* of molecules present, can be used to give a number average molecular weight result. Experimental details are available elsewhere [1].

The number average molecular weight and the molecular weight distribution of a synthetic polymer markedly affect the processing and the final properties of articles made from it. To better visualize and understand how a number average result is obtained, let us take a hypothetical sample of polymer containing an equal weight of material of 1000 molecular weight and 10,000 molecular weight (Fig. 20.1). To have the same weight, or mass, of each molecular weight polymer in the mixture, we must have

$$N_{1000} = 10 \times N_{10,000}$$

To calculate M_n we need only know the molar ratio of the two (or more) molecular weight fractions, not the actual numbers of molecules. If we let x be the number of 10,000 molecular weight molecules, then

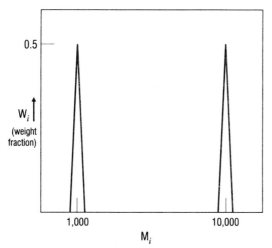

FIGURE 20.1   A representation of a hypothetical polymer sample composed of equal weights of 1000 and 10,000 molecular weight (M_i) material.

$$M_n = \frac{10x(1000 \text{ g/mol}) + x(10,000 \text{ g/mol})}{11x} = 1818 \text{ g/mol}$$

The number average molecular weight of this mixture is 1818 g/mol, or rounded, 1820 g/mol.

The other very generally useful molecular weight measure in polymer work is the weight average molecular weight M_w, $\overline{M}_w$, or $<M_w>$. It is defined as

$$M_w = \frac{\sum\limits_{i=1}^{\infty} N_i M_i^2}{\sum\limits_{i=1}^{\infty} N_i M_i} \qquad\qquad 20.8$$

Weight average molecular weight may be experimentally determined by any method in which molecular size or molar mass is the parameter being measured, e.g., light scattering or ultracentrifugation.

We can calculate weight average molecular weight in a similar way to number average molecular weight. Using the same hypothetical polymer sample used for the calculation of number average molecular weight gives us

$$M_w = \frac{10x(1000 \text{ g/mol})^2 + x\,(10,000 \text{ g/mol})^2}{10x(1000 \text{ g/mol}) + x\,(10,000 \text{ g/mol})}$$
$$= 5500 \text{ g/mol}$$

as the weight average molecular weight, or molar mass, for this sample. This markedly different value results from both the extra molecular weight term present in the numerator and denominator for the definition of M_w and from the bimodal (not "normal") distribution of molecular weights in the hypothetical sample. However, it is true that, for any ordinary synthetic polymer sample, M_w will always be greater than M_n. The "most probable" molecular weight is defined as the molecular weight of the most numerous molecular weight fraction of a normal synthetic polymer sample. Figure 20.2 gives the relationship of the most probable molecular weight to M_n and M_w for a typical molecular weight distribution synthetic polymer. Other molecular weight averages, M_z and M_{z+1}, are less commonly used. Each of these attaches one higher order of significance to molecular weight than M_w. Full details are available in standard texts [1].

Before we leave molecular weight considerations mention should be made of the heterogeneity index (HI), or polydispersity index. This is defined as the weight average molecular weight divided by the number average molecular weight:

$$HI = \frac{M_w}{M_n} \qquad\qquad 20.9$$

The result of this calculation gives us a coarse indication of the breadth of the molecular weight distribution in the polymer sample. A "normal" (statistical) polymer molecular weight distribution gives a value of about 2 for the heterogeneity index. A monodisperse (single molecular weight) polymer would

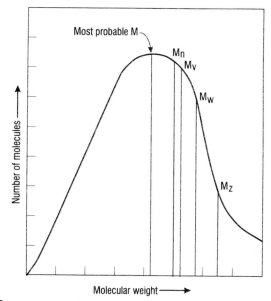

FIGURE 20.2 Relationships of the most probable (M), number average (M_n), viscosity average (M_v), weight average (M_w), and Z average (M_z) molecular weights used to define synthetic polymers.

give the same weight and number average molecular weights and so would give a heterogeneity index of 1.0. For the hypothetical example we have been considering,

$$HI = \frac{5500}{1818} = 3.03$$

This is evidently not a very "normal" distribution. Very wide bimodal or very broad molecular weight distributions can give heterogeneity indices of 5 or higher.

20.5. POLYCONDENSATION POLYMERS

A characteristic common to the monomers of all condensation, or step-growth, polymers is that they possess two (or more) moreorless reactive functional groups. The functional groups of a single monomer may be different, in which case the polymer may be produced by reaction of the single monomer with itself (Table 20.1, first two examples); or the functional groups required for condensation reactions may be on different monomers, in which case two monomers are required (Table 20.1, last three examples). Examples 1, 2, and 4 from this table are also typical of polycondensations in which a small molecule is split out, water in the first two cases and alcohol in example 4. Polyurethane formation, in which the alcohol adds across the carbon nitrogen double bond of the isocyanate function ($O=C=N-R$) to form the urethane ($-O-CO-NH-$) link without loss of a small molecule, represents

TABLE 20.1   Generalized Equations for the Formation of Some Common Types of Condensation Polymers

Polmer type	Condensation process
Polyesters	$n\ HO(CH_2)_xCO_2H \rightarrow +(CH_2)_xCO-O+_n + (n-1)\ H_2O$
Polyamides	$n\ H_2N(CH_2)_xCO_2H \rightarrow +NH(CH_2)_xCO+_n + (n-1)\ H_2O$
Polyurethanes	$n\ HO-R-OH + n\ OCNR'NCO \rightarrow +OROCO-NHR'NHCO+_n$
Polyacetals	$n\ HO-R-OH + n\ CH_2(OR')_2 \rightarrow +OROCH_2+_n + 2n\ R'OH$
Phenol-formaldehyde	$PhOH + HCHO \rightarrow$ resins of complex structure

an exception to this general recognition feature of polycondensations. Bakelite is one of several possible products from phenol-formaldehyde resins, and these may or may not involve loss of a small molecule. This more complicated process will be discussed in further detail later.

20.6. POLYCONDENSATION MECHANISMS

For efficient polycondensation to high molecular weight polymers, one generally requires monomers with a functionality of 2 (or higher), high monomer purity, stoichiometric (equimolar) presence of the monomer functional groups, relatively long reaction times, at least moderate temperatures, and sometimes a suitable catalyst. Wallace Carothers (assignor of the original nylon patent U.S. 2,071,250 of Feb. 16, 1937, to the Du Pont Company) was the chemist who first established the importance of exact stoichiometry to the preparation of high molecular weight condensation polymers. Given the right conditions, reactions between the monomer functional groups will occur along the following lines:

i. $H_2N-R-NH_2 + HOOC-R'-COOH \rightarrow$
$$H_2N-R-NH-CO-R'-COOH + H_2O$$
dimer

2. dimer + monomer → trimer + H_2O
3. trimer + monomer → tetramer + H_2O
4. trimer + dimer → pentamer + H_2O
5. trimer + trimer → hexamer + H_2O
6. trimer + tetramer → heptamer + H_2O

All kinds of possibilities exist, and in practice all will occur, i.e., any two molecular species present in a polycondensation system can react. Detailed observation of many polycondensations has established that the monomer is completely consumed, or nearly so, very early in the reaction.

To be able to describe polycondensations kinetically, and to be able to predict rates, we have to consider a large number of different component reactions. How can we do this? It was Paul Flory (Nobel Prize for polymer chemistry, 1974) who first suggested that one make an overriding assumption that the reactivity of the functional groups of all the oligomers was the same. This assumption has been shown to be a reasonable one, by experiments which compare the rates of condensation reactions of a homologous series of monomers. At trimer, and for higher oligomers, the rate constants remain unchanged regardless of molecular size. Verification was also obtained from the good agreement between experimental results and the kinetic theory derived from the use of this assumption.

To develop the kinetics, let us take an A-B type of condensation monomer, such as p-hydroxymethylbenzoic acid.

$$HOCH_2 - C_6H_4 - COOH$$

A-B Type Monomer

This monomer, if pure, makes the required 1:1 stoichiometry automatic. Let N_o be the original number of functional groups present in the system, and N the total number of unreacted functional groups left after any given reaction time, t. Then the extent of reaction,

$$p = \frac{N_0 - N}{N_0}.$$

That is, p is the fraction of the functional groups left unreacted. This expression can be rearranged to the useful form:

$$N = N_0(1 - p) \qquad 20.10$$

which is referred to as the *Carothers equation*.

The number average degree of polymerization, X_n or $\overline{X}_n$, is given by

$$\overline{X}_n = \frac{N_0}{N}$$

Substituting for N from Eq. 20.10 into this expression gives

$$\overline{X}_n = \frac{1}{(1 - p)} \qquad 20.11$$

Equation 20.11 relates the number average degree of polymerization, and ultimately the molecular weight, to the extent of reaction, p.

As an example, suppose we wish to make a polymer with 50 monomer units in the average chain. This would not give a particularly high molecular weight. How far must one carry the polycondensation reaction to obtain this? Substituting, and solving for p:

$$50 = \frac{1}{(1 - p)},$$

p = 0.98, or 98% of complete reaction would be required

When one considers the usual extent of reaction for organic preparations in relation to this requirement it is evident that this is not going to be easy. In fact, quite forceful conditions are required. For commercially useful polymer, 99.9% reaction or so is required (Fig. 20.3, 20.4). At least by using an A-B

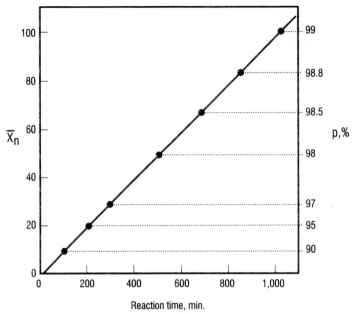

FIGURE 20.3   Relationship between $\overline{X}_n$ and extent of reaction, p, for various reaction times during a typical polycondensation.

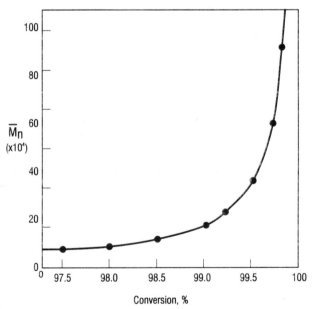

FIGURE 20.4   Asymptotic relationship between the percent conversion, as a step-growth polymerization proceeds, and molecular weight.

type of monomer, as in this example, good stoichiometry is assured. This removes one factor which might detract from obtaining a high extent of reaction. However, if the condensation polymerization is between two different monomers, i.e., a system of the A-A, B-B type, then the importance of careful mass measurement at the start to ensure good stoichiometry becomes more evident.

20.7. POLYCONDENSATION KINETICS

Let us set up a general series of steps that describe the mechanism of a polyesterification, and use this to illustrate the progress of a step polymerization. For a self-catalyzed polyesterification the initial step will be protonation of the carbonyl of one carboxyl by a proton from another, followed by the series of steps illustrated (Eq. 20.12-20.14).

$$
\underset{\text{www}}{\text{C}}\overset{\text{O}}{\overset{\|}{-}}\text{OH} + \underset{\text{www}}{\text{C}}\overset{\text{O}}{\overset{\|}{-}}\text{OH} \;\underset{k_2}{\overset{k_1}{\rightleftharpoons}}\; \underset{\text{www}}{\overset{\text{OH}}{\overset{|}{\text{C}}}}\overset{}{\underset{+}{-}}\text{OH} + \underset{\text{www}}{\text{C}}\overset{\text{O}}{\overset{\|}{-}}\text{O}^- \qquad\qquad 20.12
$$

$$
\underset{\text{www}}{\overset{\text{OH}}{\overset{|}{\text{C}}}}\overset{}{\underset{+}{-}}\text{OH} + \text{www}-\text{OH} \;\underset{k_4}{\overset{k_3}{\rightleftharpoons}}\; \underset{\underset{+}{\text{www}\,\text{OH}}}{\overset{\text{OH}}{\overset{|}{\text{C}}-\text{OH}}} \qquad\qquad 20.13
$$

$$
\underset{\underset{+}{\text{www}\,\text{OH}}}{\overset{\text{OH}}{\overset{|}{\text{www}\,\text{C}-\text{OH}}}} \;\overset{k_5}{\rightleftharpoons}\; \underset{\text{www}}{\text{C}}\overset{\text{O}}{\overset{\|}{-}}\text{O}\,\text{www} + \text{H}_2\text{O} + \text{H}^+ \qquad 20.14
$$

For this process the rate will be given by the following:

$$
\text{rate} = \frac{-d[\text{COOH}]}{dt} = k[\text{COOH}]^2[\text{OH}] \qquad 20.15
$$

To simplify this expression, and in keeping with the stoichiometry required for efficient polycondensation, we can assume the concentrations of the functional groups are the same, to give:

$$
\frac{-dc}{dt} = kc^3 \qquad 20.16
$$

where c is the initial concentration of each functional group. Integrating gives:

$$
2\,kt = \frac{1}{c^2} - \frac{1}{c_0^2}, \text{ as long as } c = c_0 \text{ at zero time}
$$

Water is removed from the system, and may be neglected.

To be kinetically useful we need to put this equation entirely in terms of c_0. Replacing the N's in the Carothers equation by c's gives $c = c_0(1 - p)$. Substituting this value for c in the integrated expression:

$$2kt = \frac{1}{c_0^2(1 - p)^2} - \frac{1}{c_0^2} \qquad\qquad 20.17$$

$$\text{or} \quad 2c_0^2kt = \frac{1}{(1 - p)^2} - 1$$

Equation 20.17 puts the kinetic expression entirely in terms of the initial concentration of monomer, c_0, and is in the form useful for following the rates of linear (only bifunctional monomers) self-catalyzed polycondensations. If $1/p^2$ is plotted against t one gets a straight line. Under self-catalyzed conditions, the reaction is third order, and the degree of polymerization (DP, or $\overline{X}_n$, is approximately proportional to $t^{1/2}$). This third-order dependency of the rate of self-catalyzed polycondensations also means that the rate of increase in the degree of polymerization slows down quite sharply as the reaction proceeds.

The rate reduction observed for self-catalyzed polycondensations may be remedied by using a strong acid catalyst. This also changes the kinetic treatment. The rate equation, Eq. 20.15, becomes

$$-\frac{d[COOH]}{dt} = k[COOH][OH][cat.] \qquad\qquad 20.18$$

$$= k[COOH][OH][H^+]$$

Since the $[H^+]$ contributed by the mineral acid will remain constant, we can simplify this rate expression by substituting k', a new constant, for $k[H^+]$ which gives

$$\frac{-d[COOH]}{dt} = k'\,[COOH][OH]$$

The concentration of carboxyls in this system equals the concentration of hydroxyls so that we can substitute c for each of these terms, which gives

$$\frac{-d[COOH]}{dt} = k'c^2$$

which is easier to integrate. Integrating this expression then gives

$$c_0k't = \frac{1}{(1 - p)} - 1 \qquad\qquad 20.19$$

From this equation we can see that a plot of $1/p$ versus t for a strong acid-catalyzed polycondensation would be expected to give a straight line. The second-order kinetics obtained gives more rapid progression toward high molecular weight material under these conditions than possible under self-catalyzed conditions, particularly toward the end of the polymerization. These second-order kinetics have been found to hold to at least a DP of 90, which is equivalent to an M_n of about 10,000.

20.8. QUANTITATIVE MOLECULAR WEIGHT CONTROL

Many useful condensation polymers employ two bifunctional monomers, one having two functional groups of one type, e.g., alcohol, and the other having two complementary functional groups, e.g., a di-acid. This type of polycondensation is often referred to as an A-A, B-B type. To develop the theory relating to molecular weight control, we need to define a few new terms. Let

N_A = moles of A functional groups present

N_B = moles of B functional groups.

Then the term "imbalance ratio,"

$$r, = \frac{N_A}{N_B}.$$

If $r = 1.000$, we have perfect balance. The imbalance ratio is always expressed as unity, or less than unity, i.e., $N_B \geq N_A$. Since each monomer molecule in a linear polycondensation possesses two functional groups, the total number of monomer molecules present,

$$N = \frac{N_A + N_B}{2}$$

Since $N_B = N_A/r$, the total number of monomer molecules present expressed in terms of just N_A, is

$$N = \frac{N_A(1 + 1/r)}{2}$$

For the purposes required here, the extent of reaction, p, is defined as the fraction of A groups reacted in a given time.

The fraction of B groups reacted is therefore $= rp$.

The fraction of unreacted A groups $= (1 - p)$.

The fraction of unreacted B groups $= (1 - rp)$. (And, if $r = 1000$, the fraction of unreacted B groups would be the same as the unreacted A groups.)

The total <u>number</u> of unreacted A groups $= N_A(1 - p)$, and the total number of unreacted B groups $= N_B(1 - rp)$.

To proceed from functional group considerations to polymer chains, the total number of polymer chain ends will be equal to the sum of the unreacted groups. Since each polymer chain has two end groups, the number of polymer chains becomes equal to

$$\frac{N_A(1 - p) + N_B(1 - rp)}{2}$$

The number average degree of polymerization is given by

$$\overline{X}_n = \frac{\text{(total no. of A-A and B-B molecules initially present)}}{\text{total number of polymer molecules}}$$

$$= \frac{\dfrac{N_A(1 + 1/r)}{2}}{\dfrac{N_A(1 - p) + N_B(1 - rp)}{2}}$$

This complex fraction can be simplified to the form

$$\overline{X}_n = \frac{1 + r}{1 + r - 2rp} \qquad\qquad 20.20$$

Equation 20.20 is the generalization of the Carothers equation which places the number average degree of polymerization in terms of the extent of reaction, p, and the imbalance ratio, r.

To obtain an idea of how Eq. 20.20 functions, let us consider two limiting cases. If the two bifunctional monomers are present in exactly stoichiometric amounts, r = 1.000, and Eq. 20.20 simplifies to

$$\overline{X}_n = \frac{2}{2 - 2rp} = \frac{1}{(1 - p)}$$

That is, the number average degree of polymerization becomes dependent only on the extent of reaction, p, as in the first version of the Carothers equation discussed.

If we consider the hypothetical situation of complete polymerization (hypothetical because in practice it could never be 100.00%), then p = 1.0000 and the generalized Carothers equation becomes

$$\overline{X}_n = \frac{1 + r}{1 + r - 2r} = \frac{(1 + r)}{(1 - r)}$$

$$\text{For } r = 0.98, \overline{X}_n = \frac{1 + 0.98}{1 - 0.98} = 99$$

$$\text{For } r = 0.99, \overline{X}_n = \frac{1 + 0.99}{1 - 0.99} = 199$$

$$\text{For } r = 1.00, \overline{X}_n = \frac{1 + 1}{1 - 1} = \frac{2}{0}, \text{ i.e. an invalid solution}$$

The invalidity of the last scenario occurs because it is never possible to have 100.00% complete polymerization, in practice. However, the extent of reaction can approach very close to 1, at which time Eq. 20.20 gives a good approximation of the number average degree of polymerization to be expected (cf. Fig. 20.4).

20.8.1. Use of the Imbalance Ratio in Practice

The imbalance ratio, r, in conjunction with the generalized Carothers equation, Eq. 20.20, may be used to control either the chain length of a linear polycondensation, or the functional groups left on the ends of the polymer chains, or both product characteristics. The initial ratio of the two bifunctional monomers present could be adjusted to be nonstoichiometric to obtain the desired molecular weight. However, it is usually more convenient to adjust the ratio of the functional groups present at the start of the polymerization by adding a third, monofunctional component with the same functionality as one of the two bifunctional components, e.g.,

$$\text{A-A + B-B + B'} \text{ (added, with single functional group)}$$

How can this be accommodated by the generalized Carothers equation? First we need to redefine the imbalance ratio to take this change into account.

$$\text{imbalance ratio, } r = \frac{N_A}{N_B + 2N_{B'}} \qquad 20.21$$

This redefinition can be explained by recognizing that one B' molecule will have the same quantitative effect on limiting the growth of a polymer chain as one excess B-B molecule. Either option will serve to stop chain growth. Only a small proportion of added monofunctional component is sufficient to have a significant effect on the DP, and ultimately on the polymer molecular weight to be expected for the product. The range of results given in Table 20.2 for extents of reaction 0.98, 0.99, and the hypothetical 1.00 indicate that to obtain commercially useful DPs in the 50 to 100 range, an extent of reaction of at least 0.990 is routinely required.

20.8.2. Estimation of Molecular Weight Control in Linear Polycondensations

Suppose we wish to prepare a nylon with an M_n of 16,000 using adipic acid ($HO_2C(CH_2)_4CO_2H$) and hexamethylene diamine ($H_2N(CH_2)_6NH_2$), which represents an A-A, B-B type of polycondensation. The common, or trade name of the product, poly(hexamethylene adipamide), $HO[OC(CH_2)_4CONH(CH_2)_6NH]_nH$, is nylon 6,6, or nylon 66. The first six refers to the number of carbon atoms between amide nitrogens for the amine component, and the second 6 refers to the number of carbon atoms separating nitrogens for the acid component. What monomer feed ratio would be required to achieve this M_n?

TABLE 20.2 **Predicted Number Average Degree of Polymerization (DP) for Different Stoichiometric Ratios, r, for an A-A, B-B Type Polycondensation**[a]

Mole % excess		Stoichiometric ratio, r	DP at p=0.980	DP at p=0.990	DP at p=1.000[c]
B-B	B'[b]				
0.20	0.10	0.9990	49	95	1999
0.50	0.25	0.9975	47	89	799
1.0	0.50	0.9950	45	80	399
2.0	1.0	0.9901	40	67	201

[a]Calculated using the generalized Carothers equation, Eq. 20.20, and Eq. 20.21.
[b]For moles of A-A = moles of B-B, in the system polymerized.
[c]Limiting case.

Polymerization in this system will stop when all the remaining unreacted groups on polymer chain ends are the same. The number average degree of polymerization, is given by

$$\overline{X}_n = \frac{\text{molecular weight of polymer}}{\text{molecular weight of monomer unit}}$$

For A-A, B-B types of condensation polymers, the molecular weight of the monomer unit is defined as being one-half of the molecular weight of the repeating unit (the portion of the formula within square brackets).

Formula of repeating unit: $C_{12}H_{22}N_2O_2$
Molecular wt. of the repeating unit = 226.32
$\therefore M_0$ (mol. wt. of monomer unit) = 113.16
$$\overline{X}_n = \frac{16000}{113.16} = 141.39$$

We need to determine the imbalance ratio, r, required to obtain an $\overline{X}_n$ of 141.39 after a reasonably attainable extent of reaction. Let's say, for this example, that p = 0.995. Then, substituting this information into the generalized Carothers equation, Eq. 20.20, we obtain

$$\overline{X}_n = 141.39 = \frac{1 + r}{1 + r - 2r(0.995)}$$

Solving for r gives r = 0.99584, or 0.9960. This could be obtained either by [COOH]/[NH$_2$] = 0.9960, i.e., 4.0×10^{-3} mol excess diamine or by 2.0×10^{-3} mol of excess monoamine per mole of diamine or diacid (using r = N_A/(N_B + 2N_B)) or by [NH$_2$]/[COOH] = 0.9960, i.e., 4.0×10^{-3} mol excess diacid or by 2.0×10^{-3} mol of excess monoacid, e.g., benzoic acid, per mole of diamine or diacid in the system.

20.9. NONLINEAR POLYCONDENSATIONS

The discussion concerning polycondensation polymers so far has been restricted to systems using only bifunctional monomers. By doing this we have seen that high purity, very close to equimolar (stoichiometric) proportions of functional groups, and nearly complete reactions are needed to obtain useful high molecular weight material. The linear products from these types of polymerizations may be more or less readily melted, or dissolved in a solvent for production of useful fibers, shapes, etc., from the crude polymer resins.

If one introduces a component into a polycondensation with a functionality of 3 or more the situation dramatically changes. Branches, branched branches, and crosslinks between polymer chains will cause an increase in molecular weight more rapidly as the polymerization proceeds, and ultimately to higher values than possible from linear polycondensations (see Section 20.2). The extent to which branches and crosslinks will occur will be proportional to the ratio of the polyfunctional to the bifunctional components present, and to the number of functionalities on the polyfunctional component. Usually 3, but up to 8, functionalities per monomer may be used for this component of polyfunctional polycondensations.

The theory for estimation of molecular weight and the "gel point," or the onset of significant crosslinking and setting or gelation during the course of a polyfunctional polycondensation, is well established [2], but is beyond the scope of this outline. The need for high-purity, stoichiometric proportions of reacting functional groups, and nearly complete reaction for useful linear polycondensation products also changes when a 3 or higher functionality component is introduced into the system. These factors all still have an effect on polyfunctional polycondensations, but the polyfunctionality introduces more resilience into the system so that these requirements are not nearly as stringent for useful products as they are for strictly linear polycondensations.

REVIEW QUESTIONS

1. (a) Calculate the average degree of polymerization expected for an equimolar mixture of adipic acid and hexamethylene tetramine for each of the following extents of reaction: 0.500, 0.800, 0.950, 0.995, 0.998.
(b) What molar ratio of benzoic acid has to be added to either one of the monomers of the system of part (a) in order to limit the molecular weight of the product to 5600, at 99.5% conversion?
(c) How else could the molecular weight of the product be controlled to 5600 while still working at 99.5% conversion? Quantitatively, how should the polymerization be adjusted to achieve this?
2. (a) A polymer blend has been prepared from 5000 and 200,000 molecular weight material in a weight ratio of 1:4. What would be $\overline{M}_n$, $\overline{M}_w$, and the heterogeneity index of this blend?
(b) Answer the same questions as in part (a) for a 1:4 mole ratio of 5000 and 200,000 molecular weight material.
(c) Comment on the normalcy of the calculated heterogeneity indexes (the polydispersities).
3. (a) A polymer blend from 40,000, 80,000, and 120,000 molecular weight material is prepared on a 1:2:1 weight ratio. What would be the $\overline{M}_n$, $\overline{M}_w$, and polydispersity of this material?
(b) How "normal" is the calculated polydispersity?
(c) Explain qualitatively the reasons for the differences from a normal polydispersity calculated for the systems of Review Question 2, as compared to this system.

FURTHER READING

F.W. Billmeyer, Jr., "Textbook of Polymer Science," 3rd ed. Wiley, New York, 1984.

M.P. Stevens, "Polymer Chemistry: An Introduction," 2nd ed. Oxford University Press, New York, 1990.

REFERENCES

1. R.B. Seymour and C.E. Carraher, Jr., "Polymer Chemistry: An Introduction," 3rd ed., Chapter 4. Dekker, New York, 1992.
2. G. Odian, "Principles of Polymerization," 3rd ed., pp. 108–111. Wiley, New York, 1991.

21

COMMERCIAL POLYCONDENSATION (STEP-GROWTH) POLYMERS

Dragline silk, used by spiders . . . exhibits a combination of strength and toughness unmatched by other high-performance synthetic fibers. [Kevlar]
— David A. Tirrell, 1996

A . . . fiber twice as strong as nylon, four times as strong as rayon, eight times as strong as cotton, and pound for pound five times as strong as steel . . . Kevlar. . . .
— Eugene Magat, 1997

21.1. EARLY DEVELOPMENT

From almost zero volume in 1910, synthetic condensation polymers are currently produced on a very large scale. The products are used both for thermosetting applications, in which the final shape of the article is decided at the time of forming the crosslinks, and for thermoplastic applications where melt-forming, and melt, or solution spinning are the dominant product processing steps. Some processors are specialists, producing a large volume of a single polymer type or a single class of polymer. Others, generally larger companies, may have extensive facilities to produce several related but different polycondensation polymers at the same time as they process vinyl-type monomers to high polymer products (cf. Chap. 22, 23).

Table 21.1 gives current production data for some of the commercially important condensation polymers [1,2]. It is evident from this information that the phenolics, and the saturated polyesters and nylons destined for fiber production, are all major players with American production levels for each of these of well over a million metric tonnes per year in 1994. Also evident, at least in the American context, is the rate of growth or decline in the listed polymers. All except rayon have shown significant growth in production in the 1984–1994 period, polyolefins (not condensation polymers) at almost 10% per year. Rayon production in the United States declined about 30% during the same period. The American proportion of the world production of modified cellulosics and noncellulosic fibers production is about 8 and 25%, respectively.

■ **TABLE 21.1 Annual Production of Some Commercially Important Condensation Polymers and Related Materials**

	Canada [1] 1994	Japan [1] 1994	Germany [1] 1994	U.S.A. [2]		World [1,2] 1993
				1984	1994	
Thermosetting resins, total				2,517	3,375	
Phenolics and related		330		1,135	1,455	
Urea-formaldehyde				543	856	
Polyesters (unsaturated)	60			559	663	
Epoxies				184	271	
Melamine resins				95.3	129	
Modified cellulosic fibers, total		196		267	225	2,630
Rayon				177	124	
Acetate				90.3	102	
Noncellulosic fibers, total		847		3,389	4,287	17,100
Polyester				1,539	1,750	
Nylon				1,094	1,246	
(Olefin)[a]				452	1,091	
(Acrylic)[a]				304	200	

[a]These vinyl-type polymers are included for comparison purposes.

21.2. POLYESTER RESINS

21.2.1. Polyesters as Fibers

Polyesters in general have less intermolecular cohesion than polyamides so that poly(ethylene terephthalate) is the only polyester which is commercially useful as a fiber. This polymer can be prepared by direct polyesterification of terephthalic acid with 1,2-ethane diol (ethylene glycol), and usually requires the help of a strong acid catalyst since the process is less favorable than direct polyamidation (Eq. 19.3, 19.4). However, thermally induced side reactions of monomer or oligomer can interfere with the direct polyesterification process under the high-temperature forcing conditions required. For this reason, and because it is a faster reaction, ester interchange is usually the method of choice (Eq. 21.1, 21.2).

First step:

$$CH_3O_2C - C_6H_4 - CO_2CH_3 + 2\ HOCH_2CH_2OH \xrightarrow{105-200°C}$$

dimethyl terephthalate   1,2-ethanediol   21.1

$$HOCH_2CH_2O_2C - C_6H_4 - CO_2CH_2CH_2OH + 2\ CH_3OH$$

bis(2-hydroxyethyl)terephthalate

Second step:

$$\text{n HOCH}_2\text{CH}_2\text{O}_2\text{C} - \text{C}_6\text{H}_4 - \text{CO}_2\text{CH}_2\text{CH}_2\text{OH} \xrightarrow{\text{ca. 260°C}}$$
$$(\text{n} - 1)\ \text{HOCH}_2\text{CH}_2\text{OH} + \qquad\qquad\qquad 21.2$$
$$\text{HOCH}_2\text{CH}_2\text{O}\big[\text{CO} - \text{C}_6\text{H}_4 - \text{CO}_2\text{CH}_2\text{CH}_2 - \text{O}\big]_\text{n}$$
$$\text{poly(ethylene terephthalate), or PET}$$

Also, when the methyl ester is used, methanol removal is easier than water removal which facilitates the first step. Removal of ethylene glycol (b.p. 198°C) to help drive the second stage of the polymerization is assisted by using reduced pressure or by sparging the hot mixture with an inert gas. There is no problem with stoichiometric imbalance since the stoichiometry is established in the second stage. The excess ethylene glycol is distilled from the polymerizing mixture. As a result the poly(ethylene terephthalate), or PET, product will be obtained with entirely alcohol end groups.

Poly(ethylene terephthalate) may also be made by Schotten-Baumann condensation of terephthaloyl chloride with 1,2-ethanediol (Eq. 21.3).

$$\text{n ClCO} - \text{C}_6\text{H}_4 - \text{COCl} + \text{n HOCH}_2\text{CH}_2\text{OH} \rightarrow \text{PET} + 2\text{n HCl} \qquad 21.3$$

Using the much more reactive acid chloride monomer allows the production of polyester at much lower temperatures than when using the ester, or free acid. In fact, the polymerization under ordinary conditions is so vigorous that interfacial polymerization of diluted solutions of the monomers in an appropriate, immiscible solvent pair may be required to control rate. With this highly reactive monomer pair the kinetics and molecular weight distributions will differ from those general cases outlined earlier. Even though Schotten-Baumann polymerization is so energetically favorable, the high cost of the acid chloride as compared to terephthalic acid or the ester makes this route commercially unattractive.

Commercial PET has a melting point of about 270°C. This makes it relatively straightforward to produce a fiber from the bulk resin by melt spinning into a dry inert gas from a melt temperature of 280–300°C (Fig. 21.1). Fabrics woven from PET resin fibers include the well known Dacron, Fortrel, Crimplene, and Trevira trade names. Although fiber usage dominates PET processing, it may also be blow molded into the familiar soft drink bottles (Fig. 21.2) and made into Mylar film.

21.2.2. Polyester in Fiberglassing Resins

Polyester-based resins are also used in bulk plastics, primarily in fiberglass reinforced fabrications such as boats and auto body repairs and for much larger structures such as industrial vessels and pipes, and mine sweeping vessels [3]. This technology uses a combination of polycondensation and vinyl-type (chain reaction) technologies to obtain the final composite product. The viscous, still fluid resin used for layup is a linear unsaturated polyester of relatively low molecular weight. Unsaturation is introduced into the backbone of the polymer by using maleic anhydride to replace a part of the phthalic anhydride (e.g., Eq. 21.4).

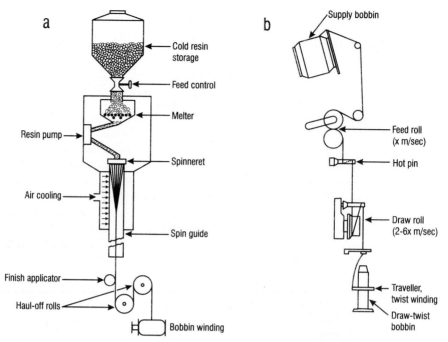

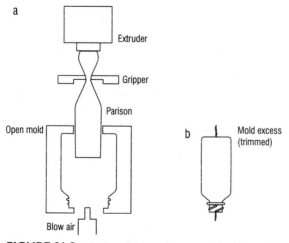

FIGURE 21.1 (a) Melt spinning of a fiber from a thermoplastic polymer. (b) The drawing process increases the degree of alignment of polymer chains into the long dimension of the fiber.

FIGURE 21.2 Outline of the machinery used for blow molding. (a) A tube of molten polymer (the parison) is clamped into the mold. Then a puff of air expands the parison to the shape of the mold, the mold is briefly chilled, and (b) the blow-molded article is ejected.

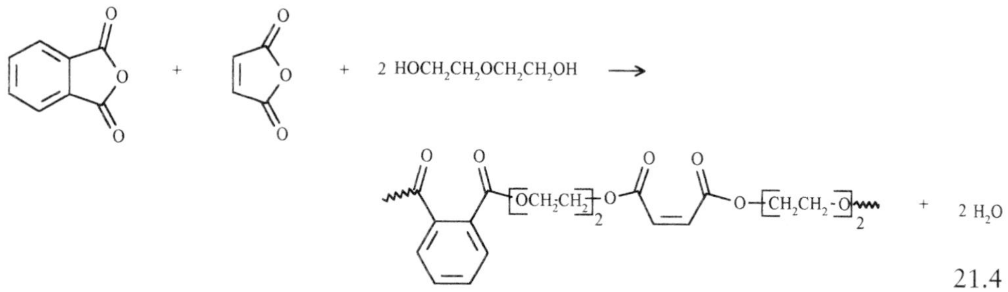

+
+ 2 HOCH2CH2OCH2CH2OH
OCH2CH2
O
CH2CH2-O
2
+ 2 H2O

21.4

The proportion of maleic anhydride used is kept low to avoid introducing too many crosslinks, which would make the product brittle. This resin, made by polycondensation, is then dissolved in styrene monomer to form the so-called "fiberglassing resin." Addition of a small proportion of peroxide initiator, the "catalyst," then initiates vinyl-type chain reaction polymerization between the styrene monomer present and the residual unsaturation from the maleic anhydride in the polyester resin (Eq. 21.5).

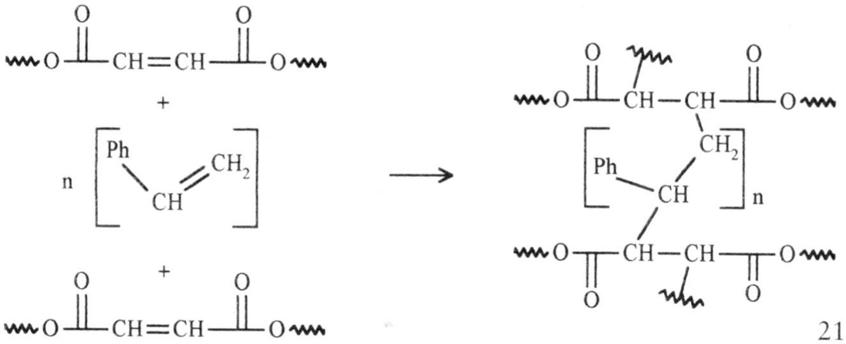

CH=CH
Ph
CH2
CH
n
CH—CH

21.5

This process very rapidly raises the molecular weight and introduces crosslinks to set the resin to a rigid, three-dimensional network polymer incorporating the glass fiber matrix. Details of the mechanism of the crosslinking process are discussed more fully in Chapter 22.

21.2.3. Polyesters in Oil-Modified Alkyds

Again employing a composite technology with some parallels to that used in the formation of fiberglassing resins, polyesters may be used to prepare durable paints and other surface coatings. In the initial step a polyol (a 3 or higher functionality alcohol), such as glycerol, or pentaerythritol is transesterified with one or more natural drying oils, e.g., linseed, tung, or cottonseed oils, or dehydrated castor oil (Eq. 21.6).

$$\begin{array}{lllll} \mathrm{CH_2OH} & \mathrm{CH_2OCOR} & & \mathrm{CH_2OCOR} & \mathrm{CH_2OH} \\ | & | & & | & | \\ \mathrm{CHOH} & +\ \mathrm{CHOCOR} & \rightarrow \text{e.g.,} & \mathrm{CHOH} & +\ \mathrm{CHOCOR} \\ | & | & & | & | \\ \mathrm{CH_2OH} & \mathrm{CH_2OCOR} & & \mathrm{CH_2OCOR} & \mathrm{CH_2OH} \\ & \text{a natural} & & & \\ & \text{drying oil} & & & \end{array} \qquad 21.6$$

A characteristic of the natural drying oils is that at least some of the carboxylic acid component of the glyceryl triesters present has to be unsaturated (Table 21.2) [4,5]. The transesterification process mixes the acids of the drying oil to form a mixture of esters among the glycerol components present and leaves some unreacted hydroxyls in the mixture.

The next stage is to take the ester mixture and react the residual alcohol functions with a di- or higher functionality carboxylic acid such as phthalic acid (or citric, or trimellitic acids; or pyromellitic dianhydride) to form a complex, already partially crosslinked polymer (Eq. 21.7).

$$
\begin{array}{cccc}
\text{phthalic} & \text{citric acid} & \text{trimellitic} & \text{pyromellitic} \\
\text{acid} & & \text{acid} & \text{dianhydride} \qquad 21.7
\end{array}
$$

This process, which involves both direct esterification and *trans*-esterification reactions, is carried to a sufficient extent of reaction to give a viscous liquid product, the oil-modified alkyd.

To formulate a paint or varnish, the oil-modified alkyd is dissolved in a solvent together with driers such as cobalt or manganese naphthenate. When applied as a coating, the driers in this solution of resin catalyze the oxidation of the residual unsaturation in the alkyd from the air, with which the resin is now in contact. This process generates radicals which serve to crosslink the

■■■■ TABLE 21.2   Typical Glyceride Content of Some Representative Drying Oils[a]

	Saturated[b]	Oleate[c]	Linoleate[c]	Linolenate[c]	Eleostearate[c]
Double bonds:	none	9,10	9,10; 12,13	9,10; 12,13; 15,16	9,10; 11,12 13,14
Geometry:	—	cis	both cis	all cis	trans, trans, cis
Cottonseed oil	25	40	35	—	—
Corn oil	11	46	43	—	—
Dehydrated castor oil	5	10	85	—	—
Linseed oil	10	18	17	55	—
Soybean oil	14	26	52	8	—
Tung oil	5	7	3	—	85

[a]Compiled from Fieser and Fieser [4] and Lange [5].

[b]Consists of varying proportions of the glycerides of the saturated n-carboxylic acids, palmitic ($C_{16}H_{32}O_2$), and stearic ($C_{18}H_{36}O_2$), which are of less importance in the film-forming functions of the oil-modified alkyds than the unsaturated acids. They serve to control the degree of crosslinking.

[c]These are all linear, 18-carbon unsaturated acids.

polymer further as the solvent dries to generate a tough, strongly adhering film. A pigment or a mixture of pigments is blended into the formulation to produce a paint coating. If a semitransparent resin is built into the oil-modified alkyd formulation instead of pigments, one obtains a varnish, a nearly transparent coating.

21.3. THE POLYAMIDES

Synthetic polyamides are almost as important as polyesters for synthetic fiber production (Table 21.1). In some respects, the structure of synthetic polyamides mimics the structure of proteins in that they, too, contain recurring amide groups. The similarity ends there, however, since proteins possess a far more complicated molecular architecture than the synthetic polyamides. Also a typical protein will be composed of various sequences of about 15 different amino acids as compared to the one or two monomer units typical of a synthetic polyamide.

The term "nylon," said to have been coined from the simultaneous announcement of this new fiber-forming material in New York and London, is a generic term used for synthetic polyamides. This name was selected as a common name for polyamides by the Du Pont Company, but apparently was never registered as a trade name. The commercialization of nylon resin and fiber formation by melt spinning grew out of the fundamental research begun by Wallace Carothers of the Du Pont Company in 1929. An early patent covered polyester, polyamide, polyanhydride, and polyacetal synthesis, and was the forerunner to a whole series of patents covering this technology [6]. Nylon 6,6 was first synthesized in 1935 and commercial production started by 1938. Initial resin production was directed entirely toward fiber production, and nylon stockings were an instant success. The high strength and abrasion resistance appealed in this application and for applications as rope, carpeting, tire cord, and fabrics. By the early 1940s moldings and engineering parts were also being produced from this resin.

The excellent properties of fiber which is melt spun from nylon 6,6 led to research by would-be competitive companies to produce similar polymers that did not infringe on the Du Pont patent. One such product, nylon 6, was developed by I.G. Farbenindustrie, a German company. Nylon 6 is a one-monomer polyamide generally made by the ring-opening cyclization of caprolactam (Eq. 21.8).

$$n\left[\begin{array}{c}\text{O}\\\\\text{N—H}\end{array}\right] \rightarrow \ \ \{(CH_2)_5CONH\}_n \qquad\qquad 21.8$$

caprolactam   nylon 6

I.G. Farbenindustrie started production of this resin in 1940. Nylon 6,6 and nylon 6 were initially the dominant commercial polyamides and remain so today.

21.3.1. Poly(hexamethylene adipamide), Nylon 6,6

For the preparation of nylon 6,6 resin, exact stoichiometry is assured by the initial preparation of the 1:1 salt, and purification of this before polymerization (Eq. 21.9).

$$\text{n } H_2N(CH_2)_6NH_2 \quad + \text{ n } HO_2C(CH_2)_4CO_2H \rightarrow$$
$$\text{hexamethylenediamine} \qquad\qquad \text{adipic acid}$$

$$\text{n} \left(\begin{array}{c} {}^-O_2C(CH_2)_4CO_2{}^- \\ + \qquad\qquad + \\ H_3N-(CH_2)_6-NH_3 \end{array} \right) \qquad\qquad 21.9$$

1:1 nylon salt

The polyamidation is sufficiently favorable under the conditions used that no catalyst is required. Initially a slurry of 60–80% of the 1:1 nylon salt in water is heated to about 80–90% reaction which is achieved at 200°C and 15 atm pressure (Eq. 21.10).

$$\text{n 1:1 nylon salt} \rightarrow H{+}NH(CH_2)_6NHCO(CH_2)_4CO{+}_nOH$$
$$\text{nylon 6,6} \qquad\qquad 21.10$$
$$+ (2n - 1) \, H_2O$$

The elevated pressure used at this stage helps to retain the presence of the water, which aids in heat transfer and mixing. To complete the polymerization the temperature is gradually raised to 270–300°C while steam is gradually released to allow the pressure to drop to normal atmospheric. This last step is referred to as melt polymerization because it is carried out at temperatures above the 250°C melting point of nylon 6,6 polymer to help drive the polyamidation to completion. The molten condition also assures continuance of good heat transfer and mixing conditions during this period.

Sometimes the production of particularly high molecular weight material is aided by evacuating the reaction vessel to assist in the removal of water. Both batch-type polymerization, in a jacketed polymerization vessel, and continuous polymerization methods, in a long tube with zones which provide the appropriate conditions, are used for polyamidation of the 1:1 salt.

21.3.2. Polycaproamide, Nylon 6

Batch and continuous processes are also employed for the preparation of nylon 6 resin. To initiate the process, caprolactam and water are heated together at about 250°C for a period of 10-12 hours in an inert atmosphere. The ring-opened caprolactam product, 6-aminohexanoic acid (Eq. 21.11),

$$n \quad \text{[caprolactam ring]} \quad N\text{—}H + H_2O \rightarrow H_2N(CH_2)_5CO_2H \qquad 21.11$$

then reacts rapidly with unchanged caprolactam to first form oligomers and then true polymers from this AB-type single monomer (Eq. 21.12).

$$n \quad \text{[caprolactam ring]} \quad N\text{—}H + H_2N(CH_2)_5CO_2H \rightarrow H\text{+}NH(CH_2)_5CO\text{+}_{n+1}OH$$

$$21.12$$

A small tendency for the 6-aminohexanoic acid to recyclize gives a crude product consisting of about 90% useful high molecular weight resin and 10% caprolactam and oligomers. To obtain optimum properties this crude product is either heated at 180–200°C in an evacuated state to form high polymer from the residual oligomers, or the mixture may be subjected to prolonged leaching with water at 80–85°C to dissolve out the somewhat soluble caprolactam and oligomers from the product. Finished nylon 6 resin has a melting point of 223°C, still high enough to be useful but somewhat lower than nylon 6,6. This lower melting point is probably because a higher degree of order is required from nylon 6 polymer chains to obtain equivalent interchain nonbonded interactions as those obtained with nylon 6,6 (try sketching H bonds).

21.3.3. Other Aliphatic Polyamides

The three other commercially important polyamides are nylon 6,10, nylon 11, and nylon 12. Nylon 6,10 is prepared from the purified 1:1 salt, hexamethylenediammonium sebacate, in a manner parallel to that described for nylon 6,6. The irregular intervals between linking amide units, 6 carbons between amide nitrogens staggered with the 10-carbon diacid units, confers a somewhat lower interchain cohesion to the product. As a result nylon 6,10 has a slightly lower melting point of 215°C.

Nylon 11 is made by the ring-opening cyclization of 11-aminoundecanoic acid at 200–220°C in a manner similar to that described for nylon 6. Removal of water in the later stages is facilitated by reduced pressure. Since the ring-chain equilibrium for the 12-membered ring lactam is much less favorable for the ring than is the case for the seven-membered lactam used to prepare nylon 6, there is only about 0.5% of the residual lactam left in the final nylon 11 product. This percentage of monomer has a minimal effect on the properties of the final resin, so is normally left in the polymer.

Nylon 12 is also produced by ring-opening polymerization, this time of the 13-membered ring dodecyl lactam. Ring opening is catalyzed by phosphoric acid and requires finishing temperatures of up to 300°C to complete the polymerization. Yields are almost quantitative so the finished polymer may be used directly. This polyamide, too, has a relatively low melting point and lower tensile strength in fibre or engineering applications than nylon 6,6 or nylon 6, like the other aliphatic polyamides described in this section (Table 21.3) [7–9]. It is interesting to note that nylon 6,6, the first of the aliphatic polyamides to be commercialized, also has the best properties for many purposes.

21.3.4. Recent Polyamide Developments

A few more recent introductions have been made to the commercial polyamide market. Nylon 4,6 is produced by replacing the hexamethylenediamine component of nylon 6,6 by 1,4-diaminobutane. This yields a regular repetition of four methylene groups between every pair of amide links, in contrast to the 4,6,4,6 alternation of methylene numbers as present in nylon 6,6 chains (Eq. 21.13).

$$\text{n } H_2N\text{—}(CH_2)_4\text{—}NH_2 + \text{n } HO_2C\text{—}(CH_2)_4\text{—}CO_2H \rightarrow \qquad 21.13$$

$$H\text{$\{$}NH\text{—}(CH_2)_4\text{—}NH\text{—}CO\text{—}(CH_2)_4\text{—}CO\text{$\}$}_nOH + (n-1)\ H_2O$$

This more regular repetition of amide functionalities together with the greater amide concentration in the polymer gives a greater interchain cohesion for nylon 4,6, raising the melting point to almost 300°C and enabling a rapid maximum crystallization rate of up to 8 per second, if desired [10] (cf. Table 21.3). Nylon 6,12, nylon 13, and blends of these resins with polyolefins, with and without fiberglass reinforcing filler were also introduced in the mid 1980s [11].

Copolymers of the homopolymer polyamides described may be obtained by heating a blend of the homopolymers above its melting point. A short

TABLE 21.3　Typical Properities of Commercial Grades of Polyamides (Nylons)[a,b]

Nylon type	Specific gravity	Melting point (°C)	Tensile strength (MPa)[c]	Elongation at break (%)
4,6	1.18	290	89	—
6,6	1.14	255	77–80	300
6	1.12–1.15	223	70–80	300
6,10	1.09	215	58–60	100–150
6,12	1.06	212	61	~300
11	1.04	185	54	330
12	1.01	175	52	250
6,6–6,10[d]	1.08	195	38	200

[a]Compiled from Brandrup and Immergut [7], Saunders [8], and Zimmerman [9].
[b]Special postspinning heat treatments and orienting by drawing can significantly alter properties.
[c]MPa ≡ N/mm². To convert to pounds per square inch, multiply by 145.
[d]A 35:65 by mass copolymer.

heating period produces mostly block copolymer. After 1–2 hr of heating, amide interchange proceeds further and random copolymer results. The same copolymers may be obtained by heating the appropriate constituent monomers. The properties of commercial 6,6–6,10 copolymer, and 6,6–6,10–6 terpolymer lie at values intermediate to those of the constituent homopolymers. This is an important factor to consider in, for example, the recycling of sometimes misidentified polyamide which may result in the accidental reprocessing of a mixture of nylon resins. This occurrence will not completely waste a batch of nylon 6,6, or nylon 6, although it will cause some lowering of the melting point and the tensile strength below the values of the nylon 6,6 or nylon 6 components. More background is given on this aspect in Section 21.7.

21.3.5. The Aromatic Polyamides or Aramids, Nomex and Kevlar

Also a relatively recent development in this technology is the use of aromatic diamines and diacid derivatives to form polyamide chains of inherently increased stiffness and order [12]. These fundamental contributions to chain geometry confer both a much higher melting point to the resin and much higher tensile strength to spun fiber than is possible with the aliphatic nylons.

Initial research in this area, by Stephanie Kwolek working at Du Pont in 1965, determined that p-aminobenzoic acid could be polymerized, and the product solution spun to produce fibers with superior properties [13]. Experience gained with this polymer system clearly established the value of the highly ordered fibers which could be produced by this means [14]. Unfortunately the high cost of the p-aminobenzoic acid monomer discoraged pursuit of this approach to the aramids.

Nomex aramid, made from m-phenylenediamine and isophthaloyl chloride, produces poly(m- phenylene isophthalamide) with a melting point of 380–390°C, and substantially avoided the problems of high monomer cost (Eq. 21.14) [15].

$$+ (n-1)HCl \qquad 21.14$$

Nomex fiber chars rather than burns at its melting point. This resistance to melting or burning has led to extensive use of Nomex fiber for the production of protective clothing for fire fighters, racing car drivers, petroleum industry workers and any others whose occupation exposes them to high fire risks [15]. It is also extensively used in a premium insulating paper for electrical equipment.

Kevlar (Du Pont) and Twaron (Enka/Akzo) are trade names for the all *para* aramid, poly(p-phenylene terephthalamide), produced by these companies (Eq. 21.15).

$$\text{n } H_2N-C_6H_4-NH_2 + \text{n } ClCO-C_6H_4-COCl \xrightarrow[\text{N-methylpyrrolidone}]{\text{CaCl}_2 \text{ in solvent}}$$

$$H\!\!+\!\!NH-C_6H_4-NH-CO-C_6H_4-CO\!\!+_n\!OH + (n-1) \text{ HCl}$$

$$21.15$$

The melting point of the product is so high that it cannot be conveniently melt spun. However, Kevlar resin may be spun from a 20% solution of the resin, in hot 100% sulfuric acid, which is solution spun into water [12,16]. The extreme conditions required for resin production and fiber spinning contribute to the high cost of the fiber. However, the ability of Kevlar cable to match steel cable, strength for strength in the same diameter and with 20% of the weight, confers the wide utility of Kevlar fiber in many demanding applications. It has a tensile strength of 3800 MPa, nearly 500 times that of the aliphatic nylons. As a result it is used in demanding applications such as cable for the anchoring of offshore drilling platforms and other shipping applications where its light weight and use without the need for grease make for easier, cleaner handling without the risk of corrosion inherent with steel. Kevlar fiber is also used for applications such as the fabrication of premium tire cord, as a reinforcing matrix for many composite resins in the production of sporting goods, bulletproof vests, aircraft components, etc. [17]. The current estimated world *para*-aramid fiber market is estimated to be 20,000 tonnes/year, split about two-thirds to Kevlar and one-third to Twaron [18].

21.4. POLYURETHANES

Initial work in the development of polyurethane technology, like nylon 6 development, was also stimulated by an interest in producing a nylon 6,6 fiber type by a polymer chemistry that did not infringe on the Du Pont patents. Fiber from the polyurethane produced from hexamethylene diisocyanate and 1,4-butanediol did have properties similar to nylon 6,6 fiber, but high monomer cost, particularly of the diisocyanate, made this commercially unattractive (Eq. 21.16).

$$\text{n } OCN(CH_2)_6NCO + \text{n } HO(CH_2)_4OH \rightarrow$$
$$HO\!\!+\!\!CONH(CH_2)_6NHCO-O(CH_2)_4O\!\!+_n\!H$$

$$21.16$$

However, during these exploratory experiments with various related monomers other polyurethanes were discovered with novel properties which were commercially attractive. Fiber, coating, molding and cushioning, and insulating applications for some of these materials are now exploited on a large scale.

The initial step in the production of cast elastomers is to terminate a low molecular weight difunctional polyether or polyester by a mixture of toluene 2,4- and 2,6-diisocyanate using an excess of the diisocyanate (Eq. 21.17).

$$21.17$$

Mixtures of isocyanates are commonly used for convenience in commercial production of the diisocyanate, since the pure toluene 2,4-diisocyanate is more expensive to produce. The resulting prepolymer is then mixed with either a glycol, such as 1,6-hexanediol, or a deactivated (sterically hindered) diamine plus pigment if required, and then promptly poured into a preheated mold of the desired shape. In a half hour or so the mixture sets to a pliable shape with stiffness and elasticity controlled by the components used and processing details [19]. Similar procedures are used to produce high-strength polyurethane fiber, e.g., Perlon U, or elastomeric fibers, e.g., Spandex and Lycra.

Rigid polyurethanes employ a 3 or higher functionality polyglycol of 2000-3000 molecular weight, e.g., made from glycerol and ethylene oxide (Eq. 21.18).

The oligomeric branches will be approximately the same length. Reacting this polyglycol with toluene diisocyanate rapidly yields a rigid, branched product. As would be expected from the background theory this approach should yield high molecular weight material more rapidly than methods employing only bifunctional components, together with a degree of crosslinking controlled by the functionality and molecular weight of the polyglycol component selected. Choice of glycerol gives a trifunctional component, as in this example. Pentaerythritol would contribute tetrafunctionality, and use of sucrose to form the polyglycol would give a functionality of 8 to this component.

21.4.1 Urethane Coatings

A mixture of intermediate molecular weight polyesters and natural drying oils (Table 21.2) is reacted with toluene diisocyanate to produce a viscous resin. A solution of this resin in solvent together with driers, and pigments if desired, comprises the basic clear or colored polyurethane coating. The driers, generally transition metal complexes, are added to the coating formulation to catalyze the crosslinking reactions of the natural drying oil component on exposure to air in a manner similar to that described for the oil- modified alkyds.

21.4.2 Polyurethane Foams

In the polyurethane applications described so far, care has to be taken to exclude moisture. Any water which is present will react with isocyanate groups to give carbon dioxide gas, which can form bubbles in the product (Eq. 21.19).

$$R\sim NCO + H_2O \rightarrow R \sim NHCOOH \rightarrow R \sim NH_2 + CO_2 \uparrow$$

isocyanate   a carbamic   primary   21.19
                              acid (unstable)   amine

This occurrence significantly affects the stoichiometry, not only from consumption of the isocyanate functionality in the process, but also by production of an amine product with two active hydrogens which adds reactive functional groups to the "glycol" component of the polyurethane reaction. For these reasons care has to be taken to exclude moisture for the polyurethane applications described so far. However, for the production of flexible and rigid polyurethane foams the reaction of isocyanate with water is exploited to produce bubbles.

The "one-shot" methods used to produce flexible polyurethane foams employ quick mixing of a (usually) triol-based polyether of fairly high molecular weight with toluene diisocyanate, catalyst, and water for gas production (Eq. 21.19). The reaction of water with the diisocyanate rapidly raises the average functionality in the polymerizing system by forming urea, as well as urethane links (Eq. 21.20).

$$OCN\sim R\sim NH_2 + OCN\sim R'\sim NCO \rightarrow \qquad\qquad 21.20$$
$$OCN\sim R\sim NHCONH\sim R'\sim NCO$$
a urea link

As a result a viscous polymer is formed which traps the carbon dioxide as it is generated. A continuing increase in molecular weight, together with a greater degree of crosslinking which becomes more important as the reaction proceeds, rapidly brings about a flexible set to the foam. By varying the details of the formulation and temperatures used, flexible foams may be obtained with densities of 15–70 kg/m^3, small or large and open or closed (low water uptake, or low gas exchange) cells, and in a variety of stiffnesses and tensile strengths [20]. Applications include furniture and crash padding, bedding and blanketing, and car seat cushioning.

Rigid polyurethane foams are produced by analogous technology to the flexible foams except for two aspects. Higher functionality, or a somewhat lower molecular weight, will be used in the polyether or polyester component. Either change effectively serves to increase the frequency of crosslinking in the product. Also, lower cost bubble production is achieved via introduction of a low boiling solvent such as Freon 11 (Cl$_3$CF, b.p. 24°C.), or carbon dioxide gas directly, rather than by the isocyanate-water reaction. This conserves the more expensive isocyanate component for the formation of crosslinked polymer to trap the gas or vapor, as it is generated. Since 97% of the insulating capability of the rigid polyurethane is the consequence of the voids present, use of a low thermal conductivity gas for this purpose greatly improves the insulating capability of the final foam. Freon-11 and carbon dioxide have about one-third and two-thirds of the thermal conductivity of air, respectively, which makes these such good candidates for foam generation. The

TABLE 21.4 Breakdown of Polyurethane Demand in Selected Areas, in Thousands of Metric Tonnes[a]

| | Canada 1981 | Latin America 1981 | United States[b] | | |
			1981	1995[b]	1996
Flexible foam	55.3	147.8	553	(844)	927[c]
Rigid foam	22.2	21.4	263	(449)	575
Elastomers	—	—	77	—	92[d]
Surface coatings	10.0	26	40	—	140
Adhesives, sealants	—	—	59	—	238[e]
Other[f]	—	—		—	118
Total	87.5	195.2	992	—	2090

[a]Data from Upjohn Polymer Chemicals [21] and Stinson [22]. The total European polyurethane market was estimated to be 1.41 million metric tonnes in 1988 with the largest use sectors being transportation, 27.8%; furniture, 21.4%; buildings, 16.6%; appliances, 5.4%; and footware, 5.2% [23].
[b]Projections for 1995 given in parentheses [24]. In the same year it was expected that a total of 669,000 tonnes of polystyrene foam products would be produced in the United States.
[c]Includes slab and molded foam products.
[d]Includes cast elastomers and fibers (Spandex).
[e]Includes binders and fillers.
[f]Includes molded thermoplastics, and automotive and non-automotive reaction injection molded products.

excellent insulating capability of these materials leads to their use in modern, thin-walled refrigerators and freezers, foamed-in-place insulation for freezer-truck bodies and housing, and for flotation billets used in floating docks, etc. A representative breakdown of polyurethane usage according to type is given in Table 21.4 [21–24].

21.5. EPOXY RESINS

As a group this class includes a number of examples of condensation polymers in which crosslinking is obtained by the reactions of epoxide groups. The products are primarily of use as highly effective adhesives for cementing of rigid materials, and as potting compounds for the physical protection or encapsulation of electronic devices and the like.

Two components are required, which are mixed just prior to use. The prepolymer, in which the crosslinkable epoxide functions reside, is made from epichlorohydrin and *bis*-phenol A. Epichlorohydrin in turn is derived from allyl chloride (Eq. 21.21).

$$CH_2=CH-CH_2Cl \xrightarrow{\text{HOCl}} HOCH_2CH(Cl)-CH_2Cl \rightarrow \qquad\qquad 21.21$$

$$\underset{\text{epichlorohydrin}}{\overset{CH_2-CH-CH_2Cl}{\underset{O}{\diagdown\diagup}}}$$

bis-Phenol A is produced from the acid-catalyzed reaction of phenol with acetone (Eq. 21.22).

$$\text{2 } C_6H_5OH + CH_3COCH_3 \xrightarrow{\text{H}^+} \text{HO}\!-\!\!\bigcirc\!\!-\!\!\underset{CH_3}{\overset{CH_3}{C}}\!\!-\!\!\bigcirc\!\!-\!\text{OH} \qquad 21.22$$

bis-phenol A, m.p. 158°

Combining an excess of epichlorohydrin with the *bis*-phenol A (4,4'-isopro-pylidenediphenol) gives the required prepolymer component of epoxy resin systems (Eq. 21.23).

$$\text{4n } CH_2\text{-}CH\text{-}CH_2Cl \quad + \quad \text{n HO}\!-\!\!\bigcirc\!\!-\!\!\underset{CH_3}{\overset{CH_3}{C}}\!\!-\!\!\bigcirc\!\!-\!\text{OH} \quad \longrightarrow \qquad\qquad 21.23$$

$$CH_2\text{-}CH\text{-}CH_2\!\!\left[\!O\!-\!\!\bigcirc\!\!-\!\!\underset{CH_3}{\overset{CH_3}{C}}\!\!-\!\!\bigcirc\!\!-\!O\text{-}CH_2\text{-}\underset{OH}{CH}\text{-}CH_2\text{-}O\!-\!\!\bigcirc\!\!-\!\!\underset{CH_3}{\overset{CH_3}{C}}\!\!-\!\!\bigcirc\!\!\right]_n\!\!O\text{-}CH_2\text{-}CH\text{-}CH_2$$

epoxide prepolymer

glycidyl ether
end group

Epoxide end groups (glycidyl ether) in the polymer are assured by the use of an excess of epichlorohydrin at this stage.

    There are two options for the other component of an epoxy resin system. Use of mono- or di-anhydrides as curing agents, usually catalyzed by a tertiary amine, causes reactions with the residual secondary hydroxyls in the repeating unit of the prepolymer, forming esters and free carboxylic acids. The carboxylic acids formed also react with the epoxide end groups forming crosslinks and further free secondary hydroxyl groups. Maleic anhydride, phthalic anhydride, or pyromellitic dianhydride are suitable for this process (Eq. 21.24).

$$\qquad\qquad\qquad\qquad 21.24$$

However, the reactions between any of these anhydrides and the prepolymer are sluggish at room temperature so that heating is required to enable the crosslinking reactions to proceed at a reasonable rate. This is relatively easy to arrange for small-scale epoxy resin castings and for encapsulation of electronic components (the "potting compounds") which are the primary end uses of this class of epoxy resins.

    For adhesive applications it is more convenient to have the two components of the epoxy resin system react at a reasonable rate at room temperature. Replacing the anhydride component by a suitable, more reactive polyfunctional amine enables this to occur at room temperature. Several aliphatic and aromatic polyfunctional amines are used for this purpose (Eq. 21.25).

$$H_2N-(CH_2)_2-NH-(CH_2)_2-NH_2$$
diethylene triamine

$$H_2N--CH_2--NH_2$$

4,4′-diaminodiphenylmethane   m-phenylenediamine

21.25

The so-called "five minute epoxy" employs a highly active amine that enables setting in five minutes and substantial strength in an hour under ambient conditions. Recent production of epoxy resins in the United States has totalled 134,000 metric tonnes per year [25].

21.6. PHENOL-FORMALDEHYDE RESINS

The constituents of phenol-formaldehyde resins owe their polymer-forming capability to the trifunctional potential of phenol and the bifunctional capability of formaldehyde. Delocalization of the electron pairs on the oxygen of phenol into the aromatic ring produces electron-rich sites at the *ortho* and *para* positions. This delocalization activates these sites, producing the potential trifunctionality of the phenol. The phenol-formaldehyde reaction is usually carried out in aqueous solution. Under these conditions formaldehyde adds water in an equilibrium reaction to form formaldehyde hydrate (Eq. 21.26).

$$H_2C{=}O \quad + H_2O \rightleftharpoons \quad HOCH_2OH$$
formaldehyde   formaldehyde hydrate

21.26

Despite the favoring of formaldehyde hydrate by this equilibrium, there is sufficient free formaldehyde available for an *ortho* or *para* position of phenol to add to the highly electrophilic carbon of formaldehyde. As formaldehyde is consumed by this process, the equilibrium is displaced to the left providing further formaldehyde for reaction until all the phenol potential functionalities are taken up or all the formaldehyde is consumed. The structures of the phenolformaldehyde polymers produced are difficult to study because the final product is infusible and insoluble. However, current thinking is that all possibilities for monomer links are thought to exist in a typical Bakelite sample (Eq. 21.27).

The phenol-formaldehyde resin system has such a high potential for crosslinking that in practice these monomers could rapidly yield an intractable mass. Strategies have to be adopted to achieve the control desired. Reagent imbalance may be used to effect control. This strategy is normally used for polymerizations carried out under acidic conditions to initially produce oligomeric linear prepolymer. Polymerization is completed at a later stage by heating the prepolymer with a component to contribute formaldehyde to the system. Or, alternatively, the reaction in bulk may be stopped by dehydration well before completion, and then completed in molds of the desired final fabrication shape.

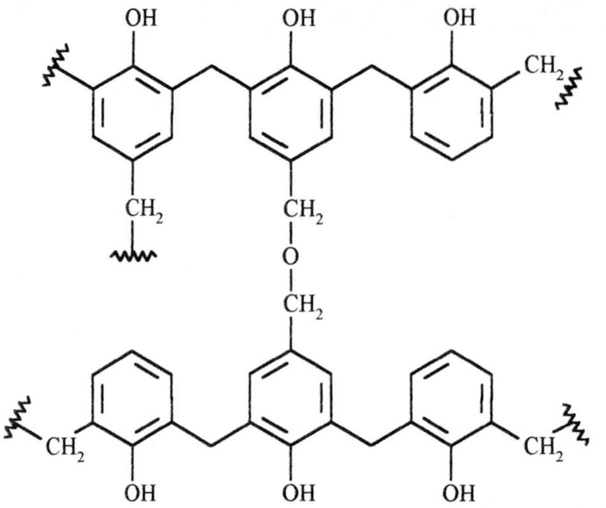

21.27

+   n H$_2$O

representative segment of phenol-formaldehyde polymer

Polymerization under acidic conditions, the first control scenario mentioned, employs an initial formaldehyde to phenol ratio of 0.75–0.85 to 1. Stoichiometric ratio would be 1.5 to 1. Catalysis is achieved by using a strong acid; 0.1–0.3%, hydrochloric acid:phenol ratio, or 0.5–2% oxalic acid to phenol ratio. The non-stoichiometry restricts polymerization to the oligomer stage and avoids crosslinking (Eq. 21.28).

$$H_2CO + H_3O^+ \rightleftharpoons H_2\overset{+}{C}OH \overset{PhOH}{\longrightarrow}$$

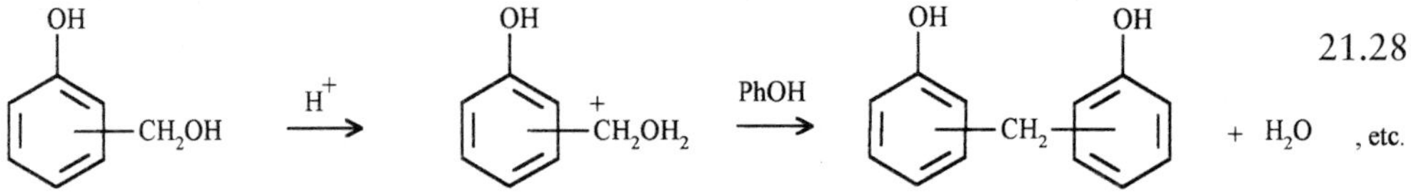

21.28

, etc.

These prepolymers, called novolacs, are dehydrated, reduced to a powder, and then blended with about 10% by weight hexamethylenetetramine ("hexa", Eq.21. 29).

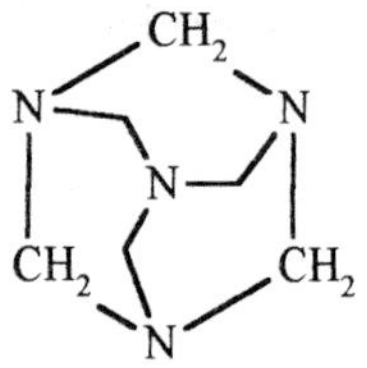

21.29

Hexa is made by reacting 4 moles of ammonia with 6 moles of formaldehyde. When the powder mixture is placed in a mold of the desired shape and heated, rapid polymerization and crosslinking of the novolac occurs by the formation of methylene (mostly) with a few benzylamine bridges.

Polymerization under alkaline conditions, using about 1% sodium hydroxide catalyst based on the weight of the phenol, proceeds via a somewhat different mechanism. Methylol phenols, and oligomers consisting of 5 or 6 phenol units connected by methylene bridges, result in the so-called resol resin prepolymer (Eq. 21.30).

methylol phenols   representation of resole prepolymer

$$21.30$$

Resole resins may be converted into high molecular weight, crosslinked materials simply by heating. For this reason the resol process is sometimes referred to as the "one-stage" process to phenol-formaldehyde resins. With this process the potential for the formation of high molecular weight, crosslinked material exists right from the start. So care has to be exercised to avoid gelation or solidification via crosslinking of the product in the primary reactor. Mechanistic details have been presented, and recent detailed characterizations of resol resins have been published [26,27].

The single largest use for the phenol-formaldehyde resins is in adhesive applications for the production of plywood, chipboard, and particle board. The resin can comprise as much as one-third of the weight of the board, particularly of particle boards, which contributes to a total demand for phenolics in the United States of over half a million metric tonnes per year. They are also used as the adhesives for the production of several types of grindstones. In combination with paper, woven cotton, glass fiber, etc., components, phenolics contribute to the production of engineering and decorative laminates in the form of rods, tubes, and sheet. The sheet products Arborite and Formica are familiar as the finished surfaces of furniture, bathroom, and kitchen counter tops and other areas where attractive patterns and water resistance are important characteristics. Molded products from phenolics are also important where heat or electrical resistance is required in applications such as saucepan and toaster handles, switches, and the printed circuit boards used in computers. Recent phenolics production in the United States for these applications has totaled over 500,000 metric tonnes per year, not including fillers [28].

21.7. ENVIRONMENTAL AND RECYCLE ASPECTS

Not long after the large-scale production of plastics, interest and motivation for reuse and recycling of these materials developed, the details of which continue to be promoted and refined [29,30]. These cover aspects relating to in-plant recycling of all types of polymer scrap (producer recycling) as well as all aspects of postconsumer recycling. They extend from applications such as drainage tile production from polyethylene milk bottles, to the conversion of used polyethylene terephthalate (PET) soft drink bottles into PET fiber for carpet production, sleeping bag insulation, and the like [31].

More specifically, a case study which outlines a vapor combustion approach to the handling of an odor problem during polyester primary production could also be usefully applied to other polymer producing facilities [32]. Recently it has been established that recycled PET resin may be reused without substantial loss of grade under the appropriate conditions, even for applications involving food contact [33,34].

Operating details for more benign batch reactors for nylon production have been studied [35]. Perhaps of greater concern, however, is the significant contribution to atmospheric nitrous oxide from the large-scale use of nitric acid to produce nylon monomers (Eq. 19.56) [36]. Nylon resins may be recycled in the form in which they are received, generally with a lowering of the grade. Thus, nylon engineering parts may be recycled into less demanding carpet fiber applications or employed in multilayer film production [37,38]. Or the resins, segregated as to type as far as possible, may be subjected to hydrolysis back to the monomers, the monomers purified, and then repolymerized to a resin which is indistinguishable from virgin material. Although hydrolysis-based recycling is possible for both nylon 6 and nylon 6,6, the two largest scale commercial polyamides, it is much more straightforward for nylon 6 because only a single monomer results.

An interest in reduced use of chlorofluorocarbon blowing agents (e.g., Freon-11, CCl_3F) for production of rigid polyurethane foams has lead to the study of suitable alternatives such as carbon dioxide [39,40]. Even though the rigid polyurethane foams are predominantly used in closed systems, eventual chlorofluorocarbon loss to the atmosphere is still almost certain in this application when the appliance is discarded. Usage precautions have been assembled [41]. Polyurethane scrap can be reusable, despite the crosslinked nature of the resin, by a variety of processes [42,43]. The polyol component may be recovered, and the diisocyanate(s) in the form of the corresponding 2,4- and 2,6-toluenediamines using hydrolytic techniques at temperatures of 190–250°C [44,45].

Destructive distillation or anaerobic pyrolysis may be used, as a last expedient, to derive fuel or feedstock values from low grades of recovered plastics. This option is more often used for vinyl-type polymers, some of which can yield fairly uniform product by this means. This option is discussed more fully in Chapter 23 with the other aspects of these polymers.

REVIEW QUESTIONS

1. (a) Draw the repeating unit obtained when the following monomers are polymerized: adipic acid and hexamethylene diamine; p-carboxybenzyl alcohol; terephthalic acid and 1,2-ethanediol; caprolactam.

(b) Name each of the polymers in part (a) using a systematic and a commercial name.

2. Titration of a solution of a 38.6-g sample of nylon 6,6 with base indicates that the sample contains 3.11×10^{-3} moles of carboxyl groups. Calculation using this information gives an $\overline{M}_n$ of 12,400 for the sample.
   (a) What is the assumption required to obtain this result?
   (b) What further experimental information is necessary in order to verify or correct this result? Explain the need and describe a feasible method.

3. (a) What would be an experimentally feasible approach to the synthesis of the following random copolymers?

$$\left(O - (CH_2)_2 - O - (CH_2)_2 O (CH_2)_2 - O \right)_x (CO - (pC_6H_4) - CO)_y$$
$$\left(NH - (pC_6H_4) - CO\right)_x - \left(NH - (CH_2)_{10} - CO\right)_y$$

   (b) How should the procedure be altered to produce the equivalent block copolymers of any particular designated block length for the structures of part (a)?

4. Drastic hydrolysis of a sample of a nylon of number average molecular weight 20,575 yields 24.07 g of 1,4-diaminobutane, 39.51 g of adipic acid, and 0.32 g of benzoic acid.
   (a) Draw the repeating unit of this polymer.
   (b) What would be the systematic and the commercial names for this polymer?
   (c) What would be the number average degree of polymerization and the extent of reaction used to prepare this polymer?
   (d) What number average molecular weight product would be expected for the same extent of reaction if the proportion of benzoic acid used was halved?

5. What are the incentives to a nylon reprocessor to keep the various types of nylon segregated in their collection and recovery system? Explain.

6. What is a probable outcome of the thermal reprocessing of a recovered mixture of nylon resins? Discuss options.

7. (a) Discuss the factors that might dictate the need to hydrolyze a polycondensation polymer to constituent monomer(s), to permit useful recycling.
   (b) Why might this be a less attractive option than cleaning and reextrusion of the resin itself?

8. Why might one monomer nylon and one monomer polyester polymer be easier to reprocess to the original monomer than those condensation polymers which contain two or more monomers?

FURTHER READING

J.L. Hedrick and J.W. Labadie, eds.,"Step-Growth Polymers for High Performance materials: New Synthetic Methods," ACS Symp. Ser., No. 624. American Chemical Society, Washington, DC, 1996.

REFERENCES

1. Polymer Production Jumpd 7.2% in 1994, *Chem. Eng. News* **73**(15), 20, April 10 (1995).
2. Facts and Figures for the Chemical Industry, *Chem. Eng. News* **73**(26), 36, June 26 (1995).
3. Platic Warship Launched, *Chem. Br.* **18**(5), 334 (1982).
4. L.F. Fieser and M. Fieser, "Advanced Organic Chemistry," p. 990, Reinhold, New York, 1961.
5. N.A. Lange, ed., "Handbook of Chemistry," 10th ed., p. 792, McGraw-Hill, New York, 1969.
6. W.H. Carothers, Linear Condensation Polymers. U.S. Pat 2,071,250 (to the Du Pont Company) (1937).
7. J. Brandrup and E.H. Immergut, eds., "Polymer Handbook," 3rd ed., pp. V/113–114, Wiley, New York, 1989.
8. K.J. Saunders, "Organic Polymer Chemistry," p. 186, Chapman & Hall, London, 1973.
9. J. Zimmerman, Polyamides. *In* "Encyclopedia of Polymer Science and Engineering," H.F. Mark *et al.,* eds., 2nd ed., Vol. 11, pp. 315–381, Wiley, New York, 1985.
10. D. O'Sullivan, Conventional Nylons Encounter Strong New Competitor in Nylon 46. *Chem. Eng. News* **62**(21), 33, May 21 (1984).
11. S. Stinson, Nylon Impact Resistance Tailored to User Need. *Chem. Eng. News* **64**(38), 39, Sept. 22 (1986)
12. D. Tanner, J.A. Fitzgerald, and B.R. Phillips, The Kevlar Story—An Adanced Materials Case Study. *Angew. Chem., Int. Ed. Engl. Adv. Mater.* **28**(5), 649 (1989).
13. S.L. Kwolek (Du Pont), Br. Pat. 1,283,064 (1972); U.S. Pat. Reissue 30,352 (1980).
14. K.A. Hodd and D.C. Turley, High Modulus Fibres. *Chem. Br.* **14**(11), 545 (1978).
15. H.J. Sanders, Flame Retardants. *Chem. Eng. News* **56**(17), 22, Apr. 24 (1978).
16. H. Short, A First Look at Aramid Processing. *Chem. Eng.* **96**, 37, Apr. (1989).
17. P.L. Layman and M.S. Reisch, DuPont, Akzo Resolve Long Hassle Over Aramid Fiber Patents. *Chem. Eng. News,* **34**(9), 10, Aug. 22 (1988).
18. Anonymous, Akzo Nobel's Bulletproof Strategy, *Chem. Ind. (London),* p. 744, Oct. 3 (1994).
19. G. Woods, "The ICI Polyurethanes Book, ICI Polyurethanes." Wiley, Chichester and New York, 1990.
20. G. Woods, "Flexible Polyurethane Foams: Chemistry and Technology." Appl. Sci. Publ., London, 1982.
21. Upjohn Polymer Chemicals, Polyurethane 81. *Chem. Eng. News* **59**(23), 55, June 8 (1981).
22. S. Stinson, Polyurethane Use Continues to Grow. *Chem. Eng. News* **75**(31), 22, Aug. 4 (1997).
23. Anonymous, Polyurethane. *Chem. Br.* **26**(4), 315 (1990).
24. "Kirk-Othmer Encyclopedia of Chemical Technology," 4th ed., Vol. 11, p. 759, Wiley, New York, 1994.
25. Key Polymers, Epoxies. *Chem. Eng. News* **59**(39), 18, Sept. 28 (1981).
26. K.J. Saunders, "Organic Polymer Chemistry," p. 282–297, Chapman & Hall, London, 1973.
27. R.A. Haupt and T. Sellers, Jr., Characterization of Phenol-Formaldehyde Resol Resins. *Ind. Eng. Chem. Res.* **33**, 693 (1994).
28. Key Polymers, Phenolics. *Chem. Eng. News* **59**(39), 16, Sept. 28 (1981).
29. Anonymous, "Recycling Plastics, A Survey and Assessment of Research and Technology," Report, 56 pages. International Research and Technology Corp., 1973.
30. R.J. Ehrig, ed., "Plastics Recycling: Products and Processes." Hanser Publishers, New York, 1992.
31. D. O'Sullivan, PET Bottles Recycled into Insulation Foam. *Chem. Eng. News* **68**, 25, Feb. 12 (1990).
32. P. Mattli, Control of Odour Nuisance from the Chemical Industry. *Chem. Ind. (London)* p. 338, May 7 (1984).
33. J.H. Schut, R-PET Cleared for Food Contact. *Plast. World* **52**, 11, Oct. (1994).
34. B. Miller, Co-injection Opens Door for Recycled PET for Foods. *Plast. World* **52**, 14, Aprl. (1994)
35. R.M. Wajge and S.K. Gupta, Multiobjective Dynamic Optimization of a Non-vaporizing Nylon 6 Batch Reactor. *Polym. Eng. Sci.* **34**, 1161 (1994).

36. M.H. Mark and W.C. Trogler, Nylon Production: An Unknown Source of Atmospheric Nitrous Oxides. *Science* **251**, 932 (1991).
37. R.Welgos, Nylon 6 and 6,6 Aren't Always the Same. *Mach. Des.* **66**, 55, Nov. 21 (1994).
38. M.M. Nir, A. Ram, and J. Miltz, Performance of Reprocessed Multilayer LDPE/Nylon-6 Film. *Polym. Eng. Sci.* **35**, 1878 (1995).
39. L.M. Sherman, Third Generation Blowing Agents Starred at Polyurethanes '95. *Plast. Technol.* **41**, 25, Oct. (1995).
40. B. Miller, Urethanes Technology is Adapting to Environmental Concerns. *Plat. World* **51**, 12, Oct. (1993).
41. Upjohn Company. "Precautions for the Proper Usage of Polyurethanes, Polyisocyanurates, and Related Materials," Tech. Bull. 107. Chemical Division, The Upjohn Company, Kalamazoo, MI, 1980.
42. E. Weigand, W. Rasshofer, and M. Herrman, Polyurethane Recycling Processes are Reaching Commercial Maturity. *Mod. Plast.* **70**, 71, Nov. (1993).
43. P. Mapelston, Bayer Process Makes Urethane Scrap Reusable, *Mod. Plast.* **68**, 20, Feb. (1991).
44. J. Braslaw and J.L. Gerlock, Polyurethane Waste Recycling. Polyol Recovery and Purification. *Ind. Eng. Chem. Process Des. Dev.* **23**, 552 (1984) (series of articles).
45. J.L. Gerlock, J. Braslaw, and L.R. Mahoney, Superheated Stream Hydrolysis of Polyurethane Foam. *175th Am. Chem. Soc. Nat. Meet.*, 1978, Anaheim, CA, Abstr., p. 12 (1978).

22
ADDITION (CHAIN REACTION) POLYMER THEORY

Our time has seen the synthesis of Polyisoprene
And many cross-linked Helices unknown
To Robert Hooke. . . .
                              —*John Updike, 1968*

22.1. ADDITION (VINYL-TYPE) POLYMERS

Synthetic polymers may be broadly classified into three major divisions on the basis of their properties as plastics, fibers, or elastomers. The main differentiating feature of these broad divisions of polymers is their modulus, or stiffness; fibers have the highest modulus and elastomers the lowest. A further subdivision of the plastics is made on the basis of whether they are melt formable, or must have the final polymerization take place in a mold of the final fabrication shape desired. The melt-formable polymers are referred to as thermoplastics, and the mold-formed or infusible polymers as thermosetting plastics. Thermosetting plastics, such as the phenol-formaldehyde resins discussed in Chapter 20, are structurally distinct from the thermoplastics in having a significant degree of crosslinking which joins adjacent polymer chains into an interlocking network. This is the feature which makes these materials incapable of melting, although they will still char or burn at sufficiently high temperatures. Thermosets comprise about 10% of the total current mass of production of all types of plastics.

Thermoplastics may be further subdivided into two broad categories on the basis of their cost and suitable end uses. "Commodity" plastics are typified by high volumes of production, good properties, and low resin cost. The four major commodity plastics are polyethylene, polypropylene, poly(vinyl chloride), and polystyrene. Their properties and low cost have led to the extensive use of these plastics in packaging applications, where they are very competitive with the more traditional paper, steel, and glass. They are also used for

TABLE 22.1   Some Representative Thermoplastic Polymers

Name	Repeating unit	Typical uses
Commodity plastics:		
Low-density polyethylene (LDPE or PE)	$+CH_2 - CH_2 +_n{}^a$	Packaging/construction film, housewares, cable insulation, flexible bottles, toys
High-density polyethylene (HDPE)	$+CH_2 - CH_2 +_n$	Drums, pipe, bottles, sheet, packaging film, cable insulation
Polypropylene (PP)	$+CH_2 - CH(CH_3) +_n$	Appliance and auto parts, carpeting, rope, film
Poly(vinyl chloride) (PVC)	$+CH_2 - CHCl +_n$	Housing, pipe, flooring, film, sheet, wire insulation
Polystyrene (PS)	$+CH_2 - CHPh +_n$	Foam and film packaging, appliance parts, housewares, foam insulation, disposable ware
Engineering plastics:		
Polyamides (nylons)	Various (Chap. 21)	Gears, bearings, latches, high-tensile cordage, rope, aircraft parts
Polycarbonate (PC, Lexan)	e.g., $+OC_6H_4C(CH_3)_2 - C_6H_4OCO +_n$	Safety helmets, bullet-proof windows, gears, auto parts, machine casings
Polyesters, e.g.,:		
PET	$+OCH_2CH_2OCOC_6H_4CO +_n$ poly(ethylene terephthalate)	High-strength moldings, electrical insulation
PBT	$+O(CH_2)_4OCOC_6H_4CO +_n$ poly(butylene terephthalate)	Engineering products, blends with PC in high-impact applications

^{a}With some branching; see Section 23.1.

appropriate less demanding applications as components of durable goods (Table 22.1). "Engineering" plastics are so called because of their significantly better mechanical properties than the commodity plastics. They are also more expensive to produce. As a class their production volume is about one-tenth that of the commodity plastics (Table 22.2) [1–3]. The repeating units of the engineering plastics also identify these as being predominantly condensation polymers, the class discussed in Chapters 20 and 21. The commodity plastics are prime examples of the utility of the vinyl-type, or chain reaction, polymers to be discussed here.

22.2. COMPARISON OF CONDENSATION AND CHAIN REACTION POLYMERIZATION

As outlined in Chapter 20, condensation polymerization can involve the re-action of any two species present in a condensation polymerization system.

TABLE 22.2 American Production of Some Large Scale Synthetic Polymers[a]

Type	Thousand metric tonnes		Average prices 1995 ($/kg)
	1985	1995	
Commodity plastics:			
Low-density polyethylene	4032	5749	0.69
High-density polyethylene	3026	5115	0.71
Polypropylene	2331	4946	0.73
Polystyrene	1839	2628	0.90
Poly(vinyl chloride)[b]	3072	5445[c]	0.77
Totals	14300	23883	
Engineering plastics:			
Polyamide, nylon[d]	181	428[e]	3.10[f]
Polyester, thermoplastic	438	1450[e]	1.25[f]
Totals	619	1878	
Synthetic elastomers:			
Styrene-butadiene	735	945	1.80
Polybutadiene	330	548	1.25[e]
Ethylene-propylene	215	314	1.10[e]
Nitrile-solid	58	73	2.20
Polychloroprene	82	75[e]	4.60
Other	289	647	—
Totals	1709	2602	

[a]Data compiled from *Chemical and Engineering News* [1], Kirschner [2], and *Chemical Marketing Reporter* [3].
[b]Includes copolymers.
[c]Includes Canadian production.
[d]Does not include resin targeted for fiber production.
[e]Data for 1994.
[f]Estimates.

The result is a generally slow progression toward moderate molecular weights if conditions are chosen to help drive the reaction to very near completion. These requirements and the results are listed for convenience of comparison with chain reaction (vinyl-type) polymerization discussed in detail here (Table 22.3).

Chain reaction polymerization involves use of a monomer with at least one carbon-carbon or carbon-heteroatom double bond. The double bond represents a masked functionality of two, which can form new single bonds linking monomer units on interaction with an initiator, which could be a radical species or a charged molecule or atom. This step represents the first of three distinct kinetic divisions recognized for chain reaction polymerization: initiation, propagation, and termination.

For a free radical initiated process with a monomer such as ethylene, the three stages can be represented as follows (Eq. 22.1–22.6). Species suitable for initiation of a free radical promoted process are normally generated by the homolytic dissociation of weak bonds of a precursor by heat or light.

■ **TABLE 22.3   Contrasting Characteristics of Condensation and Chain Reaction Polymerization**

Characteristic	Condensation (step-reaction) polymerization	Chain reaction (vinyl-type) polymerization
Chain growth	By reactions between monomers, oligomers, and polymers Usual loss of small molecule	Only by monomer addition to initiating radicals Whole monomer incorporated
Process involved	Same reaction involved with all reacting species present No initiator required (catalyst sometimes) Monomer consumed early in polymerization; molecular wt. increases slowly	Initiation, propagation, and termination mechanisms different Suitable radical, or charged (+, or −) initiator required Only monomer and high polymer present; each polymer molecule forms very rapidly
Rate	Decreases steadily as polymerization proceeds	Increases initially, then stays nearly constant until monomer consumed
Product $\overline{DP}^a$	Low to moderate for linear product	High, to very high
Heterogeneity index, $\dfrac{\overline{M}_w}{\overline{M}_n}$	ca. 2	ca. 2^b

^{a}Degree of polymerization, or $\overline{X}_n$.
^{b}Except for anionic "living" polymerizations where the value is about 1; see Section 22.5.

Initiation:

$$\underset{\substack{\text{initiator}}}{I} \xrightarrow[\text{light}]{\text{heat or}} 2\,R^{\cdot} \qquad 22.1$$

$$R^{\cdot} + CH_2 {=} CH_2 \rightarrow RCH_2 {-} \overset{\cdot}{C}H_2 \qquad 22.2$$

Propagation:

$$RCH_2\overset{\cdot}{C}H_2 + CH_2 {=} CH_2 \rightarrow RCH_2CH_2CH_2\overset{\cdot}{C}H_2 \qquad 22.3$$

$$RCH_2CH_2CH_2\overset{\cdot}{C}H_2 + n\,CH_2 {=} CH_2 \rightarrow R(CH_2CH_2)_{n+1}CH_2\overset{\cdot}{C}H_2 \qquad 22.4$$

Termination:
By combination:

$$2\,R(CH_2CH_2)_nCH_2\overset{\cdot}{C}H_2 \rightarrow$$
$$R(CH_2CH_2)_nCH_2CH_2CH_2CH_2(CH_2CH_2)_nR \qquad 22.5$$

By disproportionation:

$$2\,R(CH_2CH_2)_nCH_2\overset{\cdot}{C}H_2 \rightarrow \qquad 22.6$$
$$R(CH_2CH_2)_nCH {=} CH_2 + R(CH_2CH_2)_nCH_2CH_3$$

The initiating radical then reacts with monomer to form a propagating radical in the second phase of the initiation process. The propagating radical can then add many more monomer units in a chain reaction before intervention of a

a.

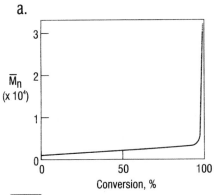

b.

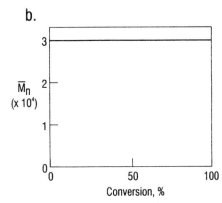

FIGURE 22.1   Relationship of the product number average molecular weight, M_n, to the percent monomer conversion for (a) a step-growth, polycondensation versus (b) a chain growth (vinyl-type) polymerization.

termination step stops the process to produce a finished (dead) polymer chain. The propagation steps consume by far the largest proportion of monomer. For a free radical-initiated process, termination can occur by combination (coupling) or by disproportionation. Coupling forms a saturated polymer product with a molecular weight equal to the sum of the molecular weights of the coupling radical species (Eq. 22.5). Termination by disproportionation occurs by abstraction of a hydrogen atom from one of the two radical species by the other, to form two stable polymer molecules of molecular weights equivalent to that of each of the initial species (Eq. 22.6). One of these is saturated and the other will have a terminal olefin. In theory, the olefin-terminated polymer could react with a new radical species and so increase in molecular weight. But in practice the olefinic terminus is so sterically hindered by the rest of the polymer chain that this molecule, too, is essentially "dead" or finished polymer.

Representative features of chain reaction polymerization are listed in Table 22.3 for comparison with the characteristics of typical condensation polymerizations. In both cases, the degree of polymerization is determined by dividing the molecular weight of the polymer by the molecular weight of the monomer unit (Section 20.8.2). However, the relationship of the number average molecular weights to the degree of conversion is quite different in the two processes (Fig. 22.1). High molecular weight material is only obtained quite near the end of the polymerization for a polycondensation, whereas the molecular weight of the product is normally independent of the degree of polymerization in chain (vinyl-type) polymerizations.

22.3. RADICAL CHAIN POLYMERIZATION METHODS

22.3.1. Bulk or Mass Polymerization

Free radical polymerization of neat monomer in the absence of solvent and with only initiator present is called bulk, or mass, polymerization. Monomer in liquid or vapor state is well mixed with initiator in a heated or cooled reactor as appropriate. The advantages of this method are that it is simple, and because of the few interacting components present there is less possibility for contamination. However, vinyl-type polymerizations are highly exothermic so that control of bulk polymerization temperatures may be difficult. Also viscosities may become very high toward the end of a polymerization which could make stirring difficult, and further complicate the removal of heat from the system. The advantages of this system, however, are sufficiently attractive for this to be used commercially for the free radical polymerization of styrene, methyl methacrylate, vinyl chloride, and also for some of the processes for the polymerization of ethylene [4].

Bulk polymerization is also, incidentally, the chief method used for commercial polycondensations. Polycondensations are not as exothermic as free radical catalyzed vinyl-type polymerizations, so thermal control is less of a problem. Bulk polycondensation also favors formation of linear polymer rather than the cyclic products that are favored by dilute solution polymerization, particularly if AB-type monomers are being used. Finally, since a high degree of polymerization (i.e., high DP, or high molecular weight) is only obtained at very near complete reaction of the functional groups, the viscosity stays low throughout most of the reaction. Hence stirring remains easy until very near the end of the polymerization. Since the high temperatures used near the end of the process to help drive a typical polycondensation to completion are often above the melting point of the polymer, high viscosity is seldom a problem with these systems.

22.3.2. Solution Polymerization

Free radical polymerization of a monomer in solution is often more versatile and more amenable to temperature control than bulk polymerization. The presence of the solvent avoids any potential viscosity or stirring problems. Also the exotherm of the polymerization is moderated by the lower monomer concentrations under these conditions, which tends to help slow down rates, and by the thermal mass contributed by the solvent. Sometimes the polymerization temperature will be conveniently controlled by the reflux temperature of the solvent used.

To optimize these advantages of solution polymerization, care has to be exercised in certain production parameters when monomer and solvent are present. Appropriate solvents for free radical polymerization need to be selected carefully, since radical chain transfer from the propagating polymer chain end to solvent molecules may occur. Hydrogen abstraction from the solvent by the propagating radical will tend to shorten the propagating chain, which will tend to decrease the average molecular weight of the polymer product obtained. The extent of transfer will depend on the choice of solvent and the conditions (concentrations, temperature, etc.) used to carry out the polymerization. Also the purity of the polymers produced by solution polymerization may be affected, either by impurities present in the solvent, or by residual traces of the solvent itself if this is difficult to remove. Nevertheless the advantages of solution polymerization have led to it being commonly used for commercial production of polymers of the acrylic esters, ethylene, vinyl acetate, and acrylonitrile. Solution polymerization of acrylonitrile in water is a particularly convenient example since the monomer is soluble in water, but the polymer is not. On completion of the polymerization the finished polymer may be recovered from the solvent and any residual monomer by simple filtration.

22.3.3. Suspension Polymerization

Water-immiscible monomers, or not more than slightly water-soluble monomers, may be polymerized in the form of a suspension of large droplets in water. The droplets are kept in suspension by agitation and by the use of stabilizers such as gelatin, talc, or bentonite clay. The free radical initiator used must be soluble in the monomer. Droplet size is 0.01–0.5 cm in diameter in typical operating modes. Polymerization in this way can be pictured as the simultaneous operation of many droplet-sized reactors, which on completion give beads, or "pearls," of polymer. In fact these are the names which are sometimes applied to this method of polymerization and to the product obtained. The polymer product is recovered by filtration, washing with water to remove any adhering stabilizer(s), and then dried with warm air.

The kinetic behavior and the methods of predicting molecular weight for suspension polymerizations are virtually the same as for bulk polymerizations. Operating in suspension mode eliminates any temperature control problems since the water phase has both a large heat capacity and a high heat transfer capability. Also there are no difficulties with high viscosities or stirring because of the large proportion of water present. The main negative factors experienced with suspension polymerization relate to the occasional difficulty with complete removal of residual suspending agents that may get incorporated into the finished polymer. There may also be a greater difficulty in controlling monomer vapor loss during the polymerization than for the corresponding bulk polymerization system. Suspension polymerization is appropriate commercially for monomers such as methyl methacrylate, styrene, and vinyl chloride.

22.3.4. Emulsion Polymerization

Polymerization of a monomer in an emulsion system is similar to suspension polymerization in that the bulk medium in both cases is water. But there the similarity ends [5–7]. Monomer and initiator are still present, but this time the initiator must be a water-soluble, monomer-insoluble type. Suitable for this purpose are the persulfate-ferrous ion or peroxide-ferrous ion systems (Eq. 22.7, 22.8).

$$S_2O_8^{2-} + Fe^{2+} \rightarrow Fe^{3+} + SO_4^{2-} + SO_4^{\overline{\cdot}} \qquad 22.7$$

$$PhC(CH_3)_2OOH + Fe^{2+} \rightarrow Fe^{3+} + HO^- + PhC(CH_3)_2O^{\cdot} \qquad 22.8$$

Surfactants are used in combination with stirring to produce a suspension of very small aggregates of monomer known as micelles. A small proportion of the total monomer will be present in these micelles, of 100–300 nm (0.1–0.3 um) diameter, and stabilized by a high concentration of surfactant molecules on the surface (Fig. 22.2). The bulk of the monomer, initially, will be present in droplets of 1500 nm in diameter and larger with no more than a few surfactant molecules present on the surface of each. Agitation performs the primary stabilizing function for the droplets. To visualize a typical emulsion polymerization, there will be up to 10^{18} micelles/cm^3 present, and up to 10^{11} droplets/cm^3. Thus the micelles present a much larger surface area exposed to the water-soluble initiator, relative to the mass of monomer, than the droplets. This factor, plus the orders of magnitude higher concentration of surfactant molecules on the surface of the micelles, focuses most of the progress of the polymerization in the micelles. As the polymerization proceeds there is a gradual transfer of monomer from the droplets, which slowly decrease in size and number while the micelles gradually increase in size.

The very small size of the micelles means that even the presence of one propagating radical in the micelle represents a very high radical concentration. For this reason there is usually only one propagating radical present per micelle. When a second radical penetrates the polar surfactant water-monomer interface, it generally terminates the process. The next radical entering initiates propagation again. Thus, the monomer in one-half of the micelles is actively being polymerized at any point in time, and the monomer in the other half is dormant. Emulsion polymerization, because of this dominant mode of progression, operates via kinetics which differ significantly from those of bulk polymerization. Molecular weight control, rate, etc., are all affected by different variables. Details of the methods used to evaluate these parameters have been developed and are outlined in standard texts.

The main advantage of emulsion polymerization is that higher DPs (degrees of polymerization) and molecular weights are possible from some mono-

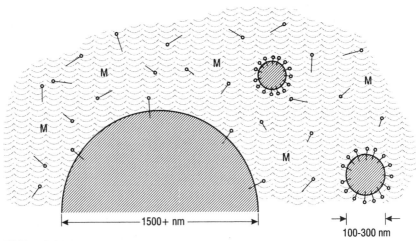

FIGURE 22.2 Pictorial representation of monomer distribution in emulsion polymerization. Dissolved monomer is represented by letters M, molecules of surfactant hydrophylic ends by open circles, connected to their hydrophobic alkyl tails represented by a solid line.

mers by this method than by use of any of the other methods discussed. In fact, in some cases molecular weights obtained may be so high that subsequent processing could be made difficult with the unmodified process. Fortunately the DPs obtained may be moderated if desired by addition of a chain transfer agent. This additive can decrease the number of monomer molecules which can add to the propagating radical before termination occurs by radical transfer to the additive. Thus the concentration of chain transfer agent used can be used to control the molecular weight of the product to the desired range.

Emulsion polymerization presents similar processing difficulties to those of suspension polymerization. The product has to be recovered, in this case usually by coagulation, and then washed and dried. Again it may be difficult to remove all traces of the surfactant, etc., used to stabilize the emulsion, as with the product from suspension polymerization. However, for some applications, such as for latex (water-based) paints and carpet adhesives, the aqueous product from emulsion polymerization may be used directly. For applications such as these it is possible to produce latices (latexes) containing in the region of 50% solids.

22.4. IONIC INITIATORS FOR VINYL MONOMERS

Vinylic monomers that are capable of forming a propagating radical on introduction of a radical species can generally be polymerized using a free radical initiator, as just discussed. However, if the double bond of the vinylic monomer has a high electron density contributed by electron-donating substituents on the double bond, such as shown by isobutylene ($CH_2 = C(CH_3)_2$) or vinyl ethers, then radical initiation fails to produce polymer. These types of monomers, however, are amenable to polymerization catalyzed by cationic species because the high electron density on the double bond can serve to stabilize a propagating carbocation (formerly called carbonium ion; proposed new name, carbenium ion), as it forms.

For analogous reasons to the monomer requirements which favor cationic initiators, vinylic monomers with electron-withdrawing substituents on the carbon-carbon double bond are amenable to polymerization under the influence of anionic catalysts, since under these conditions the electron-withdrawing substituent assists in stabilization of the propagating carbanion, as it forms. However, this class of monomer is usually still sensitive to free radical-initiated polymerization because of electronic back donation from the electron-withdrawing group to the carbon-carbon double bond (Table 22.4) [8,9].

Both cationic and anionic polymerizations proceed via chain mechanisms, like free radical polymerizations. In free radical processes the initiator is incorporated into the polymer chains, as they are formed. But in ionic polymerizations the "termination" processes which produce dead polymer chains very often also release the active initiating species again. Thus, in ionic polymerizations these are often referred to as catalysts, rather than initiators, to make this distinction.

■■■■■ **TABLE 22.4 Suitable Initiating Systems for the Chain Polymerization of Representative Vinylic Monomers**[a]

| Monomer types | Suitable initiating systems[b] | | | |
	Radical	Cationic	Anionic	Coordination
Ethylene	+	+	−	+
Propylene, monosubstituted α-olefins	−	−	−	+
Isobutylene, dialkyl olefins	−	+	−	−
Dienes	+	−	+	+
Styrene	+	+	+	+
Vinyl chloride, tetrafluoroethylene[c]	+	−	−	+
Vinylidene chloride, $CH_2 = CCl_2$[d]	+	−	+	−
Vinyl ethers, $CH_2 = CHOR$	−	+	−	+
Vinyl esters, $CH_2 = CHOCOR$	+	−	−	−
Acrylic, methacrylic esters, $CH_2 = CHCO_2R$, $CH_2 = C(CH_3)CO_2R$	+	−	+	+
Acrylonitrile	+	−	+	+
Aldehydes, ketones	−	+	+	n/a

[a]Compiled from Lenz [8] and Billmeyer [9].
[b]+, forms high polymers; − no reaction or forms only oligomers; n/a information not available.
[c]Product trade names: Teflon, Goretex.
[d]Product trade names: Saran wrap, Saran fiber.

22.4.1. Cationic Initiating Systems

Suitable catalysts for cationic polymerization may be conveniently classified into three groups. The classical protonic mineral acids such as H_2SO_4, $HClO_4$, and H_3PO_4 are often useful. Effective catalysis is obtained if the acid selected has high acid strength, plus an anion of low nucleophilicity (Eq. 22.9).

$$H^+ + XO_4^- + CH_2 = CR_2 \rightleftharpoons CH_3 - \overset{+}{C}R_2 + XO_4^- \qquad 22.9$$

It is this latter requirement that makes the halogen acids a poor choice despite their high acid strength since the halide ions are good nucleophiles and will also add to the double bond at the same time as hydrogen ion does. Some halogen acids can serve as efficient co-catalysts, however, in combination with a Lewis acid.

Friedel-Crafts reagents (Lewis acids) represent the most widely used class of cationic polymerization catalysts.

General formula, MX_n: $AlCl_3$, $TiCl_4$, BF_3, $SnCl_4$, $ZnCl_2$, etc.

They usually require a co-catalyst such as water, an organic acid, a halogen acid, or an organic halide which serves as the proton or cation donor as it associates with the Lewis acid (Eq. 22.10–22.12).

$$\text{e.g.,} \qquad AlCl_3 + RCl \rightleftharpoons R^+ + AlCl_4^- \qquad 22.10$$

$$TiCl_4 + H_2O \rightleftharpoons H^+ + TiCl_4OH^- \qquad 22.11$$

$$TiCl_4 + HCl \rightleftharpoons H^+ + TiCl_5^- \qquad 22.12$$

The Lewis acid can function in the absence of a co-catalyst only in special cases, when autoionization is possible (Eq. 22.13).

$$\text{e.g.,} \quad 2\ AlBr_3 \rightleftharpoons AlBr_2^+\ AlBr_4^- \qquad\qquad 22.13$$

A number of other materials can function as catalysts for cationic polymerization, among them iodine, tropylium salts, triphenylmethyl halides, and t-butyl halides (Eq. 22.14–22.16).

$$2\ I_2 \rightleftharpoons I^+\ I_3^- \qquad\qquad 22.14$$

$$I_2 + CH_2{=}C(CH_3)_2 \rightleftharpoons ICH_2{-}CI(CH_3)_2 \rightleftharpoons \qquad 22.15$$
$$ICH{=}C(CH_3)_2 + H^+\ I^-$$
$$\text{or,}\ ICH_2{-}CI(CH_3)_2 \rightleftharpoons ICH_2^+{-}C(CH_3)_2\ I^-$$

$$Ph_3CCl \rightleftharpoons Ph_3C^+\ Cl^- \qquad\qquad 22.16$$

Ionizing radiation of energies sufficient to cause heterolytic cleavage may also be used (Eq.22.17).

$$CH_2{=}CHR \xrightarrow[\gamma\text{-rays}]{} CH_2{=}CR^- + H^+ \qquad 22.17$$

The type and activity of a suitable catalyst are often quite closely related to the activity of the monomer to be polymerized. The relatively stable carbocation triphenylmethyl, for example, is only effective with very reactive monomers such as methyl vinyl ether. Zinc chloride plus co-catalyst will polymerize an appropriate monomer more slowly than a catalyst combination which uses one of the more active Lewis acids. Also the classical proton acids only rarely give high molecular weight polymers.

22.4.2. Example of Cationic Polymerization

While the detailed mechanistic understanding of many cationic polymerization systems has not as yet been clearly established, there are some that are reasonably well understood. Let us consider the detailed progression of the cationic polymerization of isobutylene using a $TiCl_4/H_2O$ catalyst system as an example of the mechanistic progress of cationic chain polymerizations. This system cannot be used as a general model for all cationic polymerizations unlike the generalization possible for free radical-initiated systems, since there is a wider variety of mechanisms seen for cationic, than for free radical, systems.

Initiation involves the reversible reaction of water with $TiCl_4$ to produce the active catalyst, which is then in a form capable of protonating the methylene of isobutylene (Eq. 22.18, 22.19). The active propagating species formed is a tightly bound ion pair.

Initiation:

$$TiCl_4 + \underset{\text{co-catalyst}}{H_2O} \rightleftharpoons \underset{\text{active catalyst}}{H^+(TiCl_4OH)^-} \qquad 22.18$$

$$H^+(TiCl_4OH)^- + \underset{\text{isobutylene}}{(CH_3)_2C{=}CH_2} \rightarrow \underset{\text{ion pair}}{(CH_3)_3C^+(TiCl_4OH)^-} \qquad 22.19$$

In a nonpolar, hydrocarbon-type solvent, polymerization may proceed rather slowly because of the steric hindrance of the tightly bound gegen ion, or counter ion, to the active carbocationic center. The rate of polymerization will be increased by use of a more polar solvent which will serve to increase the spacing between the gegen ion and carbocation, which in turn will increase the rate.

Propagation involves monomer insertion into the carbocationic center, and sigma bond formation via the pi electrons of the entering isobutylene (Eq. 22.20).

Propagation:

$$\underset{\text{gegen ion}}{(CH_3)_3C^+ \quad (TiCl_4OH)^-} \quad \rightarrow \quad (CH_3)_3CCH_2\overset{+}{C}(CH_3)_2(TiCl_4OH)^- +$$

$$+$$

$$CH_2\!=\!C(CH_3)_2 \qquad\qquad\qquad\qquad\qquad\qquad\qquad\qquad 22.20$$

$$(n-1)\ CH_2\!=\!C(CH_3)_2$$

$$\underset{\text{steps}}{\overset{\text{several}}{\rightarrow \rightarrow \rightarrow}}\ H\!+\!CH_2C(CH_3)_2\!+\!_nCH_2\overset{+}{C}(CH_3)_2(TiCl_4OH)^-$$

This polymerization is normally carried out at low temperatures of about 170 K ($-100°C$) and produces extremely rapid polymerization, comparable or even faster than the rates observed for many free radical-initiated polymerizations. High molecular weight product begins to appear within a few seconds of initiation.

If the cationic polymerization is carried out at room temperature or higher, both the rate and the molecular weight of the product obtained are lower. This is probably the result of an increased rate of transfer to monomer competing more effectively with the propagation process at the higher temperatures (Eq. 20.21).

Transfer:

$$H\!+\!CH_2C(CH_3)_2\!+\!_nCH_2\overset{+}{C}(CH_3)_2(TiCl_4OH)^- + CH_2\!=\!C(CH_3)_2 \rightarrow \qquad 22.21$$

$$\underset{\text{dead polymer}}{H\!+\!CH_2C(CH_3)_2\!+\!_nCH_2\!-\!C(CH_3)\!=\!CH_2} + \underset{\text{initiating species}}{(CH_3)_3C^+(TiCl_4OH)^-}$$

Hydrogen transfer from the propagating species to monomer produces "dead" polymer, which has a terminal olefin, and regenerates the close ion pair product as given in the second step of the initiation process (Eq. 20.19). This is available to start a new polymer chain all over again. Since the activation energy for this transfer process is higher than the activation energy required for propagation, the rate of transfer competes more effectively with the rate of propagation at higher temperatures than at low temperatures.

Cationic polymerizations generally have no formal termination process. Dead chains are produced by what amounts to an ionic rearrangement of the propagating species which produces an olefin-terminated chain and regenerates the first active catalyst species (Eq. 22.22).

"Termination":

$$H \{CH_2C(CH_3)_2\}_n CH_2\overset{+}{C}(CH_3)_2(TiCl_4OH)^- \rightarrow \qquad \qquad 22.22$$

$$H \{CH_2C(CH_3)_2\}_n CH_2 - C(CH_3) = CH_2 + H^+(TiCl_4OH)^-$$

dead polymer   still active

Thus, like the transfer process just discussed, this does produce dead polymer but also regenerates an active species which can continue to propagate, unlike a true termination step. Also, as with the intermolecular transfer event, this intramolecular transfer process generally requires a larger activation energy than propagation. So this type of transfer also begins to compete more effectively with propagation at higher temperatures. More details of the kinetics for at least some specific cases of cationic polymerization are accessible from standard texts (See the Further Reading Section at the end of this chapter.)

22.4.3. Suitable Catalysts for Anionic Polymerization

Progress of anionic polymerizations is most favorable with monomers which have one or more electronegative (electron-withdrawing) groups, such as nitrile or halogen, directly bonded to the vinyl group. Suitable catalysts for initiation of anionic polymerizations fall into one of two classes, according to their mode of action.

The basic class of initiators functions by the addition of the basic anion of a salt to the double bond of a suitable monomer in order to generate the propagating species (Eq. 22.23).

$$MX + CH_2 = CHZ \rightarrow X - CH_2 - \overset{..}{C}HZ^-M^+$$

e.g., where $Z = -CN, -COR$

$$22.23$$

Relatively weakly basic catalysts, such as hydroxide ion or cyanide ion, may be used to initiate polymerization of monomers which have strongly electron-withdrawing substituents on the vinyl group, such as a nitrile or carbonyl. Thus, suitable monomers would be acrylonitrile or methylmethacrylate.

At the other extreme, strongly basic anionic initiators are required for polarizable monomers or for monomers with relatively poor electron-withdrawing substituents such as styrene, isoprene, and 1,3-butadiene. With these monomers, which have no electrophilic (or hetero) atom to decrease the electron density of the vinyl group, strongly basic anions such as amide ion or alkyl anion are required for effective initiation. Sodium or potassium amide and n-butyllithium are appropriate examples.

Of intermediate utility between these two extremes of the basic initiators for anionic polymerization is alkoxide ion, which has adequate base strength for initiation of the ring-opening polymerization of epoxides (Eq. 22.24–22.26).

$$NaOCH_3 \rightleftharpoons Na^+ + CH_3O^- \qquad \qquad 22.24$$

$$CH_3O^- + CH_2 - CH_2 \rightarrow CH_3O - CH_2 - CH_2O^-$$

$$\underset{O}{\diagdown \diagup}$$

$$22.25$$

$$CH_3O-CH_2-CH_2O^- + n\,CH_2-CH_2 \rightarrow$$

$$\underset{O}{}$$

$$CH_3O \{CH_2-CH_2-O\}_n CH_2-CH_2O^- \qquad 22.26$$

Thus the relative strength of the base used is important to the effectiveness of the catalysis. A weakly basic anion will not function for a mildly activated monomer, and a strongly basic anion may not function well for a strongly activated monomer.

Donor-acceptor catalysts represent the other type of initiator useful in anionic polymerizations. In this class, the donor entity may be an electronically neutral species which gives up an electron to the monomer to form the cationic propagating species with monomer (Eq. 22.27).

$$D + M \rightarrow D^+ + M^- \qquad 22.27$$

Use of sodium metal to initiate the polymerization of methacrylonitrile is an example of this type of process (Eq. 22.28).

$$Na + CH_2{=}C(CH_3)-CN \rightarrow Na^+(\dot{\,}CH_2-\ddot{C}(CH_3)CN)^-$$
$$\text{propagating species} \qquad 22.28$$
$$\text{(may dimerize)}$$

Another donor-acceptor approach is to use a donor entity that is already charged. In the initiation process this gives up its charge to monomer to form a neutral entity and the propagating species (Eq. 22.29).

$$D^- + M \rightarrow D + M^- \qquad 22.29$$

22.4.4. Anionic Polymerization of Styrene Using Amide Ion

Polymerization of styrene in liquid ammonia at low temperatures catalyzed by potassium metal represents a good example of a base-initiated anionic polymerization. Styrene, being relatively nonpolar, requires the strongly basic amide ion for effective anionic polymerization. The catalyst is made, *in situ*, usually before the styrene is added, by the addition of small pieces of potassium metal to liquid ammonia kept at dry ice (solid carbon dioxide) temperatures (Eq. 22.30).

$$2\,K + 2\,NH_3 \rightarrow 2\,KNH_2 + H_2 \qquad 22.30$$

Initiation involves two steps, ionization of the potassium amide in the extremely polar liquid ammonia, then addition of amide anion to the styrene, which is usually added to the system at this stage (Eq. 22.31, 22.32).

Initiation:

$$KNH_2 \rightleftharpoons K^+ + H_2N{:}^- \qquad 22.31$$

$$H_2N{:}^- + CH_2{=}CHPh \rightarrow H_2N-CH_2-(Ph)\ddot{C}H^-$$
$$\text{propagating species} \qquad 22.32$$

Potassium ions are also present in this solution, but because of the high dielectric constant and high solvating capability of liquid ammonia the propagating species formed here is truly a free ion, not more than very loosely associated with the cation. The less polar the solvent used the greater would be the degree of association obtained between the propagating anion and the cation present. Thus, the polarity of the solvent chosen can affect the rate of formation and sometimes also the stereochemistry of the anionic polymerization product.

Propagation involves addition of the propagating species to styrene monomer (Eq. 22.33, 22.34). Head-to-tail addition, the normal mode, retains the negative charge at a benzylic, resonance-stabilized position.

Propagation:

$$H_2N-CH_2-(Ph)\ddot{C}H^- + CH_2=CHPh \rightarrow \qquad\qquad 22.33$$
$$H_2N-CH_2-(Ph)CH-CH_2-(Ph)\ddot{C}H^-$$

$$H_2N-M_n^- + M \xrightarrow{k_p} (H_2N-M_{n+1})^- \qquad\qquad 22.34$$

As seen with cationic polymerization, there is also no formal termination step with anionic-initiated systems. Dead chains are produced by transfer of a proton (or other positive fragment), usually from the solvent to the anion. When this happens with this particular example, it regenerates amide ion again, the primary initiating species (Eq. 22.35).

Termination:

$$H_2N\{CH_2-(Ph)CH\}CH_2-(Ph)\ddot{C}H^- + NH_3 \xrightarrow{k_{tr}} \qquad 22.35$$
$$H_2N\{CH_2-(Ph)CH\}_{n+1}H + H_2N:^-$$

This is now available to generate a new propagating species and start the process all over again. The following experimental evidence supports this portrayal of the mechanism. The rate is found to increase with increased amide ion concentration or with increased concentrations of styrene. Rate decreases with added potassium ion, presumably by increasing the extent to which the propagating species was associated with it. Also each polymer chain in the product contains only one nitrogen atom and has no residual unsaturation.

An anionic polymerization carried out under an inert atmosphere, since it has no formal termination step and uses relatively stable propagating species, is sometimes referred to as a living polymerization, and the product as a living polymer. As such, the DP of these systems becomes simply the moles of monomer present divided by the moles of initiator used (Eq. 22.36).

$$\overline{X}_n = \frac{\text{moles of monomer}}{\text{moles of initiator}} \qquad\qquad 22.36$$

The other unique features of living polymerizations are that M_n varies directly with conversion (Fig. 22.3, cf. Fig. 22.1) and the heterogeneity index (polydispersity) is 1, or very close to it.

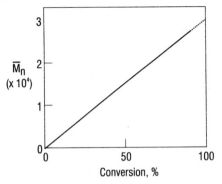

FIGURE 22.3 Relationship of number average molecular weight, M∞, to percent conversion for a living polymerization.

22.4.5. Sodium-Naphthalene Anionic Initiation

An example of the donor-acceptor initiation process is demonstrated by the sodium-naphthalene system devised by Szwarc [10]. The initiation sequence begins with the transfer of an electron from sodium metal to naphthalene. In an ether-type solvent such as THF (tetrahydrofuran) sodium naphthalide, in which the naphthalene exists as an anion radical, is formed quantitatively (Eq. 22.37).

$$Na \quad + \quad \text{(naphthalene)} \quad \xrightarrow{\text{THF}} \quad Na^+ \left[\text{(naphthalene anion radical)} \right]^- \qquad 22.37$$

green solution

The solution becomes an intense green, and an ESR (electronspin resonance) measurement at this point confirms the presence of the naphthalene radical anion. Addition of styrene at this stage causes the color of the solution to change to red and the ESR signal disappears. The styryl radical anion formed initially is correctly viewed as a resonance hybrid (Eq. 22.38, 22.39).

$$Na^+ \left[\text{(naphthalene anion radical)} \right]^- \quad + \quad PhCH{=}CH_2 \quad \rightarrow \qquad 22.38$$

green solution

$$(Ph\dot{C}H{\dddot{=}}CH_2)^- Na^+ \quad + \quad \text{(naphthalene)}$$

$$(Ph - \dot{C}H - \ddot{C}H_2 \leftrightarrow Ph\ddot{C}H - \dot{C}H_2)^- \qquad 22.39$$

The styryl radical anion species is much more reactive than the naphthalene radical anion, and rapidly couples to form a dimeric dianion, the source of the red color and the reason for the disappearance of the ESR signal (Eq. 22.40). The dimeric dianion is a double-ended anionic propagating species useful for the initiation of a number of valuable homopolymerizations, e.g., Eq. 22.41.

$$2\ (Ph\ddot{C}H \stackrel{\cdot\cdot}{=} \ddot{C}H_2)^- Na^+ \rightarrow Na^+ \bar{C}H(Ph) - CH_2CH_2 - \bar{C}H(Ph)Na^+$$
$$\textbf{1}, \text{ gives a red solution} \qquad 22.40$$

$$\mathbf{1} + n\ PhCH = CH_2 \rightarrow$$
$$\text{+}PhCH - CH_2)_{n/2}CH(Ph) - CH_2CH_2 - CH(Ph)(CH_2 - CHPh)_{n/2} \qquad 22.41$$

The same dianion is also an invaluable initiator for the block copolymerization of compatible monomers. Its ability to add monomer to both ends simultaneously doubles the rate of accumulation of blocks as compared to what is possible with a single ended propagation species (Eq. 22.42).

$$\mathbf{1}\ (=\sim R\sim) \xrightarrow[\text{monomer A}]{6A} AAA\sim R\sim AAA \xrightarrow[\text{monomer B}]{8B} \qquad 22.42$$
$$BBBBAAA\sim R\sim AAABBBB$$

22.5. BASIC POLYMER STEREOCHEMISTRY

Polymerization of ethylene itself gives essentially a high molecular weight alkane product with which there is no opportunity for differing stereostructures. However, many important vinyl monomers are monosubstituted, or 1,1-(gem)disubstituted. Most polymerizations progress via a head-to-tail addition of the propagating center to the monomer. For this reason when a substituted ethylene is polymerized the resulting product can have various stereochemical possibilities.

A monosubstituted monomer which polymerizes in a head-to-tail fashion can give two possible stereoregular polymer structures. One of these will have all chiral centers of the chain in the same configuration, and the other will have every second chiral center of the same configuration and the alternating chiral center of the opposite configuration. The terms for these configurations, originally proposed by Guilio Natta, are isotactic for the first type, and syndiotactic for the second (Fig. 22.4). The non-stereoregular structure which results from random configurations along the chain is referred to as atactic, literally "without arrangement."

A stereoregular polymer is identified by adding a prefix to the name which signifies the type of tacticity, e.g., isotactic polypropylene. If the polymer is atactic then the generic term for the polymer is used, e.g., use of polypropylene alone means atactic polypropylene. One has to be careful, however, because common usage sometimes ignores or overlooks these naming conven-

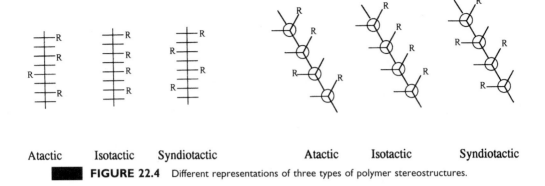

| | | | | | | |
| Atactic | Isotactic | Syndiotactic | | Atactic | Isotactic | Syndiotactic |

FIGURE 22.4   Different representations of three types of polymer stereostructures.

tions of polymers of monosubstituted ethylenes. Introduction of two, or very occasionally three, substituents on the ethylene or incorporation of a heteroatom into the backbone of a synthetic polymer can increase the number and complexity of possible stereoregular forms. Many of these more complex possibilities are discussed in detail in the text by Odian [11].

Knowledge of the existence of the various stereoregular forms and a comprehension of their respective structures adds both to the scientific interest and an appreciation of the physical properties of these materials, as well as to practical applications aspects of synthetic polymers. In general terms, atactic (heterotactic) polymers tend to be amorphous (noncrystalline) in the solid state. They will have a lower melting point than their stereoregular equivalents, and less physical strength. Some will be soft and waxy in nature. In contrast to this, the stereoregular polymers of either the isotactic or the syndiotactic form possess a stereochemistry which lends itself to more ready crystallization. The regular array of substituents in stereoregular polymer chains allows closer packing. For this reason stereoregular polymers often spontaneously crystallize to a significant extent, or may be processed in such a way as to aid crystallization during the fabrication or spinning of these materials. The highly ordered structures obtained in crystalline polymers give these resins higher melting points, higher tensile strengths, greater solvent and chemical resistance, and lowered gas permeability than that obtained for the corresponding atactic materials. So scientifically and commercially the stereochemistry of polymers is of very great interest and significance. Variations in the method used to prepare synthetic polymers, such as solvent choice or polymerization temperature, can significantly influence the stereochemistry of the polymer product obtained.

22.5.1. Stereochemistry of Free Radical-Initiated Polymerization

The configuration of an asymmetric carbon in a polymer is determined at the time of monomer addition to the propagating center. To visualize the situation for the free radical-initiated polymerization of vinyl chloride, approaching monomer may add in such a way as to produce the same or the opposite configuration for the chlorine-substituted carbon in the adding monomer as already present in the adjacent unit of the propagating center (Eq. 22.43, 22.44).

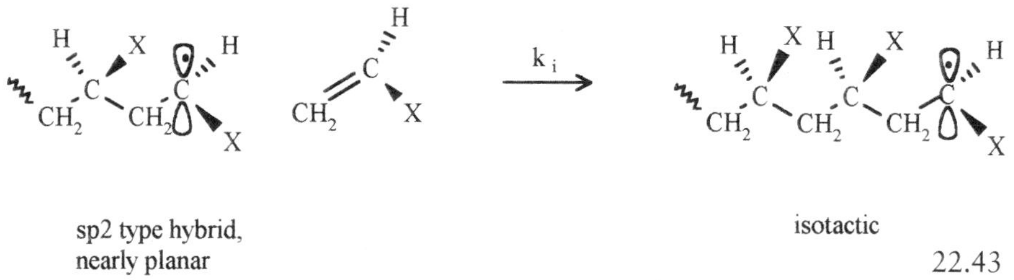

sp2 type hybrid,
nearly planar

isotactic

22.43

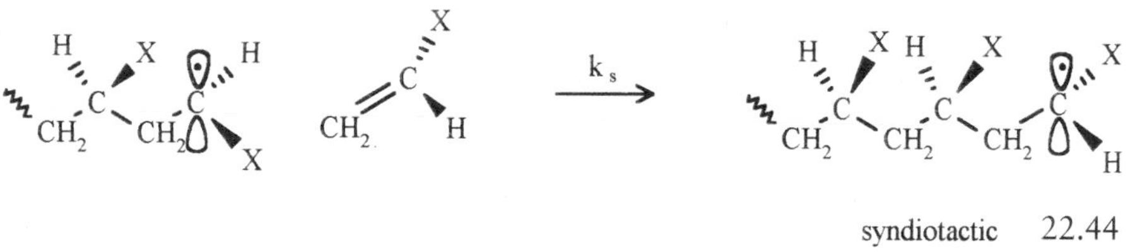

syndiotactic 22.44

Addition with the same configuration is favored if the substituent of the incoming monomer is on the same side as the substituent on the active site of the propagating center and produces an isotactic link or unit. Addition of monomer with the opposite configuration is favored if the substituent of the incoming monomer is on the opposite side to that of the propagating center and produces a syndiotactic link or unit in the polymer. Since there is generally less steric interaction at the time of bond formation for syndiotactic addition under these circumstances, syndiotactic polymer is generally the favored configuration for free radical-initiated polymerizations.

The foregoing comments are not meant to imply that the configuration of substituents in polymers produced under free radical conditions will all be syndiotactic, since the extent of this influence is significantly affected by the size and polarity of the substituent group in the monomer, as well as other factors. Also the configuration of the monomer unit just added is not finally determined until the next monomer bonds to the propagating center. The higher temperatures often used to activate typical free radical initiators may also provide sufficient collision energy for some isotactic addition to occur. So normally significant blocks of syndiotactic polymer would be observed with the occasional isotactic link. For these reasons it can be said that for free radical-initiated polymerizations syndiotactic addition is generally favored over isotactic addition for steric reasons. The preference for syndiotacticity may be increased with a bulky or a highly polar substituent and may be enhanced slightly by carrying out the polymerization at lower temperatures. This is feasible by changing from a thermally activated to a photolytically activated initiator, such as azo-*bis*-isobutyronitrile (AIBN) or a disulfide. In these ways it is possible to influence the stereoregularity of free radical-initiated polymerizations, but not to control it completely.

22.5.2. Stereocontrol in Ionic Polymerizations

In the cationic-initiated polymerization of alkyl vinyl ethers it is possible to exercise fairly rigorous control of the configuration of the product by appropriate choice of the monomer and conditions. For example, isobutyl vinyl ether polymerized by BF_3 etherate at 195 K in toluene can give isotactic polymer[12]. In this low polarity solvent, close association of the gegen ion with the cationic propagating center helps to block one mode of entry of fresh monomer (Eq. 22.45).

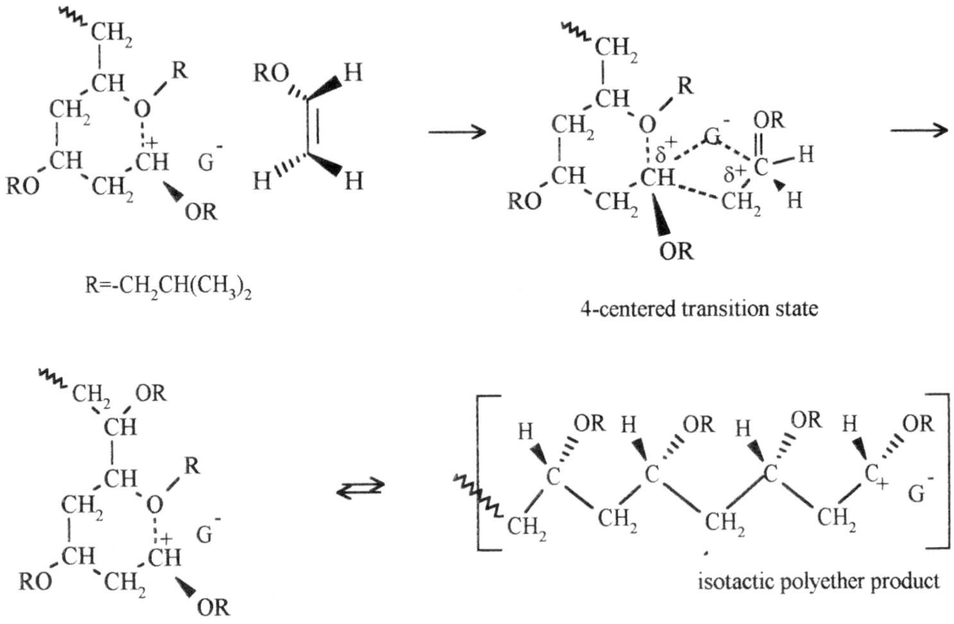

$R = -CH_2CH(CH_3)_2$

4-centered transition state

isotactic polyether product

$$22.45$$

Stereoregulation is not obtained if the isobutyl vinyl ether is replaced by ethyl vinyl ether or isopropyl vinyl ether. This makes it clear that the bulk and placement of the isobutyl group also plays a role in the stereoregulation. It is thought that a loosely associated six-membered ring helps to both stabilize the propagating carbocation and, with the help of the closely associated gegen ion, guide the approach of monomer to it in a single, reproducible orientation. Actual bond formation could occur via a four-centered cylic transition state to yield the observed isotactic product. Thus it is possible to exercise stereocontrol of cationic polymerizations but the examples available are generally special cases.

It is also possible to exercise control of the stereochemistry of vinyl polymerizations which use anionic initiators. As with the cationic systems these examples are also mostly special cases. For example, anionic polymerization of methyl methacrylate in a polar solvent such as dimethoxyethane will give a highly syndiotactic product (Table 22.5). The polar solvent provides good gegen ion separation from the propagating anion so that it behaves almost like a free ion as the polymer forms. Under these conditions, the minimum steric interaction between the propagating anion and the incoming

TABLE 22.5 **Influence of Solvent Polarity on the Tacticity of the Product from the Ionic Polymerization of Methyl Methacrylate**

Catalyst, conditions[a]	Solvent	Product stereostructure
9-Fluorenyl lithium:		
−60°C	1,2-dimethoxyethane	highly syndiotactic
−60°C	toluene	isotactic
−60°C	mixed 1,2-dimethoxyethane/toluene	mixed block copolymer, "blocky" atactic
Phenyl magnesium bromide, 0°C	toluene	100% isotactic
γ-Rays, −78°C	no solvent, i.e., bulk polymerization	96% syndiotactic

[a]All conducted under an inert atmosphere.

gem-disubstituted monomer is obtained when there is an alternation of orientation of the incoming monomer molecules at the time of bond formation. This results in syndiotactic product.

For the anionic polymerization of methyl methacrylate in toluene, a much less polar solvent, the gegen ion remains closely associated with the propagating center (Eq. 22.46).

$$22.46$$

Approach of methyl methacrylate monomer is possible in only one orientation because of the steric hindrance of the gegen ion. As a result, at the critical bond-forming step the substituents of the incoming monomer are always oriented in the same way, which results in isotactic product. The outcomes for the other examples tabulated can be rationalized in similar ways.

22.6. COORDINATION POLYMERIZATION

Until 1953 there were only three mainline methods in common use for the initiation or catalysis of the polymerization of vinylic monomers. These were based on free radical, cationic, and anionic techniques. However, in that year Karl Ziegler, working in Germany, announced his discovery that ethylene could be polymerized to a high molecular weight, easily crystallized product under relatively mild conditions. The key to this success lay in the preparation of a heterogeneous catalyst from titanium tetrachloride or titanium trichloride and an aluminum alkyl. Polyethylene prepared using this catalyst, and now referred to as high density polyethylene (HDPE), had significantly different properties than the current commercial product. This material was produced by free radical initiation and was called low density polyethylene (LDPE) to distinguish it. The HDPE resin was higher melting, more dense, and tougher than LDPE because it was substantially free of the spurious branches that are present in LDPE. Despite predictions that the new HDPE would supersede and replace the slightly older product, each has properties which are suitable for particular applications so that both continue in large-scale production today (Table 22.2). Details of the structural differences are discussed in Section 23.1.

Within 2 or 3 years of Ziegler's announcement Guilio Natta, working for Montecatini in Italy, tried Ziegler's catalyst systems for propylene. This could not be predicted to work without experiment, since the free radical systems which could be used to polymerize ethylene only yielded oligomeric, low molecular weight oils with propylene. He had some initial success, and soon realized that a stereoregular polypropylene would be required for the product to be commercially useful. Changes in the catalyst, in particular a change from $TiCl_4$ to $TiCl_3$ in the transition metal component of this, gave the desired change in activity and yielded substantially isotactic polypropylene. Commercially useful isotactic polypropylene, like high-density polyethylene, only became feasible with the coordination polymerization discoveries of Ziegler and Natta. American production alone of these two polymers now totals of the order of 5 million metric tonnes per year (Table 22.2). The Nobel prize in chemistry was awarded jointly to Otto Ziegler and Guilio Natta in 1953 in recognition of the significance of their discoveries.

22.6.1. Coordination (Ziegler-Natta) Catalysts and Mechanisms

There is a very wide range of catalyst possibilities with which it is feasible to obtain coordination polymerization, depending on the monomer to be polymerized. In the early forms they consisted of a poorly defined product obtained from the mixture of an alkyl or aryl compound of an element from Groups IA to IVA of the periodic table, with a halide or ester of a transition element from Groups IVB to VIIIB. Each catalyst produced was—and still is—highly specific for a particular monomer or monomer mixture (to produce copolymer), and there is still little theory available to enable prediction of which particular catalyst components should be combined for a new monomer substrate, or to yield a different polymer tacticity from an existing monomer. The majority of Ziegler-Natta catalysts are heterogeneous, and the stereoregularity induced in the polymer product is thought to be the consequence of this. Support for this suggestion is obtained from the discovery that vigorous mixing of the first-generation, sludgy catalyst under high shear conditions sufficient to homogenize it simultaneously decreases catalytic activity and yields only an atactic product. Homogeneous Ziegler-Natta catalysts are known, but in general these yield atactic polymers and hence are not of commercial importance.

While the exact nature of the active material present is still poorly understood, it is thought that the catalytic material produced from aluminum triethyl and titanium trichloride has a bridged structure as shown (first struc-

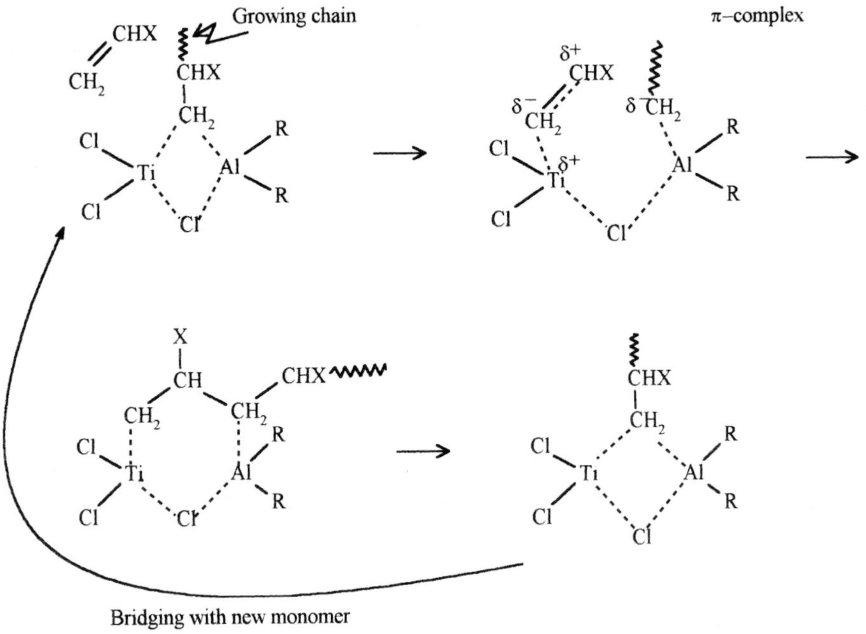

FIGURE 22.5 A bimetallic mechanism for coordination (Ziegler-Natta) polymerization.

ture, Fig. 22.5). A bimetallic mechanism, originally proposed by Natta, illustrates a probable mode of progression of a substituted vinylic monomer to stereoregular polymer.

From the utility of coordination polymerization to produce a useful, stereoregular (easily crystallized) product from propylene in the late 1950s this method has now been expanded to be useful for the polymerization of a variety of monosubstituted alpha olefins. This particular polymerization method provides the only avenue to useful products from these particular monomers (Table 22.4). Coordination polymerization also provides a useful alternative to other methods of polymerization of monomers such as ethylene, the dienes, styrene, haloalkenes, etc. Details of some of these processes are given in Chapter 23.

REVIEW QUESTIONS

1. Outline methods which may be used to experimentally determine whether a polymerization is proceeding via a step condensation or a chain reaction mechanism. Explain your suggestions.
2. (a) Describe and explain the monomer characteristics which best lend themselves to each of the following initiating systems: free radical, anionic, and cationic.
   (b) Which of the initiating systems of part (a) has the widest applicability to monomer polymerization and why?

3. (a) Explain the stereochemistry of vinyl-type addition polymerizations using each of the four available types of initiating systems.

   (b) Describe for each system of part (a) the effect of solvent (or excess monomer) polarity on the degree of stereocontrol which may be possible.

FURTHER READING

R.G. Gilbert, "Emulsion Polymerization, A Mechanistic Approach." Academic Press, San Diego, CA, 1995.

R.B. Seymour and C.E. Carraher, Jr., "Polymer Chemistry: An Introduction," 4th ed., Dekker, New York, 1996.

REFERENCES

1. Facts and Figures for the Chemical Industry, *Chem. Eng. News* **73**(26), 36–79, June 26 (1995).
2. E.M. Kirschner, Growth of Top 50 Chemicals Slowed. *Chem. Eng. News* **74**(15), 16–22, April 8 (1996).
3. "Chemical Marketing Reporter," Vol. 246, No. 26, pp. 26–33, Schnell Publ. Co., New York, June (1995).
4. M.F. Cunningham, K.F. O'Driscoll, and H.K. Mahabadi, Bulk Polymerization in Tubular Reactors, I. Experimental Observations on Fouling. *Can. J. Chem. Eng.* **69**, 630–638 (1991).
5. M. Morton, Mechanisms of Emulsion Polymerization. *Elastomerics* **120**, 2–23, Mar. (1988).
6. D.-Y. Lee, J.-F. Kuo, and J.-H. Wang, Study on the Continuous Tubular Reactor For Emulsion Polymerization of Styrene. *Polym. Eng. Sci.* **30**, 187–192 (1990).
7. D.A. Paquet and W.H. Ray, Tubular Reactors for Emulsion Polymerization (I. Experiemtns; II. Model Comparisons). *AIChE J.* **40**, 73–87, 88–96, Jan. (1994).
8. R.W. Lenz, "Organic Chemistry of Synthetic High Polymers." Wiley-Interscience, New York, 1967.
9. F.W. Billmeyer, Jr., "Textbook of Polymer Science," 3rd ed. Wiley-Interscience, New York, 1984.
10. M. Szwarc, Carbanions, "Living Polymers and Electron Transfer Procsses." Interscience, New York, 1968.
11. G. Odian, "Principles of Polymerization." 3rd ed., Wiley, New York, 1991.
12. C.E.H. Bawn and A. Ledwith, Stereoregular Additon Polymerization. *Q. Rev., Chem. Soc.* **16**, 361–434 (1962).

COMMERCIAL ADDITION (VINYL-TYPE) POLYMERS

The Federal Liberals are the Teflon Party. Nothing sticks to them.
                              —*Dave Barrett (NDP Party), 1996*

23.1. POLYETHYLENE (PE)

Polyethylene was the first of the vinylic polymers to be produced commercially and still is produced on the largest scale of any commodity plastic (Table 22.2). It was first synthesized unintentionally at the facilities of Imperial Chemical Industries (now ICI, U.K.) in 1933, when a high-pressure reaction of benzaldehyde with ethylene was found to produce a waxy, white solid product that did not contain oxygen. Closer examination revealed this unexpected product to be polyethylene, and subsequent development work led to the construction of a commercial plant that started production in 1939.

The pressures used in the polymerization vessels of the early ICI polyethylene plants were among the highest used in the chemical industry, 50–300 MPa (500–3500 atm). Until 1955 all commercial polyethylene was produced by the high-pressure process, or minor variants of this, at pressures of up to 350 MPa. In that year the first of the polyethylene plants that used the Ziegler-devised low-pressure polymerization process began production, and 2 years later an analogous process developed by Phillips Petroleum also came on stream. These production units, which used much more moderate conditions, were followed by other facilities which also used low-pressure techniques. The new low-pressure polyethylene product of all of these plants was immediately distinguishable from the original high-pressure product in that it had a density of 0.94–0.96 g/cm^3, greater than the 0.91–0.93 g/cm^3 of the high-pressure product. It also differed in other important respects

(Table 23.1). The initial product became known as low-density polyethylene (LDPE) to distinguish it from the newer, high-density polyethylene (HDPE) product.

23.1.1. Low-Density Polyethylene (LDPE)

High-pressure processes of 50–350 MPa (500–3500 atm) catalyzed using traces of oxygen or peroxides are used at temperatures of 80–300°C to convert high purity ethylene into the low-density polymer product (Eq. 23.1).

$$n\ CH_2 {=} CH_2 \xrightarrow{R\cdot} \{CH_2 {-} CH_2\}n$$
$$b.p.\ -104°\ C\ (169\ K) \tag{23.1}$$

High pressures are necessary because the propagating radicals are short lived. High monomer concentrations produced by the high pressures help to trap the short-lived radicals involved with a sufficient number of molecules of ethylene before termination to produce a high molecular weight product. When oxygen (0.03–0.1%) is used as the catalyst, conditions of 150 MPa and 190–200°C are sufficient to generate the initiating radical species *in situ*. Tubular reactors or high-pressure autoclaves are operated in continuous mode to ethylene conversions of 10–30%, depending on the process. Good heat dissipation and mixing are essential as the reaction is highly exothermic. Without these measures a hot spot could develop to give extremely rapid localized formation of carbon, hydrogen, and methane from the 99+% ethylene initially contained in the reactor. Once the desired ethylene conversion is obtained, the reaction stream passes to a high-pressure separator where the bulk of the unreacted ethylene is removed from the product and recycled (Fig. 23.1) [1]. High-pressure separation of most of the unreacted ethylene conserves compression costs in much the same way as high-pressure hydrogen recovery does during ammonia production. After high-pressure ethylene separation the hot product moves to a low-pressure separator where the rest of the unreacted

TABLE 23.1   Properties and Product Breakdown of Commercial Polyethylenes

	High-pressure	Low-pressure	
Commercial symbol	LDPE	HDPE	LLDPE[a]
Density, g/cm^3	0.91–0.93	0.94–0.96	0.918–0.94
Crystallinity	50–70%	80–90%	
Crystalline m.p, T_m, °C	110	135	123
Glass transition temperature, T_g, °C	−120	−120	
Tensile strength, MPa	10	10	30
Elongation at break, %	450	700	500
Proportion of total, %			
1988: U.S.A.	35	46	19
EEC	57	34	9

[a]Refers to linear low-density polyethylene, also produced by coordination polymerization at low pressures. It is not a homopolymer of ethylene; see text for details.

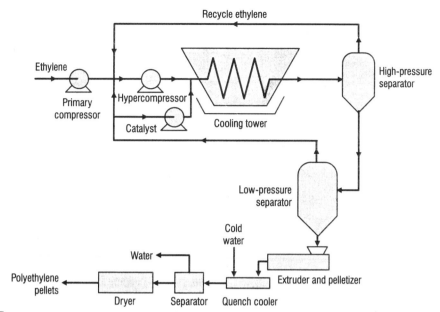

FIGURE 23.1   The high-pressure process to low-density polyethylene. (From G. T. Austin, *Shreve's Chemical Process Industries*, p. 657. Copyright © 1984, reproduced with permission of the McGraw-Hill Companies.)

ethylene is removed. The polyethylene product is then extruded into cold water and pelletized, dried, and then either bagged or shipped in bulk containers for delivery to fabricators.

The essential reaction to polymer which occurs in the high-pressure process to polyethylene is correctly represented by Eq. 23.1. However, careful examination of the high-pressure product by infrared spectroscopy shows that it has 20 to 30 methyl groups per 1000 carbon atoms of chain, i.e., far more than can be accounted for by the 1 or 2 chain end methyl groups expected per 1000 carbon atoms. This is explained by the existence of branches on the polyethylene produced by the high-pressure process. Short branches are produced by a process called "back-biting," involving a six-membered cyclic transition state (Eq. 23.2).

$$23.2$$

Depending on the mode of back-biting and the extent of subsequent propagation before termination, this can give ethyl, butyl, and occasionally longer branches. These outcomes are at least partly the consequence of the high pressure used which tends to cause the propagating chains to fold back onto themselves. In practice, the ratio of butyl to ethyl branches in high-pressure polyethylene is about 1:2, with the occasional longer branch and branched branch. The existence of branching in the product from this process is what confers the less dense packing of polymer chains in the solid, which leads to lower crystallinities and the lower density for this product.

23.1.2. High Density Polyethylene (HDPE)

Several commercial processes are used to produce high-density polyethylene. All employ more moderate pressures and most also use lower temperatures than the low-density polyethylene processes. The Ziegler-developed process uses the mildest conditions, 200–400 kPa (2–4 atm) and 50–75°C, to polymerize a solution of ethylene in a hydrocarbon solvent using a titanium tetrachloride/aluminum alkyl-based coordination catalyst. After quenching the polymerized mixture with a simple alcohol, the catalyst residues may be removed by extraction with dilute hydrochloric acid or may be rendered inert by a proprietary additive. The product is almost insoluble in the hydrocarbon solvent, so is recovered by centrifuging and drying. The final product is extruded into uniform pellets and cooled for shipping to fabricators.

The other processes all use catalysts placed on the surface of a solid support to effect the polymerization of ethylene. This may be conducted in solution in a hydrocarbon solvent, such as in the Phillips and Standard oil processes, or in the gas phase in a fluidized bed reactor, as used in the Union Carbide process [2] (Fig. 23.2). The pressures and temperatures used, 1–10 MPa and 90–300°C depending on the process, are higher in both cases than for the Ziegler process but much more moderate than those used in the

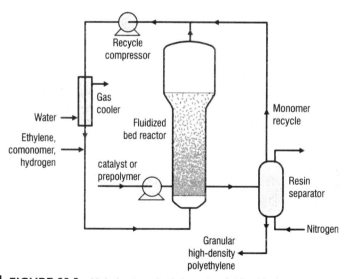

FIGURE 23.2   High-density polyethylene via a fluidized bed process.

high-pressure process. The product is recovered from the solution processes in a manner similar to that used for the Ziegler process. In the Union Carbide fluidized-bed process polyethylene granules (with traces of catalyst) form the bed material, and grow in size with time. Once the granules reach about 1-mm average diameter they are removed from the bottom of the bed, where they tend to accumulate, via a gas lock chamber. The granules are then flushed with nitrogen after which they are ready for storage or shipment without requiring further processing. The concentration of catalyst residues is so low that removal is unnecessary. Stabilizing additives may be used in some cases.

High-density polyethylene has less spurious branching than the low-density material, about 0.3 to 5 ethyl branches per 1000 carbon atoms. This much lower extent of branching in HDPE enables closer packing of the polymer chains in the solid state. Closer packing explains both the higher density and the higher crystallinities observed for this material as compared to those of low-density polyethylene.

23.1.3. Linear Low-Density Polyethylene (LLDPE)

The LLDPE product represents the outcome of a method developed to produce a low-density polyethylene but by using the more moderate, and therefore less costly, conditions employed by the processes used to produce high-density polyethylene. It is not, strictly speaking, a "polyethylene" since it is not a homopolymer of ethylene. LLDPE is actually a copolymer of ethylene which includes traces of 1-octene (Dow and Du Pont), 1-hexene (Phillips), or 1-butene (Union Carbide). This results in a polymer that has entirely short branches, and these are more uniformly spaced along the backbone than in LDPE. The spacing of the branches obtained in these cases can be closely controlled by the proportion of α-olefin to ethylene used in the feed, and the lengths by the choice of the α-olefin comonomer. Properties intermediate to those of low- and high-density polyethylene are obtained for the product (Table 23.1).

23.1.4. New Polyethylenes

It is appropriate to describe here the details of a high-tensile-strength, ultra-high-molecular-weight polyethylene fiber recently developed by Allied [3,4]. This material, produced by methods which result in a very high degree of chain orientation along the axis of the fiber, is said to be 30–40% stronger on a weight for weight basis than the aramids, its nearest competitor. An interesting application is as a thin, lightweight knitted glove to be worn under latex surgical gloves. These confer about 30 times the cut resistance of latex gloves alone, and 15 times the resistance of a medium-weight leather work glove while retaining "feel" for the surgeon.

Also worthy of mention are the new very-low-density polyethylenes (VLDPEs) which are produced using metallocene-based catalysts. These materials, which remain experimental at the moment, have a density of less than 0.9 g/cm^3.

23.1.5. Uses of Polyethylene

Applications of polyethylene depend on the resin type. An interesting early application for LDPE was as a low power factor electrical insulation for the wiring of prototype radar installations which were developed in the U.K. during World War II. The existing plasticized poly(vinyl chloride) was poor in this respect. About two-thirds of the LDPE is used to make film, mostly for packaging. A further 10% is used as a paper coating and barrier layer for paper products used in food packaging, such as milk cartons. Only about 5% of the total is used to make injection-molded products. These are formed by forcing the molten resin into a closed, stainless steel mold cavity of the desired shape followed by rapid cooling of the mold to solidify the shaped resin. The sectional mold is then opened and the article ejected from the cavity, the whole process being repeated every few seconds.

In contrast with the applications for LDPE, about two-thirds of high-density polyethylene is used in molding applications, split about 40% into blow molding of bottles and containers, and 25% into injection-molded products. A further 10% of the total HDPE resin output is consumed in each of the pipe and electrical conduit fabrication areas, and for film and sheet products.

Linear low-density polyethylene (LLDPE) is used for substantially the same applications as LDPE. As the markets for LDPE have expanded, LLDPE plants have been built to supplement the supply of resin to these markets. Very few, if any, new LDPE plants have been built in recent years; in fact, there are indications that LLDPE plants are being built to replace some aging LDPE units. Fortunately this is usually long after the original capital cost of the LDPE plant has been recovered from marketed product.

23.2. POLYPROPYLENE (PP)

Early attempts to obtain polypropylene by vinylic polymerization methods yielded atactic, low molecular weight material of little practical utility. Giulio Natta first experimentally produced isotactic polypropylene of useful molecular weight in 1954 (Eq. 23.3).

$$n\ CH_2\!=\!CHCH_3 \rightarrow isotactic\ +CH_2-CHCH_3+_n$$

$$b.p.-48°C \qquad T_m\ (m.p.)\ 175°C \qquad\qquad 23.3$$

The Montecatini Company began to produce the isotactic resin on a commercial scale in 1957. It is now produced by several slurry and gas phase processes similar to those used for high-density polyethylene. The first-generation catalysts for slurry-based commercial processes yielded about 1 kg of polypropylene (PP) per gram of titanium, with 90–95% isotacticity. Any atactic, coproduced product was soluble in the suspending solvent and could be removed with this during product recovery. Better understood second-generation commercial catalysts gave 30 kg of PP per gram of titanium in the catalyst, and a sufficiently high degree of isotacticity that atactic material did not have to be removed. Modern third-generation catalysts yield 2000 kg of essentially isotactic PP per gram of titanium by a gas phase (Unipol) process in which the residual catalyst concentration is so low that it may be left in the final polymer [5].

Isotactic polypropylene has the lowest density of the commercially dominant thermoplastics, 0.90 g/cm^3, which contributes to its high strength to weight ratio. It has a crystalline melting point (T_m) of 175°C, which gives the resin a useful upper applications temperature limit of about 120°C, and a tensile strength of 28 MPa (4000 lb/in.^{2}). Since it is priced about the same as high-density polyethylene (T_m = 135°C) it fills the application niches where its significantly higher melting point is important. Injection-molding products for the auto industry, appliances, toys and housewares, and fiber uses in carpeting, decorator fabrics, specialized outdoor wear, and lighter-than-water ropes dominate the market. About 10% of the resin is used for film, mostly for packaging applications [6].

23.3. POLY(VINYL CHLORIDE) (PVC)

Laboratory polymerization of vinyl chloride was accomplished as early as the 1870s, but the high melting point polymer product was not adopted commercially for over 50 years. The practical factor that caused the delay in commercial development of this material was the discovery that temperatures high enough to permit melt processing also caused significant thermal decomposition of the polymer. Initial measures to lower the processing temperatures involved copolymerization of vinyl chloride with vinyl acetate. But in 1930, before development of this remedy for the difficult processing got very far, it was discovered that intimate mixing of poly(vinyl chloride) (PVC) with certain high boiling point liquids called plasticizers produced a material much more amenable to processing and melt-forming techniques. This process is referred to as plasticization. Production of articles from this plasticized PVC began in both Germany and the United States in 1933, and today PVC is produced on the third largest scale of all the plastic resins. It has reached this position for reasons of its ease of processing in compounded form, its excellent physical properties, and its competitive cost relative to alternative materials.

Most commercial poly(vinyl chloride) is produced by suspension polymerization, although bulk, emulsion, and solution processes are also used. Vigorous stirring plus a suspending agent are used to form a suspension of small droplets of liquid vinyl chloride (boiling point −14°C) in water, as it is forced into the reactor under pressure using oxygen-free nitrogen. Trichloroethylene may be added as a chain transfer agent, to help control molecular weight. The contents are then heated, usually to about 50°C, to promote radical formation from the monomer-soluble initiator and start the polymerization. This also raises the pressure to about 0.5 MPa (5 atm) (Eq. 23.4).

$$\text{n CH}_2\text{=CHCl} \xrightarrow{R\cdot} \text{{+}CH}_2\text{—CHCl{+}}_n \qquad 23.4$$

As the polymerization proceeds vinyl chloride monomer is consumed and the pressure drops. At about 0.05-MPa gauge pressure (pressure above normal atmospheric) sufficient monomer has been converted to polymer. Unreacted monomer is vented and the residual vinyl chloride is removed by blowing steam through the system. Both vapor streams are led to a vinyl chloride recovery system. These steps are taken both to recover the unreacted vinyl chloride and to minimize the concentration of residual monomer left in the product. The suspension of polymer beads in water is then cooled, and the polymer separated and dried. PVC produced in this way will have a number average molecular weight in the 30,000–80,000 (or a DP or X_n of = 480–1280 vinyl chloride units) range.

The monomer chain transfer constant for vinyl chloride at 60°C is in the range $10.8–16 \times 10^{-4}$, one of the highest of the common vinyl monomers. This normally restricts the reasonably achievable number average molecular weight to the 50,000–100,000 range. Fortunately number average molecular weights for PVC of around 50,000 are high enough to be commercially useful.

Poly(vinyl chloride) always requires one or more additives before it is processed to final products because of its susceptibility to thermal decomposition. It may be stabilized for processing by the addition of one of barium, cadmium, calcium, or zinc stearates, or organic derivatives of lead or tin. In addition, a processing lubricant such as stearic acid or a metal stearate may be added. Rigid PVC products, such as pipe, sheet, and trim profiles, consume about half of the commercial resin production. For rigid pipe, a specially lubricated melt of the thermally stabilized resin is extruded through an annular die with the bore-forming disk supported on slender stays extended upstream of the annulus. PVC for use in higher temperature pipe applications may be readily chlorinated to increase its stiffness under these conditions, before it is formed into pipe. This raises the usable service temperature to about 100°C as compared to the 65°C rating for normal chlorine content rigid PVC pipe. The rigid product is also formed into siding, gutters, window frames, and other durable building products by similar extrusion methods.

The other half of commercial PVC resin is plasticized by vigorous mixing of the powdered resin at high temperatures with a suitable plasticizer. Suitable plasticizers must be resistant to extraction by water, nonflammable and stable, low in toxicity, and economical. Tricresyl phosphate was the first plasticizer used for PVC, in 1933, but this was rapidly replaced by less toxic materials such as di-octyl or by di-hexyl phthalate or the esters of adipic acid or fatty acids. Phthalate esters now fill about three-quarters of the PVC plasticizer market. Addition of 30–80% by weight plasticizer, together with varying proportions of functional filler, lubricant, and pigment, yields products which are flexible, elastic to the extent required by the final application, and attractively colored. Most pigments and some fillers also provide a degree of additional UV protection for the product, over and above that usually provided by the plasticizer.

Products from plasticized PVC include flexible sheeting ("vinyl") and tubing, floor tiles, etc. These raw products, in turn, are used to produce auto upholstery, components of medical appliances, shoe soles, and in the construction and renovation industry. User evidence of the presence of the plasticizer in "vinyl" products is obtained from the distillation of a small fraction of it from auto upholstery or lining materials onto the inner surfaces of car windows, in hot weather. Also, at one time, it was noticed that traces were extracted into blood and blood products from the plasticized PVC used in the medical tubing and bags used to handle these materials.

Poly(vinyl chloride) may also be internally plasticized by copolymerization of vinyl chloride with vinyl acetate, as mentioned earlier. It is not possible to obtain the same result from a mixture of the same proportions of poly(vinyl chloride) and poly(vinyl acetate) homopolymers. Incorporation of vinyl acetate into the polymer chains produces the intimate association of components needed to lower the temperature required for processing in much the same way as an added plasticizer does. Important end uses for poly(vinyl chloride-co-vinyl acetate) at low levels of vinyl acetate incorporation are phonograph records and floor tiles, and also for sheeting, when added filler will also be used.

23.4. POLYSTYRENE (PS)

Styrene is one of the few monomers that may be polymerized by free-radical, anionic, cationic, or coordination (Ziegler-Natta) methods. This property, common to styrene and most of its derivatives, is the consequence of the availability of a benzylic position in these monomers which is capable of stabilizing a radical, carbanionic, or carbocationic center, as well as possessing a polarizability amenable to the charge distributions required by coordination methods of polymerization.

Styrene (and derivatives) also possesses the rare monomer quality that the neat material, without initiator, may be spontaneously polymerized by simply heating to 80–100°C for 24–48 hours. It is thought that this occurs via the initial Diels-Alder dimerization of styrene to the two diasteomers a and b [7]. The two diastereomers appear to have an extremely labile hydrogen which is both doubly allylic and tertiary. However, only dimer a has the correct stereochemistry (an axial phenyl) which enables the excess styrene to abstract a hydrogen atom from it, producing two radical species (Eq. 23.5).

$$1, \text{or } 2 \longrightarrow \left[CH_2-CHPh \right]_{(n-1)} CH_2 - \overset{\bullet}{C}HPh$$

23.5

Each of the radical species produced is capable of initiating radical chain polymerization of styrene to yield useful polymer. The other diastereomer is probably consumed along with monomer during the process.

Styrene, as a monomer, also has an uncommon versatility in that bulk, solution, suspension, and emulsion techniques may be used to produce high polymer. All four methods are used commercially, though solution polymerization is the most common because it reduces the viscous material transfer problems of bulk polymerization and eliminates the higher risk of contaminated product and the need for drying required by suspension and emulsion methods. Solution polymerization also lends itself to continuous processing, unlike the batch methods usually used for bulk polymerization.

Solution polymerization of styrene employs a reactor train of polymerization vessels, usually three, each progressively hotter than the first. A stream of about 10% styrene dissolved in ethylbenzene plus a suitable thermally activated initiator, is fed into the first reactor under an inert atmosphere and kept at 110–130°C. The solution moves through each reactor in turn until it leaves the last reactor at 150–170°C, the polymerization at this point completed to the extent desired (Eq. 23.6).

$$\text{n PhCH}=\text{CH}_2 \xrightarrow{\text{R}^{\cdot}} \text{+PhCH}-\text{CH}_2\text{+}_n \qquad\qquad 23.6$$

To remove the solvent and unreacted monomer the solution is then extruded as fine strands into a devolatilization unit kept at >200°C. Solvent and monomer are recovered leaving strands of polymer containing only traces of ethylbenzene and monomer. Extruding into coarse filaments, chopping, and cooling complete the process to produce polystyrene resin pellets suitable for sale.

Commercial polystyrene has a weight average molecular weight in the 2–3 $\times\ 10^5$ g/mol range representing 2000–3000 monomer units. It is an atactic material, with significant syndiotacticity, and has a low degree of crystallinity, a density of 1.04–1.06 g/cm^3, and melting point (T_m) 240–250°C. It is possible to make isotactic polystyrene via Ziegler-Natta catalysis, but this still has a low degree of crystallinity, and is less transparent and more brittle so is of little commercial interest. Preparation of syndiotactic polystyrene is attracting research interest, but is not a commercial product as yet.

Pure polystyrene resin is a highly transparent brittle material, and relatively inexpensive to produce. About half the total production is consumed in packaging, as film and in expanded form, and in disposables (drinking glasses, etc.). Another 15% is used to produce electrical goods, such as switch cover plates, and as components of appliances such as refrigerator and freezer liners. In the latter application high-impact grades of polystyrene, described shortly, are favored for their greater toughness. Furniture components and the construction industry consume a further 10%. The chief application in construction is in expanded form, for excellent thermal insulation. Housewares, toys, and recreation and personal products consume the bulk of the remainder.

23.4.1. Polystyrene Foam Products

Expanded polystyrene is of such importance that it is of interest to briefly outline the methods used in its production. The easiest technology for a fabricator to use employs expandable polystyrene beads to produce foam products. To make the beads, styrene is suspension polymerized in the presence of *n*-pentane (b.p. 36°C), or other hydrocarbon with a low boiling point. The polystyrene beads, as they form, contain 5–6% of the *n*-pentane incorporated into the bead, the dense precursor form in which this material is shipped. At the fabricator the expandable beads are exposed to steam in a chamber which allows free expansion, which causes the polystyrene to soften and the pentane to vaporize to "puff up" the beads like popcorn, to about 40 times their original size. These once-expanded beads are then cooled in the presence of air for a few hours so that air may diffuse into the expanded beads to equalize the pressure. Then, in a mold of the desired shape for the final application such as flotation billets, television/computer packing, or disposable cups, the requisite quantity of once-foamed beads is placed in the mold and is again heated by steam [8,[9]. The secondary expansion plus slight softening of the bead surface in a limited space fuses the mass into a low-density product the shape of the mold. Cooling and then opening of the mold releases the item.

Expanded polystyrene is also produced by extrusion of the melt containing a volatile solvent, or more recently carbon dioxide, through a slit. The pressure drop caused by the emergence of the molten polymer from the slit of the extruder allows the hot solvent to vaporize, and in the process expands the extruded shape to an extent related to the proportion of solvent used. When carbon dioxide is used as the blowing agent its expansion on emergence of the melt from the slit can perform the same function.

23.4.2. Polystyrene Copolymers

High-impact polystyrenes, in which the brittle nature of the pure polystyrene resin is decreased by incorporation of a rubber, give increased durability and resistance to fracture to the resin when shock loads are expected. Simple mixing of finely divided polystyrene with *cis*-1,4-polybutadiene (a synthetic rubber) gives only slight improvement in impact strength. Optimum results are obtained by incorporation of the rubber into the polystyrene chains as a copolymer. This is accomplished by solution polymerization of styrene in a solvent in which a rubber is already dissolved. Chain transfer processes result in copolymerization of the styrene with the rubber. The product has much improved impact resistance, but at the same time has lower tensile strength and is less transparent than the pure polystyrene product. Since pigments are normally added to the high-impact resins before forming, the reduced clarity of the product is easily accommodated in commercial products.

23.5. POLYTETRAFLUOROETHYLENE (PTFE)

Discussion of polytetrafluoroethylene is included here, as one example of a high value, relatively low volume specialty polymer, to illustrate the diversity of monomers which are amenable to vinyl-type polymerization and the very special polymer properties that may be achieved by this means. PTFE technology, like the development of isotactic polypropylene, is also one of the more recently commercialized vinyl polymers. PTFE was first prepared experimentally in 1938 by Plunkett, in the United States. A pilot plant for this product came on stream in 1943 followed by a full-scale production unit in 1950, both built by the Du Pont Company. Du Pont coined the trade name Teflon for this material with such novel properties.

Pure tetrafluoroethylene monomer, under ambient conditions, is an odorless, colorless, tasteless gas with low toxicity. It has a liquefaction temperature of $-76°C$. It may be polymerized by either suspension or emulsion techniques. Both procedures require use of high pressures in an autoclave in order to maintain the monomer in liquid form. These techniques produce chemically identical product, the first a granular resin, and the second a fine powder (Eq. 23.7).

$$n\ CF_2 = CF_2 \xrightarrow{R} \ \{CF_2 - CF_2\}_n \qquad\qquad 23.7$$

If the emulsion polymerization product is retained in the water phase for use it is referred to as a PTFE dispersion. The very strong C–F bonds, plus the strengthening effect of these on the C–C bonds confers extraordinarily stable properties to this polymer. The chemical inertness, excellent electrical resistance, high heat resistance, and low coefficient of friction combine to make PTFE one of the highest performance commercial vinyl plastics produced (Table 23.2).

Unprocessed PTFE resin is a white solid with a waxy appearance and feel. It has a moderate tensile strength, much less than the nylons and polyesters. It also has a tendency to creep under compression, but this can be accommodated in bearing applications by proper engineering design of the bearing area and housing shape. The very high softening point (it doesn't form a true melt), high viscosity at the softening point, and low solubility in solvents make PTFE resins difficult to fabricate by conventional means. These difficulties slowed the early development of applications for this material. Powder sintering techniques allied to those procedures used in powder metallurgy are used to form films and coatings. Care is required in these techniques and in

TABLE 23.2 Properties of Typical Polytetrafluoroethylene (Teflon) Resin

Molecular weight, M_w, g/mol	10^6–10^7
Density, g/cm^3	2.18
Crystalline melting point, T_m, °C	327^a
Useful temperature range, °C	-100–260
K	170–530
Static coefficient of friction (against polished steel)	0.05–0.08
Tensile strength, MPa	20–28
lb/in^2	2900–4000
Solvent resistance	insoluble, ordinary conditions^b
Chemical resistance	inert, ordinary conditions^c

^{a}Holds shape of melt. Viscosity of the melt at 380°C is 10 GPa (10^{11} Pa).
^{b}But is swollen by fluorocarbon oils.
^{c}Under severe conditions reacts with alkali metals, fluorine, strong fluorinating agents, and sodium hydroxide at temperatures above 300°C.

the cold extrusion procedures used to form fabricated shapes in order to minimize shear since these very high molecular weight resins are susceptible to shear-induced chain shortening. Some of the applications of this specialty polymer are for gaskets and seals, particularly for chemically aggressive environments, low friction (bearings) and nonstick applications, high-performance tubing and pipe, and insulating wire wrappings.

Gore Tex (Gore, Inc.) membranes, with a pore size large enough to permit the passage of water vapor but small enough to exclude water droplets, are extruded from a melt under stress. The stress is applied to the extrusion by taking up the membrane on a roller which is operated at a speed significantly higher than the extruder. This stretches the membrane, as it forms, and produces the close distribution of micropores that is required for this application. Final adjustment of pore size, etc., may be made by stretching the membrane again at temperatures below the melting point, sometimes in the presence of an aromatic hydrocarbon.

23.5.1. Polytetrafluoroethylene (Teflon) Copolymer

As with the higher impact resistance achievable with polystyrene when styrene is copolymerized with a rubber, the melt-forming temperatures of PTFE resin may be conveniently lowered by copolymerization of tetrafluoroethylene with low ratios of perfluoropropyl vinyl ether (boiling point, 36°C). The polymerization may be conducted in either an aqueous medium or in an organic solvent (Eq. 23.8).

$$n\ CF_2{=}CF_2\ +\ x\ CF_3CF_2CF_2OCF{=}CF_2\ \xrightarrow{R\cdot}\ \ \ \ \ \ \ 23.8$$
$$\text{+}CF_2{-}CF_2\text{+}_n\text{+}FC(OC_3F_7){-}CF_2\text{+}_x$$

The product, commonly referred to as Teflon PFA (for perfluoro alkoxy), has a melting point of about 305°C. It retains many of the desirable properties of PTFE (Teflon) homopolymer, and is much more readily melt-formable into products than PTFE itself.

A paper by Polakoff $et\ al.$ records and considers the results of a study of urinary fluoride levels in PTFE fabricators on the line [10]. There were no markedly elevated fluoride levels found, though these did correlate with exposures.

23.6. ENVIRONMENTAL ASPECTS OF VINYLIC POLYMERS (PLASTICS)

23.6.1 Biodegradable Polymers

Medical applications originally stimulated research in the area of biodegradable polymers, but more recently accelerated litter decomposition and landfill degradation interests have provided additional incentives. These objectives have been met by a variety of strategies [11]. Among these, incorporation of a photosensitive species, such as a carbonyl group or certain metal complexes, has been used to promote accelerated photodegradation with some success.

Carbonyl incorporation can be accomplished readily, for instance, by copolymerization of carbon monoxide with ethylene, which is feasible with the low-density polyethylene processes. Photosensitization may also be achieved after polymerization by blending certain proprietary additives into a melt of the finished polymer [11]. It should be mentioned here that generally speaking far more attention is devoted to the photo*stabilization* of synthetic polymers in order to increase their useful life in outdoor applications.

Starch addition at 6–15%, or gelatinized starch addition at 40–60% by weight blended into the melt, provides an alternative strategy. Physical incorporation of a degradable constituent into the polymer in this manner facilitates breakdown by a combination of oxidative, biological, and physical processes.

Special types of purpose-built polymers can also be made with the intention of enabling more rapid direct microbiological attack and decomposition of these materials. Classes of both condensation and addition polymers among the polyesters, vinylic polymers, and polyhydroxyalkanoates typify current candidate materials being tested in these applications (e.g., Eq. 23.9), but, as yet, high costs have discouraged large-scale exploitation.

$$\begin{array}{ccc} & O & \\ & \parallel & \\ \{OCH-C\}_n & \{CH_2-CH\}_n & \{CH-CH_2-C-O\}_n \\ \mid & \mid & \mid \quad\quad \parallel \\ CH_3 & OH & CH_3 \quad\quad O \\ \text{poly(L-lactic acid)} & \text{poly(vinyl alcohol)} & \text{poly(3-hydroxybutyrate)} \end{array}$$

$$23.9$$

If further development of these biodegradable packaging options makes them economically feasible, then care is required to enable differentiation and separation of these polymers from very similar stabilized polymers so that they do not cause a loss of properties of recycle stream products.

23.6.2. Recycling of Vinylic Polymers (Plastics)

Since the plastics are produced from petrochemicals derived from hydrocarbons, the motivation to reuse, recycle, or reprocess for energy recovery is primarily driven by an interest in conservation of petroleum resources. Economic factors are also important, but the potential saving of landfill space is more a perception rather than a reality [6,7]. Most of the categories of vinylic polymers discussed in this chapter are melt-formable, that is, they are "thermoplastic" materials, rather than nonmelting or "thermosetting" as are several of the synthetic polymers discussed in Chapters 20 and 21. Thus, from a technical point of view, all of the plastics discussed here are relatively easily reprocessed and reshaped into "new" articles merely by cleaning and remelting the polymer. One notable exception to this generalization is vulcanized rubber. Although synthetic rubber is produced by a vinyl-type polymerization, which in theory should be melt-formable, the final shape of rubber products such as tires and footwear is fixed by vulcanization. This process introduces disulfide crosslinks into the rubber, which thereby stabilizes the shape of the article for its intended purpose. The presence of these crosslinks alters the reprocessing methods required for vulcanized rubber as compared to any of the other plastic materials.

23.6.3. Industrial Scrap

Fabricator scrap produced from the trimming of the sprues and gates used to direct molten polymer into the mold cavity, or from molded products which are off-specification, is usually recycled on site. The material for recovery is reduced to a uniform size and additional processing aids or stabilizers are added, if required. This suitably pretreated scrap may be directly blended with virgin material for production of the originally produced item. If blending of scrap with virgin material is unable to meet the specifications for the original product, the reprocessed producer scrap will have to be downgraded and used to form articles with less stringent specifications. This still represents an efficient form of producer recycling.

23.6.4. Postconsumer Plastic Waste Recovery

A large proportion of the vinylic plastics discussed here is used in packaging, mostly in small units. This wide dispersal in small units is what makes collection and recovery of these plastic materials difficult and costly in terms of labor and energy. Mechanical means are available to separate the major constituents of municipal waste after household collection. But the rejection capability of these processes for foreign material is imperfect, causing some nonplastic materials to enter plastic recycle streams, and some misidentified plastics to enter the wrong segregated plastics stream. Also, inevitably, some plastics are inappropriately missorted into a compostable stream. The first outcome leads to a downgraded, and therefore much lower value, recycle resin, and the second to a loss of one of the more valuable constituents of the municipal waste stream as well as contamination of the compost stream. Most plastics are highly resistant to decomposition by composting but are not harmful to the process, although they may be to the end uses of the compost. Much better differentiation of plastic waste from general refuse and segregation of the recovered plastic materials according to resin type is obtained by presorting of the waste at the householder level. As already discussed in connection with the properties of mixtures of poly(vinyl chloride) and poly(vinyl acetate), or polystyrene with rubber, crude mixtures of two or more polymers usually results in degraded properties relative to those achievable from any of the more rigorously segregated component materials. For this reason, for optimum properties from a recycle plastics stream the component polymers must be segregated as to resin type. A decade ago this would have been very difficult by mechanical means, because the densities of several of these plastics are quite similar. Separation by hand would also have been poorly feasible because of the lack of identifying marks on packaging items. Today, however, numerical codes are embossed on most types of plastic packaging which identify the type of plastic used in the package (Table 23.3). These codes, usually given together with the letter short forms commercially used for the polymer,

TABLE 23.3 Symbolism, Uses, and Properties of Several Important Packaging Plastics

Symbol for recycling	Name and (industry acronym)	Typical uses	Density (g/cm³)	Melting point, T_m (°C)
1 PETE	Poly(ethylene terephthalate) (PET)	soft drink and water bottles	1.40	240
2 HDPE	High-density polyethylene (HDPE)	milk, bleach bottles	0.94–0.96	130
3 PVC	Poly(vinyl chloride) (PVC)	water, glass cleaner bottles	1.40[a]	ca. 80
4 LDPE	Low-density polyethylene (LDPE)	plastic bags; bread, food wrap	0.91–0.93	108
5 PP	Polypropylene (PP)	yogurt, margarine containers; shampoo, syrup bottles	0.90	176
6 PS	Polystyrene (PS), often foamed	disposable picnic ware, fast food clamshells, egg cartons, meat trays	1.05[b]	ca. 100
7 OTHER	Other plastics, composites and laminates	beverage and juice boxes	various	varies

[a]Unplasticized. Decreases to 1.30–1.32, depending on the formulation used when plasticized.
[b]General purpose, not expanded.

now make handsorting relatively unambiguous, though still labor intensive. It is possible that machine-readable bar codes could supplement this information to make for rapid, relatively error-free machine sorting as a cost-effective means of providing segregated polymer streams ideal for recycling. All that is needed to return these streams into production is shredding, washing, drying, and then reextrusion to pellets suitable for all types of products. In some cases additive levels may need to be adjusted or restored. The resulting resins may either be used on their own, or may be blended with virgin material, depending on the specifications required for the product.

If downgraded specifications are acceptable it is possible to reprocess a mixture of recycled plastics to form usable articles more tolerant of minor variations in specifications such as field drainage tile and building products such as base layer flooring. One option to enable this is to thoroughly melt-blend the mixed plastic stream with a compatibilizer, such as a chlorinated polyethylene [12]. The compatibilizer aids in producing a mutual solution of all the components present in the melt so that articles formed directly from the blended resin have acceptable properties. Alternatively, hot mix machinery has been developed, for example, the Mitsubishi Reverzer, which is capable of producing an intimate mixture of plastics from a feed which is raised to a somewhat higher than normal processing temperature to obtain a lowered viscosity [13]. To avoid an extra processing step the product desired is preferably formed directly from the well-mixed melt using the extruder of this machine to force the product into molds of the desired shape.

23.6.5. Anaerobic Pyrolysis

Some segregated vinyl polymers will very readily revert to the monomer on heating the polymer out of contact with air. This process is sometimes referred to as tertiary recycling. Poly(methyl methacrylate) and poly(α-methylstyrene) can yield practically 100% monomer in this way, and polystyrene can yield 80–90% monomer [8,14,15]. Polytetrafluoroethylene, too, can yield about 95% monomer on heating to 500°C. However, polyethylene, polypropylene, and poly(vinyl chloride) are more indiscriminate in their decomposition pathways and yield a mixture of products [16,17]. Pyrolysis of a mixture of polymers, such as may be recovered from automobile shredder residues, also gives more complex fragments in the recovery system, but is nevertheless a recommended option for retrieval of usable feedstock from a mixed plastics waste stream [18,19]. Of course mixed plastics with an energy content of about twice that of paper may be burned directly for energy recovery with appropriate emission controls. This may be managed as a segregated stream or as a mixture with other combustible waste.

23.6.6. Laminates, Composites

Some of the procedures of value for processing of mixed plastic wastes may also be useful for the recovery of value from the laminates used for efficient packaging of food and beverages. A combination of heat and pressure may be employed on a recycle stream of this type of material to fuse the mixed fibrous and resin components into a dense board-like product [17,20]. This highly weather and insect resistant product, "Eco Superwood Plastic Lumber," may be sawn, nailed, drilled, etc., like wood, and it resists rotting in direct contact with the soil without the need for preservatives.

Other types of composite packaging in which a plastic and another material, or two different plastics are used, such as for a cap and a bottle, are increasingly being recognized as posing a problem for the more valuable segregated plastics recycling programs. Package designers are increasingly avoiding these problems by making a packaging unit entirely from a single resin. A further development in package design for commonplace liquid products involves initial sale of the liquid detergent, syrup, etc., in a convenient, sturdy reusable dispenser. This may or may not be made from a single plastic, a factor which has less significance in this role. Refills are offered in lightweight plastic film pouches saving the weight difference in packaging on each use of a pouch refill.

23.6.7. Vulcanized Rubber

The crosslinked rubber remaining in used tires may not be reclaimed directly by remelting. However, the rubber may be recovered by chilling the used tire carcasses and milling to produce a crumb rubber product. The fibrous plies and any steel belts are separated. Crumb rubber may be blended with an appropriate flexible polyurethane mix to bind the granules into the durable and commercially tested "Rubberloc Pavers" patio bricks [21]. Crumb rubber incorporation into asphalt has also been found to confer greater flexibility and elasticity.

Natural and butyl rubber from inner tubes and other applications which are fiber-free may be reclaimed by the addition of reclaiming oils and other additives and then subjecting the mix to temperatures of 175–205°C and high shear in a Reclaimator. These conditions are sufficiently severe to break the relatively weak bonds of the disulfide crosslinks in the feed, effectively devulcanizing the rubber in a period of 1–3 minutes under these conditions.

With waste rubber goods, as with nonsegregated plastic wastes, it is possible to simply burn the rubber and regain the considerable fuel value. Whole scrap tires have been burned in cement kilns, for example, to experience a proportionate saving in the usual oil or gas fuel supply used in these operations. At the extremely high temperatures used, the combustion is complete and any steel belts or beading are incorporated into the cement klinker without problems.

Recently developed thermoplastic rubbers, e.g., Santoprene developed by Monsanto, are taking over some of the market niches which were formerly reserved for vulcanized (crosslinked) rubbers [22]. Not only does this material lend itself to convenient processing like a thermoplastic, but once an item made from a thermoplastic rubber has outlived its usefulness it can be much more readily reprocessed. However, thermoplastic elastomers generally have lower strength, less abrasion resistance, and lower tolerance to high temperatures than vulcanized rubber so these materials will only encroach on conventional rubber applications in areas where these properties are less important.

REVIEW QUESTIONS

1. Why might isotactic polypropylene be absolutely essential for commercial utility whereas isotactic polystyrene, while possible to prepare, is not sought for commerce?
2. When poly(vinyl chloride) (PVC) is chlorinated to raise its service temperature for rigid hot water piping, chloroethyl repeating units are converted to 1,2-dichloroethyl repeat units.
   (a) What chlorine content, in percent by weight, would be expected for neat poly(vinyl chloride), as first polymerized?
   (b) What ratio of 1,2-dichloroethyl to monochloroethyl units would be present in chlorinated PVC that gave a 68% by weight chlorine content on analysis?
   (c) Would the original PVC or the chlorinated PVC be more resistant to combustion and why?
3. (a) What were the motivating factors for development of linear low-density polyethylene (LLDPE) production technology?
   (b) Why might LLDPE producing facilities be favored over LDPE technology for replacement of low-density polyethylene production facilities?
4. Compare and contrast the structures and properties of low-density polyethylene, high-density polyethylene, and linear low-density polyethylene.
5. (a) Describe the precautions necessary in a plastics recycling operation to obtain optimum properties for the reprocessed plastics.
   (b) If the procedures required for efficient recycling have not been taken, give two reasonable options for remedies that may be taken with the recycled plastics stream to conserve either material or energy values from these.
6. Efficient incineration may be practiced for energy recovery from a severely degraded waste plastic stream. What particular precaution is necessary for safe disposal of waste poly(vinyl chloride) by this method, as compared to the procedure used for most other plastics?
7. One thousand kilograms of 99.6% pure ethylene is polymerized in a heterogeneous process to 850 kg of polyethylene (HDPE) of 80,000 number average molecular weight. Ethylene (99%) totaling 148 kg is recovered.
   (a) What is the degree of polymerization of the product?
   (b) What is the percent industrial yield of polyethylene in this instance?

FURTHER READING

California Integrated Waste Management Board, "State of California, Plastics: Waste Management Alternatives Use, Recyclability, and Disposal." California Integrated Waste Management Board, Sacramento, 1992.

R.J. Ehrig, ed., "Plastics Recycling: Products and Processes." Hanser Publishers, New York, 1992.

J.R. Fried, "Polymer Science and Technology." Prentice Hall PTR, Upper Saddle River, NJ, 1995.

G.J.L. Griffin, ed., "Chemistry and Technology of Biodegradable Polymers." Chapman and Hall, Glasgow, 1994.

E.P. La Mantia, ed., "Recycling of PVC and Mixed Waste." ChemTec Publishing, Toronto, 1995.

K.J. Saunders, "Organic Polymer Chemistry," 2nd ed., Chapman & Hall, New York, 1985.

M.P. Stevens, "Polymer Chemistry: An Introduction," 2nd ed., Oxford University Press, Toronto, 1990.

REFERENCES

1. G.T. Austin, "Shreve's Chemical Process Industries," 5th ed., p. 657, McGraw-Hill, New York, 1984.

2. T. Xie, K.B. McAuley, J.C.C. Hsu, and D.W. Bacon, Gas Phase Ethylene Polymerization: Production Processes, Polymer Properties, and Reactor Modeling." *Ind. Eng. Chem. Res.* **33**, 449–479 (1994).

3. Anonymous, Allied Develops Supertough Polyethylene Fiber. *Chem. Eng. News* **63**(8) 7, Feb 25 (1985).

4. S. Borman, New High Strength Fiber Finds Innovative Use in Protective Clothing. *Chem. Eng. News* **67**(41), 23–24, Oct. 9 (1989).

5. J. Haggin, Unipol Polypropylene Process Garners Licensees. *Chem. Eng. News* **64**(13) 15–45, Mar. 15 (1986).

6. A.M. Thayer, U.S. Polypropylene Producers Confront Troubling Supply-and-Demand Scenario. *Chem. Eng. News* **72**(38) 17–18, Sept. 19 (1996)

7. W.D. Graham, J.G. Green, and W.A. Pryor, Chemistry of Methylenecyclohexadiene and the Thermal Polymerization of Styrene. *J. Org. Chem.* **44**(6), 907–914 (1979).

8. M.B. Hocking, Paper Versus Polystyrene: A Complex Choice. *Science* **251**, 504–505 (1991).

9. M.B. Hocking, Relative Merits of Polystyrene Foam and Paper in Hot Drink Cups. *Environ. Manage.* **15**(6), 731–747 (1991).

10. P.L. Polakoff, K.A. Busch, and M.T. Okawa, Urinary Fluoride Levels in Plytetrafluoroethylene Fabricators, *J. Am. Ind. Hyg. Assoc.* **35**, 99–106 (1974).

11. P.J. Hocking, The Classification, Preparation, and Utility of Degradable Polymers. *J.M.S.-Rev. Macromol. Chem. Phys.,* **C32**(1), 35–54 (1992).

12. J.N. Schramm and R.R. Blanchard, "The Use of CPE [Chlorinated Plyethylene] as a Compatibilizer for Reclamation of Waste Plastic Materials," presented at Palisades Section SPE RETEC, Plastics and Ecology, Cherry Hill Inn, Cherry Hill, NJ, 1970.

13. J. Paul, Recycling Plastics. In "Kirk-Othmer Concise Encyclopedia of Chemical Technology," p. 1007, Wiley, New York, 1985.

14. N. Grassie, Depolymerization Reactions in Vinyl Polymers. *Chem. Ind. (London),* p. 622, June 27 (1953).

15. C. Bouster, P. Vermande, and J. Vernon, Evolution of the Product Yield with Temperature and Molecular Weight in the Pyrolysis of Polystyrene. *J. Anal. Appl. Pyrolysis* **15**, 249 (1989).

16. J. Leidner, "Plastics Waste, Recovery of Economic Value." Dekker, New York, 1981.

17. M. Day, J.D. Cooney, C. Klein, and J.L. Fox, The Use of Thermal Analysis to Study Thermal Processing Effects on Polypropylene. *J. Therm. Anal.* **41**, 225 (1994).

18. B.A. Hegberg, "Technologies for Recycling Post-Consumer Mixed Plastics: Emerging Plastic Lumber Production." Office of the University of the University of Illinois Center for Solid Waste Management and Research, Chicago, 1991.

19. H. Kastner and W. Kaminsky, Recycle Plastics into Feedstocks. *Hydrocarbon Process.* **74**, 109, May (1995).

20. W.C. Fergusson, Recovery of Resources from Platics Industrial Waste. *Chem. Ind. (London),* p. 725, Sept. 4 (1976).

21. G. Rodden, From Window Scrapers to Patio "Stones," Old Tires Continue to Roll Along, *Can. Chem. News* **46**(4), 24 (1994).

22. M.S. Reisch, Thermoplastic Elastomers Target Rubber and Plastics Markets. *Chem. Eng. News* **74**(32), 10–14, Aug. 5 (1996).

▌Constants, SI Units and Multiples, and Formulas

CONSTANTS

Speed of light in a vacuum, c	$2.99793 \ 10^{10}$ cm/sec	Ideal gas constant, R:	8.2056×10^{-2} L atm/K mole
Faraday's constant, F	9.6487×10^4 C/g equiv wt.		8.3143 J/K mole
Avogadro's number	6.02252×10^{23} molecules/mole		1.9872 cal/K mole
Ice point	273.15 K, 0°C	Molar gas volume	22.4136 L/mole (at 0°C, 1 atm)

Seawater is about 3.5 wt. % dissolved solids.

Potable water contains <500 ppm dissolved solids.

One mole = Avogadro's number of atoms, molecules, ions, electrons, etc.

INTERNATIONAL SYSTEM OF UNITS AND MULTIPLES (SYSTÈME INTERNATIONAL D'UNITES, SI)

Fundamental Units:

Quantity	Unit	Abbreviation
length	meter	m
mass	kilogram	kg
time	second	s or sec
electric current	ampere	A or amp
temperature	Kelvin	K
luminous intensity	candela	cd

Prefix Factors:

Multiple or fraction	Prefix	Symbol
10^{18}	exa	E
10^{15}	peta	P
10^{12}	tera	T
10^{9}	giga	G
10^{6}	mega	M
10^{3}	kilo	k
10^{2}	hecto[a]	h
10	deka[a]	da
10^{-1}	deci[a]	d
10^{-2}	centi[a]	c
10^{-3}	milli	m
10^{-6}	micro	μ
10^{-9}	nano	n
10^{-12}	pico	p
10^{-15}	femto	f
10^{-18}	atto	a

[a]These are normally restricted to steps of 10^3 or 10^{-3}, which exclude hecto, deka, deci, and centi. However, these are in common use, so are included here.

FORMULAS

Area		Volume	
Triangle	1/2 bh	Cylinder	$\pi r^2 h$
Regular polygon[a]	$nl^2/(4 \tan\pi/n)$	Sphere	$4/3\ \pi r^3$
Circle	πr^2		
Sphere	$4\pi r^2$		

Gas Laws	Electrochemical
$pv = nRT$	watts, W = amperes $\times$ volts
	Coulombs, C = amperes $\times$ seconds
$\dfrac{p_1 v_1}{T_1} = \dfrac{p_2 v_2}{T_2}$	1 g equiv. wt. = formula weight $\div$ charge
	1 Faraday, F, deposits 1 g equiv. wt.

[a]n = number of sides, 1 = length of side.

2
Conversion Factors

From	To	Factor
Length		
miles	m	1.609×10^3
foot	m	0.3048
inches	m	2.54×10^{-2} (exact)
angstroms	m	10^{-10} (exact)
Mass		
ton, long (2240 lb)	kg	1.015×10^3
tonne, metric (1000 kg; 1 Mg)	lb	2.204×10^3
ton, short (2000 lb)	kg	9.07×10^2
pound (avdp)	kg	0.4536
ounce (avdp)	g	28.350
ounce (troy)	g	31.103
Area		
square mile	acres	640
acres	ha (hectares)	0.40469
hectare	m^2	10^4
square foot	m^2	0.0929
Volume		
barrel (bbl)	m^3	0.159
	imp. gal	34.971
	U.S. gal (liq.)	42 (exact)
cubic foot	m^3	2.832×10^{-2}
imperial gallon	L	4.5460

(continued)

Conversion Factors (continued)

From	To	Factor
U.S. gallon (liq.)	L	3.7854
imperial gallon	U.S. gal (liq.)	1.20095 (i.e., 6/5)
Pressure		
bar	atmosphere	0.98692
	MPa	0.1
	Pa (Nm^{-2})	10^5
atmosphere	bar	1.01325
	Pa (pascals)	1.013 $\times$ 10^5
	lb/in.^{2}	14.696
pound/in.^{2}	kPa	6.89
Energy, Power, work		
therm	MJ	106
British thermal unit (BTU)	kJ	1.06
calorie	J	4.1840
watt	J/sec	1
watt hour	J	3600
horsepower	W	746
horsepower-hours	kWh	0.746

■ SUBJECT INDEX